Modeling, Characterization and Production of Nanomaterials

Woodhead Publishing Series in Electronic and Optical Materials

Modeling, Characterization and Production of Nanomaterials

Electronics, Photonics and Energy Applications

SECOND EDITION

Edited by

Vinod K. Tewary
National Institute of Standards and Technology (NIST)

Yong Zhang
The University of North Carolina at Charlotte

WP
WOODHEAD
PUBLISHING

ELSEVIER   An imprint of Elsevier

Woodhead Publishing is an imprint of Elsevier
50 Hampshire Street, 5th Floor, Cambridge, MA 02139, United States
The Boulevard, Langford Lane, Kidlington, OX5 1GB, United Kingdom

Notices
Knowledge and best practice in this field are constantly changing. As new research and experience
broaden our understanding, changes in research methods, professional practices, or medical treatment
may become necessary.

Practitioners and researchers must always rely on their own experience and knowledge in evaluating
and using any information, methods, compounds, or experiments described herein. In using such
information or methods they should be mindful of their own safety and the safety of others, including
parties for whom they have a professional responsibility.

To the fullest extent of the law, neither the Publisher nor the authors, contributors, or editors, assume
any liability for any injury and/or damage to persons or property as a matter of products liability,
negligence or otherwise, or from any use or operation of any methods, products, instructions, or ideas
contained in the material herein.

ISBN: 978-0-12-819905-3 (print)
ISBN: 978-0-12-819919-0 (online)

For information on all Woodhead publications
visit our website at https://www.elsevier.com/books-and-journals

Publisher: Matthew Deans
Acquisitions Editor: Stephen Jones
Editorial Project Manager: Howi M. De Ramos
Production Project Manager: Fizza Fathima
Cover Designer: Vicky Pearson Esser

Typeset by STRAIVE, India

Working together
to grow libraries in
developing countries

www.elsevier.com • www.bookaid.org

Contents

Contributors

Nasir Ali Zhejiang Province Key Laboratory of Quantum Technology and Devices, Department of Physics, State Key Laboratory for Silicon Materials, Zhejiang University, Hangzhou, People's Republic of China

Fabio Andrijauskas School of Technology, University of Campinas—UNICAMP, Limeira; Physics Institute "Gleb Wataghin", University of Campinas—UNICAMP, Campinas, SP, Brazil

P. Balakrishna Pillai Department of Electronic and Electrical Engineering, University of Sheffield, Sheffield, United Kingdom

Mark Banash Neotericon LLC, Bedford, NH, United States

B. Cao University of New Orleans, New Orleans, LA, United States

Dominic Caracciolo Department of Chemistry, State University of New York at Binghamton, Binghamton, NY, United States

Daniel Casimir Laser Spectroscopy Laboratory, Department of Physics & Astronomy, Howard University, Washington, DC, United States

J. Chen University of New Orleans, New Orleans, LA, United States

Cristian V. Ciobanu Colorado School of Mines, Golden, CO, United States

Vitor R. Coluci School of Technology, University of Campinas—UNICAMP, Limeira, SP, Brazil

Sócrates O. Dantas Department of Physics, Institute of Exact Sciences, Federal University of Juiz de Fora, Juiz de Fora, MG, Brazil

M.M. De Souza Department of Electronic and Electrical Engineering, University of Sheffield, Sheffield, United Kingdom

Olasunbo Farinre Laser Spectroscopy Laboratory, Department of Physics & Astronomy, Howard University, Washington, DC, United States

E.J. Garboczi Applied Chemicals and Materials Division, National Institute of Standards and Technology (NIST), Boulder, CO, United States

Raul Garcia-Sanchez Laser Spectroscopy Laboratory, Department of Physics & Astronomy, Howard University, Washington, DC, United States

M.K. Harbola Indian Institute of Technology Kanpur (IITK), Kanpur, India

Ming Hu Department of Mechanical Engineering, University of South Carolina, Columbia, SC, United States

Y.H. Ikuhara Japan Fine Ceramics Center, Nagoya, Japan

Z. Ji Department of Electrical Engineering, Wright State University, Dayton, OH, United States

Sri Ranga Jai Likith Colorado School of Mines, Golden, CO, United States

Miao-Ling Lin State Key Laboratory of Superlattices and Microstructures, Institute of Semiconductors, Chinese Academy of Sciences, Beijing, China

L. Lindsay Materials Science and Technology Division, Oak Ridge National Laboratory, Oak Ridge, TN, United States

Aolin Lu Department of Chemistry, State University of New York at Binghamton, Binghamton, NY, United States

Elisabeth Mansfield Applied Chemicals and Materials Division, National Institute of Standards and Technology, Boulder, CO, United States

Prabhakar Misra Laser Spectroscopy Laboratory, Department of Physics & Astronomy, Howard University, Washington, DC, United States

A. Mookerjee S.N. Bose National Centre for Basic Sciences, Kolkata, India

J. Myers Department of Electrical Engineering, Wright State University, Dayton, OH, United States

Antardipan Pal Department of Electrical and Computer Engineering, The University of North Carolina at Charlotte, Charlotte, NC, United States

Ravindra Pandey Department of Physics, Michigan Technological University, Houghton, MI, United States

T. Pandey Department of Physics, University of Antwerp, Antwerp, Belgium

Lia Phillips Department of Physics & Astronomy, Appalachian State University, Boone, NC, United States

L. Rast Applied Chemicals and Materials Division, National Institute of Standards and Technology (NIST), Boulder, CO, United States

Richard Robinson Department of Chemistry, State University of New York at Binghamton, Binghamton, NY, United States

Nabanita Saikia School of Science, Navajo Technical University, Chinle Site, Chinle, AZ, United States

Shiyao Shan Department of Chemistry, State University of New York at Binghamton, Binghamton, NY, United States

Guojun Shang Department of Chemistry, State University of New York at Binghamton, Binghamton, NY, United States

P. Singh Ames Laboratory, US Department of Energy, Iowa State University, Ames, IA, United States

Ping-Heng Tan State Key Laboratory of Superlattices and Microstructures, Institute of Semiconductors; Center of Materials Science and Optoelectronics Engineering and CAS Center of Excellence in Topological Quantum Computation, University of Chinese Academy of Sciences, Beijing, China

Vinod K. Tewary Applied Chemicals and Materials Division, National Institute of Standards and Technology (NIST), Boulder, CO, United States

K. Vishal Department of Electrical Engineering, Wright State University, Dayton, OH, United States

K. Wang University of New Orleans, New Orleans, LA, United States

Shan Wang Department of Chemistry, State University of New York at Binghamton, Binghamton, NY, United States

Xiaoyu Wang Zhejiang Province Key Laboratory of Quantum Technology and Devices, Department of Physics, State Key Laboratory for Silicon Materials, Zhejiang University, Hangzhou, People's Republic of China

Huizhen Wu Zhejiang Province Key Laboratory of Quantum Technology and Devices, Department of Physics, State Key Laboratory for Silicon Materials, Zhejiang University, Hangzhou, People's Republic of China

Dennis D. Yau Sinai Green Lab, Cupertino, CA, United States

Shuguo Yu School of Physical Science and Technology, Xinjiang University, Urumqi, People's Republic of China

Naili Yue Department of Electrical and Computer Engineering, The University of North Carolina at Charlotte, Charlotte, NC, United States

Hongyan Zhang School of Physical Science and Technology, Xinjiang University, Urumqi, People's Republic of China

Yong Zhang Department of Electrical and Computer Engineering, The University of North Carolina at Charlotte, Charlotte, NC, United States

Zhi Zheng University of New Orleans, New Orleans, LA, United States

Chuan-Jian Zhong Department of Chemistry, State University of New York at Binghamton, Binghamton, NY, United States

Weilie Zhou University of New Orleans, New Orleans, LA, United States

Y. Zhuang Department of Electrical Engineering, Wright State University, Dayton, OH, United States

Light-effect transistors and their applications in electronic-photonic integrated circuits

Antardipan Pal[a], Yong Zhang[a], and Dennis D. Yau[b]
[a]Department of Electrical and Computer Engineering, The University of North Carolina at Charlotte, Charlotte, NC, United States, [b]Sinai Green Lab, Cupertino, CA, United States

1.1 Introduction

As demand for consumer electronic goods increase, there is an extremely high need for a high speed, low power consumption Static Random-Access Memory (SRAM) for high performance cache memories. With more and more functionality packed within a single chip (integrated circuit) to meet market demands, there is a huge constraint on the battery life of consumer electronic products (mainly portable devices like mobile phones, laptops, smart watches, etc.). Memories are the most power-hungry blocks and consume most of the area on a chip due to their high capacity. So, the performance and energy efficiency of SRAM is of paramount importance in both high performance and ultralow-power portable, battery operated electronic systems.

Integrating electronic and photonic systems on the same chip can potentially transform computing architectures and enable more powerful computers. Although photoconductive devices can potentially offer advantages in switching speed [1] and switching energy [2], one major drawback of using such devices, known as light effect transistors (LETs) [2] is the inconvenience of using the output of one LET based logic gate to directly drive the next similar logic gate without going through relatively inefficient electrical to optical energy conversion. To explore the advantages of LETs but avoid the cascading issue in computing applications [3], we seek to replace some field effect transistors (FETs) that only serve the roles of switching a circuit on and off, such as, the access transistors in a SRAM cell.

1.1.1 Various techniques and architectures to improve performance of SRAMs

SRAMs are directly interfaced with the CPU at high speeds which are not possible to attain by other memory devices and hence their performance is of paramount importance. Moreover, SRAM is a major contributor in the total energy consumption of a processor due to its high performance for even more increased computing power. One of the most crucial concerns in many ultralow-power applications is energy efficiency.

Modeling, Characterization and Production of Nanomaterials. https://doi.org/10.1016/B978-0-12-819905-3.00001-4

SRAM being one of the most critical building blocks in almost all digital systems, its packing density, speed, power consumption are all crucial performance metrics [4,5]. For a processor, on-chip caches typically consume 25%–45% of the total area and energy [6]. Moreover, in modern high performance large capacity memory arrays (mostly DRAMs) of the order of gigabytes almost 40% of the total energy is consumed due to leakage currents [7,8]. Also, leakage is the only source of static energy consumption in an idle circuit and in this regard the SRAM array is an important source of leakage, since the majority of transistors are in the idle condition when a SRAM cell is in a hold state (that is, when it just stores data). To mitigate this problem the design of a low leakage SRAM cell is desirable, which in turn can be achieved by using high threshold devices and low supply voltage. Scaling down the supply voltage also reduces the voltage swing per switching activity per cell (in both read or write operation), and thus the dynamic power consumption [4,8]. However, lowering supply voltage may result in various design issues like degradation in cell stability, noise margin, reduced on-current-to-off-current ratio, and strong sensitivity to process, voltage, temperature variations [9–14].

Aggressive scaling of complementary metal oxide semiconductor (CMOS) memory cells into nano-meter regime, especially less than 32 nm is facing many challenges and issues like degradation of the noise margin, exponential increase of the leakage current, short channel effects (SCEs) [11,12], and so on. Also, excessive scaling of device dimensions along with ultrathin gate oxide has resulted in appreciable amount of gate oxide tunneling current [9,12]. Moreover, the data storage capacity of the read and write operation is also affected (degraded noise margins and hence the stability), and the parameter fluctuation effects like line-edge roughness, random dopant fluctuation, and gate-oxide-thickness fluctuation also reduce the stability of SRAM cells [9,12]. Process variation and leakage current also limit the conventional bulk CMOS technology to go beyond 32 nm [11]. Though the sub-threshold design for low-power application is being carried out in recent years, successful operation of such memory circuits is very challenging since the performance of SRAMs degrades as very low voltages are used [9,12].

To improve the performance of SRAM cells in terms of leakage current reduction, speed enhancement and to mitigate various SCEs in highly scaled bulk MOSFETs, double gate (DG) MOSFET, tunnel FETs (TFET) and FinFETs are lucrative choices, since all of them have better gate control on the channel electrostatics, which reduces the SCEs [15–17]. As the SCEs have become alarming in highly scaled planar FETs, the most feasible device that is compatible with the standard CMOS technology and can counter the SCEs is FinFET. Due to a significant reduction in leakage currents, high carrier speed, and higher I_{on}/I_{off} ratio as compared to the planar MOSFETs, FinFETs are felicitous for design of high-performance SRAMs. Various device and circuit techniques have been implemented to enhance the performance of FinFET-based SRAM cells/arrays [15,18,19]. Different circuit techniques, such as using additional devices to improve SRAM cell stability and the usage of back gate feature of independent gate FinFETs in the access paths of the SRAM cell to reduce environmental variation and leakage, are being used to boost the overall performance of the SRAM cell [15,20,21]. Also, a tri-gated 6T SRAM cell using FinFETs, with improved write ability, read noise margin, leakage, and overall power dissipation has been reported [22]. A detailed description of the device size to obtain optimum noise margin along with write and read speed, and analysis of SRAM cell stability with respect to process temperature variations and various speed-enhancing techniques have been offered [23]. Techniques including usage of reduced gate voltage (word line [WL] voltage) for the access transistors, using negative ground under the idle condition, employing dual threshold voltage devices, and transistor stacking are being reported to minimize the leakage currents and increase speed [24]. The ways to improving read and write access time as well as leakage current have been proposed [25]. Forward body biasing technique and self-controllable-voltage-level switch techniques are used to better the performance of the SRAM cell in terms of speed improvement in the active mode, whereas a reversed body biasing technique is used to reduce the leakage in the standby mode [25].

Moreover, it has been well established that 6T SRAMs made from nanowire (NW)-based structures (e.g., gate-all-around, band-to-band tunneling FETs) have much better noise margin, lesser variability and much better leakage suppression due to lesser SCEs as compared to FinFET based structures [26–28]. It has also been reported that scaling down to more advanced technology nodes (far below 22 nm), NW based device structures offer much better SCEs, drain induced barrier lowering (DIBL), subthreshold slopes, higher switching speeds, and less variability in device performance with scaling [29], which enhances robustness of the circuits made from them.

Above-mentioned efforts primarily focus on improving the FET performance on the single device level to bring upon the improvement for the 6T SRAM cell. However, the primary factors limiting the read and write speeds and the corresponding energy consumptions are the characteristics of the access transistors (rather than the transistors of the inverters) and the capacitances of the WLs and bit lines (BLs). The hybrid FET-LET 6T SRAM discussed in this chapter can offer major improvement in performance with better energy efficiency by replacing the two access FETs with two LETs and accordingly the WL electrical wires with optical waveguides (OWGs). This hybrid structure offers a more intimate monolithic integration of the electronics and photonics at the chip level. Additionally, this application avoids the well-known energy-data rate (EDR) challenge (EDR ≤ 10 fJ/bit for on-chip communication) [30,31], because it does not require using light to address photonic devices individually, but in a group simultaneously through an optical waveguide, for example, simultaneously illuminating all the access devices in a row of the hybrid array.

1.2 Basic working mechanisms of the 6T SRAM and the role of the access transistors

In this section, the roles of access transistors in determining the overall speed and total energy consumption of a conventional 6T SRAM cell/array with 6 FETs are analyzed to highlight the needs and directions for improvement.

The three main operations of a 6T SRAM cell are writing data into the cell, reading data from the cell, and holding or storing data. As shown schematically in Fig. 1.1, a conventional 6T SRAM has two cross coupled latches formed of FETs M1-M4, along with two access devices M5 and M6 that allow to access the cross-coupled latched for read and write operations. The two access transistors along with the WL and bit lines (BL/$\overline{\text{BL}}$) play an extremely crucial role in determining the overall speed, power dissipation and stability of the cell [32–34]. The three p-FETs, encircled in red in Fig. 1.1, are the BL conditioning devices whose roles are to pre-charge and equalize the BL voltages before each read and write operation, and they also play a crucial role in the read and write processes [35].

The dynamic energy consumptions of the 6T SRAM that occur during either write or read process.is analyzed assuming that the SRAM cell is initially at logic 0 (i.e., $Q=0$, $\overline{Q}=1$), as shown in Fig. 1.2A and B. Before the 6T cell read operation the BL capacitances C_{BL} and $\overline{C}_{\text{BL}}$ are pre-charged to V_{DD} via the pre-charge circuitry; then

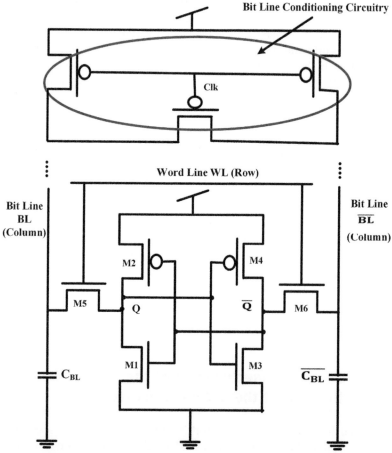

Fig. 1.1 A 6T SRAM cell.

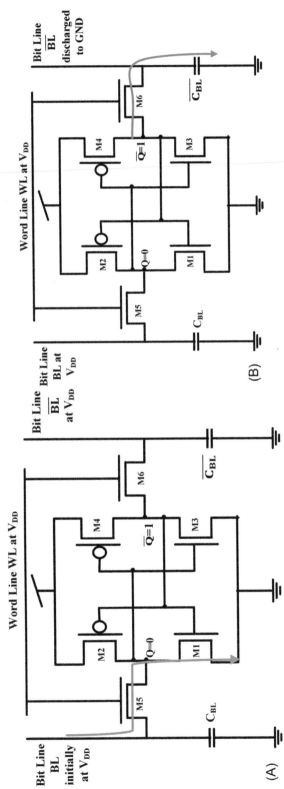

Fig. 1.2 A 6T SRAM cell showing the read operation (A) and write operation (B).

after the WL turns on the access devices M5 and M6, the bit line BL is partially discharged through the path M5 and M1 as shown in Fig. 1.2A, while bit line $\overline{\text{BL}}$ remains at V_{DD}, which allows the sense circuitry to read the state based on the voltage difference between bit lines BL and $\overline{\text{BL}}$. Similarly for write operation as shown in Fig. 1.2B, the one of the already pre-charged bit lines, $\overline{\text{BL}}$ is fully discharged via the write circuitry (not shown in figure), then node $\overline{Q}$ which was initially at logic 1 is discharged through the access device M6 such that M1 and M4 turn off while M2 and M3 turn on, and a cell flipping occurs such that $Q = 1, \overline{Q} = 0$ and hence a write-1 operation takes place. As can be seen from Fig. 1.2A and 1.2B, the bit line and the WL capacitances mainly depend on the drain/source capacitance and gate capacitance, respectively, of the access transistors, along with bit line and WL wire capacitances. Hence, as evident the access devices play a crucial role in determining the overall SRAM cell performance (read write delay and energy consumption) and in this chapter the effects of replacing the FET access devices with LETs (with very low carrier transit delay, intrinsic RC delay, and switching energy) in a 6T array are studied and analyzed elaborately.

1.3   Analytical analysis of critical capacitances, delay, and energy consumption in 6T SRAM

A regular 6T SRAM array (without considering the various peripherals and assist circuits) is shown in Fig. 1.3 and from the figure it can be seen that the WL capacitance (C_{WL}) consists of the gate capacitances (C_{gate}) of the all the access FETs in a row along with the horizontal wire capacitance and is calculated as following [32]:

$$C_{\text{WL}} = n_{\text{C}} \left(2C_{\text{gate}} + C_{\text{w\_horizontal}} \right) \tag{1.1}$$

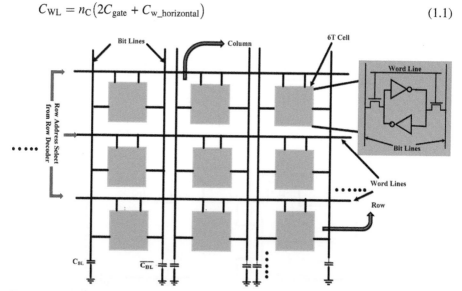

Fig. 1.3 A 6T SRAM array.

where n_C is the number of columns in the array = number of 6T cells/row, and the horizontal wire capacitance is calculated as [33]:

$$C_{w\_horizontal} = 5P_{metal}C_w \tag{1.2}$$

where P_{metal} and C_w are the technology dependent metal pitch and wire capacitance per unit length, respectively.

The bit line capacitance (C_{BL}) consists of the drain/source capacitance (C_{drain}) of the access FETs along with the vertical wire capacitance and is calculated as following:

$$C_{BL} = n_R(C_{drain} + C_{w\_vertical}) \tag{1.3}$$

where n_R is the number of rows in the array = number of 6T cells/column, and the vertical wire capacitance is calculated as [33]:

$$C_{w\_vertical} = 0.4C_{w\_horizontal} \tag{1.4}$$

The capacitance at the output node (node Q in Fig. 1.1) is the summation of output capacitance of the (M1-M2) inverter, gate capacitance of the (M3-M4) inverter and the drain capacitance of the access device M5, and is calculated as follows [34,36]:

$$C_{out} = C_Q = C_{drain,M1} + C_{drain,M2} + C_{gate,M3} + C_{gate,M4} + C_{drain,M5} \tag{1.5}$$

As evident from Eqs. (1.1), (1.3), (1.5), the critical capacitances heavily depend on the capacitances of the access transistors of the 6T SRAM.

1.3.1 Analytical expressions for read, write delay, and energy

The read delay and energy for the 6T array are calculated as follows [33]:

$$T_{read} = \frac{C_{WL}V_{DD}}{I_{WL,drive}} + \frac{C_{BL}\Delta V_{read}}{I_{read}} \tag{1.6}$$

$$E_{read} = C_{WL}V_{DD}^2 + C_{BL}V_{DD}\Delta V_{read} \tag{1.7}$$

where V_{DD} and ΔV_{read} are the supply voltage and the change of the bit line voltage at the end of read operation, respectively. $I_{WL,drive}$ and I_{read} are the WL driver current and 6T cell read current, respectively.

The write delay and energy are calculated as follows [33]:

$$T_{write} = \max\left(\frac{C_{WL}V_{WL}}{I_{WL}}, \frac{C_{BL}V_{DD}}{I_{write,ckt}}\right) + \frac{C_{out}\Delta V_{out}}{I_{write}} \tag{1.8}$$

$$E_{write} = C_{WL}V_{DD}V_{WL} + C_{BL}V_{DD}^2 + C_{out}V_{DD}\Delta V_{out} \tag{1.9}$$

where $V_{WL} = V_{DD}$ is the WL voltage when it is on, ΔV_{out} is the change of output voltage during the cell flipping operation, I_{WL}, $I_{write,ckt}$, and I_{write} are the WL current when on, write assist circuitry current, and 6T cell write current, respectively.

In the analytical relations for delay and energy, the carrier transit delay, and switching energy related to the transit delay of carriers through the FET channel is ignored as they are negligible compared to the RC delays and energies. The above relations clearly show that the read, write delay, and energy of the 6T array structure depend on the critical capacitances related to the access transistors, although the wire capacitances of the word and bit lines also play a role. Therefore, it is highly desirable to have the access FETs replaced by some gateless devices, for instance, LETs (described in Section 1.4) with a different working mechanism as compared to gated FETs.

1.3.2   Energy associated with leakage in a 6T SRAM

On the single component level, the off currents in a FET predominantly depend on the supply voltage, the threshold voltage, channel length, channel doping profile, drain and source junction depth and gate oxide thickness [37–39]. Various types of leakage currents in a conventional n-FET are shown in Fig. 1.4A [38] which includes: sub-threshold leakage current (I_{sub}), gate-induced drain and source leakage current (I_{GIDL}, I_{GISL}), punch-through leakage current ($I_{punchthrough}$), gate tunneling leakage current through the bulk (I_{GB}), source (I_{GS}) and drain(I_{GD}) summed up as, $I_{gate} = I_{GB} + I_{GS} + I_{GD}$, and p-n junction leakage currents ($I_{junction} = I_{drain,junc} + I_{source,junc}$) at the drain-substrate junction ($I_{drain,junc}$), and the source-substrate junction ($I_{source,junc}$) [38,39]. Thus, the total leakage current of an individual FET in the 6T cell may be empirically modeled as the sum of gate, subthreshold and junction leakage currents.

The leakage in a 6T cell depends on the logic state of the cell, the logic level of the WL, and the type of operation performed. In Fig. 1.4B, the 6T cell shown is in the state just before the read or write operation (reading 0 or writing 1) [37]. Both the bit lines BL and $\overline{BL}$ are precharged to V_{DD}, but the WL is still at 0 and M5, M6 are off, and hence subthreshold leakage may occur in the access devices. The arrows in Fig. 1.4B show the various leakage currents in each FET depending on their operating conditions (drain, source, gate, substrate voltages, and conduction state). For instance, M5 is off (gate is at 0), drain is at 1 and source and substrate are at 0, and hence there will be a component of the gate leakage from drain to gate (I_{GD}), sub-threshold leakage (I_{sub}) from drain to source, and a component of junction leakage ($I_{drain,junc}$) from the drain to the substrate as shown in Fig. 1.4B. Similarly, for M6, the gate is at 0, drain and source are at 1 and substrate is at 0, and hence there will be two components of the gate leakage ($I_{GD} + I_{GS}$) from the drain and source to the gate, and junction leakages ($I_{junction} = I_{drain,junc} + I_{source,junc}$) from the drain and source to the substrate as shown in Fig. 1.4B. There will be no I_{sub} between the drain and source since both are at logic 1. If these leakages are severe, it may lead to a false read or write operation and affect the reliability of the 6T cell. By analyzing Fig. 1.4B, it can be qualitatively estimated that about 40% of the total leakage in a 6T cell is in the access paths.

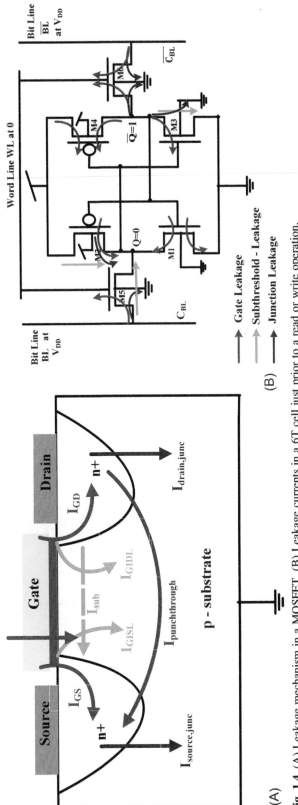

Fig. 1.4 (A) Leakage mechanism in a MOSFET. (B) Leakage currents in a 6T cell just prior to a read or write operation.

1.4   LET overview and device characteristics

A LET as shown in Fig. 1.5 is basically a nanowire-based metal semiconductor metal photodetector where the nanowire is embedded on an insulation substrate with two metal contacts at the two ends forming the source and drain [2]. The LET can emulate the current-voltage characteristics of a FET when gated by a light beam as opposed to a voltage in the FET. The LET also has a specific type of transfer characteristic, current-gate power, resembling that of the FET, current-gate voltage, when compared to a generic photoconductive device. It has various other functionalities like optical logic operation, amplifications [2,3], in this chapter, it has been used primarily as an optoelectronic switch.

A LET works on the very well-known photoconductive mechanism [40] where the source-drain conductivity is modulated by light shined on the semiconductor nanowire (SNW) which generates free carriers in the SNW due to optical absorption, and in turn modifies the electrical conductivity of the channel. The predominant advantage of an LET over a traditional FET is due to the removal of physical gate, which not only minimizes the complex gate fabrication process, but also the random dopant fluctuations in FETs [41]. Thus, due to the simple, gateless structure, the LET can be scaled down to quantum regime without the problem of SCEs which are common in most nanoscale FETs [42]. Also, it is interesting to note that an LET, when operated under multiple wavelength illuminations, can also provide many additional

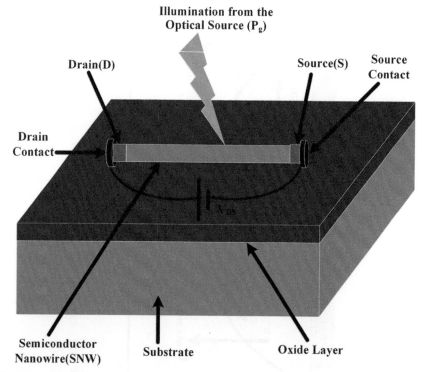

Fig. 1.5 A light-effect transistor (LET).

functionalities (e.g., optical logic), which are not achievable by FETs or photodetectors [3]. Since there is no physical gate, it can also be assumed that the LET speed is only limited by the carrier transit time or lifetime, whichever is smaller rather than gate related RC-delays. The electron saturation velocities of various semiconductors (e.g., Si, SiC, InAs, InP, GaAs, CdSe) at room temperature, are in the range of (1–10) $\times 10^7$ cm/s when the electric field is on the order of 100 kV/cm [43,44] which implies a carrier transit time of the order of (1–0.1) ps for a 100 nm long NW, which is also the ballistic transport regime where the saturation velocity can be achieved. In the non-ballistic transport regime for longer nanowires (typically >100 nm), the electron transit time depends on the electrical field, for example for Si at $E = 10$ kV/cm, the electron velocity is around 7×10^6 cm/s [43,44] and the carrier transit time (t_{LET}) can be estimated to be 4.3 ps and 7.1 ps, for a 300 nm and 500 nm long Si NW, respectively. The ultra-fast switching of 1 ps or lesser in a ballistic LET translates to ultra-small switching energy. For instance, the electrical switching energy $E_{el} (= I_{sd} \times V_{sd} \times t_{LET})$ will be of the order of 1 aJ/switch, assuming a carrier transit time, $t_{LET} = 1$ ps (for a ballistic device), an on-current of $I_{sd} = 1$ μA under $V_{sd} = 1$ V. Since the LET is an optoelectronic device, optical gating power also contributes to the total switching energy, and it can be estimated by $P_g = E_{ph} I_{sd}/(eG)$, where E_{ph} is the photon energy and G is the photo-conductive gain [2]. Assuming $E_{ph} = 2.5$ eV, $G = 10^3$, to have $I_{sd} = 1$ μA, and $t_{LET} = 1$ ps, we get optical switching energy, $E_{op} = P_g \times t_{LET} = 2.5 \times 10^{-3}$ aJ/switch $\ll E_{el}$, which leaves sufficient room allowing for below 100% light power delivery efficiency. Assuming an idealistic ballistic device with no voltage loss at the contacts and a quantum impedance of 12.9 kΩ [45], transit time of 0.1 ps, S-D current of 1 μA, the electrical switching energy can be as low as 1.3×10^{-21} J/switch at a very low V_{sd} of only 13 mV. Previous studies show that for a prototype CdSe LET structure of length 5 μm and diameter 80 nm under 532 nm illumination of 110 nW (only about 6% was actually absorbed), yielded $I_{ds} = 0.35$ μA (in dark, $I_{ds} \sim 1$ pA at $V_{ds} = 1.43$ V) [2]. Even for such a large device the total switching energy is about, $E_{tot,sw} = E_{el} + E_{op} \approx 0.06$ fJ/switch considering a typical room temperature carrier lifetime in a II–VI semiconductor in the order of 100 ps. This total switching energy is still better than typical FETs having switching energy of 0.1–1 fJ/switch [46]. Also, in case of FETs, the large gate related RC delays predominate over the transit-time delays through the conducting channel, but in the LETs the carrier transit time through the NW channel is expected to be the predominant factor for determining the switching speed and energy of a discrete LET. Moreover, the I_{on}/I_{off} ratio for a LET could be as high as 10^6 [2], which is almost an order of magnitude better than that of advanced FETs. The high I_{on}/I_{off} signifies are very low dark current which resembles the subthreshold leakage current in FETs.

1.5 Hybrid 6T SRAM cell and array structure

1.5.1 Hybrid 6T SRAM with access FETs replaced by LETs

In the hybrid 6T structure, the two access transistors (M5 and M6) in the 6T cell of Fig. 1.1 are replaced by two LETs (L1 and L2) and the WL is replaced by the OWG that illuminates the nanowire in the LETs as shown in the prototype 6T cell of Fig. 1.6.

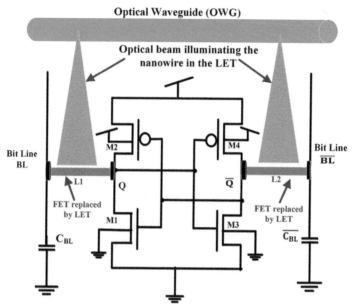

Fig. 1.6 Prototype hybrid 6T SRAM cell with LET access.

The LET has a very simple structure, with no polysilicon/metal gate embedded on a dielectric material which contributes to the oxide related part of the gate capacitance in FETs [36], also no gate-drain or gate-source overlap capacitance, which also forms a part of the total gate capacitance and drain/source capacitance in gated FETs [36]. Though a thin oxide layer maybe formed on the NW surface due to environmental oxidation, it does not contribute to any oxide related capacitance as in FETs, since there is no metallic gate electrode wrapping around the NW in a LET. Since the substrate in the LET just provides mechanical support without any electrical connection, there will also be no MS junction to substrate capacitance which is equivalent to the drain to substrate or source to substrate junction capacitance of FETs in which the substrate generally has an electrical connection. For the analytical calculations in this chapter two separate LET based on generic semiconductor nanowires are considered: one with length $(L) = 300 \, \text{nm}$, diameter $(D) = 50 \, \text{nm}$ and another with $L = 500 \, \text{nm}$ and $D = 70 \, \text{nm}$ and drive currents of $5 \, \mu\text{A}$ and $25 \, \mu\text{A}$, respectively (the plots in Fig. 1.8 are considering the smaller LET with $5 \, \mu\text{A}$ drive current).

A prototype hybrid 6T array is shown in Fig. 1.7 in which the FET access devices are replaced by LETs and electrical WLs by OWGs, keeping the core FETs and other peripherals almost unchanged. The OWGs has to be illuminated by appropriate on chip optical sources like nanoscale lasers [31] which form a part of the opto-electronic row decoding system.

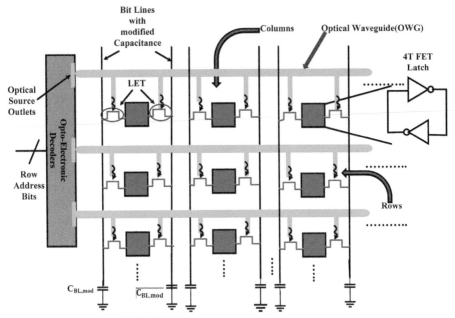

Fig. 1.7 Prototype hybrid 6T SRAM array.

1.5.2 Delay and energy analysis of the hybrid 6T array

As discussed in the previous section, the LET structure would not contribute to any MS junction capacitance and hence the bit line capacitance will consist of only the vertical wire capacitance. Therefore, Eq. (1.3) can be modified as follows:

$$C_{BL,mod} = n_R \left(C_{w\_vertical} \right) \tag{1.10}$$

The internal node capacitance given by Eq. (1.5) may be modified as follows, where the drain equivalent capacitance of M5 ($C_{drain,M5}$) would not be present due to the same reason.

$$C_{out,mod} = C_{drain,M1} + C_{drain,M2} + C_{gate,M3} + C_{gate,M4} \tag{1.11}$$

Also, the hybrid 6T structure should not have any WL capacitance (as opposed to C_{WL} described for a regular structure in Eq. (1.1)) since LETs do not have any gate capacitance and do not require a wired electrical signal to control the gates as in FETs. Instead, the light is brought in through an OWG that will contribute no practical capacitance. To incorporate the effects of the LET access devices and OWGs in the hybrid 6T structure, delay and energy Eqs. (1.6)–(1.9) are modified to Eqs. (1.12)–(1.15) as follows:

$$T_{read,mod} = t_{WG} + \max \left(\frac{C_{BL,mod} \Delta V_{read}}{I_{read,mod}}, t_{LET} \right) \tag{1.12}$$

where $C_{\text{BL, mod}}$ as shown in Eq. (1.10) is the modified bit line capacitance and $I_{\text{read,mod}}$ is the cell read current for the hybrid structure with LET access devices. The RC-word-line delay term (first term) in Eq. (1.6) is replaced by the time taken by the EM signal to propagate through the optical waveguide (t_{WG}) and illuminate the access LETs, the second term in Eq. (1.6) is modified to include the larger of the modified RC-bit line delay during read, and the LET carrier transit delay (t_{LET}) as discussed previously.

$$E_{\text{read,mod}} = \left(2n_{\text{C}}E_{\text{op}}\right) + \max\left(C_{\text{BL,mod}}V_{\text{DD}}\Delta V_{\text{read}}, E_{\text{el}}\right) \tag{1.13}$$

where the RC WL energy in case of the regular array in Eq. (1.7) will not be present because OWGs practically does not consume or dissipate any RC energy while light is being transmitted through them. The first term in Eq. (1.13) is due to the optical gating switching energy for a LET (E_{op}) described in Section 1.4 and for a whole row it is multiplied by $2n_{\text{C}}$, since there are n_{C} number of 6T cells per row and each 6T cell has two access LETs. The second term is the larger of the modified bit-line RC-energy during read and the transit time electrical switching energy (E_{el}) for the LET access device.

$$T_{\text{write,mod}} = \max\left(t_{\text{WG}}, \frac{C_{\text{BL,mod}}V_{\text{DD}}}{I_{\text{write}_{\text{ckt}}}}\right) + \max\left(\frac{C_{\text{out,mod}}\Delta V_{\text{out}}}{I_{\text{write,mod}}}, t_{\text{LET}}\right) \tag{1.14}$$

where $C_{\text{out,mod}}$ is the modified internal node capacitance and is described in Eq. (1.11), and $I_{\text{write,mod}}$, is the cell write current for the hybrid 6T cell. The RC WL delay term (first term in Eq. (1.8)) during write operation is replaced by t_{WG}, and the first term in Eq. (1.14) is the larger of the terms t_{WG} and the modified bit line delay during write, and the second term is the larger term of the modified 6T cell flipping delay during write and the LET carrier transit delay (t_{LET}).

$$E_{\text{write,mod}} = \left(2n_{\text{C}}E_{\text{op}}\right) + C_{\text{BL,mod}}V_{\text{DD}}^2 + \max\left(C_{\text{out,mod}}V_{\text{DD}}\Delta V_{\text{out}}, E_{\text{el}}\right) \tag{1.15}$$

where the first term is identical to Eq. (1.13), the second term in Eq. (1.15) is the modified bit line RC energy during write, and the last term is the larger of the hybrid 6T cell modified flipping energy during write, and the transit time electrical switching energy of the LET access device.

Using analytical relations (1.6–1.9, 1.12–1.15), a set of delays and energies for various 6T array sizes (256 bytes–512 KB) are calculated for the regular array, hybrid array, and hybrid array with ballistic LET access device. The various currents used in the analytical relations for the regular SRAM are assumed to be 25 μA [47], and $\Delta V_{\text{read}} \approx 120\,\text{mV}$ and $\Delta V_{\text{out}} \approx V_{\text{DD}}/2$ [33]. For direct comparison, the various analytical results are plotted in Fig. 1.8A–D, and also the results for the 256 KB and 512 KB arrays are summarized in Table 1.1, from where it is evident that the hybrid array is capable of giving much better performance than the regular array even at lesser energy consumption. It is noted from the various plots of Fig. 1.8, that though on a single device level the ballistic LET has much lesser carrier transit delay and switching energy as compared to the regular LET, the results for the hybrid arrays are coinciding. This is because for an array, the overall RC delay and energy will dominate over the

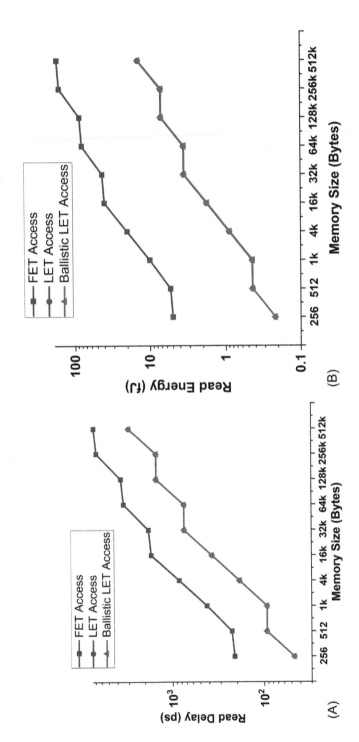

Fig. 1.8 Read and write delay and energy for various SRAM arrays with FET, LET, and ballistic LET access devices. (A) Read delay, (B) read energy, *(Continued)*

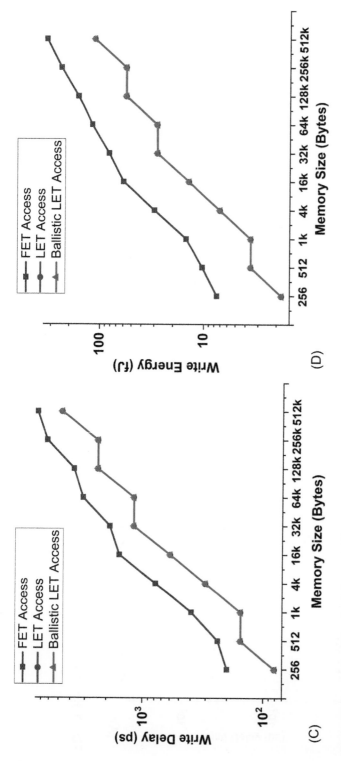

Figure 1.8, cont'd (C) write delay, and (D) write energy. The curves with LETs and ballistic LETs are indistinguishable.

Table 1.1 Comparison of various performance metrics of 256-KB and 512-KB 6T SRAM array with FET, LET, and ballistic LET access devices.

	256-KB SRAM array			512-KB SRAM array		
	FET access devices	LET access devices	Ballistic LET access devices	FET access devices	LET access devices	Ballistic LET access devices
Read delay (ps)	6768	1496.8	1495.8	7244.7	2991	2990
Write delay (ps)	6295.3	2372.7	2372.7	7552.1	4738.5	4738.5
Read energy (fJ)	160.7	7.2	7.1	172.1	14.3	14.2
Write energy (fJ)	239.1	56.32	56.2	328.8	112.5	112.4

carrier transit delay and switching energy of the individual LETs. As evident from Fig. 1.8, the read delay and energy on an average improve by factors of 3 and 16, respectively; while the write delay and write energy on an average improve by factors of 2 and 3, respectively. The figure of merit (FOM) of the SRAM array can be found from the energy-delay product (EDP) and is plotted in Fig. 1.9. Considering 50% probability of the array being accessed in a cycle, and 50% probability for each of the read and write operations [33], it can be roughly estimated that the hybrid SRAM array on an average exhibit almost one order of magnitude lesser EDP, as compared to the regular SRAM arrays.

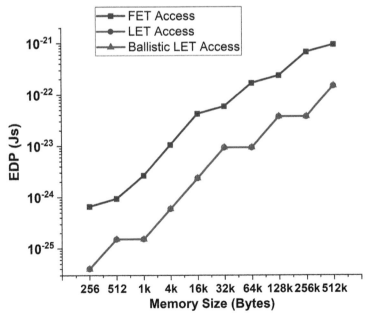

Fig. 1.9 EDP for various SRAM arrays with FET, LET, and ballistic LET access devices.

Considering a larger drive current (25 µA, 5 times the previous case) and larger device dimensions ($L = 500$ nm, $D = 70$ nm) for the access LETs in the hybrid structure, it has been found that the improvement in read delay is drastically increased (almost by a factor of 5) as compared to the case of using lesser drive current for LETs. The average improvement in case of write delay is almost the same since for large arrays it is predominantly dependent on the current from the write circuitry; while there is a very slight increase in the average read and write energies due to the increase of the transit time electrical switching energy (E_{el}) and optical gating energy (E_{op}) per LET device, and consequently the average EDP also increase slightly with larger LET drive current and LET dimensions.

By analyzing the read delay equations for the regular (Eq. 1.6) and hybrid (Eq. 1.12) arrays, it can be stated that the improvement is primarily due to the replacement of the large RC-WL delay in the regular structure by the much smaller optical waveguide delay (t_{WG}) in the hybrid structure. Also, the highest reduction achieved in the read energy is mainly due to the replacement of the large RC WL energy consumption in the regular array (Eq. 1.7) by a much smaller optical gating energy term (E_{op}) in the hybrid array (Eq. 1.13). Also as seen from plots of Fig. 1.8A and C, the improvement in write delay is lesser compared to read delay because the first max term in Eq. (1.14) shadows the effect of replacing the RC WL delay with t_{WG}. Similarly, as evident from plots of Fig. 1.8B and D, the improvement in the write energy for the hybrid structure is lesser than read energy, due to the fact that $V_{DD} > \Delta V_{out} > \Delta V_{read}$, and thus $C_{BL,mod} V_{DD}^2$, $> \max (C_{BL,mod} V_{DD} \Delta V_{read}, E_{el})$ in Eqs. (1.15), (1.13) respectively.

1.5.3 Analyzing the effects in the peripherals of the 6T array

In this section, the effects of the hybrid SRAM structure on WL driver circuitry, which is a chain of buffers of increasing size as shown in Fig. 1.10 of the 6T array, are analyzed. In the hybrid structure, since the access LETs have a different turn on

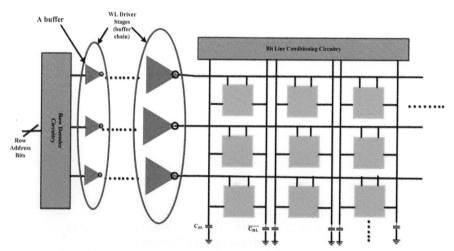

Fig. 1.10 Conventional 6T SRAM array with word line driver buffers.

mechanism, electrical WLs are replaced by OWGs and hence electrical WL drivers are not needed, which not only minimize the extra circuitry, but also reduces the WL capacitance and RC-energy consumption to almost negligible. For the regular structure, the WL driver capacitance $C_{WL,driver}$, delay $T_{WL,driver}$, and energy $E_{WL,driver}$ can be modeled as follows [33]:

$$C_{WL,driver} = n_C \left(2C_{gate} + C_{w\_horizontal}\right) + b_m \left(C_{drain,n} + C_{drain,p}\right) \tag{1.16}$$

$$T_{WL,driver} = \frac{C_{WL,driver}V_{DD}}{I_{avg,WL}} \tag{1.17}$$

$$E_{WL,driver} = C_{WL,driver}V_{DD}^2 \tag{1.18}$$

where $C_{drain,n}$ and $C_{drain,p}$ are the drain capacitances of the n and p FETs, respectively, of each buffer stage and b_m is the buffer multiplier factor and depends on the maximum size of the buffer stage used in the WL driver, $I_{avg,WL}$ is the average current of the WL driver circuitry and it is proportional to b_m. For the hybrid array, the equivalent of the WL delay and energy will be the light propagation delay (t_{WG}) through the OWG, and total optical gating energy consumption, $E_{op,tot}$, for all the LETs illuminated by the OWG and can be modeled as:

$$t_{WG} = \frac{n_C W_{6Tcell}}{v_{OWG}} \tag{1.19}$$

$$E_{op,tot} = 2n_C E_{op} \tag{1.20}$$

where W_{6Tcell} is the total width of a single hybrid 6T cell and depends on the cell layout, v_{OWG} is the group velocity of light through the OWG, and n_C and E_{op} have already been described in previous sections.

Using Eqs. (1.16)–(1.20) and considering typical values for b_m, W_{6Tcell} and v_{OWG} it can be roughly estimated that for the regular array with WL drivers, the WL delay ranges between few tens of ps to hundreds of ps over various array sizes (256 bytes–512 KB), while for the hybrid array the equivalent delay approximately ranges between one-tenth of a ps to few ps, which gives on an average almost more than two orders of magnitude improvement in the equivalent WL delay in case of the hybrid array. Similarly, the WL energy consumption with WL drivers for the regular array ranges between few fJ to tens of fJ, while for the hybrid array the equivalent optical energy consumption is in between one-hundredths of fJ to one-tenth of fJ, which gives on an average almost three orders of magnitude improvement in the equivalent WL energy consumption in the hybrid structure.

Also, it may be possible to replace the three p-FETs of the bit line conditioning circuitry as shown in Fig. 1.1 by LETs, which will reduce the bit line capacitances and hence bit line-related delays and energy consumptions, especially for large hybrid 6T arrays.

1.5.4   Possible improvements on leakage in the hybrid 6T structure

The hybrid 6T structure showing the various leakage components is shown in Fig. 1.11, where the access LETs L1 and L2 are initially off such that the 6T cell is in the hold condition. It can be seen there are almost no leakages in the access paths except the subthreshold leakage in L1. LETs have a different turn on mechanism and no SCEs as discussed previously and have minimal subthreshold leakage (e.g., an off current as low as 1 pA at 1.43 V V_{DS} and off power of 1.5 pW even for a large LET device), and hence hybrid 6T cells and arrays will have almost negligible subthreshold leakage in the access paths. Gate tunneling leakage current is one of the most critical sources of leakage in FETs as device dimensions are scaled down and ultra-thin gate oxides are used [37–39]. Gate leakages (independent of the conduction state of the device) in FETs in the access paths, as shown in Fig. 1.4B increase the total leakage energy consumption of a 6T cell and more severely in an array having a large number of 6T cells. In the hybrid structure, there will be neither any gate related nor any SCE induced leakage [37–39], in the access paths as shown in Fig. 1.11 and so the leakage power consumption in the hybrid cell will be much reduced and will be more beneficial for an array that have many such 6T cells. Also, junction leakage occurs across the p and n junctions to ground (formed between the source, drain, and the substrate regions) in FETs [38] in the access paths (as shown in Fig. 1.4B), which severely increases the leakage power consumption in arrays that have a large number of such p-n junctions to ground. Since, LETs do not have any such p-n junctions or paths to

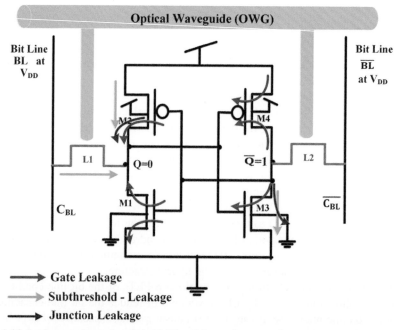

Fig. 1.11 Leakage currents in a hybrid 6T cell just prior to read or write operation.

ground (the insulating substrate does not have any electrical connection and only provides mechanical support), the hybrid 6T cell and array will have no junction leakage in the access paths as shown in Fig. 1.11. So, considering no gate and junction leakages, but still having subthreshold leakage (the dark current in the LETs) in the LETs in access paths (as shown in Fig. 1.11), there will be an overall reduction of roughly about 35% in the total leakage current components in a single hybrid 6T cell, which will be more advantageous in case of a hybrid 6T arrays.

1.5.5   Compatibility with MOSFET and scalability

In electronic-photonic integrated circuits, there are various ways for integrating an optical transport layer in a standard CMOS system, out of which possibly the most promising approach would be the use of hybrid (3D) integration of the optical layer above a complete CMOS integrated circuit, where the basic CMOS process flow would remain the same, since the optical layer can be fabricated independently [48]. Heterogeneous integration of Si electronics with electronic and photonic components/structures made from compound semiconductors and other dielectric materials has been reported extensively with CMOS compatible process flows [49–51]. For instance the necessary technologies have largely been demonstrated for different applications, like in the hybrid InGaAs/SiGe 6T SRAM [52], where the two InGaAs access transistors can be replaced with LETs, and $LiNbO_3$ photonic waveguide cavity on Si [53] which all can be readily transferred to the proposed new integration scheme.

In context of a 6T SRAM array, for a typical circuit layout, the WL spacing is generally of the order of 2 poly pitches, which is of the order of 200 nm for 22 nm technology node and 100 nm for 7 nm technology node [54,55]. OWGs fabricated on an insulating substrate can be scaled to achieve subwavelength lateral size, and can be made to have very low loss, for instance, a Si waveguide of width 400 nm for light at 1.55 μm can have loss as low as 2.8 dB/cm [56]. For the LETs operating in visible wavelengths, the OWG dimension can be significantly reduced (e.g., to around 140 nm at 532 nm wavelength), since photonic properties are scalable with wavelength. Further miniaturization is possible by using plasmonic-dielectric hybrid waveguides, though the loss in such a case may be somewhat higher [49]. Moreover, to alleviate the size mismatch issue between the light wavelength (photonic component) and the electronic components, somewhat different circuit layout strategy may be adopted in the hybrid structure, like arranging multiple LETs of the same WL in a group together such that they can be optically addressed simultaneously. This strategy not only may use the optical energy more efficiently, but also allows for more WL spacing. For cases where the space requirement is not very stringent, but SRAM performance is of paramount importance, larger waveguide spacing can easily be allowed. Since, OWGs are typically designed for interconnection in photonic circuits, in the hybrid SRAM, the optical paths are substantially shorter for on chip operation. Thus, the scalability to a few hundred or even over one thousand cells/WL may not a problem. For illuminating an OWG, the minimum light power output required from the optical decoder can be estimated to be in the order of only 5.5 μW per OWG for 1024 cells/WL, considering a propagation loss of 2.8 dB/cm, and an overestimated

OWG length of 1 mm (the actual OWG length for 1024 cells will be lesser than 1 mm, signifying lesser optical power requirement than estimated), and using the optical gating power estimate of 2.5 nW/LET given in Section 1.4. This leaves a large room for less efficient implementation or absorption of optical energy even with a large loss.

1.6   Concluding remarks

LETs offer much better performance with higher energy efficiency as compared to FETs. In gated FETs, generally RC switching delay and energy predominates over the transit delay of carriers through the channel, and hence it is much less energy efficient. In the hybrid array, the main advantage stems from the fact that the gate, source, and drain-related capacitance of the access FETs and electrical WL are no longer present, which removes the WL delay as well as energy consumption and the overall structure is much more energy efficient with better performance as compared to the regular 6T structure. In addition, due to their simple gateless structure, LETs can be expected to have much lower leakage currents than conventional FETs, and thus the hybrid 6T array will have much lesser leakage compared to regular 6T array. The use of the optical waveguide-based WL architecture in the hybrid SRAM array abolishes the need of electrical WLs and also the WL drivers, which drastically reduces the total WL capacitance, RC-delay and energy consumption to almost negligible compared to regular array structure. Furthermore, LETs may find useful applications in other peripheral and assist circuits of the SRAM array like the bit line conditioning circuit for improvement in speed and energy consumption. So, the 6T hybrid SRAM structure offers a close on chip electronic-photonic integration where both the components play active roles in the performance and efficiency of the overall photonic integrated circuit (PIC).

Acknowledgments

This work was partially supported by Bissell Distinguished Professorship Endowment Fund at UNC-Charlotte.

References

[1] S.Y. Chou, M.Y. Liu, Nanoscale tera-hertz metal-semiconductor-metal photodetectors, IEEE J. Quantum Electron. 28 (1992) 2358–2368.

[2] J.K. Marmon, S.C. Rai, K. Wang, W. Zhou, Y. Zhang, Light-effect transistor (LET) with multiple independent gating controls for optical logic gates and optical amplification, Front. Phys. 4 (2016).

[3] Y. Zhang, Light effect transistors for high speed and low energy switching and beyond, J. Phys. Conf. Ser. 1537 (2020), 012004.

[4] A. Garg, T.T. Kim, SRAM array structures for energy efficiency enhancement, IEEE Trans. Circuits Syst. Express Briefs 60 (2013) 351–355, https://doi.org/10.1109/TCSII.2013.2258247.

[5] S. Panda, N.M. Kumar, C.K. Sarkar, Power, delay and noise optimization of a SRAM cell using a different threshold voltages and high performance output noise reduction circuit,

in: 2009 4th International Conference on Computers and Devices for Communication (CODEC), 14-16 Dec. 2009, 2009, pp. 1–4.

[6] S. Gu, E.H. Sha, Q. Zhuge, Y. Chen, J.D. Hu, A time, energy, and area efficient domain wall memory-based SPM for embedded systems, IEEE Trans. Comput. Aided Des. Integr. Circuits Syst. 35 (2016) 2008–2017, https://doi.org/10.1109/TCAD.2016.2547903.

[7] Z. Liu, V. Kursun, Characterization of a novel nine-transistor SRAM cell, IEEE Trans. Very Large Scale Integr. (VLSI) Syst. 16 (2008) 488–492.

[8] Saurabh, P. Srivastava, Low power 6T-SRAM, in: 2012 International Conference on Emerging Electronics, 15-17 Dec. 2012, 2012, pp. 1–4.

[9] N. Kaur, N. Gupta, H. Pahuja, B. Singh, S. Panday, Low power FinFET based 10T SRAM cell, in: 2016 Second International Innovative Applications of Computational Intelligence on Power, Energy and Controls with their Impact on Humanity (CIPECH), 18-19 Nov. 2016, 2016, pp. 227–233.

[10] A. Pushkarna, R. Sajna, M. Hamid, Comparison of performance parameters of SRAM designs in 16nm CMOS and CNTFET technologies, in: 23rd IEEE International SOC Conference, 2010, pp. 339–342.

[11] M. Rahaman, R. Mahapatra, Design of a 32nm independent gate FinFET based SRAM cell with improved noise margin for low power application, in: 2014 International Conference on Electronics and Communication Systems (ICECS), 13-14 Feb. 2014, 2014, pp. 1–5.

[12] S. Saxena, R. Mehra, Low-power and high-speed 13T SRAM cell using FinFETs, IET Circuits Devices Syst. 11 (2017) 250–255.

[13] A. Sil, S. Ghosh, G. Neeharika, M. Bayoumi, A novel high write speed, low power, read-SNM-free 6T SRAM cell, in: 2008 51st Midwest Symposium on Circuits and Systems, 10-13 Aug. 2008, 2008, pp. 771–774.

[14] A.S. Yadav, S. Nakhate, Low power SRAM cell with reduced write PDP and enhanced noise margin, in: 2016 IEEE 1st International Conference on Power Electronics, Intelligent Control and Energy Systems (ICPEICES), 4-6 July 2016, 2016, pp. 1–4.

[15] K. Ma, H. Liu, Y. Xiao, Y. Zheng, X. Li, S.K. Gupta, Y. Xie, V. Narayanan, Independently-controlled-gate finFET 6T SRAM cell design for leakage current reduction and enhanced read access speed, in: 2014 IEEE Computer Society Annual Symposium On VLSI, 9-11 July 2014, 2014, pp. 296–301.

[16] S. Strangio, P. Palestri, D. Esseni, L. Selmi, F. Crupi, Analysis of TFET based 6T SRAM cells implemented with state of the art silicon nanowires, in: 2014 44th European Solid State Device Research Conference (ESSDERC), 22-26 Sept. 2014, 2014, pp. 282–285.

[17] X. Zhang, D. Connelly, H. Takeuchi, M. Hytha, R.J. Mears, T.K. Liu, Comparison of SOI versus bulk FinFET technologies for 6T-SRAM voltage scaling at the 7-/8-nm node, IEEE Trans. Electron Devices 64 (2017) 329–332.

[18] M. Yabuuchi, M. Morimoto, Y. Tsukamoto, S. Tanaka, K. Tanaka, M. Tanaka, K. Nii, 16 nm FinFET High-k/Metal-gate 256-kbit 6T SRAM macros with wordline overdriven assist, in: 2014 IEEE International Electron Devices Meeting, 15-17 Dec. 2014, 2014, pp. 3.3.1–3.3.3.

[19] X. Zhang, D. Connelly, P. Zheng, H. Takeuchi, M. Hytha, R.J. Mears, T.K. Liu, Analysis of 7/8-nm bulk-Si FinFET technologies for 6T-SRAM scaling, IEEE Trans. Electron Devices 63 (2016) 1502–1507.

[20] M. Chen, C. Lin, Y. Hou, Y. Chen, C. Lin, F. Hsueh, H. Liu, C. Liu, B. Wang, H. Chen, C. Chen, S. Chen, C. Wu, T. Lai, M. Lee, B. Wu, C. Wu, I. Yang, Y. Hsieh, C. Ho, T. Wang, A.B. Sachid, C. Hu, F. Yang, A 10 nm Si-based bulk FinFETs 6T SRAM with multiple fin heights technology for 25% better static noise margin, in: 2013 Symposium on VLSI Technology, 11-13 June 2013, 2013, pp. T218–T219.

[21] Z. Jakšić, R. Canal, Enhancing 6T SRAM cell stability by back gate biasing techniques for10nm SOI FinFETs under process and environmental variations, in: Proceedings of the 19th International Conference Mixed Design of Integrated Circuits and Systems—MIXDES 2012, 24-26 May 2012, 2012, pp. 103–108.

[22] R.M. Premavathi, Q. Tong, K. Choi, Y. Lee, A low power, high speed FinFET based 6T SRAM cell with enhanced write ability and read stability, in: 2016 International SoC Design Conference (ISOCC), Oct. 2016, pp. 311–312, https://doi.org/10.1109/ISOCC.2016.7799802.

[23] R. Sinha, P. Samanta, Analysis of stability and different speed boosting assist techniques towards the design and optimization of high speed SRAM cell, in: 2015 19th International Symposium on VLSI Design and Test, 26-29 June 2015, 2015, pp. 1–6.

[24] S. Ranganath, M.S. Bhat, A.C. Fernandes, Design of low leakage SRAM bit-cell and array, in: International Conference on Circuits, Communication, Control and Computing, 21-22 Nov. 2014, 2014, pp. 5–8.

[25] P.K. Sahu, Sunny, Y. Kumar, V.N. Mishra, Design and simulation of low leakage SRAM cell, in: 2016 3rd International Conference on Devices, Circuits and Systems (ICDCS), 3-5 March 2016, 2016, pp. 73–77.

[26] N.N. Mojumder, K. Roy, Band-to-band tunneling ballistic nanowire FET: circuit-compatible device modeling and design of ultra-low-power digital circuits and memories, IEEE Trans. Electron Devices 56 (2009) 2193–2201.

[27] O. Prakash, S. Maheshwaram, M. Sharma, A. Bulusu, S.K. Manhas, Performance and variability analysis of SiNW 6T-SRAM cell using compact model with parasitics, IEEE Trans. Nanotechnol. 16 (2017) 965–973.

[28] L. Wang, A. Shafaei, M. Pedram, Gate-all-around FET based 6T SRAM design using a device-circuit co-optimization framework, in: 2017 IEEE 60th International Midwest Symposium on Circuits and Systems (MWSCAS), 6-9 Aug. 2017, 2017, pp. 1113–1116.

[29] D. Nagy, G. Indalecio, A.J. García-Loureiro, M.A. Elmessary, K. Kalna, N. Seoane, FinFET versus gate-all-around nanowire FET: performance, scaling, and variability, IEEE J. Electron Devices Soc. 6 (2018) 332–340.

[30] D.A.B. Miller, Optical interconnects to electronic chips, Appl. Optics 49 (2010) F59–F70.

[31] C.-Z. Ning, Semiconductor nanolasers and the size-energy-efficiency challenge: a review, Adv. Photonics 1 (2019), 014002.

[32] H.J. David Hodges, R. Saleh, Analysis and Design of Digital Integrated Circuits, Mc Graw Hill Higher Education, 2004.

[33] A. Shafaei, H. Afzali-Kusha, M. Pedram, Minimizing the energy-delay product of SRAM arrays using a device-circuit-architecture co-optimization framework, in: 2016 53nd ACM/EDAC/IEEE Design Automation Conference (DAC), 5-9 June 2016, 2016, pp. 1–6.

[34] S.-M. Kang, Y. Leblebici, CMOS Digital Integrated Circuits, second ed., McGraw-Hill, 1996.

[35] R.J. Evans, P.D. Franzon, Energy consumption modeling and optimization for SRAM's, IEEE J. Solid State Circuits 30 (1995) 571–579.

[36] N.H.E. Weste, D.M. Harris, CMOS VLSI Design A Circuit and Systems Perspective, Pearson, Boston, MA, 2011.

[37] P.K. Bikki, P. Karuppanan, SRAM cell leakage control techniques for ultra low power application: a survey, Circuits Syst. 8 (2017) 23–52.

[38] A. Calimera, A. Macii, E. Macii, M. Poncino, Design techniques and architectures for low-leakage SRAMs, IEEE Trans. Circuits Syst. Regul. Pap. 59 (2012) 1992–2007.

[39] K. Roy, S. Mukhopadhyay, H. Mahmoodi-Meimand, Leakage current mechanisms and leakage reduction techniques in deep-submicrometer CMOS circuits, Proc. IEEE 91 (2003) 305–327.

[40] N.F. Mott, R.W. Gurney, Electronic Processes in Ionic Crystals, Oxford University Press, Oxford, London, 1948.

[41] T. Shinada, S. Okamoto, T. Kobayashi, I. Ohdomari, Enhancing semiconductor device performance using ordered dopant arrays, Nature 437 (2005) 1128–1131.

[42] H.P. Wong, Beyond the conventional transistor, IBM J. Res. Dev. 46 (2002) 133–168.

[43] M.R.S. Levinshtein, Handbook Series on Semiconductor Parameters, Vol. 1, World Scientific, Singapore [etc.], 1996.

[44] O. Madelung, Semiconductors: Data Handbook, Springer-Verlag, Berlin, Heidelberg, New York, 2004.

[45] R. Landauer, Spatial variation of currents and fields due to localized scatterers in metallic conduction, IBM J. Res. Dev. 1 (1957) 223–231.

[46] ITRS, International Technology Roadmap for Semiconductors (ITRS), 2013.

[47] E. Sicard, Introducing 10-nm FinFET Technology in Microwind, 2017. hal-01551695.

[48] K. Ohashi, K. Nishi, T. Shimizu, M. Nakada, J. Fujikata, J. Ushida, S. Torii, K. Nose, M. Mizuno, H. Yukawa, M. Kinoshita, N. Suzuki, A. Gomyo, T. Ishi, D. Okamoto, K. Furue, T. Ueno, T. Tsuchizawa, T. Watanabe, K. Yamada, S. Itabashi, J. Akedo, On-chip optical interconnect, Proc. IEEE 97 (2009) 1186–1198.

[49] J.T. Kim, S. Park, The design and analysis of monolithic integration of CMOS-compatible plasmonic waveguides for on-chip electronic–photonic integrated circuits, J. Lightwave Technol. 31 (2013) 2974–2981.

[50] P. Shen, C. Chen, C. Chang, C. Chiu, S. Li, C. Chang, M. Wu, Implementation of chip-level optical interconnect with laser and photodetector using SOI-based 3-D guided-wave path, IEEE Photonics J. 6 (2014) 1–10.

[51] T. Spuesens, J. Bauwelinck, P. Regreny, D.V. Thourhout, Realization of a compact optical interconnect on silicon by heterogeneous Integration of III–V, IEEE Photon. Technol. Lett. 25 (2013) 1332–1335.

[52] L. Czornomaz, V. Djara, V. Deshpande, E.O. Connor, M. Sousa, D. Caimi, K. Cheng, J. Fompeyrine, First demonstration of InGaAs/SiGe CMOS inverters and dense SRAM arrays on Si using selective epitaxy and standard FEOL processes, in: 2016 IEEE Symposium on VLSI Technology, 14-16 June 2016, 2016, pp. 1–2.

[53] M. Li, J. Ling, Y. He, U.A. Javid, S. Xue, Q. Lin, Lithium niobate photonic-crystal electro-optic modulator, Nat. Commun. 11 (2020) 4123.

[54] W.C. Jeong, S. Maeda, H.J. Lee, K.W. Lee, T.J. Lee, D.W. Park, B.S. Kim, J.H. Do, T. Fukai, D.J. Kwon, K.J. Nam, W.J. Rim, M.S. Jang, H.T. Kim, Y.W. Lee, J.S. Park, E.C. Lee, D.W. Ha, C.H. Park, H. Cho, S. Jung, H.K. Kang, True 7nm platform technology featuring smallest FinFET and smallest SRAM cell by EUV, special constructs and 3rd generation single diffusion break, in: 2018 IEEE Symposium on VLSI Technology, 18-22 June 2018, 2018, pp. 59–60.

[55] Nikolić, B. Advanced Digital Circuits Lecture 2—Scaling Trends. UC Berkeley.

[56] T. Tsuchizawa, K. Yamada, H. Fukuda, T. Watanabe, T. Jun-Ichi, M. Takahashi, T. Shoji, E. Tamechika, S. Itabashi, H. Morita, Microphotonics devices based on silicon microfabrication technology, IEEE J. Sel. Top. Quantum Electron. 11 (2005) 232–240.

Modeling metamaterials: Planar heterostructures based on graphene, silicene, and germanene

L. Rast* and Vinod K. Tewary
Applied Chemicals and Materials Division, National Institute of Standards and Technology (NIST), Boulder, CO, United States

2.1 Introduction

2.1.1 Overview

Nanophotonic materials allow exceptional control over light-matter interactions on the nanoscale. Such materials circumvent the traditional diffraction limit by exploiting near-field optics, making feats such as single-molecule detection possible [1], and spurring new applications including near-field spectroscopy [2] and data storage (enabling the use of big data) [3]. The engineering of more advanced nanophotonic devices, particularly those that rely on metamaterial components with emergent physical properties (novel properties that derive from complex interactions within the material), require close interplay between the new field of rational design and experimental model validation. Rational design of nanophotonic materials requires reliable and comprehensive data on material properties, efficient methods for multiscale calculation of composite properties and quasiparticle interaction, and advanced software frameworks for optimal design and model inversion. Multiscale models offer the ability to utilize accurate and efficient ab initio calculations and simultaneously efficiently bridge the gap from the atomic scale to the continuum in order to consider behavior that occurs on all physically relevant scales in nanomaterials. In this chapter, we will discuss the theoretical background and modeling techniques necessary for creating such models, including discussion of the new paradigm created by machine learning techniques.

At the heart of many nanophotonics applications is the nanoscale confinement of light by light-matter interactions. Spatial confinement of plasmons (quantized units of collective plasma oscillation) on length scales much smaller than the free-space wavelength of light greatly enhances the local electromagnetic field. The field enhancement induced by surface plasmon resonances (SPRs) in silver nanoclusters can provide up to a 10^{15} intrinsic Raman enhancement factor for molecules placed at the locations of greatest field enhancement (so-called "hot spots"), allowing for single molecule detection via surface enhanced Raman spectroscopy (SERS) [1]. In bulk

---

* Present address: The University of Alabama at Birmingham, Department of Physics, Birmingham, AL, United States

Modeling, Characterization and Production of Nanomaterials. https://doi.org/10.1016/B978-0-12-819905-3.00002-6

optoelectronic devices, when the dimensions of an optical component are near the wavelength of light, light propagation is limited by optical diffraction. In plasmonic-photonic circuits, however, the size of components such as lenses and fibers are not subject to this classical Abbe diffraction limit. This is a consequence of the small wavelength of the coupled photon-plasmon modes (surface plasmon polaritons or SPPs) relative to the free-space wavelength of light of the same frequency [4,5]. Many nanophotonic devices support plasmon resonances—this provides the important advantage of strong plasmonic coupling with light near the resonant frequency. Optical signals in plasmonic-photonic circuits may be controlled at length scales similar to the feature size of current and future electronic devices (tens of nm), and plasmonic-photonic circuit components may be much smaller in size than "traditional" optoelectronic components [6].

In addition to confinement, particle and quasiparticle interactions also determine the behavior of these materials. Plasmon oscillations are able to couple strongly with a variety of particles and quasiparticles to form other quasiparticles and collective modes. These interactions can be particularly important in determining the behavior of nanophotonic devices based on 2D materials. These effects have been studied significantly in graphene-based devices. For example, coupled plasmon-phonon oscillations have been shown to dominate the mid-IR interfacial optical response of graphene on silicon [7]. Additionally, intrinsic optical phonons have been shown to be a damping pathway for plasmon resonances in graphene, and plasmon-phonon hybridization at a graphene-SiO_2 interface has been shown to induce significant damping and modification of the plasmon dispersion in graphene [8]. Finally, it has been shown in angle-resolved photoemission spectroscopy measurements that higher-order quasiparticle interactions lead to novel quasiparticles in graphene, such as plasmarons (plasmon-electron hybrid modes), and that these modes are important in determining the electronic structure of graphene at the Dirac point [9]. Predicting the optoelectronic properties of nanophotonic devices, particularly those based on 2D materials with properties that have not been thoroughly studied, requires computational methods capable of modeling various particle and quasiparticle interactions. Beyond graphene, the capabilities of Group IV two-dimensional materials such as silicene and germanene have drawn considerable attention in recent years due to their multifunctional capabilities and band gap that is tunable via electric field [10], hydrogenation [11], strain [12], or coupling with the substrate [13]. The ability to open up a significant band gap without substantial degradation of the high carrier mobilities characteristic of such materials enables applications such as room-temperature field effect transistors [10,13]. Modern density functional theory (DFT) methods can be excellent tools for modeling such materials. A key aim of this chapter is to give the reader guidance on the use of DFT for achieving this goal.

2.1.2 Low-dimensional materials for stratified metamaterial heterostructures: Fundamental interest and applications

Much progress has been made in the production of 2D materials, and the fabrication of heterostructures with tailored electronic and optical properties is on the horizon. 2D crystal-based nanophotonic metamaterials could, for example, combine the sizable band gaps of the transition metal dichalcogenides, which can change from indirect to direct in single layers [14], with the incredible mechanical strength of graphene and the ultra-flat morphology of hexagonal boron nitride (hBN) [15]. Atomically thin materials also provide the advantage of unparalleled plasmonic and optical confinement.

Hybrid materials are also excellent candidates for low-loss plasmonic materials. As an example, devices that contain a plasmonic metal, as well as a combination of low and high refractive index dielectric materials, exhibit longer propagation distances of carriers without reduced confinement [16]. Metamaterial heterostructures have material properties that arise from the properties of their constituent materials, as well as "meta" properties that arise due to the collective interaction between component materials and various physical processes in each material layer. Interfacial phenomena are particularly important in metamaterial heterostructures. Rational design of 2D-based nanophotonic devices therefore requires well-validated data on material properties for the individual layers, efficient methods for multiscale calculation of composite properties, and advanced computational frameworks for model optimization and inversion. Further reading on computational frameworks for materials by design may be found in Section 2.4.5.

2.2 Atomistic modeling of low-dimensional materials: Modeling collective modes with DFT

2.2.1 Overview

DFT is among the most widely used tools for the calculation of excitations and collective modes in many-body systems. DFT is founded upon the Hohenburg-Kohn theorem that states that the ground-state Schrodinger equation is a unique functional of the electron density [17]. For N interacting electrons, subject to an external potential V_{ext}

$$E[n(r)] = F[n(r)] + \int n(\vec{r})V_{ext}(\vec{r})d\vec{r} \tag{2.1}$$

where F is a universal functional of the electron density $n(\vec{r})$. The minimum value of F, F_{min}, is the exact ground-state electronic energy.

The basic equations used in the modern DFT formalism were developed by [18] Kohn and Sham, who added exchange and correlation terms to the chemical potential. These are included in the DFT formulation as an additive potential term E_{xc} [18], as shown below:

$$E[n(r)] = K_s\left[n(\vec{r})\right] + \frac{1}{2}\iint \frac{n(\vec{r})n(\vec{r}')}{\left|\vec{r} - \vec{r}'\right|} \vec{dr}\,\vec{dr}' + E_{xc}\left[n(\vec{r})\right]$$

$$+ \int n(\vec{r})V_{ext}(\vec{r})d\vec{r} \tag{2.2}$$

where $K_s\left[n(\vec{r})\right]$ is the kinetic energy of a noninteracting electron gas of density $n(\vec{r})$. There are a wide variety of available exchange correlation energy functionals in the various DFT implementations. One of the most popular classes of functionals is based on the local density approximation (LDA).

$$\overset{\text{LDA}}{\underset{xc}{E}}\left[n(\vec{r})\right] = \int\left[n(\vec{r})\right]\varepsilon_{xc}\left[n(\vec{r})\right]d\vec{r} \tag{2.3}$$

where $\varepsilon_{xc}\left[n(\vec{r})\right]$ is the exchange-correlation energy density.

Used properly, DFT can yield highly accurate results for band structure, plasmon resonances, phonon modes, excitonic resonances, electron transport properties, and a host of other properties of interest for nanophotonics. However, DFT calculations are by no means a "black box." The user must carefully consider the "physics" of the system of interest, when selecting the DFT solution method and parameters, such as the exchange correlation functional, planewave cutoff energy, number of bands for convergence and grid spacing. The choice of these parameters depends upon the material systems. For example, the basic LDA formulation, while reasonably accurate for some systems, is not the most accurate approach for modeling other materials or effects. Examples include materials with strong electron-electron correlation or the calculation of bandgaps in atomically thin wide bandgap materials. In such cases, a correction factor that accounts for the self-energy of the system is needed.

2.2.2 Sensible selection of an exchange-correlation functional

Simple LDA and generalized gradient approximation (GGA) methods for electronic structure calculations may underestimate bandgaps by over 50% for some materials [19]. These errors often occur in the location of both conduction and valence bands, and this leads to errors in transition energies that are on the order of the bandgap error [19]. Methods available for corrections that bring the electronic structure in closer agreement with experimental values include the LDA+U method, LDA+GW method, as well as a variety of hybrid functionals (see Section 2.4.2). A detailed analysis of all of these methods is beyond the scope of this chapter, but it is suggested that the reader consult the references in the DFT methods section of the Suggested Reading list. Hybrid functionals are often a convenient and computationally efficient way of obtaining band structures with improved accuracy. This is true to varying degrees, and the reader is advised to do a careful search of the relevant literature for the material of interest in order to select an appropriate functional for creating an accurate representation of the behavior of the material. For example, some functionals may produce accurate electronic band structures for good conductors (such as metals), but give inaccurate results for wide bandgap semiconductors. DFT is one of the most popular methods of predicting properties of hypothetical low-dimensional materials as well as not-yet measured properties of recently discovered low-dimensional materials.

Consequentially, there has been much work done recently on improving both the accuracy and efficiency of these methods, including the development of new functionals. One such hybrid functional, the GLLBSC functional [20], reproduces bandgaps for a wide variety of materials that agree very closely with experimental values. This functional incorporates the derivative discontinuity of the exchange correlation functional explicitly at integral particle numbers, a critical feature for obtaining an accurate band structure in a DFT calculation. The GLLBSC method has also been shown to have computational costs similar to the LDA, with the advantage of accuracy comparable to methods such as the LDA-GW method [20].

2.2.3   Optical limit and electron energy loss function calculations

Special consideration is also required in selecting the method for calculation of the optical limit (also known as the long wavelength limit) and Electron Energy Loss Spectroscopy (EELS) calculations. Methods available for the accurate prediction of absorption spectra and EELS, in particular excitonic features, include time-dependent density functional theory (TDDFT) or the Bethe-Salpeter equation (BSE) [21]. The most accurate methods for calculations of spectroscopic properties of nanophotonic materials include the electronic self-energy and electron-hole exchange interactions via many-body perturbations (using, for example, the GW approximation for electron self-energy and the BSE for electron-hole interactions). The standard GW-BSE approach is as follows: Single-particle energies are calculated via a self-energy through the GW approximation, then diagonalizing the two-particle Hamiltonian yields the optical excitation energies. The GW self-energy computation can be costly. Alternatively, single-particle energies may be obtained using an efficient and accurate hybrid functional, such as the GLLBSC, and optical properties may subsequently be calculated using the BSE.

2.2.4   Available software packages

Although there are numerous DFT codes available, both open source and commercial, not all DFT packages have the capabilities for calculations of optical spectra. This section is intended to give an overview of a few of the available software packages that have integrated modules, or that may be used in combination with other packages, in order to perform optical limit or EELS calculations. Although this is by no means an exhaustive list, it should give the reader an idea of the wide variety of capabilities related to nanophotonics calculations of modern DFT software.

(1) SIESTA[a] [22]: This Fortran 95 code (available for serial or parallel calculations under MPI) is capable of band structure calculations as well as ab initio molecular dynamics modeling of molecules and solids. Since 2016, this code has been available through GPL open-source

---

[a] Disclaimer: Certain commercial equipment, instruments, or materials are identified in this paper in order to specify the experimental procedure adequately. Such identification is not intended to imply recommendation or endorsement by the National Institute of Standards and Technology, nor is it intended to imply that the materials or equipment identified are necessarily the best available for the purpose.

license. The Kohn-Sham self-consistent density functional method (using the numerical atomic orbitals basis) in the LDA-(LSD) or generalized gradient (GGA) approximations is used for the band structure calculations. Optical calculations (the imaginary part of the dielectric function) are performed using dipolar transition matrix elements between divergent eigenfunctions of the Hamiltonian. The molecular computation uses the position operator matrix elements, and for solids, the calculation is performed in momentum space.

(2) VASP (Vienna Ab-initio Simulation Package) [23,24]: VASP is a Fortran 90 code (available for serial or parallel calculations under MPI). This code is available for use through both academic and commercial licenses. The software calculates an approximate solution to the many-body Schrödinger equation, either within DFT using the Kohn-Sham equations, or within the Hartree-Fock (HF) approximation. Hybrid functionals that mix the HF approach with DFT are included in the implementation as well. A plane wave basis is utilized in VASP, and both many-body perturbation theory and GW corrections are possible with this code. A variety of optical calculations are possible with VASP, including dielectric tensor calculations that may be used to obtain absorption spectra in postprocessing. Optical calculations using the BSE are possible using this software. Available functionals include LDA, GGA, and metaGGA-based functionals, HF, and HF/DFT hybrids.

(3) Quantum ESPRESSO [25]: This open-source Fortran 90/77 and C code (available for serial or parallel calculations under MPI or OpenMP) uses a plane wave basis set and pseudopotentials to perform DFT electronic structure calculations. Capabilities for TDDFT also exist as a specialized Quantum Espresso package. Optical spectra may be calculated from TDDFT perturbation theory within Quantum ESPRESSO. Output from Quantum Espresso may also be used as input to the Yambo program [26] in order to perform calculations involving electronic excitations with the GW correction and optical properties including absorption spectra via the BSE.

(4) GPAW [27–30]: GPAW is an open-source Python/C DFT code that uses the projector-augmented wave method. It uses real-space uniform grids or multigrid approaches as well as atom-centered basis functions. The code has TDDFT and GW approximation capabilities, as well as a wide variety of functionals, including the GLLBSC functional. A variety of capabilities for absorption spectra and EELS calculations exist within GPAW, including calculations via the Bethe-Salpeter formalism.

(5) Abinit [31,32]: Abinit is an open-source Fortran DFT code that can perform electronic structure calculations using pseudopotentials and a plane wave basis set. Excited states can be computed within the many-body perturbation theory via the GW approximation and the macroscopic dielectric function can be obtained in the Bethe-Salpeter formalism. Functionals implemented in Abinit include LDA and the GGA and capabilities for corrections for delocalization in strongly correlated materials exist, including the full localized limit and around mean field methods. Output from Abinit may also be used with the code Yambo for spectroscopic calculations, in a similar manner as previously described for Quantum ESPRESSO.

2.3   Spectral properties of multilayer structures

2.3.1   Continuum models for collective modes

Nanophotonics heterostructure materials are of interest for a variety of applications, including low-loss plasmonics. Engineering novel materials for nanophotonics requires efficient methods for predicting emergent material properties. Both individual material properties and interactions between various materials must be incorporated realistically into the calculation. Using ab initio methods such as DFT for modeling an entire continuum structure can be prohibitively expensive, especially if one is interested in considering or comparing a variety of system components and geometries. However, there are hybrid techniques such as semiclassical or Green's function approaches that allow for efficient bridging of the gap between the atomistic and continuum scales. Used judiciously, these methods can be very powerful and useful as a step toward materials by design. As an example, this section describes one such semiclassical technique for modeling plasmon and phonon excitations in multilayer structures. This method has been shown to apply to phonons [33] as well as plasmon polaritons [34] in layered materials with step function-like complex dielectric functions (though the dielectric functions are required to be continuous throughout each layer) and interacting interfaces. Although, the Lambin et al. paper [33], in which Eqs. (2.7), (2.10) were derived, focused on modeling semiconductors, the expression and the formalism from which it originates are also applicable to plasmonic resonances in alternating metal-insulator layers. This method is extremely computationally efficient, as we will show in the next section. Because the method applies to both phonon and plasmon excitation in stratified structures with interacting interfaces and sharp changes in dielectric function across layer boundaries, it is convenient for modeling a variety of multilayer nanophotonic heterostructures.

2.3.2 Details of the continued fraction model

Lambin's model [33] for multilayer structures is derived from a specular reflection model. It has been previously demonstrated that for small values of the wave vector, the specular reflection model is in close agreement with the more well-known Bloch hydrodynamic model [35]. In the Lambin model, the z coordinate is normal to the surface of the material, extending from $z=0$ to $-\infty$ (as in Fig. 2.1). ω is excitation frequency and k is the wave vector of the plasmon or phonon resonance. Let $\epsilon(\omega,z)$ be the (frequency-dependent) dielectric function in the long wavelength optical limit ($k \to 0$), which is a tensor of the material along the spatial coordinate z. D and E are the electric displacement and electric field, respectively. We have the following boundary conditions for the fields: $D \perp E$ must be continuous. The general relation between D, E, and the frequency-dependent dielectric function is given by:

$$D(k, \omega, z) = \epsilon(\omega, z)E(k, \omega, z) \tag{2.4}$$

Eq. (2.4) ensures continuity of the effective dielectric function even in instances of sharp interfaces parallel to the x-y interfaces below the $z=0$ surface. The effective dielectric function $\xi(k,\omega,z)$ of the layered structure is given by:

$$\xi(k, \omega, z) = \frac{i\,D(k, \omega, z) \cdot n}{E\,(k, \omega, z) \cdot \frac{k}{k}} \tag{2.5}$$

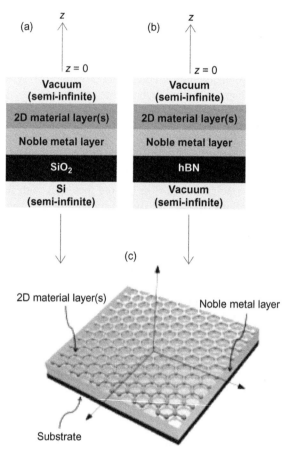

Fig. 2.1 Multilayer structure made up of a graphene top layer(s), a noble metal intermediate layer, and semiconducting substrate. EELF functions were calculated for structures of this type by the methods discussed in the text.

where $\boldsymbol{n}$ is the surface normal vector. The effective dielectric function is a solution to the following Riccati differential equation, in the long-wavelength limit ($k \to 0$), at $z = 0$, as follows:

$$\frac{1}{k}\frac{d\xi(z)}{dz} + \frac{\xi^2(z)}{\epsilon(z)} = \epsilon(z) \qquad (2.6)$$

At $z = 0$ we have for the effective dielectric function:

$$\xi_0 = a_1 - \frac{b_1{}^2}{a_1 + a_2 - \dfrac{b_2{}^2}{a_2 + a_3 - \dfrac{b_3{}^2}{a_3 + a_4 - \dots}}} \qquad (2.7)$$

where

$$a_i = \varepsilon_i \cot h(kd_i) \qquad (2.8)$$

and

$$b_i = \frac{\varepsilon_i}{\sin h(kd_i)} \qquad (2.9)$$

ε_i are the frequency-dependent dielectric functions within each layer of the stratified material. We then have for the electron energy loss function (EELF):

$$\text{EELF} = \text{Im}\left[\frac{-1}{\xi(\omega, k) + 1}\right] \qquad (2.10)$$

Applications of the above formulae to specific systems will be presented in the next section.

2.3.3 Applications

This section is intended as an example of how a semiclassical method for predicting properties of advanced nanophotonic materials may be developed based on a classical expression for a hybrid material property, such as that of Lambin. The general concept is that individual material properties are calculated on the quantum or nanoscale using an ab initio method such as DFT, then composite properties is calculated using a computational bridge such as Eq. (2.7). Such methods can be very efficient and powerful, provided they are used thoughtfully, with all of the relevant limitations of the model in mind. Such considerations should include, for example, whether the dominant physical processes in the regime of interest are accurately represented by the model. In metamaterial heterostructures, such as those described in this section, interfacial interactions are important and must be included in the model. Eq. (2.7) represents interfacial interactions classically, which is acceptable in this context because we are interested in modeling classical resonances (plasmons and phonons).

2.3.3.1 Graphene/noble metal multilayer composites

Lambin's continued fraction expression, given in Eq. (2.7), was utilized in our work on electron energy loss properties of graphene/noble metal/insulating substrate layered structures [36]. Details of the semiclassical method are as follows: Dielectric functions for the individual material layers were initially obtained. For the 2D materials of interest (graphene and h-BN), ab initio calculations were performed using the following procedure: Time-dependent DFT was employed (using the DFT code GPAW) in the LDA. These optical calculations were performed with atomically thin materials with armchair geometry and a momentum transfer value of 0.005 Å^{-1} along the Γ-M direction in the Brillouin zone.

Optical properties of other materials in the layered structure, as depicted in Fig. 2.1, were experimentally measured values obtained from the literature. With the dielectric functions of the individual layers as input, Eq. (2.7) was used to obtain the composite dielectric function for the layered system, and then Eq. (2.10) was used to obtain the EELF. The use of this procedure enables computationally efficient multiscale EELS calculations for layered structures based on many different materials. Layer thicknesses and materials may be substituted easily as desired. The CPU time needed for the EELS calculations is nearly independent of the range of wavelengths and is less than a second on a standard 2.3 GHz, dual core desktop. The graphene dielectric function was obtained in the optical limit, and the LDA calculation was executed with a Monkhorst-Pack grid of $20 \times 20 \times 1$ k-points. Johnson and Christy's empirical data [37] were used for the silver dielectric function. The Si and SiO_2 dielectric constants used in the calculation were 11.68 and 3.9, respectively. These are broadly accepted values obtained from the literature [38,39].

Fig. 2.2A–C demonstrate the effect of graphene films with varying numbers of atomic layers on a noble metal/SiO_2/Si layered structure. In Fig. 2.2A, where a silver metallic layer is used, for mono- to trilayer graphene, the lower energy surface peak near 2.6 eV remains virtually unchanged, while the bulk peak near 3.7 eV is reduced by 50% at three layers of graphene. Similar enhancement of the surface peak is also seen for gold and copper, as displayed in Fig. 2.2B and C, though the effect is less pronounced than in the case of the silver film. The imposition of a boundary layer on the noble metal surface leads to strengthened surface coupling at the expense of bulk resonance modes. The physical interpretation of this effect is similar to that of the *begrenzung* effect, which has been well studied for increasingly thin metallic thin films [40]. In the case of metallic films, the *begrenzung* effect occurs when a boundary is imposed on an infinite plasmonic metal slab, leading to a surface plasmon peak. In terms of the energy loss, this occurs at the expense of the bulk plasmon peak. At 20 graphene layers, we begin to see behavior more like that of a bulk top layer. The surface peak broadens, which indicates increased losses, as one would expect for a bulk graphite top layer.

2.3.3.2 Comparison of hexagonal boron nitride and traditional silicon substrates for plasmonics

Fig. 2.3 displays data for the combined effects of a graphene top coating (of varying numbers of graphene layers), silver middle layer, and highly insulating hBN substrate, previously calculated in [36]. The structure is displayed in Fig. 2.1B. It can be seen that the substitution of the silicon substrate with hBN produces a drastic shift toward the surface resonance. In the context of our mesoscopic model, the diminished plasmonic losses for plasmonic metals with a graphene top coating on hBN substrates is due to the following two basic physical processes:

(i) Adding a graphene boundary coating to the metal surface diminishes coupling to bulk plasmon oscillations via a process akin to the *begrenzung* effect. The *begrenzung* effect arises due to reduction in the number of degrees of freedom for resonances—additional surface confinement occurs at the expense of the bulk plasmon resonance intensity, leading to a reduction in bulk energy losses.

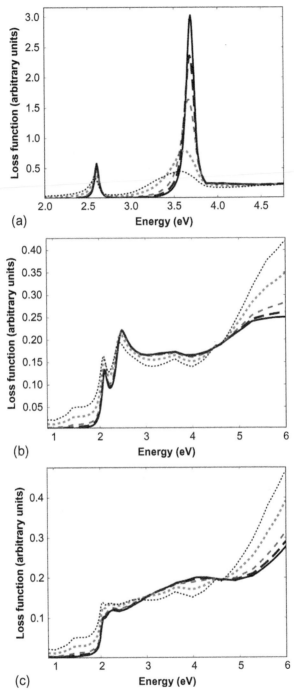

Fig. 2.2 EELF for graphene films with different numbers of atomic layers three substrate structures: (A) 34 nm Ag layer and SiO$_2$/Si substrate. (B) 34 nm Au layer and SiO$_2$/Si substrate. (C) 34 nm Cu layer and SiO$_2$/Si substrate. For each substrate, the graphene layer numbers are 0 *(solid line)*, single layer *(long dashes)*, tri-layer *(intermediate-length dashes)*, 10 layers *(short dashes)*, and 20 layers *(dotted line)*. For few-layer graphene, the lower energy surface peak near 2.6 eV remains virtually unchanged, while the bulk peak near 3.7 eV is reduced by 50% at three layers of graphene. The physical interpretation of this effect is similar to that of the *begrenzung* effect—the imposition of a boundary layer enhances coupling to surface plasmon modes and reduces coupling to bulk modes.

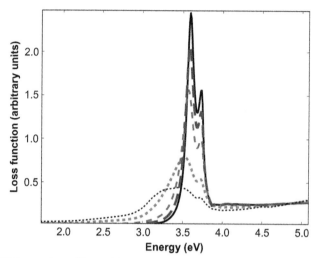

Fig. 2.3 EELF for graphene films with varying numbers of layers on a 34 nm Ag film and single-layer hBN substrate. Without graphene *(solid line)*, monolayer graphene *(long dashes)*, tri-layer graphene *(intermediate-length dashes)*, 10-layer graphene *(short dashes)*, and 20 layer graphene coating *(dotted line)*. Because hBN is highly insulating, this substrate material reduces bulk losses and strengthens surface confinement due to reduced scattering losses.

(ii) hBN is highly insulating, so this substrate material, when substituted for silicon, reduces bulk losses and strengthens surface confinement due to reduced scattering losses.

Fig. 2.4 displays EELF data for (A) free-standing Ag films of varying thickness and (B) trilayer graphene coatings on Ag layers (of various thicknesses) with a single layer hBN substrate. Comparison of Fig. 2.4A and B reveals that the spectral profile of the system with an hBN substrate (B) is strikingly similar to the case of a "suspended sample" (A).

2.3.3.3   Silicene and germanene systems

Silicene and germanene are predicted to have electron transport properties similar to graphene, with the added advantage of increased compatibility with silicon-based technology. In addition, these materials have a buckled lattice structure that leads to inversion symmetry breaking. This may be taken advantage of through the application of an external electrical field perpendicular to the plane for highly controllable bandgap tunability. These materials are therefore also of recent interest for nanophotonics applications. Fig. 2.5 displays the system considered in this section, a multilayer structure consisting of (A) silicene or germanene top layers on a semi-infinite silver substrate and (B) silicene and germanene top layer(s), Ag middle layer, and SiO_2/Si substrate. Further results and calculation details for the system described in this section may be found in our recent paper [41]. The method for EELF calculations is, in general, the same as those described for the graphene-based systems in Section 2.3.3. Eqs. (2.7), (2.10) are used for the composite dielectric function, and

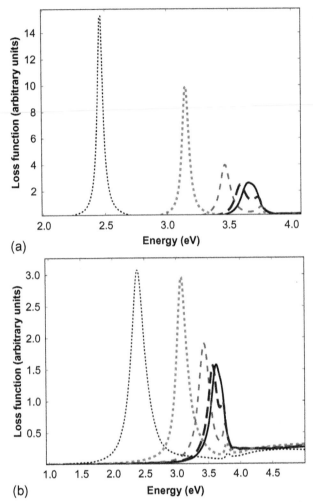

Fig. 2.4 EELF for (A) free-standing Ag films of various thickness and (B) trilayer graphene on Ag films (of varying thickness) with a monolayer hBN substrate. In both cases (A) and (B), Ag films with thicknesses of 50 nm *(solid line)*, 34 nm *(long dashes)*, 20 nm *(intermediate-length dashes)*, 10 nm *(short dashes)*, and 4 nm *(dotted line)*. The spectral profile of the system with an hBN substrate is remarkably similar to the case of a "suspended sample."

the silver optical data are again those of Johnson and Christy [37]. Both silicene and germanene dielectric functions are computed in the long wavelength optical limit with a momentum transfer of $0.005\,\text{Å}^{-1}$, along the Γ-M direction of the Brillouin zone. The k-point sampling with $20 \times 20 \times 1$ Monkhorst-Pack grid was selected for the band-structure computation.

Values of the silicene and germanene dielectric functions are obtained through ab initio DFT methods (this time using the GLLBSC functional), and the optical

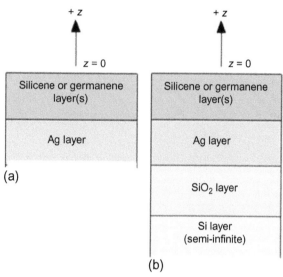

Fig. 2.5 Multilayer structure consisting of silicene and germanene multilayers (A) on semi-infinite Ag substrate, and (B) on substrate layers consisting of Ag, SiO$_2$, and semi-infinite Si. EELS functions were calculated for 000 structures of this type by the methods discussed in the text.

spectra are obtained using the BSE to include excitonic effects, which is an important spectral feature for both materials [42]. This is the highly accurate and broadly applicable method described in [20], and our the calculated optical spectra compare well with previous results in the literature [43]. The GLLBSC potential includes the derivative discontinuity of the xc-potential [18] (see Eq. 2.2) at integer particle numbers, which is important in obtaining physically meaningful band structure from a DFT simulation. A two-dimensional Coulomb cutoff method is employed in order to model the monolayer structure, which is only periodic in 2D [44]. The buckling amplitude, obtained from DFT structural calculations available in the literature, is 0.44 Å for silicene and 0.6 Å for germanene [45]. Our model utilizes dielectric functions derived from surface parallel excitations only, as the calculation of the effective dielectric function is derived from a specular reflection geometry. Complex dielectric functions for silicene and germanene are displayed in Fig. 2.6A and B, respectively. The EELF results for varying numbers of silicene and germanene layers on silver are displayed in Fig. 2.7A and B, respectively.

From the above figures, we observe that for up to three layers of silicene on silver, the bulk plasmon peak is diminished without significant broadening, as our calculations also showed for graphene. This indicates an overall reduction in bulk losses. The effect is most notable when the silver slab is coated with only one layer of silicene. The origin of the reduced bulk losses is similar to that of graphene: reduced coupling to bulk resonances and simultaneous enhancement of surface coupling. This effect could be useful for determining successful fabrication of monolayer silicene on silver through spectroscopic characterization. At 10 layers and above, the system

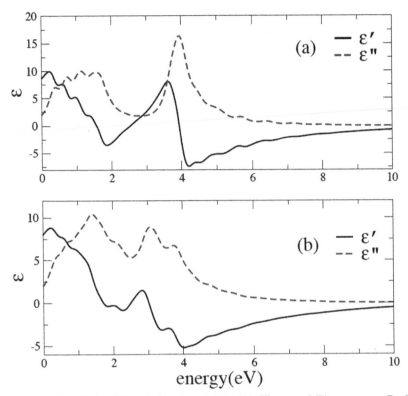

Fig. 2.6 Complex relative dielectric function $\varepsilon(\omega)$ for (A) silicene and (B) germanene. Real and imaginary parts (ε' and ε'') of the dielectric function are represented by *solid* and *dotted lines*, respectively.

approaches the expected behavior for a bulk Si/Ag system—a broad interfacial peak appears at about 2.5 eV. In the case of germanene, there is no well-defined interfacial plasmon for the germanene/silver system—in contrast to silicene, the real part of germanene's dielectric function is only momentarily slightly non-negative in the 2–4 eV regime. So, the sharp change in dielectric function needed for an enhanced surface plasmon does not occur. However, the bulk plasmon is damped in a manner similar to that of silicene with a few layers of germanene and broadens for layer thicknesses approaching the bulk. This effect may also be of use for determining whether germanene has been deposited.

In Fig. 2.8A and B, simulation results corresponding to the system shown in Fig. 2.5B are displayed. These results demonstrate the effect of decreasing the silver metallic layer thickness with single-layer silicene (Fig. 2.8A) and germanene (Fig. 2.8B) coatings. As the Ag thickness is reduced, the begrenzung effect, observed in graphene-based systems, again becomes apparent. The surface peak and negative begrunzung peak both intensify with decreasing silver layer thickness, decreasing the bulk-plasmon amplitude. Surface modes become dominant for silver thickness

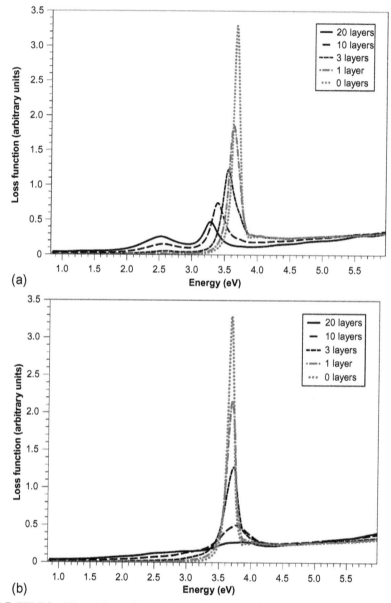

Fig. 2.7 EELF for (A) multilayer film of silicene on a semi-infinite Ag substrate; (B) multilayer film of germanene on a semi-infinite Ag substrate. The effect of differing numbers of silicene/germanene layers is demonstrated. The silicene/germanene layer numbers are 20 *(solid line)*, 10 *(long dashes)*, 3 *(short dashes)*, 1 *(dash-dot)*, and 0 *(dotted line)*. For both materials, in a manner similar to the graphene multilayer systems, the bulk plasmon is damped for a few layers and broadens for layer thicknesses approaching the bulk. This effect may be of use for confirming if few-layer silicene or germanene has been deposited experimentally.

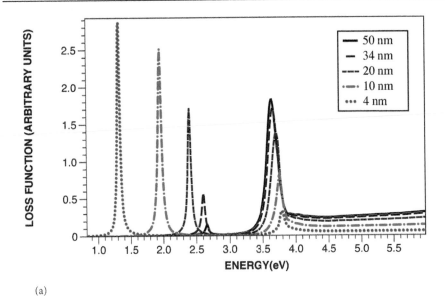

(a)

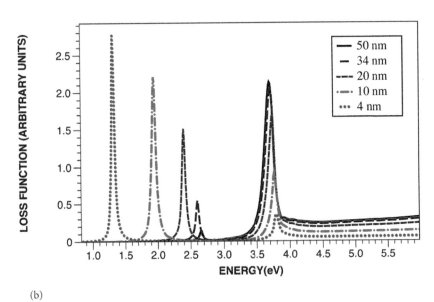

(b)

Fig. 2.8 EELS for single-layer (A) silicene and (B) germanene on Ag/SiO$_2$/Si substrate: The effect of varying thickness for the Ag layer is demonstrated. The Ag layer thicknesses are 500 Å *(solid line)*, 340 Å *(long dashes)*, 200 Å *(short dashes)*, 100 Å *(dash-dot)*, and 40 Å *(dotted line)*.

between 20 and 10 nm. These results are consistent with observations both in the well-validated and widely used empirical data by Johnson and Christy [37] and our earlier work on graphene/noble metal multilayer systems [36]. Comparison with empirical data for extremely thin silver layers provides confirmation of the validity of the model for a wide variety of silver metal layer thicknesses. Bulk and surface peak locations and relative intensities for 4 nm compare very well with experimental results for EELS of 3.4 nm silver layers [46].

2.3.4   Limits of applicability of electromagnetic continuum models

While extremely convenient, and capable of revealing a range of phenomena of interest for nanophotonics, the semiclassical method used in the work described in this chapter has a variety of limitations. The continuum model described in Section 2.2 does not include effects due to interfacial strain and doping. Superlattice heterostructures typically have lattice mismatch between adjacent layers, as well as inherent rippling and buckling in individual layers. These morphological features introduce interfacial strain as indicated by the bright spot on the top layer in Fig. 2.9. The strain induces charge distortions (leading to charge pooling, for example) to which material optical and electronic properties (such as plasmon resonances) are sensitive [47]. DFT methods or a Green's function approach should be used in order to take these effects into account.

Another limitation of the continuum model is that as boundary interactions are considered classically, it does not incorporate interlayer-hopping effects. Angle-resolved photoemission spectroscopy on graphene has shown this tunneling effect to be important for graphene's electronic structure, because it causes π band splitting [48]. This effect is more important with increasing layer numbers. For example, at four layers, the energy shift is approximately 0.7 eV. Ohta et al. calculated the graphene excitation spectrum, including interlayer hopping effects, in the Hamiltonian [48]. The authors found tunneling to be significant mainly in predicting the low energy behaviors

Fig. 2.9 Conceptual configuration of a 2D heterostructure with strain-induced charge distortions. These structures typically have a lattice mismatch between layers in addition to inherent rippling and buckling of individual layers. These morphological features lead to strain-induced charge distortions, to which spectroscopic features such as plasmon resonances are very sensitive.

($\omega \sim v_F k$) with reasonable accuracy, where v_F is the Fermi velocity. Thus, predicting material properties for low-energy photonics necessitates the inclusion of tunneling effects.

Recently, attention surrounding van der Waals heterostructures has grown, due to the ability to engineer devices with emergent opto-electronic properties not only by combining different crystals or through the arrangement of these materials within a heterostructure, but through relative rotation of the material layers. By manipulating these new degrees of freedom that are made possible through these Moiré Superlattice structures (shown conceptually in Fig. 2.10), materials like hBN, which typically serves as a highly insulating substrate for graphene, become a platform for a periodic substrate-induced moiré superlattice. A new hexagonal superstructure arises, with a pattern wavelength that can be tuned via the relative orientation of the material layers [49]. Scattering along k-vectors forbidden in graphene are possible in such structures and new Dirac points are created. DFT electronic structure and optical property calculations for these new materials require careful selection of an exchange correlation potential, including van der Waals correction and corrections for effects due to strain [49,50].

An important limitation of the model, and many of the other current methods for modeling nanophotonics devices, is that it does not consider complex morphologies. As will be discussed in the next section, many of the new systems on the horizon for nanophotonics have quite complex geometries. There is a current need for the development of methods for accurate calculations of electronic and optical properties of systems where the behavior on the quantum scale is important, but the scaling to the continuum is less ordered than simple multilayer structures. Examples of complex morphologies include quasiperiodic, aperiodic, and strongly scattering disordered

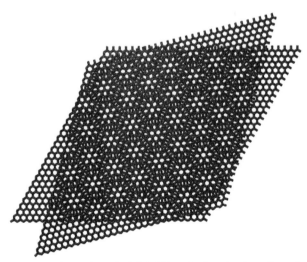

Fig. 2.10 Conceptual configuration of a Moiré Superlattice structure. In such structures, electronic structure and transport properties are sensitive to lattice mismatch and the relative orientation of the material layers.

media for nanophotonics. For example, quasiperiodic multilayers have been shown to have properties that lead to a more complex spectrum structure than periodic multi-layers, and aperiodic optical media can produce photonic systems with improved optical properties (including narrow peaks and light emission enhancement) as well as more broad "materials by design" capabilities than periodic nanophotonic systems [51]. Coupling of semiclassical methods as is described in this chapter or other multiscale methods with frameworks for uncertainty quantification, including geometric uncertainties, is one possible solution to the problems introduced by complex morphologies. Resources for further reading in this area are listed in Section 2.4.5.

The new paradigm created by artificial intelligence and big data has created unprecedented opportunities for materials-by design. Machine learning can be used in the creation of reliable and efficient phenomenological models [52] or with methods such as Monte Carlo simulation or Hartree-Fock theory [53]. The feedback loop created by incorporating machine learning algorithms into the modeling process creates exciting possibilities in terms of linking material characterization, simulation, and theory. Despite their limitations, DFT and modified DFT approaches, including the use of generalized gradient approximations and hybrid functionals, have been useful in the prediction of spectroscopic properties for many materials. However, the exchange-correlation functional presents an area of significant weakness for many systems such as strongly-correlated materials. For such systems, experimental data, including data from existing structure-property databases, offer a valuable resource that can be employed in machine learning selection of an accurate functional. Finally, the need to solve the Kohn-Sham equations can be avoided entirely through the use of density-to-energy and density-to-potential maps developed by machine learning techniques [54]. An excellent review on the subject, which includes a list of publicly available software resources for machine learning in quantum chemistry and materials science, can be found in [55].

2.4   Sources of further information

This section includes suggested reading material related to current topics in low-loss nanophotonics applications and modeling of nanophotonic materials.

2.4.1   Many-body effects in atomically thin materials

Bostwick A., Ohta T., Seyller T., Horn K., Rotenberg E. (2007) 'Quasiparticle dynamics in graphene' Nat. Phys., 3, 36–40.
Hwang E. H., Sensarma, R. and Das Sarma, S. (2010) 'Plasmon-phonon coupling in graphene' Phys. Rev. B, 82, 19, 195406.
Wei W., Dai Y., Huang B., and Jacob B. (2013) 'Many-body effects in silicene, silicane, germanene and germanane' Phys. Chem. Chem. Phys., 15, 8789–8794.

2.4.2   DFT methods

Sholl D. and Steckel J. A. (2011) Density functional theory: a practical introduction, Wiley, New Jersey.

Rinke P., A. Janotti A., M. Scheffler M., and Van de Walle C.G. (2009) 'Defect formation energies without the band-gap problem: combining density-functional theory and the GW approach for the silicon self-interstitial' Phys. Rev. Lett.,102, 026402.

Janotti A. and Van de Walle C. G. (2007) 'Native point defects in ZnO' Phys. Rev. B, 76, 165202.

Janotti A., Varley J. B., Rinke P., N. Umezawa N., G. Kresse G., and Van de Walle C. G. (2010) 'Hybrid functional studies of the oxygen vacancy in TiO_2' Phys. Rev. B, 81, 085212.

Oba F., Togo A., Tanaka I., Paier J., and Kresse G. (2008) 'Defect energetics in ZnO: A hybrid Hartree-Fock density functional study' Phys. Rev. B, 77, 245202.

2.4.3 Metal nanomaterial-functionalized two-dimensional materials

Shi Y., Huang J.K., Jin L., Hsu Y.T., Yu S.F., Li L.J., and Yang H.Y. (2013) 'Selective decoration of Au nanoparticles on monolayer MoS_2 single crystals' Sci. Rep., 31839.

Zedan A.F., Moussa S., Terner J., Atkinson G., and El-Shall M.S. (2012) 'Ultrasmall gold nanoparticles anchored to graphene and enhanced photothermal effects by laser irradiation of gold nanostructures in graphene oxide solution' ACS Nano, 71,627–636.

Zaniewski AM, Schriver M, Lee JG, Crommie MF, and Zettl A (2013) 'Electronic and optical properties of metal-nanoparticle filled graphene sandwiches' Appl. Phys. Lett. 102,023108.

2.4.4 Plasmon-assisted gain media

Liu N., Wei H., Li J., Wang Z., Tian X, Pan A, and Xu H. (2013) 'Plasmonic amplification with ultra-high optical gain at room temperature', Sci. Rep., 3, 1967.

Stockman, M. I. (2011) 'Loss compensation by gain and spasing' Phil. Trans. R. Soc. A, 369, 3510–3524.

Wuestner, S., Pusch, A., Tsakmakidis, K. L., Hamm, J. M. & Hess, O. (2011) 'Gain and plasmon dynamics in active negative-index metamaterials' Phil. Trans. R. Soc. A, 369, 3525–2550.

2.4.5 Designer nanophotonic materials: Multiscale modeling, model optimization, and model inversion

Yip, S and Short, M. P. (2013) 'Multiscale materials modelling at the mesoscale' Nat. Mater., 12, 774–777.

Owhadi H., Scovel C., Sullivan T.J., McKerns M., and Ortiz M. (2013) 'Optimal uncertainty quantification' SIAM Rev., 55, 2, 271–345.

Dienstsfrey A. and Boisvert R.F. (2011) Uncertainty quantification in scientific computing, Heidelberg, Dordrecht, London, New York. Springer.

Liu, V. and Miller, D.A.B. and Fan, S. (2013) 'Highly tailored computational electromagnetics methods for nanophotonic design and discovery' Proc. IEEE, 101, 2, 484–493.

Acknowledgments

The authors thank Christina Richey, Timothy J. Sullivan, Katie Rice, Alex Smolyanitsky, Ann Chiaramonti Debay, and Edward Garboczi for enlightening and helpful discussions.

This research was performed while one of the authors (LR) held a National Research Council Research Associateship Award at the National Institute of Standards and Technology. This work represents an official contribution of the National Institute of Standards and Technology and is not subject to copyright in the United States.

References

[1] S. Nie, S.R. Emory, Probing single molecules and single nanoparticles by surface-enhanced Raman scattering, Science 275 (5303) (1997) 1102–1106.

[2] A. Cuche, A. Drezet, Y. Sonnefraud, O. Faklaris, F. Treussart, J.-F. Roch, S. Huant, Near-field optical microscopy with a nanodiamond-based single-photon tip, Opt. Express 17 (22) (2009) 19969–19980.

[3] T. Nishida, T. Matsumoto, F. Akagi, H. Hieda, A. Kikitsu, K. Naito, T. Koda, N. Nishida, H. Hatano, M. Hirata, Hybrid recording on bit-patterned media using a near-field optical head, J. Nanophotonics 1 (1) (2007), 011597.

[4] E. Ozbay, Plasmonics: merging photonics and electronics at nanoscale dimensions, Science **311** (5758) (2006) 189–193.

[5] N. Liu, H. Wei, J. Li, Z. Wang, X. Tian, A. Pan, H. Xu, Plasmonic amplification with ultra-high optical gain at room temperature, Sci. Rep. 3 (2013) 1967.

[6] J.A. Dionne, H.A. Atwater, Plasmonics: metal-worthy methods and materials in nanophotonics, MRS Bull. 37 (8) (2012) 717–724.

[7] Z. Fei, G.O. Andreev, W. Bao, L.M. Zhang, A.S. McLeod, C. Wang, M.K. Stewart, Z. Zhao, G. Dominguez, M. Thiemens, Infrared nanoscopy of Dirac plasmons at the graphene–SiO2 interface, Nano Lett. 11 (11) (2011) 4701–4705.

[8] H. Yan, T. Low, W. Zhu, Y. Wu, M. Freitag, X. Li, F. Guinea, P. Avouris, F. Xia, Damping pathways of mid-infrared plasmons in graphene nanostructures, Nat. Photonics 7 (5) (2013) 394–399.

[9] A. Bostwick, T. Ohta, T. Seyller, K. Horn, E. Rotenberg, Quasiparticle dynamics in graphene, Nat. Phys. 3 (1) (2007) 36–40.

[10] Z. Ni, Q. Liu, K. Tang, J. Zheng, J. Zhou, R. Qin, Z. Gao, D. Yu, J. Lu, Tunable bandgap in silicene and germanene, Nano Lett. 12 (1) (2011) 113–118.

[11] T.H. Osborn, A.A. Farajian, O.V. Pupysheva, R.S. Aga, L.L.Y. Voon, Ab initio simulations of silicene hydrogenation, Chem. Phys. Lett. 511 (1–3) (2011) 101–105.

[12] T.-T. Jia, X.-Y. Fan, M.-M. Zheng, G. Chen, Silicene nanomeshes: bandgap opening by bond symmetry breaking and uniaxial strain, Sci. Rep. 6 (1) (2016) 1–8.

[13] N. Gao, J. Li, Q. Jiang, Bandgap opening in silicene: effect of substrates, Chem. Phys. Lett. 592 (2014) 222–226.

[14] Q.H. Wang, K. Kalantar-Zadeh, A. Kis, J.N. Coleman, M.S. Strano, Electronics and opto-electronics of two-dimensional transition metal dichalcogenides, Nat. Nanotechnol. 7 (11) (2012) 699.

[15] C.R. Dean, A.F. Young, I. Meric, C. Lee, L. Wang, S. Sorgenfrei, K. Watanabe, T. Taniguchi, P. Kim, K.L. Shepard, Boron nitride substrates for high-quality graphene electronics, Nat. Nanotechnol. 5 (10) (2010) 722.

[16] A. Hosseini, Y. Massoud, A low-loss metal-insulator-metal plasmonic Bragg reflector, Opt. Express 14 (23) (2006) 11318–11323.

[17] P. Hohenberg, W. Kohn, Inhomogeneous electron gas, Phys. Rev. 136 (3B) (1964) B864–B871.

[18] W. Kohn, L.J. Sham, Self-consistent equations including exchange and correlation effects, Phys. Rev. 140 (4A) (1965) A1133.

[19] A. Janotti, C.G. Van de Walle, LDA + U and hybrid functional calculations for defects in ZnO, SnO2, and TiO2, Phys. Status Solidi B 248 (4) (2011) 799–804.

[20] M. Kuisma, J. Ojanen, J. Enkovaara, T. Rantala, Kohn-Sham potential with discontinuity for band gap materials, Phys. Rev. B 82 (11) (2010) 115106.

[21] J. Yan, K.W. Jacobsen, K.S. Thygesen, Optical properties of bulk semiconductors and graphene/boron nitride: the Bethe-Salpeter equation with derivative discontinuity-corrected density functional energies, Phys. Rev. B 86 (4) (2012), 045208.

[22] J.M. Soler, E. Artacho, J.D. Gale, A. García, J. Junquera, P. Ordejón, D. Sánchez-Portal, The SIESTA method for ab initio order-N materials simulation, J. Phys. Condens. Matter 14 (11) (2002) 2745–2779.

[23] G. Kresse, J. Furthmüller, Efficient iterative schemes for ab initio total-energy calculations using a plane-wave basis set, Phys. Rev. B 54 (16) (1996) 11169–11186.

[24] G. Kresse, J. Hafner, Ab initio molecular-dynamics simulation of the liquid-metal–amorphous-semiconductor transition in germanium, Phys. Rev. B 49 (20) (1994) 14251–14269.

[25] P. Giannozzi, S. Baroni, N. Bonini, M. Calandra, R. Car, C. Cavazzoni, D. Ceresoli, G.L. Chiarotti, M. Cococcioni, I. Dabo, QUANTUM ESPRESSO: a modular and open-source software project for quantum simulations of materials, J. Phys. Condens. Matter 21 (39) (2009) 395502.

[26] A. Marini, C. Hogan, M. Grüning, D. Varsano, Yambo: an ab initio tool for excited state calculations, Comput. Phys. Commun. 180 (8) (2009) 1392–1403.

[27] J.J. Mortensen, L.B. Hansen, K.W. Jacobsen, Real-space grid implementation of the projector augmented wave method, Phys. Rev. B 71 (3) (2005), 035109.

[28] J.E. Enkovaara, C. Rostgaard, J.J. Mortensen, J. Chen, M. Dułak, L. Ferrighi, J. Gavnholt, C. Glinsvad, V. Haikola, H. Hansen, Electronic structure calculations with GPAW: a real-space implementation of the projector augmented-wave method, J. Phys. Condens. Matter 22 (25) (2010) 253202.

[29] M. Walter, H. Häkkinen, L. Lehtovaara, M. Puska, J. Enkovaara, C. Rostgaard, J.J. Mortensen, Time-dependent density-functional theory in the projector augmented-wave method, J. Chem. Phys. 128 (24) (2008) 244101.

[30] J. Yan, J.J. Mortensen, K.W. Jacobsen, K.S. Thygesen, Linear density response function in the projector augmented wave method: applications to solids, surfaces, and interfaces, Phys. Rev. B 83 (24) (2011) 245122.

[31] X. Gonze, A brief introduction to the ABINIT software package, Z. Kristallogr. Cryst. Mater. 220 (5/6) (2005) 558–562.

[32] X. Gonze, B. Amadon, P.-M. Anglade, J.-M. Beuken, F. Bottin, P. Boulanger, F. Bruneval, D. Caliste, R. Caracas, M. Côté, ABINIT: first-principles approach to material and nano-system properties, Comput. Phys. Commun. 180 (12) (2009) 2582–2615.

[33] P. Lambin, J.-P. Vigneron, A. Lucas, Electron-energy-loss spectroscopy of multilayered materials: theoretical aspects and study of interface optical phonons in semiconductor superlattices, Phys. Rev. B 32 (12) (1985) 8203–8215.

[34] A. Dereux, J.-P. Vigneron, P. Lambin, A.A. Lucas, Polaritons in semiconductor multilay-ered materials, Phys. Rev. B 38 (8) (1988) 5438–5452.

[35] R. Ritchie, A. Marusak, The surface plasmon dispersion relation for an electron gas, Surf. Sci. 4 (3) (1966) 234–240.

[36] L. Rast, T. Sullivan, V.K. Tewary, Stratified graphene/noble metal systems for low-loss plasmonics applications, Phys. Rev. B 87 (4) (2013), 045428.

[37] P.B. Johnson, R.W. Christy, Optical constants of the noble metals, Phys. Rev. B 6 (12) (1972) 4370–4379.

[38] S.P. Muraka, M. Eizenberg, A.K. Sinha, Interlayer Dielectrics for Semiconductor Technologies, Vol. 1, Elsevier, 2003.

[39] G.-C. Yi, Semiconductor Nanostructures for Optoelectronic Devices: Processing, Characterization and Applications, Springer Science & Business Media, 2012.

[40] R.H. Ritchie, Plasma losses by fast electrons in thin films, Phys. Rev. 106 (5) (1957) 874–881.

[41] L. Rast, V.K. Tewary, Electron Energy Loss Function of Silicene and Germanene Multilayers on Silver, arXiv preprint arXiv:1311.0838, 2013.

[42] W. Wei, Y. Dai, B. Huang, T. Jacob, Many-body effects in silicene, silicane, germanene and germanane, Phys. Chem. Chem. Phys. 15 (22) (2013) 8789–8794.

[43] F. Bechstedt, L. Matthes, P. Gori, O. Pulci, Infrared absorbance of silicene and germanene, Appl. Phys. Lett. 100 (26) (2012) 261906.

[44] C.A. Rozzi, D. Varsano, A. Marini, E.K. Gross, A. Rubio, Exact Coulomb cutoff technique for supercell calculations, Phys. Rev. B 73 (20) (2006) 205119.

[45] E. Scalise, Vibrational properties of silicene and germanene, in: Vibrational Properties of Defective Oxides and 2D Nanolattices, Springer, 2014, pp. 61–93.

[46] T. Nagao, S. Yaginuma, C. Liu, T. Inaoka, V. Nazarov, T. Nakayama, M. Aono, Plasmon confinement in atomically thin and flat metallic films, in: NanoScience + Engineering, SPIE, 2007.

[47] C. Sciammarella, L. Lamberti, F. Sciammarella, G. Demelio, A. Dicuonzo, A. Boccaccio, Application of plasmons to the determination of surface profile and contact strain distribution, Strain 46 (4) (2010) 307–323.

[48] T. Ohta, A. Bostwick, J.L. McChesney, T. Seyller, K. Horn, E. Rotenberg, Interlayer interaction and electronic screening in multilayer graphene investigated with angle-resolved photoemission spectroscopy, Phys. Rev. Lett. 98 (20) (2007) 206802.

[49] M. Yankowitz, Q. Ma, P. Jarillo-Herrero, B.J. LeRoy, Van der Waals heterostructures combining graphene and hexagonal boron nitride, Nat. Rev. Phys. 1 (2) (2019) 112–125.

[50] C. Zhang, C.-P. Chuu, X. Ren, M.-Y. Li, L.-J. Li, C. Jin, M.-Y. Chou, C.-K. Shih, Interlayer couplings, Moiré patterns, and 2D electronic superlattices in MoS_2/WSe_2 heterobilayers, Sci. Adv. 3 (1) (2017), e1601459.

[51] E. Macia, Exploiting aperiodic designs in nanophotonic devices, Rep. Prog. Phys. 75 (3) (2012), 036502.

[52] C. Kim, G. Pilania, R. Ramprasad, From organized high-throughput data to phenomenological theory using machine learning: the example of dielectric breakdown, Chem. Mater. 28 (5) (2016) 1304–1311.

[53] F.A. Faber, L. Hutchison, B. Huang, J. Gilmer, S.S. Schoenholz, G.E. Dahl, O. Vinyals, S. Kearnes, P.F. Riley, O.A. Von Lilienfeld, Prediction errors of molecular machine learning models lower than hybrid DFT error, J. Chem. Theory Comput. 13 (11) (2017) 5255–5264.

[54] F. Brockherde, L. Vogt, L. Li, M.E. Tuckerman, K. Burke, K.-R. Müller, Bypassing the Kohn-Sham equations with machine learning, Nat. Commun. 8 (1) (2017) 1–10.

[55] K.T. Butler, D.W. Davies, H. Cartwright, O. Isayev, A. Walsh, Machine learning for molecular and materials science, Nature 559 (7715) (2018) 547–555.

Electronic and electromechanical properties of vertical and lateral 2D heterostructures

Sri Ranga Jai Likith and Cristian V. Ciobanu
Colorado School of Mines, Golden, CO, United States

3.1   Introduction

With the recent advances in controllable growth of large-area, high-quality, two-dimensional (2D) layered materials (e.g., graphene [1–4], hexagonal boron nitride (hBN) [5–8], and transition metal dichalcogenides (TMDC)) [9,10], the field has naturally progressed toward the design and/or investigation of heterostructures based on out-of-plane [9,11–16] or in-plane [17–19] stacking of these nanomaterials. For instance, out-of-plane heterostructures of graphene on boron nitride [15,20] offer the best avenue for preserving the exotic properties of graphene and using them in nanoscale devices, because of the near-perfect epitaxial lattice match, small van der Waals (vdW) interactions between graphene and hBN, and most importantly because of the insulating nature of the underlying hBN layer(s). For applications where a bandgap is necessary and close control of that bandgap over a wide range is desired, heterostructures may offer a good avenue for creating and controlling the bandgap. New and stable heterostructures offer a way to make available a finer range of important electronic properties such as the bandgap; that range is rather sparse at the moment, with graphene being semimetallic, hBN insulator with a badgap of $\sim 5\,eV$, and MoS_2 semiconducting with a gap of $\sim 1.8\,eV$. The ideas of bandgap control by exploiting the domain width as a control parameter come from the studies of nanoribbons of graphene [21], in which a bandgap appears due to the presence of edges. The bandgap varies with the width of the nanoribbon, and, moreover, it can be induced in one spin component and not in the other—a phenomenon commonly referred to half-metallicity [22]. Like graphene nanoribbons, the hybrid domains include edges that separate the domains of each type; these edges are multiple and periodic in space, are not reactive (hence more stable), and are expected to produce similar novel electronic properties as well. Indeed, recent density functional theory computations on graphene-hBN hybrid heterostructures have revealed half-metallicity [23]. This is an important advance, since creating a bandgap in one spin component and not the other is the key to selective electronic transport based on spin (spintronics), which may lead to novel fundamental science and applications in terms of magnetic memories/storage, spin-based logic devices, and possibly others. Further, TMDC-based vertical heterostructures have received significant interest since, in

Modeling, Characterization and Production of Nanomaterials. https://doi.org/10.1016/B978-0-12-819905-3.00003-8

addition to controlling the composition of the layers, changing the relative stacking registries of the layers may have a significant effect on the properties of the heterostructures. They have also been shown to be attractive candidates for various electronic devices such as light-harvesting and detecting devices and flexible electrodes [10].

In this chapter, we present first a set of results (Sections 3.2–3.4) of first-principles studies of the electronic and piezoelectric properties [24,25] of van der Waals-layered out-of-plane TMDC heterostructures. The effects of composition as well as interlayer stacking registries are explored. We then present a brief review of the current state of the hybrid domains in terms of their synthesis, mechanical, and electronic properties, which are presented in separate sections. With excellent mechanical stability, exotic electronic properties, as well as piezoelectric properties, the bottleneck toward the large-scale usage of these heterostructures is probably the current (lack of) availability of large-scale synthesis techniques with reliable and controllable hybrid-domain dimensions. Nevertheless, given the pace at which the field is moving, it is expected that this desire of reliable, controllable, large-scale synthesis will be fulfilled—as can be inferred from very recent works on interface formation between graphene and hBN [17] and on using graphene edges as templates for the one-dimensional epitaxial growth of hBN domains [26].

3.2   Structural relaxation of vertical TMDC heterostructures

The 2H TMDC monolayers consist of one layer of transition metal atoms that are chemically bonded to and sandwiched between two layers of chalcogen atoms. The stacking of atoms within a monolayer can be denoted as bAb, where the first letter in this descriptor refers to the registry of the bottom-most layer of chalcogen atoms, and the subsequent layers are denoted in increasing order of their height along the out-of-plane lattice vector. (Lowercase letters represent chalcogen atoms occupying a given layer while uppercase letters denote a transition metal atoms.) Thus, the 2H bulk phase of TMDCs, which consists of two such monolayers in the unit cell, separated by a vdW gap, can be represented as bAb-aBa. The relative placement of the second layer can be achieved in one of six different high-symmetry stacking "sequences," i.e., bAb-aBa, bAb-aCa, bAb-cBc, bAb-cAc, bAb-bCb, and bAb-aBa. Here, we study the effects of composition and interlayer stacking registry of out-of-plane heterostructures of the form MX_2-NY_2 (where M, N: Mo, W; X, Y: S, Se, Te) as well as their parent MX_2 bulk structures.

The energies of bulk heterostructures are used to assess their relative stability. We have used density functional theory (DFT) calculations [27,28] at the level of generalized gradient approximation [29] to relax the structure and compute their electronic and piezoelectric properties. Table 3.1 lists the relative energy per formula unit for bulk MoS_2 in each of the six stacking sequences. The stacking sequence is seen to significantly influence the relative stability of the structures. Stacking sequences where the atoms of the second layer are directly above holes formed between the

Table 3.1 Computed energies per formula unit (in meV), relative to the most stable stacking registry for bulk MX$_2$.

Stacking	MoS$_2$	MoSe$_2$	MoTe$_2$	WS$_2$	WSe$_2$	WTe$_2$
bAb-aBa	0	0	0	0	0	0
bAb-aCa	0.46	3.01	12.83	2.56	5.56	16.05
bAb-cBc	0.47	3.01	12.83	2.56	5.56	16.06
bAb-cAc	16.77	29.78	78.08	20.64	35.68	89.43
bAb-bCb	71.95	77.14	102.00	70.35	75.71	100.32
bAb-bAb	132.13	163.93	259.40	131.97	164.14	264.75

atoms of the first layer, in general, are relatively more stable than the stacking sequences where the second layer of atoms is positioned directly atop the atoms of the first layer, even though the layers are separated by relatively large vdW gaps.

First, we consider the two stacking sequences where the metal atoms of the second layer are directly atop the metal atoms of the first (bottom-most) layer, i.e., bAb-cAc and bAb-bAb. Between them, we see that the bAb-bAb stacking sequence, since it has all atoms of the second layer directly atop their counterparts in the first layer, corresponds to a considerably higher final energy compared to the bAb-cAc stacking sequence. The bAb-cAc stacking sequence has the chalcogen atoms of the second layer stacked directly above the holes between chalcogen atoms of the first layer, and hence, is relatively more stable. Next, we consider the bAb-aBa and the bAb-cBc stacking sequences. Both have the metal atoms of the second layer stacked directly above the hole sites formed by the metal atoms of the first layer. In addition, both stacking sequences consist of the chalcogen atoms of the second layer stacked above hole sites of the chalcogen atoms of the first layer. These are the two most stable stacking sequences. This is true for all the heterostructures, as well as the parent bulk TMDCs. Next, we consider the bAb-aCa and the bAb-bCb stacking sequences. The bAb-bCb stacking sequence has the chalcogen atoms of the second layer stacked directly atop the chalcogen atoms of the first layer, leading to a lower stability compared to the bAb-aCa stacking sequence; even though the chalcogen atoms of the second layer are directly atop the metal atoms of the first layer, these atoms are relatively further apart compared to neighboring layers of chalcogen atoms, and hence, has a smaller effect on the stability. In fact, the bAb-aCa stacking sequence shows a higher stability compared to the bAb-cAc stacking sequence where the metal atoms of the second layer are stacked directly atop those in the first layer even though the chalcogen atoms in the bAb-cAc stacking sequence are stacked in hole sites.

An important criterion for comparing the stabilities of the heterostructures is the strain introduced during the structural relaxation. This is an artifact due to the periodic boundary conditions that are inherent to DFT. The strain in a given heterostructure is calculated as

$$\bar{\epsilon} = \frac{\epsilon_1 + \epsilon_2}{2},$$

where ϵ_i is the strain in the ith layer of the heterostructure, which are in turn given by

$$\epsilon_i = \frac{a - a_i}{a_i},$$

with a is the in-plane lattice constant of the relaxed bulk heterostructure and a_i is the in-plane lattice constant of the ith parent bulk structure. The in-plane lattice constants of the parent bulk structures are only sensitive to the chalcogen species. Since this directly affects the strain developed in a heterostructure, heterostructures with the same species of chalcogens across layers show almost zero strain, e.g., MoS_2-WS_2, $MoSe_2$-WSe_2, $MoTe_2$-WTe_2. This is because the radii of the two metals are very similar (Mo^{4+}: 79 pm, W^{4+}: 80 pm), and hence, a metal mismatch between the monolayers of a heterostructure does not lead to significant strain. A strain of about 2% is observed in heterostructures where the chalcogen species are S and Se between layers. This is because of an S-Se chalcogen mismatch between the monolayers, which stems from the difference in the chalcogens' ionic radii (S^{2-}: 184 pm, Se^{2-}: 198 pm), e.g., MoS_2-$MoSe_2$, MoS_2-WSe_2, etc. The next higher level of strain (of about 3.5%) is seen in heterostructures with Se and Te, e.g., $MoSe_2$-$MoTe_2$, $MoSe_2$-WTe_2, etc. This is the result of the larger difference in the ionic radii of Se^{2-} and Te^{2-}, which is 23 pm, compared to the difference of 14 pm between S^{2-} and Se^{2-}. The largest strains (of about 5.5%) are seen in heterostructures with an S-Te chalcogen mismatch due to the large difference in ionic radii of S^{2-} and Te^{2-}. The stacking sequence itself has little to no influence on the strain in a heterostructure.

3.3 Electronic properties of TMDC vertical heterostructures

Tables 3.2 and 3.3 show the electronic bandgaps and the corresponding transitions of all the zero-strain and the 2%-strain heterostructures, respectively. In the zero-strain structures, the Γ, K, and H points are very close in energy to each other at the top of the valence band. Hence, in some structures like WTe_2, simply changing the stacking sequence from bAb-aBa to bAb-aCa (which is another high-stability stacking sequence) can change the electronic band gap transition from K-Ψ to Γ-Ψ. Similarly, the zero-strain $MoTe_2$-WTe_2 heterostructure shows a shift from a K-Ψ transition (Fig. 3.1A) to an H-Ψ transition (Fig. 3.1B) when going from bAb-aBa to the bAb-aCa stacking sequence. In the 2%-strain heterostructures, in addition to the Γ, K, and H points being close in energy at the top of the valence band, the K and H points are also very close in energy at the bottom of the conduction band. As a result, not only does the MoS_2-WSe_2 heterostructure show a shift from a Γ-K transition to an H-K transition going from bAb-aBa to the bAb-aCa stacking sequence, it then shows a Γ-H transition in the bAb-cBc stacking sequence. This is a consequence of the K and H points being very close in energy at the bottom of the conduction band. In

Table 3.2 Bandgaps (eV) and the corresponding transitions for the zero-strain heterostructures.

	MoS_2	$MoSe_2$	$MoTe_2$	WS_2	WSe_2	WTe_2	MoS_2-WS_2	$MoSe_2$-WSe_2	$MoTe_2$-WTe_2
bAb-aBa	0.89 Γ-Φ	0.85 Γ-Φ	0.78 Γ-Ψ	1.01 Γ-Φ	0.97 Γ-Φ	0.79 K-Ψ	0.92 Γ-Φ	0.89 Γ-Φ	0.75 K-Ψ
bAb-aCa	0.90 Γ-Φ	0.86 Γ-Ψ	0.77 Γ-Ω	1.03 Γ-Φ	1.01 Γ-Φ	0.91 Γ-Ψ	0.92 Γ-Φ	0.90 Γ-Ψ	0.80 H-Ψ
bAB-cBc	0.90 Γ-Φ	0.86 Γ-Ψ	0.77 Γ-Ω	1.03 Γ-Φ	1.01 Γ-Φ	0.91 Γ-Ψ	0.95 Γ-Φ	0.92 Γ-Ψ	0.84 Γ-Ψ
bAb-cAc	0.90 Γ-Ψ	0.80 Γ-Ψ	0.60 Γ-Ω	1.02 Γ-Φ	0.95 Γ-Ψ	0.78 Γ-Ω	0.94 Γ-Φ	0.87 Γ-Ψ	0.67 Γ-Ω
bAb-bCb	1.25 Γ-Φ	1.20 Γ-Φ	0.98 H-Ψ	1.34 Γ-Φ	1.30 Γ-Φ	1.02 H-Ψ	1.25 Γ-Φ	1.19 H-Φ	0.90 H-Ψ
bAb-bAb	0.95 Γ-Φ	0.88 Γ-Φ	0.65 Σ-Ψ	1.04 Γ-Φ	0.98 Γ-Φ	0.70 K-Ψ	1.00 Γ-Φ	0.94 Γ-Φ	0.66 K-Ψ

Σ: (0.088,0.088,0).
Φ: (0.175,0.175,0).
Ψ: (0.193,0.193,0).
Ω: (0.211,0.211,0).

Table 3.3 Bandgaps (eV) and the corresponding transitions for the 2%-strain heterostructures.

	MoS_2-$MoSe_2$	WS_2-WSe_2	MoS_2-WSe_2	WS_2-$MoSe_2$
bAb-aBa	0.70	0.86	0.62	0.93
	Γ-K	Γ-K	Γ-K	Γ-K
bAb-aCa	0.64	0.82	0.56	0.89
	Γ-K	Γ-K	H-K	Γ-K
bAb-cBc	0.66	0.83	0.59	0.89
	Γ-H	Γ-H	Γ-H	Γ-H
bAb-cAc	0.67	0.85	0.60	0.86
	Γ-K	Γ-K	Γ-K	Γ-K
bAb-bCb	0.78	0.90	0.58	1.10
	H-H	H-H	H-H	H-H
bAb-bAb	0.79	0.92	0.60	0.94
	K-K	K-K	K-K	Γ-Φ

the bAb-bCb stacking sequence, all the 2%-strain heterostructures show a direct H-H band gap. All of them, except WS_2-$MoSe_2$ then shift to a K-K transition band gap in the bAb-bAb stacking sequence.

3.4 Piezoelectric properties of vertical TMDC heterostructures

The in-plane piezoelectric coefficient, e_{21}, of all the structures (where $e_{21} \geq 0.01$ C/m^2) are shown in Fig. 3.2. In the bAb-aBa, bAb-bCb, and bAb-cAc stacking sequences, the in-plane polarization of the two constituent layers develop in opposite directions, hence, the resultant piezoelectric coefficient, e_{21}, is the resultant of these polarization vectors canceling each other out. The remaining stacking sequences, i.e., bAb-aCa, bAb-bAb, and bAb-cBc, have the in-plane polarization of the constituent layers developing in the same direction. Structures in these stacking sequences, including the parent MX_2 structures, show an in-plane piezoelectric coefficient, e_{21} varying from 0.1 to 0.6 C/m^2. Within each "group" of stacking sequences, there is very little variation in piezoelectric coefficients, indicating a weak coupling between layers across the van der Waals gap. Fig. 3.3 shows the out-of-plane piezoelectric coefficients, e_{33} of all the structures where $e_{33} \geq 0.01$ C/m^2. Two stacking sequences, bAb-bAb and bAb-cAc show piezoelectric responses in the stacking direction. In these sequences, even the zero-strain (e.g., the parent MX_2 structures) and 2%-strain heterostructures (e.g., WSe_2-WS_2) show non-zero out-of-plane piezoelectric responses (e_{33}).

Sections 3.2–3.4 that illustrate how electronic and electromechanical properties can vary in vertical heterostructures. These variations occur in response to strain between the layers, and stacking sequence. The results likely to bear out in experiments are those corresponding to low strains, because any large mismatch strain between the layers will in practice be accommodated via bending, interlayer rotation

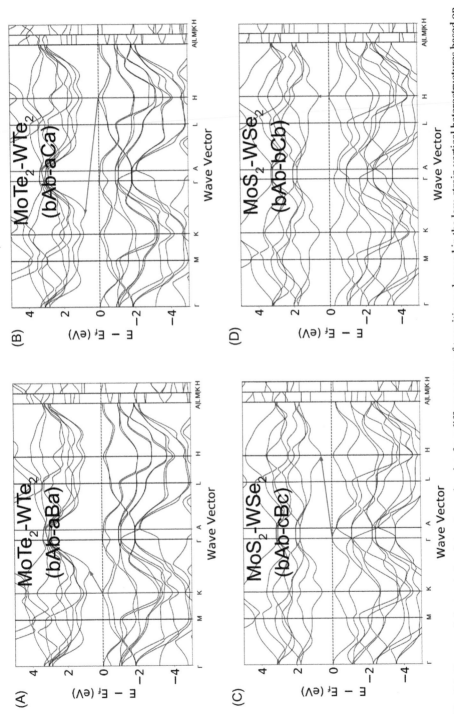

Fig. 3.1 Computed electronic band structures showing four different types of transitions observed in the low-strain vertical heterostructures based on TMDCs. The *arrows* show the lowest energy transition from valence to conduction states: (A) K-Ψ, (B) H-Ψ, (C) Γ-H, and (D) H-H.

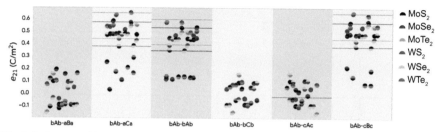

Fig. 3.2 Computed in-plane piezoelectric coefficient, e_{21}, of all the structures in the six different stacking sequences. The piezoelectric coefficients of the parent MX_2 bulk structures for a given stacking sequence are shown as *solid lines*. Structures with $|e_{21}| < 0.01$ C/m^2 are not shown.

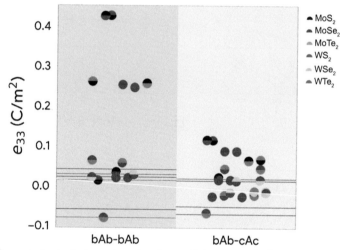

Fig. 3.3 Computed out-of-plane piezoelectric coefficient, e_{33}, of all the structures. bAb-bAb and bAb-cAc are the only stacking sequences where these structures show an out-of-plane piezoelectric response. The piezoelectric coefficients of the parent MX_2 bulk structures for a given stacking sequence are shown as solid lines. Structures with $|e_{21}| < 0.01$ C/m^2 are not shown.

or other mechanisms; large strains are computational artifacts of using unit cells in plane. The following sections deal with lateral heterostructures in different 2D materials; these are often called hybrid-domain superlattices.

3.5 Synthesis of two-dimensional hybrid-domain superlattices

The discovery of coherent and sharp in-plane boundaries between graphene and BN domains [18] practically sparked numerous investigations into the synthesis of interfaces between 2D materials, domain structures, along with the detailed characterization of the synthesis process and their electronic properties. The reason for these new investigations is the implied potential/promise that new 2D materials and combinations thereof could lead to new physical phenomena, to new ways of synthesizing such nanomaterials, and also to new ways of controlling some of their properties so as to facilitate the observation of new phenomena and their applications. Indeed, at least as far as the hybrid-domain monolayer are concerned, they have interesting electronic properties associated with the interfaces between graphene and boron nitride, such as the opening of a variable bandgap [19,23,30], magnetism [20], unique thermal transport properties [31], robust half-metallic behavior without applied electric fields [19,23], and interfacial electronic reconstructions analogous to those observed in oxide heterostructures [32]. Access to these properties depends on methods for controlling the formation of graphene-boron nitride interfaces within a single atomic layer.

The synthesis approach of Ci et al. was in essence chemical vapor deposition (CVD) performed on copper, and using methane and ammonia borane as simultaneous carbon and BN sources, respectively. The atomic percentage of C was controlled in a desired range, while the B:N ratio was always unity. Subsequent lithographic patternic was performed, along with a suite of characterization experiments (the data from some of which being shown in Fig. 3.4). By combining in-plane domains of carbon and BN, a whole range of electronic transport properties could be engineered. It is therefore not surprising that subsequent work has focused on closer control of the interfaces and domains widths. It is not the intent to review here all the subsequent work on the formation and characterization of domains and interfaces between graphene and boron nitrid; mentioning a couple of key examples of engineering graphene-BN interfaces would suffice for the purpose of illustrating the degree of control that is currently achievable in experiments.

In recent experiments, Sutter et al. (see scanning tunneling microscopy images in Fig. 3.5) [17], as well as Han et al. [33], have shown that in-plane monolayer domains of graphene and hBN can form both diffuse interfaces and sharp ones, depending on how one controls the amount of carbon adatoms remaining on the surface during the

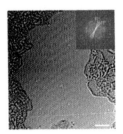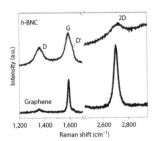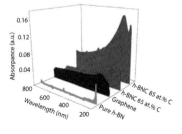

Fig. 3.4 Evidence of graphene-BN domain boundaries from Ci et al. [18]. *Left*: High resolution transmission electron images of moire pattern of hybridizer graphene-BN structure.
Middle: Raman spectrum of a hybridized domain structure in comparison to that of graphene.
Right: Adsorption spectra of pure hBN, pure graphene, and hybridized domain is varying levels of carbon.
Adapted with permission from Nature Publishing Group.

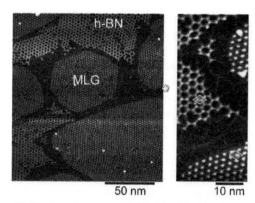

Fig. 3.5 Graphene and h-BN domains on Ru(0001), identified by the distinct moiré patterns in scanning tunneling microscopy experiments. In the region between the clearly identified graphene and hBN, any moiré patterns are absent but the substrate is not barren: in that region, the substrate is covered with a C-B-N alloy.
From P. Sutter, R. Cortes, J. Lahiri, E. Sutter, Nano Lett. 12(9) (2012) 4869–4874 with permission from American Chemical Society.

CVD growth of hBN. In both experiments, the formations of the interfaces was carefully controlled and characterized. This seems to indicate that large scale production of in-plane superlattices with alternate domains of different atomic/molecular types should not be too far away.

3.6 Mechanical properties of heterostructures

The discovery of graphene [34] has been followed by observations of a variety of other 2D monolayer crystals [35,36].in particular, hBN. The near-match of the lattice parameters of graphene and BN has triggered theoretical studies of combined graphene-BN ribbons [23,37,38], and more recently domain-hybridized graphene-BN monolayers have been fabricated [18]. Meanwhile, still-hypothetical silicon carbide nanoribbons have also been attracting growing interest [39–41], motivated by the synthesis of silicon carbide nanotubes [42]. While these studies have been focused on the electronic and magnetic properties, rigorous study on their structural and mechanical properties have only appeared later [19,43,44]; we draw our review of mechanical properties from these studies, in particular from that by Jun et al. [19] In order to tailor their physical and chemical properties for future nanoscale device applications, we need to understand their structural and mechanical properties, in particular their elastic stability.

In this section, we review the fundamental elastic properties and deformation behaviors of the pristine edges in boron nitride and silicon carbide monolayer nanoribbons (BNNR and SiCNR, respectively), and of the domain boundaries in freestanding domain hybridized graphene-BN monolayer superlattices (CBNSL). Edge and boundary energies have been reported for certain cases [18], but these can only provide us with knowledge of chemical stability. To understand their mechanical stability and deformation behavior, we start by determining the edge (or boundary in the case of heterophase domains) stresses for nanoribbons. We present below our first-principles DFT calculations of edge (boundary) energy and stress for nanoribbons (domain superlattices), and we analyze the results in comparison with previous reports on graphene nanoribbons (GNR) [43–46].

Similar to the interface energy and stress in 3D crystals [47,48], the boundary energy quantifies the cost to form a new boundary separating different domains (i.e., graphene and BN domains), while boundary stress characterizes the work necessary to stretch a pre-existing boundary along the boundary direction. As a brief definition, the boundary energy is the total excess energy (with respect to individual bulk 2D crystals), possessed by all atoms in the vicinity of the boundary, per unit boundary length. On the other hand, the boundary stress characterizes the work per unit boundary length needed to deform a boundary by elastically straining both monolayer domains by the same amount of strain along the boundary direction [48]. The relationship between boundary energy σ and boundary stress h is similar to the Shuttleworth relation [49], and is written as

$$h = \sigma + \frac{d\sigma}{de}$$

where h and the strain e are scalar quantities since we only consider one-dimensional deformation along the straight line of boundary. We note that e is the elastic strain applied to the boundary across which the lattices of both 2D crystal phases are already matched. We do not include in the boundary energy the strain that is required to match one lattice with another. We have confirmed that this mismatch strain makes less than 2% difference in the boundary-energy values because the difference between the lattice parameters of perfect graphene and BN phases is sufficiently small. The above equation also applies to the relation between edge energy and edge stress.

Boundary energies are calculated from the total-energy difference between model systems with and without boundary, and boundary stresses are obtained by numerical differentiation of the boundary energy calculated at a series of strain values within the elastic range. The total-energy calculation is thus the main numerical procedure in this work. Total-energy DFT calculations were performed using the program SIESTA [50] based on local density approximation (LDA), with pseudopotentials and basis set functions similar to those used in earlier work by Jun. [44] An energy cutoff of 250 Ry was set for the real-space integrations. The atomic positions were relaxed via conjugate gradient procedure with force tolerances set to $0.02\,\text{eV/\AA}$. We have first obtained the lattice constants of perfect graphene ($2.468\,\text{\AA}$), hBN ($2.492\,\text{\AA}$), and SiC ($3.081\,\text{\AA}$) monolayers; these values agree well with previous reports, for example, $2.465\,\text{\AA}$ for graphene [51,52], $2.49\,\text{\AA}$ for BN [36], and $3.094\,\text{\AA}$ for two-dimensional SiC [53].

We considered two types of boundary models, armchair and zigzag, depending on the way in which the two phases face each other at the C-BN boundary. Experimental fabrication of domain-hybridized graphene-BN monolayers revealed that the local atomic structure of heterophase boundaries is either armchair or zigzag, although at large length scales the boundary has an arbitrary shape [18]. Two examples of supercells with domain boundaries for armchair and zigzag graphene-BN superlattices (aCBNSL22 and zCBNSL12, respectively). The numbers in our naming scheme follow the convention for nanoribbons, and thus give the supercell width as the number of dimers in armchair models, or as the number of zigzag lines in zigzag supercells. We kept the same number of dimer (or zigzag) lines for the graphene and BN domains in a supercell, however, it is expected that varying independently the widths of the two domains will provide a valuable additional degree of control.

For the C-BN superlattice models, we computed the boundary energy as

$$\sigma = \frac{E_T - N_C E_C - N_{BN} E_{BN}}{2L}$$

where E_T is the total energy of a model supercell which has N_C number of carbon atoms and N_{BN} number of boron-nitrogen pairs. L is the length of the boundary, E_C is the energy per atom in perfect graphene, and E_{BN} is the energy per B-N pair in perfect BN monolayer. Without the term of E_{BN} (E_C), the excess energy σ becomes the edge energy of a graphene (boron nitride) nanoribbon. To calculate the energy of one carbon atom in perfect graphene, or one BN or SiC pair in perfect hBN or perfect SiC, respectively, we employed their fully periodic 4-atom unit cells, and chose $16 \times 16 \times 1$ ($24 \times 24 \times 1$ for SiC) k-point mesh by the Monkhorst-Pack scheme. This k-point sampling was scaled according to the size of supercells. Elastic properties of graphene edges have also been reported in other works [43–46].

In Fig. 3.6, we present our numerical results of edge (boundary) energy and stress for all models with respect to the width of ribbon (stripe). As the first observation, the boundary and edge energies show little dependence on the widths. The width-averaged edge (and boundary) energy and stress values are summarized in Table 3.4.

For both armchair and zigzag cases, boundary energies are substantially lower than edge energies of pristine graphene, BN, and SiC nanoribbons due to the absence of dangling bonds at the boundary. It is known that armchair edge energy of graphene nanoribbon is lower than its zigzag counterpart. We note the same trend in other ribbon edges and graphene-BN boundaries. However, the physical origins of these lower armchair energies are somewhat different. The difference of edge energy between armchair and zigzag GNRs is approximately 0.300 eV/Å. This relatively large difference comes from the fact that along armchair edge the edge-parallel C—C bond is relaxed to a shorter length than that of interior C—C bond, due to the strong pair of the sp hybridization, which results in the healing of dangling-bond nature of the armchair edge carbon dimers [54]. In contrast, zigzag edge does not change much its atomic structure after relaxation, and thus it does not substantially reduce the high edge energy caused by dangling bonds.

The edge energy difference between armchair and zigzag becomes even larger for BNNRs. After relaxation, the armchair edge in graphene still maintains the hexagonal lattice structure fairly well in spite of the shortened C—C bond. However, in the armchair BNNR (SiCNR as well), the edge hexagons are distorted so significantly that the original hexagonal symmetry is broken. Such distortion provides the armchair BN

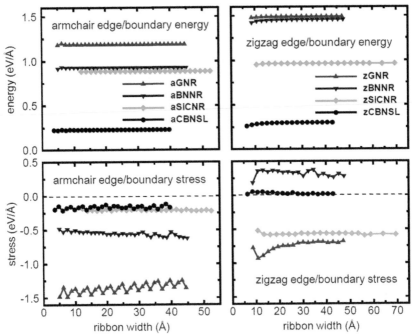

Fig. 3.6 Edge/domain-boundary energy and stress as functions of the nanoribbon/stripe width. Reproduced from S. Jun, X. Li, F. Meng, C.V. Ciobanu, Phys. Rev. B 83(15) (2011) 153407 with permission from American Physical Society.

Table 3.4 Average edge/boundary energy and stress (eV/Å).

	Edge/boundary energy (armchair)	Edge/boundary energy (zigzag)	Edge/boundary stress (armchair)	Edge/boundary stress (armchair)
GNR	1.190	1.490	−1.355	−0.743
BNNR	0.927	1.453	−0.552	+0.323
SiCNR	0.885	0.962	−0.207	−0.507
CBNSL	0.228	0.293	−0.167	+0.027

edge with the opportunity to reduce energy by further relaxing its edge structure while such relaxation is not present for zigzag BN edges. On the other hand, the difference of edge energies between armchair and zigzag SiCNRs is not as large as GNRs and BNNRs. The originally edge-parallel Si—C bond was shortened by 4.93% while those of GNR and BNNR were 12.20% and 9.91% shortened. We therefore believe that the dangling-bond healing effect is less significant in armchair SiCNR edge and that the distortion of edge hexagons is the main source of lowering the armchair edge energy of SiCNR. The boundary energy difference between armchair and zigzag boundaries is remarkably smaller, $0.065\,eV/\text{Å}$, than the above two cases of graphene and BN edges. In our superlattice models of both armchair and zigzag boundaries, we could not observe any obvious change in lattice structure and atomic positions after relaxation, even in the vicinity of boundary. Therefore, the absence of dangling bonds is the key physical reason for which the armchair and zigzag boundaries have similar energies in CBNSL.

The lower two panels of Fig. 3.6 show our results of edge (boundary) stress calculations for armchair and zigzag edges (boundaries). Determinations of BNNR and SiCNR edge stresses and CBNSL boundary stresses have not been previously reported, as only reports of GNR edge stresses are present in the literature so far [43–46]. Edge (boundary) stresses are seen to depend more sensitively on the width of the ribbon (stripe) than the edge (boundary) energies. The average stress values are given in Table 3.4. Our edge stress values for GNR are comparable with those ($-1.45\,eV/\text{Å}$ for armchair and $-0.7\,eV/\text{Å}$ for zigzag) obtained using plane-wave basis [46]. We note that the edge stress of pristine zBNNR is positive. However, all other edges have negative stress values. This means that they are in compression and that their edges tend to ripple whenever they are free to deform at a finite temperature [43]. In contrast, the bare zigzag edge of BNNR is in tension and tends to shorten relative to the interior domain. Therefore, rippling is likely to take place at the interior domain of BN ribbon, away from the zigzag edge, while the edge itself will stay straight. There are two different types of pristine zigzag edges of BNNR, B-terminated, and N-terminated edges. If one edge side is B-ended then the other side is N-ended. Since the charge densities of N- and B-terminated edges are substantially different, their edge energies and edge stresses are quite different as well, and consequently these two pristine zigzag edges may deform in distinct ways. The edge energy and stress presented for zBNNR in Table 3.4 are the average values between these two zigzag edge types. This argument also applies to the zigzag edge of SiCNR and to the zigzag boundary of CBN superlattice.

To verify the distinct behaviors of zigzag edges with different terminations, we calculate edge stresses of both sides separately. Since DFT approaches are unable to yield separate edge energies, we performed energy minimizations and classical molecular dynamics simulations using the Tersoff potential [55] for Si-C systems. Since the Tersoff potential energy is a sum over (environment-dependent) contributions of individual atoms, summing the individual atomic contributions over sufficient distance from the edge leads to edge-specific energies upon substraction of the contributions of same atoms in bulk-like (edge-free) configurations. The edge energies and stresses calculated using the Tersoff potential are virtually independent of the ribbon width. The edge energy of C-side (Si-side) zigzag edge is 0.559 (0.596) $eV/\text{Å}$, and their average value is $0.577\,eV/\text{Å}$. The edge stress of C-side (Si-side) zigzag edge is -0.587 (-0.222) $eV/\text{Å}$ while the average value of zigzag edge stress is $-0.406\,eV/\text{Å}$. Although this average edge stress value is somewhat lower than that obtained by DFT calculations ($-0.570\,eV/\text{Å}$), our empirical potential calculations clearly evidence that the C-side zigzag edge has a higher compressive edge stress value than Si-side edge. This implies that the C edge has a higher rippling tendency than the Si edge, which we have also confirmed by constant-temperature molecular dynamics

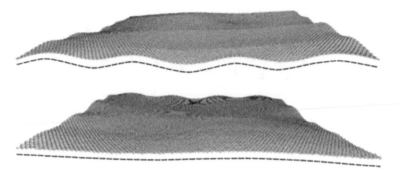

Fig. 3.7 Deformation behaviors of *(top)* C-terminated and *(bottom)* Si-terminated zigzag edges in a SiC monolayer, simulated via constant temperature molecular dynamics using the Tersoff interatomic potentials. Rippling patterns correctly reflect the edge stress values of C-terminated ($-0.587\,\text{eV/Å}$) and Si-terminated ($-0.222\,\text{eV/Å}$) edges. *Dash lines* are a guide to the eye. Reproduced from S. Jun, X. Li, F. Meng, C.V. Ciobanu, Phys. Rev. B 83(15) (2011) 153407 with permission from American Physical Society.

simulations at 300 K. A snapshot of a large SiC monolayer with 18,718 atoms (dimensions of $320.22\,\text{Å} \times 240.34\,\text{Å}$) is shown in the bottom inset of Fig. 3.7, which verifies the more frequent rippling (higher compressive stress) of the C-ended zigzag edge.

One of our findings is that the boundary stresses in graphene-BN superlattices are significantly lower than the edge stresses of the GNR and BNNR. Our results of $-0.167\,\text{eV/Å}$ for armchair and $+0.027\,\text{eV/Å}$ for zigzag boundaries are even lower than the edge stresses of hydrogen-passivated graphene edges, $-0.35\,\text{eV/Å}$ (armchair) and $+0.13\,\text{eV/Å}$ (zigzag) reported earlier [46]. This strongly suggests that C-BN superlattice boundaries experience very low stress and therefore they do not have a tendency to ripple within a hybrid domain heterostructure. We predict that the existence of domain boundaries in graphene-BN monolayer structures will cause neither structural change nor severe deformation.

Another interesting result is the oscillating behavior of boundary energy and stress in armchair CBN superlattices as functions of stripe width. It has been known that the values of armchair graphene edge energy and stress oscillate as the ribbon width is increased [46,56], and that the oscillating period is every three ribbon widths (3-family pattern) [46]. This oscillating behavior is closely related to the energy bandgap of 3-family pattern that has been found in H-passivated semiconducting armchair graphene nanoribbons and understood in the framework of topological analysis of armchair edges [22]. In Fig. 3.8, we present magnified plots for the stresses of four armchair models. Boundary stresses follow a similar 3-family pattern. As shown in the bottom panel of Fig. 3.8, we confirm that armchair graphene-hBN superlattices are semiconducting and their bandgaps depends on the stripe width following a threefold pattern.

In contrast to armchair CBNSL and GNR, the edge energies and stresses of armchair BNNR and SiCNR do not exhibit a clear 3-family pattern in their oscillations because the shapes of edge hexagons are distorted after relaxation, as shown in the insets of Fig. 3.8. Park and Louie [57] reported the oscillating energy bandgap of

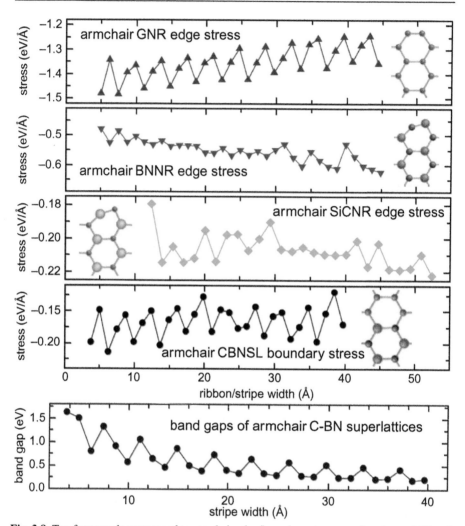

Fig. 3.8 Top four panels correspond to armchair edge/boundary stresses as functions of ribbon/stripe width. Insets show the relaxed structures at the corresponding edges or domain boundaries. The lowest panel shows results of the oscillating bandgaps of C-hBN superlattices with respect to the domain (stripe) width.
Reproduced from S. Jun, X. Li, F. Meng, C.V. Ciobanu, Phys. Rev. B 83(15) (2011) 153407 with permission from American Physical Society.

BNNRs with armchair edges, but in their work the edges were passivated and thus the original shape of the edge hexagons was preserved. However, the pristine armchair BN (SiC) edge changes its configuration significantly and the edge B-N (Si—C) bond is no longer parallel with the edge direction. Consequently, the topological analysis does not apply anymore. This broken hexagonal symmetry eliminates the 3-family pattern and results in irregular fluctuation of edge stress values of armchair BN and SiC nanoribbons.

3.7 Electronic properties of hybrid domain superlattices

The key property of these superlattices appears for those with zigzag edges, and consists in different bandstructures for the two spin components. Fig. 3.9, from a recent report of Pruneda [23], shows the variation of the bandgap for the two spin components.

If the graphene domains are large, then the (expected) semimetallic behavior is recovered. In the other limit, in which the graphene domains are narrow, the antiferromagnetic state becomes unstable and, with it, the system becomes nonmagnetic and insulating. As seen in Fig. 3.9, there is a range of domain widths in which the system is metallic in the spin-down component and semiconducting in the spin-up component. Based on DFT calculations, Pruneda has provided an explanation for the half-metallicity of the hybrid-domain superlattices: [23] the magnetic properties of the stripe edges of graphene, coupled with the polarity of the BN stripes lead to asymmetries in the spin screening, thereby inducing an electronic rearrangement ("reconstruction") at the domain edges. As seen in Fig. 3.9, the bandgap in the spin-up component can be as high as 0.3 eV while the other spin component can be metallic. This finding is anticipated to lead to very interesting phenomena at the 1D domain interfaces, phenomena that may include even superconductivity (by analogy with 2D interfaces in perovskites). Furthermore, electronic conduction is expected to be highly anisotropic, especially within the hBN domains.

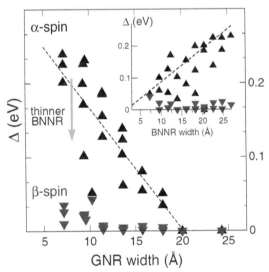

Fig. 3.9 Bandgaps for the spin-up component *(blue triangles)* and spin-down component *(red triangle)* as functions of the stripe width in the antiferromagnetic ground state for the graphene-hBN superlattices. The *green arrow* shows the dependence of the spin-up bandgap on the width of the BN strip.
From J.M. Pruneda, Phys. Rev. B 81(16) (2010) 161409 with permission of American Physical Society.

Another interesting property of this very system, i.e., graphene-BN domain super-lattice, is the very large Siebeck coefficient that can be achieved [58]. This coefficient is crucial in characterizing the efficiency of thermoelectric conversion and the primary indicator of a large thermoelectric figure of merit. While the physical origins of this increase in the Siebeck coefficient is not fully understood, the coefficient has been computed for a wide range of graphene-BN superlattices and its dependence of the stripe width is known. The key to achieve it is a low width for the hBN, which seems to be the factor that has most effect regarding the deviation of the Siebeck coefficient from the value corresponding to pure graphene.

Structurally, the domain boundaries (while not very reactive in a chemical sense) have shown promise as templates for 1D nanostructures (essentially wires or very small atomic clusters) grown on top of the graphene-BN superlattices. In a recent work [59], Haldar and coworkers have calculated the diffusion barriers of Fe atoms on such superlattices. They have shown that iron atoms diffuse much more easily on BN than on graphene, and that during only a few picoseconds at room temperature they diffuse across the C-N domain boundary to get trapped at the C-B boundaries. Furthermore, Haldar et al. showed that the magnetic exchange coupling between Fe clusters at C-B domain boundaries varies nonmonotonically as a function of the hBN stripe width. These superlattices can therefore act as templates for the spontaneous formation of magnetic nanostructures at specific interfaces.

Lastly, we mention that Bristowe et al. [60] have predicted half-metallic 1D inter-faces between two insulating domains, made, for example, several II–VI, III–V, and IV–IV compounds, whose stable bulk phase is wurtzite but which may assume a highly planar phases in atomically thin films. These authors show, expectedly, that 1D domain boundaries in these insulator-insulator superlattices are polar, having a net excess charge that determined from using the formal valence charges of the atomic species involved, irrespective of the predominant covalent character of the bonding in these materials; furthermore, they rationalize such finding on by analyzing the topol-ogy of the formal polarization lattice in the parent bulk materials. DFT calculations similar to those in Ref. [15] show an electronic compensation mechanism due to a Zener-like charge transfer between interfaces of opposite polarity. The emergence of one-dimensional electron and hole gases, which in some cases are ferromagnetic half metallic, is predicted [60].

3.8   Perspectives and concluding remarks

In conclusion, we have reviewed recent reports on the mechanical and electronic prop-erties of hybrid domain superlattices. In terms of mechanical properties, we have shown that a key quantity characterizing the stability and "flatness" of a domain boundary is the domain boundary stress. Calculating the stresses is rather straightfor-ward, with rapid meaningful conclusions for the stability of the domain edges. We have found that the C-BN boundaries experience very little stress and thus, the exis-tence of such domain boundaries will cause neither structural change nor severe defor-mation in C-BN superlattices. The oscillating values of armchair boundary stress indicate that armchair CBNSL may be semiconducting. Furthermore, it is shown that the broken hexagonal symmetry of armchair BN and SiC ribbons results in their irregularly oscillating stress values as functions of ribbon width. Next, we have shown that the two types of zigzag edges in SiC nanoribbons undergo distinctly different deformations.

In terms of electronic properties of in-plane heterostructures, three major ones stand out, all of which depend on the size of the domain-widths: (a) half metallicity, (b) edge magnetism, and (c) giant Siebeck coefficient. These properties pertain to graphene-hBN superlattices, although it is expected that conditions favorable to their occurrence can be met in superlattices made of other materials as well [60]. In addition, we present the results of first-principles studies of the structural, electronic, and piezoelectric properties of out-of-plane TMDC heterostructures. We show that out-of-plane heterostructuring of TMDC monolayers provides us with the ability to finely tune the electronic bandgap, both in magnitude and transition, by altering the composition of the monolayers as well as the interlayer registry. Lastly, we have shown that both the parent structures (MX_2) as well as heterostructures (MX_2-NY_2) can become piezoelectrically active by simply changing the interlayer registry. These exotic properties, along with their immediate "control-knobs" (domain-widths and interlayer registry, respectively), are evolving into a new niche in the field of 2D materials. Fundamental and technological advances based on these properties are expected to rapidly take off in the near future, and we expect that an enhanced understanding of the transport at the interface will lead to game-changing paradigms as far as the miniaturization of nanoelectronics devices is concerned.

Acknowledgments

We are grateful to the volume editors, Dr. Vinod Tewary and Dr. Yong Zhang for their kind invitation to write this chapter, and to the staff at Woodhead Publishing for their unmatched patience and professionalism. We are also indebted to coauthors of Ref. [19], late Prof. Sukky Jun, X. Li, and F. Meng, most of which has been included in this review. The more recent parts of this work have benefited from the financial support of the National Science Foundation through Grant No. DMREF-1534503.

References

[1] X. Li, W. Cai, J. An, S. Kim, J. Nah, D. Yang, R. Piner, A. Velamakanni, I. Jung, E. Tutuc, S.K. Banerjee, L. Colombo, R.S. Ruoff, Science 324 (5932) (2009) 1312–1314.
[2] S. Bae, H. Kim, Y. Lee, X. Xu, J.-S. Park, Y. Zheng, J. Balakrishnan, T. Lei, H.R. Kim, Y. I. Song, Y.-J. Kim, K.S. Kim, B. Ozyilmaz, J.-H. Ahn, B.H. Hong, S. Iijima, Nat. Nanotechnol. 5 (8) (2010) 574–578.
[3] Z. Sun, Z. Yan, J. Yao, E. Beitler, Y. Zhu, J.M. Tour, Nature 468 (7323) (2010) 549–552.
[4] A. Reina, X. Jia, J. Ho, D. Nezich, H. Son, V. Bulovic, M.S. Dresselhaus, J. Kong, Nano Lett. 9 (1) (2009) 30–35.
[5] K.H. Lee, H.-J. Shin, J. Lee, I.-Y. Lee, G.-H. Kim, J.-Y. Choi, S.-W. Kim, Nano Lett. 12 (2) (2012) 714–718.
[6] L. Song, L. Ci, H. Lu, P.B. Sorokin, C. Jin, J. Ni, A.G. Kvashnin, D.G. Kvashnin, J. Lou, B. I. Yakobson, P.M. Ajayan, Nano Lett. 10 (8) (2010) 3209–3215.

[7] P. Arias, A. Ebnonnasir, C.V. Ciobanu, S. Kodambaka, Nano Lett. 20 (4) (2020) 2886–2891.

[8] G. Siegel, C.V. Ciobanu, B. Narayanan, M. Snure, S.C. Badescu, Nano Lett. 17 (4) (2017) 2404–2413.

[9] A. Ebnonnasir, B. Narayanan, S. Kodambaka, C.V. Ciobanu, Appl. Phys. Lett. 105 (2014) 3.

[10] S.-Y. Kim, J. Kwak, C.V. Ciobanu, S.-Y. Kwon, Adv. Mater. 31 (2019) 20.

[11] W. Yang, G. Chen, Z. Shi, C.-C. Liu, L. Zhang, G. Xie, M. Cheng, D. Wang, R. Yang, D. Shi, K. Watanabe, T. Taniguchi, Y. Yao, Y. Zhang, G. Zhang, Nat. Mater. 12 (9) (2013) 792–797.

[12] J. Xue, J. Sanchez-Yamagishi, D. Bulmash, P. Jacquod, A. Deshpande, K. Watanabe, T. Taniguchi, P. Jarillo-Herrero, B.J. Leroy, Nat. Mater. 10 (4) (2011) 282–285.

[13] M. Wang, S.K. Jang, W.-J. Jang, M. Kim, S.-Y. Park, S.-W. Kim, S.-J. Kahng, J.-Y. Choi, R.S. Ruoff, Y.J. Song, S. Lee, Adv. Mater. 25 (19) (2013) 2746–2752.

[14] Z. Liu, L. Song, S. Zhao, J. Huang, L. Ma, J. Zhang, J. Lou, P.M. Ajayan, Nano Lett. 11 (5) (2011) 2032–2037.

[15] A. Ramasubramaniam, D. Naveh, E. Towe, Nano Lett. 11 (3) (2011) 1070–1075.

[16] Y. Murata, S. Nie, A. Ebnonnasir, E. Starodub, B.B. Kappes, K.F. McCarty, C.V. Ciobanu, S. Kodambaka, Phys. Rev. B 85 (20) (2012).

[17] P. Sutter, R. Cortes, J. Lahiri, E. Sutter, Nano Lett. 12 (9) (2012) 4869–4874.

[18] L. Ci, L. Song, C. Jin, D. Jariwala, D. Wu, Y. Li, A. Srivastava, Z.F. Wang, K. Storr, L. Balicas, F. Liu, P.M. Ajayan, Nat. Mater. 9 (5) (2010) 430–435.

[19] S. Jun, X. Li, F. Meng, C.V. Ciobanu, Phys. Rev. B 83 (15) (2011) 153407.

[20] A. Ramasubramaniam, D. Naveh, Phys. Rev. B 84 (7) (2011), 075405.

[21] M.Y. Han, B. Ozyilmaz, Y.B. Zhang, P. Kim, Phys. Rev. Lett. 98 (20) (2007) 206805.

[22] Y.-W. Son, M.L. Cohen, S.G. Louie, Phys. Rev. Lett. 97 (21) (2006) 216803.

[23] J.M. Pruneda, Phys. Rev. B 81 (16) (2010) 161409.

[24] S. Manna, P. Gorai, G.L. Brennecka, C.V. Ciobanu, V. Stevanovic, J. Mater. Chem. C 6 (41) (2018) 11035–11044.

[25] S. Manna, K.R. Talley, P. Gorai, J. Mangum, A. Zakutayev, G.L. Brennecka, V. Stevanovic, C.V. Ciobanu, Phys. Rev. Appl. 9 (3) (2018).

[26] L. Liu, J. Park, D.A. Siegel, K.F. McCarty, K.W. Clark, W. Deng, L. Basile, J.C. Idrobo, A. P. Li, G. Gu, Science 343 (6167) (2014) 163–167.

[27] G. Kresse, J. Furthmuller, Phys. Rev. B 54 (16) (1996) 11169–11186.

[28] G. Kresse, D. Joubert, Phys. Rev. B 59 (3) (1999) 1758–1775.

[29] J.P. Perdew, K. Burke, M. Ernzerhof, Phys. Rev. Lett. 77 (18) (1996) 3865–3868.

[30] J. He, K.-Q. Chen, Z.-Q. Fan, L.-M. Tang, W.P. Hu, Appl. Phys. Lett. 97 (19) (2010) 193305.

[31] J.-W. Jiang, J.-S. Wang, B.-S. Wang, Appl. Phys. Lett. 99 (4) (2011), 043109.

[32] A. Ohtomo, H.Y. Hwang, Nature 441 (7089) (2006) 120.

[33] G.H. Han, J.A. Rodriguez-Manzo, C.W. Lee, N.J. Kybert, M.B. Lerner, Z.J. Qi, E.N. Dattoli, A.M. Rappe, M. Drndic, A.T.C. Johnson, ACS Nano 7 (11) (2013) 10129–10138.

[34] K.S. Novoselov, A.K. Geim, S.V. Morozov, D. Jiang, Y. Zhang, S.V. Dubonos, I.V. Grigorieva, A.A. Firsov, Science 306 (5696) (2004) 666–669.

[35] K.S. Novoselov, D. Jiang, F. Schedin, T.J. Booth, V.V. Khotkevich, S.V. Morozov, A.K. Geim, Proc. Natl. Acad. Sci. U. S. A. 102 (30) (2005) 10451–10453.

[36] C. Jin, F. Lin, K. Suenaga, S. Iijima, Phys. Rev. Lett. 102 (19) (2009) 195505.

[37] S. Dutta, A.K. Manna, S.K. Pati, Phys. Rev. Lett. 102 (9) (2009), 096601.

[38] Y. Ding, Y. Wang, J. Ni, Appl. Phys. Lett. 95 (12) (2009).

[39] L. Sun, Y. Li, Z. Li, Q. Li, Z. Zhou, Z. Chen, J. Yang, J.G. Hou, J. Chem. Phys. 129 (17) (2008) 174114.

[40] P. Lou, J.Y. Lee, J. Phys. Chem. C 113 (29) (2009) 12637–12640.

[41] B. Xu, J. Yin, Y.D. Xia, X.G. Wan, Z.G. Liu, Appl. Phys. Lett. 96 (14) (2010) 143111.

[42] X.H. Sun, C.P. Li, W.K. Wong, N.B. Wong, C.S. Lee, S.T. Lee, B.K. Teo, J. Am. Chem. Soc. 124 (48) (2002) 14464–14471.

[43] V.B. Shenoy, C.D. Reddy, A. Ramasubramaniam, Y.W. Zhang, Phys. Rev. Lett. 101 (24) (2008) 245501.

[44] S. Jun, Phys. Rev. B 78 (7) (2008), 073405.

[45] K.V. Bets, B.I. Yakobson, Nano Res. 2 (2) (2009) 161–166.

[46] B. Huang, M. Liu, N. Su, J. Wu, W. Duan, B.-L. Gu, F. Liu, Phys. Rev. Lett. 102 (16) (2009) 166404.

[47] J.W. Cahn, F. Larche, Acta Metall. 30 (1) (1982) 51–56.

[48] R.C. Cammarata, Prog. Surf. Sci. 46 (1) (1994) 1–38.

[49] R. Shuttleworth, Proc. Phys. Soc. London, Sect. A 63 (1950) 444.

[50] J.M. Soler, E. Artacho, J.D. Gale, A. Garcia, J. Junquera, P. Ordejon, D. Sanchez-Portal, J. Phys. Condens. Matter 14 (11) (2002) 2745–2779.

[51] S. Reich, J. Maultzsch, C. Thomsen, P. Ordejon, Phys. Rev. B 66 (3) (2002), 035412.

[52] S. Reich, C. Thomsen, P. Ordejon, Phys. Rev. B 65 (15) (2002) 153407.

[53] E. Bekaroglu, M. Topsakal, S. Cahangirov, S. Ciraci, Phys. Rev. B 81 (7) (2010), 075433.

[54] S. Okada, Phys. Rev. B 77 (4) (2008), 041408.

[55] J. Tersoff, Phys. Rev. B 39 (8) (1989) 5566–5568.

[56] T. Wassmann, A.P. Seitsonen, A.M. Saitta, M. Lazzeri, F. Mauri, J. Am. Chem. Soc. 132 (10) (2010) 3440–3451.

[57] C.-H. Park, S.G. Louie, Nano Lett. 8 (8) (2008) 2200–2203.

[58] Y. Yokomizo, J. Nakamura, Appl. Phys. Lett. 103 (11) (2013).

[59] S. Haldar, P. Srivastava, O. Eriksson, P. Sen, B. Sanyal, J. Phys. Chem. C 117 (42) (2013) 21763–21771.

[60] N.C. Bristowe, M. Stengel, P.B. Littlewood, E. Artacho, J.M. Pruneda, Phys. Rev. B 88 (16) (2013).

Calculation of bandgaps in bulk and 2D materials using Harbola-Sahni and van Leeuwen-Baerends potentials

P. Singh[a], M.K. Harbola[b], and A. Mookerjee[c],*
[a]Ames Laboratory, US Department of Energy, Iowa State University, Ames, IA, United States, [b]Indian Institute of Technology Kanpur (IITK), Kanpur, India, [c]S.N. Bose National Centre for Basic Sciences, Kolkata, India

4.1 Introduction

Solving the Schrödinger equation for the many-body system of interacting valence electrons in a solid is one of the more difficult problems in condensed matter physics. To perform such calculations, several methods, such as Hartree-Fock theory, based on approximating the wavefunctions, have been proposed. However, the most successful first-principles approach to the electronic structure of solids is density functional theory (DFT) [1–4], which uses the ground-state density as the basic variable. The ideas behind DFT are quite simple and remarkably easy to implement for numerical calculations. As a result, DFT has proven to be a very powerful method for obtaining the ground-state density, total energy, and other properties associated with the ground-state of a many-electron system. Furthermore, it can handle systems ranging from a single atom to molecules, clusters, and solids with equal ease. On the other hand, one problem that has not been amenable to DFT methods is predicting the bandgap of semiconductors and insulators. Thus, interest in obtaining accurate bandgaps using DFT methods continues unabated [5–7]. This chapter describes our efforts to solve the problem by looking at it afresh. In the following, we first describe the Kohn-Sham (KS) approach to bandgap calculations and discuss [8,9] why it cannot be expected to give accurate fundamental gap, defined [10,11] as $(I - A)$, where I is the ionization potential and A is the electron affinity of a system, in solids. We then motivate the reader to regard the bandgap calculation as an excited-state problem and demonstrate that good estimates of bandgaps can be obtained by treating the exchange-correlation potential accurately.

---

* Deceased.

Modeling, Characterization and Production of Nanomaterials. https://doi.org/10.1016/B978-0-12-819905-3.00004-X

4.2   Bandgap calculations in density-functional theory and derivative discontinuity of Kohn-Sham potential

The KS scheme of obtaining the bandgaps relies on the fact that the eigenvalue ϵ_{max} of the highest occupied molecular orbital (HOMO) of the exact KS Hamiltonian for an electronic system with N electrons is equal to the negative of its ionization energy I [12,13]. Consequently, ϵ_{max} for $N+1$ electron system equals negative of the electron affinity A of the system. Furthermore, in an insulator or a semiconductor, it is expected that adding an extra electron at the bottom of the conduction band would hardly change the density of the bulk system and therefore all the eigenvalues of the KS spectrum will remain essentially unchanged. It is therefore anticipated that the difference in the KS eigenenergies corresponding to the lowest orbital of the unoccupied band (negative of the electron affinity—A) and the highest orbital of the occupied band (negative of the ionization energy—I) should give the correct fundamental gap $(I - A)$ of semiconductors and insulators [10,11]. However, the results obtained by applying the traditional exchange-correlation functionals such as the local density approximation (LDA) [14] are contrary to this expectation and grossly underestimate the bandgap with respect to experiments. This does not change even with more accurate functionals, for example, generalized gradient approximation (GGA) [15,16]. The understanding of why it happens is based on the generalization [13] of density-functional theory to fractional number of electrons. Accordingly, when the electron number changes across an integer, the chemical potential of the system changes discontinuously by $(I - A)$. This indicates that the exchange-correlation potential should also change by a constant $(\Delta > 0)$ [9,13] when a small fraction of an electron is added to a system with an integer number of electrons. For example, in the helium ion He$^+$, as the electron number is increased fractionally by δ from 1 to $1+\delta$, the exchange-correlation potential jumps by 1.1 a.u. [17]. This is known as the derivative (with respect to the electron number) discontinuity of the exchange-correlation potential. As a result of this discontinuity, the Kohn-Sham eigenenergy $\epsilon_{(N+1)}(N+1)$ for the highest occupied orbital, indicated by the subscript $(N+1)$, of the $(N+1)$ electron insulator will be higher than the eigenenergy $\epsilon_{(N+1)}(N)$ of this orbital for the N electron g system by energy Δ. Therefore, the Kohn-Sham gap $E_g^{KS} = \epsilon_{(N+1)}(N) - \epsilon_N(N)$ will always underestimate the fundamental gap by the same amount.

Let us now look at the LDA or other approximations, such as the GGA, to the exchange-correlation functional in light of our discussion above. Because all these approximations depend directly on the density or its gradient, they cannot and do not have any derivative discontinuity in the corresponding potential. It is therefore argued that the Kohn-Sham gap calculated using these approximations cannot give the correct fundamental gap of a semiconductor or an insulator. For example, in silicon the E_g^{KS} calculated with the LDA is 0.49 eV, whereas the experimental gap is 1.17 eV. Does this mean that accurate bandgaps cannot be predicted by a Kohn-Sham calculation? In this chapter we argue that this need not be the case if we look at the bandgap as transition energy between the top of the valence band and the bottom of the conduction band. This is based on the observation that the optimized potential method does give reasonably accurate bandgaps for a large number of systems [18–20]. Furthermore, it has been shown [21] in atomic systems that if the exchange-correlation potential has correct behavior in the outer regions of an atom, the difference between the eiegenenergies of the highest occupied orbital and higher unoccupied orbitals is close to experimentally observed excitation energies of these systems. Note that this is in spite of the fact that the unoccupied orbitals of a KS system have no particular significance. On similar grounds, we may expect that when exchange-correlation potential is treated exactly, the KS bandgap will be close to the experimental gap. In this chapter, we show this to be the case. Although the exact exchange (EXX) correlation potential is not known, the exchange potential alone can be calculated exactly or highly accurately. We therefore present our study employing one such exchange potential, namely, the Harbola-Sahni (HS) potential for exchange; the correlation effects are still treated approximately within the LDA.

4.3 Kohn-Sham potential in terms of the orbitals: Exact exchange and HS potential

As is well known, the KS potential of density-functional theory is a local potential that gives orbitals leading to the exact density of a system. The potential is made up of its external, classical Coulomb and the exchange-correlation potentials; in performing Kohn-Sham calculations it is the exchange-correlation potential that is approximated. The exchange-only (XO) Kohn-Sham potential is the potential whose orbitals reproduce the Hartree-Fock density of a system. Equivalently, the ground-state orbitals of the exchange-only KS potential minimize the Hartree-Fock expression for the energy of a given system. However, in contrast to Hartree-Fock theory, the exchange-potential in the XO KS theory is a local potential and is identified as the EXX potential. The question arises how such a potential can be constructed.

Sharp and Horton [22] built upon Slater's [23] idea to develop the optimized effective potential (OEP) method in 1957, but it was first applied in 1976 by Talman and Shadwick [20] to numerically generate the required local potential for atoms. We now briefly describe how this is done. In the OEP method, one solves for a local potential $v_s(\mathbf{r})$ so that the corresponding orbitals ϕ_is minimize the Hartree-Fock expression for the energy. The potential $v_s(\mathbf{r})$ comprises the external potential, the Hartree potential, and the exchange potential. Because in general, the expression for the EXX energy E_x is not known in terms of the density, the exchange potential $v_x(\mathbf{r})$ in the EXX is written in terms of the orbitals as

$$v_x([p]; \mathbf{r}) = \frac{\delta E_x}{\delta \rho(\mathbf{r})} = \sum_i^{occ} \int d\mathbf{r}' \int d\mathbf{r}'' \left[\frac{\delta E_x}{\delta \phi_i(\mathbf{r}'')} \frac{\delta \phi_i(\mathbf{r}'')}{\delta v_s(\mathbf{r}')} + c.c. \right] \frac{\delta v_s(\mathbf{r}')}{\delta \rho(\mathbf{r})} \qquad (4.1)$$

In the expression above, $\delta E_x/\delta \phi_i$ are the nonlocal potentials of Hartree-Fock theory, $\delta \phi_i/\delta v_s(\mathbf{r})$ is calculated using the first-order perturbation theory, and $\delta v_s(\mathbf{r})/\delta \rho(\mathbf{r})$ is the inverse of the noninteracting linear response function $\chi(\mathbf{r}, \mathbf{r}')$ defined via

$$\delta\rho(\mathbf{r}) = \int d\mathbf{r}' \chi(\mathbf{rr}')\delta v_s(\mathbf{r}') \tag{4.2}$$

and given as

$$\chi(\mathbf{rr}') = \sum_i^{occ} \sum_j^{unocc} \frac{\phi_i^*(\mathbf{r})\phi_j(\mathbf{r}')\phi_j^*(\mathbf{r}')\phi_i(\mathbf{r}')}{\epsilon_i - \epsilon_j} + c.c \tag{4.3}$$

where ϵ_i and ϵ_j are the orbital eigenenergies of the noninteracting system.

The first systematic study using EXX potential for a range of semiconducting materials was performed by Kotani [18,19]. These studies showed that EXX leads to bandgaps that are close to their experimental values. This led to many more studies [24,25] of semiconductors, in particular their bandgaps, using the EXX theory. These studies showed that EXX leads to bandgaps that are close to their experimental values for a large number of systems. The explanation put forth [25] for this agreement is that the derivative discontinuity of the exchange and the correlation components of the exchange-correlation potential may be in opposite directions and hence cancel each other. However, as stated above, a possible reason for this is that the accuracy of the potential leads to differences in eigenenergies of occupied and unoccupied orbitals that are close [21] to the corresponding excitation energies of a system. With this in mind, we wish to use an easy and physically motivated potential, that is, the HS exchange potential, to calculate the bandgaps of semiconductors.

The HS exchange-correlation potential [26] is derived from the exchange-correlation (XC) hole because the work is done in moving a unit charge (atomic units) in the electric field produced by the XC hole; to calculate the exchange potential, it is the exchange hole, also known as the Fermi hole, $\rho_x(\mathbf{r}, \mathbf{r}')$, that is employed to calculate the field $\varepsilon_x(\mathbf{r})$. Thus, the exchange potential is given as a line integral

$$W_{HS}(\mathbf{r}) = -\int_\infty^{\mathbf{r}} \boldsymbol{\varepsilon}_x(\mathbf{r}') \cdot d\ell' \tag{4.4}$$

where

$$\boldsymbol{\varepsilon}_x(\mathbf{r}) = \int \frac{\rho_x(\mathbf{r},,\mathbf{r}')}{|\mathbf{r} - \mathbf{r}\prime|^3}(\mathbf{r} - \mathbf{r}')d\mathbf{r}' \tag{4.5}$$

For a set of single particle orbitals $\{\phi_{i\sigma}\}$ with occupation $f_{i\sigma}$ for the ith orbital with spin σ, the Fermi hole is given [27] as

$$\rho_x(\mathbf{rr}') = \frac{1}{\rho(\mathbf{r})} \sum_\sigma \sum_{i,j} f_{i\sigma}f_{j\sigma}\phi_{i\sigma}^*(\mathbf{r})\phi_{j\sigma}^*(\mathbf{r}')\phi_{i\sigma}(\mathbf{r}')\phi_{j\sigma}(\mathbf{r}) \tag{4.6}$$

For an electron, the potential will have the opposite sign to that given in Eq. (4.6).

The HS approach provides [3,28] a fundamental understanding of KS theory. The HS potential can be derived [29] directly from the Schrödinger equation via the differential virial theorem. From this derivation, it also becomes clear that the HS potential differs from the exact KS potential by the difference in the true kinetic energy and the Kohn-Sham kinetic energy of an interacting system. However, this difference is numerically very small. From the implementation point of view, a comparison of Eqs. (4.4)–(4.6) with Eqs. (4.1)–(4.3) makes it clear that the HS approach is much easier to employ numerically than the EXX. We note that the computational complexity of EXX is somewhat simplified in the KLI approximation [27] of the EXX.

A question that can be raised is whether the HS potential is path independent [30,31] and the field $\varepsilon(\mathbf{r})$ is curl free. It is not so in general. However, for particular cases, for example, spherically symmetric densities such as those used in the atomic sphere approximations (ASAs) of tight binding linearized muffin-tin orbital (TB-LMTO) method [32–34], the field is indeed curl free [3,35,36]. On the other hand, for nonspherical charge densities, the solenoidal part of the electric field is numerically rather small [37] and is related to the difference in kinetic energies between the Hartree-Fock and the HS approaches [28].

The EXX calculations give the KS BGs [18,25] in reasonable agreement with experiments. As mentioned above, it is believed that the agreement is due to a fortuitous cancellation of errors in these systems and does not hold in general [24]. However, if we think of the BG as transition energy, the agreement can be understood as arising from the accurate treatment of the exchange potential. It is with this understanding that we perform our calculations of bandgaps employing the HS potential.

4.4   Calculation of bandgaps for bulk materials using the HS potential

We have performed our calculations for the bandgaps using the HS potential within the LMTO as follows: We calculate the potential inside the atomic sphere using Eq. (4.4) and in the empty sphere as the LDA exchange-correlation potential. We do so because in the empty sphere the density of electrons is rather small and close to being homogeneous. Thus, the LDA can be expected to be a good approximation in the ES. Furthermore, because the density is small, correlation effects would be relatively large and must be included in the potential. The HS potential inside the AS is calculated by fixing its value at the AS/ES boundary to be $-1/r$ atomic units. We present the results of our calculations in Table 4.1.

Also presented there are results obtained by employing other potentials along with the experimental values for the bandgaps of these materials.

It is clear from Table 4.1 that our predictions for BGs calculated using the HS-EX are far superior to the results of the LDA calculations and are in excellent agreement with experimental values for a majority of the systems studied. This, along with the results of the OEP/EXX, demonstrates that treating the potential accurately indeed may give accurate values of bandgaps and also that the bandgap problem may also be looked at as an excited-state problem.

Table 4.1 Bandgap of several materials obtained using different theories and their comparison with experimental numbers.

	BGs							
	BG (eV)							
System	LDA	HS-EX	vLB	GW	QPC	MBJLDA	EXX	Expt.
Ne (A1)	11.39	22.07	23.64	22.1	16.55	22.72	–	20.75
Ar (A1)	8.09	11.29	12.76	14.9	11.95	13.91	–	14.32
Kr (A1)	6.76	9.10	10.91	–	9.98	10.83	–	11.40
Xe (A1)	5.56	6.63	8.61	–	8.23	8.52	–	9.15
C (A4)	2.70	5.47	5.18	6.18	–	4.93	5.12	5.48
Si (A4)	0.49	1.24	1.21	1.41	–	1.17	1.93	1.17
MgO (B1)	4.94	6.23	6.94	9.16	–	7.17	–	7.78
CaO (B1)	3.36	7.29	7.15	–	–	–	–	7.09
LiF (B1)	8.94	9.52	12.61	15.9	–	12.94	–	13.60
LiCl (B1)	6.06	6.50	7.84	–	–	8.64	–	9.40
AlN (B3)	2.44	5.05	5.13	4.90	–	–	5.03	5.11
AlP (B3)	1.16	2.53	2.75	2.86	–	2.32	–	2.51
BP (B3)	1.51	2.22	2.09	1.90	–	2.28	–	2.00
3C-SiC (B3)	1.38	2.88	2.58	2.76	–	2.52	–	2.42

The comparison of LMTO-LB (vLB) BGs of rare-gas solids with fcc (A1), diamond (A4), rock-salt (B1), and zinc-blende (B3) crystal structure has been made with LMTO-HS-EX [38], LMTO-LDA, GW [39], quasi-particle corrections (QPC) [40], MBJLDA [41], EXX [18,25], and experiments [42–60].

We sum up this section by noting that improvement in the accuracy of the potential, calculated either via the OEP or via the HS method, gives accurate results for bandgaps. However, both these methods are based on obtaining the potential using orbitals. The question is whether the conclusion drawn will still remain valid if the potential were calculated using density alone. We now show that this is indeed the case by performing our calculations employing the van Leeuwen-Baerends (vLB) potential that is written entirely in terms of the density and its derivative.

4.5 Density-based calculations using the vLB potential

Because the LDA potential for finite systems does not have the correct asymptotic behavior, that is, it does not go as $-1/r$ far away from the system, vLB proposed [61] that a correction term be added to it that makes it go as $-1/r$ in regions far from a system and improves [61,62] the eigenvalues for the highest occupied orbitals of atoms significantly. We therefore anticipate that with this correction, the resulting potential—referred to as the vLB potential—should also lead to accurate bandgaps in all systems.

The vLB potential is given as

$$v_{xc,\sigma}^{\text{vLB}}(\mathbf{r}) = \left[v_{x,\sigma}^{\text{LDA}}(\mathbf{r}) + v_{x,\sigma}^{\text{LB}}(\mathbf{r}) \right] + v_{c,\sigma}^{\text{LDA}}(\mathbf{r}) \tag{4.7}$$

where $v_x^{\text{LDA}}(\mathbf{r})$ is the standard LDA exchange potential while $v_{c,\sigma}^{\text{LDA}}(\mathbf{r})$ is the usual LDA correlation. Here $v_{x,\sigma}^{\text{LB}}(\mathbf{r})$ is

$$v_{x,\sigma}^{\text{LB}}(\mathbf{r}) = -\beta \rho_\sigma^{1/3} \frac{x_\sigma^2}{1 + 3\beta x_\sigma \sin h^{-1}(x_\sigma)} \tag{4.8}$$

with $\beta = 0.05$ and $x = |\nabla \rho(\mathbf{r})|/\rho^{4/3}(\mathbf{r})$.

The correction vLB in Eq. (4.8) is motivated by the Becke formula [63] for exchange energy and therefore leads to correct $-1/r$ behavior of the vLB potential in the outer regions of a finite systems. The potential has been employed in the past to study [62] the effect of the correct asymptotic behavior of the potential on response properties of atoms. In the present work, we employ it to obtain the exchange-correlation potential in the AS and use only the LDA potential in the ES to perform self-consistent calculation within TB-LMTO-ASA. The results of the BGs obtained from these calculations are given in Table 4.1. As is evident from the results, the vLB potential of Eq. (4.7) gives BGs that show vast improvement over the LDA results and are in good agreement with experimental values. This shows that an improved exchange-correlation potential, even if it is given entirely in terms of the density, which is particularly accurate in the outer regions of the AS, gives significantly improved results over the LDA. In this connection, we further note that modifying the LDA potential by the Becke-Johnson potential [63] also leads to substantially improved BGs for a variety of solids [64]. For completeness, these results are also shown in Table 4.1. Having established in the above that calculating the potential accurately gives good bandgaps for bulk systems, we are now ready to employ these potentials for nanomaterials. We choose to use the vLB potential for this purpose because it is easier to implement and gives results similar to the orbital-based methods of OEP/EXX or the HS.

4.5.1 Application to clusters of graphene and hexagonal boron nitride

In this section, we apply TB-LMTO-LB with the vLB potential to the two-dimensional finite clusters of hexagonal boron nitride (h-BN) and graphene to study variation in bandgaps with increasing cluster size. As is well known, pure graphene has no bandgap [65], while h-BN shows bandgap that varies from 3.6 to 7.1 eV [66–69]. To model our system, we considered cells of orthorhombic symmetry with physical parameters

$a = 40\,\text{Å}$, $b = 25°\,\text{Å}$, and $c = 40\,\text{Å}$. We have modeled h-BN [70] and graphene [71] clusters of different sizes using experimental structure. In our calculations, we considered h-BN and graphene clusters with one, three, and seven ring sizes. The graphene cluster with one, three, and seven ring sizes has 6, 13, and 24 carbon atoms in the basis set arranged on a regular hexagonal lattice. The h-BN with 1 ring has 3 boron and 3 nitrogen atoms; with 3 rings, it has 6 boron and seven nitrogen atoms; and with 7 rings, it has 12 boron and 12 nitrogen atoms in the basis set. In the h-BN, the boron and the nitrogen are arranged on a regular hexagonal lattice. Here, we are dealing with finite-size systems, so BG is calculated as the difference of lowest unoccupied orbital (LUMO) and highest occupied orbital (HOMO). The BG in all three cases increases with increasing system size. However, we have shown here that the HOMO-LUMO gap changes substantially with an increase in the size of the clusters with the vLB-potential in compared with the LDA. The results have been quoted in Table 4.2. In Fig. 4.1, we also show the density of states, which indicate the position of the occupied and virtual levels, for the three different sizes of h-BN and graphene clusters considered. The zero in the DOS is defined to be the Fermi energy.

4.5.2 Application to two-dimensional disordered Si_xC_{1-x}

In this section, we have applied the vLB potential to the 2D disordered siliphene (Si_xC_{1-x}; $0 < x < 0.6$) [72] in order to address existing disagreement of bandgap between experiments and theory. The predictions are compared with some known experiments and other theories [73,74]. The disagreements among theoretical studies comes from the dependence on approximations and/or underlying models, e.g., crystal structure, exchange-correlation, electronic-structure method, and/or the way to handle disorder including effects of configuration fluctuations of the immediate neighborhood. The main advantage of 2D siliphene [75,76] lies in better tunability of the bandgap compared to graphene, which is an important parameter for designing nanodevices.

Ding and Wang [77], and Shi et al. [78] have shown homogenous disorder for Si at.% ≤ 50 in 2D Si_xC_{1-x}, therefore, we considered homogenous disorder [77,78]. In Table 4.3, the vLB predictions are compared with LDA (this work), other theories [73,79],

Table 4.2 The BGs of quantum dots of hexagonal-boron nitride and graphene with one, three, and seven rings within TB-LMTO-LB and TB-LMTO-LDA.

System size	BG (eV)			
	BN (vLB)	BN (LDA)	C (vLB)	C (LDA)
1	2.00	0.50	2.90	1.50
3	3.00	1.90	3.46	1.85
7	5.70	2.40	5.20	2.50

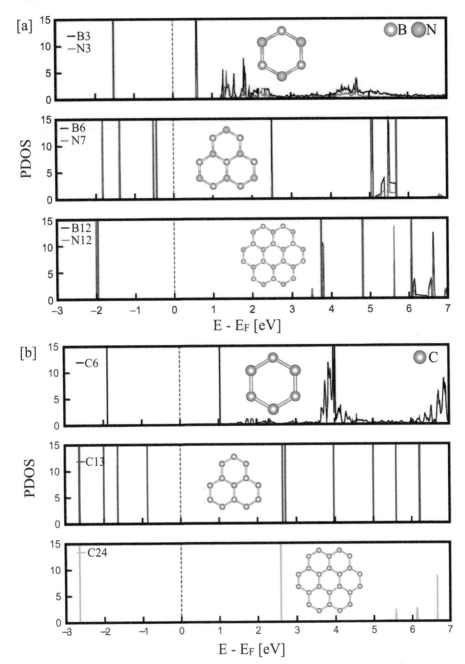

Fig. 4.1 (A and B) Density of states of boron nitride and graphene rings of varying sizes.

Table 4.3 The siliphene (50at.%Si) bandgap calculated using vLB is compared with other theory and measurements [77,78].

System/ pProperty	2D $Si_{0.50}C_{0.50}$							
	LMTO LDA	LMTO vLB	LMTO vLB (ASR[a])	NMTO vLB	VASP/ SIESTA[b]	LDA GW[b]	Expt. (TEM)	Expt. (GIXRD)
a (Å)	2.66	2.66	2.75	2.66	2.41	–	2.60	2.66
E_{gap} (eV)	2.96	3.49	3.60	3.90	2.06	3.90	2.6–2.9	2.95

The vLB shows better lattice parameters and bandgap with respect to other methods due to reduced self-interaction and the better asymptotic treatment [82].
[a] ASR = augmented space recursion [84,85].
[b] VASP/SIESTA [73,79].

and experiments [77,78]. The energy minimized lattice constants of 2D SiC using LDA and vLB show good agreement with the measured $a = 2.6 \pm 0.2 \, \mathring{A}$ but other theoretical results differ almost by $\sim 15\%$. The main source of disagreement between other theory with this work lies with the use of basis set and exchange-correlation potential [72]. The vLB correction to LDA exchange [38] improves the potential near atomic-sphere (AS) boundary (r) in the asymptotic limit, i.e., $1/r$, as $r \rightarrow$ r_AS. The vLB corrected LDA improves band positions by providing self-interaction correction [80–83] compared to other semi-local density based functionals such as LDA or GGA.

4.6 Discussion and concluding remarks

In this chapter, we have put forth a different perspective on the bandgap problem and emphasized that it be studied as a problem of calculating excitation energies. Then the accuracy of bandgap determination depends crucially on the correct treatment of exchange-correlation potential in the asymptotic region (outer regions of the atomic sphere). Keeping this in mind, we have used the HS and the vLB potentials and developed them within the TB-LMTO-ASA. We have tested this method for calculating the bandgap behavior of a range of semiconducting materials with different crystal structures. Not only are the calculated bandgaps of several bulk semiconductors in very good agreement with experiments, but the methods employed are computationally less expensive with the possibility of easy implementation in more accurate full potential methods. Finally, we have applied the vLB-based method to calculate bandgap in finite-sized clusters of graphene, hexagonal-boron nitride, and 2D siliphene. We have shown that an easy-to-implement density-based potential, which is also asymptotically correct, gives good HOMO-LUMO gaps that are that are slightly larger than the LDA results. Overall, our method gives system properties that are consistently close to experiments for a variety of systems, considered in this work.

Acknowledgments

The work at Ames Laboratory was supported by the US Department of Energy (DOE), Office of Science, Basic Energy Sciences, Materials Science, and Engineering Division. Ames Laboratory is operated for the US DOE by Iowa State University under Contract No. DE-AC02-07CH11358.

References

[1] R.O. Jones, O. Gunnarsson, The density functional formalism, its applications and prospects, Rev. Mod. Phys. 61 (1989) 689.
[2] S. Kümmel, L. Kronik, Orbital-dependent density functionals: theory and applications, Rev. Mod. Phys. 80 (2008) 3.
[3] V. Sahni, Quantal Density Functional Theory, Springer, Berlin, 2004.
[4] P. Singh, M.K. Harbola, Density-functional theory of material design: fundamentals and applications-I, Oxford Open Mater. Sci. 1 (1) (2021) itab018.

[5] R.W. Godby, M. Schlüter, L.J. Sham, Accurate exchange-correlation potential for silicon and its discontinuity on addition of an electron, Phys. Rev. Lett. 56 (1986) 2415.

[6] X. Gonze, P. Ghosez, R.W. Godby, Density-polarization functional theory of the response of a periodic insulating solid to an electric field, Phys. Rev. Lett. 74 (1995) 4035.

[7] W. Knorr, R.W. Godby, Investigating exact density-functional theory of a model semiconductor, Phys. Rev. Lett. 68 (1992) 639.

[8] J.P. Perdew, M. Levy, Physical content of the exact Kohn-Sham orbital energies: band gaps and derivative discontinuities, Phys. Rev. Lett. 20 (1983) 1884.

[9] L.J. Sham, M. Schlüter, Density-functional theory of the energy gap, Phys. Rev. Lett. 51 (1983) 1888.

[10] N.F. Mott, Metal-Insulator Transitions, Taylor and Francis, London, 1974, p. 237.

[11] N.F. Mott, L. Friedman, Metal-insulator transitions in VO2, Ti2O3 and Ti2-xVxO3, Philos. Mag. 30 (1974) 389.

[12] M. Levy, J.P. Perdew, V. Sahni, Exact differential equation for the density and ionization energy of a many-particle system, Phys. Rev. A 30 (1984) 2745.

[13] J.P. Perdew, R.G. Parr, M. Levy, J.L. Balduz, Density-functional theory for fractional particle number: derivative discontinuities of the energy, Phys. Rev. Lett. 42 (1982) 1691.

[14] U. Von Barth, L. Hedin, A local exchange-correlation potential for the spin polarized case. I, J. Phys. C 5 (1972) 1629.

[15] J.P. Perdew, Erratum: density-functional approximation for the correlation energy of the inhomogeneous electron gas, Phys. Rev. B 34 (1986) 7406.

[16] J.P. Perdew, K. Burke, M. Ernzerhof, Generalized gradient approximation made simple, Phys. Rev. Lett. 77 (1996) 3865.

[17] K.D. Sen, M.K. Harbola, Characteristic features of the electrostatic potential for negative atoms within the work formalism, Chem. Phys. Lett. 178 (1991) 347.

[18] T. Kotani, Exact exchange-potential band-structure calculations by the LMTO-ASA method: MgO and CaO, Phys. Rev. B 50 (1994) 14816.

[19] T. Kotani, Exact exchange potential band-structure calculations by the linear muffin-tin orbital–atomic-sphere approximation method for Si, Ge, C, and MnO, Phys. Rev. Lett. 74 (1995) 2989.

[20] J.D. Talman, W.F. Shadwick, Optimized effective atomic central potential, Phys. Rev. A 14 (1976) 36.

[21] A. Savin, C.J. Umrigar, X. Gonze, Relationship of Kohn–Sham eigenvalues to excitation energies, Chem. Phys. Lett. 288 (1998) 391.

[22] R.T. Sharp, G.K. Horton, A variational approach to the unipotential many-electron problem, Phys. Rev. 90 (1953) 317.

[23] J.C. Slater, A simplification of the Hartree-Fock method, Phys. Rev. 81 (1951) 385.

[24] S. Sharma, J.K. Dewhurst, C. Ambrosch-Draxl, All-electron exact exchange treatment of semiconductors: effect of core-valence interaction on band-gap and d-band position, Phys. Rev. Lett. 95 (2005), 136402.

[25] M. Stadele, M. Moukara, J.A. Majewski, P. Vogl, A. Görling, Exact exchange Kohn-Sham formalism applied to semiconductors, Phys. Rev. B 59 (1999) 10031.

[26] M.K. Harbola, V. Sahni, Quantum-mechanical interpretation of the exchange-correlation potential of Kohn-Sham density-functional theory, Phys. Rev. Lett. 62 (1989) 489.

[27] J.B. Krieger, Y. Li, G.J. Iafrate, Systematic approximations to the optimized effective potential: application to orbital-density-functional theory, Phys. Rev. A 46 (1992) 5453.

[28] V. Sahni, Physical interpretation of density-functional theory and of its representation of the Hartree-Fock and Hartree theories, Phys. Rev. A 55 (1997) 1846.

[29] A. Holas, N.H. March, Calculational scheme for exact exchange and correlation potentials based on the equation of motion for density matrix plus the perturbation theory, Phys. Rev. A 56 (1997) 3597.

[30] M. Levy, J.P. Perdew, Hellmann-Feynman, virial, and scaling requisites for the exact universal density functionals. Shape of the correlation potential and diamagnetic susceptibility for atoms, Phys. Rev. A 32 (1985) 2010.

[31] M. Rasolt, D.J.W. Geldart, Comment on "Quantum-mechanical interpretation of the exchange-correlation potential of Kohn-Sham density-functional theory", Phys. Rev. Lett. 65 (1990) 276.

[32] O.K. Andersen, O. Jepsen, Explicit, first-principles tight-binding theory, Phys. Rev. Lett. 53 (1984) 2571.

[33] O. Jepsen, O.K. Andersen, The Stuttgart TB-LMTO-ASA Program, Version 4.7, Max-Planck-Institut für Festkörperforschung, Stuttgart, Germany, 2000.

[34] H.L. Skriver, The LMTO Method: Muffin-Tin Orbitals and Electronic Structure, Springer-Verlag, New York, 1984.

[35] M.K. Harbola, V. Sahni, Harbola and Sahni reply, Phys. Rev. Lett. 65 (1990) 277.

[36] Y. Wang, J.P. Perdew, J.A. Chevary, L.D. Macdonald, S.H. Vosko, Exchange potentials in density-functional theory, Phys. Rev. A 41 (1990) 78.

[37] M. Slamet, V. Sahni, M.K. Harbola, Force field and potential due to the Fermi-Coulomb hole charge for nonspherical-density atoms, Phys. Rev. A 49 (1994) 809.

[38] P. Singh, M.K. Harbola, B. Sanyal, A. Mookerjee, Accurate determination of band gaps within density functional formalism, Phys. Rev. B 87 (2013), 235110.

[39] F. Fuchs, J. Furthmüller, F. Bechstedt, M. Shishkin, G. Kresse, Quasiparticle band structure based on a generalized Kohn-Sham scheme, Phys. Rev. B 76 (2007), 115109.

[40] N.C. Bacalis, D.A. Papaconstantopoulos, W.E. Pickett, Systematic calculations of the band structures of the rare-gas crystals neon, argon, krypton, and xenon, Phys. Rev. B 38 (1988) 6218.

[41] F. Tran, P. Blaha, Accurate band gaps of semiconductors and insulators with a semilocal exchange-correlation potential, Phys. Rev. Lett. 102 (2009), 226401.

[42] J. Heyd, J.E. Peralta, G.E. Scuseria, R.L. Martin, Energy band gaps and lattice parameters evaluated with the Heyd-Scuseria-Ernzerhof screened hybrid functional, J. Chem. Phys. 123 (2005), 174101.

[43] R.J. Magyar, A. Fleszar, E.K.U. Gross, Exact-exchange density-functional calculations for noble-gas solids, Phys. Rev. B 69 (2004), 045111.

[44] J. Paier, M. Marsman, K. Hummer, G. Kresse, I.C. Gerber, J.G. Ángyán, Screened hybrid density functionals applied to solids, J. Chem. Phys. 124 (2006), 154709.

[45] S.V. Faleev, M. van Schilfgaarde, T. Kotani, All-electron self-consistent GW approximation: application to Si, MnO, and NiO, Phys. Rev. Lett. 93 (2004), 126406.

[46] M. Shishkin, M. Marsman, G. Kresse, Accurate quasiparticle spectra from self-consistent GW calculations with vertex corrections, Phys. Rev. Lett. 99 (2007), 246403.

[47] M. Rohlfing, S.G. Louie, Electron-hole excitations and optical spectra from first principles, Phys. Rev. B 62 (2000) 4927.

[48] S. Hufner, J. Osterwalder, T. Riesterer, F. Hulliger, Photoemission and inverse photoemission spectroscopy of NiO, Solid State Commun. 52 (1984) 793.

[49] G.A. Sawatzky, J.W. Allen, Magnitude and origin of the band gap in NiO, Phys. Rev. Lett. 53 (1984) 2339.

[50] M. Marsman, J. Paier, A. Stroppa, G. Kresse, Hybrid functionals applied to extended systems, J. Phys. Condens. Matter 20 (2008), 064201.

[51] U. Rossler, Electron and exciton states in solid rare gases, Phys. Status Solidi B 42 (1970) 345.

[52] U. Rössler, in: M.L. Klein, J.A. Venables (Eds.), Rare-Gas Solids, Academic Press, New York, 1975, p. 545.

[53] O. Madelung, R. Poerschke (Eds.), Semiconductor: Other than Group IV Elements and III-V Compounds, Springer, Berlin, 1992.

[54] M.L. Cohen, J.R. Chelikowsky, Electric Structure and Optical Properties of Semiconductors, Springer, New York, 1998.

[55] M.S. Hybersten, S.G. Louie, Electron correlation in semiconductors and insulators: band gaps and quasiparticle energies, Phys. Rev. B 34 (1986) 5390.

[56] M.E. Levinshtein, S.L. Rumyantsev, M.S. Shur, Properties of Advanced Semiconductor Materials, Wiley, New York, 2001.

[57] Ambacher, O., (Private communication) n.d.

[58] L.I. Berger, Semiconductor Materials, CRC Press, New York, 1990.

[59] D.J. Stukel, Energy-band structure of BeS, BeSe, and BeTe, Phys. Rev. B 1 (1970) 12.

[60] K.H. Hellwege, et al., Numerical Data and Functional Relationships in Science and Technology, New Series, Group III, vol. 17, Springer-Verlag, New York, 1982.

[61] R. van Leeuwen, E.J. Baerends, Exchange-correlation potential with correct asymptotic behavior, Phys. Rev. A 49 (1994) 2421.

[62] A. Banerjee, M.K. Harbola, Density-functional-theory calculations of the total energies, ionization potentials, and optical response properties with the van Leeuwen–Baerends potential, Phys. Rev. A 60 (1999) 3599.

[63] A.D. Becke, Density-functional exchange-energy approximation with correct asymptotic behavior, Phys. Rev. A 38 (1988) 3098.

[64] A. Becke, E.R. Johnson, A simple effective potential for exchange, J. Chem. Phys. 124 (2006), 221101.

[65] H. Jiang, Band gaps from the Tran-Blaha modified Becke-Johnson approach: a systematic investigation, J. Chem. Phys. 138 (2013), 134115.

[66] W. Baronian, The optical properties of thin boron nitride films, Mater. Res. Bull. 7 (1972) 119.

[67] A.H. CastroNeto, F. Guinea, N.M.R. Peres, K.S. Novoselov, A.K. Geim, The electronic properties of graphene, Rev. Mod. Phys. 81 (2009) 109.

[68] V.A. Fomichev, X-ray M 2, 3 emission bands of transition metals of the first long period, Fizika Tverdogo Tela (Sov. J. Solid State Phys.) 13 (1971) 907.

[69] M.B. Khusidman, Some peculiarities of hexagonal boron nitride, Fizika Tverdogo Tela (Sov. J. Solid State Phys.) 14 (1972) 3287.

[70] L.G. Carpenter, P.Y. Kirby, The electrical resistivity of boron nitride over the temperature range 700 degrees C to 1400 degrees C, J. Phys. D 15 (1982) 1143.

[71] V.L. Solozhenko, G. Will, F. Elf, Isothermal compression of hexagonal graphite-like boron nitride up to 12 GPa, Solid State Commun. 96 (1995) 1.

[72] B. Sadhukhan, P. Singh, A. Nayak, S. Datta, D.D. Johnson, A. Mookerjee, Band-gap tuning and optical response of two-dimensional SixC1-x: a first-principles real-space study of disordered two-dimensional materials, Phys. Rev. B 96 (2017), 054203.

[73] M.S. Azadeh, A. Kokabi, M. Hosseini, M. Fardmanesh, Tunable bandgap opening in the proposed structure of silicon doped graphene, Micro Nano Lett. 6 (2011) 582.

[74] S. Lin, Light-emitting two-dimensional ultrathin silicon carbide, J. Phys. Chem. C 116 (2012) 3951.

[75] H. Nakano, T. Mitsuoka, M. Harada, K. Horibuchi, H. Nozaki, N. Takahashi, T. Nonaka, Y. Seno, H. Nakamura, Soft synthesis of single-crystal silicon monolayer sheets, Angew. Chem. 118 (2006) 6451.

[76] P. Vogt, P. DePadova, C. Quaresima, J. Avila, E. Frantzeskakis, M.C. Asensio, A. Resta, B. Ealet, G. LeLay, Silicene: compelling experimental evidence for graphene like two-dimensional silicon, Phys. Rev. Lett. 108 (2012), 155501.

[77] Y. Ding, Y. Wang, Geometric and electronic structures of two-dimensional SiC3 compound, J. Phys. Chem. C 118 (2014) 4509.

[78] Z. Shi, Z. Zhang, A. Kutana, B.I. Yakobson, Predicting two-dimensional silicon carbide monolayers, ACS Nano 9 (2015) 9802.

[79] E. Bekaroglu, M. Topsakal, S. Cahangirov, S. Ciraci, First-principles study of defects and adatoms in silicon carbide honeycomb structures, Phys. Rev. B 81 (2010), 075433.

[80] S. Datta, P. Singh, C.B. Chaudhuri, M.K. Harbola, D.D. Johnson, A. Mookerjee, Simple correction to bandgap problems in IV and III-V semiconductors: an improved, local first-principles density functional theory, J. Phys. Condens. Matter 31 (2019), 495502.

[81] S. Datta, P. Singh, D. Jana, C.B. Chaudhuri, D. Jana, M.K. Harbola, D.D. Johnson, A. Mookerjee, Exploring the role of electronic structure on photo-catalytic behavior of carbon-nitride polymorphs, Carbon 168 (2020) 125–134.

[82] P. Singh, M.K. Harbola, M. Hemanadhan, A. Mookerjee, D.D. Johnson, Better band gaps with asymptotically corrected local exchange potentials, Phys. Rev. B 93 (8) (2016) 085204.

[83] P. Singh, M.K. Harbola, D.D. Johnson, Better band gaps for wide-gap semiconductors from a locally corrected exchange-correlation potential that nearly eliminates self-interaction errors, J. Phys. Condens. Matter 29 (2017), 424001.

[84] P. Singh, et al., Magnetic behaviour of AuFe and NiMo alloys, Pramana 76 (2011) 639–656.

[85] P. Singh, M. Rahaman, A. Mookerjee, Magnetic transitions in Ni1-xMox and Ni1-xWx disordered alloys, J. Magn. Magn. Mater. 323 (2011) 2478–2482.

Multiscale Green's functions for modeling graphene and other Xenes [☆]

Vinod K. Tewary and E.J. Garboczi

Applied Chemicals and Materials Division, National Institute of Standards and Technology (NIST), Boulder, CO, United States

5.1 Introduction

Nanomaterials represent a giant new step in the science and technology of materials. A large number of papers have already been published on the development and fabrication of new nanomaterials. This number has been growing exponentially over the last several years and many new applications of nanomaterials have been either developed or suggested (see, for example, the papers quoted in this volume and also in its previous edition [1]). However, to advance the nanotechnology in a systematic way, it is vital that robust and reliable mathematical models are available.

In fact, materials modeling has now emerged as a new branch of knowledge that has tremendous scientific and industrial applications. These models help in the visualization of materials and understanding of the physical processes at length and time scales that are difficult and expensive to explore by direct experimental methods. The models are, therefore, useful for scientific research as well as pedagogical purposes. Moreover, they are indispensable tools for industry for designing and testing of new materials because they can provide quick and inexpensive answers to "what if" type questions and can give an estimate of the lifetime and reliability of the materials for different applications and under different operating conditions.

Modern 2D (two-dimensional) materials such as graphene and other Xenes form a special class of nanomaterials. Their thickness is of atomistic dimensions, but their length and breadth can be of macro dimensions. The word Xenes collectively refers to monoelemental 2D graphene-like materials: borophene, silicene, germanene, stanene, phosphorene, arsenene, antimonene, bismuthene, and tellurene [2]. However, any reference to Xenes in this chapter will be restricted to Xenes of group 4 elements: graphene, silicene, and germanene. Silicene and germanene have a dual planar lattice structure, in which atoms are located in two different planes, in contrast to graphene, in which all atoms are on a single plane.

In the literature, the term Xenes does not necessarily include graphene. The nomenclature is still evolving. In this chapter, any reference to Xenes will include graphene.

[☆] Contribution of the National Institute of Standards and Technology, an agency of the US Federal Govt. Not subject to copyright in the USA.

Modeling, Characterization and Production of Nanomaterials. https://doi.org/10.1016/B978-0-12-819905-3.00005-1
2023 Published by Elsevier Ltd.

In addition to the advantage of brevity, it will underline an important commonality between graphene and other Xenes in the context of the multiscale modeling. All Xenes, including graphene, have the same Bravais lattice structure, which determines the primary characteristics of the multiscale Green's functions, as described in this chapter.

Among Xenes, graphene is special because it is exactly one atom thick, apart from intrinsic ripples [3]. Neglecting the quantum and the relativistic effects, we can treat graphene as a limiting case of Xenes, in which the separation between the two planes reaches zero in the limit. Hence, for the purpose of modeling, graphene can be treated as exactly planar within the limits of the current measurements. It, therefore, serves as a prototype for modeling 2D lattices. Further, many newer and even more exciting possible industrial applications have been identified for graphene, as described in the next section. This explains the strong contemporary interest in graphene.

In this chapter, we describe a Green's function (GF) method for multiscale modeling of Xenes with special emphasis on graphene. Green's functions are a well-established technique for solving a variety of problems in science and engineering [4,5]. The application of the GFs to phonons in a lattice is given in the classic treatise by Maradudin et al. [6]. The formulation of the static lattice GF, which is the zero frequency limit of the phonon GF, has been described in [7] for 3D solids. The GF techniques have been extensively used in the continuum model of a solid, e.g., Pan and Cheng [8] and Ting [9]. At NIST, we have been developing GF-based techniques for a variety of problems in materials physics. These techniques have contributed to the multiscale Green's function (MSGF) method described in this chapter for 2D materials.

The plan of the chapter is as follows: The present section, Section 5.1 is the introduction. Section 5.2 gives the preliminaries and discusses the meaning of multiscale modeling in the present context, the need for bridging length and time scales, and its application to Xenes, specifically to graphene. The basic definition of the GF and its role in measurement science is discussed in Section 5.3. A brief review of the discrete lattice model of a solid, as relevant for lattice GFs, is given in Section 5.4 and the lattice statics Green's function (LSGF) is described in Section 5.5. The MSGF method for multiple length scales is given in Section 5.6.

The MSGF technique is illustrated by applying it to Xenes in Section 5.7. In this section, we calculate the static lattice distortion due to a point defect, such as a vacancy, in graphene [10] and silicene [11]. For perfect graphene without defects, we use the set of force constants published earlier [12], which gives a good fit between the calculated and the measured phonon dispersion data. The defect-host interaction is expressed in terms of arbitrary parameters by defining an effective force. For silicene, we simply assume an idealized model potential for the limited purpose of illustrating the application of the MSGF method to Xenes other than graphene.

The causal GF method for temporal problems and its application to simulate the propagation of elastic waves in graphene is described in Section 5.8. These calculations show that the use of causal GF can accelerate the temporal convergence, at least in this idealized case, by several orders of magnitude [13]. Finally, conclusions and suggestions for future work are given in Section 5.9.

5.2 Preliminaries

5.2.1 Multiple scales in solids: Continuum and lattice models

The term "multiscale model" has been used in the literature with multiple meanings and in different contexts. In the context of this paper, a multiscale model should be valid at different length scales and should be able to link the length scales seamlessly.

A mathematical model of nanomaterials must be applicable to perfect lattices as well as those containing lattice defects. The effect of defects can be identified at three broad length scales: (i) the core region containing the lattice defects at the atomistic scale (sub-nano regime), where nonlinear and quantum effects may be important, (ii) outer region away from the defect (nanoregion), and (iii) free surfaces and interfaces in case of nanocomposites (nano or microregion). There may also be an intermediate mesoscopic region. Thus, a mathematical model of a nanomaterial needs to be an integrated model, in which each length scale seamlessly merges into the next scale.

Lattice defects can be classified into two broad categories: potential defects and structural defects. The potential defects are those that can be simulated by a change in the potential energy or the Hamiltonian of the crystal. Examples of potential defects are point defects, such as vacancies and interstitials or their aggregates such as voids and hillocks. In contrast, structural defects or extended defects consist of major rearrangements of atoms in the lattice and cannot be represented by a change in the crystal Hamiltonian alone. Examples of structural defects are dislocations [14], free surfaces, and interfaces. In this chapter, our interest will be confined to potential defects, specifically vacancies and substitutional defects. Further, we will not explicitly discuss any quantum effects, except to the extent that they are phenomenologically included in the interatomic potentials.

For ordinary 3D macroscopic solids, lattice defects can be adequately modeled by using anisotropic continuum theory, which can be used to calculate the stresses and strains in the bulk solid. These two are the most important measurable parameters that characterize the mechanical behavior of a solid and are mutually related through its elastic constants. See, for example, the excellent treatise by Anderson et al. [14].

It is generally accepted that continuum theory is inadequate to describe the response of a crystal lattice in the core and the nanoregions close to a defect [7]. This region, in which continuum models are not valid, extends from sub-nanometer to a few nanometers. It then becomes necessary to model the discrete atomistic structure of the lattice in this region. On the other hand, at least in a 3D case, continuum theory is adequate to model the response of the solid at sufficiently large distances from the defects. It may also be adequate to model averaged physical characteristics of extended defects, such as free surfaces and interfaces that are not too close to the point defects.

At 3D macroscopic scales, the continuum theory is a highly developed theory that has been very successful in modeling the macroscopic response of solids and for interpreting measurements of stresses and strains in terms of the elastic constants. The problem is that the physical processes that determine the macroscopic stresses and strains, in particular, the energetics of interatomic displacements, occur at the atomistic or sub-nano scales. Hence, a multiscale model is needed to relate the physical processes to macroscopic parameters for the purpose of design and interpretation of measurements that are sensitive to atomic-scale phenomena.

Now, consider the effect of a lattice defect on an otherwise perfect lattice. The presence of a defect perturbs the translation symmetry of the lattice and thus distorts the lattice by changing the location of the atoms. A defect can thus be characterized by two main parameters—its strength and range. The strength is sensitive to atomistic structure near the defect, which is an atomistic scale problem. The range is a macroscopic quantity because it extends over the whole crystal, and the continuum model should be applicable at large distances.

However, for 2D lattices, there is an additional problem. In the ideal case, a 2D lattice is assumed to have zero thickness. If z is the space variable in the direction perpendicular to the lattice plane, the derivative of the displacement field with respect to z does not exist. Hence, the z components of the strain are not defined. We must use a discrete atomistic model for the lattice distortion in the Z-direction. Actually, strain is a continuum model parameter, which is defined in terms of the derivatives of the displacement field. Hence, it cannot be uniquely defined for a discrete lattice model of a solid because the displacement field is a discrete variable. For a discrete lattice, the corresponding physical quantity is lattice distortion, represented by a set of numbers giving the change in atomic positions at all the lattice sites.

5.2.2   Modeling temporal physical processes in solids

A large class of material modeling problems deals with the solution and understanding of time-dependent processes. Examples include diffusion, phonon transport, thermal conduction, crack propagation, radiation damage, and many more. Presently, various Xenes such as silicene and graphene are being considered as adjuncts or additions to complementary metal oxide semiconductor technology and other semiconductor and quantum applications [15]. For many applications, these processes must be modeled using the discrete lattice structure of the solid. To appreciate the problem of temporal modeling and the need to bridge the time scales, consider radiation damage in solids. Although we will not discuss radiation damage as such in this chapter, we have selected this example, because it clearly shows the importance of temporal modeling. Because of the obvious difficulties associated with the purely experimental investigation of radiation damage, particularly at nanoscales, it is necessary to develop reliable mathematical models to study the effect of radiation on nanomaterials. Such models are needed to estimate the reliability of nanomaterials-based solid-state devices in outer space and other environments exposed to high energy radiation.

Radiation damage causes atomistic displacements in the solid [16], which can result into creation of vacancies and interstitials in the lattice (primary event). These lattice defects can agglomerate into clusters and, under certain conditions, form voids, and hillocks/dislocations (secondary and higher events). Structural changes in the solid, caused by the radiation damage, can be, in principle, measured by studying the width of the Bragg peaks [17]. The charge distribution in the crystal is strongly perturbed due to the lattice defects, which may cause solid-state devices to fail. In order to model the reliability and performance of the devices when exposed to high-energy radiation, it is, therefore, necessary to track the movement of vacancies and interstitials and determine the conditions under which they aggregate or settle randomly.

Radiation damage is just one of many physical processes, where bridging time scales is crucial. A precise knowledge of atomic locations is also needed for calculating charge distribution and other processes such as transport of hydrogen in graphene and nanotubes that are important for estimating the reliability of those materials. The most challenging aspect of modeling these processes is that the primary event occurs over femtoseconds, whereas the accumulation of defects and other secondary events occur over several nano or even microseconds. It is, therefore, necessary to develop models that can bridge the time scales from femto to microseconds or even more.

Another example is modeling the phonon transport and thermal conductivity of solids for energy and heat management applications. The thermal conductivity is sensitive to low frequency phonons, and, therefore, to long-time response of the solid, which must be accounted for by the model. In case of the ordinary 3D solids, this difficulty is partially mitigated by using the continuum model at large distances from the source of the primary event. This is not possible in case of nanomaterials because of the limited validity of the continuum model and the necessity to include a large number of atoms in the discrete atomistic model of the solid.

Presently, atomistic modeling of materials and physical processes is mostly done by using molecular dynamics (MD) and its variations [18]. A major stumbling block in classical as well as ab initio MD is the extremely limited time scale of the MD. Convergence requirements limit the time step in MD analysis to a few femtoseconds in most materials. In some idealized cases, the time step can be extended to at most a few picoseconds by using very elegant techniques [19]. Still, it would require 10^6 to 10^9 time integration steps in the MD analysis to model physical processes at time scales of practical interest (up to microseconds) in nanomaterials. Hence, temporal modeling of physical processes using MD is CPU intensive and not very convenient for "what if" type design questions. Extending the temporal scales in materials has, therefore, been a long-sought goal. We will describe herein the CGFMD technique that is a hybrid of causal GF and MD. It has been shown to accelerate the temporal convergence by several orders of magnitude [13], at least in an idealized model of graphene.

5.2.3 Applications

As an illustration of the MSGF method, we apply it to Xenes for bridging the length scales and to graphene for temporal scales. Although many new 2D materials have been fabricated or identified, graphene [20,21] remains a material of strong topical interest. It is widely regarded as having a strong potential for application in revolutionary new devices [22]. To develop 2D material-based devices for industrial applications, an important requirement is to control the local curvature of the solid films. This would require an understanding of the mechanical deflection of the films and to develop techniques for its multiscale modeling ranging from the atomistic to the device level.

Furthermore, the MSGF can be used to model the mechanical strength of Xenes, which should be useful for developing, for example, flexible electronic devices. Temporal modeling is useful for developing new methods of characterizing the samples and understanding processes like propagation of ripples in graphene [3].

Despite its short history, so much work has been published on graphene that it is not treated as a new material anymore. In fact, many 2D materials have now been identified that may be even more promising than graphene such as boron nitride, molybdenum disulfide, and other Xenes such as silicene and phosphorene [23,24]. It is apparent that a whole new class of materials is emerging that has the potential to revolutionize the materials industry. Still, graphene remains the main prototype of this new class of materials because of the simplicity of its lattice structure and fascinating electronic and mechanical properties. The importance of understanding graphene, therefore, cannot be overemphasized. This explains our emphasis on graphene in this chapter.

5.3   Green's function and its role in measurement science

In any scientific measurement process, we measure the response of a system to a probe. The mathematical function that defines the response is called the response function and is a characteristic of the system. We need to know the physical process that determines this function. The physical process is represented in terms of an equation or a mathematical relationship which may be an algebraic, differential, or integral equation. The equation needs to be solved to obtain the response function. In general, the objective of modeling is to calculate this function, which will enable us to predict the response of the system to a probe. The GF is essentially the response function of a system. Obviously, the choice of the GF would depend upon the nature of the response that we try to model.

We can write symbolically the following equation, which is the master equation for all scientific measurements:

$$\text{Response} = \text{Response function} * \text{Probe} \tag{5.1}$$

In the so-called forward problem, the response function and the probe are known and we find the response. In the inverse problem, the probe and the response are known, and we obtain the response function.

For example, consider a simple experiment on a particle A of mass m attached at the end of a massless spring of spring constant ϕ. The spring is attached to a wall as shown in Fig. 5.1. We apply a force $f(t)$ on the spring and measure $u(t)$, the displacement of A, with t being the time variable.

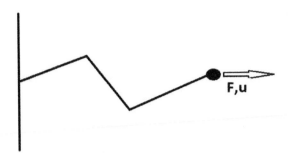

Fig. 5.1 A particle attached to a spring. F is the applied force and u is the displacement of the particle.

The displacement of A is given by the Newtonian equation, written below as a linear operator equation

$$Pu(t) = f(t) \tag{5.2}$$

where

$$P = \left(\phi + m \, \partial^2/\partial t^2 \right) \tag{5.3}$$

The formal solution of Eq. (5.2) is given by

$$u(t) = P^{-1} f = Gf(t) \tag{5.4}$$

We identify $G = P^{-1}$ as the GF that is the inverse of the operator P. It is the particular solution of the equation

$$P_t \, G(t,t') = \delta(t - t'), \tag{5.5}$$

where t' is also a time variable in the same space as t but is independent of t, and $\delta(t)$ is the Dirac delta function, which is zero for a nonzero argument. The delta function in Eq. (5.5) is defined by the following well-known integral relationship:

$$Z(t) = \int \delta(t - t') \, Z(t') \, dt', \tag{5.6}$$

where $Z(t)$ is any arbitrary but integrable function of t and the integration is over all t-space. The subscript for P in Eq. (5.5) indicates that it operates only on the variable t and not on t'. We can now write the formal solution given by Eq. (5.4) in the following general form

$$u(t) = \int G(t - t') \, f(t') \, dt', \tag{5.7}$$

where the integration is over all t-space. It can be easily verified that Eq. (5.7) is the solution of Eq. (5.2) by substitution and using Eqs. (5.5), (5.6). Note that we have written G as a function of a single argument $t - t'$, which assumes that G depends upon t and t' only through their difference.

We derive an expression for G and solve Eq. (5.7) using the Fourier representation as follows:

$$f(t) = (1/2\pi) \int f(\omega) \exp{(\imath\omega t)}d\omega, \tag{5.8}$$

$$u(t) = (1/2\pi) \int u(\omega) \exp{(\imath\omega t)}d\omega, \tag{5.9}$$

and

$$G(t) = (1/2\pi) \int G(\omega) \exp{(\imath\omega t)}d\omega, \tag{5.10}$$

where ω is the variable conjugate to t and $\imath = \sqrt{-1}$. The integration in Eqs. (5.8)–(5.10) is over all ω space. We now specify that both the variables t and ω range from $-\infty$ to $+\infty$.

The delta function has the following Fourier representation [5]

$$\delta(t) = (1/2\pi) \int_{-\infty}^{\infty} \exp{(\imath w t)}\, d\omega. \tag{5.11}$$

From Eq. (5.11), we obtain the inversion relationship for the Fourier transforms. For example

$$G(\omega) = \int_{-\infty}^{\infty} G(t) \exp{(-\imath\omega t)}\, d\omega. \tag{5.12}$$

Similar relationships can be derived for the Fourier transforms of other variables. Substitution of Eq. (5.10) in Eq. (5.5) and using Eqs. (5.3), (5.11) gives the following result

$$G(\omega) = 1/M(\omega_0{}^2 - \omega^2), \tag{5.13}$$

where

$$\omega_0 = \sqrt{(\phi/M)}, \tag{5.14}$$

is the natural frequency of the spring. Eq. (5.13) gives the desired GF in the frequency space. The GF in the time space can then be calculated from Eq. (5.10). Finally, we obtain by using Eqs. (5.7)–(5.11)

$$u(\omega) = G(\omega)f(\omega). \tag{5.15}$$

Eq. (5.15) along with Eq. (5.9) yields Eq. (5.7), the desired particular solution of Eq. (5.2).

The temporal GF $G(t)$ is defined as a retarded or advanced GF for positive and negative times to account for causality [6]. The causal GFs that we consider in this chapter must obey the causality condition that the response cannot precede the cause. Hence

$$G(t) = 0 \text{ for } t < 0. \tag{5.16}$$

Since $G(\omega)$ has a singularity on the real axis (Eq. 5.13), one has to take an appropriate contour for the integration to ensure causality.

For positive values, the variable ω can be identified as the frequency. In Eqs. (5.8)–(5.10), we have used the same symbol for the functions and their Fourier transforms for notational brevity, the identifying feature being the functional dependence on t or ω.

Although the example given above was rather trivial, it does bring out some of the primary characteristics of the GF, summarized below:

i. GF is an operator, being the inverse of another operator, as defined by Eq. (5.5).
ii. GF is linear so the solutions are additive, which can be verified from Eq. (5.7). It is, however, possible to include nonlinear effects in the probe.
iii. GF has poles at the natural frequencies of the system. These poles can be identified as resonances.
iv. GF is a characteristic of the system and is independent of the probe.
v. The GF can be used to obtain the solution (or the response) to any probe by using Eq. (5.7). This feature of the GF makes it a very powerful computational technique.

In the static case, $G, f,$ and u are independent of time. The above expressions reduce to the static case in the zero-frequency limit, which is apparent from Eqs. (5.8)–(5.10). Note that the static GF is independent of the mass of the particle. It is simply given by

$$G = [\phi]^{-1}. \tag{5.17}$$

5.4 Discrete atomistic model of a solid

In this section, we describe the main features of the lattice model of a solid that explicitly accounts for its discrete atomistic structure. We will include only those features that are directly relevant to the GF method. For details, please consult any textbook on solid-state physics or lattice dynamics such as the excellent classics by Kittel [25],

Born and Huang [26], or Maradudin et al. [6]. The discussion given in this section is formally applicable to lattices of any dimensions.

An important input to all lattice calculations is the interatomic potential. All atomistic defect calculations are based upon minimizing the free enthalpy of the solid that consists of an ionic part, which gives the elastic contribution, and an electronic part. These are, of course, coupled. When a defect such as a vacancy is introduced in the lattice, the relaxed configuration depends upon the charge states of the defect [27]. The same applies to extended defects like a quantum dot or a quantum well. A rigorous calculation of the relaxed configuration would require an ab intio quantum mechanical modeling of the coupled ion-electron system in the whole lattice. Such calculations are limited to very small model crystallites consisting of only a few hundred atoms [28]. At the other extreme is the continuum model, in which the electron effects are totally neglected. The continuum model reproduces the bulk mechanical characteristics of the defect and has been extensively used for a long time. It has the advantage of computational convenience but obviously has limited validity.

The intermediate approach is to use models in which the effect of the electrons is included in an empirical and phenomenological manner by using an effective interatomic potential [29–34]. This approximation has been used in almost all lattice-statics/lattice-dynamics/MD defect calculations. See, for example, the review article by Stangl et al. [35] and the monographs by Harrison [36] and Bimberg et al. [37] for application of phenomenological potentials to semiconductors. Such model potentials have been used in many atomistic calculations in Ge/Si and other semiconductors using MD. See, for example, the papers by Makeev and Madhukar [38], and Swadener et al. [39], which also give other references.

As discussed by Harrison [36], the inherent assumptions in all these calculations are: (i) tight binding approximation, which allows us to treat atoms as separate entities, (ii) adiabatic approximation, which assumes that the electrons respond adiabatically to the ionic displacements, and (iii) the independent electron approximation. These assumptions result into a separation of the crystal Hamiltonian into a part that corresponds to ionic interactions and another part that gives the energy of the electrons. Of course, the ionic interactions are also affected by the electrons. This contribution is included in a parametric model potential. Such models [29–32,34,39–42] give correct values for many observable parameters including the energy of vacancies and other defects, which lends credence to the validity of the model potential. The potentials [29–32,41] to which we refer in this paper belong to this class of models. Some other potentials are available in the literature [40,43–47], which have comparable advantages. We have used the Tersoff potential for some applications to graphene because of its computational convenience. It reproduces the correct energy of the defect and has been widely used in defect calculations on covalent solids. Its optimized version has been developed in an excellent paper by Lindsay and Broido [48].

The GF method described in this chapter is based upon the Born von Karman (BvK) model [6,25,26] of a lattice. This model fully accounts for the discrete structure of the lattice. The basic assumptions of the BvK model are:

i. Adiabatic approximation: As described above, it implies that the contribution of the electrons to the crystal Hamiltonian is additive. We therefore consider only the ionic interactions in the lattice model. The electronic energies can be calculated separately and added to the ionic energies to obtain the total energy of the crystal.

ii. Cyclic boundary conditions: The whole crystal is assumed to be divided into identical supercells. This enables us to neglect the surface effects in a model lattice. This is, therefore, applicable only to large crystals in the regions far away from the surfaces. This assumption is obviously not valid for nanocrystals. In such cases, the surfaces are modeled as defects in the ideal BvK lattice.

iii. Harmonic approximation: The atomic displacements are assumed to be small enough such that their cubic and higher powers can be neglected in expressing the potential energy of the crystal in terms of atomic displacements.

iv. Equilibrium and stability: The crystal is assumed to be in equilibrium in the absence of external forces so the net force on any atom in the lattice is zero.

The lattice structure of a solid can be described in terms of the primitive real-space lattice vectors of its Bravais lattice, which define the vector space of the lattice. In a 2D lattice, like an Xene, each Bravais lattice vector is a 2D vector. The corresponding reciprocal lattice vectors (see, for example, [25]), which define the reciprocal space of the lattice are also 2D. The primitive lattice vectors in the real space define a unit cell. Each unit cell in the real space may contain one or more atoms. These atoms are usually called the basis atoms. Their position vectors are defined with respect to the origin of their unit cell. We will refer to them as the basis vectors.

A lattice in which each cell contains only one atom is called a monatomic lattice. All Xenes have 2 atoms in a unit cell. The Wigner-Seitz cell in the reciprocal space is called the first Brillouin zone [25] of the lattice and plays an important role in the transformation of various lattice functions including the GF. The Brillouin zone of a lattice is a characteristic of its Bravais lattice and does not depend upon the basis vectors or the basis atoms.

We assume a Cartesian frame of reference with its origin located at a Bravais lattice site, as shown in Fig. 5.2. The X- and Y-axes are parallel to the crystallographic axes. With this choice of axes, the Bravais lattice vectors of the Xenes are confined to the XY-plane. These lattice sites are labeled as B-sites and are shown as stars in Fig. 5.2. The Z-axis is assumed to be normal to the plane of the paper. The non-Bravais sites, where the basis atoms ($\kappa \neq 0$) are located, are shown as solid ellipses in Fig. 5.2. These are labeled as A-sites in Fig. 5.2. For all Xenes, except graphene, the A-sites have non-zero Z-components, so that Fig. 5.2 only shows their projection on the XY-plane.

We denote the lattice sites of the origin of the unit cells by the indices L, L', etc., and the indices α, β, and γ denote the Cartesian coordinates x, y, and z. The atoms inside each unit cell are labeled by the index κ. The atom at the origin of each unit cell is indexed as $\kappa = 0$ and other atoms as $\kappa = 1, 2, p - 1$. For Xenes considered in the present paper, $p = 2$. In what follows, symbols in bold font will denote vectors or matrices, whereas scalars will be denoted by plain font.

We can index each atom by the index doublet $L\kappa$ and write its position vector as follows:

$$\mathbf{R}(L\kappa) = \mathbf{R}(L0) + \mathbf{b}(\kappa), \tag{5.18}$$

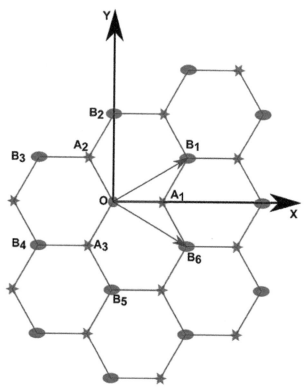

Fig. 5.2 Two-dimensional hexagonal lattice structure of Xenes, showing the lattice sites and the Cartesian coordinate axes. The Z-axis is normal to the plane of the paper. The primitive lattice vectors, each of length 2a, are OB_1 and OB_6, where 2a is the lattice constant. These vectors define the underlying Bravais lattice and the unit cell. The atoms B are all at the Bravais lattice sites. Each unit cell contains two basis atoms one each of type A and B. For graphene, all the atoms A and B are on the same plane, which is the XY- plane. For silicene, the atoms A and B are on different planes with the A atoms shifted in the Z-direction by $\Delta z = +0.44$ Å. The defect is at the origin which is a B site. Its three nearest-neighbors are A_1-A_3 and its six second-neighbors are B_1–B_6.

where $\mathbf{R}(L0)$ is the position vector of the origin of the Lth unit cell and $\mathbf{b}(\kappa)$ is a basis vector, which is the position vector of the κth atom relative to the origin of its own unit cell. For brevity, we will omit 0 in $\mathbf{R}(L0)$, unless required for clarity. The plane containing $\mathbf{R}(L0)$ is assumed to be the XY-plane. Thus, the Z-components of $\mathbf{R}(L0)$ and $\mathbf{b}(\kappa=0)$ are 0. This is a characteristic of the 2D lattice. In general, for $\kappa \neq 0$, $\mathbf{b}(\kappa)$ and therefore $\mathbf{R}(L\kappa)$ have non-zero Z-components.

The translation symmetry of the perfect solid implies that $\mathbf{b}(\kappa)$ is same for all unit cells. Thus, atoms in all unit cells for the same value of κ lie on the same plane. Atoms with different κ are, in general, on different planes. Since $\mathbf{b}(\kappa)$ is the same for all L, these planes will be parallel to each other. Hence, a Xene lattice consists of 2 planes parallel to each other. This is referred to as the dual planar structure of Xenes.

We can now treat all lattice vectors as 3D vectors in the standard Euclidean space of the lattice vectors. The 2D Bravais lattice vectors and the reciprocal lattice vectors are treated as 3D vectors with zero Z-components. This enables us to develop a unified formulation of the Green's functions for the Xenes, including graphene, in the lattice space. This space is obviously 3pN dimensional, where N is the number of the unit cells in the underlying Bravais lattice. Thus, the GF and the corresponding matrices in this representation, will be $6\,N \times 6\,N$.

For graphene, the Z-component of $\mathbf{b}(\kappa = 1)$ is also zero. This implies that all atoms lie on the XY-plane. Graphene can then be visualized as a limiting case of Xenes in which the separation between the two planes is zero. At equilibrium, each atom is assumed to be located at its lattice site and the energy at equilibrium is assumed to be zero. This defines the ground state from which the energies in the system caused by atomic displacements are measured.

We now introduce the lattice distortion. We denote the displacement of the atom Lκ by $\mathbf{u}(L\kappa)$. In time dependent problems, $\mathbf{u}(L\kappa)$ will also depend upon t. Its Fourier transform over time will be a function of ω, the conjugate of t. However, for brevity of notation, we will not show the explicit dependence of $\mathbf{u}(L\kappa)$ on t or ω unless needed for clarity.

We write the potential energy of the lattice in the form of a Taylor series as follows

$$
\begin{aligned}
W = & -\sum_{L\kappa\alpha} f_\alpha(L\kappa)u_\alpha(L\kappa) \\
& + (1/2)\sum_{L\kappa\alpha}\sum_{L'\kappa'\beta}\left[\phi_{\alpha\beta}(L\kappa, L'\kappa')u_\alpha(L\kappa)u_\beta(L'\kappa')\right],
\end{aligned} \tag{5.19}
$$

where $-\mathbf{f}$ and $\boldsymbol{\phi}$ are the Taylor coefficients. They are defined by

$$
f_\alpha(L\kappa) = -\partial W/\partial u_\alpha(L\kappa), \tag{5.20}
$$

and

$$
\phi_{\alpha\beta}(L\kappa, L'\kappa') = \partial^2 W/\partial u_\alpha(L\kappa)\partial u_\beta(L\kappa'), \tag{5.21}
$$

where the derivatives are evaluated at zero displacement. To clarify the notation, $\mathbf{f}$ (Lκ), $\mathbf{R}(L\kappa)$, and $\mathbf{b}(\kappa)$, and $\mathbf{u}(L\kappa)$ are 3D vectors. Their three Cartesian components are identified by the index α. Physically, $f_\alpha(L\kappa)$ denotes the α component of the force at the atom Lκ. Similarly, $\boldsymbol{\phi}(L\kappa, L'\kappa')$ is a 3×3 matrix, which is called the force-constant matrix between atoms at Lκ and L'κ'. Its nine elements are denoted by $\phi_{\alpha\beta}(L\kappa, L'\kappa')$.

The force and the force-constant matrices can be obtained from the interatomic potential [6]. These matrices must obey the constraints imposed by the translation and rotation symmetry of the lattice, which determine the number of independent elements in the matrices [6]. In addition, the equilibrium condition of zero net force on the solid imposes the following constraints:

i. For all α

$$\Sigma_{L\kappa} \, f_\alpha(L\kappa) = 0, \tag{5.22}$$

ii. For all L, κ, α, and β

$$\Sigma_{L'\kappa'} \, \phi_{\alpha\beta}(L\kappa, L'\kappa') = 0. \tag{5.23}$$

At equilibrium W must be at its minimum. By minimizing the right-hand side of Eq. (5.19), we obtain the following equation for the displacements.

$$f_\alpha(L\kappa) = \Sigma_{L'\kappa'\beta} \, \phi_{\alpha\beta}(L\kappa, L'\kappa')u_\beta(L'\kappa'). \tag{5.24}$$

Eq. (5.24) is the steady state part of Eq. (5.2) for a many-particle system in which **f** is the probe and **u** is the response. For time-dependent problems, the Newtonian term has to be added to Eq. (5.24) as in Eq. (5.3). For a finite system, one can write a set of coupled equations for pN atoms as in Eq. (5.24) and solve it numerically for all **u**. This is essentially the technique used in MD. As mentioned in Section 5.1, solving Eq. (5.24) directly may become computationally too expensive for large N—say a few million or billion atoms. As we shall see in this section, the GF method offers a computationally efficient method for solving Eq. (5.24).

Using Eqs. (5.19), (5.24), we can write the equation of motion for the atom $L\kappa$ for a perfect lattice as follows:

$$M(L\kappa)\partial^2 u(L\kappa)/\partial t^2 = f_\alpha(L\kappa) - \sum_{L'\kappa'\beta}\phi_{\alpha\beta}(L\kappa, L'\kappa')u_\beta(L'\kappa'). \tag{5.25}$$

For modeling of phonons, we take the Fourier transform of the displacement over time as given in Eq. (5.9). This will make **u**($L\kappa$) a function of the frequency ω, which can be identified as the phonon frequency. The nonlinear terms in the potential, if any, will add terms on the right of Eq. (5.25) that will depend upon u. Alternatively, nonlinear effects can be included by treating **f** and ϕ as functions of **u** and hence t. In the harmonic approximation, **f** and ϕ are independent of **u** and t and phonon frequencies are independent of **f**. If **f** is nonzero, the equilibrium location of atoms will change and they will simply vibrate about their new position of equilibrium.

We define the discrete Fourier transform over space of the displacement vector as follows

$$u_\alpha(L\kappa) = (1/N)\Sigma_k \, u_\alpha(k; \kappa) \, \exp[\imath k.R(L\kappa)], \tag{5.26}$$

where **k** is a 3D vector in the reciprocal space of the lattice. The number of allowed **k** vectors in the first Brillouin zone is N, equal to the number of unit cells in a supercell. For notational brevity, we follow the common convention and denote the matrix in real space and the Fourier (reciprocal) space by the same symbol. The distinguishing feature between the two is that the functional dependence on $\mathbf{k}$ will be shown explicitly for a matrix in reciprocal space. An equation similar to Eq. (5.26) can be written down for the force vector defined by Eq. (5.20). Instead of Eq. (5.26), an alternative definition of the Fourier transform has been used in the literature in which $\mathbf{R}(L0)$ is used in the exponent instead of $\mathbf{R}(L\kappa)$. The two definitions differ only by a phase factor $\exp[\imath\mathbf{k}.\mathbf{b}(\kappa)]$.

Because of the orthogonality of the reciprocal and the direct-lattice vectors, the vector $\mathbf{k}$ can take only certain allowed discrete values [6,7] and can be confined to the first Brillouin zone of the lattice. The inverse Fourier transform of Eq. (5.26) can be obtained by using the following relation [6]

$$\Sigma_L \exp[\imath\mathbf{k}.\mathbf{R}(L0)] = \delta_d(\mathbf{k}), \tag{5.27}$$

and

$$\Sigma_k \exp[\imath\mathbf{k}.\mathbf{R}(L0)] = \delta_d(L), \tag{5.28}$$

where δ_d is the discrete analogue of the Dirac delta function defined by Eq. (5.6). It is 0 for non-zero value of the argument and equal to N if the argument is 0. Strictly speaking, the sum in Eq. (5.27) is equal to N not just for $\mathbf{k}=0$ but whenever $\mathbf{k}$ is equal to a reciprocal lattice vector. However, in this chapter we will be concerned only with those cases when $\mathbf{k}$ is confined to the first Brillouin zone of the lattice. Hence $\mathbf{k}=0$ is the only relevant reciprocal lattice vector.

Using Eqs. (5.27), (5.28) in Eq. (5.26), we find the following equation for the inverse transform of the displacement vector

$$u_\alpha(\mathbf{k};\kappa) = \Sigma_L u_\alpha(L\kappa) \exp[-\imath\mathbf{k}.\mathbf{R}(L\kappa)]. \tag{5.29}$$

Note the absence of the factor (1/N) in Eq. (5.29) as compared to Eq. (5.26).

The above equations are valid for a perfect lattice as well as a lattice containing defects. A perfect infinite lattice or a perfect lattice with periodic boundary conditions has full translation symmetry. Combined with the condition of invariance of the potential energy against rigid body translations, this leads to the following additional constraints [6]:

i. The Taylor coefficients $f_\alpha(L\kappa)=0$ for all $L\kappa$. Physically, this condition implies that for a perfect lattice in equilibrium without any external forces, there cannot be any force on any lattice site. Of course, if external forces are applied, then $\mathbf{f}$ can represent the applied external forces.

ii. The force-constant matrix $\phi_{\alpha\beta}(L\kappa,L'\kappa')$ depends upon L and L' only through their difference. This is a consequence of the translation symmetry. Since all atoms on one sublattice are equivalent, any atom can be chosen as the origin. The force-constant matrix can, therefore, be labeled by a single L index as $\phi_{\alpha\beta}(0\kappa,L\kappa')$ that gives the force constant between any two atoms for which the unit cells are separated by $\mathbf{R}(L\kappa')-\mathbf{R}(0\kappa)$. This enables us to write the force-constant matrix in terms of its discrete Fourier transform as defined below:

$$\phi_{\alpha\beta}(0\kappa,L\kappa') = (1/N)\Sigma_k \phi_{\alpha\beta}(\mathbf{k};\kappa,\kappa') \exp[\imath\mathbf{k}.\{\mathbf{R}(L\kappa')-\mathbf{R}(0\kappa)\}], \tag{5.30}$$

where $\boldsymbol{\phi}(\mathbf{k})$ is the Fourier transform of the force-constant matrix. Its matrix elements are $\phi_{\alpha\beta}(\mathbf{k};\kappa,\kappa')$. As in Eq. (5.29), the inverse transform of $\boldsymbol{\phi}(\mathbf{k})$ is given by

$$\phi_{\alpha\beta}(\mathbf{k};\kappa,\kappa') = \Sigma_L\ \phi_{\alpha\beta}(0\kappa, L\kappa')\ \exp\left[-\imath\mathbf{k}.\{\mathbf{R}(L\kappa') - \mathbf{R}(0\kappa)\}\right]. \tag{5.31}$$

Since κ and κ' can take p different values, the matrix $\boldsymbol{\phi}(\mathbf{k})$ is a $3p \times 3p$ square matrix. By generalizing the derivation given in Section 5.2 to 3D many particle systems, we can construct the dynamical matrix [6,26] of the lattice in terms of $\boldsymbol{\phi}(\mathbf{k})$ as follows

$$D_{\alpha\beta}(\mathbf{k};\kappa,\kappa') = (M_\kappa M_{\kappa'})^{-1/2}\ \phi_{\alpha\beta}(\mathbf{k};\kappa,\kappa'), \tag{5.32}$$

where M_κ is the mass of the atom κ. If ω denotes the phonon frequency, the eigenvalues of $\mathbf{D}(\mathbf{k})$ can be shown to be ω^2. In view of the translation symmetry in a perfect lattice for which Eq. (5.32) is valid, M_κ is independent of L. The dynamical matrix $\mathbf{D}(\mathbf{k})$ is used extensively in lattice dynamical calculations. Lattice statics for static problems is independent of the atomic masses. The relevant matrix for lattice statics is $\boldsymbol{\phi}(\mathbf{k})$.

The matrices $\mathbf{D}(\mathbf{k})$ and $\boldsymbol{\phi}(\mathbf{k})$ have 3p eigenvalues. From Eq. (5.32), their eigenvalues are related through a factor depending upon the atomic masses. By using Eq. (5.23), it can be shown that the determinant of $\mathbf{D}(\mathbf{k})$ is zero at $k=0$ and at least three of its eigenvalues vary as k^2 near $k=0$. These are called the acoustic modes. The remaining $3p-3$ eigenvalues are, in general, nonzero at $k=0$. These are called the optical modes. For monatomic lattices, there are no optical lattice modes. In the case of graphene, one of the acoustic modes varies as k^4 near $k=0$. In fact, it does have a k^2 term, but its coefficient is zero [49]. This mode is polarized perpendicular to the plane of graphene. It has significant implications in the calculation of the flexural rigidity of graphene.

5.5   Lattice statics Green's function (LSGF)

First, we consider a perfect lattice. The phonon GF of the lattice is defined as follows [6]:

$$\mathbf{G}\left(\mathbf{k},\omega^2\right) = \left[-\omega^2\mathbf{I} + \boldsymbol{\phi}(\mathbf{k})\right]^{-1}, \tag{5.33}$$

where $\mathbf{I}$ is the unit matrix. In steady state, the displacements are independent of time so that the steady state corresponds to $\omega=0$. We define the lattice statics Green's function (LSGF) as the zero frequency limit of the phonon GF [7]. In the Fourier $\mathbf{k}$-space, the LSGF is given by:

$$\mathbf{G}(\mathbf{k}) = [\boldsymbol{\phi}(\mathbf{k})]^{-1}. \tag{5.34}$$

Eq. (5.34) is the equivalent of Eq. (5.17) for a lattice. Since the determinant of $\phi(\mathbf{k})$ is zero at $k=0$, $\mathbf{G(k)}$ is singular at $k=0$, which is consistent with the discussion in Section 5.2. Analogous to $\phi(\mathbf{k})$, the matrix elements of the 3p × 3p matrix $\mathbf{G(k)}$ are labeled as $G_{\alpha\beta}(\mathbf{k};\kappa,\kappa')$. The LSGF in the real space is defined by its inverse Fourier transform in analogy with Eq. (5.30) as follows

$$G_{\alpha\beta}(0\kappa, L\kappa') = (1/N)\Sigma_{\mathbf{k}}\, G_{\alpha\beta}(\mathbf{k}; \kappa, \kappa')\, \exp\left[\imath\mathbf{k}.\{\mathbf{R}(L\kappa') - \mathbf{R}(0\kappa)\}\right]. \qquad (5.35)$$

Eq. (5.35) provides a convenient and computationally efficient semi-analytic representation for calculation of the LSGF. The representation is semi-analytic in the sense that analytical expressions are derived in terms of GFs but the GF itself may require extensive numerical work. In the present case, for a solid containing N atoms, Eq. (5.35) requires inversion of only 3p × 3p matrices.

For real space calculations, it is convenient to define a 3pN × 3pN vector space in terms of the three coordinates of the pN atoms. In this representation, we denote the real-space force constant matrix by $\boldsymbol{\Phi}$ and the GF matrix by $\boldsymbol{\Gamma}$. These matrices will be 3pN × 3pN square matrices. The matrix elements of $\boldsymbol{\Phi}$ are $\phi_{\alpha\beta}(L\kappa, L'\kappa')$ and the matrix elements of $\boldsymbol{\Gamma}$ are $G_{\alpha\beta}(L\kappa, L'\kappa')$. Similarly, in the same vector space, we define ηpN dimensional column matrices $\mathbf{U}$ and $\mathbf{F}$ for the displacement and the force vectors with matrix elements $U_{\alpha}(L\kappa)$ and $f_{\alpha}(L\kappa)$, respectively. Using Eq.(5.34), it can be easily shown that in this representation

$$\boldsymbol{\Gamma} = \boldsymbol{\Phi}^{-1}. \qquad (5.36)$$

In the same representation, Eq. (5.19) can be written in the following compact form

$$\mathbf{W} = -\mathbf{F}.\mathbf{U} + (1/2)\mathbf{U}^{\mathrm{T}}.\boldsymbol{\Phi}.\mathbf{U} \qquad (5.37)$$

If defects are introduced into the lattice then its translation symmetry is broken. Consequently, even in the absence of any external force, $\mathbf{f}$ becomes nonzero for at least some L and κ [7]. Further, ϕ and G depend upon L and L' separately and not merely on their difference [6]. Hence, it is not possible to define their Fourier transform in terms of a single vector $\mathbf{k}$ as in Eqs. (5.30), (5.35). In such cases, the defect LSGF, the GF for the lattice with defects, is calculated as follows [7,50].

We denote the force constant matrix for the defective lattice and the defect LSGF by $\boldsymbol{\Phi}^*$ and $\boldsymbol{\Gamma}^*$, respectively. In the representation of lattice sites, $\boldsymbol{\Gamma}^*$ and $\boldsymbol{\Phi}^*$ are 3 Np × 3 Np matrices. The GF is formally given by, as in Eq. (5.36),

$$\boldsymbol{\Gamma}^* = \left[\boldsymbol{\Phi}^*\right]^{-1}. \qquad (5.38)$$

We write

$$\boldsymbol{\Phi}^* = \boldsymbol{\Phi} - \Delta\boldsymbol{\Phi}, \qquad (5.39)$$

where $\mathbf{\Delta\Phi}$ denotes the change in the force-constant matrix $\mathbf{\Phi}$ caused by the defect(s). From Eqs. (5.36), (5.38), and (5.39), we obtain the following equation.

$$\mathbf{\Gamma}^* = \mathbf{\Gamma} + \mathbf{\Gamma}\,\mathbf{\Delta\Phi}\,\mathbf{\Gamma}^*. \qquad (5.40)$$

Eq. (5.40) is the Dyson equation for calculation of the defect GF. This equation shows that in order to calculate $\mathbf{G}^*$, we have to first define a reference state and calculate $\mathbf{G}$ and the corresponding $\mathbf{\Delta\Phi}$. The change $\mathbf{\Delta\Phi}$ is obtained by the interatomic potential between the defect and the host atoms. In quantum mechanical problems, $\mathbf{\Delta\Phi}$ gives the change in the system Hamiltonian. Methods for solving the Dyson equation have been discussed, for example, in references [6, 7, 51].

In the present case, we have taken the perfect lattice with full translational symmetry as the reference state. In practice, any state can be chosen to be the reference state for which $\mathbf{G}$ can be calculated and $\mathbf{\Delta\Phi}$ can be conveniently defined [51–53]. Although $\mathbf{G}$ for the reference state is calculated using the harmonic or the linear approximation, $\mathbf{\Delta\Phi}$ can include non-linear terms [54,55].

For a defect lattice, Eq. (5.37) is modified as follows:

$$\mathbf{W}^* = -\mathbf{F}.\mathbf{U} + (1/2)\,\mathbf{U}^{\mathrm{T}}.\mathbf{\Phi}^*.\mathbf{U}, \qquad (5.41)$$

where $\mathbf{U}$ and $\mathbf{F}$ are now identified as the displacement and the force vectors in the defect lattice. The solution of Eq. (5.41) gives the displacement field:

$$\mathbf{U} = \mathbf{\Gamma}^*\mathbf{F}. \qquad (5.42)$$

A very useful form of Eq. (5.42) has been derived in [7]. It is shown that Eq. (5.42) is exactly equivalent to

$$\mathbf{U} = \mathbf{\Gamma}\,\mathbf{F}^*, \qquad (5.43)$$

where

$$\mathbf{F}^* = \mathbf{F} + \mathbf{\Delta\Phi}\,\mathbf{U}. \qquad (5.44)$$

Eqs. (5.42) or (5.43) are the main equations of the LSGF method. They are used to calculate the atomic displacements, which give the lattice distortion, or lattice relaxation, caused by the presence of the defect. In this method, the defect is characterized by $\mathbf{\Delta\Phi}$. Eq. (5.42) gives the displacement in terms of the defect GF and the forces on the atoms at the lattice sites of the reference state. Eq. (5.43), the alternative form of Eq. (5.42), gives the displacement in terms of the perfect-lattice GF and an effective force denoted by $\mathbf{F}^*$, as defined by Eq. (5.44).

The effective force $\mathbf{F}^*$ is called the Kanzaki force [7] in the harmonic approximation. From Eq. (5.44), we can identify it as the force due to the defect on the relaxed lattice sites, in contrast to $\mathbf{F}$, that denotes the force at the original lattice site.

5.6   Multiscale Green's function (MSGF)

The LSGF method provides a convenient and computationally efficient mathematical technique for multiscale modeling of nanomaterials. We need multiscale models of nanomaterials for bridging length scales as well as temporal scales. In this section, we will discuss only the length scales. The problem of multi-time scales will be discussed in the following section.

Linking between the lattice and the continuum models in the GF method is achieved by using Born's method of long waves [26]. This method establishes the correspondence between the force constants and the elastic constants. Using Born's method, it can be shown [7] that for large R, the LSGF varies as $1/R$ for 3D solids. and reduces asymptotically to the continuum GF (CGF). In the 2D solids case, the CGF has a logarithmic behavior and shows effects of lattice size. This will be discussed in Section 5.7.

To establish the correspondence between the LSGF and CGF, we make $\mathbf{R(L\kappa)}$ and $\mathbf{k}$ continuous variables and replace the summation in Eq. (5.35) by integration over the reciprocal space. In conformity with the continuum model notation, we replace $\mathbf{R(L\kappa)}$ by the continuous variable $\mathbf{r}$ for large $R(L\kappa)$, and the discrete wave vector $\mathbf{k}$ by the continuous wave vector $\mathbf{q}$, which spans all space from $-\infty$ to ∞.

$$\mathbf{G(r)} \cong (1/2\pi)^3 \int \mathbf{G(q)}\ exp\ (\imath \mathbf{q.r})\ \mathbf{dq}, \tag{5.45}$$

We obtain the asymptotic limit of $\mathbf{G(r)}$ by using Duffin's lemma [7]. In the limit $r \to \infty$, smaller values of q make more significant contributions to the integral in Eq. (5.45). Accordingly, we expand $\mathbf{G(q)}$ for $q \to 0$ in powers of q as given below.

$$Lim_{q\to 0}\ \mathbf{G(q)} = Lim_{q\to 0}\ [\phi(\mathbf{q})]^{-1} \tag{5.46}$$

For low values of q, $\phi(\mathbf{q})$ has the following behavior [6] for the acoustic modes

$$\phi(q) \sim O(q^2) + O(q^4). \tag{5.47}$$

Using the expansion as in Eq. (5.47), we can carry out the integral in Eq. (5.45) term by term which gives the asymptotic expansion of the LSGF. The leading term in $\phi(\mathbf{q})$ as $q \to 0$ is the q^2 term. This term is identical to the Christoffel matrix in the continuum model [7,50]. Keeping only this term, we can show [7] that

$$Lim_{q\to 0}\ \mathbf{G(q)} = Lim_{q\to 0}\ [\boldsymbol{\Phi}(\mathbf{q})]^{-1} = [\boldsymbol{\Lambda}(\mathbf{q})]^{-1}, \tag{5.48}$$

where $\boldsymbol{\Lambda}$ is the well-known Christoffel matrix of the continuum model [9]. It is defined in terms of $\mathbf{c}$, the elastic constant tensor, as follows

$$\Lambda_{ij}\ (\mathbf{q}) = c_{ikjl}\ q_k\ q_l, \tag{5.49}$$

and i,j,k,l are Cartesian components in the continuum model. Summation over repeated Cartesian indices is assumed. The CGF $\mathbf{G_c}(\mathbf{q})$ in q space is defined as the inverse of the Christoffel matrix. Thus, we get

$$\text{Lim}_{q \to 0}\, \mathbf{G}(\mathbf{q}) = \mathbf{G_c}(\mathbf{q}) = [\mathbf{\Lambda}(\mathbf{q})]^{-1} = O(1/q^2) \tag{5.50}$$

In view of Eq. (5.50), the integrand in Eq. (5.45) has a $1/q^2$ singularity at q = 0. It is integrated out in 3D because the integral element in 3D is proportional to q^2. It can be shown [7] that the integral is proportional to $1/r$, which is the continuum GF in 3D. In 2D solids, the behavior of the GF is qualitatively different. This is because the integral element in 2D is proportional to q and the singularity in Eq. (5.50) does not get integrated out. This leads to the logarithmic dependence of the GF in 2D and makes the GF size dependent [56].

The subscript c on $\mathbf{G}$ in Eq. (5.45) denotes the GF for the continuum model. It has been extensively used in modeling solids using boundary element analysis. It provides a powerful technique for solving the boundary value problems for solids with and without defects [57–60].

It is important to note that the asymptotic relation given by Eq. (5.45) is valid only for the perfect lattice GF $\mathbf{G}$. It is generally not valid for the defect GF $\mathbf{G}^*$ defined by Eq. (5.40) unless the term containing $\mathbf{\Delta\Phi}$ is negligible [61]. In most cases of practical interest, it is not negligible. For example, in the case of a vacancy $\mathbf{\Delta\Phi} = \mathbf{\Phi}$. Moreover, if the effect of $\mathbf{\Delta\Phi}$ is negligible, then the information about the defect is lost. Calculations in this region are therefore not of interest for studying the properties of the defect.

Eq. (5.43) is the master equation of our MSGF method. The displacement of the atom at Lκ, which is a matrix element of $\mathbf{U}$, is given by

$$u_\alpha(L\kappa) = \Sigma_{L'\kappa'\beta}\, G_{\alpha\beta}(L\kappa, L'\kappa')\, f^*_{\beta}(L'\kappa') \tag{5.51}$$

where f^* is the Kanzaki force, which is the corresponding matrix element of $\mathbf{F}^*$ given by Eq. (5.44). For Lκ close to the defect, we use $\mathbf{G}$ to be the LSGF, whereas for large values $|\mathbf{R}(L\kappa) - \mathbf{R}(L'\kappa')|$ we choose $\mathbf{G} = \mathbf{G_c}$, the CGF. Since $\mathbf{G_c}$ is the asymptotic limit of $\mathbf{G}$ for large $|\mathbf{R}(L\kappa) - \mathbf{R}(L'\kappa')|$, the linkage is seamless.

It must be emphasized that even with $\mathbf{G_c}$, we use the discrete lattice value of f^* (or $\mathbf{F}^*$) as defined by Eq. (5.44) near the defect in terms of $\mathbf{\Delta\Phi}$. The Kanzaki force retains all the characteristics of the defect through $\mathbf{\Delta\Phi}$. Thus Eq. (5.51) is the multiscale representation of the lattice distortion or strains due to the defect since it relates the discrete lattice parameters through f^* to the continuum model parameters through $\mathbf{G_c}$.

The advantage of writing the displacement in the form of Eq. (5.43) is now obvious. We can use the full power of the continuum mechanics by using the continuum-model GF $\mathbf{G_c}$ for $\mathbf{G}$ where needed, while retaining the discrete lattice effects and all the characteristics of the defect exactly in $\mathbf{F}^*$.

Further, nonlinear effects associated with the defect host interaction can be incorporated in the MSGF method through $\mathbf{\Delta\Phi}$ while still retaining the linear simplicity of the GFs. A convenient technique for including nonlinear effects is to calculate the atomic displacements in a zone near the defect by using MD and use these values to calculate $\mathbf{f}^*$. MD accounts for the nonlinear forces. The displacements outside the zone are then calculated by using Eq. (5.51) by using LSGF or CGF for $\mathbf{G}$ depending upon the value of $|\mathbf{R}(\mathbf{L}\kappa) - \mathbf{R}(\mathbf{L}'\kappa')|$. This method has been applied to model realistic size Ge quantum dots in Si and Au islands in Cu [54,55]. A somewhat different version of the MSGF method has been applied to quantum nanostructures in semiconductors [53,62].

One important advantage of the MSGF method is that the continuum parameters like strain and stresses can be defined in this method asymptotically or by expressing the derivatives in terms of their Fourier transforms. In a purely discrete lattice model, the atomic displacements are defined at discrete lattice points. In many applications, one needs elastic strains that can be measured experimentally. As mentioned earlier, strain is a continuum parameter defined in terms of the derivatives of the displacement field and cannot be defined as such for discrete values of the displacements. In practice, one has to assume some averaging scheme in order to define the derivatives. Although elegant techniques [63] have been developed for this purpose, the averaging process is not unique and requires careful attention to various conservation laws. The MSGF method provides one averaging process, which makes the variable $\mathbf{r}$ in Eq. (5.45) continuous in the asymptotic limit. In this region, the derivatives of $\mathbf{u}$ with respect to x and y are uniquely defined.

5.7 Application of MSGF method to calculation of strain field in Xenes

5.7.1 Graphene

In this sub-section, we discuss graphene in some detail. It will serve as the prototype for the MSGF method for other Xenes. As we have seen in Section 5.5, a multiscale model of a solid can be constructed if we can establish a correspondence between its lattice and continuum GFs. Such a correspondence has been rigorously proved for normal 3D solids in Section 5.5. Unfortunately, that does not apply to 2D graphene or any solid that is strictly 2D. The reason for this discrepancy will be apparent from Eq. (5.45). The integration volume element is proportional to q^2 in 3D, which integrates out the q^2 singularity at $q=0$. The corresponding element in 2D is proportional to q. Hence, the singularity in the integral survives as $1/q$. This is responsible for the long-range logarithmic behavior of the GF and size effect in the static as well as dynamic response of graphene [49,64].

However, it is still possible to set up a correspondence between lattice and continuum models for a 2D graphene film in the deflection mode. The displacement field in the deflection mode is normal to the plane of graphene. For small displacements, in the harmonic approximation, the out-of-plane displacements are not coupled with the in-plane displacements. Consequently, the GF for the deflection mode is a scalar.

Moreover, the q^2 term in Eq. (5.47) is identically zero [49] in the deflection mode for graphene. Hence, the leading term in $\phi(q)$ in the deflection mode is $O(q^4)$ for graphene.

The integral for the $O(q^4)$ term in Eq. (5.45) can be carried out analytically. This shows that the continuum limit of the LSGF of a graphene sheet, in the deflection mode, corresponds to the GF for an elastically stable Kirchhoff plate but not the GF for two-dimensional Christoffel equations. This correspondence demonstrates the mechanical stability of graphene in deflection and is necessary for relating its mechanical parameters to its lattice parameters. Using this approach, an explicit expression has been derived [49,65,66] for relating the continuum flexural rigidity to the force constants of graphene. This relationship can be used to measure flexural rigidity of graphene directly from experimentally observed phonon dispersion. However, a more rigorous calculation or measurement of rigidity will involve some more contributions.

Now we consider the force constants for graphene, which are the most important parameters for the lattice GFs. For multiscale applications, force constants give the elastic constants. This relation ensures that the LSGF is seamlessly linked with the CGF. The force constants can be obtained from the interatomic potential. A popular choice of potential for graphene is the Tersoff potential and its variations. This potential, which is essentially a bond-order potential, extends up to second neighbor distances. It is convenient for MD simulations and has been used for calculations of strains, friction, and many other physical characteristics of graphene [67–73] and many more solids.

The second neighbor Tersoff type potential does not seem to be a good choice for phonon calculations and those properties of graphene that are sensitive to details of the phonon spectrum. An optimized Tersoff Brenner potential seems to be more suitable for modeling of phonons in graphene [48]. The main point is that to obtain a good fit between the calculated and the experimental values of the phonon dispersion for graphene, interatomic interactions have to be included at least up to fourth neighbor atoms [65,66,74–76]. However, if only the harmonic GF is needed, a detailed knowledge of the potential is not required. The force constants can be obtained by parametric fitting [6] of the calculated phonon dispersion with measured values.

For phonon applications, we give the 3×3 force constant matrices for graphene as defined by Eq. (5.21). A graphene unit cell contains two non-equivalent atoms O and A1 as shown in Fig. 5.2. In the harmonic approximation, the atomic displacements graphene in the XY plane and the Z-direction are not coupled. This is reflected by the fact that the XZ, ZX, YZ, and ZY elements of the force constant matrices are zero.

The force constant matrices $\phi(0;\mathbf{l})$ between the atom at the origin and its first four neighbors are given below, where $\mathbf{l}$ is the position vector of the atom. The coordinates of the atoms are given in the units of a where 2a is the lattice constant. The form of the matrices is general for the hexagonal symmetry of 2D graphene:

$$\phi(0; 2s, 0, 0) = - \begin{pmatrix} \alpha_1 & 0 & 0 \\ 0 & \beta_1 & 0 \\ 0 & 0 & \delta_1 \end{pmatrix}, \tag{5.52}$$

$$\phi(0;0,2,0) = -\begin{pmatrix} \alpha_2 & \gamma_2 & 0 \\ -\gamma_2 & \beta_2 & 0 \\ 0 & 0 & \delta_2 \end{pmatrix}, \tag{5.53}$$

$$\phi(0;-4s,0,0) = -\begin{pmatrix} \alpha_3 & 0 & 0 \\ 0 & \beta_3 & 0 \\ 0 & 0 & \delta_3 \end{pmatrix}, \tag{5.54}$$

$$\phi(0;5s,1,0) = -\begin{pmatrix} \alpha_4 & \gamma_4 & 0 \\ \gamma_4 & \beta_4 & 0 \\ 0 & 0 & \delta_4 \end{pmatrix}, \tag{5.55}$$

where $s = 1/\sqrt{3}$. The form of the force constant matrices between O and other atoms in the same neighbor shell and also between A1 and its neighbors can be obtained from symmetry.

In the multiscale method, based upon Born's method of long waves as described in Section 5.5, we compare the phonon frequencies with the frequencies given by the Christoffel matrix as in Eq. (5.47). This yields the following relations between the force constants and the elastic constants for graphene:

$$\begin{aligned}
c11 = C_u[2\alpha_1^2 &+ \alpha_1(6\alpha_2 + 19\alpha_3 + 35\alpha_4 + 54\alpha_5 + 6\beta_1 + 18\beta_2 + 3\beta_3 + 33\beta_4 + 18\beta_5 \\
&+ 14\sqrt{3\gamma_4}) + \alpha_2(6\alpha_3 + 12\alpha_4 + 6\beta_1 + 6\beta_3 + 12\beta_4) + 8\alpha_3^2 + \alpha_3(83\alpha_4 + 54\alpha_5 \\
&+ 9\beta_1 + 18\beta_2 + 24\beta_3 + 21\beta_4 + 18\beta_5 + 2\sqrt{3\gamma_4}) + 53\alpha_4^2 + \alpha_4(108\alpha_5 + 51\beta_1 \\
&+ 36\beta_2 + 27\beta_3 + 162\beta_4 + 36\beta_5 + 40\sqrt{3\gamma_4}) + \alpha_5(54\beta_1 + 54\beta_3 + 108\beta_4) \\
&+ \beta_1(18\beta_2 + 9\beta_3 + 9\beta_4 + 18\beta_5 + 6\sqrt{3\gamma_4}) + \beta_2(18\beta_3 + 36\beta_4) + \beta_3(45\beta_4 \\
&+ 18\beta_5 + 18\sqrt{3\gamma_4}) + 9\beta_4^2 + 36\beta_4\beta_5 - 12\gamma_4^2],
\end{aligned} \tag{5.56}$$

$$\begin{aligned}
c66 = C_u[2\beta_1^2 &+ \beta_1(6\beta_2 + 19\beta_3 + 35\beta_4 + 54\beta_5 + 6\alpha_1 + 18\alpha_2 + 3\alpha_3 + 33\alpha_4 + 18\alpha_5 \\
&- 14\sqrt{3\gamma_4}) + \beta_2(6\beta_3 + 12\beta_4 + 6\alpha_1 + 6\alpha_3 + 12\alpha_4) + 8\beta_3^2 + \beta_3(83\beta_4 + 54\beta_5 \\
&+ 9\alpha_1 + 18\alpha_2 + 24\alpha_3 + 21\alpha_4 + 18\alpha_5 - 2\sqrt{3\gamma_4}) + 53\beta_4^2 + \beta_4(108\beta_5 + 51\alpha_1 \\
&+ 36\alpha_2 + 27\alpha_3 + 162\alpha_4 + 36\alpha_5 - 40\sqrt{3\gamma_4}) + \beta_5(54\alpha_1 + 54\alpha_3 + 108\alpha_4) \\
&+ \alpha_1(18\alpha_2 + 9\alpha_3 + 9\alpha_4 + 18\alpha_5 - 6\sqrt{3\gamma_4}) + \alpha_2(18\alpha_3 + 36\alpha_4) + \alpha_3(45\alpha_4 \\
&+ 18\alpha_5 - 18\sqrt{3\gamma_4}) + 9\alpha_4^2 + 36\alpha_4\alpha_5 - 12\gamma_4^2],
\end{aligned} \tag{5.57}$$

where

$$C_u = 1/\left[4c\sqrt{3}(\alpha_1 + \alpha_3 + 2\alpha_4 + \beta_1 + \beta_3 + 2\beta_4)\right], \tag{5.58}$$

and $c = 3.355$ Å is the interplanar separation in graphite and $2a = 2.462$ Å is the lattice constant of graphene. In deriving the above equations, we have assumed that the volume per atom is equal to $a^2 c \sqrt{3}$.

The numerical values of the force constants, which give a good fit between the measured and calculated phonon dispersion, are given below in units of N/m [65,66]:

$$\alpha_1 = 409.705; \beta_1 = 145.012; \delta_1 = 98.920;$$

$$\alpha_2 = -40.8; \beta_2 = 74.223; \gamma_2 = -9.11; \delta_2 = -8.191$$

$$\alpha_3 = -33.203; \beta_3 = 50.10; \delta_3 = 5.802;$$

$$\alpha_4 = 10.539; \beta_4 = 4.993; \gamma_4 = 2.184; \delta_4 = -5.213.$$

(5.59)

The values of the two elastic constants calculated by using Eqs. (5.56), (5.57) are: $c_{11} = 1060$ GPa and $c_{66} = 440$ GPa, which fit exactly with the experimental values for graphite given by Blakslee et al. [77]. Because the interplanar interaction in graphite is much weaker than the intraplanar interactions, c_{11} and c_{66} of graphite should be good estimates for the corresponding elastic constants of graphene.

Using the method described earlier in this section, the flexural rigidity of graphene for the present fourth-neighbor interaction model is given by the following expression [66]:

$$D = -\left(\sqrt{3}/36\right)(\delta_1 + 18\delta_2 + 16\delta_3 + 98\delta_4 + 162\delta_5)a^2 = 2.13 \text{ eV}. \qquad (5.60)$$

This value is in the range of values as reported in the literature [78,79]. There is some confusion in the literature about the actual value of the rigidity and a wide range of values have been reported. In any case, the derivation as given here, shows that the rigidity is very sensitive to the range of the interatomic potential. The contribution of the farther neighbors is comparable to the contribution of the nearer neighbors. The rigidity may also be size dependent. Hence, it is difficult to assess the validity of the calculated value.

Now we use the MSGF equations to calculate the lattice distortion due to a substitutional point defect in graphene. A substitutional defect can be a foreign atom or a vacancy. The lattice distortion in the continuum model is represented as strain in the solid, which is defined in terms of the space derivatives of the displacement field as given below:

$$E_{\alpha\beta} = \partial u_\alpha / \partial x_\beta. \qquad (5.61)$$

We use the standard symmetrized form of $\mathbf{E}$, so we have three independent strain tensor components, E_{xx}, E_{yy}, and E_{xy}.

Note from Eq. (5.61) that strain is essentially a continuum model macroscale parameter because the derivatives are rigorously defined only in continuous space. The corresponding quantity for a discrete lattice is the atomic displacement field [61] given by Eq. (5.51). As discussed earlier in Section 5.6, the discrete displacement field at large distances from the defect merges into a continuum. The strain in this asymptotic limit is then defined in the usual continuum representation for macroscales. This is similar to the correspondence principle in quantum mechanics.

Thus, we see that the use of the MSGF for the displacement field in Eq. (5.51) gives the lattice distortion in the near field region and then using Eq. (5.61) gives the standard continuum strain in the asymptotic limit. The use of the MSGF in Eq. (5.51) ensures that the discrete displacement field, defined at the atomistic scales, merges seamlessly into the continuum in the asymptotic limit corresponding to the macroscales.

The MSGF technique as described above has been used by Tewary and Garboczi [10] to calculate the lattice distortion due to a general point defect such as a vacancy in graphene. Eq. (5.51) is applicable to any point defect. The parameter which characterizes the defect is the effective force $\mathbf{F}^*$ in Eq. (5.44). The force constants used in [10] for the calculation of the MSGF are the same as given in Eq. (5.59) and should be quite reliable [65,66]. The effective force, however, is arbitrarily restricted to the first and second neighbors of the defect. Their values, chosen arbitrarily, are given by

$$f^*(2s,0,0) = (f1,0,0) \text{ and } f^*(0,2,0) = (0,f2,0), \tag{5.62}$$

where $f1 = 1$, which simply defines the units, and $f2 = -0.2$.

The strain tensor for atoms of type B in the (0,1,0) direction for graphene, as calculated by Tewary and Garboczi [10], is shown in Fig. 5.3.

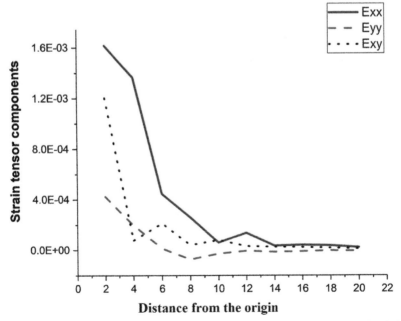

Fig. 5.3 showing variation of the strain field in the $<0,1,0>$ direction due to a point defect at the origin in graphene.

5.7.2 Silicene

In this subsection, we demonstrate the application of the MSGF technique to general Xenes using silicene as a specific example. We here the formulation and the results of Tewary and Garboczi [11] for the strain field due to a vacancy. The lattice structure of silicene is also represented by Fig. 5.2 with the difference that in silicene the A and B atoms are on different planes. The plane of the atoms A is displaced along the Z-axis by an amount $\Delta z = +0.44$ Å. In graphene, $\Delta z = 0$. For silicene, therefore, the stars in Fig. 5.2 represent the projection of atoms A on the XY plane. The effect of non-zero Δz is that the planar components of the atomic displacements get coupled with their Z-components, even in the harmonic approximation.

We use a particularly simple model for the force constants for calculating the GFs. We assume the interatomic potential to be central (axially symmetric), extending only up to the second-nearest neighbors. This potential is obviously unrealistic, in view of the covalent bonding of the atoms in silicene. However, it does fulfill our limited objective of demonstrating the application of the MSGF, without clouding it with unnecessary numerical details. It gives the following values of the force constants for a perfect silicene lattice without defects:

Units: R in units of a, force constants in eV/a^2, F* in eV/a; where $2a = 3.82$ Å is the lattice constant of silicene.

Atom A_1: $R(L\kappa) = (2s, 0, 0.23)$, where $s = 1/\sqrt{3}$.

$$\phi(0; 2s, 0, 0.23) = - \begin{pmatrix} \alpha_1 & 0 & \tau_1 \\ 0 & \beta_1 & 0 \\ \tau_1 & 0 & \delta_1 \end{pmatrix}, \tag{5.63}$$

where $\alpha_1 = 39.66$, $\beta_1 = 12.98$, $\delta_1 = 14.05$, and $\tau_1 = 5.32$.

Atom B_2: $R(L\kappa) = (0, 2, 0)$,

$$\phi(0; 0, 2, 0) = - \begin{pmatrix} \alpha_2 & \gamma_2 & 0 \\ -\gamma_2 & \beta_2 & 0 \\ 0 & 0 & \alpha_2 \end{pmatrix}, \tag{5.64}$$

where $\alpha_2 = -0.46$ and $\beta_2 = 4.90$. In Eq. (5.64), $\gamma_2 = 0$ due to the assumption of a central potential and not because of symmetry.

Now if we create a vacancy at the origin, the force constants between the vacancy and other atoms will be zero. This gives $\Delta\mathbf{\Phi} = \mathbf{\Phi}$ from Eq. (5.39). The presence of a lattice defect disrupts the equilibrium of the nearby atoms and induces a force on the atoms. The displacements can be calculated by solving the Dyson's equation [7]. In these calculations, we avoid these details and use the following values of the effective Kanzaki forces, as defined in Eq. (5.43):

Atom A_1 : $R(L\kappa) = \left(2/\sqrt{3}, 0, 0.23\right)$; f* $= \left(f_x^*, 0, f_z^*\right)$ (5.65)

$f_x^* = 1.0$; $f_z^* = 0.2$ ($f_x^* = 1$ defines the units)

$$\text{Atom B}_2 : R(L\kappa) = (0, 2, 0); f_y^* = -0.06; \tag{5.66}$$

To calculate the strain field, we use Eq. (5.51) for the displacement field, the symmetrized form of Eq. (5.61) for the components of the strain tensor, and Eqs. (5.65), (5.66) for the forces. In practice, we select the real part of the physical quantities for the final result. As expected, the strain field in silicene is qualitatively very similar to that for graphene shown in Fig. 5.3. The actual results are given in the paper by Tewary and Garboczi [11] and, therefore, not quoted here.

As described above, the nature of the strain field in graphene and silicene due to a point defect are qualitatively similar. The field has a discrete behavior at short distances from the defect and gradually becomes continuous asymptotically. This behavior is consistent with the multiscale nature of the MSGF.

5.8 Green's function method for molecular dynamics

In this section, we show that the use of the causal Green's functions in MD can significantly accelerate temporal convergence. We refer to this hybrid method as the CGFMD method. At least in one idealized case [13], it has been shown that CGFMD can model time scales over 6 to 9 orders of magnitude at the atomistic level. So far, this technique has been applied to only a few idealized cases [13,80], but it clearly has the potential for much wider and realistic applications in all physical, chemical, and biological systems where MD is used.

The primary difference between the basic MD and the CGFMD method is that in the conventional MD, the crystal potential energy is expanded only up to the first power of the atomic displacements. The apparent incentive for neglecting the quadratic and the higher powers is that the equations of motion can be solved exactly for the linear term. Iterations are then used to account for the neglected terms in the potential energy, which slows down the temporal convergence. Various additional techniques are then used to accelerate (see, for example, [19]) the temporal convergence of the MD. In the CGFMD method, we retain up to quadratic terms in the atomic displacements in the expansion of the crystal potential energy. The cubic and the higher terms are then included by iteration. The power of the GF method is that the equations of motion can still be solved exactly by using the normal mode transformation [6] and the Laplace transform technique [13]. This results in a substantial improvement of the temporal convergence. A more quantitative estimate of the relative convergence will be given later in this section. The use of the Laplace transform automatically ensures causality.

The physical significance of the quadratic terms in the potential energy is that they represent the phonons in the crystal. For that reason, these terms are arguably the most important terms for simulating the thermomechanical characteristics of solids, such as their thermal conductivity, elastic behavior, and even some optical and dielectric properties. The fact that the CGFMD method includes the quadratic terms exactly, makes it especially suitable for modeling phonons for calculating the thermomechanical response of solids. The cubic and the higher order terms are the anharmonic terms, that are included in the CGFMD method by iteration. Thus, the physical significance of these terms is that they represent interaction between the phonons. Recall that the phonons do not interact in the harmonic approximation and have infinite lifetimes.

In a certain class of problems in which atoms vibrate about an equilibrium site, CGFMD gives exact results in the harmonic approximation. Even for nonlinear vibrational problems, which depend upon the cubic and the higher terms in the potential energy, CGFMD has been shown to accelerate the MD by up to eight orders of magnitude [13]. This enables modeling the physical processes even up to microseconds. Examples of such problems include phonon transport and thermal conduction. In problems involving itinerant atoms, such as diffusion or crystal growth, CGFMD has not yet been used. However, we expect that in those cases also the CGFMD should be able to substantially accelerate MD computations.

The CGFMD method has been described in detail in [13,80]. Here we will only give the main features of this technique. Consider the equation of motion for the atom $L\kappa$ as given by Eq. (5.25). The symbols in this section will have a slightly more general meaning. The time-dependent displacement and the velocity vectors of the atom $L\kappa$ will be denoted by $\mathbf{u}(L\kappa,t)$ and $\mathbf{c}(L\kappa,t)$, respectively. We assume that at time $t=0$, $\mathbf{u}(L\kappa,0)$ and $\mathbf{f}(L\kappa)$, respectively, denote the values of displacement and force for the atom $L\kappa$. As in classical MD, we need to solve the following equation for $\mathbf{u}$:

$$
\begin{aligned}
M_{L\kappa}\frac{\partial^2 u_\alpha(L\kappa,t)}{\partial t^2} &= -\frac{\partial W}{\partial u_\alpha(L\kappa,t)} \\
&= f_\alpha(L\kappa) - \sum_{L'\beta}\phi_{\alpha\beta}(L\kappa,L'\kappa')u_\beta(L'\kappa',t) + \Delta f_\alpha(L\kappa),
\end{aligned}
\tag{5.67}
$$

where $\Delta \mathbf{f}$ represents higher than harmonic terms in the expansion of W. This term depends upon $\mathbf{u}$ and hence on t.

As in Eq. (5.41), we define 3pN dimensional vectors $\mathbf{U}(t)$, $\mathbf{C}(t)$, $\mathbf{F}$, and $\Delta\mathbf{F}(t)$, and 3pN x 3pN matrices $\mathbf{\Phi}$ and $\mathbf{D}$. However, in the present time dependent case, for brevity of notation, their elements are denoted by the corresponding lower-case quantities weighted with atomic masses. For example, the $(\alpha,L\kappa)$ element of $\mathbf{U}(t)$ is $\sqrt{M_{L\kappa}}u_\alpha(L\kappa,t)$ and of $\mathbf{F}$ is $(1/\sqrt{M_{L\kappa}})f_\alpha(L\kappa)$. The formal solution of Eq. (5.67) in the operator form is then given by

$$
\mathbf{U}(t) = \left(\mathbf{I}\frac{\partial^2}{\partial t^2} - \mathbf{D}\right)^{-1}\mathbf{F}_{eff}(t),
\tag{5.68}
$$

where $\mathbf{I}$ is the unit matrix,

$$
\mathbf{F}_{eff}(t) = \mathbf{F} + \Delta\mathbf{F}(t).
\tag{5.69}
$$

and

$$\mathbf{D} = \mathbf{M}^{-1/2}\mathbf{\Phi}\mathbf{M}^{-1/2}, \tag{5.70}$$

and $\mathbf{M}$ is a diagonal matrix with atomic masses as its elements.

The inverse operator in Eq. (5.68) is the GF. We solve Eq. (5.68) by using the Laplace transform that gives the causal GF [6], which is 0 for $t < 0$, to ensure causality. Using the numerical approximation $\mathbf{F}_{\text{eff}}(t) = \mathbf{F}$, we obtain the exact result [13]:

$$\mathbf{U}(t) = \mathbf{V}\mathbf{U}^*(t), \tag{5.71}$$

where

$$U_i^*(t) = \left(F_i^*/E_i^2\right)\left[\cos\left(E_it\right) - H(t)\right] + \left(C_{0i}^*/E_i\right)\sin\left(E_it\right) + U_{0i}^*\cos\left(E_it\right), \tag{5.72}$$

($\mathbf{U}^*$, $\mathbf{F}^*$, $\mathbf{C}_0^*$, and $\mathbf{U}_0^*$) are equal to $\mathbf{V}^{\mathrm{T}}(\mathbf{U}, \mathbf{F}, \mathbf{C}_0, \text{ and } \mathbf{U}_0)$, $\mathbf{V}$ is the matrix of eigenvectors of $\mathbf{D}$, E_i ($i = 1 \ldots 3\mathrm{pN}$) is an eigenvalue of $\mathbf{D}$, and $H(t)$ is the Heaviside step function, which is 0 for $t < 0$, and 1 for $t > 0$.

Eq. (5.71) can be used as stated for phonons in systems where the harmonic approximation is valid but analytical lattice dynamics cannot be used due to lack of translational symmetry. To account for anharmonic effects in a multiparticle system, we use the MD type iterative approach. We expand W locally at each time step and calculate U(t) from Eq. (5.71) in steps of Δt. We keep Δt small so that $\Delta\mathbf{F}(t)$ is negligible at each step. This introduces a constraint on the step size Δt, which is much less severe than that in the conventional MD. An estimate of the relative convergence of the CGFMD and MD is given below.

In the most basic formulation of MD, only the first term on the right of Eq. (5.67) is retained, whereas CGFMD retains up to the second term during each time interval. In the CGFMD, an exact solution of the temporal equation is obtained for any value of t by using the Laplace transform. The constraint on t or Δt arises because of the necessity to keep $\Delta\mathbf{F}(t)$ small. The basic MD can of course be accelerated somewhat by using more refined numerical techniques that partly account for the second and higher order terms by iteration [18,19].

We can compare the convergence of the CGFMD and the MD techniques by estimating the errors in the two techniques as follows. We denote the displacement variable in the expansion of W in Eq. (5.19) by ξ. Suppose the actual normalized value of a displacement is unity, which we calculate in S_{MD} steps in the MD and S_{CGFMD} steps in the GFMD technique. Then $\xi_{\mathrm{MD}} = 1/S_{\mathrm{MD}}$ for MD and $1/S_{\mathrm{CGFMD}}$ for CGFMD. In the conventional MD, second and higher powers of ξ_{MD} are neglected so the error in each step is of the order of $(\xi_{\mathrm{MD}})^2$. The corresponding error in CGFMD is $(\xi_{\mathrm{CGFMD}})^3$. For MD to be at least as accurate as CGFMD, we must have.

$$\left(\xi_{\mathrm{MD}}\right)^2 = \left(\xi_{\mathrm{CGFMD}}\right)^3 \tag{5.73}$$

or

$$S_{MD} \approx (S_{CGFMD})^{3/2}. \tag{5.74}$$

Considering that in practical cases $S_{CGFMD} \approx 10^6$, Eq. (5.74) shows that the number of iterations needed in the conventional MD to attain the same accuracy as the CGFMD technique is greater by about a factor of 1000.

Note that the u^2 term in Eq. (5.19), the potential energy of the lattice in the form of a Taylor series in u, is included exactly in the CGFMD technique. As discussed at the beginning of this section, it is this term that defines the phonon and elastic response of a solid [6]. In the conventional MD, the u^2 term is included only through additional iterations, which is a relatively inefficient process. The CGFMD method has recently been generalized by Coluci et al. [80] to enable simulations within the canonical ensemble.

Finally, we quote the result [13] obtained by using CGFMD method for propagation of elastic pulses or ripples in graphene. Propagation of pulses is an important characteristic of a material and is useful in understanding its elastic response and phonon transport. Such a calculation for a finite lattice has not been done analytically even in the harmonic approximation.

The model used in [13] is particularly simple and only serves the purpose of illustration. The origin and the system of axes is the same as given in Fig. 5.2. The model consists of about 1100 carbon atoms located at the equilibrium graphene lattice sites at time $t = 0$. As in [49], the outer atoms within the second neighbor distance of the outermost vibrating atoms are constrained so that their displacement is zero at all times. The size of the active lattice along the X-axis is about 5.52 nm. Propagation is initiated by imposing an initial displacement d in the Z-direction on the central atom.

We consider only the atomic displacements in the Z-direction. Only the Z-components f_z, Δf_z, and ϕ_{zz} in Eq. (5.67) contribute to these displacements. To account for the anharmonic effects in the Z-direction, the MD type iterative approach is used. The quantities f_z and ϕ_{zz} are calculated for each atom at each time step. These components change at each step due to anharmonicity. A major approximation in these calculations is that the displacements in the Z-direction and in the XY plane are not coupled. This is strictly valid only in the harmonic approximation. It is, therefore, important to keep d small so that the coupling between the ZZ and the planar components of ϕ can be neglected at all times. In these calculations, d was taken to be 0.01a and the results are normalized to this initial value.

The results are illustrated in Fig. 5.4, which shows a snapshot of the ripples or the instantaneous displacements of all atoms in the lattice at about 20 μs. It is important to realize that, as stated earlier, the CGFMD approach can reproduce the rippling dynamics of graphene in the case of infinitesimally small displacements.

An important test of the numerical convergence of the model is the invariance of the total energy of the system at all times. It was found that the change in the energy of the system at each time step was less than 10^{-4}%. This shows that the results had converged even up to microseconds, so that the CGFMD extended and bridged the time

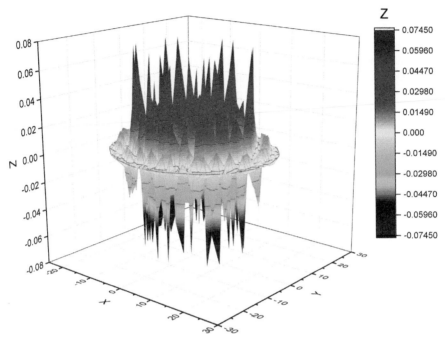

Fig. 5.4 Snapshot of normalized atomic displacements in graphene at about 20 microseconds. Coordinates X and Y are in units of half lattice constant.

scales, at least in this model problem, by eight orders of magnitude—from femtoseconds to microseconds.

In the basic MD, the lack of energy conservation results in an increase of the crystal temperature. This necessitates regular quenching of the temperature, which may introduce errors in the displacements. This problem should be substantially reduced in the CGFMD calculations because of better energy conservation. The example discussed here is of course highly idealized. In real cases the planar and the Z-components of the displacements will be coupled, and the initial displacement may be larger.

5.9   Summary of conclusions and future work

In conclusion, we note that MSGF is a powerful technique for multiscale modeling of the strains in Xenes. Fig. 5.3 shows the seamless linking between the atomistic and the continuum scale calculations of the strain field in graphene. Results for silicene are qualitatively similar. Some similarity between the strain curves for graphene and silicene should not be surprising because the Bravais lattice structure and hence the Brillouin zones of the two lattices are identical. Eq. (5.46), which is the basic equation of the MSGF, shows that the asymptotic behavior of the MSGF is determined by the behavior of the dynamical matrix near the center of the Brillouin zone. That is a characteristic of the Bravais lattice and is not sensitive to the detailed lattice structure of the solid.

For temporal problems, the convergence of the CGFMD technique should be much better than the basic MD as shown in the preceding section. The MSGF and the CGFMD techniques are both computationally very efficient, because of the power of the Green's functions. The next step is to combine them into a unified technique that can be applied to static as well as dynamic problems. These techniques have so far been applied to only symmetric lattices. They need to be developed for unsymmetric, disordered, and finite systems like amorphous solids, polymers, and liquids. One interesting application in which these techniques will be useful is to model plasmonics in graphene, which are sensitive to mechanical strains [81]. One reason why the use of MSGF and the CGFMD techniques are not very widespread is, probably, the lack of efficient software. It would be useful, therefore, to develop efficient and freely available software for using these techniques in real cases.

Finally, the CGFMD technique should be useful for solving time-dependent problems that are very difficult to solve by using existing methods. So far, this technique has been applied only to some idealized cases [13,80]. This technique should be applicable to modeling time-dependent processes in a wide class of physical, chemical, and biological systems, for which presently MD is used. For example, this technique should be useful in the simulation of protein folding and unfolding, which is of great topical interest [82].

Acknowledgments

The authors thank Drs. Newell Moser, Lauren Rast, David Read, and Alex Smolyanitsky for reading the manuscript and for their valuable comments.

References

[1] V.K. Tewary, Y. Zhang (Eds.), Modeling, Characterization and Production of Nanomaterials, Elsevier, Amsterdam, 2015, p. 536.

[2] C. Grazianetti, C. Martella, The rise of the xenes: from the synthesis to the integration processes for electronics and photonics, Materials (Basel) 14 (15) (2021) 4170.

[3] A. Fasolino, J.H. Los, M.I. Katsnelson, Intrinsic ripples in graphene, Nat. Mater. 6 (2007) 858–861.

[4] P.M. Morse, H. Feshbach, Methods of Theoretical Physics Vol. 1, first ed., Vol. 1, McGraw-Hill Publishing Company, New York, 1939, p. 1953.

[5] G. Barton, Elements of Green's Functions and Propagation, Clarendon Press, Oxford, 1989.

[6] A.A. Maradudin, E.W. Montroll, G.H. Weiss, I.P. Ipatova, Theory of lattice dynamics in the harmonic approximation, in: H. Ehrenreich, F. Seitz, D. Turnbull (Eds.), Solid State Physics, second ed., Vol. Supp 3, Academic Press, New York, 1971.

[7] V.K. Tewary, Green-function method for lattice statics, Adv. Phys. 22 (1973) 757–810.

[8] E. Pan, W. Chen, Static Green's Functions in Anisotropic Media, Cambridge University Press, New York, 2015, p. 338.

[9] T.C.T. Ting, Anisotropic Elasticity: Theory and Applications, Oxford University Press, New York, 1996.

[10] V.K. Tewary, E.J. Garboczi, Lattice Green's function for multiscale modeling of strain field due to a vacancy or other point defects in graphene, MRS Adv. 5 (2020) 2717–2725.

[11] V.K. Tewary, E.J. Garboczi, Multiscale Green's function for silicene and its application to calculation of the strain field due to a vacancy, MRS Adv. 6 (2021), https://doi.org/10.1557/s43580-021-00125-x.

[12] V.K. Tewary, Chapter 2: Multiscale Green's functions for modeling of nanomaterials, in: V. K. Tewary, Z. Yong (Eds.), Modeling, Characterization and Production of Nanomaterials, Elsevier, Amsterdam, 2015, pp. 55–85.

[13] V.K. Tewary, Extending the time scale in molecular dynamics simulations: propagation of ripples in graphene, Phys. Rev. B 80 (2009) 161409.

[14] P.M. Anderson, J.P. Hirth, J. Lothe, Theory of Dislocations, third ed., Cambrridge University Press, New York, 2017.

[15] X. Liu, M.C. Hersam, 2D materials for quantum information science, Nat. Rev. Mater. 4 (2019) 669–684.

[16] J.B. Gibson, A.N. Goland, M. Milgram, G.H. Vineyard, Dynamics of radiation damage, Phys. Rev. 120 (1960) 1229–1253.

[17] E.J. Garboczi, Elastic softening versus amorphization in a simple model of ion-induced radiation damage, Phys. Rev. B 39 (1989) 4.

[18] D.C. Rapaport (Ed.), The art of molecular dynamics simulation, second ed., Cambridge University Press, Cambridge, UK, 2004.

[19] A.F. Voter, F. Montalenti, T.C. Germann, Extending the time scale in atomistic simulation of materials, Annu. Rev. Mater. Res. 32 (2002) 321–346.

[20] A.K. Geim, K.S. Novoselov, The rise of graphene, Nat. Mater. 6 (2007) 183–191.

[21] M.I. Katsnelson, The Physics of Graphene, second ed., Cambridge University Press, Cambridge, UK, 2020.

[22] K.S. Novoselov, V.I. Fal'ko, L. Colombo, P.R. Gellert, M.G. Schwab, K. Kim, A roadmap for graphene, Nature 490 (2012) 192–200.

[23] K.S. Novoselov, D. Jiang, F. Schedin, T.J. Booth, V.V. Khotkevich, S.V. Morozov, A.K. Geim, Two-dimensional atomic crystals, Proc. Natl. Acad. Sci. U. S. A. 102 (2005) 10451–10453.

[24] R. Mas-Balleste, C. Gomez-Navarro, J. Gomez-Herrero, F. Zamora, 2D materials: to graphene and beyond, Nanoscale 3 (2011) 20–30.

[25] C. Kittel, Introduction to Solid State Physics, seventh ed., John Wiley, New York, 1996.

[26] M. Born, K. Huang, Dynamical Properties of Crystal Lattices, Oxford University Press, London, 1954.

[27] M. Lannoo, J. Bourgoin, Point Defects in Semiconductors I. Springer Series in Solid-State Sciences, Vol. 22, Springer Verlag, New York, 1981.

[28] R.G. Parr, W. Yang, Density-functional theory of atoms and molecules, in: R. Breslow, et al. (Eds.), International Series of Monographs on Chemistry, Oxford Science Publications, Oxford, 1989.

[29] J. Tersoff, Empirical interatomic potential for silicon with improved elastic properties, Phys. Rev. B 38 (1988) 9902–9905.

[30] J. Tersoff, Modeling solid-state chemistry—interatomic potentials for multicomponent systems, Phys. Rev. B 39 (1989) 5566–5568.

[31] M.I. Baskes, Modified embedded-atom potentials for cubic materials and impurities, Phys. Rev. B 46 (1992) 2727–2742.

[32] M.I. Baskes, Determination of modified embedded atom method parameters for nickel, Mater. Chem. Phys. 50 (1997) 152–158.

[33] D.W. Brenner, O.A. Shenderova, D.A. Areshkin, J.D. Schall, S.J.V. Frankland, Atomic modeling of carbon-based nanostructures as a tool for developing new materials and technologies, Comput. Model. Eng. Sci. 3 (2002) 643–673.

[34] V.K. Tewary, Phenomenological interatomic potentials for silicon, germanium and their binary alloy, Phys. Lett. A 375 (2011) 3811–3816.

[35] J. Stangl, V. Holy, G. Bauer, Structural properties of self-organized semiconductor nanostructures, Rev. Mod. Phys. 76 (2004) 725–783.

[36] P. Harrison, Quantum Wells, Wires, and Dots: Theoretical and Computational Physics, John Wiley, New York, 2002.

[37] D. Bimberg, M. Grundmann, N. Ledentsov, Quantum Dot Heterostructures, John Wiley, New York, 1999.

[38] M.A. Makeev, A. Madhukar, Simulations of atomic level stresses in systems of buried Ge/Si islands, Phys. Rev. Lett. 86 (2001) 5542–5545.

[39] J.G. Swadener, M.I. Baskes, M. Nastasi, Stress-induced platelet formation in silicon: a molecular dynamics study, Phys. Rev. B 72 (2005) 201202.

[40] H. Balamane, T. Halicioglu, W.A. Tiller, Comparative-study of silicon empirical interatomic potentials, Phys. Rev. B 46 (1992) 2250–2279.

[41] F. Cleri, V. Rosato, Tight-binding potentials for transition-metals and alloys, Phys. Rev. B 48 (1993) 22–33.

[42] J. Tersoff, C. Teichert, M.G. Lagally, Self-organization in growth of quantum dot superlattices, Phys. Rev. Lett. 76 (1996) 1675–1678.

[43] W.S. Cai, Y. Lin, X.G. Shao, Interatomic potential function in cluster research, Prog. Chem. 17 (2005) 588–596.

[44] D. Cheng, S. Huang, W. Wang, The structure of 55-atom cu-au bimetallic clusters: Monte Carlo study, Eur. Phys. J. D 39 (2005) 41–48.

[45] Y. Mishin, M.R. Sorensen, A.F. Voter, Calculation of point-defect entropy in metals, Philos. Mag. A 81 (2001) 2591–2612.

[46] A. Yavari, M. Ortiz, K. Bhattacharya, A theory of anharmonic lattice statics for analysis of defective crystals, J. Elast. 86 (2007) 41–83.

[47] L.V. Zhigilei, A.M. Dongare, Multiscale modeling of laser ablation: applications to nanotechnology, Comput. Model. Eng. Sci. 3 (2002) 539–555.

[48] L. Lindsay, D.A. Broido, Optimized Tersoff and Brenner empirical potential parameters for lattice dynamics and phonon thermal transport in carbon nanotubes and graphene, Phys. Rev. B 81 (2010) 205441.

[49] B. Yang, V.K. Tewary, Multiscale Green's function for the deflection of graphene lattice, Phys. Rev. B 77 (2008) 245442.

[50] R. Thomson, S.J. Zhou, A.E. Carlsson, V.K. Tewary, Lattice imperfections studied by use of lattice green-functions, Phys. Rev. B 46 (1992) 10613–10622.

[51] B. Yang, V.K. Tewary, Continuum Dyson's equation and defect Green's function in a heterogeneous anisotropic solid, Mech. Res. Commun. 31 (2004) 405–414.

[52] B. Yang, V.K. Tewary, Efficient Green's function modeling of line and surface defects in multilayered anisotropic elastic and piezoelectric materials, Comput. Model. Eng. Sci. 15 (2006) 165–177.

[53] B. Yang, V.K. Tewary, Green's function-based multiscale modeling of defects in a semi-infinite silicon substrate, Int. J. Solids Struct. 42 (2005) 4722–4737.

[54] D.T. Read, V.K. Tewary, Multiscale model of near-spherical germanium quantum dots in silicon, Nanotechnology 18 (2007) 105402.

[55] V.K. Tewary, D.T. Read, Integrated Green's function molecular dynamics method for multiscale modeling of nanostructures: application to Au nanoisland in Cu, Comput. Model. Eng. Sci. 6 (2004) 359–371.

[56] V.K. Tewary, E.J. Garboczi, Semi-discrete Green's function for solution of anisotropic thermal/electrostatic Boussinesq and Mindlin problems: application to two-dimensional material systems, Eng. Anal. Bound. Elem. 110 (2020) 56–68.

[57] E. Pan, Mindlin's problem for an anisotropic piezoelectric half-space with general boundary conditions, Proc. R. Soc. A: Math. Phys. Eng. Sci. 458 (2002) 181–208.

[58] B. Yang, V.K. Tewary, Formation of a surface quantum dot near laterally and vertically neighboring dots, Phys. Rev. B 68 (2003) 6.

[59] V.K. Tewary, Elastostatic Green's function for advanced materials subject to surface loading, J. Eng. Math. 49 (2004) 289–304.

[60] J.R. Berger, V.K. Tewary, Greens functions for boundary element analysis of anisotropic bimaterials, Eng. Anal. Bound. Elem. 25 (2001) 279–288.

[61] V.K. Tewary, Multiscale Green's-function method for modeling point defects and extended defects in anisotropic solids: application to a vacancy and free surface in copper, Phys. Rev. B 69 (2004) 13.

[62] B. Yang, V.K. Tewary, Multiscale modeling of point defects in Si-Ge(001) quantum wells, Phys. Rev. B 75 (2007) 144103.

[63] J. Cormier, J.M. Rickman, T.J. Delph, Stress calculation in atomistic simulations of perfect and imperfect solids, J. Appl. Phys. 89 (2001) 99–104.

[64] V.K. Tewary, B. Yang, Singular behavior of the Debye-Waller factor of graphene, Phys. Rev. B 79 (2009) 125416.

[65] V.K. Tewary, B. Yang, Parametric interatomic potential for graphene—erratum (vol 79, 075442, 2009), Phys. Rev. B 81 (2010), 075442.

[66] V.K. Tewary, B. Yang, Parametric interatomic potential for graphene, Phys. Rev. B 79 (2009) 075442.

[67] Z. Deng, A. Smolyanitsky, Q.Y. Li, X.Q. Feng, R.J. Cannara, Adhesion-dependent negative friction coefficient on chemically modified graphite at the nanoscale, Nat. Mater. 11 (2012) 1032–1037.

[68] A. Smolyanitsky, J.P. Killgore, Anomalous friction in suspended graphene, Phys. Rev. B 86 (2012) 125432.

[69] A. Smolyanitsky, J.P. Killgore, V.K. Tewary, Effect of elastic deformation on frictional properties of few-layer graphene, Phys. Rev. B 85 (2012) 035412.

[70] A. Smolyanitsky, V.K. Tewary, Manipulation of graphene's dynamic ripples by local harmonic out-of-plane excitation, Nanotechnology 24 (2013) 055701.

[71] A. Smolyanitsky, V.K. Tewary, Simulation of lattice strain due to a CNT-metal interface, Nanotechnology 22 (2011) 085703.

[72] A. Smolyanitsky, V.K. Tewary, Atomistic simulation of a graphene-nanoribbon-metal interconnect, J. Phys. Condens. Matter 23 (2011) 355006.

[73] A. Smolyanitsky, S.Z. Zhu, Z. Deng, T. Li, R.J. Cannara, Effects of surface compliance and relaxation on the frictional properties of lamellar materials, RSC Adv. 4 (2014) 26721–26728.

[74] L. Wirtz, A. Rubio, The phonon dispersion of graphite revisited, Solid State Commun. 131 (2004) 141–152.

[75] J. Zimmermann, P. Pavone, G. Cuniberti, Vibrational modes and low-temperature thermal properties of graphene and carbon nanotubes: minimal force-constant model, Phys. Rev. B 78 (2008) 104426.

[76] M. Mohr, J. Maultzsch, E. Dobardzic, S. Reich, I. Milosevic, M. Damnjanovic, A. Bosak, M. Krisch, C. Thomsen, Phonon dispersion of graphite by inelastic x-ray scattering, Phys. Rev. B 76 (2007) 035439.

[77] O.L. Blakslee, D.G. Proctor, E.J. Seldin, G.B. Spence, T. Weng, Elastic constants of compression-annealed pyrolytic graphite, J. Appl. Phys. 41 (1970) 3373.

[78] N. Lindahl, D. Midtvedt, J. Svensson, O.A. Nerushev, N. Lindvall, A. Isacsson, E.E.B. Campbell, Determination of the bending rigidity of graphene via electrostatic actuation of buckled membranes, Nano Lett. 12 (2012) 3526–3531.

[79] Y.J. Wei, B.L. Wang, J.T. Wu, R.G. Yang, M.L. Dunn, Bending rigidity and Gaussian bending stiffness of single-layered graphene, Nano Lett. 13 (2013) 26–30.

[80] V.R. Coluci, S.O. Dantas, V.K. Tewary, Generalized Green's function molecular dynamics for canonical ensemble simulations, Phys. Rev. E 97 (2018) 053310.

[81] L. Rast, T.J. Sullivan, V.K. Tewary, Stratified graphene/noble metal systems for low-loss plasmonics applications, Phys. Rev. B 87 (2013) 045428.

[82] J.E. Shea, C.L. Brooks, From folding theories to folding proteins: a review and assessment of simulation studies of protein folding and unfolding, Annu. Rev. Phys. Chem. 52 (2001) 499–535.

Modeling phonons in nanomaterials [*]

L. Lindsay[a] and T. Pandey[b]
[a]Materials Science and Technology Division, Oak Ridge National Laboratory, Oak Ridge, TN, United States, [b]Department of Physics, University of Antwerp, Antwerp, Belgium

6.1 Introduction

"Graphene" in particular and "nanotechnology" in general are terms that have pervaded our modern culture and which nearly every condensed matter scientist ponders, fundamentally and for applications. New low dimensional nanomaterials are being hypothesized and synthesized regularly, highlighted by novel discoveries including superconductivity in twisted bilayer graphene [1], high carrier mobilities in 2D materials [2–4], ultrahigh thermal conductivity in graphene and carbon nanotubes [5–7], high thermoelectric figures of merit in SiGe nanowires [8], and optical control of electronic properties in MoS_2 [9]. Many of the interesting properties found in nanomaterials are governed by, or limited by, the vibrational behaviors of the atoms on the underlying structure—including stability, adhesion, and electronic and thermal transport. In particular, phonons (normal modes of vibration of a crystal) determine the heat capacity and thermal conductivity in semiconducting and insulating nanomaterials, and interact with electronic degrees of freedom to play a role in electronic transport, optical properties, and thermoelectricity.

Theoretical models play a critical role in understanding the various physical behaviors observed in nanomaterials, particularly since nanoscale measurements are challenging and measured observables are typically integrated quantities, giving little information of the underlying microscopic behaviors. This chapter discusses calculations of phonon properties of nanoscale materials, those related to thermal transport in particular, while highlighting theoretical and experimental contributions in the literature and discussing the state-of-the-art and related challenges. In Section 6.2, phonons and their calculation in 1D and 2D crystal structures are introduced. Low-dimensional thermal conductivity calculations and related challenges are discussed in Section 6.3. A summary is given in Section 6.4.

[*] This manuscript has been authored by UT-Battelle, LLC under Contract No. DE-AC05-00OR22725 with the US Department of Energy. The United States Government retains and the publisher, by accepting the article for publication, acknowledges that the United States Government retains a non-exclusive, paid-up, irrevocable, world-wide license to publish or reproduce the published form of this manuscript, or allow others to do so, for United States Government purposes. The Department of Energy will provide public access to these results of federally sponsored research in accordance with the DOE Public Access Plan (http://energy.gov/downloads/doe-public-access-plan).

Copyright © 2023 Elsevier Ltd. All rights reserved.

6.2 Phonons, constraints, and symmetries

6.2.1 Phonons

The concept of the phonon was first introduced by Debye [10] and has since become a cornerstone of condensed matter physics having provided a foundation for understanding measured vibrational frequencies [11,12], heat capacity [10], thermal expansion [13,14], and lattice thermal conductivity [15] in crystalline solids. Phonons are built from the equations of motion of the atoms described as simple harmonic oscillators combined to construct the so-called dynamical matrix [16,17]:

$$D_{\alpha\beta}^{kk'}\left(\vec{q}\right) = \frac{1}{\sqrt{m_k m_{k'}}} \sum_{l'} \Phi_{\alpha\beta}^{0k,\,l'k'} e^{i\vec{q}\,\cdot\,\vec{R}_{l'}} \tag{6.1}$$

for a particular wavevector $\vec{q}$ and diagonalized to determine the phonon frequencies, $\omega_{\vec{q}j}$, for each polarization j. These depend on the atomic masses, m_k, where k labels a unit cell atom, and on the harmonic interatomic force constants (IFCs), $\Phi_{\alpha\beta}^{0k,\,l'k'} = \partial^2 V / \partial u_\alpha^{0k} \partial u_\beta^{l'k'}$, which describe how the atoms interact. These are second-order derivatives of the interatomic potential, V, with respect to small atomic displacement u_α^{lk} of the kth atom in the $l=0$ unit cell in the αth direction and the k'th atom in the l'th cell in the βth direction at equilibrium. Fig. 6.1A gives the calculated phonon dispersion for monolayer WS_2 using density functional theory (DFT) [23,24]. Specific details of the WS_2 calculations are presented in the caption of Fig. 6.1. The three lowest frequency curves correspond to vibrations for which the unit cell atoms oscillate in-phase and are called "acoustic" modes as they are responsible for sound propagation—at low frequency. These phonon branches (i.e., polarizations) have $\omega_{\vec{q}j} \to 0$ as $\vec{q} \to 0$ corresponding to translational invariance of the crystal in three dimensions (these will be discussed in more detail in the next section). The higher frequency curves correspond to vibrations for which the unit cell atoms oscillate out-of-phase for the zone center phonons and are called "optic" modes as they are excited by light—particularly as measured by Raman and infrared measurements. The number of polarizations corresponds to the number of degrees of freedom (N unit cell atoms times three dimensions). Here, WS_2 has three atoms in its unit cell, each allowed to vibrate in x, y, and z Cartesian directions corresponding to nine polarizations.

Much can be learned of phonons from a simple linear diatomic chain model (atoms in a chain with alternating masses (m_{light}, m_{heavy}) and a spring constant (k) between nearest neighbors), for which frequencies and eigenvectors can be solved analytically [25]. In this case, the zone boundary optic frequencies behave as $\omega_{op} \sim \sqrt{k/m_{light}}$, while the zone boundary acoustic frequencies go as $\omega_{ac} \sim \sqrt{k/m_{heavy}}$. The heavier masses govern acoustic vibrations and the lighter masses govern optic vibrations, while both types of modes depend on the interatomic bonding. Thus, mass variation between constituent atoms in a material can open a frequency gap between acoustic and optic phonon branches. This can be seen in the WS_2 dispersion shown in Fig. 6.1A and partial phonon density of states ($pDOS_k$):

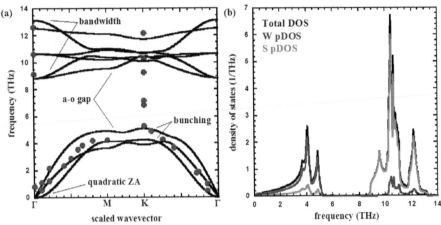

Fig. 6.1 (A) Calculated phonon dispersion for monolayer WS$_2$ (*black curves*) compared with measured data for bulk WS$_2$ (*red circles*, [18,19]). The low frequency branches correspond to acoustic vibrations, while the high frequency branches correspond to optic vibrations. (B) Calculated partial phonon density of states (pDOS) for W vibrations (*red curve*), S vibrations (*green curve*), and total DOS (*black curve*). Harmonic IFCs were derived from numerical derivatives of the interatomic forces in 243 atom supercells using the Quantum Espresso package [20], PBEsol exchange/correlations [21] with Grimme D2 for vdW corrections [22], Marzari-Vanderbilt electron Gaussian smearing of 0.008 Ry, 70 Ry plane-wave energy cutoff, 1e^{-10} Ry energy convergence threshold [20], 14 Å vacuum between layers, and considering interactions up to 13th nearest neighbors of the unit cell atoms. Energy relaxation (80 Ry plane-wave energy cutoff and 1e^{-12} Ry energy convergence threshold) gave lattice constant $a = 3.142$ Å and distance between S layers $d_S = 3.132$ Å.

$$pDOS_k(\omega) = \sum_{\vec{q}j} \left| \hat{\varepsilon}_{\vec{q}j}^k \right|^2 \delta\left(\omega - \omega_{\vec{q}j} \right) \qquad (6.2)$$

which is given in Fig. 6.1B for WS$_2$. Here, $\hat{\varepsilon}_{\vec{q}j}^k$ is the eigenvector of the kth unit cell atom for phonon mode $\vec{q}j$. Phonon frequencies and density of states (and lifetimes to some degree) can be probed by various experimental spectroscopic and scattering methods. The most prominent being laser-based Raman and infrared measurements [11], and inelastic neutron and X-ray scattering methods [12].

A description of the interatomic potential is required to make calculations of the harmonic and anharmonic IFCs and construct the phonon band structure of a given material. Historically, empirical potentials built to describe experimentally measured structural properties of bulk materials were used to describe lattice dynamics of their lower dimensional counterparts. Examples include the Tersoff [26] and Brenner (REBO) [27] empirical potentials used to describe phonons in carbon nanotubes, graphene, and other lower dimensional carbon systems. Other potentials were built with lower dimensional phonons in mind [28,29]. Such potentials typically have a simple form, thus provide an efficient means to obtain harmonic and anharmonic IFCs and are often used in molecular dynamics simulations. However, such potentials are not easily transferable to other systems, have limited predictive capabilities, and were not parameterized for lower dimensional systems. Some attempts have been made to vary empirical parameters to fit vibrational properties to available measured data of widely studied materials such as graphene [30] or to data derived from DFT calculations [31]. State-of-the-art calculations now employ DFT supercell calculations [32] or density functional perturbation theory methods [33] to obtain IFCs. These tend to be more accurate and transferable, but at a larger computational cost, too prohibitive for large scale molecular dynamics simulations.

6.2.2 Symmetries and invariance constraints

Symmetry plays a fundamentally important role in the behavior of lattice vibrations and resulting material properties, particularly in lower dimensional systems. Physically relevant IFCs should conform to the variety of point group operations (translations, rotations, inversions) that particular crystal symmetries define. Furthermore, IFCs should be invariant to derivative permutations (e.g., $\Phi_{\alpha\beta}^{0k,l'k'} = \Phi_{\beta\alpha}^{l'k',0k}$) and collectively should ensure that the interatomic potential is unchanged relative to rigid translations and rotations of the crystal lattice. For phonons, translational and rotational invariances can be maintained by ensuring that harmonic IFCs satisfy [34–36]:

$$\sum_{l'k'}\Phi_{\alpha\beta}^{0k,l'k'} = 0 \tag{6.3}$$

$$\sum_{k'l'}\Phi_{\alpha\beta}^{0k,l'k'}\left(X_\gamma^{l'k'} - X_\gamma^{0k}\right) = \sum_{k'l'}\Phi_{\alpha\gamma}^{0k,l'k'}\left(X_\beta^{l'k'} - X_\beta^{0k}\right) \tag{6.4}$$

where X_α^{lk} is the αth Cartesian position of the kth atom in the lth unit cell. The above invariance constraints are not satisfied in typical DFT calculations of harmonic IFCs due to truncation of interatomic interactions to finite distance and numerical artifacts. Thus, to obtain "physically meaningful" phonon dispersions [37] in many cases, a very large interaction cutoff distance coupled with stringent DFT parameters must be used. This can become numerically prohibitive. Alternatively, harmonic IFCs can be processed after calculation to enforce the above invariance conditions with care taken to satisfy point group operations and derivative permutations. Various schemes have been introduced to achieve this, including χ^2 minimization [38,39], Lagrange multipliers [40], and singular value decomposition with quadratic programming [41,42].

Additionally, Born and Huang put forth a set of invariance constraints for the harmonic IFCs of a crystal lattice in equilibrium and without stress [34]:

$$[\alpha\beta,\gamma\lambda] = [\gamma\lambda,\alpha\beta] \tag{6.5}$$

$$[\alpha\beta, \gamma\lambda] = -\sum_{kk'}\sum_{l'}\Phi_{\alpha\beta}^{0kl'k'}\left(X_{\gamma}^{l'k'} - X_{\gamma}^{0k}\right)\left(X_{\lambda}^{l'k'} - X_{\lambda}^{0k}\right) \tag{6.6}$$

The relation of the atomic positions to the harmonic IFCs given by Eq. (6.6) and derived in the context of elastic constants assumes that the crystal is free of stress. Thus, application of these constraints to calculated harmonic IFCs, even with residual stress (not fully relaxed), will simulate a relaxed, stress free system. Note that Eq. (6.5) is not valid for systems away from equilibrium (simulated stress or pressure) and technically applicable to non-polar materials only, though has been applied to polar systems (e.g., monolayer MoS_2 [43]) to improve the low frequency phonon dispersions.

These invariance constraints can be particularly important for describing the low frequency behavior of acoustic vibrations of 1D and 2D materials. While in bulk materials the three acoustic branches are linear ($\omega \sim q$) near the Brillouin zone center, 2D systems have two linear in-plane acoustic branches (transverse (TA) and longitudinal (LA)) and a quadratic ($\omega \sim q^2$) out-of-plane acoustic branch (ZA) and 1D materials have two linear acoustic branches (longitudinal vibrations (LA) along the chain or tube axis and twisting vibrations (TA) around the symmetry axis) and two quadratic flexure vibrations (FA) perpendicular to the axis of symmetry.

Fig. 6.2A demonstrates that application of the conditions given by Eqs. (6.3)–(6.6) gives physically meaningful harmonic IFCs and ZA dispersions for a variety of 2D materials: silicene [44,45], monolayer hexagonal BN [46,47], monolayer GaSe [48,49], and graphene [47,50]. Fig. 6.2B shows the calculated low frequency dispersions for three 1D materials: Single-chain polyethylene [51,52], single-chain Ba_3N [53], and a (5,5) single-walled carbon nanotube [28,54–58]. All calculations presented here are original to this work and structural descriptions of the materials can be found in the cited literature. For these 1D and 2D materials the harmonic IFCs determined without enforcing Eqs. (6.4)–(6.6) give unphysical flexure phonon branches with imaginary modes or linear dispersions.

In flat 2D materials like graphene and monolayer hBN, reflection symmetry across the crystal plane gives the conditions: $\Phi_{xz}^{0k,l'k'} = \Phi_{yz}^{0k,l'k'} = \Phi_{zx}^{0k,l'k'} = \Phi_{zy}^{0k,l'k'} = 0$, decoupling the in-plane (TA, LA, transverse optic (TO), and longitudinal optic (LO)) vibrations from the out-of-plane (ZA and out-of-plane optic (ZO)) vibrations [59,60]. The potential energy of the crystal can be written as a Taylor expansion around small atomic displacements:

$$V = V_0 + \sum_{lk}\sum_{\alpha}\Phi_{\alpha}^{lk}u_{\alpha}^{lk} + \frac{1}{2}\sum_{lk,l'k'}\sum_{\alpha\beta}\Phi_{\alpha\beta}^{lk,l'k'}u_{\alpha}^{lk}u_{\beta}^{l'k'}$$

$$+ \frac{1}{6}\sum_{lk,l'k',l''k''}\sum_{\alpha\beta\gamma}\Phi_{\alpha\beta\gamma}^{lk,l'k',l''k''}u_{\alpha}^{lk}u_{\beta}^{l'k'}u_{\gamma}^{l''k''} + \cdots. \tag{6.7}$$

where V_0 is a constant energy, Φ_{α}^{lk} are interatomic forces assumed zero at equilibrium, $\Phi_{\alpha\beta\gamma}^{lk,l'k',l''k''}$ are third-order anharmonic force constants discussed in the next section. Reflection symmetry in flat 2D materials mandates that each term in the expansion be unchanged if atoms are perturbed above or below the plane. That is, these atoms see the

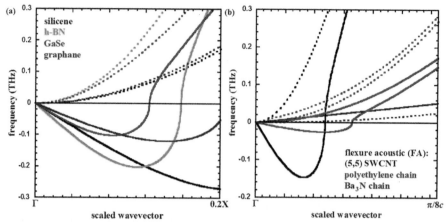

Fig. 6.2 (A) Calculated low frequency ZA dispersion along the $\Gamma \rightarrow X$ direction for silicene (*black curves*), single-layer h-BN (*green curves*), single-layer GaSe (*red curves*), and graphane (*purple curves*). *Solid curves* correspond to harmonic IFCs with only translational invariance enforced (Eq. 6.3). *Dashed curves* correspond to harmonic IFCs with Eqs. (6.3)–(6.6) enforced. For harmonic IFCs with all conditions satisfied, the ZA branch gives the expected quadratic behavior. For these calculations harmonic IFCs were only considered to eighth nearest neighbors. (B) Calculated low frequency FA dispersions for single-chain polyethylene (*red curves*), single-chain Ba$_3$N (*purple curves*), and a (5,5) single-walled carbon nanotube (SWCNT) (*black curves*). Here, c is the lattice constant along the chain/tube direction. Note that the two FA branches for the Ba$_3$N chain and the SWCNT are degenerate, while these are not so for the polyethylene chain. For harmonic IFCs with all conditions satisfied, the two FA branches in each system give the expected quadratic behavior.

same potential landscape. Thus, terms like $\Phi_{xz}^{lk,l'k'} u_x^{lk} u_z^{l'k'} = \Phi_{xz}^{lk,l'k'} u_x^{lk} u_{-z}^{l'k'} = -\Phi_{xz}^{lk,l'k'} u_x^{lk} u_z^{l'k'}$, which implies $\Phi_{xz}^{lk,l'k'} = -\Phi_{xz}^{lk,l'k'} = 0$. For any terms with an odd number of displacements the coupling constant is zero. This also applies to higher order anharmonic terms and has significant repercussions for thermal transport in 2D materials [61], as will be discussed in the next section.

In 1D materials, screw operators (combined translations and rotations) are another symmetry aspect that can play a particularly powerful role in describing phonons and deriving understanding of the behavior of phonon interactions [53–55]. As an example, a (30,30) armchair single-walled carbon nanotube has 120 atoms in its translational unit cell corresponding to 360 phonon polarizations [57], which presents a formidable challenge for describing the phonon dynamics and interactions. However, the dynamical matrix given by Eq. (6.1) that describes this system can be written in terms of smaller two atom unit cells (similar to that of graphene) that are related by rotations and translations [53–55,58]:

$$D_{\alpha\beta}^{kk'}(q_z\ell) = \frac{1}{\sqrt{m_k m_{k'}}} \sum_{l'} \sum_{\gamma} \Phi_{\alpha\gamma}^{0k,l'k'} S_{\gamma\beta}^{l'} e^{i(q_z R_{l'z} + \ell\theta_{l'})} \tag{6.8}$$

where q_z is a continuous 1D momentum along the tube axis (designated z direction here), k labels the atoms in the two atom cell, the lattice vector $\vec{R}_l = (R_{lz}, \theta_l)$ gives the location of the lth two atom cell along the tube axis (R_{lz}) and around the axis (θ_l), and $S_{\alpha\beta}^l$ is a rotation matrix corresponding to the angle of rotation around the tube axis of the lth unit cell. In this description, the phonon dynamics for each q_z is given by sixty 6×6 dynamical matrices, each labeled by a new quantum number ℓ rather than a single 360×360 dynamical matrix. Fig. 6.3 compares the phonon dispersions of a (30,30) carbon nanotube (colored curves) and graphene (black curves) using an optimized Tersoff empirical potential [30]. Essentially, each ℓ of the nanotube dynamics corresponds to a slice through the Brillouin zone of graphene (in this case continuous along q_y of graphene with quantized slices along q_x). The colors correspond to different branch indices; for instance, all red curves are $j = 1$ for dynamical matrices labeled by different ℓ. These correspond to nanotube "breathing" modes [57] and can be correlated with the ZA vibrations in graphene.

As will be discussed in the next section, the quantum number ℓ corresponds to a pseudo angular momentum that is conserved in phonon interactions, similar to the conservation of translational momentum q_z. This restricts the phase space available for phonon interactions which influences thermal conductivity. This dynamical description has been applied to study thermal transport in carbon nanotubes [54,55], boron nitride nanotubes [62], and Ba$_3$N chains [53].

6.3 Thermal conductivity

Increasing attention continues to be given to phonon thermal transport in 1D and 2D materials, partly due to their potential for technological applications and partly due to interesting behaviors that arise in lower dimensions, some of which will be discussed

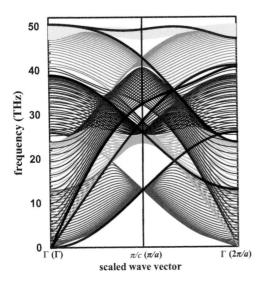

Fig. 6.3 Calculated phonon dispersions for a (30,30) single-walled carbon nanotube (*colored curves*) and graphene (*black curves*) using an optimized Tersoff empirical potential [30]. Here, a and c are the lattice constants of graphene and the nanotube, respectively. The colors correspond to the 6 j values (1 = *red*, 2 = *green*, 3 = *blue*, 4 = *purple*, 5 = *orange*, and 6 = *yellow*) for each of the 30 quantum numbers ℓ for the (30,30) nanotube. For this achiral armchair nanotube, another 180 modes are degenerate with these. The phonon branches are correlated with the six polarizations of graphene and represent slices through graphene's Brillouin zone.

here. To this end, in the past several years, many computational methods have been used to predict lattice thermal conductivity (κ) in a variety of low dimensional materials, including: Peierls-Boltzmann transport methods and molecular dynamics (MD) simulations. Of these, the Peierls-Boltzmann transport equation coupled with DFT has proven a powerful predictive tool for examining transport in lower dimensional systems, particularly for perfect, or nearly so, crystals with relatively small unit cells ($\sim$25 or less). MD methods are particularly useful for exploring large and defective systems where application of DFT-based methods is computationally prohibitive. In the next subsection we focus on low-dimensional thermal transport calculations based on the Peierls-Boltzmann transport equation, with particular focus on those derived from DFT.

6.3.1   Peierls-Boltzmann transport and intrinsic scattering

In 1929 Peierls described phonon transport in a crystalline solid following Boltzmann's theory of gas transport [15–17,63]:

$$v_{\vec{q}j\alpha} \nabla_\alpha T \left(\partial n_{\vec{q}j} / \partial T \right) = \left(\partial n_{\vec{q}j} / \partial t \right)_{\text{scatter}} \tag{6.9}$$

where $v_{\vec{q}j\alpha} = d\omega_{\vec{q}j}/dq_\alpha$ is the αth component of the phonon group velocity, $\nabla_\alpha T$ is the temperature gradient in the αth direction, and $n_{\vec{q}j}$ is the phonon distribution (not necessarily the Bose-Einstein distribution). The right-hand side of Eq. (6.9) is the set of all scattering mechanisms considered, e.g., three-phonon interactions, electron–phonon couplings, phonon-defect scattering, and phonon-boundary scattering. Solving Eq. (6.9), a complex set of coupled integro-differential equations, gives the unknown distributions that are then related to the lattice thermal conductivity:

$$\kappa_{\alpha\beta} = \sum_{\vec{q}j} C_{\vec{q}j} v_{\vec{q}j\alpha} v_{\vec{q}j\beta} \tau_{\vec{q}j\beta} \tag{6.10}$$

via Fourier's law:

$$J_\alpha = -\kappa_\alpha \nabla_\alpha T = \frac{1}{V} \sum_{\vec{q}j} \hbar \omega_{\vec{q}j} v_{\vec{q}j\alpha} n_{\vec{q}j} \tag{6.11}$$

where $C_{\vec{q}j}$ is phonon volumetric specific heat, $\tau_{\vec{q}j\alpha}$ is the phonon lifetime with the temperature gradient applied along the αth direction, J_α is the thermal current density, and V is the crystal volume. Here, a material's thermal conductivity can depend on the direction α of the applied temperature gradient. Often the thermal conductivity tensor given by Eq. (6.10) is diagonal with degenerate terms depending on the symmetry of the material. When discussing thermal conductivity in general we will drop the subscripts.

To determine the intrinsic κ as limited by anharmonicity, the only inputs to this formalism are harmonic and anharmonic IFCs. When these are calculated from DFT, this approach is essentially *parameter-free*, and thus *predictive*. Harmonic IFCs determine $\omega_{\vec{q}j}$, $v_{\vec{q}j}$, $C_{\vec{q}j}$, and eigenvectors, while anharmonic IFCs determine phonon scattering matrix elements as dictated by quantum perturbation theory (Fermi's golden rule) [16,17,64]:

$$W_{i \to f} = \frac{2\pi}{\hbar} |\langle f | V_{\text{anharm}} | i \rangle|^2 \delta(E_f - E_i) \qquad (6.12)$$

where E_f and E_i are the final and initial phonon energies and V_{anharm} are higher order anharmonic corrections to the potential energy that act as a perturbation to the harmonic Hamiltonian. Typical κ calculations only consider anharmonic perturbations to lowest order built from $\Phi_{\alpha\beta\gamma}^{0k,\ l'k',\ l''k''} = \partial^3 V / \partial u_\alpha^{0k} \partial u_\beta^{l'k'} \partial u_\gamma^{l''k''}$ and describing three-phonon interactions [16,17]. More recently, four phonon scattering [16,65–72] and anharmonic phonon renormalization [71] have been shown important for calculation of κ in some materials. For example, in monolayer graphene four phonon scattering was found to reduce the κ contributions of ZA phonons from 70% to 30% [65]. We note that κ calculations involving fourth order anharmonicity comes with a significant computational cost; thus, only a handful of studies have explored this.

In Eq. (6.12), the energy Dirac delta function ensures conservation of energy, while conservation of momentum is implicitly built into the scattering matrix elements [16,17,73]. For three phonon interactions, there are two types of scattering transitions: (i) two lower frequency phonons combine (coalesce) to produce a higher frequency phonon and (ii) one high frequency phonon decays into two lower frequency modes, with conservation conditions given by:

$$\omega_{\vec{q}j} \pm \omega_{\vec{q}'j'} = \omega_{\vec{q}''j''} \quad \text{and} \quad \vec{q} \pm \vec{q}' = \vec{q}'' + \vec{G} \qquad (6.13)$$

where $\vec{G}$ is a reciprocal lattice vector that maps a phonon scattered outside the first Brillouin zone back in. $\vec{G} = 0$ is called a normal (N) process, while $\vec{G} \neq 0$ is called an umklapp (U) process (folding back in) [15–17]. While the U processes generate thermal resistance, the N processes play an important role by redistributing momentum and energy among the modes [15]. Due to the complexity of solving Eq. (6.9), often the relaxation time approximation (RTA) is invoked, which considers a single mode distribution to be independent of the other phonon distributions (these are assumed to be in equilibrium). However, this considers both N and U processes to be strictly resistive, an approximation that is valid when U processes are relatively strong. The RTA has been shown to give general agreement with measured κ for a variety of materials [39,74], but fails for high κ bulk [39,75] and nanoscale [45,47,54,61,76,77] materials where N processes are important. For example, RTA underpredicts the κ of graphene by nearly nine times at room temperature [76,77]. The transport regime where N scattering is strong relative to U scattering and other resistances can be described by hydrodynamics [78]. Recently, hydrodynamic behaviors (e.g., second sound) have been predicted in nanoscale systems near room temperature [47,77,79–83] and even observed in graphite above 100 K [84].

As with the harmonic IFCs described in the previous section, supercell based finite displacement methods are widely used for obtaining anharmonic IFCs, particularly from DFT calculations [41,85]. For a given interaction distance there are many more anharmonic IFCs than harmonic and these calculations can be quite expensive. Thus, invoking the various symmetry properties to reduce the number of calculations is important, and typically anharmonic IFCs are truncated at smaller interaction distances. Therefore, application of translational invariance is important and convergence of κ with respect to interaction distance should be tested [39], particularly for low dimensional materials [86]. Currently, many open-source Peierls-Boltzmann transport based computational packages are available for prediction of κ using IFCs obtained from DFT calculations, including Sheng/almaBTE [87,88], phono3py [89], TDEP [90], ALAMODE [91], and PhonTS [92].

6.3.2 Other scattering resistance

6.3.2.1 Boundaries and substrates

In contrast to typical bulk materials, κ of low-dimensional materials can be very sensitive to the material environment (e.g., supporting substrates, polymer residues) [93–95] and sample size [96]. For example, transport measurements examining graphene on a SiO_2 substrate demonstrated a $\sim$5-fold reduction of κ [93] compared to measured κ values of suspended samples [5]. This was explained via strong interactions of ZA modes with the supporting substrate and phonons leaking into it. Unfortunately, realistic modeling of these effects is very difficult, particularly based on first principles methods. For finite sample size, phonon-boundary scattering is often modeled within the RTA by some variant of the empirical relation $1/\tau_{\vec{q}j}^{\text{boundary}} = 2|v_{\vec{q}j\alpha}|/L_\alpha$, where L_α is the system size along the transport direction [16]. For low dimensional materials, sample sizes are typically restricted to $\sim$1–100 μm where phonon-boundary scattering becomes comparable to intrinsic resistance and κ scales with the system size. This is demonstrated in Fig. 6.4, which shows the calculated κ of graphene as a function of size. Other methods have also been applied to describe transport in nanoribbons [96], across grain boundaries in polycrystalline graphene [97] and substrate supported graphene [98].

6.3.2.2 Point defects

All crystal samples contain defects, which, in some cases, can provide strong thermal resistance due to phonon-defect scattering. However, modeling phonon-defect interactions and resulting defect-limited κ from first principles methods is challenging as these break periodicity, locally distort structures, and alter the interatomic forces. Perhaps the least challenging defect resistance to model is the mass perturbations presented by different isotopes of the constituent elements comprising a material.

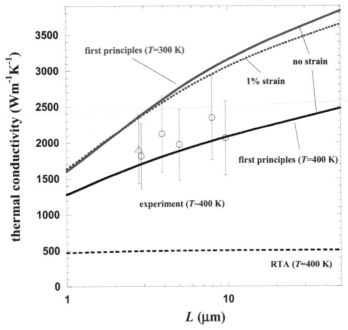

Fig. 6.4 Calculated (*curves*) thermal conductivity of graphene for selected temperatures and strains as a function of system size via the Peierls-Boltzmann transport in the ~1–50 μm range [76] compared with measurements (symbols). Details of the calculations and references for the measured data can be found in the corresponding reference.
Reproduced from L. Lindsay, W. Li, J. Carrete, N. Mingo, D.A. Broido, T. Reinecke, Phonon thermal transport in strained and unstrained graphene from first principles, Phys. Rev. B 89 (2014) 155426 with permission.

Phonon-isotope scattering can also be realistically modeled as an elastic scattering via quantum perturbation theory [99,100]:

$$1/\tau_{\vec{q}j}^{\text{isotope}} = \frac{\pi}{2N}\omega_{\vec{q}j}^2 \sum_k g_k \sum_{\vec{q'}j'} \left| \widehat{e}_{\vec{q}j}^{k*} \cdot \widehat{e}_{\vec{q'}j'}^{k*} \right|^2 \delta\left(\omega_{\vec{q}j} - \omega_{\vec{q'}j'}\right) \tag{6.14}$$

where N is the number of unit cells in the material, $\widehat{e}_{\vec{q}j}^{k}$ is the eigenvector for the kth atom of the $\vec{q}j$ phonon mode, and g_k is a mass variance parameter that depends on the masses and concentrations of the different isotopes of the kth atom [76,99,100]. Measurements demonstrate that κ is sensitive to varying isotope concentrations in graphene [101], monolayer MoS_2 [102], and hBN samples [103]. However, calculations of κ in these systems predict a significantly smaller isotope effect [47,76,102–105]; the reasons for these discrepancies are still unclear. Phonon-isotope scattering and resulting κ have also been calculated and examined in monolayer hBN [106], boron nitride nanotubes [62,107], and other nanoscale materials [108–110].

This perturbative mass variance method [99,100] has also been used to approxi-mate phonon scattering from other types of point defects, e.g., vacancies and substi-tutional atoms; however, this ignores force constant variance and local structural changes. Recently, Green's function methods have demonstrated the importance of considering such features when calculating phonon-defect interactions [111–113]. In this formalism, a so-called T-matrix is built from the Green's function describing the perfect system (G_0) and the difference in harmonic IFCs ($V_d = H_d - H_0$) from sup-ercells of the perfect system (H_0) and those with a relaxed defect site (H_d) [111–115]:

$$T_d = [I - V_d G_0]^{-1} V_d \tag{6.15}$$

where I is the identity matrix. This is related to phonon-defect scattering rates via:

$$1/\tau_{\vec{q}j}^{\text{defect}} = N_d \frac{\Omega}{\omega_{\vec{q}j}} \text{Im} \left\{ \left\langle \vec{q}j \middle| T_d \middle| \vec{q}j \right\rangle \right\} \tag{6.16}$$

where N_d is the concentration of randomly distributed defects and Ω is the volume over which the mode is normalized. Regarding defect-limited κ in lower dimensional materials, this method (and others) has been coupled with Peierls-Boltzmann transport and DFT to describe isotope clusters in graphene [114], vacancies and substitutional atoms in graphene [42], and vacancies and Sulfur adatoms in monolayer MoS_2 [43,116]. Fig. 6.5 gives size-dependent calculations of room temperature mono-layer MoS_2 with and without Sulfur vacancies compared with available measured data. The gray area gives the κ calculated using the experimentally observed vacancy densities for a number of MoS_2 samples [117].

6.3.2.3   Electrons

Electron–phonon coupling can become important for determining both electronic and phonon transport behaviors in heavily doped semiconductors and metals. The techno-logical applications of low dimensional materials in transistors, optoelectronics, and energy-harvesting devices often involve doping/gating to optimize the carrier concen-tration, making electron–phonon scattering a tuning knob to control thermal transport. Electron–phonon scattering rates for phonons can be computed from quantum pertur-bation theory [16,118]:

$$1/\tau_{\vec{q}j}^{el-ph} = \frac{2\pi}{\hbar} \sum_{\vec{k}mn} \left| g_{mnj}\left(\vec{k}, \vec{q}\right) \right|^2 \left(f_{\vec{k}n} - f_{\left(\vec{k}+\vec{q}\right)m} \right) \delta\left(\varepsilon_{\left(\vec{k}+\vec{q}\right)m} - \varepsilon_{\vec{k}n} - \hbar\omega_{\vec{q}j} \right)$$

$$\tag{6.17}$$

where $g_{mnj}\left(\vec{k}, \vec{q}\right)$ is the scattering matrix element for an electron initially in band n with wavevector $\vec{k}$ and scattered to band m and wave vector $\vec{k} + \vec{q}$. Here, $f_{\vec{k}m}$ is the Fermi Dirac distribution function and $\varepsilon_{\vec{k}m}$ is the electron energy. Numerical

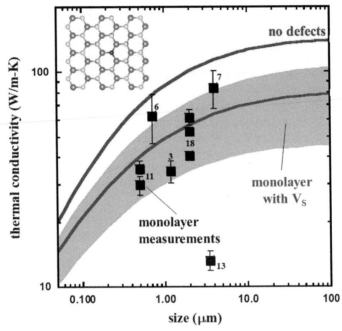

Fig. 6.5 Calculated κ of monolayer MoS$_2$ compared with room temperature measurements (*numbers label* references that can be found in Ref. [43]) as a function of size. The *dark gray curve* corresponds to the calculated monolayer MoS$_2$ κ including 1.2×10^{13} cm^{-2} Sulfur vacancies (average estimated density in Ref. [117]), while the *gray area* represents the range of κ with Sulfur vacancy densities between 0.5×10^{13} and 3.5×10^{13} cm^{-2} (range from Ref. [117]). The inset shows the defect structure: Molybdenum atoms (*purple*), Sulfur atoms (*yellow*), and Sulfur vacancy (*red*).
Reproduced from C.A. Polanco, T. Pandey, T. Berlijn, L. Lindsay, Defect-limited thermal conductivity in MoS2, Phys. Rev. Mater. 4 (2020) 014004 with permission.

procedures to calculate these scattering rates by utilizing Wannier interpolation [118,119] have been implemented in DFT codes such as Quantum Espresso [20] and EPW [120]. Theoretical studies on silicene [121] and phosphorene [121,122] have demonstrated that electron–phonon scattering at carrier concentrations $\sim 10^{13}$ cm^{-2} can result in $\sim 40\%$ reduction in room temperature κ as shown in Fig. 6.6.

6.3.3 Strain

Two dimensional materials possess interesting mechanical properties, which can be manipulated by application of external strain to tune optical, electronic, and magnetic properties [123–129]. 2D materials usually contain residual strain induced by lattice mismatch during growth processes, which plays an important role in stabilizing some materials. For example, monolayers of germanane [130] and gallenene [131] are stable only under tensile strain on supporting substrates. Strain also alters the phonon dispersion and can therefore modify κ [132–134], depending on the nature of the applied strain (uniaxial, biaxial, compression, or tensile). Due to its popularity and potential for applications, the strain dependence of κ of graphene has been examined in great detail [76,77,105,135–137]. Fig. 6.7A shows the calculated phonon dispersion of graphene with varying tensile strain. With applied tensile strain, the dispersion of the flexural branch becomes linear [137].

The strain-dependent κ behavior in graphene and other materials is sensitive to sample size, strain magnitude, and strain direction [77,135,139]. For example, κ of monolayer MoS_2 was found sensitive to applied biaxial strain, with 2%–4% strain resulting in 10%–20% reduction in κ [139]. Other studies have also investigated the strain dependence of κ of monolayer transition metal dichalcogenides [140–144]. Several first-principles calculations have demonstrated that the competition of variations in phonon lifetimes and group velocities under tensile strain can give rise to unusual non-monotonic behaviors in 2D materials [45,145,146]. Fig. 6.7B depicts the nonlinear strain dependence of the calculated κ of a graphene/hBN superlattice from a non-equilibrium molecular dynamics simulation. This nonlinear strain dependence originates from competition between phonon group velocities and phonon–phonon scattering. For smaller strain values the enhancement in κ is driven by increased flexural phonon group velocities. For larger strain values the phonon–phonon scattering becomes dominant and results in decreasing κ with further strain [138]. Likewise, despite having similar lattice structure, the κ of penta-graphene

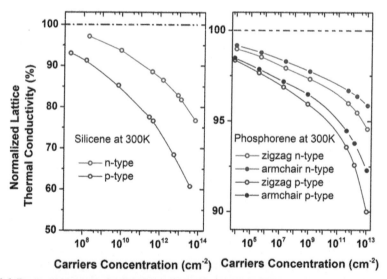

Fig. 6.6 Room temperature lattice thermal conductivity of silicene and phosphorene (along zigzag and armchair directions) at 300 K as a function of carrier concentration.
Reproduced from S.-Y. Yue, R. Yang, B. Liao, Controlling thermal conductivity of two-dimensional materials via externally induced phonon-electron interaction, Phys. Rev. B 100 (2019) 115408 with permission.

(a)

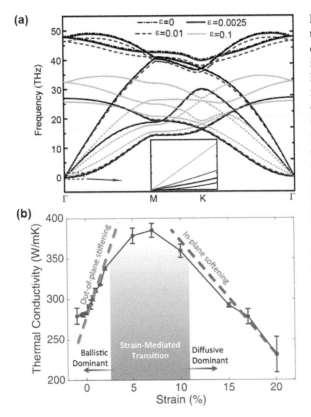

Fig. 6.7 (A) Effect of isotropic tensile strain on the phonon dispersion of graphene. The inset displays the quadratic to linear crossover of the low frequency ZA phonon dispersion with strain. (B) Illustrated effect of isotropic tensile strain on the calculated thermal conductivity of a monolayer graphene/hBN superlattice [138].
Reproduced from Y. Kuang, L. Lindsay, S. Shi, X. Wang, B. Huang, Thermal conductivity of graphene mediated by strain and size, Int. J. Heat Mass Transf. 101 (2016) 772; T. Zhu, E. Ertekin, Resolving anomalous strain effects on two-dimensional phonon flows: the cases of graphene, boron nitride, and planar superlattices, Phys. Rev. B 91 (2015) 205429 with permission, respectively.

monotonically reduces with increasing strain, while κ of penta-SiC$_2$ displays non-monotonic behavior [145].

6.3.4 Scattering and symmetry

6.3.4.1 2D materials

As mentioned in the previous section, reflection symmetry plays a critical role in dictating phonon interactions and κ in flat, monolayer materials. This symmetry mandates that only even numbers of out-of-plane modes can interact because anharmonic IFCs follow $\Phi_{xyz}^{lk,l'k',l''k''} = 0 = \Phi_{zzz}^{lk,l'k',l''k''}$ (and like terms). This limits phonon–phonon interactions in flat 2D materials, leading to a number of important consequences. First, ZA phonons, once thought to carry little heat [147], have actually been found to dominate thermal conduction in graphene [61,76,93,98,148], monolayer hBN [106,149], and other flat monolayer boron-based systems [149]. Since the out-of-plane vibrations are especially susceptible to this scattering selection rule, which supplements the conservation conditions of Eq. (6.13), they find less phase space for scattering and have longer lifetimes. Thus, despite having lower group velocities, these modes can carry a lot of heat (see fig. 3 in Ref. [76]).

Since κ values of flat materials are sensitive to this scattering rule, related systems with broken reflection symmetry tend to have lower κ values. The curvature of single-walled carbon and boron nitride nanotubes breaks reflection symmetry. Thus, κ values of moderate-diameter nanotubes (large enough so that quantization effects do not dominate) are smaller than the corresponding flat materials [55,62]. As the nanotube diameter increases, the curvature decreases, the reflection scattering rule is better approximated, and κ increases, approaching that of the flat materials in the large diameter limit. More specifically, only the out-of-tube vibrations (related to the ZA vibrations in flat materials) were found to be sensitive to nanotube diameter in moderate diameter nanotubes [55,62]. Similarly, bilayer, multilayer, and bulk BN and carbon systems break reflection symmetry and have lower in-plane κ [62,150,151]. Buckling of the monolayer backbone of materials such as silicene and germanene [45,146,152], and non-flat bonding environment of compound materials such as graphane [47,50], MoS_2 [43,102,153,154], and borophene [37,155] also give broken symmetry and generally lower κ values. Fig. 6.8 demonstrates that artificially applying reflection symmetry scattering rules to silicene and germanene give much higher κ values, comparable to their bulk cubic counterparts.

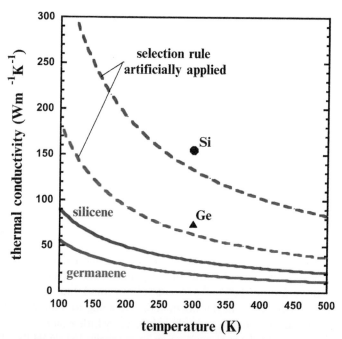

Fig. 6.8 Calculated thermal conductivities of monolayer buckled silicene (*red*) and germanene (*blue*) as a function of temperature. *Solid curves* give the normal calculations, while *dashed curves* have artificially imposed reflection-symmetry scattering rules. Calculated room temperature values for bulk Si and Ge are given for comparison.

6.3.4.2 1D materials

Screw symmetry operations in 1D chains and nanotubes not only provide insight into the dynamical description of phonons presented by Eq. (6.8), but also dictate how phonons interact. Considering the third-order anharmonic perturbation term in Eq. (6.7), the atomic displacements are written in the language of raising $\left(a_{\vec{q}j}^{\dagger}\right)$ and lowering $(a_{\vec{q}j})$ operators [64,156] and in terms of the screw symmetries described in the previous section [53,54,58]:

$$
u_{\alpha}^{lk} = \sqrt{\frac{\hbar}{2Nm_k}} \sum_{\vec{q}j\beta} \frac{\left(a_{\vec{q}j} + a_{-\vec{q}j}^{\dagger}\right)}{\sqrt{\omega_{\vec{q}j}}} S_{\alpha\beta}^{l} \varepsilon_{\vec{q}j\beta}^{k} e^{i(q_z R_{lz} + \ell\theta_l)}
\tag{6.18}
$$

As described for phonons above, lattice vectors $\vec{R_l} = (R_{lz}, \theta_l)$ are two-dimensional in this chiral description locating unit cells along the 1D axis (R_{lz}) and around the axis (θ_l) and N is the number of these cells in the crystal. The wave vectors $\vec{q} = (q_z, \ell)$ are described by a continuous 1D translational momentum (q_z) and a quantized pseudo angular momentum (ℓ). Just as crystal momentum must be conserved for each phonon–phonon interaction, so too must angular momentum. For three-phonon interactions this gives the new condition $\ell \pm \ell' = \ell'' + \ell_G$ which augments the conservation conditions in Eq. (6.13). Here, ℓ_G is an integer described in Refs. [54, 55] with $\ell_G = 0$ for N processes and $\ell_G \neq 0$ for U processes.

This additional conservation condition provides limits on phonon–phonon interactions and thus longer-lived phonons and higher κ in 1D materials than in bulk counterparts where coupling can break this description [53]. This description made possible the calculation of κ for nanotubes with very large translational unit cells (>200 atoms) with diameters up to 75 Å and for somewhat arbitrary chirality [55,62]. This pseudo angular momentum may also dictate how phonons interact with polarized light, electrons, and other quasiparticles [157].

6.4 Summary

In summary, nanoscale materials and heterostructures offer a variety of unique vibrational and transport characteristics of interest fundamentally and for a range of applications. This chapter discussed the calculation of phonons and thermal transport properties in 1D and 2D materials, primarily grounded in DFT and the Peierls-Boltzmann transport equation. All discussions were developed within the context of state-of-the-art theoretical methods and related experimental observables. Basic features of nanoscale phonon dispersions (e.g., polarizations, frequency scales, energy gaps) and their relations to symmetry and interactions were discussed. In particular, low frequency flexure branches in 1D and 2D materials are sensitive to invariance conditions (e.g., rotations) and special scattering selections rules can be derived from the symmetries of flat 2D materials and 1D chiral and achiral systems. Thermal conductivity of nanoscale materials was discussed in the context of the Peierls-Boltzmann transport equation and with regards to a variety of sources of thermal resistance. The role of defects, strain, size, substrates, and electron–phonon interactions were discussed with respect to the wide body of literature on nanoscale thermal transport.

Acknowledgments

This research was supported by the US Department of Energy, Office of Science, Basic Energy Sciences, Materials Sciences and Engineering Division with computational resources from the National Energy Research Scientific Computing Center (NERSC), a DOE Office of Science User Facility supported by the Office of Science of the US Department of Energy under Contract No. DE-AC02-05CH11231.

References

[1] Y. Cao, V. Fatemi, S. Fang, K. Watanabe, T. Taniguchi, E. Kaxiras, P. Jarillo-Herrero, Unconventional superconductivity in magic-angle graphene superlattices, Nature 556 (2018) 43.
[2] S.V. Morozov, K.S. Novoselov, M.I. Katsnelson, F. Schedin, D.C. Elias, J.A. Jaszczak, A.K. Geim, Giant intrinsic carrier mobilities in graphene and its bilayer, Phys. Rev. Lett. 100 (2008), 016602.
[3] B. Radisavljevic, A. Kis, Mobility engineering and a metal-insulator transition in monolayer MoS_2, Nat. Mater. 12 (2013) 815.
[4] D.J. Perello, S.H. Chae, S. Song, Y.H. Lee, High-performance n-type black phosphorus transistors with type control via thickness and contact-metal engineering, Nat. Commun. 6 (2015) 7809.
[5] A.A. Balandin, S. Ghosh, W. Bao, I. Calizo, D. Teweldebrhan, F. Miao, C.N. Lau, Superior thermal conductivity of single-layer graphene, Nano Lett. 8 (2008) 902.
[6] P. Kim, L. Shi, A. Majumdar, P.L. McEuen, Thermal transport measurements of individual multiwalled nanotubes, Phys. Rev. Lett. 87 (2001) 215502.
[7] X. Xu, L.F.C. Pereira, Y. Wang, J. Wu, K. Zhang, X. Zhao, S. Bae, C.T. Bui, R. Xie, J.T. L. Thong, B.H. Hong, K.P. Loh, D. Donadio, B. Li, B. Özyilmaz, Length-dependent thermal conductivity in suspended single-layer graphene, Nat. Commun. 5 (2014) 3689.
[8] E.K. Lee, L. Yin, Y. Lee, J.W. Lee, S.J. Lee, J. Lee, S.N. Cha, D. Whang, G.S. Hwang, K. Hippalgaonkar, A. Majumdar, C. Yu, B.L. Choi, J.M. Kim, K. Kim, Large thermoelectric figure-of-merits from SiGe nanowires by simultaneously measuring electrical and thermal transport properties, Nano Lett. 12 (2012) 2918.
[9] K.F. Mak, K. He, J. Shan, T.F. Heinz, Control of valley polarization in monolayer MoS_2 by optical helicity, Nat. Nanotechnol. 7 (2012) 494.
[10] P. Debye, Zur Theorie der Spezifischen Wärmen, Ann. Phys. 344 (1912) 789.
[11] R.A. Cowley, The theory of Raman scattering from crystals, Proc. Phys. Soc. 84 (1964) 281.
[12] R.A. Cowley, E.C. Svensson, W.J.L. Buyers, Interference between one- and multiphonon processes in the scattering of neutrons and X rays by crystals, Phys. Rev. Lett. 23 (1969) 525.
[13] J.S. Dugdale, D.K.C. MacDonald, The thermal expansion of solids, Phys. Rev. 89 (1953) 832.

[14] A.A. Maradudin, Thermal expansion and phonon frequency shifts, Phys. Status Solidi B 2 (1962) 1493.

[15] R.E. Peierls, On the kinetic theory of thermal conduction in crystals, Ann. Phys. 3 (1929) 1055.

[16] J.M. Ziman, Electrons and Phonons: The Theory of Transport Phenomena in Solids, Clarendon Press, Oxford, 1960.

[17] G.P. Srivastava, The Physics of Phonons, Taylor and Francis Group, New York, 1990.

[18] C. Sourisseau, M. Fouassier, M. Alba, A. Ghorayeb, O. Gorochov, Resonance Raman, inelastic neutron scattering and lattice dynamics studies of 2H-WS$_2$, Mater. Sci. Eng. B 3 (1989) 119.

[19] C. Sourisseau, F. Cruege, M. Fouassier, M. Alba, Second-order Raman effects, inelastic neutron scattering and lattice dynamics in 2H-WS$_2$, Chem. Phys. 150 (1991) 281.

[20] P. Giannozzi, S. Baroni, N. Bonini, M. Calandra, R. Car, C. Cavazzoni, D. Ceresoli, G.L. Chiarotti, M. Cococcioni, I. Dabo, A.D. Corso, S. de Gironcoli, S. Fabris, G. Fratesi, R. Gebauer, U. Gerstmann, C. Gougoussis, A. Kokalj, M. Lazzeri, L. Martin-Samos, N. Marzari, F. Mauri, R. Mazzarello, S. Paolini, A. Pasquarello, L. Paulatto, C. Sbraccia, S. Scandolo, G. Sclauzero, A.P. Seitsonen, A. Smogunov, P. Umari, R. M. Wentzcovitch, QUANTUM ESPRESSO: a modular and open-source software project for quantum simulations of materials, J. Phys. Condens. Matter 21 (2009) 395502.

[21] J.P. Perdew, A. Ruzsinszky, G.I. Csonka, O.A. Vydrov, G.E. Scuseria, L.A. Constantin, X. Zhou, K. Burke, Restoring the density-gradient expansion for exchange in solids and surfaces, Phys. Rev. Lett. 100 (2008) 136406.

[22] S. Grimme, J. Antony, S. Ehrlich, S. Krieg, A consistent and accurate ab initio parameterization of density functional dispersion correction (dft-d) for the 94 elements H-Pu, J. Chem. Phys. 132 (2010) 154104.

[23] P. Hohenberg, W. Kohn, Inhomogeneous electron gas, Phys. Rev. 136 (1964) B864.

[24] W. Kohn, L.J. Sham, Self-consistent equations including exchange and correlation effects, Phys. Rev. 140 (1965) A1133.

[25] C. Kittel, Introduction to Solid State Physics, seventh ed., John Wiley and Sons, Inc., Hoboken, 1996.

[26] J. Tersoff, Empirical interatomic potential for carbon, with applications to amorphous carbon, Phys. Rev. Lett. 61 (1988) 2879.

[27] D.W. Brenner, O.A. Shenderova, J.A. Harrison, S.J. Stuart, B. Ni, S.B. Sinnott, A second-generation reactive empirical bond order (REBO) potential energy expression for hydrocarbons, J. Phys. Condens. Matter 14 (2002) 783.

[28] G.D. Mahan, G.S. Jeon, Flexure modes in carbon nanotubes, Phys. Rev. B 70 (2004), 075405.

[29] V.K. Tewary, B. Yang, Parametric interatomic potential for graphene, Phys. Rev. B 79 (2009), 075422.

[30] L. Lindsay, D.A. Broido, Optimized Tersoff and Brenner empirical potential parameters for lattice dynamics and phonon thermal transport in carbon nanotubes and graphene, Phys. Rev. B 81 (2010) 205441.

[31] A. Rohskopf, H.R. Seyf, K. Gordiz, T. Tadano, A. Henry, Empirical interatomic potentials optimized for phonon properties, NPJ Comput. Mater. 3 (2017) 27.

[32] A. Togo, I. Tanaka, First principles phonon calculations in materials science, Scr. Mater. 108 (2015) 1.

[33] S. Baroni, S. de Gironcoli, A.D. Corso, P. Giannozzi, Phonons and related crystal properties from density-functional perturbation theory, Rev. Mod. Phys. 73 (2001) 515.

[34] M. Born, K. Huang, Dynamical Theory of Crystal Lattices, Oxford University Press, Oxford, 1954.

[35] G. Leibfried, W. Ludwig, Theory of anharmonic effects in crystals, Solid State Phys. 12 (1961) 275.

[36] D.C. Gazis, R.F. Wallis, Conditions for rotational invariance of a harmonic lattice, Phys. Rev. 151 (1966) 578.

[37] J. Carrete, W. Li, L. Lindsay, D.A. Broido, L.J. Gallego, N. Mingo, Physically founded phonon dispersions of few-layer materials and the case of borophene, Mater. Res. Lett. 4 (2016) 204.

[38] W.H. Press, S.A. Teukolsky, W.T. Vetterling, B.P. Flannery, Numerical Recipes in Fortran, Cambridge University Press, Cambridge, UK, 1992.

[39] L. Lindsay, D.A. Broido, T.L. Reinecke, Ab initio thermal transport in compound semiconductors, Phys. Rev. B 87 (2013) 165201.

[40] N. Mingo, D.A. Stewart, D.A. Broido, L. Lindsay, W. Li, Ab intio thermal transport, in: S.L. Shinde, G.P. Srivastava (Eds.), Length-Scale Dependent Phonon Interactions, vol. 128, Springer, New York, NY, 2014.

[41] K. Esfarjani, H.T. Stokes, Method to extract anharmonic force constants from first principles calculations, Phys. Rev. B 77 (2008) 144112.

[42] C.A. Polanco, L. Lindsay, Ab initio phonon point defect scattering and thermal transport in graphene, Phys. Rev. B 97 (2018), 014303.

[43] C.A. Polanco, T. Pandey, T. Berlijn, L. Lindsay, Defect-limited thermal conductivity in MoS_2, Phys. Rev. Mater. 4 (2020), 014004.

[44] P. Vogt, P. De Padova, C. Quaresima, J. Avila, E. Frantzeskakis, M.C. Asensio, A. Resta, B. Ealet, G. Le Lay, Silicene: compelling experimental evidence for graphene like two-dimensional silicon, Phys. Rev. Lett. 108 (2012) 155501.

[45] Y.D. Kuang, L. Lindsay, S.Q. Shi, G.P. Zheng, Tensile strains give rise to strong size effects for thermal conductivities of silicene, germanene and stanene, Nanoscale 8 (2016) 3760.

[46] R.-J. Chang, X. Wang, S. Wang, Y. Sheng, B. Porter, H. Bhaskaran, J.H. Warner, Growth of large single-crystalline monolayer hexagonal boron nitride by oxide-assisted chemical vapor deposition, Chem. Mater. 29 (2017) 6252.

[47] A. Cepellotti, G. Fugallo, L. Paulatto, M. Lazzeri, F. Mauri, N. Marzari, Phonon hydrodynamics in two-dimensional materials, Nat. Commun. 6 (2015) 6400.

[48] X. Li, M.-W. Lin, A.A. Puretzky, J.C. Idrobo, C. Ma, M. Chi, M. Yoon, C.M. Rouleau, I. I. Kravchenko, D.B. Geohegan, K. Xiao, Controlled vapor phase growth of single crystalline, two-dimensional GaSe crystals wit high photoresponse, Sci. Rep. 4 (2014) 5497.

[49] T. Pandey, D.S. Parker, L. Lindsay, Ab initio phonon thermal transport in monolayer InSe, GaSe, GaS, and alloys, Nanotechnology 28 (2017) 455706.

[50] L. Lindsay, Y. Kuang, Effects of functional group mass variance on vibrational properties and thermal transport in graphene, Phys. Rev. B 95 (2017) 121404.

[51] X. Wang, M. Kaviany, B. Huang, Phonon coupling and transport in individual polyethylene chains: a comparison study with the bulk crystal, Nanoscale 9 (2017) 18022.

[52] N. Shulumba, O. Hellman, A.J. Minnich, Lattice thermal conductivity of polyethylene molecular crystals from first-principles including nuclear quantum effects, Phys. Rev. Lett. 119 (2017) 185901.

[53] T. Pandey, C.A. Polanco, V.R. Cooper, D.S. Parker, L. Lindsay, Symmetry-driven phonon chirality and transport in one-dimensional and bulk Ba_3N-derived materials, Phys. Rev. B 98 (2018) 241405.

[54] L. Lindsay, D.A. Broido, N. Mingo, Lattice thermal conductivity of single-walled carbon nanotubes: beyond the relaxation time approximation and phonon-phonon scattering selection rules, Phys. Rev. B 80 (2009) 125407.

[55] L. Lindsay, D.A. Broido, N. Mingo, Diameter dependence of carbon nanotube thermal conductivity and extension to the graphene limit, Phys. Rev. B 82 (2010) 161402.

[56] S.-Y. Yue, T. Ouyang, M. Hu, Diameter dependence of lattice thermal conductivity of single-walled carbon nanotubes: study from ab initio, Sci. Rep. 5 (2015) 15440.

[57] S. Reich, C. Thomsen, J. Maultzsch, Carbon Nanotubes: Basic Concepts and Physical Properties, Wiley-VCH, Weinheim, Germany, 2004.

[58] V.N. Popov, V.E. Van Doren, M. Balkanski, Elastic properties of single-walled carbon nanotubes, Phys. Rev. B 61 (2000) 3078.

[59] D.R. Nelson, L. Peliti, Fluctuations in membranes with crystalline and hexatic order, J. Phys. 48 (1987) 1085.

[60] E. Mariani, F. von Oppen, Flexural phonons in free-standing graphene, Phys. Rev. Lett. 100 (2008), 076801.

[61] L. Lindsay, D.A. Broido, N. Mingo, Flexural phonons and thermal transport in graphene, Phys. Rev. B 82 (2010) 115427.

[62] L. Lindsay, D.A. Broido, Theory of thermal transport in multilayer hexagonal boron nitride and nanotubes, Phys. Rev. B 85 (2012), 035436.

[63] L. Boltzmann, Further studies on the balance among gas molecules, Wiener Berichte 66 (1872) 275.

[64] R. Shankar, Principles of Quantum Mechanics, second ed., Springer US, New York, 1994.

[65] T. Feng, X. Ruan, Four-phonon scattering reduces intrinsic thermal conductivity of graphene and the contributions from flexural phonons, Phys. Rev. B 97 (2018), 045202.

[66] T. Feng, L. Lindsay, X. Ruan, Four-phonon scattering significantly reduces intrinsic thermal conductivity of solids, Phys. Rev. B 96 (2017) 161201.

[67] J.S. Kang, M. Li, H. Wu, H. Nguyen, Y. Hu, Experimental observation of high thermal conductivity in boron arsenide, Science 361 (2018) 575.

[68] F. Tian, B. Song, X. Chen, N.K. Ravichandran, Y. Lv, K. Chen, S. Sullivan, J. Kim, Y. Zhou, T.H. Liu, M. Goni, Z. Ding, J. Sun, G.A.G.U. Gamage, H. Sun, H. Ziyaee, S. Huyan, L. Deng, J. Zhou, A.J. Schmidt, S. Chen, C.-W. Chu, P.Y. Huang, D. Broido, L. Shi, G. Chen, Z. Ren, Unusual high thermal conductivity in boron arsenide bulk crystals, Science 361 (2018) 582.

[69] N.K. Ravichandran, D. Broido, Non-monotonic pressure dependence of the thermal conductivity of boron arsenide, Nat. Commun. 10 (2019) 827.

[70] Y. Xia, Revisiting lattice thermal transport in PbTe: the crucial role of quartic anharmonicity, Appl. Phys. Lett. 113 (2018), 073901.

[71] N.K. Ravichandran, D. Broido, Unified first-principles theory of thermal properties of insulators, Phys. Rev. B 98 (2018), 085205.

[72] X. Yang, T. Feng, J. Li, X. Ruan, Stronger role of four-phonon scattering than three-phonon scattering in thermal conductivity of III-V semiconductors at room temperature, Phys. Rev. B 100 (2019) 245203.

[73] R.A. Cowley, The lattice dynamics of an anharmonic crystal, Adv. Phys. 12 (1963) 421.

[74] Z. Tian, J. Garg, K. Esfarjani, T. Shiga, J. Shiomi, G. Chen, Phonon conduction in PbSe, PbTe, and $PbTe_{1-x}Se_x$ from first-principles, Phys. Rev. B 85 (2012) 184303.

[75] A. Ward, D.A. Broido, D.A. Stewart, G. Deinzer, Ab initio theory of the lattice thermal conductivity in diamond, Phys. Rev. B 80 (2009) 125203.

[76] L. Lindsay, W. Li, J. Carrete, N. Mingo, D.A. Broido, T. Reinecke, Phonon thermal transport in strained and unstrained graphene from first principles, Phys. Rev. B 89 (2014) 155426.

[77] G. Fugallo, A. Cepellotti, L. Paulatto, M. Lazzeri, N. Marzari, F. Mauri, Thermal conductivity of graphene and graphite: collective excitations and mean free paths, Nano Lett. 14 (2014) 6109.

[78] R.A. Guyer, J.A. Krumhansl, Thermal conductivity, second sound, and phonon hydrodynamic phenomena in nonmetallic crystals, Phys. Rev. 148 (1966) 778.

[79] S. Lee, D.A. Broido, K. Esfarjani, G. Chen, Hydrodynamic phonon transport in suspended graphene, Nat. Commun. 6 (2015) 6290.

[80] A. Cepellotti, N. Marzari, Thermal transport in crystals as a kinetic theory of relaxons, Phys. Rev. X 6 (2016), 041013.

[81] S. Lee, L. Lindsay, Hydrodynamic phonon drift and second sound in a (20,20) single-wall carbon nanotube, Phys. Rev. B 95 (2017) 184304.

[82] Z. Ding, J. Zhou, B. Song, V. Chiloyan, M. Li, T.-H. Liu, G. Chen, Phonon hydrodynamic heat conduction and Knudsen minimum in graphite, Nano Lett. 18 (2018) 638.

[83] X. Li, S. Lee, Role of hydrodynamic viscosity on phonon transport in suspended graphene, Phys. Rev. B 97 (2018), 094309.

[84] S. Huberman, R.A. Duncan, K. Chen, B. Song, V. Chiloyan, Z. Ding, A.A. Maznev, G. Chen, K.A. Nelson, Observation of second sound in graphite at temperatures above 100 K, Science 364 (2019) 375.

[85] W. Li, L. Lindsay, D.A. Broido, D.A. Stewart, N. Mingo, Thermal conductivity of bulk and nanowire $Mg_2Si_xSn_{1-x}$ alloys from first principles, Phys. Rev. B 86 (2012) 174307.

[86] G. Qin, M. Hu, Accelerating evaluation of converged lattice thermal conductivity, NPJ Comput. Mater. 4 (2018) 3.

[87] W. Li, J. Carrete, N.A. Katcho, N. Mingo, ShengBTE: a solver of the Boltzmann transport equation for phonons, Comput. Phys. Commun. 185 (2014) 1747.

[88] J. Carrete, B. Vermeersch, A. Katre, A. van Roekeghem, T. Wang, G.K. Madsen, N. Mingo, almaBTE: a solver of the space-time dependent Boltzmann transport equation for phonons in structured materials, Comput. Phys. Commun. 220 (2017) 351.

[89] A. Togo, L. Chaput, I. Tanaka, Distributions of phonon lifetimes in Brillouin zones, Phys. Rev. B 91 (2015), 094306.

[90] O. Hellman, I.A. Abrikosov, Temperature-dependent effective third-order interatomic force constants from first principles, Phys. Rev. B 88 (2013) 144301.

[91] T. Tadano, Y. Gohda, S. Tsuneyuki, Anharmonic force constants extracted from first-principles molecular dynamics: applications to heat transfer simulations, J. Phys. Condens. Matter 26 (2014) 225402.

[92] A. Chernatynskiy, S.R. Phillpot, Phonon transport simulator (PhonTS), Comput. Phys. Commun. 192 (2015) 196.

[93] J.H. Seol, I. Jo, A.L. Moore, L. Lindsay, Z.H. Aitken, M.T. Pettes, X. Li, Z. Yao, R. Huang, D. Broido, N. Mingo, R.S. Ruoff, L. Shi, Two-dimensional phonon transport in supported graphene, Science 328 (2010) 213.

[94] W. Jang, Z. Chen, W. Bao, C.N. Lau, C. Dames, Thickness-dependent thermal conductivity of encased graphene and ultrathin graphite, Nano Lett. 10 (2010) 3909.

[95] I. Jo, M.T. Pettes, L. Lindsay, E. Ou, A. Weathers, A.L. Moore, Z. Yao, L. Shi, Reexamination of basal plane thermal conductivity of suspended graphene samples measured by electro-thermal micro-bridge methods, AIP Adv. 5 (2015), 053206.

[96] S. Mei, L.N. Maurer, Z. Aksamija, I. Knezevic, Full-dispersion Monte Carlo simulation of phonon transport in micron-sized graphene nanoribbons, J. Appl. Phys. 116 (2014) 164307.

[97] Z. Aksamija, I. Knezevic, Lattice thermal transport in large-area polycrystalline graphene, Phys. Rev. B 90 (2014), 035419.

[98] Z.-Y. Ong, E. Pop, Effect of substrate modes on thermal transport in supported graphene, Phys. Rev. B 84 (2011), 075471.

[99] S.-I. Tamura, Isotope scattering of dispersive phonons in Ge, Phys. Rev. B 27 (1983) 858.

[100] S.-I. Tamura, Isotope scattering of large-wave-vector phonons in GaAs and InSb: deformation-dipole and overlap-shell models, Phys. Rev. B 30 (1984) 849.

[101] S. Chen, Q. Wu, C. Mishra, J. Kang, H. Zhang, K. Cho, W. Cai, A.A. Balandin, R.S. Ruoff, Thermal conductivity of isotopically modified graphene, Nat. Mater. 11 (2012) 203.

[102] X. Li, J. Zhang, A.A. Puretzky, A. Yoshimura, X. Sang, Q. Cui, Y. Li, L. Liang, A.W. Ghosh, H. Zhao, Isotope-engineering the thermal conductivity of two-dimensional MoS_2, ACS Nano 13 (2019) 2481.

[103] C. Yuan, J. Li, L. Lindsay, D. Cherns, J.W. Pomeroy, S. Liu, J.H. Edgar, M. Kuball, Modulating the thermal conductivity in hexagonal boron nitride via controlled boron isotope concentration, Compr. Physiol. 2 (2019) 43.

[104] P. Jiang, X. Qian, R. Yang, L. Lindsay, Anisotropic thermal transport in bulk hexagonal boron nitride, Phys. Rev. Mat. 2 (2018), 064005.

[105] L.F.C. Pereira, D. Donadio, Divergence of the thermal conductivity in uniaxially strained graphene, Phys. Rev. B 87 (2013) 125424.

[106] L. Lindsay, D.A. Broido, Enhanced thermal conductivity and isotope effect in single-layer hexagonal boron nitride, Phys. Rev. B 84 (2011) 155421.

[107] D.A. Stewart, I. Savić, N. Mingo, First-principles calculation of the isotope effect on boron nitride nanotube thermal conductivity, Nano Lett. 9 (2009) 81.

[108] S. Srinivasan, U. Ray, G. Balasubramanian, Thermal conductivity reduction in analogous 2D nanomaterials with isotope substitution: graphene and silicene, Chem. Phys. Lett. 650 (2016) 88.

[109] M. Raeisi, S. Ahmadi, A. Rajabpour, Modulated thermal conductivity of 2D hexagonal boron arsenide: a strain engineering study, Nanoscale 11 (2019) 21799.

[110] B. Peng, H. Zhang, H. Shao, Y. Xu, G. Ni, R. Zhang, H. Zhu, Phonon transport properties of two-dimensional group-IV materials from ab initio calculations, Phys. Rev. B 94 (2016) 245420.

[111] N. Katcho, J. Carrete, W. Li, N. Mingo, Effect of nitrogen and vacancy defects on the thermal conductivity of diamond: an ab initio Green's function approach, Phys. Rev. B 90 (2014), 094117.

[112] A. Katre, J. Carrete, B. Dongre, G.K. Madsen, N. Mingo, Exceptionally strong phonon scattering by B substitution in cubic SiC, Phys. Rev. Lett. 119 (2017), 075902.

[113] C.A. Polanco, L. Lindsay, Thermal conductivity of InN with point defects from first principles, Phys. Rev. B 98 (2018), 014306.

[114] N. Mingo, K. Esfarjani, D.A. Broido, D.A. Stewart, Cluster scattering effects on phonon conduction in graphene, Phys. Rev. B 81 (2010), 045408.

[115] E.N. Economou, Green's Functions in Quantum Physics, third ed., Springer, Berlin, 2006.

[116] B. Peng, Z. Ning, H. Zhang, H. Shao, Y. Xu, G. Ni, H. Zhu, Beyond perturbation: role of vacancy-induced localized phonon states in thermal transport of monolayer MoS_2, J. Phys. Chem. C 120 (2016) 29324.

[117] J. Hong, Z. Hu, M. Probert, K. Li, D. Lv, X. Yang, L. Gu, N. Mao, Q. Feng, L. Xie, J. Zhang, D. Wu, Z. Zhang, C. Jin, W. Ji, X. Zhang, J. Yuan, Z. Zhang, Exploring atomic defects in molybdenum disulphide monolayers, Nat. Commun. 6 (2015) 6293.

[118] F. Giustino, M.L. Cohen, S.G. Louie, Electron-phonon interaction using Wannier functions, Phys. Rev. B 76 (2007) 165108.

[119] G.D. Mahan, Many-Particle Physics, Springer Science and Business Media, 2013.

[120] S. Poncé, E.R. Margine, C. Verdi, F. Giustino, EPW: Electron-phonon coupling, transport and superconducting properties using maximally localized Wannier functions, Comput. Phys. Commun. 209 (2016) 116.

[121] S.-Y. Yue, R. Yang, B. Liao, Controlling thermal conductivity of two-dimensional materials via externally induced phonon-electron interaction, Phys. Rev. B 100 (2019) 115408.

[122] B. Liao, J. Zhou, B. Qiu, M.S. Dresselhaus, G. Chen, Ab initio study of electron-phonon interaction in phosphorene, Phys. Rev. B 91 (2015) 235419.

[123] A.P. Nayak, T. Pandey, D. Voiry, J. Liu, S.T. Moran, A. Sharma, C. Tan, C.-H. Chen, L.-J. Li, M. Chhowalla, J.-F. Lin, A.K. Singh, D. Akinwande, Pressure-dependent optical and vibrational properties of monolayer molybdenum disulfide, Nano Lett. 15 (2014) 346.

[124] D. Akinwande, C.J. Brennan, J.S. Bunch, P. Egberts, J.R. Felts, H. Gao, R. Huang, J.-S. Kim, T. Li, Y. Li, K.M. Liechti, N. Lu, H.S. Park, E.J. Reed, P. Wang, B.I. Yakobson, T. Zhang, Y.-W. Zhang, Y. Zhou, Y. Zhu, A review on mechanics and mechanical properties of 2D materials—graphene and beyond, Extreme Mech. Lett. 13 (2017) 42.

[125] Z. Dai, L. Liu, Z. Zhang, Strain engineering of 2D materials: issues and opportunities at the interface, Adv. Mater. 31 (2019) 1805417.

[126] C. Si, Z. Sun, F. Liu, Strain engineering of graphene: a review, Nanoscale 8 (2016) 3207.

[127] S. Manzeli, D. Ovchinnikov, D. Pasquier, O.V. Yazyev, A. Kis, 2D transition metal dichalcogenides, Nat. Rev. Mater. 2 (2017) 17033.

[128] Y. Ma, Y. Dai, M. Guo, C. Niu, Y. Zhu, B. Huang, Evidence of the existence of magnetism in pristine VX_2 monolayers (X = S, Se) and their strain-induced tunable magnetic properties, ACS Nano 6 (2012) 1695.

[129] B. Mortazavi, O. Rahaman, T. Rabczuk, L.F.C. Pereira, Thermal conductivity and mechanical properties of nitrogenated holey graphene, Carbon 106 (2016) 1.

[130] E. Bianco, S. Butler, S. Jiang, O.D. Restrepo, W. Windl, J.E. Goldberger, Stability and exfoliation of germanane: a germanium graphane analogue, ACS Nano 7 (2013) 4414.

[131] V. Kochat, A. Samanta, Y. Zhang, S. Bhowmick, P. Manimunda, S.A.S. Asif, A.S. Stender, R. Vajtai, A.K. Singh, C.S. Tiwary, P.M. Ajayan, Atomically thin gallium layers from solid-melt exfoliation, Sci. Adv. 4 (2018) e1701373.

[132] W.-P. Hsieh, B. Chen, J. Li, P. Keblinski, D.G. Cahill, Pressure tuning of the thermal conductivity of the layered muscovite crystal, Phys. Rev. B 80 (2009) 180302.

[133] S. Bhowmick, V.B. Shenoy, Effect of strain on the thermal conductivity of solids, J. Chem. Phys. 125 (2006) 164513.

[134] A. Jain, A.J.H. McGaughey, Strongly anisotropic in-plane thermal transport in single-layer black phosphorene, Sci. Rep. 5 (2015) 8501.

[135] Y. Kuang, L. Lindsay, B. Huang, Unusual enhancement in intrinsic thermal conductivity of multilayer graphene by tensile strains, Nano Lett. 15 (2015) 6121.

[136] N. Wei, L. Xu, H.-Q. Wang, J.-C. Zheng, Strain engineering of thermal conductivity in graphene sheets and nanoribbons: a demonstration of magic flexibility, Nanotechnology 22 (2011) 105705.

[137] Y. Kuang, L. Lindsay, S. Shi, X. Wang, B. Huang, Thermal conductivity of graphene mediated by strain and size, Int. J. Heat Mass Transf. 101 (2016) 772.

[138] T. Zhu, E. Ertekin, Resolving anomalous strain effects on two-dimensional phonon flows: the cases of graphene, boron nitride, and planar superlattices, Phys. Rev. B 91 (2015) 205429.

[139] L. Zhu, T. Zhang, Z. Sun, J. Li, G. Chen, S.A. Yang, Thermal conductivity of biaxial-strained MoS_2: sensitive strain dependence and size-dependent reduction rate, Nanotechnology 26 (2015) 465707.

[140] A. Shafique, Y.-H. Shin, Strain engineering of phonon thermal transport properties in monolayer 2H-$MoTe_2$, Phys. Chem. Chem. Phys. 19 (2017) 32072.

[141] Z. Ding, Q.-X. Pei, J.-W. Jiang, Y.-W. Zhang, Manipulating the thermal conductivity of monolayer MoS_2 via lattice defect and strain engineering, J. Phys. Chem. C 119 (2015) 16358.

[142] K. Yuan, X. Zhang, L. Li, D. Tang, Effects of tensile strain and finite size on thermal conductivity in monolayer WSe_2, Phys. Chem. Chem. Phys. 21 (2019) 468.

[143] X. Meng, T. Pandey, J. Jeong, S. Fu, J. Yang, K. Chen, A. Singh, F. He, X. Xu, J. Zhou, W.-P. Hsieh, A.K. Singh, J.-F. Lin, Y. Wang, Thermal conductivity enhancement in MoS_2 under extreme strain, Phys. Rev. Lett. 122 (2019) 155901.

[144] D. Qin, X.-J. Ge, G.-Q. Ding, G.-Y. Gao, J.-T. Lü, Strain-induced thermoelectric performance enhancement of monolayer $ZrSe_2$, RSC Adv. 7 (2017) 47243.

[145] H. Liu, G. Qin, Y. Lin, M. Hu, Disparate strain dependent thermal conductivity of two-dimensional penta-structures, Nano Lett. 16 (2016) 3831.

[146] H. Xie, Large tunability of lattice thermal conductivity of monolayer silicene via mechanical strain, Phys. Rev. B 93 (2016), 075404.

[147] D.L. Nika, S. Ghosh, E.P. Pokatilov, A.A. Balandin, Lattice thermal conductivity of graphene flakes: comparison with bulk graphite, Appl. Phys. Lett. 94 (2009) 203103.

[148] D. Singh, J.Y. Murthy, T.S. Fisher, Spectral phonon conduction and dominant scattering pathways in graphene, J. Appl. Phys. 110 (2011), 094312.

[149] H. Fan, H. Wu, L. Lindsay, Y. Hu, Ab initio investigation of single-layer high thermal conductivity boron compounds, Phys. Rev. B 100 (2019), 085420.

[150] L. Lindsay, D.A. Broido, N. Mingo, Flexural phonons and thermal transport in multilayer graphene and graphite, Phys. Rev. B 83 (2011) 235428.

[151] D. Singh, J.Y. Murthy, T.S. Fisher, Mechanism of thermal conductivity reduction in few-layer graphene, J. Appl. Phys. 110 (2011), 044317.

[152] X. Gu, R. Yang, First-principles prediction of phononic thermal conductivity of silicene: a comparison with graphene, J. Appl. Phys. 117 (2015), 025102.

[153] W. Li, J. Carrete, N. Mingo, Thermal conductivity and phonon linewidths of monolayer MoS_2 from first principles, Appl. Phys. Lett. 103 (2013) 253103.

[154] A.N. Gandi, U. Schwingenshclögl, Thermal conductivity of bulk and monolayer MoS_2, Europhys. Lett. 113 (2016) 36002.

[155] H. Xiao, W. Cao, T. Ouyang, S. Guo, C. He, J. Zhong, Lattice thermal conductivity of borophene from first principle calculation, Sci. Rep. 7 (2017) 45986.

[156] N.W. Ashcroft, N.D. Mermin, Solid State Physics, Thomson Learning Inc, 1976.

[157] H. Zhu, J. Yi, M.-Y. Li, J. Xiao, L. Zhang, C.-W. Yang, R.A. Kaindl, L.-J. Li, Y. Wang, X. Zhang, Observation of chiral phonons, Science 359 (2018) 579.

Computational modeling of thermal transport in bulk and nanostructured energy materials and systems

Ming Hu
Department of Mechanical Engineering, University of South Carolina, Columbia, SC, United States

7.1   Introduction

Nanostructured materials or nanomaterials are materials with characteristic size of structural elements on the order or less than several hundreds of nanometers at least in one dimension. Examples of nanostructured materials include, but not limited to, nanocrystalline materials, nanofibers, nanotubes, and nanoparticle reinforced nanocomposites, multilayered systems with submicron thickness of the layers. Recent advances in nanotechnology have provided a variety of nanostructured materials with highly controlled and interesting properties—from exceptionally high strength in terms of enhanced mechanical ability to target drugs in biophysics, from high performance heat dissipation in thermal science to unique optical properties. By controlling structure at the nanoscale dimensions, one can control and tailor the properties of nanostructures, such as semiconductor nanocrystals and metal nanoshells, in a very predictable manner to meet the needs of a specific application. These materials can bring new and unique capabilities to a variety of engineering and technological applications ranging from diagnosis of diseases to novel energy nanotechnologies. In fact, the past decades have witnessed rapid advances in synthesis of nanostructured materials combined with reports of their enhanced or unique properties that have created a new active area of materials research. Due to the nature of the nanoscopic size of the structural elements in nanomaterials, the interfacial regions, which represent an insignificant volume fraction in traditional materials with coarse microstructures, start to play the dominant role in defining the physical and chemical properties of nanostructured materials. This implies that the behavior of nanomaterials cannot be understood and predicted by simply applying scaling arguments from the structure-property relationships developed for conventional polycrystalline, multiphase, and composite materials. New models and fundamentally new structure-property relationships, therefore, are needed for an adequate description of the behavior and properties of nanomaterials.

Modeling, Characterization and Production of Nanomaterials. https://doi.org/10.1016/B978-0-12-819905-3.00007-5

Computer modeling is playing a prominent role in the development of the theoretical understanding of the connections between the atomic-level structure and the effective (macroscopic) properties of nanomaterials. Atomistic modeling has been at the forefront of computational investigation of nanomaterials and has revealed a wealth of information on structure and properties of individual structural elements (various nanolayers, nanoparticles, nanofibers, nanowires, and nanotubes) as well as the characteristics of the interfacial regions and modification of the material properties at the nanoscale. Generally speaking, atomistic modeling is based on atoms as elementary units in the models, thus providing the atomic-level resolution in the computational studies of materials structure and properties. The main atomistic methods in material research are (1) molecular dynamics technique that yields "atomic movies" of the dynamic material behavior through the integration of the equations of motion of atoms and molecules, (2) metropolis Monte Carlo (MC) method that enables evaluation of the equilibrium properties through the ensemble averaging over a sequence of random atomic configurations generated according to the desired statistical-mechanics distribution, and (3) kinetic MC method that provides a computationally efficient way to study systems where the structural evolution is defined by a finite number of thermally activated elementary processes.

Rapid progress in the synthesis and processing of materials with structure on nanometer length scales has created an urgent demand for greater scientific understanding of thermal transport in nanoscale devices, individual nanostructures, and nanostructured materials. On one hand, as transistor power density has been exponentially going up in recent years, efficient heat dissipation across interfaces is one of the crucial challenges that limits the development of innovative microelectronic device technologies. Due to the unique architecture of the nanoelectronic devices, interfacial thermal transport has become the dominant factor in the performance of electronic cooling system. On the other hand, thermoelectric materials find important applications in the direct conversion of thermal energy to electric power and in solid-state cooling. Although thermoelectric devices possess unique advantages such as high reliability, lack of moving parts, and the ability to be scaled down to small sizes, the energy conversion efficiencies of these devices remain a generally poor factor that severely limits their competitiveness and range of employment. To materialize wide spread use of thermoelectrics at levels that would impact global energy issues the material and device efficiencies existing currently will need to be improved significantly. The energy conversion efficiency of thermoelectric devices is characterized by the figure-of-merit: $ZT = S^2 \sigma T / \kappa$, where S, σ, κ are the Seebeck coefficient, electrical conductivity, and thermal conductivity of the material, respectively, and T is the absolute temperature. In general, development schemes to improve thermoelectric conversion efficiency are driven by the need to maximize the Seebeck coefficient and to balance the competing requirements of high electrical conductivity and low thermal conductivity. To this end, considerably large amount of scientific effort is dedicated to reducing the thermal conductivity. Nanostructuring of existing thermoelectrics (nanowires, superlattices, composites, etc.) has emerged as a promising pathway to improve thermoelectric performance by manipulating phononic transport. The development of advanced thermoelectric materials calls for emergent understanding phonon behavior in various structures, especially in nanostructured materials.

However, the physics of heat transfer at the nanoscale can differ distinctly from that predicted by classical laws, primarily due to the significance of surface, boundary, and size effects. When characteristic length scales of the device or structure are comparable to the mean free path of heat carriers (mainly phonons for semiconductors and insulators), conventional heat transfer theory is no longer valid. Tremendous progress has been made in the past two decades to understand and characterize heat transfer in nanostructures. From numerical modeling point of view, computational methods including lattice dynamics (LD) based on force constants from empirical force fields or first-principles calculations and in combination with Boltzmann transport equation (BTE), nonequilibrium Green's function (NEGF), classical molecular dynamics (MD) simulations, and MC simulations have been used to study nanoscale thermal transport. These will be introduced in sequence below.

7.2 Overview of computational methods

7.2.1 Anharmonic lattice dynamics

Anharmonic lattice dynamics (ALD), combined with Boltzmann transport equation (BTE) [1], has been able to predict phonon properties and thermal conductivity with unprecedented accuracy and without the need of any empirical input, when using first-principles calculations [2,3]. Evaluating force constants from first-principles calculations has been made possible thanks to the aggressive improvements in computational hardware and software, such as ShengBTE [4], PhonTS [5], and Phono3py [6]. The coupled ALD/BTE method has been successfully used to predict the lattice thermal conductivity of perfect bulk crystals and some compounds [7–19]. However, ALD/BTE method becomes more challenging if inhomogeneity has to be taken into account, which involves significantly large supercell that requires unbearable computational demands. In addition, the method has difficulty in studying the effect of free surface, interface, and boundary.

7.2.2 Atomistic Green's function

A Green's function is a mathematical method for solving differential equations such as the Schrodinger equation. In an atomistic Green's function approach, the system is represented at the molecular level by atomistic potential models. Heat current in the system subject to a small temperature difference is related to the interatomic force constants. This heat current is expressed in terms of Green's function, and the phonon transmission as a function of phonon frequency is calculated [20–23]. The Green's function method can provide frequency dependent transmission, but the disadvantage is that it can only deal with ballistic transport so far.

7.2.3   Classical molecular dynamics simulations

Classical molecular dynamics simulations trace the time-dependent trajectories of all atoms in the simulation domain based on Newton's second law of motion and interatomic potentials [24–26]. Nonequilibrium molecular dynamics (NEMD) and equilibrium molecular dynamics (EMD) simulations are the two major methods to calculate thermal transport properties like thermal conductivity [27] and interfacial thermal conductance [28] with their respective advantages and disadvantages. Molecular dynamics has also been developed to study detailed phonon properties such as density of state, dispersion relation, relaxation time [29], and transmission across interfaces [30]. The serious disadvantage of MD simulation is that it relies on accurate interatomic potential. Hu group recently developed a computational scheme based on time Fourier transform of atomistic heat current, called the frequency domain direct decomposed method (FDDDM) [31,32], to analyze the contributions of frequency dependent thermal conductivity in NEMD simulations. The MD simulations and/or FDDDM method have been successfully applied to various bulk systems, nanostructures, and interfaces [33–39], and even partial-crystalline partial-liquid systems [40].

7.2.4   Boltzmann transport equation

As an alternative method to classical molecular dynamics, a phonon Boltzmann transport equation, a first-order partial differential equation for a phonon distribution function, has been used for small time and space scale. The distribution function is a scalar quantity in the six-dimensional phase space, namely, three space coordinates and three wave-vector coordinates. Generally the BTE is very difficult to solve so in practical relaxation time approximation is used. Monte Carlo simulations are often used to solve the Boltzmann equation for arbitrary structures [41–45]. However, when frequency-dependent phonon mean free paths need to be considered, the computational time of MC simulation becomes very large. Although recently researchers have developed some method of sampling the free path of phonon–phonon scattering with a Poisson distribution and added boundary scattering by launching phonons in a random position in the structure [46], the widely used Matthiessen rule were found to be large especially when the structure sizes have a large variation and such a treatment can significantly overestimate thermal conductivity when applied to more complex structures.

7.2.5   Multi-scale simulation

Although atomistic simulations can handle systems with heterogeneity on the order of nanometer or up to micrometers scale, this size is far not large enough to consider the heterogeneous effects occurring in realistic systems. To this end, multi-scale simulation tools are needed to bridge the length-scale and time-scale gaps between atomistic simulations and real materials and devices, as the traditional Fourier heat conduction model is inapplicable at length scales below the mean free path and the atomistic simulation methods are inadequate for simulating large systems. The phonon Boltzmann equation has been an important multi-scale tool for thermal transport studies. The information that is required as input for Boltzmann transport equation can be obtained from phonon relaxation time calculation for bulk materials and interfacial phonon transmission for interfaces using the technique such as wave packet molecular dynamics simulation.

The comparison of the main advantage and disadvantage between the above approaches is summarized in Table 7.1. Generally speaking, among the above methods MD simulation is the most popular approach in simulating nanoscale heat transfer, primarily due to its ease for coding (less effort to implement in a code, high parallel computing efficiency, etc.) and also its strong ability to deal with relatively large length scale and long time scale problems, as compared with first-principles calculations [24]. However, since MD simulation is principally based on empirical interatomic potential and classical Newton's second law, the interatomic potential used directly determines the quality of an MD simulation. Even worse, MD simulation may have severe problem when the interatomic potential for a specific material is not available, e.g., the novel nanostructures and their chemical modifications. Another significant drawback of classical MD simulation is the lack of quantum effect that should be taken into account when the system temperature is well below the Debye temperature. In these cases, first-principles based LD/BTE and NEGF methods are more suitable, as they do not need any fitting parameters.

Table 7.1 Advantage and disadvantage of typical computational approaches for nanoscale heat transfer.

	Advantage	Disadvantage
ALD/ BTE	Very accurate when combining with first-principle calculations; no need for any input parameters; can obtain phonon mode specific conductivity	High computation demand for large supercells; cannot simulate free surface and boundary
NEGF	Accurate (can combine with first-principle calculations as well); can obtain frequency dependent transmission coefficient; can simulate free surface, boundary, and interface	Only calculate ballistic transport so far; lack of accurate interfacial interaction model
Classical MD	Fast; ease for coding; high computing efficiency; can tackle various systems including free surface and boundary, especially when inhomogeneity occurs at large length scale	Empirical approach; rely on accurate interatomic potential; lack of quantum effect
MC/BTE	Fast; can simulate length scale beyond microns	Need input parameters for inhomogeneous system

7.3   Selected topics of modeling nanomaterials for energy nanotechnology

7.3.1   Low thermal conductivity of nanostructures for efficient thermoelectrics

7.3.1.1   Research motivation

Meeting the constantly increasing global demand for energy is one of the most important challenges facing humanity in the 21st century. One way to improve the sustainability of our energy base, and especially our electricity supply, is through the direct conversion of heat to electricity and by the scavenging of waste heat with thermoelectric generators. Home heating, automotive exhaust, and industrial processes all generate an enormous amount of unused waste heat that could be converted to electricity by using thermoelectrics. The thermoelectric effect is the direct conversion of temperature differences to electric voltage, and vice versa, by taking advantage of the so-called "Seebeck effect." Since thermoelectric generators are solid-state devices with no moving parts, they are small, light-weight, silent, reliable, scalable, and environmentally friendly, making them ideal for small, distributed power generation and even for tasks in harsh environments such as automobiles, incinerators, and spacecraft applications. Actually, efforts are already underway to replace the alternators in cars with thermoelectric generators mounted on the exhaust stream, thereby improving fuel efficiency.

The energy conversion efficiency of a thermoelectric material is given by figure of merit, ZT, $ZT = S^2 \sigma T / (\kappa_p + \kappa_e)$, where S is the Seebeck coefficient, σ is the electrical conductivity, T is the absolute temperature, and κ_p and κ_e are the lattice (phonon) and electronic contributions to thermal conductivity, respectively. An ideal thermoelectric with high ZT should be an electrical conductor, and also a thermal insulator. This conflicting requirement poses a material challenge. In general, development schemes to improve thermoelectric conversion efficiency in the past decades were guided by the concept of "phonon glass – electron crystal" (Fig. 7.1) (i.e., reducing the lattice contribution to the thermal conductivity as closely as possible to an amorphous state, while keeping relatively high electrical conductivity and Seebeck coefficient (high thermopower)). Along this line, a considerably large amount of scientific effort has been dedicated to reducing the lattice thermal conductivity. Initially, a "phonon glass – electron crystal" material features cages or tunnels in its crystal structure inside which reside massive atoms that are small enough relative to the cage to "rattle." This situation produces a phonon-damping effect that can result in the dramatic reduction of the lattice thermal conductivity. In the "phonon glass – electron crystal" picture, a glass-like thermal conductivity can, in principle, coexist with charge carriers of high mobility. The "phonon glass – electron crystal" approach has stimulated a significant amount of new research and has led to significant increases in ZT for several compounds such as clathrates, skutterudites, and half-heusler intermetallic compounds (see "Single Phases" in Fig. 7.1). Later on, due to the gained ability to create nanostructured materials, the nanostructuring of existing thermoelectric materials of

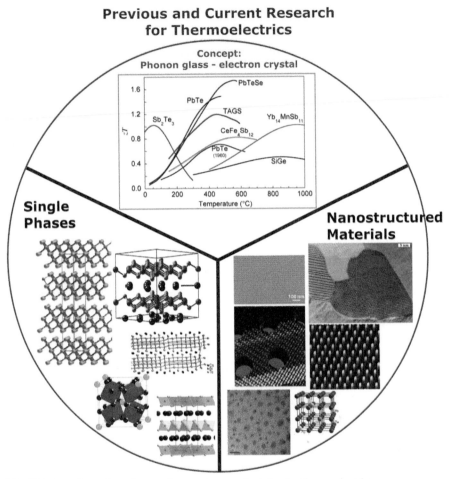

Fig. 7.1 Overview of previous and current research status for thermoelectrics.

interest has emerged as a promising pathway to greatly reduce the lattice thermal conductivity to values lower than were previously thought possible and, as a result, improve thermoelectric performance. Typical examples in this route include low-dimensional nanostructures such as superlattices, quantum dots, nanowires along with subsequent structure modulation, nanocomposites, and bulk nanostructured materials. By exploiting nanoscale effects, such as strong boundary or interfacial phonon scattering, and by taking advantage of the nanoconfinement effect, the nanostructured materials are able to achieve decent ZT values at room temperature and record-high ZT values in the range of 1.5–2.0 at medium and high temperatures (see "Nanostructured Materials" in Fig. 7.1), which cannot be achieved in traditional bulk materials. The above nanostructuring routes will be explained in detail in sequence.

7.3.1.2 Bulk (three-dimensional) materials

Among various types of nanostructured materials, bulk nanostructured materials have shown the most promise for commercial use because, unlike many other nanostructured materials (in particular low-dimensional materials), they can be fabricated in large quantities and in a form that is compatible with existing thermoelectric device configurations. The generation of these bulk nanostructured materials is currently being developed for commercialization, but still requires a fundamental understanding of carrier transport in these complex materials which is presently lacking for next generation high performance thermoelectrics.

In the past decade, significantly large amount of bulk nanostructured materials have experimentally demonstrated a higher ZT than that of their bulk counterparts by reducing the thermal conductivity to values lower than were previously thought possible, such as n-type $Si_{80}Ge_{20}$, p-type $Si_{80}Ge_{20}$, p-type $Bi_xSb_{2-x}Te_3$. Despite tremendous successes in experiments, fundamental understanding of transport in these complex materials and how their nanoscale structure affects bulk properties is still lacking. In particular, the precise way in which interfaces affect phonon transport to reduce thermal conductivity in nanocomposites remains poorly understood. Obviously, pure atomistic simulation such as molecular dynamics is not enough to capture the real physics of phonon transport in bulk-like nanostructured materials, such as nanocrystalline. This is because the heterogeneity, e.g., grains and grain boundaries which are essentially the dominant scattering process in nanocomposites and nanocrystallines, usually occurs at large length scale and the size of the grains can vary from a few nanometers up to several microns. This situation makes the explicit atomistic simulations almost impossible to obtain a reliable thermal conductivity result. To this end, solving Boltzmann transport equation at large length scales with input information obtained from atomistic simulations might be an efficient and promising method to tackle the phonon transport problem in bulk nanostructured materials. The numerical modeling which helps reveal the fundamental physics of phonon transport is key to further reducing the thermal conductivity. It is expected that with a better understanding of phonon transport, further lattice thermal conductivity reductions should be possible.

7.3.1.3 Two-dimensional materials

Graphene and grapheme-based heterostructures

Two-dimensional (2-D) materials have historically been one of the most intensely studied materials, due to the tremendously large amount of unusual physical phenomena that occur when charge and heat transport is confined to a plane. The discovery of graphene in 2004 by Novoselov and Geim [47,48] has shown that it is not only possible to isolate stable, single atom thick 2-D layers from van der Waals solids, but that these materials can exhibit unique and fascinating physical properties, which have led to novel routes for many applications. Graphene is only one-atom thick layer of sp^2-bonded carbon in a honeycomb crystal structure and is probably the most investigated system in materials science during the last decade. The recent developments in the "science of graphene" prompted an unprecedented surge of activity and demonstration of new physical phenomena. These developments also triggered a renewed emphasis on the interdisciplinary nature of nanoscience and created new opportunities in materials science, physics, chemistry, and electrical engineering. Two most striking physical properties of graphene that have already been explored extensively are: (1) In single layer graphene's band structure, the conduction π band and the valence π band of graphene meet exactly at the corners of the hexagonal first Brillouin zone called the Dirac points. Around the Dirac points, the dispersion relation of the π bands is linear with the separation distance, opposite to the other semiconductors, which chiefly exhibit a parabolic dispersion at the Fermi energy. Originating from this, electrons in graphene mimic the behavior of photons or other ultra-relativistic particles, with an energy-independent Fermi velocity that is only 300 times smaller than the speed of light. The linear dispersion at the K point gives rise to novel phenomena, such as the anomalous room-temperature quantum Hall effect [49], and has opened up a new category of "Fermi-Dirac" physics. (2) Even at single atom thick, high quality graphene is very strong, light, nearly transparent, and an excellent conductor of heat. The experimentally measured thermal conductivity of suspended single layer graphene at room temperature is reported to be as high as $\sim(4.84 \pm 0.44) \times 10^3$ to $(5.30 \pm 0.48) \times 10^3$ W/mK [50], which is almost the highest thermal conductivity among the existing materials.

Qiu and Ruan [51] addressed the problem of relative contributions of ZA phonons to thermal transport in the framework of the equilibrium MD simulations. Their conclusion was that in suspended single layer graphene out-of-plane ZA phonons are coupled with in-plane phonons due to the third-order and higher-order anharmonic interactions, which results in about 25%–30% contribution of ZA phonons to the thermal conductivity of graphene. In supported single layer graphene the contribution of all acoustic and ZO phonon branches are reduced owing to the SLG–substrate interface scattering and breakdown of the symmetries for both in-plane and out-of-plane phonons. The contributions of ZA phonons to thermal conductivity are suppressed more strongly than the contributions of TA and LA phonons. However, Ong and Pop [52] examined thermal transport in graphene supported on SiO$_2$ using MD simulations with reactive empirical bond order (REBO) potential. They found from their calculations that increasing the strength of the graphene–substrate interactions further can enhance the heat flow and effective thermal conductivity along the supported graphene. The authors attributed this result to the coupling of graphene ZA modes to the substrate Rayleigh waves, which linearizes the phonon dispersion, increases the group velocity of the hybridized modes and, thus, enhances the thermal flux.

In addition to pristine single layer graphene, the chemical modifications of graphene and graphene-based heterostructures have been proposed for a host of applications ranging from transparent conductors to thermal interface materials [53] to field effect transistors [54]. Furthermore, as single layer graphene is entirely surface area, its properties and reactivity profoundly depend on the chemical modifications, e.g., surface functionalization [55,56], and mechanical deformations.

Graphene analogs (silicene)

The graphene's success created a new era in materials science, especially in 2-D materials. It has shown the promise that it is possible to create stable, single or few-atom-thick layers out of other materials, and also that these materials can exhibit fascinating and technologically useful properties. It was not until the recent surge of intense research on graphene that the general potential of 2-D materials became apparent. In fact, two strongest motivation of exploring other graphene-like 2-D materials stems from the facts that: First, intrinsic graphene is a semi-metal or zero band-gap semiconductor. However, the presence of a sizable gap in the electronic band is an essential requirement for the realization of common electronic devices like field effect transistors, and finding the best way to induce such a gap in graphene has been a very active research topic in the past few years. Second, notwithstanding its phenomenal properties, the integration of graphene with the actual silicon-based technologies has proven a quite challenging task, whose solution would probably require the complete redesign of all electronic devices. As our present technology is primarily based on silicon (Si) semiconductors, a natural question is raised: could silicon, which is another element of the same IV-group of the periodic table as carbon, also forms graphene-like structures? Fortunately, the answer is affirmative.

Structurally, the silicon counterpart of graphene is called silicene, i.e., the Si atoms can also form one-atom thick honeycomb crystal structure. Silicene was first mentioned in a theoretical study by Takeda and Shiraishi [57] and then reinvestigated by Guzman-Verri and Lew Yan Voon [58], who named it silicene. Experimentally, in recent years silicence and silicene nanoribbons have been successfully grown on various metal substrates, such as Ag [59,60], Ir, and ZrB_2. Free standing silicene is still difficult to be synthesized until now. The experimental synthesis stimulated tremendous theoretical investigation of its physical properties. Among these, a significant effort has been invested on electronic property. For instance, it has been shown theoretically that the band structure of silicene resembles that of graphene, featuring Dirac-type electron dispersion in the vicinity of the corners of its hexagonal Brillouin zone [61,62]. As a result, the Fermi velocity of silicene is predicted to be approximately half of that for graphene. A very recent study demonstrated that silicene decorated with certain 3d transition metals (Vanadium) can sustain a stable quantum anomalous Hall effect. Moreover, silicene has been shown theoretically to be metastable as a free standing 2-D crystal [61], implying that it is possible to transfer silicene onto an insulating substrate and gate it electrically. Despite significant progress has been made in theoretical investigation of electrical transport property of silicene, however, thermal transport (mainly phonons according to our preliminary calculation) property has not been cultivated yet. For almost all nanoelectronics applications, thermal conductivity is always one of the most important physical parameters in electronics design and operation. Thermal transport can play crucial role in many applications such as heat dissipation, thermal management, and electronic packaging. Therefore, there is an emerging need of characterizing thermal transport properties of silicene based nanostructures for the development of relevant micro/nanoelectronics.

We have optimized Stillinger-Weber (SW) parameters for silicene, and the new parameters successfully reproduce the low buckling structure of silicene and the phonon dispersion curves from ab initio calculations [63]. We then calculated the thermal conductivity of silicene by both EMD and NEMD simulations, as well as the anharmonic lattice dynamics (ALD) method. The mode-specific contribution to lattice thermal conductivity is analyzed using the ALD method, and we found that the thermal conductivity of silicene is mainly contributed by phonons with MFP smaller than 10 nm, which is quite different from bulk silicon. Moreover, by qualifying the relative contributions of lattice vibrations in different directions, we found that the longitudinal phonon modes dominate the thermal transport in silicene, which is fundamentally different from graphene, despite the similarity of their two-dimensional honeycomb lattices. We then calculated thermal conductivity of monolayer silicene from first-principles method [64]. At 300 K, the thermal conductivity of monolayer silicene is found to be 9.4 W/mK and much smaller than bulk silicon. The contributions from in-plane and out-of-plane vibrations to thermal conductivity are quantified, and the out-of-plane vibration contributes less than 10% of the overall thermal conductivity, which is different from the results of the similar studies on graphene. The difference is explained by the presence of small buckling, which breaks the reflectional symmetry of the structure. The flexural modes are thus not purely out-of-plane vibration and have strong scattering with other modes. We also conducted investigation of thermal transport in a single-layer silicene sheet under uniaxial stretching [65] and hybrid graphene/silicene structures [66] with NEMD simulations.

7.3.1.4   One-dimensional materials

Single component nanowires

Experiments have demonstrated that, rough Si nanowires that are 20–300 nm in diameter have Seebeck coefficient and electrical resistivity values that are the same as doped bulk Si, but those with diameters of about 50 nm exhibit 100-fold reduction in thermal conductivity, yielding $ZT \sim 0.6$ at room temperature. For such nanowires, the lattice contribution to thermal conductivity approaches the amorphous limit for Si, which cannot be explained by current theories. Motivated by experiments, Wang et al. [67] calculated the thermal conductivity of silicon nanowires using the NEMD method with the Stillinger–Weber potential model and the Nose–Hoover thermostat. The dependence of the thermal conductivity on the wire length, cross-sectional area, and temperature was investigated. The surface along the longitudinal direction was set as a free boundary with potential boundaries in the other directions. The cross-sectional areas of the nanowires ranged from about 5 to 19 nm^2 with lengths ranging from 6 to 54 nm. The thermal conductivity dependence on temperature agrees well with the experimental results. The reciprocal of the thermal conductivity was found to be linearly related to the nanowire length. These results quantitatively show that decreasing the cross-sectional area reduces the phonon mean free path in nanowires. Martin et al. [68] modeled thermal properties of artificially rough Si, Ge, and GaAs NWs based on a perturbative treatment of the interaction between lattice vibrations and surface asperities. This approach, based on a full phonon dispersion in each material, accurately accounts for the frequency dependence of phonon scattering processes resulting from surface roughness, isotope, boundary, and anharmonic decays. The model predicts that for the same diameter and roughness, GaAs and Ge NWs are expected to have approximately 5 times and 10 times lower thermal conductivity than Si NWs, respectively, at room temperature. Later on, we performed NEMD simulations to investigate schemes for enhancing the energy conversion efficiency of Si nanowires [69], including (1) roughening of the nanowire surface, (2) creating nanoparticle inclusions in the nanowires, and (3) coating the nanowire surface with other materials. The enhancement in energy conversion efficiency was inferred from the reduction in thermal conductivity of the nanowire, which was calculated by imposing a temperature gradient in the longitudinal direction. Compared to pristine nanowires, our simulation results show that the schemes proposed above lead to nanocomposite structures with considerably lower thermal conductivity (up to 82% reduction), implying $\sim 5 \times$ enhancement in the ZT coefficient. This significant effect appears to have two origins: (1) increase in phonon-boundary scattering and (2) onset of interfacial interference. The results suggest new fundamental–yet realizable ways to improve markedly the energy conversion efficiency of nanostructured thermoelectrics.

Nanowire based heterostructures

Recently we proposed two new concepts to significantly reduce the thermal conductivity of Si-based nanowires in terms of enhancing thermoelectric properties: (1) Si-core Ge-shell nanowire [70,71] and (2) Si/Ge superlattice nanowire [72]. First, we investigated the effect of germanium (Ge) coatings on the thermal transport properties of silicon (Si) nanowires using NEMD simulations. Our results showed that a simple deposition of a Ge shell of only 1–2 unit cells in thickness on a single crystalline Si nanowire can lead to a dramatic 75% decrease in thermal conductivity at room temperature compared to an uncoated Si nanowire. By analyzing the vibrational density states of phonons and the participation ratio of each specific mode, we demonstrated that the reduction in the thermal conductivity of Si/Ge core-shell nanowire stems from the depression and localization of long-wavelength phonon modes at the Si/Ge interface and of high frequency non-propagating diffusive modes. After that, we presented main results of an investigation with NEMD simulations on thermal transport in Si/Ge superlattice nanowires aiming at taking advantage of the inherent one dimensionality and the combined presence of surface and interfacial phonon scattering to yield ultra-low values for their thermal conductivity. Our calculations revealed that the thermal conductivity of a Si/Ge superlattice nanowire varies nonmonotonically with both the Si/Ge lattice periodic length and the nanowire cross-sectional width. The optimal periodic length corresponds to an order of magnitude (92%) decrease in thermal conductivity at room temperature, compared to pristine single-crystalline Si nanowires. We also identified two competing mechanisms governing the thermal transport in superlattice nanowires, responsible for this nonmonotonic behavior: interface modulation in the longitudinal direction significantly depressing the phonon group velocities and hindering heat conduction, and coherent phonons occurring at extremely short periodic lengths countering the interface effect and facilitating thermal transport.

Our results showed trends for superlattice nanowire design for efficient thermoelectrics.

7.3.2 Low interfacial thermal resistance for high performance electronic cooling

7.3.2.1 Research motivation

Large heat generation and insufficient heat removal mechanisms due to inherent limitations in manufacturing methods and materials employed in electronics is one of the crucial challenges hindering the development of disruptive high power microelectronic device technologies. In most modern electronics systems (supercomputers, thermal tower, high power amplifiers, etc.), the electronic device is the warmest element in the system, and waste heat is removed by conduction, spreading, and convection to an appropriate working fluid (e.g., air, water, or a refrigerant) with gradual reductions in the temperature as heat travels through a series of thermal resistances from the source to the ambient. Although large air fans and/or liquid-based cooling solutions have been applied and can dissipate more than 100 W of total power, thermal resistances at multiple interfaces from the die through the heat spreader to the outside heat sink remain a bottleneck. Therefore, fundamental understanding of thermal transport across interfaces (including solid-solid and solid-liquid) is crucial to design and generate high efficiency thermal management systems.

7.3.2.2 Thermal transport across single interface

The majority of previous research is focused on studying how phonons transport from one material to the other, i.e., a single interface in between. To facilitate heat transfer across interfaces, interfacial nanoengineering has shown to be a very promising way in both experiments and numerical simulations [73–76]. For solid-solid interfaces, molecular simulations were used to evaluate thermal resistance between crystalline silicon and a vertically oriented carbon nanotube (CNT) [77]. Without chemical bonds between CNT and Si, the thermal resistance is high and its values are consistent with that measured in experiment on vertical CNT arrays. With chemical bonds the thermal resistance is reduced by two orders of magnitude demonstrating significant potential of CNT arrays for thermal management applications. The underlying mechanism for the very large effect of chemical bonding is revealed by simulations of individual phonon scattering across the interface and understood within an analytical solution of a simple spring-mass chain model. For solid-liquid interfaces, later on we proposed two mechanisms that enhance heat dissipation at solid-liquid interfaces investigated from the atomistic point of view using NEMD simulation [78]. The mechanisms include surface functionalization, where –OH terminated headgroups and self-assembled monolayers (SAMs) with different chain lengths are used to recondition and modify the hydrophilicity of silica surface, and vibrational matching between crystalline silica and liquid water, where three-dimensional nanopillars are grown at the interface in the direction of the heat flux with different lengths to rectify the vibrational frequencies of surface atoms. The heat dissipation is measured in terms of the thermal conductance of the solid-liquid interface and is obtained by imposing a one-dimensional heat flux along the simulation domain. A comparison with reported numerical and experimental thermal conductance measurements for similar interfaces indicates that the thermal conductance is enhanced by 1.8–3.2 times when the silica surface is reconditioned with hydrophilic groups. The enhancement is further promoted by SAMs, which results in a 20% higher thermal conductance compared with that of the fully hydroxylated silica surface. Likewise, the presence of nanopillars enhances the interface thermal conductance by 2.6 times compared with a bare surface (without nanopillars). Moreover, for different nanopillar densities, the conductance increases linearly with the length of the pillar and saturates at around 4.26 nm. Changes in the vibrational spectrum of surface atoms and water confinement effects are found to be responsible for the increase in conductance. The modification of surface vibrational states provides a tunable path to enhance heat dissipation, which can also be easily applied to other fluids and interfaces.

7.3.2.3 Thermal transport across multi-interfaces

At nano scale, sometimes it is possible to incorporate a third material between two different materials to reduce overall thermal resistance. This is realized by taking advantage of phonon confinement and phonon bridge effect. We first introduced this idea in solid-liquid interfaces. We studied the effect of water nanoconfinement on the thermal transport properties of two neighbor hydrophilic quartz interfaces [55,56]. A significant increase and a nonintuitive, nonmonotonic dependence of the overall interfacial thermal conductance between the quartz surfaces on the water layer thickness were found. By probing the interfacial structure and vibrational properties of the connected components, we demonstrated that the mechanism of the peak occurring at submonolayer water originates from the freezing of water molecules at extremely confined conditions and the excellent match of vibrational states between trapped water and hydrophilic headgroups on the two contact surfaces. Our results showed that incorporation of polar molecules into hydrophilic interfaces is very promising to enhance the thermal transport through thermally smooth connection of these interfaces.

Later on, for solid-solid interfaces, we studied a model system of SiC/GaN interface [79,80]. A thin AlN layer was introduced in the GaN–SiC gap to serve as a phonon bridge. The vibrational density of states (VDOS) of Al and N atoms in the low frequency region (<8 THz) falls between that of Ga and Si atoms. This feature enables the N...Al ensemble to serve as a vibrational bridge between the GaN and SiC surfaces. The matched VDOS between N...Al and Ga/Si facilitates the thermal transport across the GaN/N...Al/SiC interface. Our simulation results indicated that for both epitaxial and nonepitaxial crystal AlN with thickness up to 12.5 nm the overall thermal conductance across the interface is at least 45%–55% higher than that for a bare surface. $Al_xGa_{1-x}N$ ($0 < x < 1$) heterostructures were also considered to thermally bridge the GaN–SiC interface. It was found that for both epitaxial and nonepitaxial growths the overall thermal conductance increases monotonically with Al content from the value for pure GaN to that for pure AlN. Furthermore, the conductance for 1 nm thick $Al_xGa_{1-x}N$ only depends on the Al content and is independent of the Al distribution in the heterostructure. However, it is important to stress that the interfacial thermal conductance in possible future experiments is expected to be much lower than that in all theoretical cases studied herein, due to the presence of impurities, dislocations, etc.

We also proposed a junction structure that the carbon nanotube is bonded with a monolayer graphene, which could potentially enhance the interface thermal conductance [81]. NEMD simulations showed that the interface thermal conductance can be enhanced by at least 40% compared to direct carbon nanotube and silicon interface with strong covalent bonding, while for weak van der Waals bonding the conductance can be enhanced by almost one order of magnitude. The enhancement of thermal conductance is attributed to the efficient thermal transport between carbon nanotube and graphene, as well as the good contact between graphene and silicon surface.

7.4 Summary and perspective

The continuous decrease in the size of devices and structures, the increase in their operating speeds and frequencies, and the ever more aggressive thermal conditions imposed upon them requires sophisticated understanding and control of thermal transport at the nanoscale. Nowadays thermal performance itself is a key metric for applications such as thermal barrier coatings, nuclear fuels, and materials for thermoelectric and photovoltaic energy conversion. Fundamental understanding of nanoscale thermal transport has resulted from significant advances in experiment, theory, and simulation over the past decades. Modeling and simulation of thermal transport in nanostructures or nanostructured materials and interfacial heat transport has been the focus of numerous computational efforts in the past decade. The most prominent atomistic approach is to use molecular dynamics simulations where atoms and molecules follow classical dynamics based on the numerical solution of the Newton's second law of motion. The forces between atoms and molecules are derived from interatomic potentials. Equilibrium and non-equilibrium MD thermal conductivity techniques have shown the power in characterizing either macroscopic or interfacial behavior "integrated" over all vibrations. In recent years, direct first-principles calculations have been combined with anharmonic LD to evaluate phonon properties of complex compounds. With powerful computer, one can now quantify thermal transport process in very complicated structures with ab initio level accuracy.

Another promising strategy for advancing computational thermal science and materials physics is to use the artificial intelligence algorithms, such as machine learning techniques, which are transforming the materials science studies from property prediction to inverse materials design. Indeed, big data and deep learning have already brought transformative revolution in computer vision, autonomous cars, and speech recognition. However, in computational material science especially thermal science field, their potential has yet to be fully implemented [82–84] due to the inherent difference between materials data and image/audio data, and lack of sufficient high quality (first-principles level) data. Machine learning method is also very promising for uncovering inherent physics that was missing in the past from huge amount of materials data and/or atomic configurations and/or dynamics (such as correlation between moving or transporting ions and interaction between mobile ions and propagating phonons). In this line, we anticipate that more robust, fast, and efficient machine learning methods will be developed in the near future in terms of revealing the detailed phonon-phonon and phonon-impurity scattering in large-scale structures and systems. It is also expected that advanced machine learning algorithms can be very useful for extracting or interpreting complex phonon transport phenomena in highly inhomogeneous systems.

References

[1] A. Majumdar, Microscale heat conduction in dielectric thin films, J. Heat Transf. 115 (1993) 7–16.

[2] D. Broido, M. Malorny, G. Birner, N. Mingo, D. Stewart, Intrinsic lattice thermal conductivity of semiconductors from first principles, Appl. Phys. Lett. 91 (2007) 231922.

[3] K. Esfarjani, G. Chen, H.T. Stokes, Heat transport in silicon from first-principles calculations, Phys. Rev. B 84 (2011) 085204.

[4] W. Li, J. Carrete, N.A. Katcho, N. Mingo, ShengBTE: a solver of the Boltzmann transport equation for phonons, Comput. Phys. Commun. 185 (2014) 1747–1758.

[5] A. Chernatynskiy, S.R. Phillpot, Phonon transport simulator (PhonTS), Comput. Phys. Commun. 192 (2015) 196–204.

[6] A. Togo, L. Chaput, I. Tanaka, Distributions of phonon lifetimes in Brillouin zones, Phys. Rev. B 91 (2015) 094306.

[7] J. Garg, G. Chen, Minimum thermal conductivity in superlattices: a first-principles formalism, Phys. Rev. B 87 (2013), 140302.

[8] Y. Han, J.-Y. Yang, M. Hu, Unusual strain response of thermal transport in dimerized three-dimensional graphene, Nanoscale 10 (2018) 5229.

[9] T. Ouyang, M. Hu, Competing mechanism driving diverse pressure dependence of thermal conductivity of X Te (X = Hg, Cd, and Zn), Phys. Rev. B 92 (2015), 235204.

[10] T. Ouyang, H. Xiao, C. Tang, X. Zhang, M. Hu, J. Zhong, First-principles study of thermal transport in nitrogenated holey graphene, Nanotechnology 28 (2016), 045709.

[11] G. Qin, X. Zhang, S.-Y. Yue, Z. Qin, H. Wang, Y. Han, M. Hu, Resonant bonding driven giant phonon anharmonicity and low thermal conductivity of phosphorene, Phys. Rev. B 94 (2016), 165445.

[12] G. Qin, Z. Qin, H. Wang, M. Hu, Anomalously temperature-dependent thermal conductivity of monolayer GaN with large deviations from the traditional 1/T law, Phys. Rev. B 95 (2017), 195416.

[13] G. Qin, Z. Qin, H. Wang, M. Hu, Lone-pair electrons induced anomalous enhancement of thermal transport in strained planar two-dimensional materials, Nano Energy 50 (2018) 425.

[14] G. Qin, K.-R. Hao, Q.-B. Yan, M. Hu, G. Su, Exploring T-carbon for energy applications, Nanoscale 11 (2019) 5798.

[15] G. Qin, H. Wang, Z. Qin, L. Zhang, M. Hu, Giant effect of spin-lattice coupling on the thermal transport in two-dimensional ferromagnetic CrI3, J. Mater. Chem. C 8 (2020) 3520.

[16] J.-Y. Yang, G. Qin, M. Hu, Nontrivial contribution of Fröhlich electron-phonon interaction to lattice thermal conductivity of Wurtzite GaN, Appl. Phys. Lett. 109 (2016), 242103.

[17] J.-Y. Yang, L. Cheng, M. Hu, Unravelling the progressive role of rattlers in thermoelectric clathrate and strategies for performance improvement: concurrently enhancing electronic transport and blocking phononic transport, Appl. Phys. Lett. 111 (2017), 242101.

[18] K. Yuan, X. Zhang, D. Tang, M. Hu, Anomalous pressure effect on the thermal conductivity of ZnO, GaN, and AlN from first-principles calculations, Phys. Rev. B 98 (2018), 144303.

[19] S.-Y. Yue, X. Zhang, G. Qin, J.-Y. Yang, M. Hu, Insight into the collective vibrational modes driving ultra-low thermal conductivity of perovskite solar cells, Phys. Rev. B 94 (2016), 115427.

[20] N. Mingo, L. Yang, Phonon transport in nanowires coated with an amorphous material: an atomistic Green's function approach, Phys. Rev. B 68 (2003) 245406.

[21] N. Mingo, Anharmonic phonon flow through molecular-sized junctions, Phys. Rev. B 74 (2006) 125402.

[22] J.S. Wang, J. Wang, J.T. Lü, Quantum thermal transport in nanostructures, Eur. Phys. J. B 62 (2008) 381–404.

[23] T. Yamamoto, K. Watanabe, Nonequilibrium Green's function approach to phonon transport in defective carbon nanotubes, Phys. Rev. Lett. 96 (2006) 255503.

[24] Y. He, I. Savic, D. Donadio, G. Galli, Lattice thermal conductivity of semiconducting bulk materials: atomistic simulations, Phys. Chem. Chem. Phys. 14 (2012) 16209–16222.

[25] P.K. Schelling, S.R. Phillpot, P. Keblinski, Comparison of atomic-level simulation methods for computing thermal conductivity, Phys. Rev. B 65 (2002) 144306.

[26] X. Zhang, M. Hu, D. Poulikakos, A low-frequency wave motion mechanism enables efficient energy transport in carbon nanotubes at high heat fluxes, Nano Lett. 12 (2012) 3410–3416.

[27] M. Hu, Y. Jing, X. Zhang, Low thermal conductivity of graphyne nanotubes from molecular dynamics study, Phys. Rev. B 91 (2015), 155408.

[28] X. Zhang, M. Hu, D. Tang, Thermal rectification at silicon/horizontally aligned carbon nanotube interfaces, J. Appl. Phys. 113 (2013), 194307.

[29] J.A. Thomas, J.E. Turney, R.M. Iutzi, C.H. Amon, A.J.H. McGaughey, Predicting phonon dispersion relations and lifetimes from the spectral energy density, Phys. Rev. B 81 (2010) 081411(R).

[30] Y. Chalopin, S. Volz, A microscopic formulation of the phonon transmission at the nanoscale, Appl. Phys. Lett. 103 (2013) 051602.

[31] Y. Zhou, X. Zhang, M. Hu, Quantitatively analyzing phonon spectral contribution of thermal conductivity based on nonequilibrium molecular dynamics simulations. I. From space Fourier transform, Phys. Rev. B 92 (2015) 195204.

[32] Y. Zhou, M. Hu, Quantitatively analyzing phonon spectral contribution of thermal conductivity based on nonequilibrium molecular dynamics simulations. II. From time Fourier transform, Phys. Rev. B 92 (2015) 195205.

[33] Y. Sun, Y. Zhou, J. Han, W. Liu, C. Nan, Y. Lin, M. Hu, B. Xu, Strong phonon localization in PbTe with dislocations and large deviation to Matthiessen's rule, NPJ Comput. Mater. 5 (2019) 97.

[34] Y. Zhou, M. Hu, Record low thermal conductivity of polycrystalline Si nanowire: breaking the Casimir limit by severe suppression of propagons, Nano Lett. 16 (2016) 6178.

[35] Y. Zhou, Y. Chen, M. Hu, Strong surface orientation dependent thermal transport in Si nanowires, Sci. Rep. 6 (2016) 24903.

[36] Y. Zhou, Y. Yao, M. Hu, Boundary scattering effect on thermal conductivity of nanowires, Semicond. Sci. Technol. 31 (2016), 074004.

[37] Y. Zhou, X. Zhang, M. Hu, An excellent candidate for largely reducing interfacial thermal resistance: a nano-confined mass graded interface, Nanoscale 8 (2016) 1994–2002.

[38] Y. Zhou, M. Hu, Full quantification of frequency-dependent interfacial thermal conductance contributed by two- and three-phonon scattering processes from nonequilibrium molecular dynamics simulations, Phys. Rev. B 95 (2017), 115313.

[39] Y. Zhou, X. Zhang, M. Hu, Nonmonotonic diameter dependence of thermal conductivity of extremely thin Si nanowires: competition between hydrodynamic phonon flow and boundary scattering, Nano Lett. 17 (2017) 1269–1276.

[40] Y. Zhou, S. Xiong, X. Zhang, S. Volz, M. Hu, Thermal transport crossover from crystalline to partial-crystalline partial-liquid state, Nat. Commun. 9 (2018) 4712.

[41] Q. Hao, G. Chen, M.-S. Jeng, Frequency-dependent Monte Carlo simulations of phonon transport in two-dimensional porous silicon with aligned pores, J. Appl. Phys. 106 (2009), 114321.

[42] M.-S. Jeng, R. Yang, D. Song, G. Chen, Modeling the thermal conductivity and phonon transport in nanoparticle composites using Monte Carlo simulation, J. Heat Transf. 130 (2008), 042410.

[43] A.J. Minnich, Determining phonon mean free paths from observations of quasiballistic thermal transport, Phys. Rev. Lett. 109 (2012) 205901.

[44] J.-P.M. Peraud, N.G. Hadjiconstantinou, Efficient simulation of multidimensional phonon transport using energy-based variance-reduced Monte Carlo formulations, Phys. Rev. B 84 (2011), 205331.

[45] W. Tian, R. Yang, Thermal conductivity modeling of compacted nanowire composites, J. Appl. Phys. 101 (2007) 054320.

[46] A.J.H. McGaughey, A. Jain, Nanostructure thermal conductivity prediction by Monte Carlo sampling of phonon free paths, Appl. Phys. Lett. 100 (2012) 061911.

[47] A.K. Geim, K.S. Novoselov, The rise of graphene, Nat. Mater. 6 (2007) 183–191.

[48] K.S. Novoselov, A.K. Geim, S.V. Morozov, D. Jiang, Y. Zhang, S.V. Dubonos, I.V. Grigorieva, A.A. Firsov, Electric field effect in atomically thin carbon films, Science 306 (2004) 666–669.

[49] K.S. Novoselov, Z. Jiang, Y. Zhang, S.V. Morozov, H.L. Stormer, U. Zeitler, J.C. Maan, G.S. Boebinger, P. Kim, A.K. Geim, Room-temperature quantum hall effect in graphene, Science 315 (2007) 1379.

[50] A.A. Balandin, S. Ghosh, W.Z. Bao, I. Calizo, D. Teweldebrhan, F. Miao, C.N. Lau, Superior thermal conductivity of single-layer graphene, Nano Lett. 8 (2008) 902–907.

[51] B. Qiu, X.L. Ruan, Reduction of spectral phonon relaxation times from suspended to supported graphene, Appl. Phys. Lett. 100 (2012) 193101.

[52] Z.-Y. Ong, E. Pop, Effect of substrate modes on thermal transport in supported graphene, Phys. Rev. B 84 (2011) 075471.

[53] A.A. Balandin, Thermal properties of graphene and nanostructured carbon materials, Nat. Mater. 10 (2011) 569–581.

[54] X. Jia, J. Campos-Delgado, M. Terrones, V. Meuniere, M.S. Dresselhaus, Graphene edges: a review of their fabrication and characterization, Nanoscale 3 (2011) 86–95.

[55] J. Hu, S. Schiffli, A. Vallabhaneni, X. Ruan, Y.P. Chen, Tuning the thermal conductivity of graphene nanoribbons by edge passivation and isotope engineering: a molecular dynamics study, Appl. Phys. Lett. 97 (2010), 133107.

[56] M. Hu, J.V. Goicochea, B. Michel, D. Poulikakos, Water nanoconfinement induced thermal enhancement at hydrophilic quartz interfaces, Nano Lett. 10 (2010) 279–285.

[57] K. Takeda, K. Shiraishi, Theoretical possibility of stage corrugation in Si and Ge analogs of graphite, Phys. Rev. B 50 (1994) 14916–14922.

[58] G.G. Guzman-Verri, L.C. Lew Yan Voon, Electronic structure of silicon-based nanostructures, Phys. Rev. B 76 (2007), 075131.

[59] B. Feng, Z. Ding, S. Meng, Y. Yao, X. He, P. Cheng, L. Chen, K. Wu, Evidence of silicene in honeycomb structures of silicon on Ag(111), Nano Lett. 12 (2012) 3507–3511.

[60] P. Vogt, P. De Padova, C. Quaresima, J. Avila, E. Frantzeskakis, M.C. Asensio, A. Resta, B. Ealet, G. Le Lay, Silicene: compelling experimental evidence for graphene like two-dimensional silicon, Phys. Rev. Lett. 108 (2012) 155501.

[61] S. Cahangirov, M. Topsakal, E. Akturk, H. Sahin, S. Ciraci, Two- and one-dimensional honeycomb structures of silicon and germanium, Phys. Rev. Lett. 102 (2009), 236804.

[62] C.C. Liu, W. Feng, Y. Yao, Quantum spin hall effect in silicene and two-dimensional germanium, Phys. Rev. Lett. 107 (2011) 076802.

[63] X. Zhang, H. Xie, M. Hu, H. Bao, S. Yue, G. Qin, G. Su, Thermal conductivity of silicene calculated using an optimized Stillinger-Weber potential, Phys. Rev. B 89 (2014) 054310.

[64] H. Xie, M. Hu, H. Bao, Thermal conductivity of silicene from first-principles, Appl. Phys. Lett. 104 (2014), 131906.

[65] M. Hu, X. Zhang, D. Poulikakos, Anomalous thermal response of silicene to uniaxial stretching, Phys. Rev. B 87 (2013), 195417.

[66] Y. Jing, M. Hu, L. Guo, Thermal conductivity of hybrid graphene/silicon heterostructures, J. Appl. Phys. 114 (2013), 153518.

[67] S.-C. Wang, X. Liang, X.-H. Xu, T. Ohara, Thermal conductivity of silicon nanowire by nonequilibrium molecular dynamics simulations, J. Appl. Phys. 105 (2009), 014316.

[68] P.N. Martin, Z. Aksamija, E. Pop, U. Ravaioli, Reduced thermal conductivity in nanoengineered rough Ge and GaAs nanowires, Nano Lett. 10 (2010) 1120–1124.

[69] X. Zhang, M. Hu, K.P. Giapis, D. Poulikakos, Schemes for and mechanisms of reduction in thermal conductivity in nanostructured thermoelectrics, J. Heat Transf. 134 (2012), 102402.

[70] M. Hu, K.P. Giapis, J.V. Goicochea, X. Zhang, D. Poulikakos, Significant reduction of thermal conductivity in Si/Ge core-shell nanowires, Nano Lett. 11 (2011) 618–623.

[71] M. Hu, X. Zhang, K.P. Giapis, D. Poulikakos, Thermal conductivity reduction in core-shell nanowires, Phys. Rev. B 84 (2011), 085442.

[72] M. Hu, D. Poulikakos, Si/Ge superlattice nanowires with ultralow thermal conductivity, Nano Lett. 13 (2013) 5487–5494.

[73] M. Hu, S. Shenogin, P. Keblinksi, N. Raravikar, Thermal energy exchange between carbon nanotube and air, Appl. Phys. Lett. 90 (2007), 231905.

[74] M. Hu, P. Keblinski, B.W. Li, Thermal rectification at silicon-amorphous polyethylene interface, Appl. Phys. Lett. 92 (2008), 211908.

[75] M. Hu, J.V. Goicochea, B. Michel, D. Poulikakos, Thermal rectification at water/functionalized silica interfaces, Appl. Phys. Lett. 95 (2009), 151903.

[76] M. Hu, P. Keblinski, P.K. Schelling, Kapitza conductance of silicon–amorphous polyethylene interfaces by molecular dynamics simulations, Phys. Rev. B 79 (2009) 104305.

[77] M. Hu, P. Keblinski, J.-S. Wang, N. Raravikar, Interfacial thermal conductance between silicon and a vertical carbon nanotube, J. Appl. Phys. 104 (2008) 083503.

[78] J.V. Goicochea, M. Hu, B. Michel, D. Poulikakos, Surface functionalization mechanisms of enhancing heat transfer at solid-liquid interfaces, J. Heat Transf. 133 (2011) 082401.

[79] M. Hu, X. Zhang, D. Poulikakos, C.P. Grigoropoulos, Large "near junction" thermal resistance reduction in electronics by interface nanoengineering, Int. J. Heat Mass Transf. 54 (2011) 5183–5191.

[80] M. Hu, D. Poulikakos, Graphene mediated thermal resistance reduction at strongly coupled interfaces, Int. J. Heat Mass Transf. 62 (2013) 205–213.

[81] H. Bao, C. Shao, S. Luo, M. Hu, Enhancement of interfacial thermal transport by carbon nanotube-graphene junction, J. Appl. Phys. 115 (2014) 053524.

[82] Y. Dan, Y. Zhao, X. Li, S. Li, M. Hu, J. Hu, Generative adversarial networks (GAN) based efficient sampling of chemical composition space for inverse design of inorganic materials, NPJ Comput. Mater. 6 (2020) 84.

[83] A. Rodriguez, Y. Liu, M. Hu, Spatial density neural network force fields with first-principles level accuracy and application to thermal transport, Phys. Rev. B 102 (2020), 035203.

[84] Y. Zhao, K. Yuan, Y. Liu, S.-Y. Louis, M. Hu, J. Hu, Predicting elastic properties of materials from electronic charge density using 3D deep convolutional neural networks, J. Phys. Chem. C 124 (2020) 17262.

Further reading

M. Hu, Z. Yang, Perspective on multi-scale simulation of thermaltransport in solids and interfaces, Phys. Chem. Chem. Phys. 23 (2021) 1785–1801.

P.K. Schelling, S.R. Phillpot, P. Keblinski, Phonon wave-packet dynamics at semiconductor interfaces by molecular-dynamics simulation, Appl. Phys. Lett. 80 (2002) 2484–2486.

Modeling thermal conductivity with Green's function molecular dynamics simulations

Vitor R. Coluci[a], Fabio Andrijauskas[a,b], and Sócrates O. Dantas[c]
[a]School of Technology, University of Campinas—UNICAMP, Limeira, SP, Brazil, [b]Physics Institute "Gleb Wataghin", University of Campinas—UNICAMP, Campinas, SP, Brazil, [c]Department of Physics, Institute of Exact Sciences, Federal University of Juiz de Fora, Juiz de Fora, MG, Brazil

8.1 Introduction

Understanding the main properties of materials is a tremendous task. Currently, computer atomistic simulations have played an important role toward this task. Altogether with the traditional ways of doing science—by developing theories and doing experiments—computer simulations have helped describe the behavior of the atoms that compose a material. The simulations have speeded up the understanding of main properties and even revealed new ones, especially in extreme physical conditions, where carrying out experiments is difficult. By quoting Richard Feynman [1]

> *if we were to name the most powerful assumption of all, which leads one on and on in an attempt to understand life, it is that all things are made of atoms, and that every-thing that living things do can be understood in terms of the* jigglings *and* wigglings *of atoms*

The main aim of atomistic simulations is to reproduce these "jigglings" and "wigglings" of atoms.

Among the various types of simulations, molecular dynamics (MD) simulations are especially designed to study the atomic behavior in chemical, biological, and physical systems. Because the jigglings and wigglings of atoms occur so fast, in the order of 1 femtosecond (fs), the time-step size used in MD simulations by typical numerical integrators should be smaller than 1 fs to accurately describe the system evolution. This small-time step requires a large number of integration steps to study the systems in real-time scales (μs–s). Such a large number can significantly accumulate numerical errors and make impossible probing this long-time behavior.

In this chapter, we present a technique aimed to tackle the small step issue in MD simulations. Proposed by Vinod Tewary from NIST in 2009, the so-called Green's function molecular dynamics (GFMD) was able to extend the time in MD simulations by six to eight orders of magnitude (pico to microseconds) in certain idealized nonlinear elastic problems such as the propagation of ripples in graphene [2].

Modeling, Characterization and Production of Nanomaterials. https://doi.org/10.1016/B978-0-12-819905-3.00008-7

First, we describe the main aspects of a typical MD simulation. Second, we derive the main GFMD equations and present the advantages and drawbacks of this technique. We also show how high-performance computing can be used to tackle one of the GFMD drawbacks. Third, we present the GFMD performance on treating vibrational problems in a model one-dimensional (1D) system. Finally, we discuss the possibility of using GFMD to calculate thermal conductivity.

8.2   Classical molecular dynamics simulations

Classical MD simulations are applied to systems when classical approximation is reasonable. For liquids and monoatomic gases, this happens when the de Broglie's length (which depends on the mass of the particle and on the temperature) is much less than the average distance between the particles. For molecular systems, the classical approximation is reasonable when the thermal energy is lesser than the energy associated with molecular vibrations.

The core of a classical MD simulation is the interaction potentials. They are approximations or classical representations of quantum potentials. By using these potentials, MD simulations can treat the contribution of electrons to the atomic motion in an implicit way. Because the nuclei are much heavier than the electrons, the motion of the nuclei can be treated independently of the motion of the electrons (Born-Oppenheimer approximation). Thus, the trajectory of the particles (nuclei) is determined by the energy landscape created by the electrons. The energy landscape is usually described by analytical expressions, which depend on the interaction type. For example, the various types of interaction in a molecular system are described by inter- and intramolecular interactions through analytical expressions in the total interaction potential. These interactions include, for instance, bond stretching, angular and torsion deformations for bonded atoms, and van der Waals and electrostatic interactions for nonbonded atoms.

Having determined the most appropriate interaction potential V to describe the physical system, the trajectories of the particles of the system are obtained by numerically solving Newton's equations of motion for each particle i:

$$m_i \frac{\partial^2 r_{i,\alpha}}{\partial t^2} = -\frac{\partial V}{\partial r_{i,\alpha}}, \tag{8.1}$$

where m_i is the mass of the particle i and $\alpha = x, y, z$ represent the Cartesian components of the position vectors $r_i(t)$.

Different algorithms have been developed to solve Eq. (8.1). One of the most used is Verlet's algorithm [3]. By using the Taylor's expansion for the positions of the system [4] for $t \pm \Delta t$ around $r_i(t)$ (Δt is the time-step size), the positions in the time $t + \Delta t$ are obtained by adding these two expansions

$$r_i(t + \Delta t) = 2r_i(t) - r_i(t - \Delta t) + \frac{\Delta t^2}{m_i} F_i(t) + O(\Delta t^4), \tag{8.2}$$

where F_i is the force on atom i. Similarly, the subtraction of the Taylor's expansions for $r_i(t + \Delta t)$ and $r_i(t - \Delta t)$ yields to the velocity

$$v_i(t) = \frac{r_i(t + \Delta t) - r_i(t - \Delta t)}{2\Delta t}. \qquad (8.3)$$

From Eqs. (8.2), (8.3), a mathematically equivalent version of the Verlet algorithm—velocity-Verlet [5]—can be obtained. The velocity-Verlet algorithm is self-starting, minimizes roundoff errors, and is given by

$$r_i(t + \Delta t) = r_i(t) + v_i(t)\Delta t + \frac{\Delta t^2}{2m_i} F_i(t), \qquad (8.4)$$

$$v_i(t + \Delta t) = v_i(t) + \frac{\Delta t}{2m_i} \left(F_i(t + \Delta t) - F_i(t) \right). \qquad (8.5)$$

From the atomic positions and velocities, properties of interest such as energy, temperature, and pressure can be obtained. For systems where temperature/pressure needs to keep constant (NVT/NPT ensembles), thermostats and barostats can be applied [6]. A pseudo-code for the main steps of a typical MD simulation is shown in Algorithm 8.1.

8.3 Green's function molecular dynamics

8.3.1 Main equations

We present here a way to solve Eq. (8.1) using the casual Green's function for a system with N interacting particles of mass m_i ($i = 1, ..., N$) in a thermal bath of temperature T. In this condition, Eq. (8.1) is modified and the instantaneous force on the particle i is then given by

Algorithm 8.1 MD pseudo-code.

1 Set up system parameters (e.g., geometry, initial particle positions and velocities, and atomic masses);
2 Set up simulation parameters (e.g., time step size Δt, number of integration steps N_{steps}, and the interaction potential V);
3 Initialize parameters (e.g., $t \leftarrow 0$, velocities);
4 **while** $step < N_{\text{steps}}$ **do**
5   Calculate the force on each particle $\vec{F}_i = -\nabla_i V(\{\vec{r}\})$ for the current configuration;
6   Update positions and velocities (e.g., using Eqs. 4 and 5);
7   Update time $t \leftarrow t + \Delta t$;
8   Calculate properties (e.g., kinetic and potential energies, temperature) using the updated configuration;
9 **end**

$$m_i \frac{\partial^2 u_{i,\alpha}(t)}{\partial t^2} = -\frac{\partial V}{\partial u_{i,\alpha}} - \frac{m_i}{\tau_i}\frac{\partial u_{i,\alpha}(t)}{\partial t} + \eta_{i,\alpha}(t), \tag{8.6}$$

where V is the total potential energy, $u_{i,\alpha}$ is the α cartesian component of the displacement of the particle i with respect to its equilibrium position, and τ_i is the characteristic viscous damping time for the atom i (with friction coefficient $\xi_i \equiv m_i/\tau_i$). Eq. (8.6) is used to formulate Langevin's thermostat [6], where the degrees of freedom of the thermal bath are omitted and a set of stochastic differential equations are used. The thermal bath acts on the system through a random force $\eta_{i,\alpha}(t)$, obeying the fluctuation-dissipation theorem, described by a white noise

$$\langle \eta_{i\alpha}(t) \rangle_\eta = 0,$$
$$\langle \eta_{i\alpha}(t)\eta_{j\beta}(t') \rangle_\eta = 2\xi_i k_b T \delta_{ij}\delta_{\alpha\beta}\delta(t - t'), \tag{8.7}$$

where k_b is Boltzmann's constant and $\langle \ldots \rangle_\eta$ is the average over many realizations of the random function $\eta(t)$.

The total potential V can be expanded in a Taylor's series in atomic displacements and given in a matrix notation as

$$V = V_0 - F \cdot u + \frac{1}{2} u^T \phi u + \Delta F(u). \tag{8.8}$$

The constant term V_0 can be taken to be zero. $\Delta F(u)$ includes $O(u^3)$ and higher-order nonlinear terms. The force term $F_{i,\alpha}(t)$ is the first term in the Taylor's expansion of the crystal potential energy in the Born-von Karman model. Physically, it represents the negative of the force on the atom i. This term vanishes for a lattice in equilibrium. If the lattice at equilibrium is subjected to a perturbation, such as by introducing a defect, a force is induced on each atom and the second- and the higher-order terms (the force constants) change. The force term is treated as an effective force, which includes the linear $(F_{i,\alpha})$ and all nonlinear terms $\left(\Delta F_{i,\alpha} = -\frac{\partial \Delta V(u_{0i})}{\partial u_{i,\alpha}} \right)$ in the potential except the quadratic term. In a pure harmonic model, $\Delta F_{i,\alpha}$ is zero and F is constant. On the other hand, in the anharmonic case, $\Delta F_{i,\alpha}$ is a function of u and, therefore, t. The quadratic term is the force constant matrix $\phi_{u_{i,\alpha},u_{j,\beta}} = \frac{\partial^2 V(u_{0i})}{\partial u_{i,\alpha}\partial u_{j,\beta}}$, $u_{0i,\alpha}$ is the displacement at $t = 0$.

Using the transformation $u_{i,\alpha}(t) \rightarrow \sqrt{m_i}u_{i,\alpha}(t)$, Eq. (8.6) can be written in the following matrix form:

$$\left(\mathbb{I}\frac{\partial^2}{\partial t^2} + \mathbb{X}\frac{\partial}{\partial t} + \mathbb{D} \right)\mathbb{U} = \mathbb{F}(t) + \Delta\mathbb{F}(t) + \mathbb{F}^R(t), \tag{8.9}$$

where $\mathbb{U}$, $\mathbb{F}$, $\Delta\mathbb{F}$, and $\mathbb{F}^R$ are vectors whose components are $\sqrt{m_i}u_{i,\alpha}$, $(1/\sqrt{m_i})F_{i,\alpha}$, $(1/\sqrt{m_i})\Delta F_{i,\alpha}$, and $(1/\sqrt{m_i})\eta_{i,\alpha}$, respectively. $\mathbb{X}$ is a diagonal matrix with the elements $1/\tau_i$, $\mathbb{D}$ is the system's dynamical matrix with the elements $(1/\sqrt{m_i m_j})\phi_{i,\alpha,j,\beta}$, and $\mathbb{I}$ is the unit matrix.

The formal solution of Eq. (8.9) is

$$U(t) = \left(\mathbb{I}\frac{\partial^2}{\partial t^2} + \mathbb{X}\frac{\partial}{\partial t} + \mathbb{D} \right)^{-1} \mathbb{F}_{\text{eff}}(t), \tag{8.10}$$

where $\mathbb{F}_{\text{eff}}(t) \equiv \mathbb{F}(t) + \Delta F(t) + \mathbb{F}^R(t)$.

The inverse operator in Eq. (8.10) is the causal Green's function $\mathbb{G}(t - t')$, defined as a solution for $t > t'$, of

$$\left(\mathbb{I}\frac{\partial^2}{\partial t^2} + \mathbb{X}\frac{\partial}{\partial t} + \mathbb{D} \right)\mathbb{G}(t - t') = \mathbb{I}\delta(t - t'). \tag{8.11}$$

The Laplace transform $\mathcal{L}$ of Eq. (8.11) is

$$\left[s^2\mathbb{I} + s\mathbb{X} + \mathbb{D} \right]\mathcal{L}[\mathbb{G}] - (s\mathbb{I} + \mathbb{X})\mathbb{G}(0) - \mathbb{G}'(0) = \mathbb{I}\mathcal{L}[\delta(t - t')], \tag{8.12}$$

where s is the Laplace variable conjugate to t. For the initial conditions $\mathbb{G}(0) = \mathbb{G}'(0) = 0$, the Laplace transform of the Green's function takes the form

$$\mathcal{L}[\mathbb{G}] = \frac{\mathbb{I}}{s^2\mathbb{I} + s\mathbb{X} + \mathbb{D}} \tag{8.13}$$

and the Laplace transform of Eq. (8.9) is

$$\left[s^2\mathbb{I} + s\mathbb{X} + \mathbb{D} \right]\mathcal{L}[\mathbb{U}] - (s\mathbb{I} + \mathbb{X})\mathbb{U}(0) - \mathbb{C}(0) = \mathcal{L}[\mathbb{F}_{\text{eff}}], \tag{8.14}$$

or

$$\mathcal{L}[\mathbb{U}] = \mathcal{L}[\mathbb{G}]\mathcal{L}[\mathbb{F}_{\text{eff}}] + L[H]U(0) + \mathcal{L}[\mathbb{G}]\mathbb{C}(0), \tag{8.14}$$

where

$$\mathcal{L}[\mathbb{H}] \equiv \frac{s\mathbb{I} + \mathbb{X}}{s^2\mathbb{I} + s\mathbb{X} + \mathbb{D}}. \tag{8.15}$$

In cases where $\mathbb{D}$ is a real, symmetric matrix, we can write $\mathbb{V}^T\mathbb{D}\mathbb{V} = \mathbb{E}^2$, where $\mathbb{V}$ is orthogonal with the eigenvectors of $\mathbb{D}$ in the columns and $\mathbb{E}^2$ is the diagonal containing the eigenvalues E_i^2 of $\mathbb{D}$. Multiplying Eq. (8.14) by $\mathbb{V}^T$ and using $\mathbb{I} = \mathbb{V}\mathbb{V}^T$, Eq. (8.14) is written as

$$\mathbb{U}^{L,*} = \mathbb{K}_G \, \mathbb{F}_{\text{eff}}^{L,*} + \mathbb{K}_H \, \mathbb{U}^{L,*}(0) + \mathbb{K}_G \, \mathbb{C}^{L,*}(0), \tag{8.16}$$

where $\mathbb{U}^{L,*} \equiv \mathbb{V}^T\mathcal{L}[\mathbb{U}]$, $\mathbb{K}_G \equiv \mathbb{V}^T\mathcal{L}[\mathbb{G}]\mathbb{V}$, $\mathbb{F}_{\text{eff}}^{L,*} \equiv \mathbb{V}^T\mathcal{L}[\mathbb{F}_{\text{eff}}]$, $\mathbb{K}_H \equiv \mathbb{V}^T\mathcal{L}[\mathbb{H}]\mathbb{V}$, $\mathbb{U}^{L,*}(0) \equiv \mathbb{V}^T\mathbb{U}(0)$, and $\mathbb{C}^{L,*}(0) \equiv \mathbb{V}^T\mathbb{C}(0)$, with the superscript L standing for the

Laplace transform, and $*$ standing for the multiplication by $\mathbb{V}^T$. We used the term "projected" for the quantities with $*$ ($\mathbb{U}^*$, $\mathbb{C}^*$, and $\mathbb{F}^*$).

Finally, the inverse Laplace transform of Eq. (8.16) is

$$\mathbb{U}^* = \mathcal{L}^{-1}[\mathbb{K}_G \ \mathbb{F}_{\text{eff}}^{L,*}] + \mathcal{L}^{-1}[\mathbb{K}_H \ \mathbb{U}^{L,*}(0)] + \mathcal{L}^{-1}[\mathbb{K}_G \ \mathbb{C}^{L,*}(0)].$$
(8.17)

If we assume the same viscous damping time τ for all atoms then $\mathbb{X} = \tau^{-1}\mathbb{I}$, and through Eq. (8.15), Eq. (8.17) can be simplified to

$$\mathbb{U}^* = \mathcal{L}^{-1}[\mathbb{K}_G \ \mathbb{F}_{\text{eff}}^{L,*}] + \mathcal{L}^{-1}[(s + \tau^{-1})\mathbb{K}_G \ \mathbb{U}^{L,*}(0)] + \mathcal{L}^{-1}[\mathbb{K}_G \ \mathbb{C}^{L,*}(0)],$$
(8.18)

where $\mathbb{K}_G$ is now a diagonal matrix with elements $K_i = (s^2 + s/\tau + E_i^2)^{-1}$. The elements of matrix in Eq. (8.18) are written as [7]

$$
\begin{aligned}
U_i^*(t) &= \frac{(F_i^* + F_i^{R*})}{E_i^2}\left[1 - \frac{e^{-t/(2\tau)}}{2\omega_i\tau}\left(2\tau\dot{S}(t) + S(t)\right)\right] \\
&+ \frac{C_i^*(0)S(t)}{\omega_i}e^{-t/(2\tau)} \\
&+ \frac{U_i^*(0)}{2\omega_i\tau}\left(2\tau\dot{S}(t) + S(t)\right)e^{-t/(2\tau)},
\end{aligned}
$$
(8.19)

where $S(t) \equiv \sin(\omega_i t)$ with $\omega_i \equiv (2\tau)^{-1}\sqrt{4E_i^2\tau^2 - 1}$ for the case where $4E_i^2\tau^2 > 1$, and $S(t) \equiv \sinh(\omega_i t)$ with $\omega_i \equiv (2\tau)^{-1}\sqrt{1 - 4E_i^2\tau^2}/2\tau$ for $4E_i^2\tau^2 < 1$, and

$$
\begin{aligned}
U_i^*(t) &= (F_i^* + F_i^{R*})\tau^2\left[4 - 2e^{-t/(2\tau)}\left(2 + \frac{t}{\tau}\right)\right] \\
&+ C_i^*(0)te^{-t/(2\tau)} + U_i^*(0)e^{-t/(2\tau)}\left(1 + \frac{t}{2\tau}\right)
\end{aligned}
$$
(8.20)

for the case where $4E_i^2\tau^2 = 1$. The velocities of the particles are obtained from $U_i^*(t)$ by $C_i^*(t) = \frac{dU_i^*(t)}{dt}$. Eqs. (8.19), (8.20) are, therefore, the GFMD equations for canonical ensemble simulations. If the action of the thermal bath occurs rarely ($\tau \to \infty$), the temperature of the system is no longer conserved and the microcanonical ensemble holds. In this case, Eq. (8.19) becomes the equation originally derived by Tewary [2]:

$$U_i^*(t) = \frac{-F_i^*}{E_i^2}\left[\cos(E_i t) - H(t)\right] + \frac{C_i^*(0)}{E_i}\sin(E_i t) + C_i^*(0)\cos(E_i t),$$
(8.21)

where $H(t)$ is the Heaviside function.

By discretizing the time in time steps Δt as $t_n = n\Delta t$, the displacement $U_i(t + \Delta t)$ is calculated from the previous step t as, for instance, for $4E_i^2 \tau^2 \neq 1$,

$$
\begin{aligned}
U_i^*(t + \Delta t) &= \frac{(F_i^* + F_i^{R*})}{E_i^2} \left[1 - \frac{e^{-\Delta t/(2\tau)}}{2\omega_i \tau} \left(2\tau \dot{S}(\Delta t) + S(\Delta t) \right) \right] \\
&+ \frac{C_i^*(t) S(\Delta t)}{\omega_i} e^{-\Delta t/(2\tau)} \\
&+ \frac{U_i^*(t)}{2\omega_i \tau} \left(2\tau \dot{S}(\Delta t) + S(\Delta t) \right) e^{-\Delta t/(2\tau)},
\end{aligned}
\tag{8.22}
$$

with F_i^* and F_i^{R*} calculated at time t.

A pseudo-code for a typical GFMD simulation is shown in Algorithm 8.2. GFMD requires more steps when compared with the pseudo-code of a conventional MD simulation (Algorithm 8.1). These steps include the diagonalization of $\mathbb{D}$ and the calculation and updating of the projected quantities ($\mathbb{U}^*$, $\mathbb{C}^*$, $\mathbb{F}^*$), which increase the computational cost. As we will see in the following section, the computational cost

Algorithm 8.2 GFMD pseudo-code.

1 Set up system parameters (geometry, initial conditions $\mathbb{U}(0)$, $\mathbb{C}(0)$, and atomic masses);
2 Set up simulation parameters (time-step size Δt, number of integration steps N_{steps}, τ, etc.);
3 Initialize parameters (e.g., $t \leftarrow 0$);
4 Set $\mathbb{D}(0)$ and $\mathbb{F}(0)$ matrices from initial configuration;
5 Diagonalize $\mathbb{D}$ to obtain E_i^2 and $\mathbb{V}$;
6 Calculate the initial projected displacements $\mathbb{U}^*(0) \leftarrow \mathbb{V}^T \mathbb{U}(0)$;
7 Calculate the initial projected velocities $\mathbb{C}^*(0) \leftarrow \mathbb{V}^T \mathbb{C}(0)$;
8 Calculate the initial projected forces $\mathbb{F}^*(0) \leftarrow \mathbb{V}^T \mathbb{F}(0)$;
9 *Integration loop:*
10 **while** *step $< N_{\text{steps}}$* **do**
11   Calculate the 1st Taylor's coefficients $\mathbb{F}(t)$ and the force constants $\mathbb{D}(t)$ for the current atomic configuration;
12   Diagonalize $\mathbb{D}(t)$ to obtain E_i^2 and $\mathbb{V}$;
13   Calculate the projected displacements $\mathbb{U}^*(t) \leftarrow \mathbb{V}^T \mathbb{U}(t)$;
14   Calculate the projected velocities $\mathbb{C}^*(t) \leftarrow \mathbb{V}^T \mathbb{C}(t)$;
15   Calculate the projected 1st Taylor's coefficients $\mathbb{F}^*(t) \leftarrow \mathbb{V}^T \mathbb{F}(t)$;
16   Update projected displacements $\mathbb{U}^*(t + \Delta t)$ using Eq.(22);
17   Update projected velocities $\mathbb{C}^*(t + \Delta t)$;
18   Update displacements $\mathbb{U}(t + \Delta t) \leftarrow \mathbb{V} \mathbb{U}^*(t + \Delta t)$;
19   Update velocities $\mathbb{C}(t + \Delta t) \leftarrow \mathbb{V} \mathbb{C}^*(t + \Delta t)$;
20   Update atomic positions from the displacements $\mathbb{U}(t + \Delta t)$;
21   Reset the displacements $\mathbb{U}(t + \Delta t) \leftarrow 0$;
22   Update time $t \leftarrow t + \Delta t$;
23   Calculate properties (kinetic and potential energies, temperature, etc.) using the updated atomic configuration ($\mathbb{U}(t + \Delta t)$, $\mathbb{C}(t + \Delta t)$);
24 **end**

of GFMD is one of its drawbacks, which has been reduced by using high-performance computing techniques.

8.3.2 Advantages and drawbacks

As pointed out in the previous section, the time-step size can be set larger for GFMD simulations. However, it cannot be increased arbitrarily. The limit is related to the numerical convergence and accuracy in real solids for nonlinear terms in the potential energy. In GFMD, the time-step size limit arises from the "space" part of the MD equations, as the approximation used for the total interaction potential. On the other hand, for the "time" part, the time step can be arbitrarily large because the solution is exact in time. However, in order to keep negligible the cubic and higher-order terms in displacements in potential energy, the displacements must be small, which constrains the time-step size. This constraint, in general, is less severe than in the conventional MD [2].

GFMD treats the linear ($F_{i,\alpha}$) and nonlinear ($\Delta F_{i,\alpha}$) terms (the effective force) by iteration [2]. At each iteration, the effective force is calculated in terms of the displacements obtained in the previous iteration. Effectively, ΔF_{i_α} is calculated using the displacements obtained in the previous iteration and absorbed into $F_{\text{eff } i,\alpha}$. During the new iteration, the effective force is taken constant and the displacements are obtained from the GFMD. For the harmonic case, the effective force is indeed constant so no iterations are necessary. New iterations are performed until the effective force is small enough when compared to a prescribed value. This iteration technique is similar to the conventional MD. The main difference is that in the GFMD the terms up to the quadratic term in the potential at each iteration are retained, whereas in the conventional MD only the first-order term is retained.

The inclusion of the quadratic term improves the convergence of the GFMD by several orders of magnitude [2]. Consider the calculated value of u after one iteration, u_c. If $\mathcal{N}_{\text{MD}}$ is the number of iterations necessary to provide a correct value for u_a, then $u_c \sim u_a/\mathcal{N}_{\text{MD}}$ for linear convergence. Because the leading neglected term is $O(u^2)$ in conventional MD, the error is $O(u^2) \sim O[1/(\mathcal{N}_{\text{MD}})^2]$. On the other hand, the error in GFMD is $O(u^3) \sim O[1/(\mathcal{N}_{\text{GFMD}})^3]$. Imposing the same error in conventional MD and GFMD ($O[1/(\mathcal{N}_{\text{MD}})^2] \sim O[1/(\mathcal{N}_{\text{GFMD}})^3]$), the estimated convergence improvement factor in GFMD, defined as $\mathcal{N}_{\text{MD}}/\mathcal{N}_{\text{GFMD}}$, is $(\mathcal{N}_{\text{GFMD}})^{1/2}$. Thus, for $\mathcal{N} \sim 10^{6-8}$, the improvement will be $\sim 10^{3-4}$.

The elastic characteristics of a solid is defined by the second-order terms in displacements. These terms are neglected in conventional MD. The limitation of the conventional MD becomes even more severe in large systems when the total potential energy is a function of many variables, generally exhibiting local minima and saddle points, where the first-order terms are vanishingly small. Differently from conventional MD, GFMD includes the second-order term in potential energy and exactly determines the phonon frequency and the elastic response of the solid.

One drawback of the GFMD, however, is that it is applicable to only stable solids, where the atoms vibrate about their equilibrium positions (i.e., no diffusion). This is because $\mathbb{D}$ is assumed to be positive definite and, consequently, the eigenvalues E_i^2 are

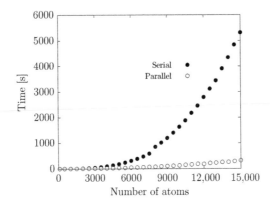

Fig. 8.1 Elapsed time after 20,000 simulation steps of a GFMD run ($\Delta t = 1$ fs) using the serial and parallel versions as functions of the system size (number of atoms in the linear chain). The serial simulations were carried out in a single core of an Intel(R) Xeon(R) CPU X5650 of 2.67 GHz. The parallel simulations were carried out using 12 cores and one Nvidia Tesla M2070.

real and positive, the condition for stable solids. Negative eigenvalues indicate unstable phonons and diffusion. The derivation of GFMD equations for this case will allow that other phenomena can be studied with GFMD.

Another drawback is that GFMD may be computationally more costly than the conventional MD. This is because GFMD involves a $O(N^3)$ process of the diagonalization of $\mathbb{D}$ for the whole lattice. The computational cost, however, should be partly offset because of the need for much fewer iterations. A possible approach is to divide the whole system into smaller subsystems. For each subsystem, a smaller $\mathbb{D}$ is assigned to it, which will lead to an $O(N)$ process.

Another way to reduce the computational cost of GFMD is by using high-performance computing techniques. Matrix diagonalization and multiplications are the main costly operations in GFMD. We combined shared memory techniques, many-core architectures, and linear algebra libraries to accelerate those operations. Parallel versions of GFMD were implemented to use CPU multithreading and graphic processing units. We find that relatively simple implementations, using routines from well-known linear algebra libraries (Intel MKL, Magma, and ESSL libraries) can reduce the overall time of a GFMD simulation run. To benchmark our parallel GFMD implementation, we applied it to simulate the atomic motion of a linear chain considering only nearest neighbors interaction through Morse's potential. Fig. 8.1 shows that our GFMD parallel version is able to significantly reduce the time of the GFMD run. For example, for a 15,000-atom chain, the elapsed time of a parallel GFMD run is 6% of the time of a serial run.

8.4 Performance

To show GFMD is able to produce the correct trajectory of the atoms in a system for larger time steps than the conventional MD, we determine the evolution of an 1D

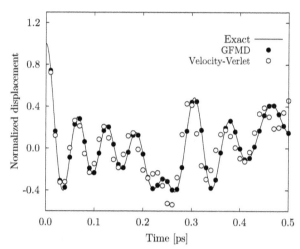

Fig. 8.2 Evolution of the normalized displacement of the central atom ($u_0(t)/d_0$) obtained from the GFMD and velocity-Verlet methods within the microcanonical ensemble. The exact solution for a linear chain with 23 atoms is also shown for comparison [2]. The following parameters were used: $m = 12$ amu, $V_0 = 1.72435$ eV, $\Delta t = 10$ fs, and $\gamma = 1.0\text{Å}^{-1}$.

atomic chain with fixed boundary conditions. The N atoms of the chain have the same mass and are initially separated by a distance $l_0 = 1\text{Å}$. The atoms interact only with their nearest neighbors through Morse's potential $V(x) = V_0[\exp[-2\gamma(x - l_0)] - 2\exp[-\gamma(x - l_0)]]$, where x is the distance between the atoms.

Fig. 8.2 shows the evolution of the normalized displacement of the central atom of the chain for $\tau \to \infty$ (microcanonical ensemble) calculated with the GFMD and with the velocity-Verlet method for a time-step size of 10 fs. In this case, all the initial velocities and displacements were set to zero except the displacement of the central atom, chosen to be $d_0 = 10^{-4}\text{Å}$. Such a small initial displacement allows the comparison with the exact evolution for a linear chain within the harmonic approximation [2]. Both velocity-Verlet and GFMD are able to reproduce the exact behavior of the system for a time step of 1 fs (not shown in Fig. 8.2) but only GFMD reproduces the correct behavior for the larger time step of 10 fs.

We now consider the atomic chain in contact with a thermal bath of temperature T. One way to probe the properties of the system such as phonon density of states and the diffusion coefficient is through the normalized velocity autocorrelation function (VACF). The VACF $C_{v_0}(\zeta)$ was calculated as [8]

$$C_{v_0}(\zeta) = \lim_{t \to \infty} \frac{\langle v_0(t + \zeta)v_0(t)\rangle}{\langle v_0^2(t)\rangle}, \tag{8.23}$$

where the $\langle\,\rangle$ is the average over different time intervals ζ and v_0 is the velocity of the central atom. For a linear chain of equal masses in contact with a thermal bath

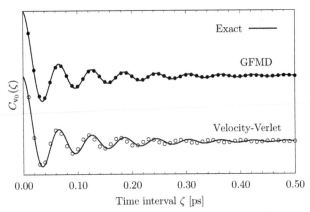

Fig. 8.3 Normalized velocity autocorrelation function obtained from GFMD and velocity-Verlet for $\Delta t = 10$ fs ($N = 203$, $\tau = 100$ fs, $T = 30$ K, and total simulation time of 10 ns).

within the harmonic approximation, the exact VACF can be obtained [7] and is given by

$$C_{v_0}(\zeta) = \frac{e^{-\zeta/2\tau}}{N} \sum_{k_i} f(\omega_{k_i}, \zeta, \tau), \tag{8.24}$$

where

$$f(\omega_{k_i}, \zeta, \tau) = \cos\left[\omega_{k_i}\zeta\right] - \frac{\sin\left[\omega_{k_i}\zeta\right]}{2\omega_{k_i}\tau} \tag{8.25}$$

if $2\omega_o(k_i)\tau < 1$, and

$$f(\omega_{k_i}, \zeta, \tau) = \cosh\left[\omega_{k_i}\zeta\right] - \frac{\sinh\left[\omega_{k_i}\zeta\right]}{2\omega_{k_i}\tau} \tag{8.26}$$

if $2\omega_o(k_i)\tau > 1$, where $\omega_0^2(k_i) = (2K/m)[1 - \cos(k_i)]$, $\omega_{k_i}^2 \equiv \frac{1}{4\tau^2} - \omega_o^2(k_i)$, and K is the nearest-neighbor spring constant.

Fig. 8.3 shows the VACF obtained from GFMD and velocity-Verlet for a time step of 10 fs. Whereas GFMD agrees with the correct VACF behavior for the whole range of the time intervals, velocity-Verlet fails to reproduce it, especially for $\zeta \gtrsim 0.3$ ps. This region on VACF is related to the low-frequency phonons that are important on thermal conductivity as we will see in the following section.

8.5 Calculation of thermal conductivity using GFMD

Most of the heat produced by modern solid-state devices at the nanoscale is wasted. For this reason, energy harvesting and thermal dissipation are currently strong research topics in materials science. A crucial property for designing suitable materials in these topics is thermal conductivity [9, 10].

Considered one of the most difficult transport coefficients to calculate (p. 150 in [11]), thermal conductivity [12]—the proportionality between a heat flux and a temperature gradient—impacts the designing of efficient thermal devices. For example, thermoelectric devices provide a method of recycling the wasted heat produced in solid-state devices and should exhibit a high figure of merit $S^2 \sigma T / \kappa$, where S is the Seebeck coefficient, σ is the electrical conductivity, T is the absolute temperature, and κ is the thermal conductivity. Therefore, whereas solids with low thermal conductivity are especially useful for thermoelectric devices since their figure of merit is inversely proportional to thermal conductivity, solids with high thermal conductivity are needed for heat dissipation.

The overall thermal conductivity of a material is influenced by electronic and lattice contributions. In semiconductors and insulators, lattice vibrations—phonons—dominate thermal transport. Vacancies, defects, allowing, and interfaces play a role in phonon scattering and propagation and can be explored to reduce the lattice thermal conductivity [13]. Decreasing the material size by nanostructuring processes has also been used to design materials with reduced lattice thermal conductivity since the vast number of interfaces in nanostructures act as scatters to phonon propagation.

Different methods have been used to predict the thermal conductivity of materials. These methods involve phenomenological models based on the Fourier law and Boltzmann transport equation [14–16], harmonic and quasiharmonic lattice dynamics [17], and MD simulations [18]. For nanomaterials, for instance, anisotropy of the thermal conductivity needs to be considered. This makes the thermal conductivity calculation difficult even in the continuum approximation and, therefore, the discrete atomistic structure of the material has to be considered. For this atomistic modeling, MD simulations are the standard numerical technique since they naturally allow the inclusion of anharmonic effects and the analysis of thermal transport at the atomic level.

In MD simulations, the thermal conductivity can be predicted, for instance, by using the Green-Kubo formalism [19, 20]. The net flow of heat in a solid is described by the heat current vector $\mathbf{J}$ and, at thermal equilibrium, $\mathbf{J}$ fluctuates about zero. In the Green-Kubo formalism, thermal conductivity is related to how long it takes to these fluctuations to dissipate [18] and is calculated as [21, 22]

$$\kappa_{ij} = \frac{V}{k_b T^2} \int_0^\infty \langle (J_i(0) - \langle J_i \rangle)(J_j(t) - \langle J_j \rangle) \rangle \, dt, \tag{8.27}$$

where V is the volume, T is the temperature, k_b is the Boltzmann constant, J_i is the heat flux in the i-direction, and $\langle \cdot \rangle$ is the average over the equilibrium ensemble. From the trajectories obtained from an equilibrium MD simulation, the heat flux $\mathbf{J}$ is calculated from [22–24]

$$\mathbf{J} = \frac{1}{V} \sum_\alpha (\epsilon_\alpha \mathbf{I} + \mathbf{v}_\alpha^T) \mathbf{v}_\alpha, \tag{8.28}$$

where ϵ_α is the per-atom energy of atom α, and ν_α and v_α are the virial and the velocity for the atom α, respectively.

Following Jones and Mandadapu [22], the thermal conductivity can be expressed generically as

$$\kappa = \lambda \int_0^\infty \langle \chi(0)\chi(t)\rangle dt = \lambda \int_0^\infty C(t)dt, \tag{8.29}$$

where χ is the fluctuation in the heat flux, λ is a constant containing k_b, T, and V, and $C(t)$ is the autocorrelation function of χ. $C(t)$ is calculated as a time average from a single trajectory by using the ergodic hypothesis as

$$C(t) = \lim_{\tau_0 \to \infty} \frac{1}{\tau_0} \int_0^{\tau_0} \langle \chi(s)\chi(s+t)\rangle ds = \lim_{\tau_0 \to \infty} C_{\tau_0}(t). \tag{8.30}$$

Because it is impossible to achieve the infinite correlation time τ_0 in MD simulations, the averaging in Eq. (8.32) is approximate. This inaccuracy is associated with the so-called *averaging* error. Another type of error in calculating the thermal conductivity is the *truncation* error due to the finite integration limit in Eq. (8.31), associated with the long-time behavior of the tail of $C_{\tau_0}(t)$. Thus, the estimate for κ is

$$\kappa_{\tau_0}(\tau_c) = \lambda \int_0^{\tau_c} C_{\tau_0}(t)dt. \tag{8.31}$$

Therefore, the accuracy of the thermal conductivity depends on the convergence of κ about τ_0 and τ_c. The convergence can be faster if, instead of using a single trajectory, one averages $\kappa_{\tau_0}(\tau_c)$ over N_r different trajectories of the same energy but with different initial conditions. Thus,

$$\kappa_{\tau_0}(\tau_c) = \lambda \frac{1}{N_r \tau_0} \sum_{r=1}^{N_r} \int_0^{\tau_c} \int_0^{\tau_0} \langle \chi_r(s)\chi_r(s+t)\rangle dt \, ds. \tag{8.32}$$

Finally, because the equations of motions in MD are numerically solved for discrete times, discretization of Eq. (8.34) gives the following approximation for $\kappa_\tau(\tau_c = N_c\Delta t)$:

$$\kappa_{\tau_0}(N_c\Delta t) \simeq \sum_{j=0}^{N_c} \frac{\lambda}{N_r N_{\tau_0}} \sum_{r=1}^{N_r} \sum_{i=1}^{N_{\tau_0}} \chi_r(i\Delta t)\chi_r((i+j)\Delta t) \, \Delta t \, w_j, \tag{8.33}$$

where w_j is the quadrature weight for the numerical integration (e.g., $w_j = 1$ except for $w_0 = 1/2$ for the trapezoidal rule), and N_c is the number of time steps necessary to guarantee that τ_c is much greater than the characteristic decay time of the autocorrelation for the corresponding system.

The discretization in time introduces a third error in the calculation of the thermal conductivity, the *time-integration* error, which is due to inaccuracies in the MD trajectories. Small time-step sizes Δt are typically used to avoid this error on the computed thermal conductivity. However, such small time-step sizes, as we have seen in the previous sections, greatly impact on the computational cost of the MD simulations carried out to calculate the thermal conductivity because of the large number of time steps necessary to reach the required long simulations (much greater than N_c). The limited time scale provided by MD simulations, therefore, impacts directly the accuracy of thermal conductivity calculations since long averaging times are required to properly describe the slower decay in the autocorrelation function.

The problem of convergence is even more crucial for two-dimensional (2D) solids because of the size effect. The frequency response of 2D solids is sensitive to the size of the material [25]. Hence, it becomes necessary to model a large solid and include a large number of atoms in the model, as compared to three-dimensional (3D) solids. The size effect arises because the response of the solid has a $1/k^2$ type singularity at $k = 0$, where k is the wave vector. The thermal conductivity of a material involves the integral of its frequency spectrum over the entire range of frequencies or over the entire Brillouin zone. For a 3D solid, the singularity is integrated out because the volume element in the k-space is proportional to k^2 near $k = 0$, which is multiplied to the singularity $1/k^2$. In a 2D solid, the volume element is proportional to k, so the integrand is $\sim 1/k$ near $k = 0$ and consequently singular. The region $k \approx 0$ corresponds to long waves (low-frequency phonons, THz–GHz) which make a significant contribution to thermal conductivity.

Different methods have been proposed to manage the errors on thermal conductivity calculations from MD simulations using the Green-Kubo approach [22, 26–29]. Chen et al. [26] studied the uncertainty in determination of the truncation time and proposed a scheme to directly integrate the autocorrelation function and a correction to the fitting of that function to describe the contribution from low-frequency phonons. Jones and Mandadapu [22] developed an on-the-fly technique to determine the thermal conductivity to within a preset error tolerance taking into account data from the initial conditions, truncation of the integral of the autocorrelation function, and the finite averaging time. de Sousa Oliveira and Greaney [27] tackled the problem of the need of long averaging time to reduce remnant noise in the autocorrelation function by modeling the integrated noise as a random walk, being able to choose integration conditions to minimize the uncertainty in the thermal conductivity calculations. Ercole et al. [28] developed a data-analysis protocol based on full sample power spectrum of the heat current to compute thermal conductivity from short simulations (up to few hundred picoseconds) with accuracy of the order of 10% on the estimated thermal conductivities. Khaled et al. systematically studied the effect of time step, simulation time, correlation time, and system size and determined the values, which provide reasonable agreement for the thermal conductivity of silicon [29].

Due to larger time-step sizes that can be used in GFMD simulations and the large convergence improvement factor, we can expect improvements in the convergence of thermal conductivity calculations. GFMD is also expected to impact directly on the *time-integration* error since more accurate trajectories can be obtained. In addition, we also expect effects on the *averaging* and *truncation* errors as more accurate long-time behavior can be achieved, possibly reducing the necessary number of trajectories to be averaged.

8.6   Concluding remarks

GFMD is a promising technique for materials science simulations. By extending the GFMD original formulation [2], we showed that GFMD can also be used to simulate systems in contact with thermal baths within the canonical ensemble. We tackled the GFMD high computational cost by applying high-performance computing techniques into the most time-consuming operations in GFMD. Even with affordable computational resources, significant reduction of the GFMD runtime has been achieved, which allows applying GFMD to more realistic systems. These results have been obtained only for 1D model systems. Thus, we have ongoing projects to apply GFMD in 2D and 3D systems to study its reliability in those systems. Based on the advantages presented by GFMD, we hypothesize that it can improve the calculations of thermal conductivity by reducing the main errors (averaging, truncation, and time-integration errors) related to these calculations.

References

[1] R.P. Feynman, R.B. Leighton, M. Sands, The Feynman Lectures on Physics, New Millennium ed., Basic Books, New York, NY, 2010 (originally published 1963–1965, https://cds.cern.ch/record/1494701).

[2] V.K. Tewary, Extending the time scale in molecular dynamics simulations: propagation of ripples in graphene, Phys. Rev. B 80 (16) (2009) 161409, https://doi.org/10.1103/PhysRevB.80.161409.

[3] L. Verlet, Computer "experiments" on classical fluids. I. Thermodynamical properties of Lennard-Jones molecules, Phys. Rev. 159 (1) (1967) 98–103, https://doi.org/10.1103/PhysRev.159.98.

[4] M.P. Allen, D.J. Tildesley, Computer Simulation of Liquids, Clarendon Press, Oxford, 1987.

[5] W.C. Swope, H.C. Andersen, P.H. Berens, K.R. Wilson, A computer simulation method for the calculation of equilibrium constants for the formation of physical clusters of molecules: application to small water clusters, J. Chem. Phys. 76 (1) (1982) 637–649, https://doi.org/10.1063/1.442716.

[6] P.H. Hünenberger, Thermostat algorithms for molecular dynamics simulations, in: C. Holm, K. Kremer (Eds.), Advanced Computer Simulation: Approaches for Soft Matter Sciences I, Springer, Berlin, Heidelberg, 2005, pp. 105–149, https://doi.org/10.1007/b99427.

[7] V.R. Coluci, S.O. Dantas, V.K. Tewary, Generalized Green's function molecular dynamics for canonical ensemble simulations, Phys. Rev. E 97 (5) (2018) 053310, https://doi.org/10.1103/PhysRevE.97.053310.

[8] M.A. Despósito, A.D. Viñales, Subdiffusive behavior in a trapping potential: mean square displacement and velocity autocorrelation function, Phys. Rev. E 80 (2) (2009) 021111, https://doi.org/10.1103/PhysRevE.80.021111.

[9] J.G. Stockholm, Applications in thermoelectricity, Mater. Today Proc. 5 (4, Pt 1) (2018) 10257–10276, https://doi.org/10.1016/j.matpr.2017.12.273.

[10] Y. Wang, Z. Xu, The critical power to maintain thermally stable molecular junctions, Nat. Commun. 5 (1) (2014) 4297, https://doi.org/10.1038/ncomms5297.

[11] D.J. Evans, G. Morriss, Statistical Mechanics of Nonequilibrium Liquids, second ed., Cambridge University Press, Cambridge, 2008, https://doi.org/10.1017/CBO9780511535307.

[12] M. Ausloos, Thermal conductivity, in: K.H.J. Buschow, R.W. Cahn, M.C. Flemings, B. Ilschner, E.J. Kramer, S. Mahajan, P. Veyssière (Eds.), Encyclopedia of Materials: Science and Technology, Elsevier, Oxford, 2001, pp. 9151–9155, https://doi.org/10.1016/B0-08-043152-6/01650-8.

[13] T. Hori, J. Shiomi, Tuning phonon transport spectrum for better thermoelectric materials, Sci. Technol. Adv. Mater. 20 (1) (2019) 10–25, https://doi.org/10.1080/14686996.2018.1548884.

[14] J. Callaway, Model for lattice thermal conductivity at low temperatures, Phys. Rev. 113 (4) (1959) 1046–1051, https://doi.org/10.1103/PhysRev.113.1046.

[15] P.G. Klemens, Theory of lattice thermal conductivity: role of low-frequency phonons, Int. J. Thermophys. 2 (1) (1981) 55–62, https://doi.org/10.1007/BF00503574.

[16] A.J.H. McGaughey, M. Kaviany, Quantitative validation of the Boltzmann transport equation phonon thermal conductivity model under the single-mode relaxation time approximation, Phys. Rev. B 69 (9) (2004) 094303, https://doi.org/10.1103/PhysRevB.69.094303.

[17] J.E. Turney, E.S. Landry, A.J.H. McGaughey, C.H. Amon, Predicting phonon properties and thermal conductivity from anharmonic lattice dynamics calculations and molecular dynamics simulations, Phys. Rev. B 79 (6) (2009) 064301, https://doi.org/10.1103/PhysRevB.79.064301.

[18] A.J.H. McGaughey, M. Kaviany, Phonon transport in molecular dynamics simulations: formulation and thermal conductivity prediction, in: G.A. Greene, J.P. Hartnett, A. Bar-Cohen, Y.I. Cho (Eds.), Advances in Heat Transfer, 39, Elsevier, 2006, pp. 169–255, https://doi.org/10.1016/S0065-2717(06)39002-8. vol.

[19] M.S. Green, Markoff random processes and the statistical mechanics of time-dependent phenomena. II. Irreversible processes in fluids, J. Chem. Phys. 22 (3) (1954) 398–413, https://doi.org/10.1063/1.1740082.

[20] R. Kubo, Statistical-mechanical theory of irreversible processes. I. General theory and simple applications to magnetic and conduction problems, J. Phys. Soc. Jpn 12 (6) (1957) 570–586, https://doi.org/10.1143/JPSJ.12.570.

[21] D.A. McQuarrie, Statistical Mechanics, Harper's Chemistry Series, HarperCollins Publishing, Inc., New York, 1976.

[22] R.E. Jones, K.K. Mandadapu, Adaptive Green-Kubo estimates of transport coefficients from molecular dynamics based on robust error analysis, J. Chem. Phys. 136 (15) (2012) 154102, https://doi.org/10.1063/1.3700344.

[23] J.H. Irving, J.G. Kirkwood, The statistical mechanical theory of transport processes. IV. The equations of hydrodynamics, J. Chem. Phys. 18 (6) (1950) 817–829, https://doi.org/10.1063/1.1747782.

[24] K.K. Mandadapu, R.E. Jones, P. Papadopoulos, Generalization of the homogeneous nonequilibrium molecular dynamics method for calculating thermal conductivity to multibody potentials, Phys. Rev. E 80 (4) (2009) 047702, https://doi.org/10.1103/PhysRevE.80.047702.

[25] V.K. Tewary, B. Yang, Singular behavior of the Debye-Waller factor of graphene, Phys. Rev. B 79 (12) (2009) 125416, https://doi.org/10.1103/PhysRevB.79.125416.

[26] J. Chen, G. Zhang, B. Li, How to improve the accuracy of equilibrium molecular dynamics for computation of thermal conductivity? Phys. Lett. A 374 (23) (2010) 2392–2396, https://doi.org/10.1016/j.physleta.2010.03.067.

[27] L. de Sousa Oliveira, P.A. Greaney, Method to manage integration error in the Green-Kubo method, Phys. Rev. E 95 (2) (2017) 023308, https://doi.org/10.1103/PhysRevE.95.023308.

[28] L. Ercole, A. Marcolongo, S. Baroni, Accurate thermal conductivities from optimally short molecular dynamics simulations, Sci. Rep. 7 (1) (2017) 15835, https://doi.org/10.1038/s41598-017-15843-2.

[29] M.E. Khaled, L. Zhang, W. Liu, Some critical issues in the characterization of nanoscale thermal conductivity by molecular dynamics analysis, Model. Simul. Mater. Sci. Eng. 26 (5) (2018) 055002, https://doi.org/10.1088/1361-651x/aabd3d.

Static Green's function for elliptic equations formulated using a partial Fourier representation and applied to computing the thermostatic/electrostatic response of nanocomposite materials[☆]

9

Vinod K. Tewary and E.J. Garboczi
Applied Chemicals and Materials Division, National Institute of Standards and Technology
(NIST), Boulder, CO, United States

9.1   Introduction

Modeling the thermal and electrostatic response of solids, and in particular semiconductors, is a subject of strong contemporary interest because of widespread diverse applications in modern solid-state devices. The applications [1,2] include thermoelectric energy conversion, heat management, electronic, and sensing. The problem of heat management is especially acute in nanomaterials. As the power generated in the devices increases, while their dimensions shrink, the power density in the devices becomes extremely high. This, if uncontrolled, would cause a huge rise in the temperature of the device. This is noticeable even in ordinary laptop computers. High temperatures reduce the semiconducting efficiency of the devices, cause material damage, and can be hazardous.

In this chapter, we describe a computationally efficient technique for calculating the continuum Green's functions (GFs) for thermal and electrostatic response of solids by solving the Laplace/Poisson equations. These equations are needed to calculate the effective or average values of material parameters such as the thermal or electrical conductivity of nanocomposite materials. These parameters are useful for characterizing the materials and simulating their performance in operating conditions [3–6]. The technique is based upon the use of partial Fourier transforms (PFTs). It is computationally efficient and is applicable to anisotropic materials and those that may contain more than one component or phases of materials. Both 3D (three-dimensional) and emerging 2D systems (graphene and beyond) are of interest.

[☆] Contribution of the National Institute of Standards and Technology, an agency of the US Federal Govt. Not subject to copyright in the USA.

Modeling, Characterization and Production of Nanomaterials. https://doi.org/10.1016/B978-0-12-819905-3.00009-9
2023 Published by Elsevier Ltd.

This chapter is primarily pedagogical. Over the last several years, we have been developing GF-based techniques for modeling and simulation of different kinds of materials. The main application of these techniques is in developing measurement and characterization techniques for new materials of industrial interest, ranging from the simplest cubic solids to modern quantum material systems. Here we describe the techniques for calculating GFs for nanomaterials and composites and quote our own numerical results for illustrating the techniques. We describe one specific technique, that is, PFT for calculating the GFs in enough detail so that the readers will be well prepared to self-learn that technique in depth. They should then be able to incorporate and adapt the technique in their own work. No attempt is made here to review the literature on solving Laplace/Poisson equation or modeling of thermal/electrostatic response of composites. A reader interested in that area can start with the references given in [2,3] and other references quoted there.

It is sometimes stated that the GF gives only the particular solution of an operator equation. We point out that the GF also provides very effective mathematical techniques for constructing homogeneous solutions, that is, solutions of the homogeneous part of the operator equation. Homogeneous solutions are essential for solving boundary value problems, which is the primary application of the GF. This also enables us to construct GFs in modules, each module corresponding to one homogeneous solution. This gives the computational advantage that numerical results obtained in one module can be reused in different problems. Ability to write homogenous, as well as particular solutions, in the same representation is what makes the GF method a powerful technique for solving a diverse variety of problems in physical and engineering sciences.

Here we will describe the general formalism of the GF method for a 3D anisotropic material and then give its adaptation to modern 2D materials. This will show the interesting mathematical transition from 3D to 2D systems (see, e.g., [7]). The GFs for the 3D and 2D systems are qualitatively different [8–10], but the mathematical transformation from 3D to 2D is intuitive and straightforward. We will illustrate our GF technique by applying it to 2D isotropic graphene and 2D anisotropic phosphorene without and with metallic inclusions. Metallic inclusions are potentially useful for tuning the conductivity of nanomaterials. In the isotropic continuum limit, phosphorene becomes identical to graphene. It is a material of strong topical interest, because of its unusual properties (see, e.g., [11–14]) and its strong potential for revolutionary new applications in semiconducting and heat management devices.

Note that the GFs, as described in this chapter, are for the continuum model of solids. In contrast, the GFs described in Chapter 5 are for discrete lattices [10,15]. In a discrete lattice, the phonons and other quantum effects are included explicitly [16,17]. The continuum GFs are applicable to solids at length scales in which the lattice structure can be smeared into a continuum. In the continuum model, all the field variables are continuous variables and the response of the solid is simulated in terms of differential or integral equations [8,9]. This is in contrast to discrete lattice models, in which the field variables are discrete, and the response is simulated using difference equations. It can be shown mathematically that the continuum GF is the asymptotic limit of the discrete GF [10]. A more detailed theory of thermal conductivity needs to include atomistic scale lattice dynamical effects and corresponding time scales [18–20], which are smeared out in the continuum model.

Section 9.2 gives the preliminary equations and defines the notation. The GF is defined in Section 9.3, which gives the virtual source and the delta function methods for obtaining the homogeneous solutions of the Laplace and the Poisson equations. The GF for material systems is described in Section 9.4, which also introduces the PFT for representing and calculating GFs in 3D and for free infinite space. The calculation of the GF for finite spaces and for a two-component composite is described in Section 9.5. In Section 9.6, we discuss the GF for 2D composites, its application to calculate the effective conductivity of graphene and phosphorene containing square-shaped metallic inclusions, and quote our own numerical results published earlier [21]. A summary and the conclusions are presented in Section 9.7. Finally, Section 9.8 suggests some possible topics of interest for future work.

9.2 Preliminary equations and definitions

We assume the Cartesian frame of reference in which Z and X axes are in the plane of the paper and the Y axis is normal to the paper. The coordinates along these axes will be numerically labeled as 3, 1, and 2, respectively. These coordinates will be denoted by the lower case Roman letters i, j, and k. We will follow the usual notation of denoting vectors, matrices, and operators in bold font. The same symbol in non-bold font will denote the corresponding scalar. For example, $\mathbf{R}$ is a vector whose magnitude is R.

We will denote the position vectors by space variables $\mathbf{R}, \mathbf{R}'$. The X, Y, and Z components of $\mathbf{R}$ will be given by x_1, x_2, and x_3. Thus

$$\mathbf{R} = (x_1, x_2, x_3), \tag{9.1}$$

$$R = \left(x_1^2 + x_2^2 + x_3^2\right)^{1/2}, \tag{9.2}$$

or

$$R^2 = x_i \, x_i. \tag{9.3}$$

In Eq. (9.3), and henceforth in this chapter, we have followed Einstein's convention of summation over repeated indices, unless otherwise mentioned.

Our main interest in this chapter is elliptic equations for 3D and 2D spaces. Strictly speaking, elliptic equations are defined in the 2D space but this term is also used for the corresponding equations in the 3D case. We treat each equation as an operator equation in terms of an appropriately defined differential operator L and find a GF for that operator. For heat diffusion or electrostatic potential problems requiring the solution of the Laplace/Poisson equation, L is the generalized Laplacian as given below

$$L = \nabla_i \, K_{ij} \, \nabla_j \tag{9.4}$$

where $\mathbf{K}$ is the (thermal or electrical) conductivity tensor of the second rank and ∇ is the gradient operator. Eq. (9.4) is the general form of the operator valid for anisotropic cases, which we will use in this chapter.

For isotropic cases, the tensor $\mathbf{K}$ is diagonal with all equal elements, or a scalar K^0, such that

$$K_{ij} = K^0 \, \delta_{ij} \text{ (isotropic)} \tag{9.5}$$

where $\boldsymbol{\delta}$ is the Dirac delta tensor, which is 1 for $i = j$ and zero otherwise. For such cases, Eq. (9.4) reduces to the usual Laplacian:

$$L = \frac{\partial^2}{\partial_i \partial_i} = \nabla^2, \text{(isotropic)} \tag{9.6}$$

where K^0 has been absorbed in the units.

We remark in passing that L in Eq. (9.4) is a special case of the second rank Christoffel tensor operator K^e for elastic equilibrium, which is defined in the Cartesian system as given below:

$$K_{ij}^e = c_{ikjl} \frac{\partial^2}{\partial_k \partial_l} \tag{9.7}$$

where $\mathbf{c}$ is the fourth rank elastic constant tensor. It is apparent that the Christoffel operator in Eq. (9.7) is essentially a more general tensorial form of the Laplacian in Eq. (9.4) or (9.6). Hence, the method developed for finding the GF for Eq. (9.4) can be generalized to find the GF for Eq. (9.7). In this chapter, however, our interest is limited to the thermostatic and electrostatic problems.

The treatment given in this chapter is fully applicable to both electrical as well as thermal problems. However, for the sake of definiteness, we will refer mainly to the electrostatic problems. For example, a reference to conductivity will be for electrical conductivity. For other operator equations of physical interest, and GFs for the operators, see, for example, Refs. [8, 22, 23].

Consider the operator equation

$$L_R \, G \, (\mathbf{R}, \mathbf{R}') = -\delta \, (\mathbf{R} - \mathbf{R}') \tag{9.8}$$

where L is an elliptic second-degree differential operator given by Eq. (9.4) in the general case, $\mathbf{R}$ and $\mathbf{R}'$ are space variables, $G(\mathbf{R}, \mathbf{R}')$ is the GF to be determined, and $\delta \, (\mathbf{R}, \mathbf{R}')$ is the Dirac delta function, which is zero for $\mathbf{R} \neq \mathbf{R}'$ with the following integral property:

$$\int \delta(\mathbf{R} - \mathbf{R}')F(\mathbf{R}')d\mathbf{R}' = F(\mathbf{R}) \tag{9.9}$$

In Eq. (9.9), the integration is over all **R**-space. The subscript **R** in Eq. (9.8) indicates that L operates only on the variable **R**. The physical meaning of Eq. (9.8) is that G(**R**, **R**') gives the response of the material at the point **R** to a stimulus applied at the point **R**'. We will refer to **R** and **R**' as the field and the source points, respectively. Henceforth, for brevity, we will not write the suffix **R** on L, which will be apparent by the context.

Once the GF solution for Eq. (9.8) is obtained, it can be used to solve any linear elliptic equation involving the same operator L by using Eq. (9.9). For example, consider the following general equation involving the operator L:

$$L\,\Phi(\mathbf{R}) = -\rho(\mathbf{R}) \tag{9.10}$$

where ρ is the inhomogeneity and Φ is the unknown function that is a solution of Eq. (9.10).

The homogeneous solution or the solution of the homogeneous part of Eq. (9.10) is needed for satisfying the boundary conditions. As we shall see later, the GF can also be used to obtain the solution of the homogeneous equation, which is given below:

$$L\,\Phi(\mathbf{R}) = 0. \tag{9.11}$$

From Eqs. (9.8), (9.9), the particular solution of Eq. (9.10) can be written as

$$\Phi(\mathbf{R}) = \int G(\mathbf{R}, \mathbf{R}')\,\rho(\mathbf{R}')d\mathbf{R}', \tag{9.12}$$

where the integration is over the entire **R**-space. The **R**-space, or the so-called solution space, consists of the entire region spanned by the vector **R** in which the solution can be constructed. It can be infinite or finite for bounded solids. It can be verified by direct substitution that Eq. (9.12) is indeed the particular solution of Eq. (9.10). Substituting for Φ from Eq. (9.12) in Eq. (9.10) and using Eq. (9.8), we obtain

$$L\,\Phi(\mathbf{R}) = \int L\,G(\mathbf{R}, \mathbf{R}')\rho(\mathbf{R}')d\mathbf{R}',$$

$$= \int \delta(\mathbf{R} - \mathbf{R}')\rho(\mathbf{R}')d\mathbf{R}' = \rho(\mathbf{R}) \tag{9.13}$$

where we have used Eq. (9.9) in the last step.

Note that the GF used in Eq. (9.12) does not depend upon the inhomogeneity ρ— the RHS of Eq. (9.10). The same function G(**R**, **R**') can, therefore, be used to obtain solutions of Eq. (9.10) for different ρ. This results in a considerable economy of the computational effort. Thus, the main computational effort consists of solving Eq. (9.8) for the GF.

The formal solution of Eq. (9.8) can be written for G in the operator form as follows:

$$G = L^{-1} \qquad (9.14)$$

or, equivalently,

$$G\,L = L\,G = I, \qquad (9.15)$$

where I is the identity operator.

A representation is needed for obtaining the inverse of an operator. In general, the orthogonal set of eigenfunctions of an operator provides a useful representation for obtaining the inverse of that operator. Let $\Psi(\mathbf{Q}, \mathbf{R})$ denote the set of orthonormal eigenfunctions of L with the eigenvalue $E(\mathbf{Q})$ such that

$$L\,\Psi(\mathbf{Q}, \mathbf{R}) = E(\mathbf{Q})\Psi(\mathbf{Q}, \mathbf{R}) \qquad (9.16)$$

where $\mathbf{Q}$ is a vector conjugate to $\mathbf{R}$. It is, therefore, of the same dimensionality as $\mathbf{R}$. Here $\mathbf{Q}$ is used only for labeling the eigenfunctions and the eigenvalues of L. In Section 9.3, we will identify $\mathbf{Q}$ as a Fourier vector in the reciprocal space. The reciprocal space is also referred to as the Fourier space.

The eigenfunctions are assumed to be orthonormal in the vector space of the $\mathbf{R}$ as well as its conjugate space. Thus

$$\int \Psi(\boldsymbol{Q}, \boldsymbol{R})\, \overline{\Psi}(\boldsymbol{Q}', \boldsymbol{R})\, d\boldsymbol{R} = \delta(\boldsymbol{Q} - \boldsymbol{Q}') \qquad (9.17)$$

and

$$\int \Psi(\boldsymbol{Q}, \boldsymbol{R})\overline{\Psi}\,(\boldsymbol{Q}, \boldsymbol{R}')\, d\boldsymbol{Q} = \delta(\boldsymbol{R} - \boldsymbol{R}') \qquad (9.18)$$

where the overhead bar denotes complex conjugate, and the integration is over the entire $\mathbf{Q}$-space.

Now, we write a solution of Eq. (9.8) as an expansion in the eigenfunctions as follows:

$$G(\boldsymbol{R}, \boldsymbol{R}') = \int \Gamma(\boldsymbol{Q}, \boldsymbol{R}')\Psi(\boldsymbol{Q}, \boldsymbol{R})d\boldsymbol{Q} \qquad (9.19)$$

where $\Gamma(\boldsymbol{Q}, \boldsymbol{R}')$ is the coefficient in the expansion of the GF and is yet to be determined. Substituting this expression in Eq. (9.8), and using the orthonormality relations, we obtain the following well-known expression for the GF:

$$G(\boldsymbol{R}, \boldsymbol{R}') = \int \overline{\Psi}(\boldsymbol{Q}, \boldsymbol{R})\, \Psi(\boldsymbol{Q}, \boldsymbol{R}')\, E(\boldsymbol{Q})^{-1}\, d\boldsymbol{Q} \qquad (9.20)$$

Eq. (9.20) provides a computationally convenient representation of the continuum GF and has been used in an earlier paper [24]. We will use it here to develop a technique for calculating the GF for bounded composite materials.

9.3 Green's function for boundary value problems

Eq. (9.12), or Eq. (9.14) in the operator form, gives the particular solution of the original operator equation. In order to construct a solution that satisfies the prescribed boundary conditions, we need to add homogeneous solutions to the particular solution. The GF also provides a convenient technique for constructing a homogeneous solution. It can be done in the following two ways.

9.3.1 Virtual source method

In this method, we apply virtual sources just outside the boundary of the solution domain [21,24–27]. If the virtual sources are denoted by the unknown function f($\mathbf{R}$), their distribution is such that

$$f(\mathbf{R}) = 0 \text{ for } \mathbf{R} < \mathbf{R}_b, \tag{9.21}$$

and

$$f(\mathbf{R}) = f(\mathbf{R}_b) \text{ for } \mathbf{R} = \mathbf{R}_b + \varepsilon \tag{9.22}$$

where $\mathbf{R}_b$ scans the boundary of the solution domain and $\varepsilon = 0^+$ in the limit. The positive value of ε denotes that $\mathbf{R}$ approaches the boundary from outside and there are no virtual sources inside the solution domain. The function f is assumed to be analytic with a unique limit as $\mathbf{R} \to \mathbf{R}_b$ from outside the boundary but is otherwise arbitrary.

Using Eq. (9.12), the solution $\Phi_0(\mathbf{R})$ of Eq. (9.10) corresponding to the source term f($\mathbf{R}$) of Eq. (9.22) can be written as

$$\Phi_0(\mathbf{R}) = \oint G(\mathbf{R}, \mathbf{R}_{b+}) f(\mathbf{R}_{b+}) d\mathbf{R}_{b+} \tag{9.23}$$

where the plus subscript on $\mathbf{R}_b$ denotes that integration is over the boundary of the solution domain in the limit when $\mathbf{R}$ approaches $\mathbf{R}_b$ from outside. It can be verified that Eq. (9.23) indeed gives the homogeneous solution by using Eq. (9.13). This equation shows that for $\mathbf{R}$ in the solution domain:

$$L \; \Phi_0(\mathbf{R}) = f(\mathbf{R}) = 0 \text{ for } \mathbf{R} \leq \mathbf{R}_b, \tag{9.24}$$

where we have used Eq. (9.21).

The total solution of Eq. (9.10) is written as

$$\Phi(\mathbf{R}) = \Phi_P(\mathbf{R}) + \Phi_0(\mathbf{R}), \qquad (9.25)$$

where $\Phi_P(R)$ is given by Eq. (9.12) or, equivalently, by Eq. (9.14).

By applying the prescribed boundary values at $\mathbf{R} = \mathbf{R}_b$, we obtain the following integral equation for the unknown virtual force function $f(\mathbf{R}_b)$:

$$\oint G(\mathbf{R}_b, \mathbf{R}_{b+}) f(\mathbf{R}_{b+}) d\mathbf{R}_{b+} = \Phi_b(\mathbf{R}_b) - \Phi_P(\mathbf{R}_b), \qquad (9.26)$$

where $\Phi_b(\mathbf{R}_b)$ are the prescribed values at the boundary of the solution domain.

Eq. (9.26) is the Fredholm integral equation of the first kind that needs to be solved for $f(\mathbf{R}_b)$. This equation has been derived for Dirichlet boundary conditions, but a similar equation can be derived for Neumann or mixed boundary conditions. In general, Eq. (9.26) has to be solved numerically by discretizing the boundary. It would require inversion of an N × N matrix, where N is the number of discrete points on the boundary over which the boundary values are specified, and the virtual source function is calculated. The computational economy occurs because in the GF method, only the boundary needs to be discretized and not the entire solid, as in the conventional methods like the FEM (finite element method). So, the order of the matrix to be inverted is much less in the GF method. This computational advantage can be significant because the matrix inversion requires $O(N^3)$ computational steps. The advantage is partly offset in favor of the FEM, because the matrix to be inverted in the FEM is a sparse matrix, which requires less effort than a dense matrix in the GF method.

9.3.2 Delta-function representations

Eq. (9.20) can be used to derive a delta function representation for the GF that gives the particular solution as well as a homogeneous solution for constructing the GF for boundary value problems. It is apparent that the RHS of Eq. (9.20) is singular in Q-space at $\mathbf{Q} = \mathbf{Q}_s$ where

$$E(\mathbf{Q}_s) = 0. \qquad (9.27)$$

In general $E(\mathbf{Q}_s)$ is real for stable solids. To account for the singularity on the real axis, we write Eq. (9.20) in the complex space and introduce a small imaginary part in the denominator as given below

$$G(\mathbf{R}, \mathbf{R}') = \lim_{\epsilon \to 0} \int \Psi(\mathbf{Q}, \mathbf{R}) \, \Psi(\mathbf{Q}, \mathbf{R}') \, [E(\mathbf{Q}) - \iota\epsilon]^{-1} \, d\mathbf{Q}, \qquad (9.28)$$

where $\iota = \sqrt{-1}$.

For the Laplace operator, $\Psi(Q, R)$ and $E(\mathbf{Q})$ are scalars even though $\mathbf{Q}$ and $\mathbf{R}$ may be 2D or 3D vectors. We mention in passing that for the Christoffel equation for elastic equilibrium, $E(Q)$ will be a 3×3 diagonal matrix. In this chapter, we consider only the Laplace operator. The case of the Christoffel operator case has been discussed in Ref. [24].

We separate the real and imaginary parts of the RHS of Eq. (9.28) and write it as follows in the limit $\epsilon \to 0^+$.

$$\text{Re } G(\mathbf{R}, \mathbf{R}') = P\left(\int \Psi(\mathbf{Q}, \mathbf{R}) \, \Psi(\mathbf{Q}, \mathbf{R}') \, [E(\mathbf{Q})]^{-1} \, d\mathbf{Q} \right), \qquad (9.29)$$

and

$$\text{Im } G(\mathbf{R}, \mathbf{R}') = \int \Psi(\mathbf{Q}, \mathbf{R}) \, \Psi(\mathbf{Q}, \mathbf{R}') \, \delta[E(\mathbf{Q})] \, d\mathbf{Q}, \qquad (9.30)$$

where P denotes the principal value and Re and Im denote the real and the imaginary parts, respectively. In writing Eqs. (9.29), (9.30), we have used the following representation of the delta function of any real variable ξ,

$$\lim_{\epsilon \to 0} \frac{1}{\xi - \iota\epsilon} = P\left(\frac{1}{\xi} \right) + \iota\pi\delta(\xi) \qquad (9.31)$$

We now show that Im $G(\mathbf{R}, \mathbf{R}')$ is the solution of the homogeneous equation. Operating by L on both sides of Eq. (9.30), we obtain

$$L[\text{Im } G(\mathbf{R}, \mathbf{R}')] = \int L[\, \Psi(\mathbf{Q}, \mathbf{R})\,]\Psi(\mathbf{Q}, \mathbf{R}') \, \delta[E(\mathbf{Q})] \, d\mathbf{Q}$$

$$= \int E(\mathbf{Q}) \, \Psi(\mathbf{Q}, \mathbf{R}) \, \Psi(\mathbf{Q}, \mathbf{R}') \, \delta[E(\mathbf{Q})] \, d\mathbf{Q} = 0 \qquad (9.32)$$

where we have used Eq. (9.16) and the following property of the delta function of a real variable ξ:

$$\xi \, \delta(\xi) = 0, \qquad (9.33)$$

for all real values of ξ.

Eq. (9.32) is valid separately for all $\mathbf{Q}$. Thus, we can write the following as a solution of the homogeneous part of Eq. (9.12) or (9.10):

$$\Phi_0(\mathbf{R}) = \int f(\mathbf{Q})\Psi(\mathbf{Q}, \mathbf{R}) \, \Psi(\mathbf{Q}, \mathbf{R}')\delta[E(\mathbf{Q})] \, d\mathbf{Q} \qquad (9.34)$$

for any arbitrary function f. Note that Eq. (9.32) is valid for all values of $\mathbf{Q}$ separately. Hence, as in the previous case, Eq. (9.25) along with Eq. (9.34) gives an integral equation for the determination of the unknown functions f($\mathbf{Q}$) that would satisfy the prescribed boundary conditions. One interesting aspect of this technique is that, unlike the case of virtual sources, the location of the unknown function in the real space does not need to be specified.

We will now use the techniques described above to derive equations for the GF of a material system.

9.4   Green's function for a material system

Our objective in this section is to solve the Laplace/Poisson equation in a material system and derive expressions for the GF, subject to appropriate boundary conditions. Later, in Section 9.6, we will use the GF to calculate the effective value of the electrical or thermal conductivity of the material. It is a physical problem of measurement interest for the purpose of material characterization.

Fig. 9.1 shows a two-component composite material, along with the Cartesian frame of reference and the origin O. This figure represents both 2D and 3D solids that we consider in this chapter. In 2D, it represents the solid in the base plane. In case of a 3D, the figure shows the projection of the solid on the base plane. We have assumed the solid to be a cuboid, that is, that the solid is rectangular with its edges perpendicular to the base plane.

The material system consists of the host labeled in the figure as H and an inclusion labeled as I. The X, Y, and Z coordinates will be denoted by x_1, x_2, and z (or x_3), respectively. A 2D vector on the X–Y plane, which is perpendicular to the plane of the paper, will be denoted by **x**. Thus, a general 3D position vector **R** in the solid that has the coordinates (x_1, x_2, z) can be written as follows:

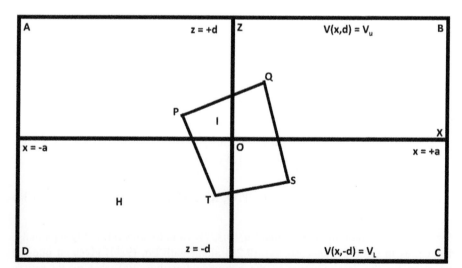

Fig. 9.1 A cuboid solid H containing an inclusion I.

$$\mathbf{R}\ (\mathbf{x},\ \mathbf{z})\ (\mathbf{x}1,\ \mathbf{x}2,\ \mathbf{z}) \tag{9.35}$$

The magnitude of the vector $\mathbf{R}$ is related to the magnitude of $\mathbf{x}$ as follows:

$$R^2 = x^2 + z^2 \tag{9.36}$$

and

$$x^2 = x_1^2 + x_2^2 \tag{9.37}$$

Obviously, the vector $\mathbf{x}$ is the component of $\mathbf{R}$ on the X–Y plane.

In Fig. 9.1, the host solid H is bounded by ABCD and the inclusion I has the boundary PQST. The host in the figure is shown to be a rectangle, whereas the inclusion can be of arbitrary shape. In the Z-direction, the solid extends from $Z = -d$ to $Z = +d$. Consider a thought experiment that an electrostatic potential V_U is applied on the upper plane $z = d$ and V_L at the lower plane $z = -d$. A measurement of the current across the solid in the Z-direction will then give the effective electrical conductivity of the composite solid. It would require solution of the Laplace/Poisson equation with appropriate boundary conditions. The corresponding thermal problem is similar in which the faces at $Z = d$ and $Z = -d$ are maintained at temperatures T_U and T_L, respectively. Measurement of the heat flow then gives the effective thermal conductivity. As mentioned earlier, the technique used for solving the Laplace/Poisson equation can be generalized for solving the Christoffel equation for elastostatic problems. The GF technique described in this section is a 3D generalization of the virtual force method described in Ref. [27].

9.4.1 Basic equations for a composite material

The electric potential in a medium is the solution of the Poisson equation:

$$\nabla.[K\nabla\Phi(\mathbf{R})] = -\rho(\mathbf{R}) \tag{9.38}$$

where $\Phi\ (\mathbf{R})$ is the potential at $\mathbf{R}$, $\mathbf{K}$ is the conductivity tensor of the material, and ρ is the charge density at $\mathbf{R}$. In a homogeneous isotropic material, $\mathbf{K}$ is a constant scalar. We consider the solid separately in the two regions shown in Fig. 9.1: host and inclusion labeled, respectively, as H and I. In such cases, Eq. (9.38) reduces to the familiar Poisson's equation:

i. Host region; $\mathbf{K} = \mathbf{K^H}$

$$\mathbf{L^H}\Phi_H\ (\mathbf{R}) \equiv \nabla.\left[\mathbf{K^H}\nabla\Phi_H\ (\mathbf{R})\right] = -\rho_H(\mathbf{R}) \tag{9.39}$$

where the dot between two vectors denotes their scalar (or dot) product.

ii. Inclusion, $\mathbf{K} = \mathbf{K}^I$

$$L^I \Phi_I\,(\mathbf{R}) \equiv \nabla . \left[\mathbf{K}^I \nabla \Phi_I\,(\mathbf{R}) \right] = -\rho_I(\mathbf{R}) \qquad (9.40)$$

Using Eq. (9.4), we can write Eq. (9.39) for the host in the following form:

$$K_{ij}^H\,\nabla_i \nabla_j\,\Phi_H\,(\mathbf{R}) = -\rho_H(\mathbf{R}) \qquad (9.41)$$

In many materials, the off-diagonal elements of the conductivity tensor can be assumed to be zero. In addition, for materials with cubic symmetry, the three diagonal elements are equal and the material is isotropic. However, we consider the slightly more general case when the material is only laterally isotropic. We also assume that the conductivity tensor is constant in the host and in the inclusion with two different tensors. That means that the conductivity tensor can be taken outside the gradient operators. We assume the plane of isotropy to be the X–Y plane. This applies to hexagonal materials with their c-axis along the Z-direction and also to many modern 2D solids like Xenes [21,28].

For such cases, $K_{22} = K_{11}$ and the off-diagonal elements are assumed to be zero. We then obtain the following simplified form of Eq. (9.41):

$$\left[K^H_{11} \left(\nabla_1{}^2 + \nabla_2{}^2 \right) + K^H_{33} \nabla_3{}^2 \right] \Phi_H(\mathbf{R}) = -\rho_H(\mathbf{R}) \qquad (9.42)$$

and the corresponding form for Eq. (9.40) with H replaced by I:

$$\left[K^I_{11} \left(\nabla_1{}^2 + \nabla_2{}^2 \right) + K^I_{33} \nabla_3{}^2 \right] \Phi_I(\mathbf{R}) = -\rho_I(\mathbf{R}) \qquad (9.43)$$

We prescribe the following boundary and continuity conditions:

A. Outer boundaries:
  i. Applied potential V_U at $z = d$ and V_L at $z = -d$

$$\Phi_H(\mathbf{x},\, z{=}d\,) = V_U \qquad (9.44)$$

$$\Phi_H(\mathbf{x},\, z{=}-d\,) = V_L \qquad (9.45)$$

For such boundary conditions, it is convenient to include V_U and V_L in the inhomogeneity on the right of Eq. (9.38) and specify homogeneous boundary conditions at both boundaries for the unknown function Φ. We will use this technique in writing the final solution in Section 9.5.

  ii. Periodic boundary conditions on the XY plane:

$$\Phi_H(x_1 + 2a_1, x_2, z) = \Phi_H\,(x_1, x_2, z) \qquad (9.46)$$

$$\Phi_H\,(x_1, x_2 + 2a_2, z) = \Phi_H\,(x_1, x_2, z) \qquad (9.47)$$

It is apparent that the volume of the whole cuboid composite (host plus inclusion) corresponding to the above conditions is $4a_1a_2d$. Although our treatment is valid for $a_1 \neq a_2$, in actual calculations we will further assume that

$$a_1 = a_2 = a \tag{9.48}$$

This assumption does not make any significant change in our results because the results for $a_1 \neq a_2$ are qualitatively similar. In any case, it is straightforward to generalize this treatment to the case when $a_1 \neq a_2$.

B. Interfacial continuity conditions between H and I
   i. Continuity of the function

$$\Phi_H(\mathbf{R}_b) = \Phi_I(\mathbf{R}_b) \tag{9.49}$$

   ii. Continuity of the flux

$$\left[\mathbf{K}^H \nabla\Phi_H(\mathbf{R}_b)\right]_n = \left[\mathbf{K}^I \nabla\Phi_I(\mathbf{R}_b)\right]_n \tag{9.50}$$

where the suffix n denotes the normal component of the corresponding vector.

We will also consider the following two limiting cases. These cases are of interest because an important application of nanocomposites is heat management. This requires functionalization of materials by introducing metallic or insulator inclusions.

I. Metallic inclusion (infinite conductivity): The potential must be constant at all points in a metal, assuming its conductivity to be infinite. Since the quantity of physical interest is the potential difference rather than the actual value of the potential, we can assume without loss of generality that the whole of the inclusion is at zero potential. It only sets the zero point of the potential. We, therefore, impose the following interfacial condition instead of Eq. (9.49):

$$\Phi_H(\mathbf{R}_b) = \Phi_I(\mathbf{R}_b) = 0 \tag{9.51}$$

II. Insulating inclusion (infinite resistance): In this case, no current can flow into or out of the inclusion. Hence, instead of Eq. (9.50), we impose the following interfacial condition (normal current from either side of host-inclusion boundary is zero):

$$\left[\mathbf{K}^H \nabla\Phi_H(\mathbf{R}_b)\right]_n = \left[\mathbf{K}^I \nabla\Phi_I(\mathbf{R}_b)\right]_n = 0 \tag{9.52}$$

We solve the coupled Eqs. (9.39), (9.40), by first solving them separately in their respective domains using GFs in the Fourier space, and then couple them using the interfacial boundary conditions given by Eqs. (9.49), (9.50) or one of the two Eqs. (9.51), (9.52). Note that only one interfacial condition is needed in the two limiting cases.

9.4.2 Partial Fourier transform of a general function

In this subsection, we shall derive the PFT of a general function and then apply it to obtain the PFT of the GF. The PFT representation is especially useful for solving coupled equations. We can write the Fourier transform of any integrable function $P(\mathbf{R})$ of the continuous vector variable $\mathbf{R}$ as follows:

$$P(\mathbf{R}) = \left(\frac{1}{2\pi}\right)^3 \int_{-\infty}^{+\infty} \widehat{P}(\mathbf{Q}) \exp(\imath \mathbf{Q}.\mathbf{R}) \, d\mathbf{Q} \qquad (9.53)$$

where $\widehat{P}(\mathbf{Q})$ is the Fourier transform of $P(\mathbf{R})$ over $\mathbf{R}$, and $\mathbf{Q}$ is a reciprocal vector, conjugate to $\mathbf{R}$. For a 3D geometry, both $\mathbf{R}$ and $\mathbf{Q}$ are 3D vectors. We will refer to the vector space of the position vectors $\mathbf{R}$ as the direct space and that of $\mathbf{Q}$ as the reciprocal or the Fourier space.

For free infinite space,

$$-\infty \leq Q_i, R_i \leq \infty \qquad (9.54)$$

Using the orthonormality of the exponential functions, $\widehat{P}(\mathbf{Q})$ is given by the following inverse relation:

$$\widehat{P}(\mathbf{Q}) = \int_{-\infty}^{+\infty} P(\mathbf{R}) \exp(-\imath \mathbf{Q}.\mathbf{R}) \, d\mathbf{R} \qquad (9.55)$$

Eqs. (9.53), (9.55) give the standard definition of the Fourier transform of a function of a single 3D variable. For the boundary value problem, shown in Fig. 9.1, we use the technique of PFTs, described below.

In PFT, we use the fact that a 3D Fourier function of Cartesian coordinates is fully separable in three separate 1D Fourier functions that are each orthonormal in their own space. Hence, we can write the full Fourier transform of any function of a 3D variable in terms of PFTs over the separate variables. In the present case, we define a hybrid representation of a function. It is a hybrid of 2D Fourier space and 1D direct space. The Fourier space part is the PFT over the 2D X–Y plane and in the direct space it will be a function of the z coordinate.

Corresponding to Eq. (9.35) for the 3D space variable $\mathbf{R}$, we express the 3D reciprocal space vector $\mathbf{Q}$ as a 2D vector $\mathbf{q}$ and k, as follows:

$$\mathbf{Q} = (\mathbf{q}, k) \qquad (9.56)$$

where k denotes the Z component of $\mathbf{Q}$, and the 2D vector q is conjugate to $\mathbf{x}$ such that

$$\mathbf{q}.\mathbf{x} = q_1 x_1 + q_2 x_2 \qquad (9.57)$$

$$q^2 = q_1{}^2 + q_2{}^2 \qquad (9.58)$$

and

$$Q^2 = q^2 + k^2 \tag{9.59}$$

We immediately note one important advantage of writing a function in terms of its Fourier transform. From Eq. (9.53), we see that

$$\nabla_i P(\mathbf{R}) = \iota \left(\frac{1}{2\pi}\right)^3 \int_{-\infty}^{+\infty} Q_i \, \widehat{P}(\mathbf{Q}) \exp\left(\iota \mathbf{Q}.\mathbf{R}\right) d\mathbf{Q} \tag{9.60}$$

and, by repeated application,

$$\nabla_i \nabla_j P(\mathbf{R}) = -\left(\frac{1}{2\pi}\right)^3 \int_{-\infty}^{+\infty} Q_i \, Q_j \, \widehat{P}(\mathbf{Q}) \exp\left(\iota \mathbf{Q}.\mathbf{R}\right) d\mathbf{Q} \tag{9.61}$$

From Eqs. (9.60), (9.61),

$$\text{if } \mathcal{F}\, P(\mathbf{R}) = \widehat{P}(\mathbf{Q}) \tag{9.62}$$

$$\text{then } \mathcal{F}\nabla_i\nabla_j\, P(\mathbf{R}) = -Q_i \, Q_j \, \widehat{P}(\mathbf{Q}) \tag{9.63}$$

where $\mathcal{F}$ denotes the Fourier transform from R-space to the reciprocal space.

We see from Eq. (9.60) that the Fourier representation allows us to represent a differential operator as a multiplicative operator in the reciprocal space. This makes it convenient to solve operator equations by converting them into algebraic equations. Note that $\mathbf{Q}$ in Eq. (9.61) has components q and k. The integration over $\mathbf{Q}$ is 3D and is over k and two components of $\mathbf{q}$.

Using Eqs. (9.42), (9.61)

$$\mathcal{F}\left[L^H \Phi_H(\mathbf{R})\right] = -K^H{}_{33}\left[\beta_H{}^2 q^2 + k^2\right] \widehat{\Phi}_H(\mathbf{Q}), \tag{9.64}$$

where $\widehat{\Phi}_H(\mathbf{Q})$ is the Fourier transform of $\Phi(\mathbf{R})$ and β_H is the anisotropy parameter of H, defined by

$$\beta_H{}^2 = K^H{}_{11}/K^H{}_{33} \tag{9.65}$$

Similarly, for the domain I,

$$\mathcal{F}\left[L^I \Phi_I(\mathbf{R})\right] = -K^I{}_{33}\left[\beta_I{}^2 q^2 + k^2\right] \widehat{\Phi}_I(\mathbf{Q}) \tag{9.66}$$

where

$$\beta_I{}^2 = K^I{}_{11}/K^I{}_{33} \tag{9.67}$$

We write the Fourier transform equations (9.53), (9.55) as follows:

$$P(\mathbf{x}, z) = \left(\frac{1}{2\pi}\right) \int_{-\infty}^{+\infty} \widetilde{P}(\mathbf{x}; k) \exp(\imath kz) \, dk \qquad (9.68)$$

where $(\mathbf{x},z)$ denotes the vector $\mathbf{R}$, and $\widetilde{P}(\mathbf{x}; k)$, the PFT of $P(\mathbf{R})$ over $\mathbf{x}$ is given by the inverse relation:

$$\widetilde{P}(\mathbf{x}; k) = \int_{-\infty}^{+\infty} P(\mathbf{x}, z) \exp(-\imath kz) \, dz \qquad (9.69)$$

or, the equivalent relation,

$$\widetilde{P}(\mathbf{x}; k) = \left(\frac{1}{2\pi}\right)^2 \int_{-\infty}^{+\infty} \exp\imath(q_1 x_1) dq_1 \int_{-\infty}^{+\infty} \widehat{P}(\mathbf{Q}) \exp\imath(q_2 x_2) dq_2 \qquad (9.70)$$

In Eq. (9.69) and elsewhere in the definition of PFTs, we separate the list of real and reciprocal space variables like $\mathbf{x}$ and k, by a semicolon in contrast to other functions where the arguments are separated by a comma.

Similarly, we can define the PFT of $P(\mathbf{R})$ over $\mathbf{x}$ as given below:

$$\widetilde{P}(\mathbf{q}; z) = \int_{-\infty}^{+\infty} \int_{-\infty}^{+\infty} P(\mathbf{R}) \exp(-\imath \mathbf{q}.\mathbf{x}) dx_1 dx_2 \qquad (9.71)$$

or, for consistency with Eqs. (9.53), (9.68),

$$\widetilde{P}(\mathbf{q}; z) = \left(\frac{1}{2\pi}\right) \int_{-\infty}^{+\infty} \widehat{P}(\mathbf{Q}) \exp(\imath kz) dk \qquad (9.72)$$

Finally, $P(\mathbf{x},z)$ is given by the following expression in terms of its PFT over $\mathbf{x}$.

$$P(\mathbf{x}, z) = \frac{1}{(2\pi)^2} \int_{-\infty}^{+\infty} \exp\imath(q_1 x_1) dq_1 \int_{-\infty}^{+\infty} \widetilde{P}(\mathbf{q}; z) \exp\imath(q_2 x_2) dq_2 \qquad (9.73)$$

Now we impose the next set of boundary conditions—the periodic boundary conditions on the X–Y plane, given by Eqs. (9.46), (9.47). For this purpose, we introduce a semi-discrete model. In this model, the coordinate variables x_1 and x_2 are discretized but not the coordinate variable z. Thus, in our semi-discrete model, the position vector $\mathbf{R}$ in Eq. (9.35) is expressed in terms of a 2D discrete position vector $\mathbf{x}$ and a continuous variable z. The ranges of their components, as shown in Fig. 9.1, are

$$-a \leq x_{1,2} \leq a \qquad (9.74)$$

and

$$-d \leq z \leq d \qquad (9.75)$$

In order to enforce periodic boundary conditions on the X–Y plane, we follow Born's method of cyclic boundary conditions as in the Born von Karman model (see, e.g., [15]). This model on the X–Y plane is similar to a discrete Born von Karman square lattice with the following primitive 2D lattice vectors:

$$\mathbf{A_1} = a\,(1, 0) \text{ and } \mathbf{A_2} = a\,(0, 1) \tag{9.76}$$

Any linear combination of these two vectors will be a lattice site of the above square lattice. A general lattice site on the X–Y plane can then be written as

$$\mathbf{A}(n) = n_1\mathbf{A_1} + n_2\mathbf{A_2} \tag{9.77}$$

where n_1 and n_2 can be any integers—positive, negative, or zero. The X and Y coordinates of the nth lattice site can be simply written as:

$$x_1(n) = n_1 a \text{ and } x_2(n) = n_2 a \tag{9.78}$$

Eqs. (9.76)–(9.78) correspond to a square lattice that form the grid of the base of a cubic 3D solid. This equation can be generalized to hexagonal and other symmetries by defining an appropriate set of 2D lattice vectors in the Born von Karman lattice model.

Following the Born von Karman method, we also define two primitive reciprocal lattice vectors as follows:

$$\mathbf{B_1} = (2\pi/a)\,(1, 0) \text{ and } \mathbf{B_2} = (2\pi/a)\,(0, 1) \tag{9.79}$$

As in Eq. (9.77), a general reciprocal lattice vector is given by

$$\mathbf{B}(m) = m_1\mathbf{B_1} + m_2\mathbf{B_2} \tag{9.80}$$

where m_1 and m_2 can be any integers—positive, negative, or zero. As in Eq. (9.78), the X and Y coordinates of the mth reciprocal lattice site can be simply written as:

$$Q_1(m) = (2\pi/a)m_1 \text{ and } Q_2(m) = (2\pi/a)m_2 \tag{9.81}$$

The GFs as well as all other space-dependent functions are written in terms of their Fourier transforms as in Eqs. (9.70), (9.71). We see from the above equations that the periodic boundary conditions prescribed in Eqs. (9.46), (9.47) will be satisfied if we restrict $\mathbf{q}$ to the following discrete values:

$$-\frac{\pi}{a} \leq q_{1,2} \leq \frac{\pi}{a} \tag{9.82}$$

The range of the allowed values of $\mathbf{q}$ defines the Brillouin zone of the discrete structure. The total number of $\mathbf{q}$ vectors in the first zone is $\mathbf{N}$. The exponential functions are orthonormal with respect to discrete summations on the lattice sites in the direct and the conjugate Fourier space. Analogous to Eqs. (9.17), (9.18), the discrete orthogonality relations are given below:

$$\frac{1}{(2N)^2} \sum_q \exp\left(\iota \mathbf{q}.\mathbf{x}\right) = \delta\left(\mathbf{x}\right) \tag{9.83}$$

and

$$\frac{1}{(2N)^2} \sum_x \exp\left(\iota \mathbf{q}.\mathbf{x}\right) = \delta\left(\mathbf{q}\right) \tag{9.84}$$

where the delta function in the above equations is the discrete delta function. Like the delta function of a continuous variable, it is zero for non-zero values of the argument. Unlike its continuum analog, it is unity for the zero value of the argument. Strictly speaking, $\delta\left(\mathbf{q}\right)$ is unity whenever q = a reciprocal lattice vector defined by Eq. (9.80). However, in the present case, we restrict $\mathbf{q}$ in the first Brillouin zone, which contains only one lattice vector $q = 0$.

In the semi-discrete model, all integrals in the X–Y plane are replaced by sums over discrete variables $\mathbf{x}$ and $\mathbf{q}$ using the following correspondence formula:

$$\frac{1}{(2N)^2} \sum_q = \left(\frac{1}{2\pi}\right)^2 \int d\mathbf{q} \tag{9.85}$$

where the integral is over the entire 2D Brillouin zone and $(2\pi)^2$ is its area.

To summarize, in the semi-discrete model, x_1 and x_2 assume the following discrete values within a cyclic supercell (see Eqs. (9.78), (9.81)). A supercell refers to the unit cell over which the crystal is made periodic by restricting the allowed values of $\mathbf{x}$ and $\mathbf{q}$ in accordance with Eqs. (9.86), (9.87). This makes the space-dependent functions periodic over the supercell.

$$x_1 = n_1 a \text{ and } x_2 = n_2 a \tag{9.86}$$

$$aq1 = \frac{\pi m_1}{N} \text{ and } aq2 = \frac{\pi m_2}{N} \tag{9.87}$$

where $2N$ is the number of allowed points in the X or Y directions. The ranges of the integers n and m are given by

$$-N \le n_{1,2} \text{ and } m_{1,2} \le N \tag{9.88}$$

9.4.3 PFT of the Green's function

Before calculating the GF for the composite, we illustrate the PFT technique by applying it to a simple case. We calculate the known free-space GF (also called fundamental solution) for an isotropic, homogeneous, infinite medium on which no boundary conditions are imposed. It is the solution of the following isotropic equation:

$$\nabla^2 G(\mathbf{R}, \mathbf{R'}) = -\delta(\mathbf{R} - \mathbf{R'}) \qquad (9.89)$$

The particular solution, without any boundary conditions, obtained by using the Fourier transform technique, is given in A.

Note that the anisotropic case, described in the previous section, reduces to the present isotropic case for $K_{11} = K_{22} = K_{33}$ for both the H and I domains. The conductivity tensors of the H and I domains become identical. The operator L for both the domains in Eqs. (9.42), (9.43) reduces to ∇^2. These equations then reduce to Eq. (9.89) for the GF as in Eq. (9.8).

The free-space GF, which is the solution of Eq. (9.89), is well known [9,29] and is given by

$$G(\mathbf{R}, \mathbf{R'}) = 1/[4\pi|(\mathbf{R}-\mathbf{R'})|] \ (\text{free} - \text{space GF in 3D}) \qquad (9.90)$$

Eq. (9.90) gives the free-space GF in the real vector space. It does not obey the periodic boundary conditions in the XY plane. For satisfying the periodic boundary conditions, it is more convenient to use the Fourier representation of the GF, as used in, for example, Ref. [27]. One major difference, between the present method and the method used in those references, is that here we will use a discrete Fourier transform. Specifically, for the geometry shown in Fig. 9.1, we use a PFT of the GF. The advantage of using a discrete Fourier transform here is that by a suitable choice of the Fourier functions, we can automatically satisfy the periodic boundary conditions given by Eqs. (9.46), (9.47). It also provides a simple and convenient method for including anisotropic effects, as described below.

Now we calculate the GF of the same system by using the PFT method, which will satisfy the periodic boundary conditions on the XY plane. We shall use the semi-discrete model described in the preceding subsection. We use the same Fourier representation for the free-space GF for an infinite solid and write it in terms of its Fourier transform as follows:

$$G(\mathbf{R}, \mathbf{R'}) = \left(\frac{1}{2\pi}\right)^3 \int_{-\infty}^{+\infty} \widehat{G}(\mathbf{Q}) \exp \imath[\mathbf{Q}.(\mathbf{R} - \mathbf{R'})] \, d\mathbf{Q} \qquad (9.91)$$

The exponential functions are eigenfunctions of the Laplace operator introduced in Eq. (9.16), such that

$$\Psi(\mathbf{Q}, \mathbf{R}) = \left(\frac{1}{2\pi}\right)^{\frac{3}{2}} \exp \imath(\mathbf{Q}.\mathbf{R}) \qquad (9.92)$$

The exponential functions obey the orthonormality relation given in Eqs. (9.17), (9.18).

Substituting for $\mathbf{G}$ from Eq. (9.91) into Eq. (9.89), and using Eq. (9.59), we obtain the following expression for $\widehat{G}(\mathbf{Q})$:

$$\widehat{G}\,(\mathbf{Q}) = 1/Q^2 \text{ (isotropic case)} \qquad (9.93)$$

Note that the relation given in Eq. (9.91) is valid only for those functions that depend upon the single variable $\mathbf{R} - \mathbf{R}'$, as in the case of the GF for a perfect homogenous solid. For such solids, any point can be chosen as the origin. Hence, without loss of generality, we choose $\mathbf{R}' = 0$. For a solid containing defects or discontinuities, the GF will depend upon R and $\mathbf{R}'$ separately. For such functions, we need to use a double Fourier transform [15].

It can be verified that Eq. (9.93) along with Eq. (9.91) gives Eq. (9.90) for the free-space GF after carrying out the Q-integration in the continuum model. Recall that in the continuum model, we assume that $\mathbf{Q}$ and $\mathbf{R}$ are continuous variables. Eqs. (9.93), (9.91) are also consistent with Eq. (9.20) because $E(Q) = Q^2$ is an eigenvalue of ∇^2 with $\Psi(\mathbf{Q},\mathbf{R})$ as the corresponding eigenfunction (apart from signs and multiplicative factors).

From Eqs. (9.72), (9.93), we obtain the following integral expression for the PFT of the semi-discrete GF:

$$\widetilde{G}(\mathbf{q}; z) = \left(\frac{1}{2\pi}\right)\int_{-\infty}^{+\infty}\frac{1}{q^2 + k^2}\,\exp\,(\imath\,kz)dk \qquad (9.94)$$

In our semi-discrete model, $\mathbf{q}$ in Eq. (9.94) is a real discrete vector variable, whereas k and z are continuous variables. Accordingly, integration over $\mathbf{q}$ is replaced by a summation over the allowed $\mathbf{q}$ vectors that ensure the periodic boundary condition. Thus, finally, we obtain the following expression for the GF using Eqs. (9.73), (9.85), and (9.141) in Appendix A:

$$G(\mathbf{x}, z) = \frac{1}{(2N)^2}\sum_{\mathbf{q}}\widetilde{G}(\mathbf{q}; z)\exp\,(\imath\mathbf{q}.\mathbf{x}) \qquad (9.95)$$

where, after carrying out the integration in Eq. (9.94),

$$\widetilde{G}(q, z) = \frac{\exp\,(-q|z|)}{2q} \qquad (9.96)$$

In writing Eq. (9.95), we have chosen $\mathbf{R}' = 0$, which amounts to selecting the origin of coordinates. For a different origin, the variables $\mathbf{x}$ and z should be interpreted as $(\mathbf{x},z)$ for the vector $\mathbf{R} - \mathbf{R}'$.

Eq. (9.95) is computationally efficient for the GF because only a 2D sum over discrete values of q needs to be done numerically. The summand in Eq. (9.95) decreases exponentially with increasing q as well as z. The sum is, therefore, nicely convergent. It is singular at $q = 0$, which is the characteristic singularity of the GF. In 3D, it is compensated by the volume element which varies as q^2. The case of 2D is discussed in Section 9.6.

The integral over k has been obtained analytically in Eq. (9.141) of Appendix A, which is exact. The calculated GF obeys periodic boundary conditions on the X–Y plane, as prescribed by Eqs. (9.46), (9.47). Another characteristic of the PFT method, which makes it computationally attractive, is that the coordinate z is a continuous variable. It can, therefore, simulate any shape as precisely as desired, without any need to make approximations at the corners. With minor modifications, this equation can be generalized to less symmetric systems.

9.5 GF for a two-component composite material

Now we apply the techniques described in this chapter to calculate the GF for the two-component composite material shown in Fig. 9.1 by solving the coupled Eqs. (9.42), (9.43). As mentioned at the end of Section 9.4.1, we solve these equations separately and then couple their solutions using the interfacial condition. First, we calculate the GF for the H component only over the finite space from $Z = -d$ to $Z = d$ in Fig. 9.1.

9.5.1 GF for the homogeneous, finite host without inclusion

The GF given by Eq. (9.90) in the previous section is valid for a free infinite space. Similarly, Eq. (9.95) gives GF of a homogeneous solid in the partial Fourier representation for an infinite space subject to periodic boundary conditions in the X–Y plane but infinite in the Z-direction. We now "move" toward the derivation of the GF for the composite solid in the space bounded in the Z-direction as shown in Fig. 9.1. This will show an important advantage of the GF method in that the GF for a material system can be constructed in modules by imposing the boundary conditions successively. For any new addition to the material system, we do not need to recalculate the GF for the previous module.

In the present case, our starting module is the solid H without the inclusion and extending over all space. In the previous section, we obtained its GF, subject only to the periodic boundary conditions in the XY plane given by Eqs. (9.46), (9.47). Next, we will impose the boundary conditions given by Eqs. (9.44), (9.45), which confine the host solid H between the planes $Z = -d$ and $Z = +d$. For this purpose, we use the homogeneous solution using the delta function representation, as described in Section 9.3.2.

We continue with the isotropy approximation. The results for anisotropic case are obtained from the results given here by replacing q by βq in accordance with Eq. (9.64) [see the paragraph following Eq. (9.89)]. We will discuss the changes due to anisotropy in Section 9.5.3.

We obtain the solution of Eq. (9.39) for a charge given by $F(\mathbf{R}')$ located at $\mathbf{R} = \mathbf{R}'$. So far the $\mathbf{R}'$ is not defined except that it is a point in the vector space of $\mathbf{R}$. Different values of $\mathbf{R}'$ along with $\mathbf{F}(\mathbf{R}')$ will then give the distribution of charges. Our objective is to find the potential at all points due to an arbitrary charge distribution. The Poisson equation for a point charge at $\mathbf{R}'$ is written below:

$$\nabla^2 \Phi_H \left(\mathbf{R}, \mathbf{R}' \right) = -F(\mathbf{R}')\delta(\mathbf{R} - \mathbf{R}') \qquad (9.97)$$

Note that $\mathbf{R}'$ in Eq. (9.97) is a vector in the same space as $\mathbf{R}$ but it is not a variable for the operation of ∇^2. It is written only to label a solution corresponding to the location of the charge.

Its components, in the same vector space as $\mathbf{R}$, are given by

$$\mathbf{R}' = (\mathbf{x}', \mathbf{z}') \qquad (9.98)$$

where $z' = x_3'$,

$$\mathbf{x}' = (x_1', x_2') \qquad (9.99)$$

obeying the same periodicity as $\mathbf{x}$ in Eq. (9.77).

Eq. (9.97) corresponds to Eq. (9.10) with the following charge distribution:

$$\rho(\mathbf{R}, \mathbf{R}') = F(\mathbf{R}')\, \delta(\mathbf{R} - \mathbf{R}') \qquad (9.100)$$

Using Eqs. (9.12), (9.95), and the GF in the region of the host solid, the particular solution of Eq. (9.97) for a charge $F(\mathbf{R}')$ at $\mathbf{R}'$ is given by

$$\Phi_{HP}\left(\mathbf{x}-\mathbf{x}', z-z'\right) = \frac{1}{(2N)^2} \left[\sum_q \widetilde{\Phi_{HP}}(\mathbf{q}; z-z') \exp\left\{\iota\mathbf{q}.(\mathbf{x}-\mathbf{x}')\right\} \right] F(\mathbf{R}')$$

$$(9.101)$$

where

$$\widetilde{\Phi_{HP}}(\mathbf{q}; z-z') = \frac{\exp\left(-q|z-z'|\right)}{2q} \qquad (9.102)$$

The subscript H in the above equations labels the host solid and should not be confused with the homogeneous solution, which is identified by the subscript 0.

Eq. (9.101), being the particular solution, does not obey any boundary conditions. We, therefore, need a homogeneous solution for satisfying the boundary conditions. Let $f(\mathbf{q}, k)$ be an arbitrary but integrable function in the Fourier space where $(\mathbf{q}, k)$ denotes a vector in the Fourier space that has components $\mathbf{q}$ in the X–Y plane and k in the Z-direction. The vector $\mathbf{q}$ is a 2D vector in the reciprocal space, which is conjugate to the 2D vector in the real space, defined by Eq. (9.37). Using Eq. (9.34), we write the following expression for the homogeneous part of the solution:

$$\Phi_{H0}(\mathbf{R}) = \frac{1}{(2N)^2} \sum_q \int_{-\infty}^{+\infty} f\,(\mathbf{Q})\delta[E(\mathbf{Q})] \exp\left(\iota\mathbf{q}.\mathbf{x} + \iota\,kz\right) dk \qquad (9.103)$$

where the vector $\mathbf{Q}$ and its magnitude are defined in Eqs. (9.56), (9.59). The k-integral is carried out by using the properties of the delta function in Eq. (9.103). This gives

$$\Phi_{H0}(\mathbf{R}) = \frac{1}{(2N)^2} \sum_q f(\mathbf{q}) \exp{(\imath \mathbf{q}.\mathbf{x} + \imath kz)} \qquad (9.104)$$

where in view of the delta function of $E(\mathbf{Q})$, $Q_3 = k$ is a function of q, which satisfies the following equation:

$$q^2 + k^2 = 0 \qquad (9.105)$$

Hence $f(\mathbf{q}, k)$ can be written as a function of $\mathbf{q}$ alone. Further, Eq. (9.105) gives the following two permissible values of k

$$k = \pm \imath q \qquad (9.106)$$

and, therefore, two permissible values of $f(\mathbf{q})$. We need both because we need to prescribe the boundary conditions on two planes as in Eqs. (9.44), (9.45). Denoting the two functions by $f_+ (\mathbf{q})$ and $f_- (\mathbf{q})$, and using Eq. (9.106), we write the homogeneous solution as

$$\Phi_{H0} (\mathbf{R}) = \Phi_{0+} (\mathbf{R}) + \Phi_{0-} (\mathbf{R}) \qquad (9.107)$$

where

$$\Phi_{0+}(\mathbf{R}) = \frac{1}{(2N)^2} \sum_q f_+(\mathbf{q}) \exp{(\imath \mathbf{q}.\mathbf{x} - qz)} \qquad (9.108)$$

and

$$\Phi_{0-}(\mathbf{R}) = \frac{1}{(2N)^2} \sum_q f_-(\mathbf{q}) \exp{(\imath \mathbf{q}.\mathbf{x} + qz)} \qquad (9.109)$$

Note the similarity between the delta function representation leading to the homogeneous solution in Eq. (9.107) and the contour integration leading to the particular solution in Eq. (9.141). This is consistent with Eq. (9.31) for the representation of the delta function. It can be verified by direct substitution that $\Phi_{H0}(\mathbf{R})$ is a solution of the homogeneous equation (9.11) for $L = \nabla^2$ and for any $f_+(\mathbf{q})$ and $f_-(\mathbf{q})$.

For solids of finite size, it is more convenient to write the z-dependence of f_+ and f_- in terms of hyperbolic sine and hyperbolic cosine. In this case, the arbitrary functions will be linear combinations of f_+ and f_-. We denote them by f_1 and f_2 and write the complete solution of Eq. (9.97) in the following form:

$$\Phi_H(\mathbf{R}, \mathbf{R}') = \Phi_{HP} (\mathbf{x} - \mathbf{x}', z - z') + \Phi_{H0} (\mathbf{R}, \mathbf{R}') \qquad (9.110)$$

where

$$\Phi_{H0}(\mathbf{R}, \mathbf{R}') = \frac{1}{(2N)^2} \sum_q \left[f_1(\mathbf{q}; \mathbf{R}')(\cosh qz)/(\cosh qd) \right. \tag{9.111}$$

$$\left. + f_2(\mathbf{q}; \mathbf{R}')(\sinh qz)/(\sinh qd) \right] \exp(\iota \mathbf{q}.\mathbf{x})$$

We recall that $\mathbf{R}'$ in Eq. (9.97) is only a label for the solution, which identifies the location of the charge $F(\mathbf{R}')$ and is not a variable of the original equation.

The functions f_1 and f in Eq. (9.111) will be linear combinations of $f+$ and $f-$, which are still undetermined. We determine them by imposing the boundary conditions given by Eqs. (9.44), (9.45). This is easily done for fixed values of z by using the PFT in the X–Y plane and using the orthonormality relations given by Eqs. (9.83), (9.84). After some algebraic manipulation, we obtain the following result for the potential distribution in a bounded solid (without inclusions) that satisfies Eq. (9.97) for any charge $F(\mathbf{R}')$ located at the arbitrary point $\mathbf{R}'$. It also satisfies the boundary conditions at the two outer surfaces (see Fig. 9.1) at $z = -d$ and $z = d$ and is periodic in the X–Y plane, which can be verified by direct substitution:

$$\Phi_H(\mathbf{R}, \mathbf{R}') = \left(\frac{1}{2N}\right)^2 \left[\sum_q \tilde{\Phi}_H(\mathbf{q}; z - z') \exp\{\iota \mathbf{q}.(\mathbf{x} - \mathbf{x}')\} \right] F(\mathbf{R}')$$
$$+ V_U(z+d)/2d - V_L(z-d)/2d \tag{9.112}$$

where

$$\widetilde{\Phi}_H(\mathbf{q}; z - z') = \widetilde{\Phi}_{HP}(\mathbf{q}; z - z')$$
$$- \left[\Phi_p(\mathbf{q}; z') \frac{\cosh qz}{\cosh qd} + \Phi_n(\mathbf{q}; z') \frac{\sinh qz}{\sinh qd} \right] \tag{9.113}$$

$$\Phi_p(\mathbf{q}; z') = \left(\frac{1}{2}\right) \left[\widetilde{\Phi}_{HP}(\mathbf{q}; d - z') + \widetilde{\Phi}_{HP}(\mathbf{q}; -d - z') \right] \tag{9.114}$$

and

$$\Phi_n(\mathbf{q}; z') = \left(\frac{1}{2}\right) \left[\widetilde{\Phi}_{HP}(\mathbf{q}; d - z') - \widetilde{\Phi}_{HP}(\mathbf{q}; -d - z') \right] \tag{9.115}$$

A very useful property of the GF method for satisfying the boundary conditions arises from the fact that Φ_H is the valid solution for all $\mathbf{R}'$ and all $F(\mathbf{R}')$. It can, therefore, serve as the starting GF for any additional boundary or interfacial conditions by using the virtual source method as described in the next subsection.

Eq. (9.112) gives the solution for the host in Fig. 9.1 assuming it to be homogeneous, that is, without any inclusion. It satisfies the boundary conditions specified by Eqs. (9.44)–(9.47). To model the composite in Fig. 9.1, we also need it to satisfy the interfacial conditions. If H has no free charges, then $F(\mathbf{R}')$ is 0 for all $\mathbf{R}'$ inside H.

For anisotropic H, Eqs. (9.102), (9.113) become, respectively,

$$\widetilde{\Phi}_{HP}(\mathbf{q}; z - z') = (1/K_{33}^H) \frac{\exp(-\beta_H \, q|z - z'|)}{2\beta_H q} \tag{9.116}$$

$$\widetilde{\Phi}_H(\mathbf{q}; z - z') = \widetilde{\Phi}_{HP}(\mathbf{q}; z - z')$$
$$- \left[\Phi_p(\mathbf{q}; z') \frac{\cosh \beta_H \, qz}{\cosh \beta_H qd} + \Phi_n(\mathbf{q}; z') \frac{\sinh \, \beta_H qz}{\sinh \, \beta_H qd} \right] \tag{9.117}$$

9.5.2 GF for the inclusion

We write the GF and the solution for the inclusion. Since we will need to couple it with the host, it will be convenient to write it also in the PFT in the same Q-space, which ensures periodicity on the X–Y plane. The solution for the domain I does not have any boundary conditions on the Z-coordinates. So, we only need a homogeneous solution with enough free parameters to satisfy the interfacial boundary conditions given by Eqs. (9.49), (9.50). We find these free parameters by using the virtual source method described in Section 9.3.1.

The GF equation for the inclusion is, from Eq. (9.66),

$$\mathcal{F}\left[L^I\Phi_I(\mathbf{R})\right] = -K^I_{33}\left[\beta_I{}^2 q^2 + k^2\right]\widehat{\Phi}_I(\mathbf{Q}) \tag{9.118}$$

From Eq. (9.142) of Appendix A, the particular solution for Eq. (9.118) is

$$\widetilde{\Phi}_{IP}(\mathbf{q}; z) = (1/K^I_{33}) \frac{\exp(-\beta_I \, q|z|)}{2\beta_I q} \tag{9.119}$$

As in Eq. (9.97) for H, we assume a distribution of sources located at $\mathbf{R}''$, each with the source strength $T(\mathbf{R}'')$. We have not yet specified the number, location, and strength of these sources, which we will do in the next subsection when we enforce the interfacial conditions. We follow the same steps as used for deriving Eq. (9.113) except that we do not need the terms in the second bracket containing hyperbolic functions. These terms were needed because of the boundary conditions on the Z-coordinate arising due to the finite size of the domain H. In case of the domain I, its size is specified by the interfacial conditions. Thus, we obtain the following solution for the domain I:

$$\Phi_I(\mathbf{R}, \mathbf{R}'') = \left(\frac{1}{2N}\right)^2 \left[\sum_q \widetilde{\Phi}_{IP}(\mathbf{q}; z - z'') \exp\{iq.(\mathbf{x} - \mathbf{x}'')\}\right] T(\mathbf{R}'') \tag{9.120}$$

9.5.3 Coupling of the inclusion and the host

In this section, we couple the solutions Φ_H and Φ_I through the interfacial conditions. Recall that $F(\mathbf{R}')$ and $T(\mathbf{R}'')$, along with their numbers and locations $\mathbf{R}'$ and $\mathbf{R}''$, are not yet specified. As shown in Section 9.3.1, if we choose all $\mathbf{R}'$ outside the domain of H, then the solution given by Eq. (9.117) will be the homogeneous solution for the domain H for all values of $F(\mathbf{R}')$. Similarly, if all $\mathbf{R}''$ are outside the domain I, the solution given by Eq. (9.120) will be the homogeneous solution for the domain I for all $T(\mathbf{R}'')$.

We have to choose a number of points on the interface at which to satisfy the interfacial conditions, specified in Eqs. (9.49), (9.50). Suppose we choose N_B points denoted by $\mathbf{R}_B$ on the interface denoted by PQST in Fig. 9.1. Next, we choose the same number of points denoted by $\mathbf{R}'_{HB}$ for domain H and $\mathbf{R}''_{IB}$ for domain I such that

$$\mathbf{R}'_{HB} = \mathbf{R}_B - \xi_{HB} \qquad \text{(in domain I)} \qquad (9.121)$$

and

$$\mathbf{R}''_{IB} = \mathbf{R}_B + \xi_{IB} \qquad \text{(in domain H)} \qquad (9.122)$$

where $|\xi|$ are small numbers to be chosen judiciously. The signs of ξ_{HB} and ξ_{IB} in the above equations are symbolic. They only mean that $\mathbf{R}'$ locations are in domain I to ensure that they are outside H. Similarly, $\mathbf{R}''$ must be in domain H to ensure that they are outside I. In addition, we need to ensure that $\mathbf{R}'$ or $\mathbf{R}''$ are near enough $\mathbf{R}$ but never equal to it because the GF will be singular at $\mathbf{R} = \mathbf{R}'$ and $\mathbf{R} = \mathbf{R}''$.

The remaining steps are simply algebraic with no new GF concepts. We calculate the solutions for domains H as well as I at N_B preselected points $\mathbf{R}_B$ on PQST using the formulae given in the preceding two subsections in terms of $2N_B$ unknowns $F(\mathbf{R}_B')$ and $T(\mathbf{R}''_B)$. For each of the N_B points, we have two equations: Eqs. (9.49), (9.50). This gives $2N_B$ equations from which we determine unique values of $F(\mathbf{R}_B')$ and $T(\mathbf{R}''_B)$, and hence the complete solution at all points in H and I. The solution will, obviously, satisfy all the prescribed boundary and interfacial conditions. We will illustrate this method by choosing a 2D example in the next section.

The major numerical effort in this method is in the inversion of a $2N_B \times 2N_B$ matrix for the determination of $F(\mathbf{R}_B')$ and $T(\mathbf{R}_B'')$. What makes the GF method numerically attractive is the value of N_B is proportional to the relative surface area PQST, which is much less than the number of points at which the solution needs to be discretized in the conventional purely numerical methods. Since the numerical effort in matrix inversion is of $O(N_B^3)$, the saving in computational effort is substantial. The actual computation of matrix elements in this method is also not numerically intensive because the integration over one variable k is done analytically even for anisotropic cases.

9.6 Effective conductivity of a 2D composite

Modern 2D materials are of strong topical interest because they represent a revolutionary new development in the science and technology of materials. Examples of these materials are graphene, silicene, and phosphorene. All materials except graphene have a puckered layered structure in which the atoms in a unit cell are arranged in two layers, both parallel to the X–Z plane with reference to the coordinate system defined in Fig. 9.1. The spacing between the two layers is only a few Angstroms, which can be neglected on the scales at which the Laplace and Poisson equations are applicable. Hence, in the continuum model for elliptic equations, all the 2D materials can be assumed to be exactly planar.

In this section, we shall apply our technique to solve the Laplace/Poisson equation in a 2D composite and illustrate the technique by applying it to calculate the effective conductivity of the composite. This problem is of practical interest because the conductivity measurements are useful in characterization of materials. It will also illustrate our method numerically for calculating the GF and show the useful correspondence between 3D and 2D modeling in the Fourier representation. Specifically, we will consider a metallic inclusion in graphene and in anisotropic phosphorene. The shape of the inclusion will be assumed to be square and that of the host as rectangular. The work in this section is a brief review of our two earlier papers [21,25], where more details are available.

In our GF method, all the 3D expressions, given in the preceding sections, reduce to the 2D case by making the following simple changes:

i. Neglect the Y-dependence of all field quantities like potential, field, displacements, etc. by assuming

$$\partial F / \partial y \text{ or } \partial F / \partial x_2 = 0 \tag{9.123}$$

where F is any field quantity.

ii. In view of Eq. (9.123), all vectors on the plane are 2D and tensors like conductivity are represented by 2×2 matrices. This is easily achieved in all the previous equations of this chapter by simply suppressing the x_2 component. The coordinate vectors $\mathbf{x}$ and $\mathbf{q}$ are now 1D scalars. For example, Eq. (9.57) simply reduces to qx and Eq. (9.58) to q^2 or x^2. The PFT is still with respect to the x coordinate but $\mathbf{Q}$ and $\mathbf{R}$ are 2D vectors with components (q,k) and (x,z), respectively. Eqs. (9.116), (9.120) are valid as such. The Fourier transform on the X–Y plane is now only with respect to x, and q is defined by the following equations:

$$\frac{1}{2N} \sum_q \exp\left(\imath qx\right) = \delta\left(x\right) \tag{9.124}$$

and

$$\frac{1}{2N} \sum_x \exp\left(\imath qx\right) = \delta\left(q\right) \tag{9.125}$$

The most significant difference in the 3D and 2D GFs occurs in the summation or integration over all values of Q, which are equivalent in view of Eq. (9.85). The integrand in these cases is singular at $Q = 0$. In 3D, the volume element is proportional to $Q^2 dq$, which mitigates the $1/Q^2$ singularity, leading to Eq. (9.90). On the other hand, the area element in 2D integration is proportional to QdQ, which does not balance the singularity. This leads to the following expression for GF in 2D [29,30]:

$$G(\mathbf{R}, \mathbf{R}') = (1/2\pi) \left[\ln |(\mathbf{R} - \mathbf{R}')| \right] \text{ (free − space GF in 2D)} \qquad (9.126)$$

The logarithmic dependence of the GF on $|\mathbf{R} - \mathbf{R}'|$ gives rise to a size effect in 2D, as discussed in [31] in the context of the Debye-Waller factor.

We divide our numerical results for 2D into two cases—one for the isotropic case given by Eq. (9.6) and the other for the general anisotropic case given by Eq. (9.4). The isotropic case corresponds to that of graphene and to other 2D materials in appropriate units. In case of phosphorene, it is known that its thermal conductivity is highly aniso-tropic. Its thermal conductivity has been estimated to be 110 W/m K along the arm-chair direction and 36 W/m K along the zigzag direction. For our calculations, we will assume that the ratio of these two values is a good estimate of β_H^2 for phosphorene [21] in Eq. (9.116). Thus, we will take $\beta_H^2 \sim 0.3$ for a composite in which phosphorene is the host material.

Fig. 9.2 gives a schematic illustration of the 2D composite that we model in this section. It is almost the same as shown in Fig 9.1. The only differences are in the shape of the inclusion and in the dimensionality of the solid. Fig. 9.1 gives a projection of the 3D solid on the XZ plane whereas Fig. 9.2 is the actual shape of the 2D solid, which exists in the XZ plane.

Consider the thought experiment described in Section 9.4, following Fig. 9.1. Our first object is to find Φ_H for the host solid. We will use the same equations given in the preceding sections. They are valid for the 2D case as such except for the changes described earlier in this section. We choose N_B points on the tilted square PQST,

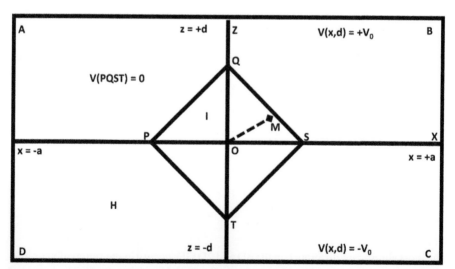

Fig. 9.2 A rectangular solid H containing a metallic inclusion I. See also Fig. 9.1. The point M is just inside the inclusion and shows the location of the virtual charge (see text for details).

and the same number of points just inside I, making sure that the points are sufficiently far from the interfacial line. This will ensure that the GF has no singularities.

The points just inside I will be labeled by b' and their position vectors will be denoted by $R_{b'}$ Thus $\mathbf{R_b} = (x_b, z_b)$ and $\mathbf{R_{b'}} = (x_{b'}, z_{b'})$ with the constraint $\mathbf{R_b} \neq \mathbf{R_{b'}}$, where b and b' are running indices from 1 to N_B or $N_{B'}$. The total number of points $N_{B'}$ inside I is equal to N_B, the number of points on the boundary, and none of the points R_b is equal to any $R_b{}'$.

In the present case, we do not have to calculate the GF for the inclusion because the boundary condition given by Eq. (9.51) is sufficient to determine the GF for H. This is a consequence of the fact that the inclusion has infinite conductivity and so the potential at the interface is constant. We choose the constant to be 0.

Let $F_{b'}$ denote the virtual charge at the point $\mathbf{R_{b'}}$, which, as explained in (9.121), is just inside I, and, therefore, just outside H. Using Eq. (9.112), we can write the following $N_B \times N_B$ matrix equation

$$\mathbf{\Gamma\,F=V} \tag{9.127}$$

or

$$\mathbf{F = \Gamma^{-1}V} \tag{9.128}$$

where $\mathbf{\Gamma}$ is a $N_B \times N_B$ square matrix and $\mathbf{F}$ and $\mathbf{V}$ are N_B dimensional column vectors. From Eq. (9.112), their matrix elements labeled by the indices b and b' are given by

$$\Gamma_{b,b'} = \frac{1}{2N}\left[\sum_q \tilde{\Phi}_H(q;z_b - z'_{b'})\exp\left\{\imath q\left(x_b - x'_{b'}\right)\right\}\right] \tag{9.129}$$

$$V_b = V_U(z_b + d)/2d - V_L(z_b - d)/2d \tag{9.130}$$

R_b and $R'_{b'}$ are the judiciously selected points on PQST and the sum over q is on all allowed values of q in the Brillouin zone as prescribed in Eq. (9.87), except that here q and x have only one component. The value of F, calculated by using Eq. (9.128), gives the potential V(x,z) at all points in the domain H.

The judicious selection of parameters involves the choice of N_B, R_B, and ξ_{IB} in Eq. (9.121). All these parameters affect numerical convergence and the computational cost. For example, if N_B is too small and/or the points R_B are not properly distributed on PQST, the boundary condition will not be representative of the actual physical condition. Another crucial parameter is ξ_{IB}. If it is too small, the matrix Γ in Eq. (9.129) may become too ill-conditioned for inversion. Similar considerations govern the choice of other parameters. Note that the symmetry of the system can be exploited to reduce the computational cost substantially. We recall that the matrix inversion is $O(N_B^3)$, so any increase or decrease in N_B significantly affects the total computational cost [32].

Finally, we calculate the effective conductivity of the material system using the methodology given by Garboczi [32]. Refer to the same thought experiment, given in Section 9.4. We can write the current flowing in the Z-direction at a point (x,z) in Fig. 9.2 as follows:

Using the values of F in Eq. (9.112), we get the potential V(x,z) at all points in H. The total current through a line at $z = z_0$ parallel to the X axis is given by [32]

$$J_z = K_{33} \sum_x E_z (x, z = z_0) \tag{9.131}$$

where E_z is the field in the Z-direction at the point (x, z). It is given by

$$E_z = [\partial V(x, z)/\partial z]_{z0} \tag{9.132}$$

The sum in Eq. (9.131) is over all the allowed values of x of the points on the line $z = z_0$. If there are no sources or sinks in H, J_z will be independent of z_0, that is, the location of the line. This property can be used as a numerical check on the numbers.

A numerically useful property of the PFT representation of the GF can be used in the evaluation of the sum in Eq. (9.132). From Eq. (9.125), we note that the sum over all values of x leads to $\delta(q)$ in the result. This means the sum over x of all the terms in E_z is simply given by the $q = 0$ limit of the Fourier transform after inversion.

We calculate the effective value of the conductivity as follows: the measured average macroscopic field in the solid (see Fig. 9.2) is given by

$$E_0 = V_0/d \tag{9.133}$$

If we K* denotes the effective conductivity, then

$$J_z = K^*_{33} E_0 \tag{9.134}$$

Comparing Eqs. (9.134), (9.131) gives the required value of K^*_{33}. Following Garboczi [32], we express the effective conductivity in terms of the fractional area v of the inclusion as follows:

$$K^*_{33} (v)/K_{33} = 1 + \Omega (v) v \tag{9.135}$$

where Ω is usually referred to as the Enhancement factor. The effective conductivity K^*_{33}, and hence Ω will obviously be a function of size, shape, and nature of the defect. It is a very useful single parameter for characterizing the composite. The region where Ω is nearly constant implies that effective conductivity is a linear function of the area fraction. This region is the usually the low concentration limit of $\Omega(v)$.

Fig. 9.3 shows the variation of $\Omega(v)$ with respect to v for graphene and phosphorene as calculated in Ref. [21] using Eq. (9.135). The low concentration limit of $\Omega(v)$ is called the intrinsic conductivity. For isotropic solids and simple inclusion shapes, this limit can be derived analytically. For a square, the intrinsic conductivity has been

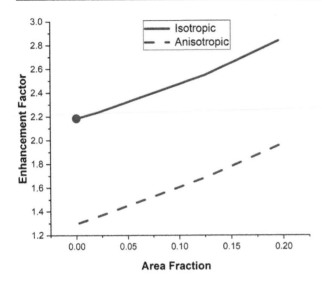

Fig. 9.3 The variation of the enhancement factor with the area concentration of the inclusion for the isotropic case *(solid line)* and one anisotropic case *(dotted line)*. The isotropic case is graphene, whereas the anisotropy parameter $=0.3$ for the *dotted curve* corresponds to phosphorene (see text). The *thick dot* at the end of the isotropic curve is the analytical value.

determined numerically but very accurately. In Fig. 9.3, this limiting value is shown as a thick point at the $v = 0$ end of the isotropic curve. Note that it shows an almost exact agreement between our calculated value (2.187 and the known value of 2.188).

9.7 Summary and conclusions

We have described a static GF-based method for the solution of 3D and 2D Laplace/ Poisson equation for anisotropic composites and applied it to compute the electro-static/thermostatic response of modern two-component nanocomposites consisting of a host and an inclusion. We calculate the GF by using the method of PFT over a plane (X–Y plane, the PFT plane) in 3D solids or over a line (X-axis, PFT line) in 2D solids. It enables us to treat the third variable (Z) analytically. This results in con-siderable economy of the computational cost because the solid needs to be numeri-cally digitized on a plane for 3D cases and a line for 2D cases. This is in contrast to the conventional methods, which require digitization in all the coordinates. We have illustrated the method by applying it to a general solid, specifying periodic boundary conditions on the X–Y plane for the 3D case or the X-axis in the 2D case, and the Dirichlet boundary condition in the Z-direction. In addition, we assume the standard interfacial conditions. In 2D cases, we have applied the method to calculate the effective conductivity of graphene (isotropic) and phosphorene (anisotropic) con-taining a metallic inclusion. Numerical results are presented for the variation of the conductivity enhancement factor as a function of the area fraction of the metallic inclusion. Our result agrees almost exactly with the analytical result in the limit of low area fraction, which is the only analytical result available.

9.8   Possible future work of interest

i. Application to more complex and nonsymmetric shapes of inclusions of arbitrary conductivities.

ii. GF for Laplace equation for bilayers for more realistic simulation of phosphorene and other Xenes beyond graphene.

iii. It will be interesting to explore the possibilities of linking this methodology with that aimed at improving the computational efficiency at macroscales and mesoscales, for application to materials engineering and machine intelligence (see, e.g., Refs. [33, 34]).

Appendix A

In this appendix, we obtain the solution of Eq. (9.89) for free-space GF for the anisotropic case. Expressing the equation in the Fourier representation, we obtain from Eq. (9.64):

$$\left(\frac{1}{2\pi}\right)^3 \int_{-\infty}^{+\infty} A(\mathbf{Q})\widehat{G}(\mathbf{Q}) \exp \iota [\mathbf{Q}.(\mathbf{R} - \mathbf{R}')] \, d\mathbf{Q} = \delta(\mathbf{R}-\mathbf{R}') \qquad (9.136)$$

where the integral is a 3D integral over all components of $\mathbf{Q}$ and

$$A(\mathbf{Q}) \;\; = K_{33}\left[\beta^2 q^2 + k^2\right] \qquad (9.137)$$

In Eq. (9.137), we have removed the suffix H for the sake of generality. Using Eq. (9.18), we obtain

$$\widehat{G}(\mathbf{Q}) = A(\mathbf{Q})^{-1} \qquad (9.138)$$

This gives

$$G(\mathbf{R}, \mathbf{R}') = (1/K_{33}) \, (1/2\pi)^3 \int \left[1/\left(\beta^2 q^2 + k^2\right)\right] \exp\left[\iota \mathbf{Q}.(\mathbf{R}-\mathbf{R}')\right] d\mathbf{Q}, \qquad (9.139)$$

where the integration is over all 3D Q-space and we have used Eq. (9.56) for $\mathbf{Q}$. For the isotropic case, $\beta = 1$ and Eq. (9.139) reduces to Eq. (9.93), which gives the well-known result quoted in Eq. (9.90) for $K_{33} = 1$, which only redefines the units of the GF.

We now show the evaluation of the 1D integral in Eq. (9.94) for the PFT, which is needed for 3D as well as 2D solids. We write it in a slightly more general form, as given below:

$$B(\mathbf{q}; z) = \left(\frac{1}{2\pi}\right) \int_{-\infty}^{+\infty} \frac{1}{\beta^2 q^2 + k^2} \, \exp\left(-\iota \, kz\right) dk \qquad (9.140)$$

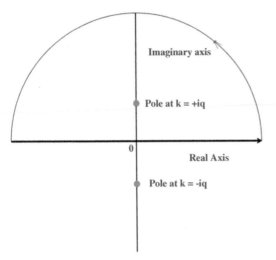

Fig. A.9.1 Semicircular Cauchy contour for evaluation of the integral in Eq. (9.94) or (9.140). We choose the contour in the UHP (upper half plane) for z <0 and in the LHP for z <0.

where $\beta >0$ and $q >0$. For $q <0$, the integral can be evaluated using similar steps. For the semi-discrete model discussed in this chapter, q is 1D or a 2D discrete vector. The integral in Eq. (9.140) is over k, the Z-component of $\mathbf{Q}$, for constant q.

First, we consider the case when $\beta = 1$. This corresponds to the isotropic case when the operator L is simply ∇^2 as in Eq. (9.94). To evaluate the integral in Eq. (9.140) or Eq. (9.94), we choose the semicircular Cauchy contour given in Fig. A.9.1. The integrand has two simple poles: one at $k = +\imath q$ in the UHP (upper half plane) and one at $k = -\imath q$ in the LHP. The choice of the contour, and hence the value of the integral, depends upon the sign of z. We choose the contour in the UHP for z <0 and in the LHP for z >0. This gives

$$B(\mathbf{q}; z) = \frac{\exp\left(-q|z|\right)}{2q} \tag{9.141}$$

Eq. (9.141) gives the PFT of the GF for the L operator in the isotropic case. In the anisotropic case, considered in this chapter, the L operators for the host and the inclusion domains are defined by Eqs. (9.64), (9.66). Notice that the integrand in this case is same as in Eq. (9.140) with q replaced by βq with $\beta = \beta_H$ or β_I. The result can, therefore, be obtained from the result for the isotropic case with q replaced by the appropriate βq, as quoted below:

$$B(\mathbf{q}; z) = \frac{\exp\left(-\beta q|z|\right)}{2\beta q} \cdot (\beta > 0) \tag{9.142}$$

Acknowledgments

The authors thank Drs. Orion Kafka and Newell Moser for their useful suggestions and comments, and Ms. Vani Murarka for help with the diagrams.

References

[1] F. Bonaccorso, L. Colombo, G.H. Yu, M. Stoller, V. Tozzini, A.C. Ferrari, R.S. Ruoff, V. Pellegrini, Graphene, related two-dimensional crystals, and hybrid systems for energy conversion and storage, Science 347 (2015), 1246501.

[2] D.G. Cahill, Extremes of heat conduction-pushing the boundaries of the thermal conductivity of materials, MRS Bull. 37 (2012) 855–863.

[3] W.F. Brown, Solid mixture permittivities, J. Chem. Phys. 23 (1955) 1514–1517.

[4] E.J. Garboczi, J.G. Berryman, Elastic moduli of a material containing composite inclusions: effective medium theory and finite element computations, Mech. Mater. 33 (2001) 455–470.

[5] E.J. Garboczi, J.F. Douglas, Intrinsic conductivity of objects having arbitrary shape and conductivity, Phys. Rev. E 53 (1996) 6169–6180.

[6] E.J. Garboczi, J.F. Douglas, Elastic moduli of composites containing a low concentration of complex-shaped particles having a general property contrast with the matrix, Mech. Mater. 51 (2012) 53–65.

[7] D. Tomanek, Interfacing graphene and related 2D materials with the 3D world, J. Phys. Condens. Matter 27 (2015), 133203.

[8] G. Barton, Elements of Green's Functions and Propagation, Clarendon Press, Oxford, 1991.

[9] E. Pan, W. Chen, Static Green's Functions in Anisotropic Media, Cambridge University Press, New York, 2015.

[10] V.K. Tewary, Green-function method for lattice statics, Adv. Phys. 22 (1973) 757–810.

[11] K. Duan, L. Li, Y.J. Hu, X.L. Wang, Interface mechanical properties of graphene reinforced copper nanocomposites, Mater. Res. Express 4 (2017), 115020.

[12] Z. Hu, G. Tong, D. Lin, C. Chen, H. Guo, J. Xu, L. Zhou, Graphene-reinforced metal matrix nanocomposites—a review, Mater. Sci. Technol. 32 (2016) 930–953.

[13] D. Kuang, W.B. Hu, Research progress of graphene composites, J. Inorg. Mater. 28 (2013) 235–246.

[14] G. Zhang, Y.W. Zhang, Thermal properties of two-dimensional materials, Chin. Phys. B 26 (2017), 034401.

[15] A.A. Maradudin, E.W. Montroll, G.H. Weiss, I.P. Ipatova, Theory of Lattice Dynamics in the Harmonic Approximation, Academic Press, New York, 1971.

[16] X.K. Gu, R.G. Yang, First-principles prediction of phononic thermal conductivity of silicene: a comparison with graphene, J. Appl. Phys. 117 (2015), 025102.

[17] G.W. Hanson, Dyadic Green's functions for an anisotropic, non-local model of biased graphene, IEEE Trans. Antennas Propag. 56 (2008) 747–757.

[18] V.R. Coluci, S.O. Dantas, V.K. Tewary, Generalized Green's function molecular dynamics for canonical ensemble simulations, Phys. Rev. E 97 (2018), 053310.

[19] V.R. Coluci, S.O. Dantas, V.K. Tewary, Accelerated causal Green's function molecular dynamics, Comput. Phys. Commun. 277 (2022) 108378.

[20] M. Hu, X. Zhang, D. Poulikakos, Anomalous thermal response of silicene to uniaxial stretching, Phys. Rev. B 87 (2013) 195417.

[21] V.K. Tewary, E.J. Garboczi, Green's function formulation of conductivity of anisotropic two-dimensional materials containing metallic inclusions: application to phosphorene, Phys. Lett. A 384 (2020) 126851.

[22] M. Philip, Morse and Herman Feshbach, Methods of Theoretical Physics, Vol. 1, McGraw-Hill Publishing Company, New York, 1953.

[23] M. Philip, Morse and Herman Feshbach, Methods of Theoretical Physics, Vol. 2, McGraw-Hill Publishing Company, New York, 1953.

[24] V.K. Tewary, Elastostatic Green's function for advanced materials subject to surface loading, J. Eng. Math. 49 (2004) 289–304.

[25] V.K. Tewary, E.J. Garboczi, Semi-discrete Green's function for solution of anisotropic thermal/electrostatic Boussinesq and Mindlin problems: application to two-dimensional material systems, Eng. Anal. Bound. Elem. 110 (2020) 56–68.

[26] V.K. Tewary, R.H. Wagoner, J.P. Hirth, Elastic Green-function for a composite solid with a planar crack in the interface, J. Mater. Res. 4 (1989) 124–136.

[27] V.K. Tewary, R.H. Wagoner, J.P. Hirth, Elastic Green-function for a composite solid with a planar interface, J. Mater. Res. 4 (1989) 113–123.

[28] V.K. Tewary, E.J. Garboczi, Semi-discrete Green's function for solution of anisotropic thermal/electrostatic Boussinesq and Mindlin problems: application to two-dimensional material systems, Eng. Anal. Bound. Elem. 110 (2020) 56–68.

[29] G. Barton, Elements of Green's Functions and Propagation, Clarendon Press, Oxford, 1989.

[30] V.K. Tewary, R.C. Quardokus, F.W. DelRio, Green's function modeling of response of two-dimensional materials to point probes for scanning microscopy, Phys. Lett. A 380 (2016) 1750–1756.

[31] V.K. Tewary, B. Yang, Singular behavior of the Debye-Waller factor of graphene, Phys. Rev. B 79 (2009) 125416.

[32] E.J. Garboczi, Technical Report, IR 6269, NIST, Gaithersburg, MD, USA, 1998.

[33] Y. Cheng, O.L. Kafka, W.K. Liu, Self-consistent clustering analysis for multiscale modeling at finite strains, Comput. Methods Appl. Mech. Eng. 349 (2019) 339–359.

[34] Y. Cheng, O.L. Kafka, W.K. Liu, Multiresolution clustering analysis for efficientmodeling of hierarchical material systems, Comput. Mech. 67 (2021) 1293–1306.

Atomistic simulation of biological molecules interacting with nanomaterials

Nabanita Saikia[a] and Ravindra Pandey[b]
[a]School of Science, Navajo Technical University, Chinle Site, Chinle, AZ, United States,
[b]Department of Physics, Michigan Technological University, Houghton, MI, United States

10.1 Introduction

Combining biomolecules with nanomaterials yields new functional materials with a vast array of properties and potential applications. The fact that nanomaterials and biological molecules share a comparable size domain with at least one dimension on the order of 1–100 nm (Fig. 10.1), makes nanomaterials suitable as biological labels for in vivo imaging without introducing interference in the cellular machinery [1–3]. The similarity in size and intrinsic properties of biomolecules can yield novel biomolecule-nanoparticle hybrids with synergistic characteristics and function [2,4]. Research on the nano-bio interface is critical in the design and development of safe and effective strategies in nanomedicine, drug delivery, pathological site targeting, metabolism, and biocompatibility [5,6]. Nanomaterials such as metal oxide nanoparticles, fullerenes, nanotubes, and nanowires have been extensively studied in molecular imaging, diagnosis of diseases, biosensing, biochips, and bioelectronic device applications [7,8]. Graphene-based nanomaterials have been utilized for nanosensor fabrication and detection of biomolecules as the conjugated graphene surface can promote electron transfer and facilitate the immobilization of molecules [9,10]. One-dimensional nanomaterials like carbon nanotubes (CNTs) and boron nitride nanotubes (BNNTs) have also been explored in biomedical research including biosensing [11], cancer therapy [12,13], and drug delivery [14,15]. The directional property of nanotubes along with the ability to functionalize through both covalent or noncovalent methods have proven to be an efficient alternative for targeting therapeutic molecules, peptides, proteins, and nucleic acids to cells in a controlled mediated manner [16,17]. There are many excellent reviews on the potential application of graphene-based nanomaterials for biomolecular imaging and medical therapy [18–21].

Studies on the interaction of biological molecules such as peptides, proteins, enzymes, DNA/RNA oligonucleotides, small ligands, lipids, and cell membranes are thought to be beneficial in designing protein-nanoparticle conjugates [22–24]. The interaction of biological systems with engineered nanomaterials (ENMs) can promote the formation of a "nanoparticle-protein corona" that can influence the biological reactivity of conjugated nanoparticles [25–27]. ENMs encounter several

Modeling, Characterization and Production of Nanomaterials. https://doi.org/10.1016/B978-0-12-819905-3.00010-5

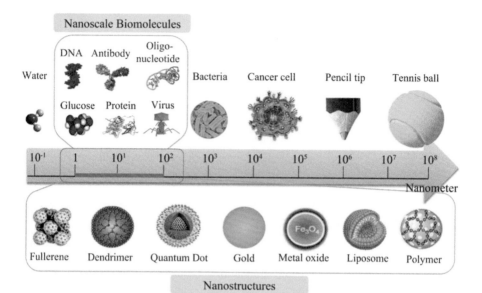

Fig. 10.1 A comparison of the length scale of nanostructures and biomolecules in the range of 1–100 nm. The nanometer length scale is a thousand times smaller than bacteria and cancer cells and a million times smaller than the tip of a pencil.

Reproduced with permission from S. Saallah, I.W. Lenggoro, Nanoparticles carrying biological molecules: recent advances and applications, KONA Powder Part. J. 35 (2018) 89–111. Copyright 2018. Published by Hosokawa Powder Technology Foundation.

interactions with biological molecules including vdW force, electrostatic interaction, H-bonding, ligand-receptor, stereospecific interactions, and membrane curvature effects (Fig. 10.2). The molecular interactions are mainly determined by the peptide composition, charge, and amino acid sequence. Since most nanoparticles are hydrophobic, to render them compatible for biological applications, surface functionalization by chemical modifications using hydrophilic ligands or small cross-linking molecules can stabilize nanoparticles for aqueous dispersion [28]. Scientific interest in this interesting intersection of research is based on the discernment that nanotechnology can offer new biological insights and that biology, in turn, can offer nanotechnology the access to new types of functional nanosystems [29].

The nano-bio interface represents a juxtaposition of interfacial properties that is determined by the biophysical interactions including vdW, electrostatic, hydrophobic, and hydrogen bonding [30]. Often, these outlined parameters are interdependent as they are determined by the synthesis procedure. Multiscale modeling algorithms can address these molecular-level interactions through the design and surface modification of nanomaterials. Computational methods within the domain of quantum mechanics (QM) and molecular dynamics (MD) simulations have accelerated atomic-level resolution of the complex biophysical interactions under complex molecular environments that have been largely elusive by experimental methods. With the

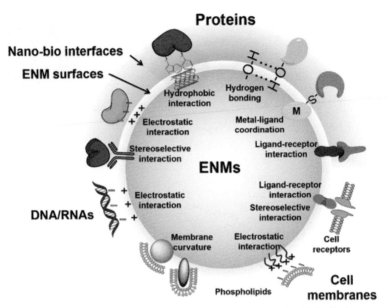

Fig. 10.2 The interaction of engineered nanomaterials (ENMs) with biological systems. Reproduced from Y. Wang, R. Cai, C. Chen, The nano-bio interactions of nanomedicines: understanding the biochemical driving forces and redox reactions, Acc. Chem. Res. 52 (2019) 1507–1518. Copyright (2019) American Chemical Society.

ability to extend the length and time-scales of multiscale simulations, it is conceivable to match computational observables with experiments and validate computational methods [19]. One major caveat with experimental measurements is that it is ensemble-based which makes it challenging to gain specific details at the single-molecule level. QM calculations have limitations on a system size of the order of 1000 atoms making it computationally expensive to scale to a larger system size (Fig. 10.3).

On the other hand, atomistic simulations using empirical force fields can scale to around 100,000 atoms at the expense of accuracy or time scale/sampling and obtain a reasonable conformational sampling of the phase space of the molecular systems [31]. In this regard, benchmarking atomistic force fields based on quantum chemical calculations such as *first-principles* density functional theory (DFT) is a prerequisite to accurately characterize the binding properties, molecular interactions at the interface, and efficient sampling of the conformational space. Enhanced sampling techniques such as coarse-grained (CG) MD, Kinetic Monte Carlo simulations, mesoscale dynamics, and continuum models have realized modeling nanomaterial-biological interface at longer time scales [32–35]. CG modeling relies on grouping several atoms into larger "pseudo-atoms" or beads thereby increasing computational efficiency by reducing the number of degrees of freedom [36]. In the case of proteins, the polypeptide chain is represented by two beads, H for hydrophobic and P for polar

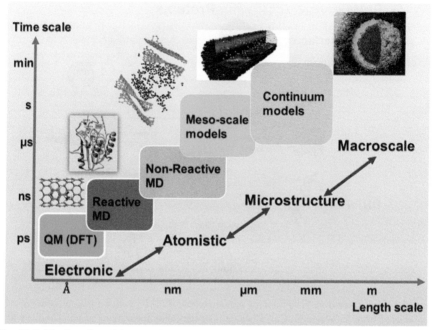

Fig. 10.3 The length and time scale of different simulation techniques: quantum mechanics (QM), reactive MD including molecular mechanics (MM), all-atom molecular dynamics (AA-MD) simulations, implicit solvent, coarse-grained MD, mesoscale dynamics, and continuum models.

residues [37]. This minimalistic approximation has been successful in sampling the conformational free energy landscape of protein folding and dynamics [36,38]. While CG-MD simulation and continuum methods are suitable for modeling bulk properties of materials, accurate calculation of molecular interaction energies from a single peptide molecule requires an all-atom representation [39].

In light of the advancements in scientific research on the amalgamation between biological molecules and nanomaterials and molecular-level understanding of the interfacial properties, this chapter focuses on the emergent approaches to the application of multiscale atomistic simulation methods such as MD, Monte Carlo, and coarse-grained MD to study the interaction of biomolecules such as amino acids, peptides, proteins, DNA oligonucleotides, and small molecules with nanomaterials. Since the intermolecular interactions in biomolecule-nanomaterial conjugates are primarily driven through electrostatic, hydrophobic, H-bonding, and vdW forces, we cover the scalability of multiscale modeling techniques and focus on integrative approaches that combine molecular modeling with experiments to uncover the opportunities and challenges in this field. We realize that a comprehensive study of the application of nanomaterial-biological conjugates is beyond the scope of this chapter and summarize selected examples to highlight the important advances in this research field.

10.2   Emergent approaches to integrating biological molecules with nanomaterials

Peptides and proteins can bind to the nanomaterial surface such that solvent exposure of nanomaterials occurs in a controlled manner without resulting in nanoparticle agglomeration [40]. Xia et al. [41], suggested surface adsorption energy to be the major driving force behind the dynamic changes of nanomaterials within biological molecules. The interaction of graphene with two peptides, cecropin P1 and MSI-78 (C1) was mediated by both planar and hydrophilic residues [42], accompanied by peptide – graphene, peptide–peptide, and peptide – solution interactions [43]. The competitive interaction between vdW stacking and hydrophilic residues on graphene determined the orientation of peptides on the graphene surface. Using Titanium (Ti) as a ubiquitous surface with strong cell adhesion characteristics, Schneider et al. [44], investigated the minimum titanium-binding motif, minTBP-1 with peptide sequence RKLPDA employing atomistic simulations. The R residue was mainly responsible for stabilizing peptide anchoring to the surface. Charged residues K and D were responsible for the equilibrium adsorption and D residue was shown to promote a tightly bound, flat geometry of the peptide on Ti surface. The variations in local solvent density were detected by the peptide side chains and alternation between hydrophilic and hydrophobic residues could match with the solvent density oscillations, thereby mediating specific recognition in biomolecular complexes. In a recent study on the design, development, and potential application of phosphorene, a monolayer of black phosphorus and its interaction with biological systems was reviewed [45]. The instability of phosphorene to water and oxygen made it prone to rapid dissolution, structural instability, and generation of surface defects. Although this characteristic property is perhaps beneficial for medical applications, its rapid degradation can lead to oxidative stress and other health concerns, thereby limiting its potential medical integration.

Fig. 10.4 elucidates some of the emergent approaches for signaling biomolecular interactions and assembly of proteins and DNA oligonucleotides on Au nanoparticles and 2D nanomaterials. Shao and Hall suggested that the molecular mechanism underlying nanoparticle influence on protein structure was regulated by the allosteric effect of nanoparticles and altered protein structure and function [46]. The allosteric effects are considered important when designing nanoparticles for medical and biological applications as the allosteric effect of nanoparticles can distort the structure and flexibility of protein regions distant from the binding site. MD simulations indicated that variation in local protein domains and changes in protein structure and flexibility can occur when the proteins remain folded (Fig. 10.4A). MXenes, a new family of 2D materials consisting of transition metal carbides and carbonitrides was investigated for interaction with dsDNA using fluorescence spectroscopy and MD simulations (Fig. 10.4D) [48]. The weak interaction of dsDNA allowed for a kinetically-dynamic system, making MXenes unique among other layered/2D materials for hybridization-based biosensing applications. A recent study by Luan and Zhao designed a novel DNA sequencing technique using 2D in-plane heterostructure comprising of graphene

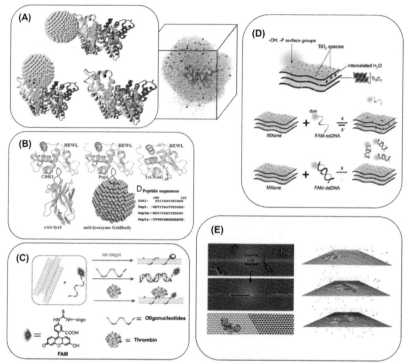

Fig. 10.4 (A) The binding configurations of human serum albumin (HSA) protein on a 4.0 nm Au nanoparticle [46]. The subdomains of HSA protein are shown in different colors. The inset shows a snapshot of the initial configuration of an HSA–nanoparticle complex in explicit solvent from atomistic MD simulations. The HSA protein is shown in cartoon model to emphasize its secondary structure and Au nanoparticle is shown in the vdW model.
(B) Schematic of the design of an anti-lysozyme Goldbody. (C) Scheme for signaling biomolecular interactions using an assembly between SWCNT and dye-labeled ssDNA [47]. (D) Representation of MXene interacting with ssDNA and dsDNA. Noncovalent binding of ssDNA to the surface of nanomaterials was based on vdW, hydrogen bonds, and π-stacking interactions, involving the phosphate backbone and/or the nucleobases [48]. (E) An illustration of ssDNA dynamics on graphene, h-BN, and their in-plane heterostructure, and how ssDNA can be spontaneously stretched on an in-plane graphene/h-BN/graphene heterostructure [49,50]. The figure illustrates the hydroxide-assisted ball milling method for h-BN exfoliation and edge functionalization and high-resolution TEM and UV–vis spectrum of basal edge exposed and –OH terminated BN porous sheets.

Panel A: Adapted from Q. Shao, C.K. Hall, Allosteric effects of gold nanoparticles on human serum albumin, Nanoscale, 9 (2017) 380–390, from The Royal Society of Chemistry; Panel B: Adapted from G.-H. Yan, K. Wang, Z. Shao, L. Luo, Z.-M. Song, J. Chen, et al., Artificial antibody created by conformational reconstruction of the complementary-determining region on gold nanoparticles. Proc. Natl. Acad. Sci. U. S. A. 115 (2018) E34–E43. Copyright (2017) National Academy of Sciences; Panel C: Reproduction from W. Zhong, Nanomaterials in fluorescence-based biosensing, Anal. Bioanal. Chem. 394 (2009) 47–59, Published by SpringerLink; Panel D: Reproduction from C.L. Manzanares-Palenzuela, A.M. Pourrahimi, J. Gonzalez-Julian, Z. Sofer, M. Pykal, M. Otyepka, et al., Interaction of single- and double-stranded DNA with multilayer MXene by fluorescence spectroscopy and molecular dynamics simulations, Chem. Sci. 10 (2019) 10010–10017, Published by The Royal Society of Chemistry; Panel E: Reproduction from Q. Weng, X. Wang, X. Wang, Y. Bando, D. Golberg, Functionalized hexagonal boron nitride nanomaterials: emerging properties and applications, Chem. Soc. Rev. 45 (2016) 3989–4012, Published by The Royal Society of Chemistry.

and hexagonal boron nitride (h-BN) [49]. This technique facilitated the efficient stretching of ssDNA in a linear conformation on a h-BN nanostrip sandwiched between two adjacent graphene domains (also referred to as nanochannels) and by applying a biasing voltage, the stretched ssDNA can be electrophoretically transported along the nanochannel (Fig. 10.4E). These examples underline the broad repertoire in designing strategies for integrating nanomaterials with biological systems by combining experimental and state-of-the-art theoretical approaches. These capabilities proffer the potential for engineering new functional materials and systems with unprecedented precision—that is, the ability to assemble structures that approach the dimensions of single molecules or even single atoms [51].

10.3 Single-walled carbon nanotubes as nanovehicles in drug targeting and delivery

Functionalized single-walled carbon nanotubes (SWCNTs) have been developed as novel drug delivery systems to improve the pharmacological and therapeutic profile of drugs [52] with an ability to cross the cell membrane upon administration [53]. The most important characteristic of functionalized SWCNTs as a drug delivery system is its ability to cross the cell membrane [54] and facilitate cellular internalization and trafficking with the cell cytoplasm [16,55]. Kam et al., showed that fluorescein probe molecules attached to the SWCNT-biotin complex were easily detected in the endosomes, suggesting a cellular uptake through endocytosis with the ability of SWCNTs to penetrate cell cytoplasm, nucleus, and fibroplasts [56]. Although the potential application of SWCNTs in a real diagnostic scenario is yet to be fully ascertained, the development of such systems with a capability for cellular recognition, imaging, and target release of bioactive agents is foreseen especially in the treatment of cancer, bacterial, and viral infections [57]. Co-tethering fluorescent probe molecules like fluorescein dyes can assist in the cellular trafficking and uptake of nanomaterials within the body which can be monitored till the drug is released to the active region (Fig. 10.5). Noncovalent functionalization of CNTs using polyethylene glycol, PEGylated–phospholipid chains are another effective means of high surface loading of drug and biomolecules at the free end of PEG functionalized nanotubes. Major studies have focused on integrating PEG terminated nanotubes as carriers for the attachment of therapeutics such as cisplatin, doxorubicin, carboplatin, including peptides, proteins, and fluorescent probe molecules.

10.4 Hybrids of DNA oligonucleotides and nanomaterials

The hybridization of single-strand DNA (ssDNA) with nanomaterials exemplifies another archetypal example wherein, the inherent electronic properties of nanomaterials along with the natural amphiphilic nature of ssDNA promote the interfacing of molecular nanostructures; distinct in properties yet unique in applications [63–65]. The complementary structure of DNA is stabilized by vdW π-stacking between the

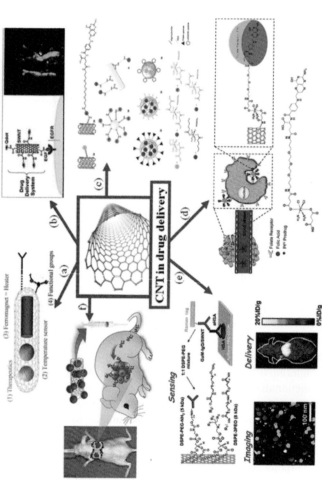

Fig. 10.5 SWCNTs as potential drug delivery modules in transdermal therapy. (A) Functionalized CNTs with different filling and bio-functional groups @ http://www.carbio.eu/research.html. (B) Cisplatin and epidermal growth factor (EGF) attached to SWCNT to specifically target squamous cancer cells [58]. (C) Metal-based prodrugs based on SWCNTs, Au NPs, and polymeric nanoparticles for the delivery of Pt(IV) anticancer drugs [59]. (D) Folate receptor-mediated targeting and SWCNT-mediated delivery of cisplatin anticancer drug [60]. (E) CNTs in vitro and in vivo detection, imaging, and drug delivery [61]. (F) A schematic depiction of conjugated drug injection [62].

Panel B: Adapted from A.A. Bhirde, V. Patel, J. Gavard, G. Zhang, A.A. Sousa, A. Masedunskas, et al., Targeted killing of cancer cells in vivo and in vitro with EGF-directed carbon nanotube-based drug delivery, ACS Nano 3 (2009) 307–316. Copyright 2009, American Chemical Society; Panel C: Adapted from N. Graf, S. J. Lippard, Redox activation of metal-based prodrugs as a strategy for drug delivery, Adv. Drug Del. Rev. 64 (2012) 993–100. Copyright 2012 Elsevier B.V.; Panel D: Adapted from S. Dhar, Z. Liu, J. Thomale, H. Dai, S.J. Lippard, Targeted single-walled carbon nanotube mediated Pt(IV) prodrug delivery using folate as a homing device, J. Am. Chem. Soc. 130 (2008) 11467–11476. Copyright 2008, American Chemical Society; Panel F: Adapted from S.-W. Kim, J.-Y. Park, S. Lee, S.-H. Kim, D. Khang, Destroying deep lung tumor tissue through lung-selective accumulation and by activation of caveolin uptake channels using a specific width of carbon nanodrug, ACS Appl. Mater. Interfaces 10 (2018) 4419–4428. Copyright 2018, American Chemical Society.

nucleobases and the backbone phosphate group enhances the stability of DNA/CNT hybrids in aqueous solution [66]. DNA, because of its complementary base pairing sequence, helps in nanotube assembly. Recent studies on the helical wrapping of ssDNA around SWCNTs are motivated by this inherent mobility to disperse nanotubes for DNA sensing and biomedical applications [67–70]. Rodger and co-workers investigated the interaction of CNT with DNA strands using the Linear Dichroism (LD) method [71]. DNA molecules upon interaction with SWCNT showed higher LD signals, higher than the sum of the LD spectrum of DNA and nanotube taken independently. Direct coupling of functional quantum dot (QD)-DNA hybrids allowed specific detection of nanomolar unlabeled complementary DNA at low DNA probe/QD copy numbers via a "signal-on" FRET response [72]. This technique reduced the donor-acceptor distance and enhanced the fluorescence resonance energy transfer (FRET) efficiency at low (1:1) DNA/QD. The specific electronic and optical detection of DNA hybridization was studied using biotin-streptavidin as an interlinker between SWCNT and 20-mer DNA oligonucleotides [73]. Streptavidin was shown to prevent the nonspecific adsorption of biomolecules onto nanotubes and prevent the aggregation of DNA strands to the source and drain electrodes that can modify the Schottky barrier. The wrapping of streptavidin with CNTs allowed electrical and fluorescence detection of DNA hybridization.

Real-time detection of ssDNA hybridization on graphene was monitored using surface plasmon resonance spectroscopy (SPR) [74]. The interaction of ssDNA with graphene was mediated by hydrophobic and π-stacking interactions, and a noticeable change in the SPR signal was detected due to this interaction. The SPR signal amplification was observed using Au nanostar and Au nanostar modified ssDNA as Au nanostar interacted strongly with graphene nanosheet and enabled the integration of ssDNA. The hybridization of DNA on Au nanoparticles and graphene oxide using polyethylene glycol (PEG) polymer to mimic the excluded volume effect of cellular proteins was employed for designing intracellular biosensors and drug delivery systems [75]. PEG enhanced the rate of DNA hybridization to DNA-functionalized AuNPs by 50%–100% and this rate enhancement was attributed to the surface blocking effect by PEG instead of the macromolecular crowding [75]. Similar observations were made with graphene oxide which suggested PEG to be an effective blocking agent for both hydrophilic Au NP and graphene oxide. These examples highlight some of the important aspects of conjugated nanomaterials with DNA and the scalability of DNA to assemble nanomaterials.

The emergent applications proffered by DNA nanotechnology has motivated research interest in the synthetic analogs of DNA, termed as the peptide nucleic acids (PNAs) [76,77]. In PNA, the chiral sugar-phosphate backbone is mimicked by N-(2-aminoethyl) glycine polyamine, with the nucleobases attached via methyl carbonyl linker (see Fig. 10.6). The lack of a charged backbone eliminates electrostatic repulsion during hybridization thereby rendering high stability to PNA. Similar to DNA, PNA can form duplexes by Watson-Crick base pairing, and adopt a helical (P-type) structure with 18 bases/turn, and has a diameter of 28 Å [79]. Exciting new developments in the application of PNA for molecular recognition, gene expression, biosensing, and antisense hybridization are foreseen owing to the sequence-specific

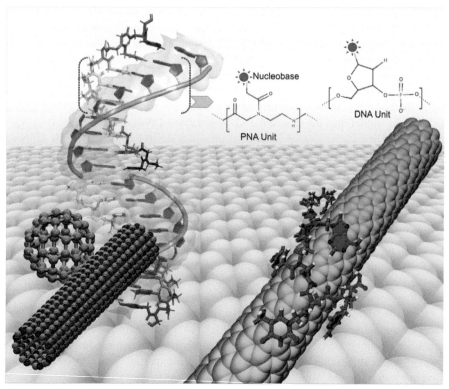

Fig. 10.6 Schematic of the biological interaction of PNA-DNA nanohybrids with functional nanomaterials such as graphene, fullerene, SWCNT, and BNNT [78].

recognition of complementary DNA/RNA [80–83]. Although, both PNA and ssDNA demonstrate similar dynamics in aqueous solution, with the structural transition to a random coil-like aggregate, the dynamics of PNA is independent of the salt concentration due to the lack of a charged backbone. PEG polymer was reported to impede PNA adsorption on Au nanoparticle in a molecular weight dependent manner (rate of PNA adsorption in absence of PEG was $2.2\,\text{min}^{-1}$ and in the presence of PEG 20000 it was $0.31\,\text{min}^{-1}$) [75]. Contrary to DNA, the energy barrier for PNA adsorption on Au nanoparticle was increased by PEG but its small absolute value was from the lack of electrostatic interaction. In our study [78], using atomistic simulations we investigated the self-assembly of single-stranded PNA on SWCNT and BNNT of varying diameter and show that in contrast to DNA, the uncharged pseudopeptide backbone of PNA eliminates electrostatic repulsion and delineates the dynamics at the biomolecular solid/liquid interface. In aqueous solution, PNA provides definitive contrasts in self-recognition of the nanotube inherent electronic properties: PNA wraps along the circumferential direction of SWCNT, whereas it binds along the axial direction adopting an extended configuration on BNNT. The binding free energy of PNA toward the lateral translocation on the tubular surface was higher of BNNT compared to SWCNT, which suggests that the semi-ionic nature of the B—N bond render additional stability for the PNA/BNNT hybrid. Our study also highlights that the physiochemical interactions that stabilize the intermolecular nano-bio interface are mainly dominated by vdW, hydrophobic, and hydrogen bonding interactions [78].

10.5 Overview of experimental methods and single-molecule detection techniques

To date, characterizing complex surface–biomolecule interactions and surface properties of nanomaterials remains challenging both experimentally and at computational levels [84]. There have been significant efforts over the past decade to resolve the single-molecule structure of biomolecules with nanomaterials and understand the conformational states at near atomic-level resolution [85]. Theoretical and experimental studies have prompted the use of nanomaterials with biomolecules for application in biomechanics, optical, thermal, biomedical, biosensors, and nanoscience. A myriad of experimental techniques such as atomic force microscopy (AFM), isothermal titration calorimetry (ITC), ellipsometry, flow cytometry, gel filtration, surface plasmon resonance spectroscopy, fluorescence spectroscopy, UV/vis spectroscopy, circular dichroism (CD) spectroscopy, Fourier transform infrared spectroscopy (FTIR), Raman spectroscopy, nuclear magnetic resonance (NMR) spectroscopy, dynamic light scattering (DLS), zeta potential measurements, X-ray photoelectron spectroscopy (XPS), and high-resolution powder X-ray diffraction (HR-XRD) have been used to characterize the intermolecular interactions, binding selectivity, thermodynamics, molecular stability of nanomaterial-biomolecule systems [31,86–88]. Single-molecule detection is a rapidly developing area that can characterize the dynamic properties of individual molecules in ensemble-average experiments [85]. Raman spectroscopy can detect changes in the molecular bond force constants, pointing to the subtle interactions at the nano-bio interface. FTIR spectroscopy can detect conformational changes in the vibrational bands of amide (or peptide) bonds. CD spectra can identify structural changes in proteins upon interaction with nanomaterials. DLS can measure the autocorrelation function of scattered light intensity from which the diffusion coefficient of the particles and hence the apparent particle size can be calculated. Small-angle neutron scattering (SANS) can provide details on the interaction through form factor and structure factor of the scattering intensity [89]. Flow cytometry can detect the adsorption of biological molecules on the nanoparticle surface by using intense fluorescent reporter binders, such as antibodies conjugated to QD [90].

Sethi and Knecht reported the interaction and surface aggregation of Arginine residue with citrate capped Au NPs using UV–vis spectroscopy, TEM, and DLS measurements [40]. Arginine can partially replace the citrate stabilizer and self-assemble on the Au surface. The Au NPs assembled in an organized linear chain in presence of Arginine as confirmed by both UV–vis and DLS measurements. The electrostatic interactions between zwitterionic Arginine in close proximity and citrate stabilizers result in surface segregation on Au NPs. Raman spectroscopy, XRD, and electron microscopy measurements on the interaction of ZnO NPs and adenosine triphosphate

(ATP) at neutral and acidic solution conditions suggest that ATP forms chelate complexes with Zn^{2+} through the N7 atom in adenine nucleobase [91]. Satzer et al., investigated the adsorption of myoglobin and bovine serum albumin (BSA) on spherical silica NPs of the range of 30 to 1000 nm using the optimized CD method [92]. Both myoglobin and BSA proteins showed conformational changes such as a decrease in alpha-helical structure upon adsorption on silica NPs bigger than 150 nm. Although the kinetics of conformational change was different for both proteins, it showed a nanoparticle size dependence and the conformational changes were only visible for particles above 200 nm. The interaction of anionic silica NPs of the dimension of 10, 18, and 28 nm, respectively with a cationic lysozyme and an anionic BSA was studied using UV–vis spectroscopy, DLS, and SANS method [89]. The adsorption of lysozyme on silica NPs followed an exponential growth behavior with concentration dependence. BSA protein, on the other hand did not show any adsorption on the NPs. The nonadsorption of BSA and adsorption of lysozyme led to two-phase (turbid) and one-phase (clear) transformation in the NP-protein complexes. The size-dependent phase behavior for lysozyme was due to the dominance of protein-mediated short-range attractive interaction over the long-range repulsion between the NPs as confirmed from SANS measurement.

Peptide binding to SWCNTs was probed using CD, fluorescence, NMR spectroscopy, and molecular modeling methods [93], to characterize the structural features of the peptide-CNT complex. Two newly isolated phage, UW-1, and UW-4 showed stronger binding affinity to CNT. The CD spectra of UW-1/peptide-CNT complex exhibited a β-sheet-like conformation at low pH. At neutral pH, free-standing UW-1 peptide adopted a random coil conformation and changed into a β-turn (or β-hairpin) structure when bound to CNT. The study emphasized that different nanotube-binding peptides exhibit distinct and different conformations relevant to their binding selectivity, and provide a molecular basis for understanding the interaction of biomolecules with CNTs. High-resolution solution NMR, including nuclear Overhauser effect spectroscopy (NOESY) and saturation-transfer difference NMR techniques, were used to determine the structure and orientation of peptides bound to SiO_2 and TiO_2 NP surfaces [94]. A 12-mer titania binding peptide (TBP12, RKLPDAPGMHTW) identified from phage-display experiment was shown to be a strong binder to TiO_2 ($K_d = 13.2 \mu M$), SiO_2 ($K_d = 11.1 \mu M$), and silver. High-resolution solution NMR, NOESY, and saturation-transfer difference NMR were used to determine the orientation and structure of the peptide, TBP6 bound to SiO_2 and TiO_2 NPs. 2D NOESY spectrum of TBP6 in the presence and absence of 25 nm titania nanoparticles (1 mg/mL) showed cross-peaks for TBP6 in the absence of nanoparticles that are opposite in sign from the diagonal, as expected for a low molecular weight peptide [94]. The cross-peaks had the same sign as the diagonal in presence of nanoparticles, indicating slow dynamics. Saturation-transfer difference NMR implied that ionic interactions between the amino acid and nanoparticle surface mediated peptide binding on the surface.

Fullerenes C_{60} and C_{70} nanomaterials was shown to inhibit the aggregation of $A\beta_{1-40}$ protein in vitro with high specificity to the KLVFF recognition motif [95]. The mechanism of amyloid disassembly mediated by fullerenes C_{60} and C_{70} dispersed in 1-methyl-2-pyrrolidinone (NMP) on insulin and lysozyme amyloid fibrils was reported using a combination of biophysical and biochemical techniques including spectrofluorometry, AFM, and SANS [96]. Thioflavin-T (ThT) fluorescence assays were used to monitor protein fibrillation in solutions. AFM and SANS allowed the detection of changes in the morphology of $A\beta_{1-40}$ fibrils interacting with fullerenes. Both hydrophobic and hydrophilic interactions enhanced fullerene binding to amyloid fibrils and hydrophilic sites in fullerene had additional interactions with the polar residues of $A\beta_{1-40}$ amyloids.

Experimental techniques have thus been indispensable in providing valuable information on how biological molecules can be integrated with nanomaterials. Nevertheless, experimental techniques are often limited in providing atomic-level details of the nano-bio interactions and insights on the conformational changes within a biomolecule on its adsorption on the nanomaterial surface. This is mainly due to the high intricate complexity and conformational heterogeneity of biological systems. Computational methods can complement existing experiment techniques in providing molecular-level insights on the nature of protein-surface interactions and detailed description of the structure, free energy landscape, and intermolecular interactions. Present research initiatives are therefore directed to integrate experiments with computational methods that would help us further our knowledge of biomolecule–surface interactions at various spatiotemporal levels.

10.6   Quantum mechanical approaches

Computational modeling of biomolecular systems can provide molecular-level insights to physiochemical events and predict the behavior of complex systems that are often limited or not measured in experiments. QM methods broadly are classified into ab initio, semiempirical, and first-principles DFT. Ab initio methods require no a priori empirical information about the molecular system for the calculation of molecular properties but uses approximations to solve the Schrödinger's equation. The semiempirical QM method considers only the valence electrons thus considerably reducing the computational cost of the calculations. DFT calculations are based on the electron density of the molecular system and can be applied to a system involving hundreds of atoms. High-level QM methods such as Møller–Plesset (MP2) can accurately account for the electron-correlation effect between nonbonded atoms. MP2 methods are computationally expensive and can only be applied to systems with a limited number of atoms. QM calculations can accurately describe the electronic, optical, and mechanical properties of subatomic particles, atoms, molecules, and molecular systems in the sub-nanometer dimension. It is important to mention that QM methods have limitation in terms of system size which makes them largely unsuitable for simulating large protein–nanomaterial systems. Nevertheless, QM calculations can precisely characterize both intra- and inter-molecular interactions without requiring any empirical information, thereby making them extremely valuable to parameterize empirical force fields for protein-nanomaterial interactions in solvent media.

10.7 Electronic structure theory: Time-independent Schrödinger equation

Schrödinger's approach involves partial differential equations to describe the correlation between the quantum state of a physical system with time (time evolution of wave function) [97]. The time-independent Schrödinger equation is given by:

$$\frac{d^2\psi(x)}{dx^2} + \frac{[2m\{E - V(x)\}]}{\hbar^2} = 0$$

$$-\frac{\hbar^2}{2m}\frac{d^2\psi(x)}{dx^2} + V(x)\psi(x) = E\psi(x)$$

(10.1)

The Schrödinger equation extended in three dimensions is given by:

$$-\frac{\hbar^2}{2m}\nabla^2\psi(r) + V(r)\psi(r) = E\psi(r)$$

(10.2)

where ∇^2 is the Laplacian operator given by the equation $\nabla^2 = \frac{d^2}{dx^2} + \frac{d^2}{dy^2} + \frac{d^2}{dz^2}$ and $V(r)$ is the potential energy of the system. In analogy to the time-independent Schrödinger wave equation, the time-dependent equation is given by:

$$i\hbar\frac{\delta\psi(r,t)}{\delta t} = -\frac{\hbar^2}{2m}\nabla^2\psi(r,t) + V(r)\psi(r,t)$$

(10.3)

The permitted solutions of Eq. (10.3) are called eigenfunctions and the corresponding energy values the eigenvalues. Each permitted solution corresponds to a definite energy state and is known as orbital. The electron orbitals in atoms are called atomic orbitals while those in a molecule are called molecular orbitals. The Schrödinger equation can be resolved into two parts solving for the electronic (for a fixed nuclear geometry) and nuclear wave function.

The nonrelativistic Hamiltonian for a molecular system consisting of M nuclei and N electrons in absence of an external magnetic or electric field is given by:

$$\widehat{H} = \frac{-\hbar^2}{2m}\sum_{i=1}^{N}\nabla_i^2 - \sum_{A=1}^{M}\frac{\hbar^2}{2M_A}\nabla_A^2 - \sum_{i=1}^{N}\sum_{A=1}^{M}\frac{Z_A e^2}{4\pi\varepsilon_0 r_{iA}}$$

$$+ \sum_{i=1}^{N}\sum_{j>i}^{N}\frac{e^2}{4\pi\varepsilon_0 r_{ij}} + \sum_{A=1}^{M}\sum_{B>A}^{M}\frac{Z_A Z_B e^2}{4\pi\varepsilon_0 R_{AB}}$$

(10.4)

where i, j refers to the electrons and A, B to the nuclei. In atomic units, Eq. (10.4) reduces to,

$$\hat{H} = \frac{-1}{2} \sum_{i=1}^{N} \nabla_i^2 - \frac{1}{2} \sum_{A=1}^{M} \frac{\nabla_A^2}{M_A} - \sum_{i=1}^{N} \sum_{A=1}^{M} \frac{Z_A}{r_{iA}} + \sum_{i=1}^{N} \sum_{j>i}^{N} \frac{1}{r_{ij}}$$

$$+ \sum_{A=1}^{M} \sum_{B>A}^{M} \frac{Z_A Z_B}{R_{AB}} \tag{10.5}$$

The Hamiltonian can be simplified to the generalized form:

$$\hat{H} = \hat{T}_N(R) + \hat{T}_e(r) + \hat{V}_{Ne}(r,R) + \hat{V}_{ee}(r) + \hat{V}_{NN}(R) \tag{10.6}$$

where "R" spans on nuclear coordinates and "r" on electronic coordinates. $\hat{T}_N$, $\hat{T}_e$ represent the kinetic energy operators for nuclei and electron, $\hat{V}_{Ne}$ the attractive electrostatic interaction between nuclei and electron, $\hat{V}_{ee}$ the electron–electron repulsion term, and $\hat{V}_{NN}$ the nuclear-nuclear repulsion terms.

$$\hat{T}_e = -\frac{1}{2} \sum_{i=1}^{N} \nabla_i^2$$

$$\hat{T}_N = -\frac{1}{2} \sum_{A=1}^{M} \frac{1}{M_A} \nabla_A^2$$

$$\hat{V}_{Ne} = -\sum_{i=1}^{N} \sum_{A=1}^{M} \frac{Z_A}{r_{iA}} \tag{10.7}$$

$$\hat{V}_{ee} = \sum_{i=1}^{N} \sum_{j>1}^{N} \frac{1}{r_{ij}}$$

$$\hat{V}_{NN} = \sum_{A=1}^{M} \sum_{B>A}^{M} \frac{Z_A Z_B}{R_{AB}}$$

The nuclear and electronic terms in the Schrödinger equation is given by:

$$\hat{H}_{elect} \Psi_{elect} = E_{elec} \Psi_{elect} \tag{10.8}$$
$$\psi(R,r) = \psi_N(R)\psi_e(r;R)$$

10.8 Hartree–Fock (HF) approximation and self consistent field (SCF) approach

An important unsolved problem in QM is how to deal with indistinguishable, interacting particles in particular electrons that determine the behavior of almost every object in nature [98]. For H and H-like atoms the exact solution of the Schrödinger equation is possible by including the interatomic distances in the variational function.

However, for atoms with higher atomic numbers like He and Li, the variational method is not accurate enough to solve for the exact wave function and the best-adopted approach to solve for the wave function lies in calculating an approximate wave function using the HF procedure. The HF theory is considered to be the cornerstone for all wave function-based quantum chemical methods and much of the electronic structure theory which approximates the N electron wave function by an antisymmetric product of N–one-electron wave function $\chi_i\left(\vec{x}_i\right)$ termed as the *Slater determinant*. The HF approximation is a primary example of central field approximation in which the Coulomb electron–electron repulsion is considered by integrating the repulsion term. According to Hartree's SCF model, the motion of an electron in an effective field of N–1 electrons is governed by the one-particle Schrödinger equation [99]. The general procedure of solving the HF equation is to make the orbitals self-consistent with the potential field and is achieved through an iterative SCF method.

The HF theory is approximated as an antisymmetric product of N orthonormal spin orbitals $\chi_i\left(\vec{x}_i\right)$, which is a product of the spatial orbital $\phi_i(r)$ and one of the two spin functions $\alpha(s)$ or $\beta(s)$ such that,

$$\chi\left(\vec{x}\right) = \phi(r)\sigma(s) \tag{10.9}$$

where $\sigma=\alpha$, β. The spatial orbital $\phi_i(r)$ is a function of position vector "r" and describes the spatial distribution of an electron, and $|\phi(r)|^2 dr$ denotes the probability density of finding an electron within the volume element "dr." However, to completely describe an electron it is necessary to incorporate the spin functions into the wave function which consists of two orthonormal functions $\alpha(s)$ and $\beta(s)$ corresponding to up (↑) and down (↓) spins, respectively. The two Hartree products are distinguishable between electrons 1 and 2 and we obtain a wave function which satisfies the antisymmetric principle by considering the linear combination of the two Hartree products,

$$\psi(x_1,x_2) = \frac{1}{\sqrt{2}}\left[\chi_i(x_1)\chi_j(x_2) - \chi_j(x_1)\chi_i(x_2)\right] \tag{10.10}$$

where $\frac{1}{\sqrt{2}}$ is the normalization factor and the minus sign signifies $\psi(x_1,x_2)$ to be antisymmetric with respect to the interchange of the coordinates of the two electrons,

$$\psi(x_1,x_2) = -\psi(x_2,x_1) \tag{10.11}$$

The antisymmetric wave function can be expressed in the determinant form as:

$$\psi(x_1,x_2) = \frac{1}{\sqrt{2}}\begin{vmatrix} \chi_i(x_1) & \chi_j(x_1) \\ \chi_i(x_2) & \chi_j(x_2) \end{vmatrix} \tag{10.12}$$

which is the Slater determinant. For an *N* electron system, the Slater determinant can be generalized in the determinant form as:

$$\psi_{HF}(x_1, x_2....x_N) = \frac{1}{\sqrt{N!}} \begin{vmatrix} \chi_1(\vec{x}_1) & \chi_2(\vec{x}_1) & \cdots & \chi_N(\vec{x}_1) \\ \chi_1(\vec{x}_2) & \chi_2(\vec{x}_2) & & \chi_N(\vec{x}_2) \\ \vdots & \vdots & & \vdots \\ \chi_1(\vec{x}_N) & \chi_2(\vec{x}_N) & \cdots & \chi_N(\vec{x}_N) \end{vmatrix} \qquad (10.13)$$

The elements of the determinant are termed as spin orbitals. Introducing a short-hand notation for the normalized Slater determinant which includes the diagonal elements, the Slater determinant can be expressed as,

$$\psi_{HF}(x_1, x_2, x_3..........,x_N) = \frac{1}{\sqrt{N!}} \det\{\chi_1(x_1)\chi_2(x_2).........\chi_N(x_N)\} \qquad (10.14)$$

There are two basic techniques involved in constructing the HF wave function for systems with paired and unpaired electrons. For a molecule having paired spins in a close-shell system, singlet state orbital spatial functions can describe the α and β spins in each pair (restricted HF method, RHF). In the unrestricted HF (UHF) method, two separate sets of orbitals are assigned for α and β spins and the two electrons have different spatial distributions. Apart from RHF and UHF methods, another possibility of constructing the wave function for open-shell systems is the restricted open HF method (ROHF), wherein the paired electrons share the same spatial orbitals. The advantage of the ROHF method is that there is no spin contamination.

10.9 Basis set [100]

A basis set is a set of mathematical functions (known as basis functions) used to describe the shape of the orbitals in an atom. Basis sets are used to create molecular orbitals by expanding as a linear combination of atomic orbitals. Two types of basis sets are commonly used in electronic structure calculation: Slater Type Orbitals (STO) and Gaussian Type Orbitals (GTO). The STO is represented by the functional form:

$$\chi_{\zeta,n,l,m}(r, \theta, \varphi) = N Y_{l,m}(\theta, \varphi) r^{n-1} e^{-\xi r} \qquad (10.15)$$

where N is the normalization constant, $Y_{l,m}$ are the spherical harmonics corresponding to the angular momentum, and n, l, m the principal, angular momentum, and magnetic quantum numbers, respectively. The STOs do not have any radial nodes and exhibit an exponential dependence on the distance between the nucleus and electrons and show rapid convergence with an increasing number of functions. For hydrogen-like systems, the STO is given by:

$$STO = \left[\frac{\alpha^3}{\pi}\right]^{0.5} e^{-\alpha r},$$

where α is the Slater orbital exponent.

The Gaussian type functions for the atomic orbitals in an LCAO wave function replaces the $e^{-\alpha r}$ exponential term of STO with $e^{-\alpha r^2}$. GTOs are expressed in terms of polar and Cartesian coordinates:

$$\chi_{\zeta,n,l,m}(r,\theta,\varphi) = NY_{l,m}(\theta,\varphi)r^{2n-2-l}e^{-\xi r^2}$$
$$\chi_{\xi,l_x,l_y,l_z}(x,y,z) = Nx^{l_x}y^{l_y}z^{l_z}e^{-\xi r^2} \tag{10.16}$$

where the sum of l_x, l_y, and l_z determine the type of orbitals (e.g., $l_x+l_y+l_z=1$ for p-orbital).

The simplified form of the GTO is given by: $GTO = \left[\frac{2\alpha}{\pi}\right]^{0.5}e^{-\alpha r^2}$.

10.9.1 Types of basis sets

10.9.1.1 Minimal basis sets

The minimal basis set contains one basis function for each atomic orbital to describe a free atom. The simplest and common of all the minimal basis sets is the STO–nG basis sets devised by Pople and his group, which involves a linear combination of "n" GTOs fitted to each STO. The individual GTOs are termed primitive orbitals while the combined functions are called contracted functions. E.g., in the STO–3G basis set (elements H—Xe) each basis function is a contraction of three primitive Gaussians.

10.9.1.2 Split valence, double and triple–zeta basis sets

The split Valence basis set uses one function for orbitals that are not in the valence shell and two functions for orbitals lying in the valence shell. The basis sets beyond minimal basis sets are termed as "extended" basis set the most common improvement over the minimal basis set is the Double–Zeta (DZ) basis set which uses two contracted Gaussian functions in place of each minimal basis function. The core electrons are represented by a single contracted Gaussian per atomic orbital, whereas valence electrons are represented by two contracted Gaussians. The D95 basis sets of Dunning and coworkers are DZ-type basis sets that use 9 s-type primitive Gaussians to describe the 1 s and 2 s atomic orbitals and 5 p-type primitives for the atomic 2p orbitals. This type of basis set is available for H—Cl. The triple-zeta (TZ) basis set uses three basis functions instead of one. The Valence double-zeta basis sets include 3-21G, 4-31G, and 6-31G. The first number is the number of primitive Gaussians in the core contracted Gaussian functions (CGF), the next is the number of two valence CGFs, and the last number (1 in these bases) is the number of primitives in the outer contracted Gaussian. The smallest split-valence basis set is denoted 3-21G (available for H—Xe). It uses a three-primitive expansion for the 1 s orbital and then splits the valence orbitals into a two basis function, the inner function being a contraction of two Gaussians and the outer function being just a single Gaussian.

In a basis set, the polarization function represents angular momentum values higher than valence orbitals of free atoms, i.e., it uses p- and d-type basis functions to describe the distortion in s and p orbitals. The inclusion of polarization functions permits more flexibility to the molecular wave function to distort away from spherical symmetry. In "Pople nomenclature," the 6-31G(d) (or 6-31G*) basis sets add a single set of (Cartesian) d-type functions to the basis sets of all nonhydrogen atoms; the 6-31G(d,p) (or 6-31G**) basis sets [4] add a set of p-type polarization functions on H as well. In general, polarization functions improve the description of molecular geometries (bond lengths and angles) as well as relative energies. The number of basis functions and integrals involved in computation, the dimension of matrices, etc. increase rapidly with the increase in the number of functions in the basis set. A single star ("*") indicates that polarization functions have been added to the "heavy" atoms—but not H. Two stars indicate that, in addition to the heavy atom polarization, p functions have been added to H. Example: 6-31G**. In addition to polarization functions, diffuse functions (represented by "+") are used which considers a larger spatial extent, a single "+" corresponds to p orbitals and "++" to both s and p orbitals.

10.10 Density functional theory (DFT)

The electron density $\rho(r)$ is defined as the number of electrons present per unit volume at a given point r in space such that the integration of $\rho(r)$ over the region dr gives the total number of electrons N,

$$\int \rho(r)dr = N \tag{10.17}$$

The relation between electron density and wave function is given by:

$$\rho(r) = N \int \int \ldots\ldots \int |\Psi_e(r\sigma_1, x_2, \ldots\ldots x_N)|^2 d\sigma_1 dx_2 \ldots\ldots dx_N \tag{10.18}$$

which is similar to the normalization equation $\langle \Psi_e | \Psi_e \rangle = \int \int \ldots\ldots \int |\Psi_e|^2 dx_1 dx_2 \ldots\ldots dx_N = 1$, but without one of the spatial orbitals. Since the wave functions are normalized and follow the antisymmetric principle the integral of the electron density should be equal to the total number of electrons N. The external potential V_{ext} can be described in terms of the electron density as:

$$V_{ext} = \int \int \ldots\ldots \int \sum_{i=1}^{N} |\Psi_e|^2 v(r_i) dx_1 dx_2 \ldots\ldots dx_N$$

$$= \frac{1}{N} \sum_{i-1}^{N} \int \rho(r)v(r_i)dr_i = \int \rho(r)v(r)dr \tag{10.19}$$

The total energy is given by:

$$E[\rho] = T_0[\rho] + \int \left\{ \hat{V}_{ext}(r) + \hat{U}_{el}(r) \right\} \rho(r)dr + E_{xc}[\rho] \tag{10.20}$$

where $T_0[\rho]$ is the kinetic energy of the system of noninteracting electrons, $\widehat{V}_{ext}(r)$ is the external potential of the electrons under the effect of the nuclei and is given by:

$$\widehat{V}_{ext}(r) = \sum_a \frac{-Z}{|R-r|} \tag{10.21}$$

$\widehat{U}_{el}(r)$ is the Coulomb electron–electron self-interaction given by:

$$\widehat{U}_{el}(r) = \int \frac{\rho(r')}{|r'-r|} dr' \tag{10.22}$$

$$E_{el}[\rho] = \int \int \frac{\rho(r)\rho(r')}{|r'-r|} drdr' \tag{10.23}$$

and $E_{xc}[\rho]$ the exchange–correlation energy. The chemical potential is given by:

$$\mu = \frac{\delta E[\rho(r)]}{\delta \rho(r)} = \frac{\delta T_0[\rho(r)]}{\delta \rho(r)} + \widehat{V}_{ext}(r) + \widehat{U}_{el}(r) + \frac{\delta E_{xc}[\rho(r)]}{\delta \rho(r)},$$

$$= \frac{\delta T_0[\rho(r)]}{\delta \rho(r)} + \widehat{V}_{eff}(r) \tag{10.24}$$

where, $\widehat{V}_{eff}(r) = \widehat{V}_{ext}(r) + \widehat{U}_{el}(r) + \widehat{V}_{xc}(r)$ such that $\left[-\frac{1}{2}\nabla_i^2 + \widehat{V}_{eff}(r)\right]\phi_i^{KS}(r) = \varepsilon_i \phi_i^{KS}(r)$ which is similar to the HF equation with the nonlocal exchange potential replaced by the local exchange correlation potential $\widehat{V}_{xc}(r)$. The KS orbital $\phi_i^{KS}(r)$ is a function of r and can be derived to obtain the total electron density:

$$\rho(r) = \sum_i^N |\phi_i^{KS}(r)|^2 \tag{10.25}$$

Thus, the electron density term can be used to predict the total energy, where the kinetic energy term can be determined from the corresponding orbitals,

$$T_0[\rho] = \frac{1}{2} \sum_{i=1}^N \langle \phi_i^{KS}|\nabla_i^2|\phi_i^{KS}\rangle \tag{10.26}$$

10.11   Local density approximation (LDA)

The LDA forms the basis of all approximate exchange-correlation functionals the central idea being that of the uniform electron gas, wherein the electrons move on a positive background charge distribution such that the total ensemble is neutral. In the LDA, it is assumed that the exchange-correlation energy per particle is dependent on the local density, $\rho(r)$. The exchange–correlation energy is given by:

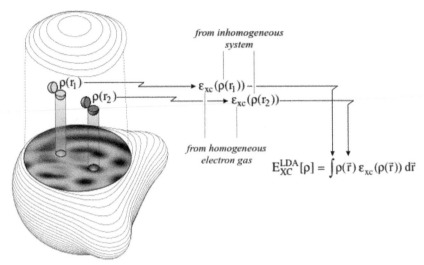

Fig. 10.7 A model depiction of local density approximation (LDA). Reproduced from W. Koch, M.C. Holthausen, A Chemist's Guide to Density Functional Theory, second ed. Wiley-VCH Verlag GmbH, ISBNs: 3-527-30372-3 (Softcover); 3-527-60004-3 (Electronic). Copyright 2001 Wiley-VCH Verlag GmbH.

$$E_{xc}^{LDA}[\rho] = \int \rho(r)\varepsilon_{xc}\{\rho(r)\}dr \tag{10.27}$$

where $\varepsilon_{xc}\{\rho(r)\}$ is the exchange-correlation energy per particle in a uniform electron gas weighed with a probability $\rho(r)$ at any particular position. The LDA can be interpreted with the help of Fig. 10.7. In this model of an open shell system, $\rho(r_1)$ and $\rho(r_2)$ denote the two spin densities at position r. The two density terms are inserted in the exchange–correlation energy expression to obtain $E_{xc}(r)$. The spin densities can be associated with the exchange and correlation energies and mapped for all the points in space and summed up to give the total exchange–correlation LDA energy, $E_{xc}^{LDA}[\rho(r)]$.

10.12   Generalized gradient approximation (GGA)

In the GGA method, the gradient of the electron density is considered to account for the inhomogeneity of the true electron density. The exchange-correlation energy in GGA is defined as:

$$E_{xc}^{GGA} = \int \rho(r)\varepsilon_{xc}^{GGA}\{\rho, \nabla\rho\}dr \tag{10.28}$$

where $\nabla\rho$ is the gradient term.

10.13 Applications of QM calculations to describe the interaction of nanomaterials with biological systems

QM calculations have been successfully employed to study the interaction of amino acids and small peptides with finite-size nanoparticles and nanoclusters [35,101,102]. The suitability of SWCNT and BNNT in the biosensing of DNA/RNA nucleobases were probed, as interaction with nucleobases can affect the electronic transport properties of the nanomaterials [103]. In the same study, the conductivity (I/V characteristics) in BNNTs was shown to be sensitive to the adsorbed molecule with G nucleobase offering an additional conduction channel near the Fermi energy of the bioconjugated system. Instead, high background current associated with the metallic SWCNT made a small variation in current due to the molecular adsorption difficult to detect. Noncovalent adsorption of nucleobases, A, G, C, T, and U on a high curvature (5,0) BNNT showed that, except for G nucleobase, A, C, T, U exhibited similar binding energy values on BNNT sidewall [104]. The binding energy not only depended on the polarizability of bases but also marginally on the degree of hybridization of electronic states with BNNT. A relatively higher binding energy of G nucleobase was predicted on a (5,0) SWCNT and followed the order: $G > A > T > C > U$ [105]. In the equilibrium configuration, A, T, and U bases adopted a configuration reminiscent of the Bernal's AB stacking of two adjacent graphene layers in graphite. The G and C bases, on the other hand, showed a lesser degree of resemblance to AB stacking. The deviation in molecular geometry of the bases on SWCNT sidewall was attributed to the presence of both nitrogen and carbon in the rings and the different side groups containing CH_3, NH_2, or O. Molecular polarizability of the bases along with the nanotube curvature dependence determined the interaction strength in the bioconjugated system [105]. Molecular polarizability of nucleobases also drive the physisorption on graphene that allowed a weakly attractive dispersion force to be induced in the bioconjugated system [106]. The calculated binding energy for the physisorption on graphene followed the order: $G > A \approx T \approx C > U$.

Using an integrative theoretical and experimental study, Das et al. [107], investigated the interaction of DNA nucleobases with (5,5) SWCNT using both DFT and classical force field methods. DFT calculations showed that the binding energy of nucleobases onto SWCNT was governed mainly by vdW forces and followed the order: $C > G > A > T$. The noncovalent stacking of DNA/RNA nucleobases onto zigzag (7,0) SWCNT sidewall showed a binding energy trend: $G > C > A > T > U$ [108]. Nanotubes with small diameter due to the low curvature, exhibited low binding whereas nanotubes with high diameter had higher interaction energy with the nucleobases [108]. DFT study on the noncovalent binding of DNA and RNA nucleobases, adenine (A), guanine (G), cytosine (C), thymine (T), and uracil (U) with a $(ZnO)_{12}$ nanocluster (Fig. 10.8A) followed the order: $G < T < U < A < C$ [109]. The covalent and electrostatic and/or vdW interactions dominated the bonding characteristics of the conjugates, where the degree of hybridization between Zn-d with N-p orbitals determined the interaction strength of the individual nucleobases and

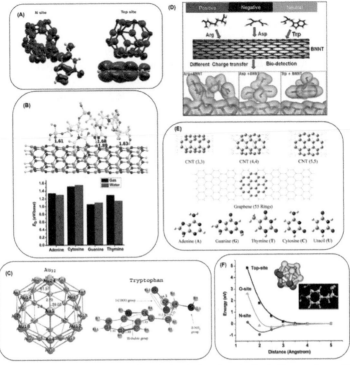

Fig. 10.8 QM approaches to functionalize biological molecules including amino acids, small peptides, and DNA nucleobases with nanomaterials and nanoclusters. Charge interactions along with hydrophobic, vdW, and H-bonding stabilizes the interaction in the bioconjugates.
(A) Noncovalent binding of DNA and RNA nucleobases, adenine (A), guanine (G), cytosine (C), thymine (T), and uracil (U) with a $(ZnO)_{12}$ nanocluster [109]. (B) Binding of DNA nucleobases with chalcogenide QDs and BNNT, studied using dispersion corrected DFT calculations [110]. (C) Interaction of tryptophan (Trp) amino acid with Au_{32} nanocluster [111]. (D) Interaction of Trp, aspartic acid (Asp), and arginine (Arg) amino acids with BNNT [112]. (E) Physisorption of nucleobases, A, T, G, C, and U on SWCNT and graphene nanostructures [113]. (F) Interaction of metallic nanoclusters of Mn_{13}, Ag_{13}, and Al_{13} with dopamine molecule [114].
Panel A: Adapted from V. Shewale, P. Joshi, S. Mukhopadhyay, M. Deshpande, R. Pandey, S. Hussain, S.P. Karna, First-principles study of nanoparticle–biomolecular interactions: anchoring of a (ZnO)12 cluster on nucleobases, J. Phys. Chem. C 115 (2011) 10426–10430. Copyright 2011, American Chemical Society; Panel B: Adapted from Z. Wang, H. He, W. Slough, R. Pandey, S.P. Karna, Nature of interaction between semiconducting nanostructures and biomolecules: chalcogenide QDs and BNNT with DNA molecules, J. Phys. Chem. C 119 (2015) 25965–25973. Copyright 2015, American Chemical Society; Panel C: Adapted from P. Joshi, V. Shewale, R. Pandey, R.V. Shanker, S. Hussain, S.P. Karna, Tryptophan–gold nanoparticle interaction: a first-principles quantum mechanical study, J. Phys. Chem. C 115 (2011) 22818–22826. Copyright 2011, American Chemical Society; Panel D: Adapted from S. Mukhopadhyay, R.H. Scheicher, R. Pandey, S.P. Karna, Sensitivity of boron nitride nanotubes toward biomolecules of different polarities, J. Phys. Chem. Lett. 2 (2011) 2442–2447. Copyright 2011, American Chemical Society; Panel E: Adapted from D. Umadevi, G.N. Sastry, Quantum mechanical study of physisorption of nucleobases on carbon materials: graphene versus carbon nanotubes, J. Phys. Chem. Lett. 2 (2011) 1572–1576. Copyright 2011, American Chemical Society; Panel F: Adapted from C. Liu, H. Haiying, R. Pandey, S. Hussain, S.P. Karna, Interaction of metallic nanoparticles with a biologically active molecule, dopamine, J. Phys. Chem. B 112 (2008) 15256–1525. Copyright 2008, American Chemical Society.

$(ZnO)_{12}$ cluster. The study finds importance in understanding ZnO-biomolecule inter-actions at the atomic level and the potential biomedical application of ZnO nanoparticles. Wang, et al., investigated the binding of DNA nucleobases with chal-cogenide QDs and BNNT using dispersion corrected DFT calculations [110]. The chalcogenide QDs were shown to bind at their metal centers through electrostatic interactions with the O- and N-site of DNA nucleobases. A comparison of the average binding energy per DNA base showed BNNTs to have a stronger binding than QDs and cytosine had the strongest binding strength (Fig. 10.8B).

QM calculations on the interaction of tryptophan (Trp) amino acid with Au_{32} nanocluster was performed at the level of DFT to predict the equilibrium configura-tions, interaction energies, the density of states, molecular orbitals, and charge density of the complexes (Fig. 10.8C) [111]. The Trp-Au conjugate was stabilized by salt-bridge COO^-, charge solvent interaction involving the indole ring and amine group interaction involving the NH_2 group. The π–π stacking along with ionic interactions contributed to the formation of stable Trp–Au complex and modified the electronic states of Trp by introducing hybrid states associated with Au as low-lying excited states. The interaction of Trp, aspartic acid (Asp), and arginine (Arg) amino acids with BNNT at the LDA–DFT level of theory reported that for the Trp–BNNT complex, the interaction was stabilized by vdW forces while for the polar amino acids, electrostatic interactions were predominant in stabilizing the conjugates (Fig. 10.8D) [112]. The stronger binding of Arg and conceivably other positively charged amino acids with the BNNT surface might explain the experimentally observed natural affinity of a pro-tein toward BNNTs and thereby enable BNNTs to immobilize proteins.

For the noncovalent adsorption of nucleobases on SWCNT and graphene nanostructures, the curvature effect determined the binding strength of DNA–nanostructure complexes (Fig. 10.8E) [113]. The effect of nanotube curvature on the physisorption of DNA/RNA nucleobases on CNTs studied at the level of B3LYP-D/6-31G*//ONIOM(M06-2X/6-31G*: AM1) level with BSSE correction showed that adsorption of nucleobases on CNT followed the order: $G > T \sim A > C > U$ and $G > A > T > C > U$ for monolayer graphene. An inverse relationship between nanostructure curvature and binding affinity was reported and the least curved graphene exhibited the highest binding energy in excellent agreement with experi-ment and previous theoretical calculations. The aromaticity of nucleobases increased significantly upon binding for both SWCNTs and graphene projecting graphene with superior DNA sequencing properties compared to SWCNTs. DFT calculations on the interaction of metallic NPs of Mn_{13}, Ag_{13}, and Al_{13}, with a biologically active mol-ecule, dopamine showed a higher interaction strength for Mn cluster over Ag or Al clusters [114]. The nanoclusters exhibited site-specific interaction toward dopamine molecule, Mn_{13} cluster preferred to bind at the top-site (Fig. 10.8F), whereas Ag_{13} and Al_{13} preferred the N-site of dopamine. The site selectivity of interaction was con-trolled by the hybridization of the atomic orbitals of metals with the molecular orbitals in the dopamine-nanocluster complex.

In a previous study on the covalent functionalization of therapeutic drugs with (5,5) SWCNTs using azomethine ylide functional group, we showed that covalent func-tionalization of nanotube sidewall was thermodynamically favored with negative

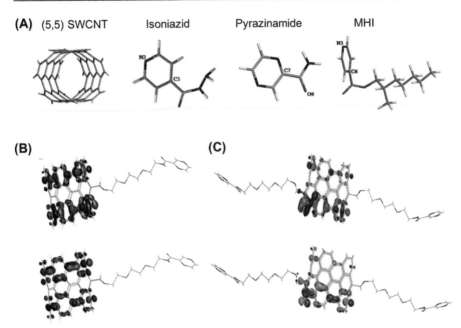

Fig. 10.9 (A) Optimized geometries of (5,5) SWCNT, isoniazid, pyrazinamide, and MHI antitubercular drug molecules. (B) Charge distribution of HOMO and LUMO orbitals of functionalized (5, 5)-SWNT loaded with one pyrazinamide molecule. (C) HOMO and LUMO orbitals of functionalized (5,5)-SWCNT loaded with two pyrazinamide molecules.
Adapted from N. Saikia, R.C. Deka, Theoretical study on pyrazinamide adsorption onto covalently functionalized (5,5) metallic single-walled carbon nanotube, Chem. Phys. Lett. 500 (2010) 65–70. Copyright 2010, Elsevier.

binding energy and decreased HOMO–LUMO energy gap (Fig. 10.9) [115,116]. The decrease in global hardness and HOMO–LUMO energy gap of the drug–nanotube complex with an increase in functionalization connote decreased stability of the complex with increase in sidewall functionalization. The increased functionalization of nanotube sidewall induced structural deformation within the nanotubes thereby modulating its electronic properties. We also established the curvature effect on the binding energy of the PZA–SWCNT complex, nanotubes with narrow diameter were thermodynamically more favorable compared to tubes with higher diameter. The covalent functionalization of SWCNTs perturbed the electronic, structural properties, and frontier orbitals compared to pristine nanotubes.

We also extended the study to a new class of graphene-based 2D nanomaterials namely BN sheet, silicene, SiC, and phosphorene (Fig. 10.10A) [117]. Using a finite cluster model, at the level of first principles DFT with vdW dispersion correction, we showed that pyrazinamide (PZA) selectively binds to the 2D clusters and the extent of interaction was governed solely by site-specificity and intrinsic electronic properties of 2D clusters. PZA was chemisorbed on silicene and SiC with a strongly localized puckering along the absorption site mediated by Si—O covalent bond (Fig. 10.10B).

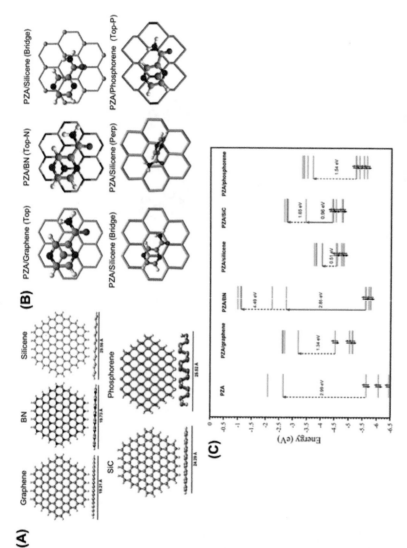

Fig. 10.10 (A) Equilibrium geometries of graphene, BN, silicene, SiC, and phosphorene nanoclusters were calculated at DFT(PBE)–D2 level of theory. (B) Equilibrium configurations of PZA–conjugated complexes of graphene, BN, SiC, silicene (perpendicular), and phosphorene clusters. (C) Molecular orbital diagram of PZA, PZA/graphene, PZA/BN, PZA/silicene, PZA/SiC, and PZA/phosphorene. The *red dashed arrows* correspond to the energy gap in PZA and PZA/2D nanoclusters. The *pink arrows* correspond to midgap states within the energy gap of PZA/BN and PZA/SiC complexes introduced by the PZA molecule.

Adapted from N. Saikia, M. Seel, R. Pandey, Stability and electronic properties of 2D nanomaterials conjugated with pyrazinamide chemotherapeutic: a first-principles cluster study, J. Phys. Chem. C 120 (2016) 20323–20332. Copyright 2016, American Chemical Society.

The BN and SiC nanoclusters exhibited significant changes in the calculated density of states (DOS) around the Fermi level where the adsorbed PZA introduced midgap states (Fig. 10.10C). Of the studied cluster models, silicene and SiC clusters formed the most stable payloads for PZA favoring covalent functionalization over noncovalent interactions.

10.14 QM calculations on the molecular self-assembly of noncanonical DNA bases on graphene

The molecular self–assembly of noncanonical guanine nucleobases $(Gua)_n$ on graphene monolayer within the framework of dispersion-corrected DFT was shown to be mediated by the interplay between base–base intermolecular and bifurcated H–bonds (Fig. 10.11) [118]. At the M06–2X level of theory, $(Gua)_n$ bases stabilized

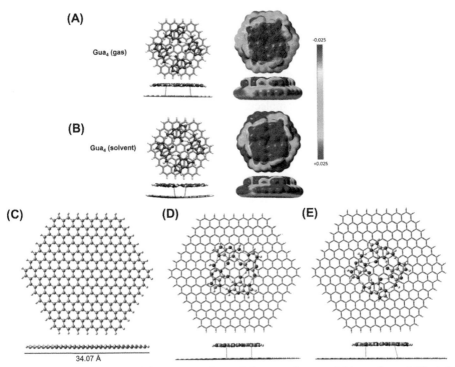

Fig. 10.11 (A and B) Equilibrium geometries and electrostatic potential isosurface of $(Gua)_4$/graphene complex in gas and solvent phases. (C) Equilibrium configuration of graphene nanoflakes consisting of 336 atoms, (D and E) equilibrium configuration of $(Gua)_4$ physisorbed on graphene nanoflake in gas and solvent phases.
Adapted from N. Saikia, S.P. Karna, R. Pandey, Theoretical study of gas and solvent phase stability and molecular adsorption of noncanonical guanine bases on graphene, Phys. Chem. Chem. Phys. 19 (2017) 16819–16830, with permission from the PCCP Owner Societies.

through bifurcated H–bonds yield nearly planar geometries while bases stabilized by intermolecular H–bonds resulted in puckered geometries. The substrate–induced effects on the assembly of $(Gua)_n$ bases were associated with the coupling of base–base and base–substrate intermolecular interactions. The presence of graphene substrate stabilized the adsorption of free-standing $(Gua)_n$ bases by lowering the inherent polarity of $(Gua)_n$ bases. The subtle influence on the nature of adsorption and the overall stability of $(Gua)_n$ complexes was also determined by the way we modeled the 2D substrate as shown in Fig. 10.11C–E. For the graphene cluster considered, DFT–D2 calculations indicated an overall stability of $(Gua)_n$ complexes in gas and solvent phases. Our study formed a basis for atomistic MD simulations of free–standing and interacting $(Gua)_n$ bases in gas and solvent phases. We determined the dominant interaction modes namely the base–base vs. base–substrate π-stacking in governing the supramolecular self–assembly at the solid/liquid interface.

10.15   Overview of molecular mechanics method

All-atom MD simulations are based on molecular mechanics (MM), a method to accurately determine a priori structure and energies of molecules. MM considers atoms as hard spheres and the bonds between the atoms to be linked by mechanical springs, which control the bond distance, angles, rotation around the bonds, and so on. The potential energy of an arbitrary molecule is given as a superposition of two body, three body, and four body interactions and is expressed as a sum of valence or bonded and nonbonded interactions [119]. In the MM method the energy of the molecule is denoted as a parametric function of the nuclear coordinates [2] approximated as:

$$
E_{total} = \overset{bond}{\sum} E_{stretch} + \overset{bond-angle}{\sum} E_{bend}
$$
$$
+ \overset{torsion-angle}{\sum} E_{torsion} + \overset{out-of-plane}{\sum} E_{out-of-plane}
$$
$$
\overset{non-bonded}{\sum} E_{non-bonded} + \sum E_{cross-term} \qquad (10.29)
$$

where E_{total} is the total energy of a system, $E_{stretch}$ is the energy associated with bond stretching or compression from an ideal bond length, E_{bend} is the bond angle bending, $E_{torsion}$ is the dihedral torsional energy associated with rotation around a bond, $E_{nonbonded}$ corresponds to nonbonded interactions (vdW and electrostatic interactions), and $E_{cross-term}$ denotes the coupling between the first three terms. The above equation together with the parametric functions required to describe the different atoms and bonds is termed as a force field. The force field is defined as a mathematical function that determines the energy of a system as a function of its conformation. With the introduction of MM2 in 1977, a relatively good prediction of molecular structure and energies is possible. An optimized force field can accurately gauge the accuracy of simulation in terms of enhanced conformational sampling and extensive parameterization. Sampling refers to how closely the conformations represent a Boltzmann-weighted ensemble average and measures the confidence in thermodynamic accuracy, while parametrization assigns values for atomic properties such as atomic charge or bonded parameters to atoms [39,120].

Empirical force fields for molecular simulations are broadly classified as class I force fields (AMBER, CHARMM, OPLS, and GROMOS) and class II force fields (MM2–4, CFF, PCFF, and COMPASS). Class I force fields are typically used for the simulation of proteins, DNA, RNA, carbohydrates in aqueous solution, and class II force fields are typically used for the simulation of small molecules in vacuum and bulk materials such as polymers, metals, and ceramics [120,121]. The energy expression of the class II force field is much more complex than the class I force field as it includes higher-order terms for the energy contributions for the bond stretch, bend, and dihedral rotations, and cross terms for simulating the effect of bond stretching on bending and dihedral rotations. The Lennard-Jones terms are particularly subject to error because they are generally developed for individual atom types, with arithmetic or geometric combining rules then used to provide the equilibrium separation distance and well-depth characteristics for pairing one atom type with another [120]. To this end, accurate description of noncovalent interactions has become critical for obtaining equilibrium geometries, relative energies, and binding affinity of noncovalent complexes. The semi-experimental equilibrium geometries of noncovalent complexes (small rigid molecules and ab initio geometries of complexes) were benchmarked to create a reliable database of structural parameters based on experimental data [122]. For a detailed review on developing a nonpolarizable molecular model for peptide–surface interactions, fitting of model parameters as well as validation of force field with experimental data, we would direct to an earlier study by Martin et al. [39].

10.15.1 Molecular dynamics simulation

The MD simulation is a technique of generating atomic trajectories by numerical integration of Newton's equation of motion for a specific interatomic potential within certain initial and periodic boundary conditions [123]. Due to their remarkable resolution in space (single atom), time (femtosecond), and energy, MD simulations represent a powerful complement to experimental techniques, providing mechanistic insight into experimentally observed processes [124]. MD simulations are broadly classified into classical MD (where molecules are treated classically as ball and stick models and laws of classical mechanics determines the dynamics) and quantum (ab initio) MD which is based on the quantum nature of the chemical bond [125].

The basic idea of MD is the time evolution of a particle in an N-body system obtained by solving the Newton's equation of motion:

$$m_i \frac{d^2 r_i(t)}{dt^2} = F_i \quad \text{for } i = 1, 2, \ldots\ldots N \tag{10.30}$$

where $r_i(t)$ denotes the position vector of the ith particle, m_i is the mass, and F_i is the force acting on the ith particle. Two types of boundary conditions are used in MD:

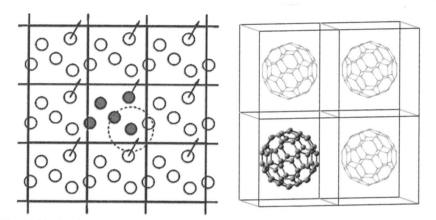

Fig. 10.12 The depiction of a periodic boundary condition (PBC) model. As a particle moves out of the simulation box, an image particle moves in to replace its position.
The figure in the left panel is adapted from M.P. Allen, Introduction to Molecular Dynamics Simulation, Computational Soft Matter: From Synthetic Polymers to Proteins, Lecture Notes, N. Attig, K. Binder, H. Grubmuller, K. Kremer (Eds.), John von Neumann Institute for Computing, Julich, NIC Series, vol. 23, 2004, pp. 1–28.

isolated boundary condition (IBC) and periodic boundary condition (PBC). In the IBC model, the system is suspended in a vacuum isolated from the surrounding with no interactions on the outside except for some well-defined "external forcing." In the PBC model, the motion of N particles is confined in a periodic supercell wherein the supercell is surrounded by replicating periodic images as illustrated in Fig. 10.12.

10.15.2 Force calculation using Newton's equation of motion

The numerical integration of Newton's equation of motion solves for an expression that defines position $r(t+\delta t)$ at a time $(t+\delta t)$ in terms of already known positions [17]. The position of each atomic coordinate is thus expressed as an expansion of Taylor series given by:

$$r(t + \delta t) = r(t) + v(t)\delta t + \frac{1}{2}a(t)(\delta t)^2 + \cdots\cdots \tag{10.31}$$

$$v(t + \delta t) = v(t) + a(t)\delta t + \frac{1}{2}b(t)(\delta t)^2 + \cdots\cdots \tag{10.32}$$

$$a(t + \delta t) = a(t) + b(t)\delta t \tag{10.33}$$

where "r" represents the velocity and "a" the acceleration. The high order terms in the Taylor series are neglected if acceleration does not fluctuate during the time step. The time step in MD simulation should be small enough such that the gradients do not fluctuate significantly at a particular time step, δt.

10.15.3   Ewald summation [126,127]

Ewald summation named after Paul Peter Ewald is a method for computing the inter-action energies (particularly electrostatic energy) in periodic systems [98]. The sum-mation of interaction energies in real space is replaced by an equivalent summation in Fourier space. The Coulomb (electrostatic) energy for a system of N particles in a cubic box of size L within the PBC is given by:

$$U = \frac{1}{2} \sum_{i,j,n} \frac{q_i q_j}{|r_i - r_j + nL|} \tag{10.34}$$

where ½ factor cancels double-counting of charges q_i, q_j on atoms i, j, r_i, r_j are the corresponding coordinates and nL the displacement vector. To gain further insight we consider the electrical potential field which is defined as the electrostatic energy acting on an atom having +1 charge when placed at position r given by the expression:

$$\phi(r) = \sum_{i,n} \frac{q_i}{|r_i - r + nL|} \tag{10.35}$$

Similarly, for position r_i, the potential $\phi(r_i)$ is given by:

$$\phi(r_i) = \sum_{j,n}' \frac{q_i}{|r_i - r_j + nL|} \tag{10.36}$$

where the prime symbol in equation excludes double counting of atom i from the potential. The total electrostatic energy reduces to:

$$U = \frac{1}{2} \sum_i q_i \phi(r_i) \tag{10.37}$$

In the Ewald sum, the charge is split into two components, one that decays rapidly with r and the other that is dominated by the long-range interaction. The electrostatic and total Coulomb interaction energy is given by:

$$E = \frac{1}{2} \sum_{i=1}^{N} q_i \phi_i^S(r_i) + \frac{1}{2} \sum_{i=1}^{N} q_i \phi^L(r_i) - \frac{1}{2} \sum_{i=1}^{N} q_i \phi_i^L(r_i) \equiv E^S + E^L - E^{self} \tag{10.38}$$

$$E = \frac{1}{4\pi\varepsilon_0} \frac{1}{2} \sum_{i,j,n} \frac{q_i q_j}{|r_i - r_j + nL|} erfc\left(\frac{|r_i - r_j + nL|}{\sqrt{2}\sigma}\right)$$

$$+ \frac{1}{2V\varepsilon_0} \sum_{k \neq 0} \frac{e^{-\sigma^2 k^2/2}}{k^2} |S(k)|^2 - \frac{1}{4\pi\varepsilon_0} \frac{1}{\sqrt{2\pi}\sigma} \sum_i q_i^2 \tag{10.39}$$

10.15.4   Solvation effect

The solvation effect in empirical force field simulations is, in general, explicitly represented using an accurate water model. The benefit of using explicit solvent in interfacial simulations is that water molecules can specifically interact with the functional groups of amino acid residues of peptide/protein and the nanomaterial surface. These interactions can compete with the interactions between water molecules and peptide–surface or protein–surface interactions. One caveat of implementing explicit solvent simulations is that a large number of water molecules must be used to appropriately represent the solvation environment. Accordingly, over 90% of the computational time is spent simulating the behavior of bulk water as opposed to peptide–water or peptide–surface interactions. To circumvent this problem, implicit solvent simulation has been developed that considers the effect of the aqueous environment by incorporation a mean-field approximation that is directly integrated into the force field equation. Although implicit solvent simulations significantly reduces the degrees of freedom of the molecular system, thereby reducing the computational requirements and extending the simulation to longer timescales, it comes at the expense of decreased accuracy.

10.15.5   Atomistic simulation of biological molecules with nanomaterials

Using all-atom atomistic MD simulation, we probed the supramolecular self-assembly of noncanonical DNA nucleobases, guanine, and cytosine on monolayer graphene and BN nanomaterial [128–130]. The molecular assembly was shown to depend on the base–base and base-substrate intermolecular interactions. The base-substrate interactions determined the formation of self-assembled monolayer on graphene, whereas the base–base interactions governed the nature of the aggregation of nucleobases on the nanomaterial surface [128]. The substrate-induced effect was important in mediating the monolayer assembly of the nucleobases. In the gas phase, guanine bases adopted random dispersion on graphene, rendered from the inherent polarity and the availability of donor/acceptor sites enabling myriad polymorphic arrangements. In solvent media, guanine bases assimilated in disordered aggregates with no well-defined arrays on graphene that was in excellent agreement with previous AFM studies on a highly-ordered pyrolytic graphite surface. In the case of a h-BN monolayer, we showed that the distinctive patterns in the self-assembly of noncanonical cytosine and guanine nucleobases on the h-BN surface can serve as fingerprints for biomolecular recognition at the solid/liquid interface [130]. The polarity of nucleobases determined the molecular ordering and growth patterns on h-BN: cytosine into short-to medium-order 1D linear arrays and interconnected molecular chains and guanine into a 2D aggregated network as shown in Fig. 10.13A and B. The base–base interactions via the intermolecular hydrogen bonding were crucial in the self-assembly while the base-surface interactions facilitated surface recognition and adsorption on h-BN. Cytosine demonstrated well-defined periodicity in the self-assembly at both intermediate and saturated coverage with medium-range molecular ordering (Fig. 10.13A and C).

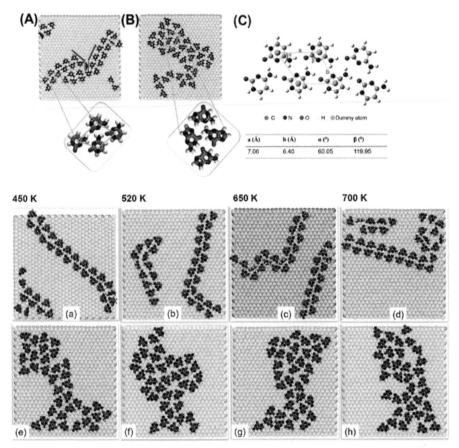

Fig. 10.13 (Top panel) Gas-phase (intermediate coverage) snapshots of cytosine and guanine self-assemblies on h-BN: (A) C_{36}/h-BN and (B) G_{36}/h-BN. (C) Unit cell parameter of cytosine. (Bottom panel) Snapshots of C_{36}/h-BN and G_{36}/h-BN complexes at elevated temperatures from 450 to 700 K. (–a–d) C_{36}/h-BN and (–e–h) G_{36}/h-BN.
Adapted from N. Saikia, R. Pandey, Polarity-induced surface recognition and self-assembly of noncanonical DNA nucleobases on h-BN monolayer, J. Phys. Chem. C 122 (2018) 3915–3925. Copyright 2018, American Chemical Society.

The growth patterns of both the nucleobases were supported by the vdW corrected DFT calculations which illustrate the charge density distribution over the assembled molecular arrays.

Karachevtsev et al. [131], integrated Raman spectroscopy along with DFT and MD simulations to investigate the interface interaction of 1-pyrenebutanoic acid N-hydroxysuccinimide ester (PSE) with SWCNTs and enzyme glucoseoxidase (GOX). Raman spectroscopy monitored the shifting of the tangential G mode (1500–$1650\,\mathrm{cm}^{-1}$) upon the formation of CNT/PSE hybrids and quantum chemical calculations at the level of M05-2X/6–31++G(d,p) confirmed that the hybrids were stabilized by π-stacking interactions between the pyrene fragment of PSE and the nanotube surface [131]. MD simulations showed that the total interaction energy was about 35 and 25 kcal/mol, respectively, for CNT/PSE and between pyrene fragment and nanotube surface. A single pyrene effectively anchored the GOX enzyme to the nanotube surface. In a recent study, Samieegohar et al. [132], employed ReaxFF MD simulations to study gold nanoparticles modified with different short-chain peptides consisting of amino acid residues of cysteine and glycine (CGCG and CGGG) in different grafting densities. The peptide chains had higher preferential adsorption on Au(111) facet compared to Au(100) and Au(110). In addition to the stable thiol interaction with the Au nanoparticle, polarizable oxygen and nitrogen atoms showed strong interactions with gold surface. Due to the thiol – Au bonding, the peptide chains were tightly adsorbed on AuNP surfaces. Furthermore, proton transfer from thiol groups of cysteine residues to water or carboxyl groups was detected over the course of simulation, in good agreement with previous quantum simulations and experiments. Atomistic simulation of collagen – like peptides with Au nanosurface (AuNS) and AuNP suggested a rapid association of collagen peptide with the surface due to the presence of charged layers above Au nanosurface [133]. The interaction between gold atoms and the protein backbone led to the rapid unfolding of collagen helices on nanosurface and nanoparticles. This unfolding of collagen was attributed to the interaction of protein side chains (e.g., OH and NH_3^+) and the backbone of collagen triple helices.

ssDNA and oligopeptides have been engineered to spontaneously wrap onto solid surfaces of 2D layered materials, including graphite, MoS_2, WSe_2, WS_2, $MoSe_2$, and hBN through hydrophobic vdW π–stacking interactions [66]. The organized assembly of oligonucleotide structures can distinctively modify the electronic states and optical properties of nanomaterials [134]. Spontaneous helical wrapping of highly charged, conjugated polymer poly-[p-phenylene]-ethynylene (PPE) on SWCNT in the absence and presence of aqueous environment showed that in vacuum simulations, PPE maintained vdW contact with the nanotube surface but had no persistent global helical structure [135]. In aqueous simulations the helical structure of PPE was formed spontaneously. The relative stabilities of different helical configurations of the PPES/ SWNT system obtained using the potential of mean force calculations and the experimentally measured helical pitch in aqueous solution was found to be a global free energy minimum. The side-chain conformations of the aliphatic segments were dictated by the vdW contact with the tubular surface while maintaining solvation of the sulfonate groups of PPE. Helical wrapping of BNNT by polynucleotides in aqueous solution was investigated by MD simulations to understand the interaction between biomolecules and nanomaterials and the dispersion mechanism of BNNT [136]. BNNT was shown to wrap along the sidewall through the hydrophobic nucleobases and hydrophilic phosphate groups with the bases to closely interact with the BNNT surface and the phosphate atoms to interact with water molecules. The phosphate groups facilitated the dispersion of BNNT and hindered its self-aggregation in aqueous solution. Further nanotube diameter could also affect the dispersion of BNNT in aqueous solution by changing the interaction strength between nucleotides and the BNNT surface [136]. Binding and unzipping of double-stranded xylonucleic acid (dsXNA) containing xylose, a stereoisomer of ribose sugar was assisted by SWCNT and XNA underwent faster unzipping compared to dsRNA of the same sequence [137]. Atomistic simulations performed to unravel this unzipping mechanism showed

XNA to adopt an extended helical geometry with a severely strained phosphate backbone. The screening of negative ionic charges at higher salt concentration decreased the binding on the SWCNT sidewall as compared to charge neutral cases causing slower unzipping. The thermal fluctuations at relatively higher temperatures were more favorable for duplex unwinding. The findings have relevance in extending the integration of SWCNTs with biological systems as an efficient transporter for nucleic acids and drugs into human cells.

Atomistic simulations on the assembly of gellan gum with clay mineral were studied in atomic resolution using MD simulation and AFM method [138]. The agglomeration of gellan was shown to be pH-dependent. The polysaccharide backbone experienced rapid conformation equilibria between extended and coiled states in pH values from 2 to 5 or higher due to ionization and dynamic intramolecular ion bridge as shown in Fig. 10.14. The incorporation of clay nanomaterial enhanced the adsorption of gellan along the edges with direct contact via the formation of H-bonds. Strong clay–polymer attraction increased the overall conformational flexibility and reduced chain stiffness at intermediate clay surface charge and high mobility of interlayer cations along with negatively charged gellan backbones at a pH value of 9 [138].

10.16   Challenges in atomistic simulations at the nano-bio interface

Despite the diverse potential applications presented by nanomaterials, the integration with biological systems is one of the most challenging research areas due to the high structural diversity exhibited by nanomaterials coupled with molecular heterogeneity, high surface to volume (aspect) ratio, low dissolution in common solvents, and molecular crowding effects [28,139,140]. While nanomaterials have their inherent properties, it is generally necessary to impart additional functionality through noncovalent or chemical modifications to increase their biocompatibility. In biological media that closely mimics the intracellular environment, the increasing influence of ions, small molecules, or biological molecules can strongly interfere with the nano-bio interactions [141]. Since noncovalent forces mainly govern the interaction of nanomaterials with biological systems, the solvent effect on the intermolecular interactions cannot be completely overlooked. Compelling evidence points that electrostatic interactions are not the primary determinant in driving the interfacial interactions rather the ability of amino acids to sense the solvent density variations at the solid/liquid interface directs surface recognition [44]. Despite the remarkable advancements in the design and development of nanomaterials, not much is known about its interaction with biological systems. While in vitro studies have made great advancements in unraveling the effect of nanoparticles, there is a pressing need to understand how nanoparticles interact and adhere to cell membranes in vivo, the mechanism of cellular entry, endocytosis, and membrane disruption [142]. Such knowledge is critical in the design of nanomaterials with surface morphologies that do not induce adverse side effects to living organisms and achieve the desired functionality [142]. The dynamic changes of nanomaterials in presence of biological molecules, aggregation mechanism, surface adsorption, and interfacial properties are still speculative at the molecular

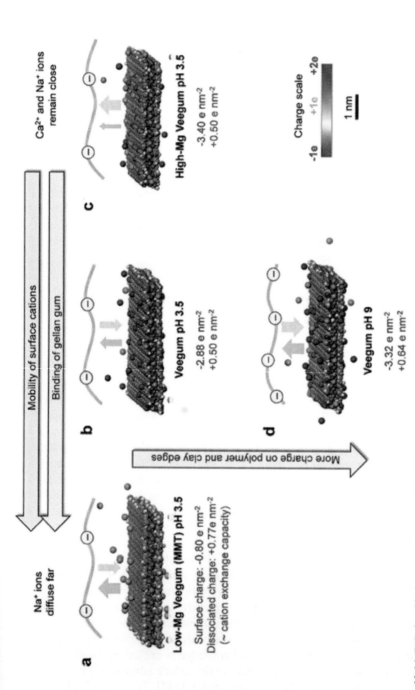

Fig. 10.14 Relationship between clay–water interfaces and carbohydrate-binding. (A–C) The clay composition determines the total surface charge (red color), number of dissociated ions (blue color), and resulting polymer attraction via ion pairs and close contacts (orange arrows). High surface charge binds surface cations tightly and creates dense hydration layers that do no attract gellan gum. High-Mg Veegum pH 3.5 edges and the gellan backbone exhibits twice the number of ionic sites to interact with the clay mineral, leading to stronger binding. (D) At pH 9, more cations dissociate from the clay mineral. Reproduced from T. Jamil, J.R. Gissinger, A. Garley, N. Saikia, A.K. Upadhyay, H. Heinz, Dynamics of carbohydrate strands in water and interactions with clay minerals: influence of pH, surface chemistry, and electrolytes, Nanoscale 11 (2019) 11183–11194, with permission from The Royal Society of Chemistry.

level [41,139]. Also nanomaterials can interfere with fluorescence and degrade fluorophore dye labels [143,144]. This knowledge gap limits the rational development of methods/models for peptides with the specific binding propensity to nanomaterials and the quantitative measurements of binding affinity, structure–function relationship, and other interfacial properties.

10.17 Conclusions and general remarks

Nanobiotechnology represents a broad area of research at the intersection of biology and nanomaterials that focuses on utilizing the unique physical, electronic, optical, and mechanical properties of nanomaterials with a biological molecule [28]. The improved solubility, high volume-to-surface (aspect) ratio, quantum size, and tunneling effect, coupled with its multifunctionality have rendered nanomaterials suitable for a wide range of applications, including life science, engineering, biotechnology, and medicine [145,146]. The higher free energy of nanomaterials enable surface functionalization and adsorption of proteins or biomolecules. While the entropic change associated with protein adsorption is either due to dehydration of uncharged surfaces or removal of the electric double layer from charged surfaces, the enthalpic changes are either from hydrogen bonding or formation of donor–acceptor coordination between the protein and nanomaterial surface [147,148]. The integration of biological molecules with nanomaterials is pursued in the design of nanocomposites with enhanced mechanical properties, molecular electronic devices, and in the biomedical field [149,150].

Although the inherent electronic and optical properties of nanomaterials can be tailored to improve its functionality, the heterogeneity of biomolecular interaction, structural integrity of the conjugates on binding and interfacial properties of nanomaterial surface–bound biomolecules remain elusive both experimentally and at the computational levels. Concomitant with recent experimental techniques, computational methods have facilitated the molecular-level understanding of biomolecular systems interacting with nanomaterials. In this chapter, we cover the applications of QM calculations and atomistic simulations to study the interaction of amino acids, peptides, proteins, and DNA oligonucleotides with functional nanomaterials at varying length and timescales. We discuss the scalability of multiscale modeling techniques and state-of-the-art simulation methods to account for biological molecules interacting with nanomaterials and provide a new perspective in the application of integrative methods that combine computation with experiments at various spatiotemporal levels.

Acknowledgments

NS would like to thank Prof. Ramesh Ch. Deka of the Department of Chemical Sciences at Tezpur University and Prof. Max Seel of the Department of Physics at Michigan Technological University for the helpful discussions.

References

[1] O.V. Salata, Applications of nanoparticles in biology and medicine, J. Nanobiotechnol. 2 (2004) 3.

[2] F. Wang, W.B. Tan, Y. Zhang, X. Fan, M. Wang, Luminescent nanomaterials for biological labelling, Nanotechnology 17 (2006) R1–R13.

[3] T.A. Taton, Nanostructures as tailored biological probes, Trends Biotechnol. 20 (2002) 277–279.

[4] S. Saallah, I.W. Lenggoro, Nanoparticles carrying biological molecules: recent advances and applications, KONA Powder Part. J. 35 (2018) 89–111.

[5] X. Tian, Y. Chong, C. Ge, Understanding the nano-bio interactions and the corresponding biological responses, Front. Chem. 8 (2020) 446.

[6] A.A. Shemetov, I. Nabiev, A. Sukhanova, Molecular interaction of proteins and peptides with nanoparticles, ACS Nano 6 (2012) 4585–4602.

[7] A. Zhang, C.M. Lieber, Nano-bioelectronics, Chem. Rev. 116 (2015) 215–257.

[8] X.Q. Zhang, X. Xu, N. Bertrand, E. Pridgen, A. Swami, O.C. Farokhzad, Interactions of nanomaterials and biological systems: implications to personalized nanomedicine, Adv. Drug Deliv. Rev. 64 (2012) 1363–1384.

[9] J. Peña-Bahamonde, H.N. Nguyen, S.K. Fanourakis, D.F. Rodrigues, Recent advances in graphene-based biosensor technology with applications in life sciences, J. Nanobiotechnol. 16 (2018) 75.

[10] C.N.R. Rao, A.K. Sood, K.S. Subrahmanyam, A. Govindaraj, Graphene: the new two-dimensional nanomaterial, Angew. Chem. Int. Ed. Engl. 48 (2009) 7752–7777.

[11] A. Star, E. Tu, J. Niemann, J.-C.P. Gabriel, S.C. Joiner, C. Valcke, Label-free detection of DNA hybridization using carbon nanotube network field-effect transistors, Proc. Natl. Acad. Sci. U. S. A. 24 (103) (2006) 921–926.

[12] W. Zhang, Z. Zhang, Y. Zhang, The application of carbon nanotubes in target drug delivery systems for cancer therapies, Nanoscale Res. Lett. 6 (2011) 555.

[13] A.M. Elhissi, W. Ahmed, I.U. Hassan, V.R. Dhanak, A. D'Emanuele, Carbon nanotubes in cancer therapy and drug delivery, J. Drug Deliv. 2012 (2012), 837327.

[14] N. Saikia, Functionalized carbon nanomaterials in drug delivery: emergent perspectives from application, in: G.Z. Kyzas, A.C. Mitropoulos (Eds.), Novel Nanomaterials—Synthesis and Applications, IntechOpen, 2017, pp. 231–255.

[15] N. Saikia, A.N. Jha, R.C. Deka, Dynamics of fullerene-mediated heat-driven release of drug molecules from carbon nanotubes, J. Phys. Chem. Lett. 4 (2013) 4126–4132.

[16] A. Bianco, K. Kostarelos, M. Prato, Applications of carbon nanotubes in drug delivery, Curr. Opin. Chem. Biol. 9 (2005) 674–679.

[17] N.W.S.K. Kam, H. Dai, Carbon nanotubes as intracellular protein transporters: generality and biological functionality, J. Am. Chem. Soc. 127 (2005) 6021–6026.

[18] G. Hong, S. Diao, A.L. Antaris, H. Dai, Carbon nanomaterials for biological imaging and nanomedicinal therapy, Chem. Rev. 115 (2015) 10816–10906.

[19] C.J. Murphy, A.M. Vartanian, F.M. Geiger, R.J. Hamers, J. Pedersen, Q. Cui, C.L. Haynes, E.E. Carlson, R. Hernandez, R.D. Klaper, G. Orr, Z. Rosenzweig, et al., Biological responses to engineered nanomaterials: needs for the next decade, ACS Cent. Sci. 1 (2015) 117–123.

[20] V.C. Sanchez, A. Jachak, R.H. Hurt, A.B. Kane, Biological interactions of graphene-family nanomaterials: an interdisciplinary review, Chem. Res. Toxicol. 25 (2012) 15–34.

[21] C. Cheng, S. Li, A. Thomas, N.A. Kotov, R. Haag, Functional graphene nanomaterials based architectures: biointeractions, fabrications, and emerging biological applications, Chem. Rev. 117 (2017) 1826–1914.

[22] S.R. Saptarshi, A. Duschl, A.L. Lopata, Interaction of nanoparticles with proteins: relation to bio-reactivity of the nanoparticle, J. Nanobiotechnol. (2013) 11–26.

[23] C. Ge, J. Du, L. Zhao, L. Wang, Y. Liu, D. Li, Y. Yang, R. Zhou, Y. Zhao, Z. Chai, C. Chen, Binding of blood proteins to carbon nanotubes reduces cytotoxicity, Proc. Natl. Acad. Sci. U. S. A. 108 (2011) 16968–16973.

[24] W. Ma, A. Saccardo, D. Roccatano, D. Aboagye-Mensah, M. Alkaseem, M. Jewkes, F. Di Nezza, M. Baron, M. Soloviev, E. Ferrari, Modular assembly of proteins on nanoparticles, Nat. Commun. 9 (2018) 1489.

[25] T. Lima, K. Bernfur, M. Vilanova, T. Cedervall, Understanding the lipid and protein Corona formation on different sized polymeric nanoparticles, Sci. Rep. 10 (2020) 1129.

[26] T. Cedervall, I. Lynch, S. Lindman, T. Berggård, E. Thulin, H. Nilsson, K.A. Dawson, S. Linse, Understanding the nanoparticle–protein corona using methods to quantify exchange rates and affinities of proteins for nanoparticles, Proc. Natl. Acad. Sci. U. S. A. 104 (2007) 2050–2055.

[27] E. Casals, T. Pfaller, A. Duschl, G.J. Oostingh, V. Punte, Time evolution of the nanoparticle protein Corona, ACS Nano 4 (2010) 3623–3632.

[28] K.E. Sapsford, W.R. Algar, L. Berti, K.B. Gemmill, B.J. Casey, E. Oh, M.H. Stewart, I.L. Medintz, Functionalizing nanoparticles with biological molecules: developing chemistries that facilitate nanotechnology, Chem. Rev. 113 (2013) 1904–2074.

[29] G.M. Whitesides, The 'right' size in nanobiotechnology, Nat. Biotechnol. 21 (2003) 1161–1165.

[30] Y. Wang, R. Cai, C. Chen, The Nano-bio interactions of nanomedicines: understanding the biochemical driving forces and redox reactions, Acc. Chem. Res. 52 (2019) 1507–1518.

[31] M.J. Limo, A. Sola-Rabada, E. Boix, V. Thota, Z.C. Westcott, V. Puddu, C.C. Perry, Interactions between metal oxides and biomolecules: from fundamental understanding to applications, Chem. Rev. 118 (2018) 11118–11193.

[32] E. Heikkilä, H. Martinez-Seara, A.A. Gurtovenko, M. Javanainen, H. Häkkinen, I. Vattulainen, J. Akola, Cationic Au nanoparticle binding with plasma membrane-like lipid bilayers: potential mechanism for spontaneous permeation to cells revealed by atomistic simulations, J. Phys. Chem. C 118 (2014) 11131–11141.

[33] T. Casalini, V. Limongelli, M. Schmutz, C. Som, O. Jordan, P. Wick, G. Borchard, G. Perale, Molecular modeling for nanomaterial-biology interactions: opportunities, challenges, and perspectives, Front. Bioeng. Biotechnol. 7 (2019) 268.

[34] R.E. Amaro, A.J. Mulholland, Multiscale methods in drug design bridge chemical and biological complexity in the search for cures, Nat. Rev. Chem. 2 (2018) 0148.

[35] Q. Cui, Perspective: quantum mechanical methods in biochemistry and biophysics, J. Chem. Phys. 145 (2016), 140901.

[36] J. Roel-Touris, A. Bonvin, Coarse-grained (hybrid) integrative modeling of biomolecular interactions, Comput. Struct. Biotechnol. J. 18 (2020) 1182–1190.

[37] K.A. Dill, S. Bromberg, K. Yue, K.M. Fiebig, D.P. Yee, P.D. Thomas, H.S. Chan, Principles of protein folding—a perspective from simple exact models, Prot. Sci. 4 (1995) 561–602.

[38] A.R. Dinner, A. Šali, M. Karplus, The folding mechanism of larger model proteins: role of native structure, Proc. Natl. Acad. Sci. U. S. A. 93 (1996) 8356–8361.

[39] L. Martin, M.M. Bilek, A.S. Weiss, S. Kuyucak, Force fields for simulating the interaction of surfaces with biological molecules, Interface Focus 6 (2016) 20150045.

[40] M. Sethi, M.R. Knecht, Experimental studies on the interactions between Au nanoparticles and amino acids: bio-based formation of branched linear chains, ACS Appl. Mater. Interfaces 1 (2009) 1270–1278.

[41] X.R. Xia, N.A. Monteiro-Riviere, S. Mathur, X. Song, L. Xiao, S.J. Oldenberg, B. Fadeel, J.E. Riviere, Mapping the surface adsorption forces of nanomaterials in biological systems, ACS Nano 5 (2011) 9074–9081.

[42] X. Zou, S. Wei, J. Jasensky, M. Xiao, Q. Wang, C.L. Brooks, Z. Chen, Molecular interactions between graphene and biological molecules, J. Am. Chem. Soc. 139 (2017) 1928–1936.

[43] J.M. MacLeod, F. Rosei, Molecular self-assembly on graphene, Small 10 (2014) 1038–1049.

[44] J. Schneider, L.C. Ciacchi, Specific material recognition by small peptides mediated by the interfacial solvent structure, J. Am. Chem. Soc. 134 (2012) 2407–2413.

[45] G. Qu, T. Xia, W. Zhou, X. Zhang, H. Zhang, L. Hu, J. Shi, X.F. Yu, G. Jiang, Property-activity relationship of black phosphorus at the nano-bio interface: from molecules to organisms, Chem. Rev. 120 (2020) 2288–2346.

[46] Q. Shao, C.K. Hall, Allosteric effects of gold nanoparticles on human serum albumin, Nanoscale 9 (2017) 380–390.

[47] W. Zhong, Nanomaterials in fluorescence-based biosensing, Anal. Bioanal. Chem. 394 (2009) 47–59.

[48] C.L. Manzanares-Palenzuela, A.M. Pourrahimi, J. Gonzalez-Julian, Z. Sofer, M. Pykal, M. Otyepka, M. Pumera, Interaction of single- and double-stranded DNA with multilayer MXene by fluorescence spectroscopy and molecular dynamics simulations, Chem. Sci. 10 (2019) 10010–10017.

[49] B. Luan, R. Zhou, Spontaneous ssDNA stretching on graphene and hexagonal boron nitride in-plane heterostructures, Nat. Commun. 10 (2019) 4610.

[50] Q. Weng, X. Wang, X. Wang, Y. Bando, D. Golberg, Functionalized hexagonal boron nitride nanomaterials: emerging properties and applications, Chem. Soc. Rev. 45 (2016) 3989–4012.

[51] https://www.nano.gov/sites/default/files/pub_resource/nni_materials_by_design.pdf.

[52] K. Kostarelos, Rational design and engineering of delivery systems for therapeutics: biomedical exercises in colloid and surface science, Adv. Colloid Interf. Sci. 106 (2003) 147–168.

[53] D. Pantarotto, J.P. Briand, M. Prato, A. Bianco, Translocation of bioactive peptides across cell membranes by carbon nanotubes, Chem. Commun. (2004) 16–17.

[54] N.W.S. Kam, T.C. Jessop, P.A. Wender, H. Dai, Nanotube molecular transporters: internalization of carbon nanotube-protein conjugates into mammalian cells, J. Am. Chem. Soc. 126 (2004) 6850–6851.

[55] D. Pantarotto, R. Singh, D. McCarthy, M. Erhardt, J.-P. Briand, M. Prato, K. Kostarelos, A. Bianco, Functionalized carbon nanotubes for plasmid DNA gene delivery, Angew. Chem. Int. Ed. Engl. 43 (2004) 5242–5246.

[56] N.W.S. Kam, Z. Liu, H. Dai, Carbon nanotubes as intracellular transporters for proteins and DNA: an investigation of the uptake mechanism and pathway, Angew. Chem. 44 (2005) 1–6.

[57] M. Ferrari, Cancer nanotechnology: opportunities and challenges, Nat. Rev. Cancer 5 (2005) 161–171.

[58] A.A. Bhirde, V. Patel, J. Gavard, G. Zhang, A.A. Sousa, A. Masedunskas, R.D. Leapman, R. Weigert, J.S. Gutkind, J.F. Rusling, Targeted killing of cancer cells in vivo and in vitro with EGF-directed carbon nanotube-based drug delivery, ACS Nano 3 (2009) 307–316.

[59] N. Graf, S.J. Lippard, Redox activation of metal-based prodrugs as a strategy for drug delivery, Adv. Drug Deliv. Rev. 64 (2012) 993–1004.

[60] S. Dhar, Z. Liu, J. Thomale, H. Dai, S.J. Lippard, Targeted single-walled carbon nano-tube mediated Pt(IV) prodrug delivery using folate as a homing device, J. Am. Chem. Soc. 130 (2008) 11467–11476.

[61] Z. Liu, S. Tabakman, K. Welsher, H. Dai, Carbon nanotubes in biology and medicine: in vitro and in vivo detection, imaging, and drug delivery, Nano Res. 2 (2009) 85–120.

[62] S.-W. Kim, J.-Y. Park, S. Lee, S.-H. Kim, D. Khang, Destroying deep lung tumor tissue through lung-selective accumulation and by activation of caveolin uptake channels using a specific width of carbon nanodrug, ACS Appl. Mater. Interfaces 10 (2018) 4419–4428.

[63] M. Zheng, A. Jagota, M.S. Strano, A.P. Santos, P. Barone, S.G. Chou, B.A. Diner, M.S. Dresselhaus, R.S. McLean, G.B. Onoa, et al., Structure-based carbon nanotube sorting by sequence-dependent DNA assembly, Science 302 (2003) 1545–1548.

[64] Z. Gao, C. Zhi, Y. Bando, D. Golberg, T. Serizawa, Noncovalent functionalization of boron nitride nanotubes in aqueous media opens application roads in nanobiomedicine, Nanobiomedicine 1 (2014) 7.

[65] S.W. Shin, S.Y. Ahn, S. Yoon, H.S. Wee, J.W. Bae, J.H. Lee, W.B. Lee, S.H. Um, Dif-ferences in DNA probe-mediated aggregation behavior of gold nanomaterials based on their geometric appearance, Langmuir 34 (2018) 14869–14874.

[66] M. Zheng, A. Jagota, E.D. Semke, B.A. Diner, R.S. McLean, S.R. Lustig, R.E. Richard-son, N.G. Tassi, DNA-assisted dispersion and separation of carbon nanotubes, Nat. Mater. 2 (2003) 338–342.

[67] H. Gao, Y. Kong, Simulation of DNA-nanotube interactions, Annu. Rev. Mater. Res. 34 (2004) 123–150.

[68] C. Hu, Y. Zhang, G. Bao, Y. Zhang, M. Liu, Z.L. Wang, DNA functionalized single-walled carbon nanotubes for electrochemical detection, J. Phys. Chem. B 109 (2005) 20072–20076.

[69] B. Cai, S. Wang, L. Huang, Y. Ning, Z. Zhang, G.-J. Zhang, Ultrasensitive label-free detection of PNA DNA hybridization by reduced graphene oxide field-effect transistor biosensor, ACS Nano 8 (2014) 2632–2638.

[70] B. Gigliotti, B. Sakizzie, D.S. Bethune, R.M. Shelby, J.N. Cha, Sequence-independent helical wrapping of single-walled carbon nanotubes by long genomic DNA, Nano Lett. 6 (2) (2006) 159–164.

[71] J. Rajendra, M. Baxendale, L.D.G. Rap, A. Rodger, Flow linear dichroism to probe bind-ing of aromatic molecules and DNA to single-walled carbon nanotubes, J. Am. Chem. Soc. 126 (2004) 11182–11188.

[72] D. Zhou, L. Ying, X. Hong, E.A. Hall, C. Abell, D. Klenerman, A compact functional quantum dot-DNA conjugate: preparation, hybridization, and specific label-free DNA detection, Langmuir 24 (2008) 1659–1664.

[73] M.T. Martínez, Y.-C. Tseng, M. González, J. Bokor, Streptavidin as CNTs and DNA linker for the specific electronic and optical detection of DNA hybridization, J. Phys. Chem. C 116 (2012) 22579–22586.

[74] O. Zagorodko, J. Spadavecchia, A.Y. Serrano, I. Larroulet, A. Pesquera, A. Zurutuza, R. Boukherroub, S. Szunerits, Highly sensitive detection of DNA hybridization on com-mercialized graphene-coated surface plasmon resonance interfaces, Anal. Chem. 86 (2014) 11211–11216.

[75] X. Zhang, P.J. Huang, M.R. Servos, J. Liu, Effects of polyethylene glycol on DNA adsorption and hybridization on gold nanoparticles and graphene oxide, Langmuir 28 (2012) 14330–14337.

[76] P.E. Nielsen, Peptide nucleic acid (PNA) from DNA recognition to antisense and DNA structure, Biophys. Chem. 68 (1997) 103–108.

[77] P.E. Nielsen, M. Egholm, An introduction to peptide nucleic acid, Curr. Issues Mol. Biol. 1 (1999) 89–104.

[78] N. Saikia, M. Taha, R. Pandey, Molecular insights on the dynamic stability of peptide nucleic acid functionalized carbon and boron nitride nanotubes, Phys. Chem. Chem. Phys. 23 (2021) 219–228.

[79] A. Paul, R.M. Watson, P. Lund, Y. Xing, K. Burke, Y. He, E. Borguet, C. Achim, D. Waldeck, Charge transfer through single-stranded peptide nucleic acid composed of thymine nucleotides, J. Phys. Chem. C 112 (2008) 7233–7240.

[80] J.C. Hanvey, N.J. Peffer, J.E. Bisi, S.A. Thomson, R. Cadilla, J.A. Josey, D.J. Ricca, C.F. Hassman, M.A. Bonham, K.G. Au, et al., Antisense and antigene properties of peptide nucleic acids, Science 258 (1992) 1481–1485.

[81] R. Chakrabarti, A.M. Klibanov, Nanocrystals modified with peptide nucleic acids (PNAs) for selective self-assembly and DNA detection, J. Am. Chem. Soc. 125 (2003) 12531–12540.

[82] V.K. Singh, R.R. Pandey, X. Wang, R. Lake, C.S. Ozkan, K. Wang, M. Ozkan, Covalent functionalization of single-walled carbon nanotubes with peptide nucleic acid: nanocomponents for molecular level electronics, Carbon 44 (2006) 1730–1739.

[83] D. Bonifazi, L.-E. Carloni, V. Corvaglia, A. Delforge, Peptide nucleic acids in materials science, Artif. DNA PNA XNA 3 (2012) 112–122.

[84] S. Schwaminger, S.A. Blank-Shim, M. Borkowska-Panek, P. Anand, P. Fraga-Garcia, K. Fink, W. Wenzel, S. Berensmeier, Experimental characterization and simulation of amino acid and peptide interactions with inorganic materials, Eng. Life Sci. 18 (2018) 84–100.

[85] Y. Li, L. Zhao, Y. Yao, X. Guo, Single-molecule nanotechnologies: an evolution in biological dynamics detection, ACS Appl. Bio Mater. 3 (2019) 68–85.

[86] J. Klein, Probing the interactions of proteins and nanoparticles, Proc. Natl. Acad. Sci. U. S. A. 104 (2007) 2029–2030.

[87] S. Mourdikoudis, R.M. Pallares, N.T.K. Thanh, Characterization techniques for nanoparticles: comparison and complementarity upon studying nanoparticle properties, Nanoscale 10 (2018) 12871–12934.

[88] K. Nienhaus, H. Wang, G.U. Nienhaus, Nanoparticles for biomedical applications: exploring and exploiting molecular interactions at the nano-bio interface, Mater. Today Adv. 5 (2020).

[89] I. Yadav, V.K. Aswal, J. Kohlbrecher, Size-dependent interaction of silica nanoparticles with lysozyme and bovine serum albumin proteins, Phys. Rev. E 93 (2016), 052601.

[90] M.C.L. Giudice, L.M. Herda, E. Polo, K.A. Dawson, In situ characterization of nanoparticle biomolecular interactions in complex biological media by flow cytometry, Nat. Commun. 7 (2016) 13475.

[91] A. Bhaumik, A.M. Shearin, R. Delong, A. Wanekaya, K. Ghosh, Probing the interaction at the nano-bio interface using Raman spectroscopy: ZnO nanoparticles and adenosine triphosphate biomolecules, J. Phys. Chem. C Nanomater. Interfaces 118 (2014) 18631–18639.

[92] P. Satzer, F. Svec, G. Sekot, A. Jungbauer, Protein adsorption onto nanoparticles induces conformational changes: particle size dependency, kinetics, and mechanisms, Eng. Life Sci. 16 (2016) 238–246.

[93] Z. Su, T. Leung, J.F. Honek, Conformational selectivity of peptides for single-walled carbon nanotubes, J. Phys. Chem. B 110 (47) (2006) 23623–23627.

[94] P.A. Mirau, R.R. Naik, P. Gehring, Structure of peptides on metal oxide surfaces probed by NMR, J. Am. Chem. Soc. 133 (2011) 18243–18248.

[95] J.E. Kim, L. Minyung, Fullerene inhibits β-amyloid peptide aggregation, Biochem. Biophys. Res. Commun. 303 (2003) 576–579.

[96] K. Siposova, V.I. Petrenko, O.I. Ivankov, A. Musatov, L.A. Bulavin, M.V. Avdeev, O.A. Kyzyma, Fullerenes as an effective amyloid fibrils disaggregating nanomaterial, ACS Appl. Mater. Interfaces 12 (2020) 32410–32419.

[97] K.I. Ramachandran, G. Deepa, K. Namboori, Computational Chemistry and Molecular Modeling, Principles and Application, Springer-Verlag Berlin Heidelberg, 2008, https://doi.org/10.1007/978-3-540-77304-7.

[98] http://www2.ph.ed.ac.uk/~gja/qp/qp10.pdf.

[99] S.M. Blinder, Basic concepts of self-consistent-field-theory, Am. J. Phys. 33 (1965) 431–443.

[100] C.J. Cramer, Essentials of Computational Chemistry: Theory and Models, John Wiley & Sons, New York, 2002, pp. 165–202.

[101] M.H. Abdalmoneam, K. Waters, N. Saikia, R. Pandey, Amino-acid-conjugated gold clusters: interaction of alanine and tryptophan with Au_8 and Au_{20}, J. Phys. Chem. C 121 (2017) 25585–25593.

[102] https://hal.archives-ouvertes.fr/hal-01275043/document.

[103] X. Zhong, S. Mukhopadhyay, S. Gowtham, R. Pandey, S.P. Karna, Applicability of carbon and boron nitride nanotubes as biosensors: effect of biomolecular adsorption on the transport properties of carbon and boron nitride nanotubes, Appl. Phys. Lett. 102 (2013), 133705.

[104] S. Mukhopadhyay, S. Gowtham, R.H. Scheicher, R. Pandey, S.P. Karna, Theoretical study of physisorption of nucleobases on boron nitride nanotubes: a new class of hybrid nano-bio materials, Nanotechnology 21 (2010), 165703.

[105] S. Gowtham, R.H. Scheicher, R. Pandey, S.P. Karna, R. Ahuja, First-principles study of physisorption of nucleic acid bases on small-diameter carbon nanotubes, Nanotechnology 19 (2008), 125701.

[106] S. Gowtham, R.H. Scheicher, R. Ahuja, R. Pandey, S.P. Karna, Physisorption of nucleobases on graphene: density-functional calculations, Phys. Rev. B 76 (2007), 033401.

[107] A. Das, A.K. Sood, P.K. Maiti, M. Das, R. Varadarajan, C.N.R. Rao, Binding of nucleobases with single-walled carbon nanotubes: theory and experiment, Chem. Phys. Lett. 453 (2008) 266–273.

[108] M. Shukla, M. Dubey, E. Zakar, R. Namburu, Z. Czyznikowska, J. Leszcznski, Interaction of nucleic acid bases with single–walled carbon nanotubes, Chem. Phys. Lett. 480 (2009) 269–272.

[109] V. Shewale, P. Joshi, S. Mukhopadhyay, M. Deshpande, R. Pandey, S. Hussain, S.P. Karna, First-principles study of nanoparticle–biomolecular interactions: anchoring of a (ZnO)12 cluster on nucleobases, J. Phys. Chem. C 115 (2011) 10426–10430.

[110] Z. Wang, H. He, W. Slough, R. Pandey, S.P. Karna, Nature of interaction between semi-conducting nanostructures and biomolecules: chalcogenide QDs and BNNT with DNA molecules, J. Phys. Chem. C 119 (2015) 25965–25973.

[111] P. Joshi, V. Shewale, R. Pandey, V. Shanker, S. Hussain, S.P. Karna, Tryptophan–gold nanoparticle interaction: a first-principles quantum mechanical study, J. Phys. Chem. C 2011 (115) (2011) 22818–22826.

[112] S. Mukhopadhyay, R.H. Scheicher, R. Pandey, S.P. Karna, Sensitivity of boron nitride nanotubes toward biomolecules of different polarities, J. Phys. Chem. Lett. 2 (2011) 2442–2447.

[113] D. Umadevi, G.N. Sastry, Quantum mechanical study of physisorption of nucleobases on carbon materials: graphene versus carbon nanotubes, J. Phys. Chem. Lett. 2 (2011) 1572–1576.

[114] C. Liu, H. Haiying, R. Pandey, S. Hussain, S.P. Karna, Interaction of metallic nanoparticles with a biologically active molecule, Dopamine, J. Phys. Chem. B 112 (2008) 15256–15259.

[115] N. Saikia, R.C. Deka, Theoretical study on pyrazinamide adsorption onto covalently functionalized (5,5) metallic single-walled carbon nanotube, Chem. Phys. Lett. 500 (2010) 65–70.

[116] N. Saikia, R.C. Deka, Density functional calculations on adsorption of 2-methylheptylisonicotinate antitubercular drug onto functionalized carbon nanotube, Comput. Theor. Chem. 964 (2011) 257–261.

[117] N. Saikia, M. Seel, R. Pandey, Stability and electronic properties of 2D nanomaterials conjugated with pyrazinamide chemotherapeutic: a first-principles cluster study, J. Phys. Chem. C 120 (2016) 20323–20332.

[118] N. Saikia, S.P. Karna, R. Pandey, Theoretical study of gas and solvent phase stability and molecular adsorption of noncanonical guanine bases on graphene, Phys. Chem. Chem. Phys. 19 (2017) 16819–16830.

[119] A.K. Rappi, C.J. Casewit, K.S. Colwell, W.A. Goddard, W.M. Skiff, UFF, a full periodic table force field for molecular mechanics and molecular dynamics simulations, J. Am. Chem. Soc. 114 (1992) 10024–10039.

[120] R.A. Latour, Perspectives on the simulation of protein–surface interactions using empirical force field methods, Colloids Surf. B Interfaces 124 (2014) 25–37.

[121] A.D. Mackerell Jr., Empirical force fields for biological macromolecules: overview and issues, J. Comput. Chem. 25 (2004) 1584–1604.

[122] P. Kraus, D.A. Obenchain, I. Frank, Benchmark-quality semiexperimental structural parameters of van der Waals complexes, J. Phys. Chem. A 122 (2018) 1077–1087.

[123] S. Yip (Ed.), Chapter 2.8 Basic molecular dynamics, in: Handbook of Materials Modeling, Springer, 2005, pp. 565–588.

[124] P.H. Hünenberger, Thermostat algorithms for molecular dynamics simulations, Adv. Polym. Sci. 173 (2005) 105–149.

[125] J. Meller, Molecular Dynamics, Encyclopedia of Life Science, Nature Publishing Group, 2001, pp. 1–8.

[126] A.Y. Toukmaji, J.A. Board Jr., Ewald summation techniques in perspective: a survey, Comput. Phys. Commun. 95 (1996) 73–92.

[127] P.P. Ewald, The calculation of optical and electrostatic grid potential, Ann. Phys. (Leipzig) 64 (1921) 253–287.

[128] N. Saikia, K. Waters, S.P. Karna, R. Pandey, Hierarchical self-assembly of noncanonical guanine nucleobases on graphene, ACS Omega 2 (2017) 3457–3466.

[129] N. Saikia, F. Johnson, K. Waters, R. Pandey, Dynamics of self-assembled cytosine nucleobases on graphene, Nanotechnology 29 (2018), 195601.

[130] N. Saikia, R. Pandey, Polarity-induced surface recognition and self-assembly of noncanonical DNA nucleobases on h-BN monolayer, J. Phys. Chem. C 122 (2018) 3915–3925.

[131] V.A. Karachevtsev, S.G. Stepanian, A.Y. Glamazda, M.V. Karachevtsev, V.V. Eremenko, O.S. Lytvyn, L. Adamowicz, Noncovalent interaction of single-walled carbon nanotubes with 1-pyrenebutanoic acid Succinimide ester and Glucoseoxidase, J. Phys. Chem. C 115 (2011) 21072–21082.

[132] M. Samieegohar, F. Sha, A.Z. Clayborne, T. Wei, ReaxFF MD simulations of peptide-grafted gold nanoparticles, Langmuir 35 (2019) 5029–5036.

[133] M. Tang, N.S. Gandhi, K. Burrage, Y. Gu, Interaction of gold nanosurfaces/nanoparticles with collagen-like peptides, Phys. Chem. Chem. Phys. 21 (2019) 3701–3711.

[134] Y. Hayamizu, C.R. So, S. Dag, T.S. Page, D. Starkebaum, M. Sarikaya, Bioelectronic interfaces by spontaneously organized peptides on 2D atomic single-layer materials, Sci. Rep. 6 (2016) 33778.

[135] C.D. Von Bargen, C.M. MacDermaid, O.-S. Lee, P. Deria, M.J. Therien, J.G. Saven, Origins of the helical wrapping of phenyleneethynylene polymers about single-walled carbon nanotubes, J. Phys. Chem. B 117 (2013) 12953–12965.

[136] L. Liang, W. Hu, Z. Zhang, J.W. Shen, Theoretic study on dispersion mechanism of boron nitride nanotubes by polynucleotides, Sci. Rep. 6 (2016) 39747.

[137] S. Ghosh, R. Chakrabarti, Spontaneous unzipping of xylonucleic acid assisted by a single-walled carbon nanotube: a computational study, J. Phys. Chem. B 120 (2016) 3642–3652.

[138] T. Jamil, J.R. Gissinger, A. Garley, N. Saikia, A.K. Upadhyay, H. Heinz, Dynamics of carbohydrate strands in water and interactions with clay minerals: influence of pH, surface chemistry, and electrolytes, Nanoscale 11 (2019) 11183–11194.

[139] W.R. Algar, D.E. Prasuhn, M.H. Stewart, T.L. Jennings, J.B. Blanco-Canosa, P.E. Dawson, I.L. Medintz, The controlled display of biomolecules on nanoparticles: a challenge suited to bioorthogonal chemistry, Bioconjug. Chem. 22 (2011) 825–858.

[140] R. Bhattacharya, P. Mukherjee, Biological properties of "naked" metal nanoparticles, Adv. Drug Deliv. Rev. 60 (2008) 1289–1306.

[141] M. Kopp, S. Kollenda, M. Epple, Nanoparticle-protein interactions: therapeutic approaches and supramolecular chemistry, Acc. Chem. Res. 50 (2017) 1383–1390.

[142] C.W. Yong, Study of interactions between polymer nanoparticles and cell membranes at atomistic levels, Philos. Trans. R. Soc. Lond. Ser. B Biol. Sci. 370 (2015) 20140036.

[143] S.P. Schwaminger, D. Bauer, P. Fraga-García, F.E. Wagner, S. Berensmeier, Oxidation of magnetite nanoparticles: impact on surface and crystal properties, CrystEngComm 19 (2017) 246–255.

[144] S.A. Blank-Shim, S.P. Schwaminger, M. Borkowska-Panek, P. Anand, P. Yamin, P. Fraga-Garcia, K. Fink, W. Wenzel, S. Berensmeier, Binding patterns of homo-peptides on bare magnetic nanoparticles: insights into environmental dependence, Sci. Rep. 7 (2017) 14047.

[145] T. Nagamune, Biomolecular engineering for nano-bio/bionanotechnology, Nano Converg. 4 (2017) 9.

[146] A.J. Sposito, A. Kurdekar, J. Zhao, I. Hewlett, Application of nanotechnology in biosensors for enhancing pathogen detection, WIREs Nanomed. Nanobiotechnol. (2018) 10-e1512.

[147] A. Mukhopadhyay, S. Basu, S. Singha, H.K. Patra, Inner-view of nanomaterial incited protein conformational changes: insights into designable interaction, Res.: A Sci. Partner J. 2018 (2018) 9712832.

[148] C.D. Walkey, W.C. Chan, Understanding and controlling the interaction of nanomaterials with proteins in a physiological environment, Chem. Soc. Rev. 41 (2012) 2780–2799.

[149] Z. Hazarika, A.N. Jha, Computational analysis of the silver nanoparticle – human serum albumin complex, ACS Omega 5 (2020) 170–178.

[150] A. Epanchintseva, P. Vorobjev, D. Pyshnyi, I. Pyshnaya, Fast and strong adsorption of native oligonucleotides on citrate-coated gold nanoparticles, Langmuir 34 (2018) 164–172.

Carbon nanotubes and graphene: From structural to device properties

P. Balakrishna Pillai and M.M. De Souza
Department of Electronic and Electrical Engineering, University of Sheffield,
Sheffield, United Kingdom

11.1 Introduction

The technological achievements of the electronics sector are a result of miniaturization and improved performance derived from scaling [1]. As geometrical scaling continues, to reduce the lateral electric field (electric field $E \propto V_{DD}/L_G$) and the power density, $V_{DD} \times I_{ON}/A$ (where A is the area of the device), the supply voltage should also be scaled. Ultimately, new strategies are required to keep the gate overdrive sufficiently high to avoid high OFF current and switching delay. As silicon cannot sustain these requirements, the search is therefore on for a replacement material and device for the semiconductor industry. Among the various graphitic nanostructures, single-wall carbon nanotubes invite special attraction for digital logic due their structural simplicity and tunable electronic properties arising from a finite semiconducting bandgap. Carbon-based devices are also being touted for radio frequency (RF) applications: where the transistor on/off ratio is not of relevance, the transistor here is never switched off, a small signal is superimposed on the transistor characteristics at a predesigned bias condition. Good RF performance requires high cut-off frequency, achieved via a high mobility of carriers. High mobility of carriers in graphene as compared to nanotubes/graphene nanoribbon (GNR) makes it more suitable for RF applications.

11.2 Outline

In this chapter, fundamental details of the atomic and electronic structure of carbon nanotubes are introduced from the nearest neighbor tight binding approximation. The importance of 1-D electrostatics in various carbon-based device geometries is discussed. An overview of state of the art developments in CNTFET technology is presented and the performance compared with rivaling III–V technologies. Technological issues in implementing these materials in the mainstream CMOS are briefly reviewed. Using non-equilibrium Green's function-based quantum mechanical device simulations, the performance of Carbon Nanotube (CNT) and GNR tunnel FETs is benchmarked using the requirements for ITRS-2024 and an advantageous diameter

Modeling, Characterization and Production of Nanomaterials. https://doi.org/10.1016/B978-0-12-819905-3.00011-7

and width range for a CNT and a GNR to meet the proposed International Technology Roadmap for Semiconductors (ITRS) requirements are identified.

11.3   Electronic structure of graphite

Graphite is a well-known allotropic form of carbon composed of layers of sp^2-bonded atoms in a honeycomb arrangement. The electronic structure of individual layers of graphite called "graphene" was first discussed by P.R. Wallace in 1947 [2]. Single-atom thick graphite sheets (graphene) and their seamlessly rolled up form (CNTs) [3] consist of a hexagonal structure with each atom forming three σ bonds with each of its nearest neighbors (Fig.11.1).

These covalent bonds are nearly equivalent to those holding diamond together, giving graphene similar mechanical and thermal properties as diamond. The fourth valence electron does not participate in covalent bonding. It lies in the $2P_z$ orbital oriented perpendicular to the sheet of graphite and forms a conducting π band. Overlap of the P_z orbitals of neighboring atoms results in the formation of delocalized π-bonds which determine the electronic properties of the material [4].

A graphene unit cell composed of two atoms per unit cell is fully described by a wave function corresponding to the individual atoms and a coupling term which denotes their interaction with nearest neighbors [5]. The dispersion relation can be obtained by solving the general time independent Schrodinger equation as follows

$$E\{\phi(r)\} = \left(\frac{-\hbar^2}{2m}\nabla^2 + V(r)\right)\{\phi(r)\} \tag{11.1}$$

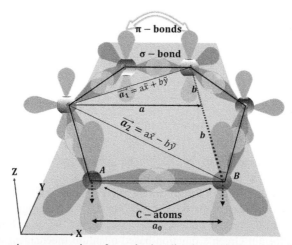

Fig. 11.1 Schematic representation of σ and π-bonding in graphene responsible for its elastic and electronic properties where each corner of the hexagon represents a carbon atom. The vector distances between two identical nearest neighbors of an atom are represented as $\vec{a_1} = a\tilde{x} + b\tilde{y}$ and $\vec{a_2} = a\tilde{x} - b\tilde{y}$, where the distances a and b are connected to a_0 as $a = \frac{3}{2}a_0$ and $b = \frac{\sqrt{3}a_0}{2}$.

here $\frac{-\hbar^2}{2m}\nabla^2 + V = H$ is the Hamiltonian, $\phi(r)$ corresponds to the wave function and $V(r)$ is the confining potential due to the $2P_z$ orbitals of each carbon atom. For a periodic solid, the wave function can be expressed by the sum of basis functions as $\phi(r) = \sum_m \phi_m U_m(r)$, where $U_m(r)$ is the basis function at point m in the lattice and ϕ_m is the coefficient of the basis function at m. Thus one can convert the Schrodinger equation into its matrix form as

$$E\begin{bmatrix} \phi_1 \\ \phi_2 \\ \vdots \\ \phi_N \end{bmatrix} = [H] \begin{bmatrix} \phi_1 \\ \phi_2 \\ \vdots \\ \phi_N \end{bmatrix} \tag{11.2}$$

The above matrix represents N algebraic equations and for an arbitrary unit cell n,

$$E\{\phi_n\} = \sum_m [H_{nm}]\{\phi_m\} \tag{11.3}$$

here $\{\phi_n\}$ is a matrix which includes the basis functions of the unit cell at point n, which consists of two basis functions corresponding to the "A" and "B" lattice sites in that unit cell. and H_{nm} is the Hamiltonian at the nth unit cell due to the neighboring m unit cells.

Considering the periodicity of the lattice, one can express the wave function of the nth unit cell as $\{\phi_n\} = \{\phi_1\} \, e^{i\vec{k}\cdot\vec{d_n}}$ where d_n is the distance to the nth unit cell from the first unit cell, "k" is the reciprocal lattice vector and ϕ_1 is the wave function of the first unit cell. Using this expression for $\{\phi_n\}$ in Eq. (11.3) results in

$$E\{\phi_1\} = [h]\{\phi_1\} \tag{11.4}$$

where $h = \sum_m [H_{nm}] e^{i\vec{k}\cdot\left(\vec{d_m}-\vec{d_n}\right)}$

In the case of graphene, the position of every other unit cell from a given unit cell can be easily obtained by defining the vector distances as a linear combination of $\vec{a_1}$ and $\vec{a_2}$ given by $n\vec{a_1} + m\vec{a_2}$ (Fig. 11.1). The Hamiltonian corresponding to any arbitrary unit cell and the derivation of the energy dispersion relation from there (shown in Eq. 11.16) have been discussed by S. Datta et al. in [5]. Datta's approach begins with Hamiltonian of the unit cell as

$$h\left(\vec{k}\right) = \begin{bmatrix} \varepsilon & t \\ t & \varepsilon \end{bmatrix} + \begin{bmatrix} 0 & 0 \\ t & 0 \end{bmatrix} e^{i\vec{k}\cdot\vec{a_1}} + \begin{bmatrix} 0 & 0 \\ t & 0 \end{bmatrix} e^{i\vec{k}\cdot\vec{a_2}} + \begin{bmatrix} 0 & t \\ 0 & 0 \end{bmatrix} e^{-i\vec{k}\cdot\vec{a_1}}$$

$$+ \begin{bmatrix} 0 & t \\ 0 & 0 \end{bmatrix} e^{-i\vec{k}\cdot\vec{a_2}} \tag{11.5}$$

where ε is the self-energy term and t is the nearest neighbor coupling term in graphene. The first term arises from the unit cell under consideration and the subsequent four terms from each of its nearest neighbors. Adding all the terms yields,

$$\left|h\left(\vec{k}\right)\right| = \left[t\left(1 + e^{i \cdot k^{\to} \cdot a_1^{\to}} + e^{i \cdot k^{\to} \cdot a_2^{\to}}\right) \quad \begin{matrix} \varepsilon \\ \\ t\left(1 + e^{-i \cdot k^{\to} \cdot a_1^{\to}} + e^{-i \cdot k^{\to} \cdot a_2^{\to}}\right) \\ \varepsilon \end{matrix} \right] \tag{11.6}$$

The eigenvalue of this matrix is given by

$$E = \varepsilon \pm |h_0| \tag{11.7}$$

where

$$h_0 = t\left(1 + e^{i \cdot \vec{k} \cdot \vec{a_1}} + e^{i \cdot \vec{k} \cdot \vec{a_2}}\right) \tag{11.8}$$

From Fig. 11.1, the following vectors can be expressed in their component form as

$$\vec{k} = \tilde{x}k_x + \tilde{y}k_y \tag{11.9}$$

$$\vec{a_1} = \tilde{x}\frac{3a_0}{2} + \tilde{y}\frac{\sqrt{3}a_0}{2} \tag{11.10}$$

$$\vec{a_2} = \tilde{x}\frac{3a_0}{2} - \tilde{y}\frac{\sqrt{3}a_0}{2} \tag{11.11}$$

Assuming the distance $\frac{3a_0}{2} = a$ and $\frac{\sqrt{3}a_0}{2} = b$, the expression for h_0 yields

$$h_0 = t\left(1 + 2e^{ik_x a}\cos k_y b\right) \tag{11.12}$$

here $|h_0|^2 = h_0 h_0^* = t^2(1 + 4\cos k_x a \cos k_y b + 4\cos^2 k_y b)$ and hence

$$\left|h_0\left(\vec{k}\right)\right| = t\sqrt{1 + 4\cos k_x a \cos k_y b + 4\cos^2 k_y b} \tag{11.13}$$

This energy eigenvalue E can be expressed as the dispersion relation for valence and conduction bands from Eq. (11.7) as

$$E^{\pm} = \varepsilon \pm t\sqrt{1 + 4\cos k_x a \cos k_y b + 4\cos^2 k_y b} \tag{11.14}$$

where the + and − sign represent the conduction and valence band energies respectively.

The energy dispersion curve has minima in reciprocal space when

$$(k_x, k_y) = \left(0, \frac{2\pi}{3b}\right), \left(0, -\frac{2\pi}{3b}\right), \left(\frac{\pi}{a}, \frac{\pi}{3b}\right), \left(\frac{\pi}{a}, -\frac{\pi}{3b}\right), \left(-\frac{\pi}{a}, \frac{\pi}{3b}\right) \& \left(-\frac{\pi}{a}, -\frac{\pi}{3b}\right)$$

which corresponds to the valley points of the Brillouin zone in k-space.

The value of the function $h_0\left(\vec{k}\right) = t(1 + 2e^{ik_x a}\cos k_y b)$ around one of the valley points $(k_x a, k_y b) = \left(0, \frac{2\pi}{3}\right)$ can be evaluated by considering the Taylor series expansion of $h_0(k_x, k_y) = t(1 + 2e^{ik_x a}\cos k_y b)$ as

$$h_0\left(k_x, k_y\right) = h_0\left(0, \frac{2\pi}{3}\right) + \left[\frac{\partial h_0}{\partial k_x}\right]_{\left(0, \frac{2\pi}{3}\right)}(k_x - 0) + \left[\frac{\partial h_0}{\partial k_y}\right]_{\left(0, \frac{2\pi}{3}\right)}\left(k_y - \frac{2\pi}{3}\right) \quad (11.15)$$

where $\left[\frac{\partial h_0}{\partial k_x}\right]_{\left(0, \frac{2\pi}{3}\right)} = -iat$ and $\left[\frac{\partial h_0}{\partial k_y}\right]_{\left(0, \frac{2\pi}{3}\right)} = -\sqrt{3}tb = -at$, taking $\sqrt{3}\,b = a$

so $h_0\left(k_x, k_y\right) = -iatk_x - at\left(k_y - \frac{2\pi}{3}\right) = -iat\left[k_x - i\beta_y\right]$ where $\beta_y = k_y - \frac{2\pi}{3}$

Therefore, the energy eigenvalue around the valley point is given by

$$E\left(k_x, k_y\right) = \varepsilon \pm |h_0| = \varepsilon \pm at\sqrt{k_x^2 + \beta_y^2} \quad (11.16)$$

This is the energy dispersion relation is graphene around the valley points $(k_x, k_y) = \left(0, \frac{2\pi}{3b}\right)$ and the energy dispersion relation over the entire Brillouin zone can be represented as shown in Fig. 11.2B. The conduction (π^*) band and the valence (π) bands meet at these six points—known as Dirac points—giving the Fermi level in undoped graphene at these six points [2–4]. Each valley is shared between three adjacent Brillouin zones and effectively only one-third belongs to the first Brillouin zone.

11.4 Types of carbon nanotubes

CNTs are labeled in terms of a pair of integers (n, m) defining their chiral vector $C_h = na_1 + ma_2$ that describes the circumference ($C_h = \pi d_{CNT}$), where a_1 and a_2 are the unit vectors of the graphene honeycomb lattice. The vector distances are given by Eqs. (11.10) and (11.11), respectively, which can be represented in simple terms as, $\vec{a_1} = a\tilde{x} + b\tilde{y}$ and $\vec{a_2} = a\tilde{x} - b\tilde{y}$. The periodic boundary condition along the circumference forces the boundary condition $\vec{k} \cdot \vec{C_h} = 2\pi\vartheta$ where, $\vec{k} = k_x\tilde{x} + k_y\tilde{y}$. Hence $\vec{k} \cdot \vec{C_h} = k_x a(m + n) + k_y b(m - n) = 2\pi\vartheta$. At the valley point $(k_x, k_y) = \left(0, \frac{2\pi}{3b}\right)$, $\vec{k} \cdot \vec{C_h} = \frac{2\pi}{3}(m - n) = 2\pi\vartheta \Rightarrow \frac{(m-n)}{3} = \vartheta$, an integer. In a nanotube, a series of lines which follow the equation $\vec{k} \cdot \vec{C_h} = k_x a(m + n) + k_y b(m - n) = 2\pi\vartheta$ are called its subbands [4,5]

The chiral angle θ, i.e., the angle between C_h and a_1 is estimated as

$$\cos\theta = \frac{C_h \cdot a_1}{|C_h||a_1|} = \frac{2n + m}{2\sqrt{n^2 + m^2 + nm}} \quad (11.17)$$

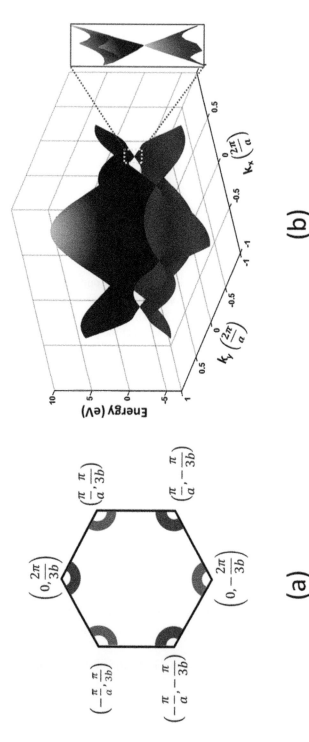

Fig. 11.2 (A) Illustration of the six Brillouin valley points of graphene. As each vertex is shared between two other neighboring Brillouin zones, an effective contribution of only 1/3 per corner point can be considered. To represent the states belonging to the first Brillouin zone, $13 * 6 = 2$ valley points are adequate [6]. (B) 3-D electronic band dispersion of graphene with valence and conduction bands shown in the first Brillouin zone. A zoom of the bands near the K point or Dirac point where the two bands meet is highlighted.

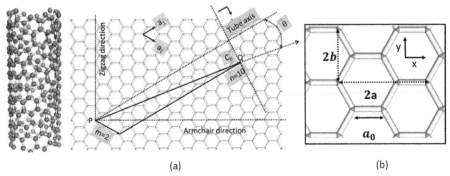

(a)   (b)

Fig. 11.3 (A) Illustration of geometrical construction of (10,2) nanotube from a 2D sheet of graphene. A rolled up (10,2) tube is shown on the *right*. (B) The minimum distance corresponds to periodic boundary conditions for tube wrapping along x and y directions. The vectors "$2a$" and "$2b$" connect equivalent points of an nanotube and hence periodic boundary conditions apply so that $k_\vartheta \cdot 2am = 2\pi\vartheta$ (along x) or $k_\vartheta \cdot 2bm = 2\pi\vartheta$ (along y) where m is the chiral index and ϑ, an integer.

The hexagonal symmetry of graphene restricts the value of θ to the range $0 \le |\theta| \le 30°$. In armchair tubes, wrapped along the x direction, the circumferential vector $\vec{C} = 2am$, where m is an integer (the equivalent points along the x-direction of the nanotube are separated by a distance "$2a$" as shown in Fig. 11.3B). These types of tubes are of type (n, n) $(\theta = 30°)$ and their bonds are perpendicular to the tube axis. Their periodic boundary conditions result in $k_\vartheta \cdot 2am = 2\pi\vartheta$, $\Rightarrow k_\vartheta = \frac{2\pi}{2am}\vartheta$.

Every value of the integer ϑ defines a k_ϑ at which the dispersion relation becomes

$$\epsilon_\vartheta \left(k_y \right)^\pm = \pm at \sqrt{k_\vartheta^2 + \left(k_y - \frac{2\pi}{3b} \right)^2} \tag{11.18}$$

At $\vartheta = 0$,

$$\epsilon_0 \left(k_y \right)^\pm = \pm at \left| k_y - \frac{2\pi}{3b} \right| \tag{11.19}$$

For tubes wrapped along the x-direction (arm-chair tubes), a linear dispersion relation is always valid and results in metallic nanotubes.

For tubes wrapped along the y direction, the circumferential vector $\vec{C} = 2bm$ and such nanotubes are called zig-zag. Their indices are described as $(n, 0)$ $(\theta = 0°)$ and they display bonds parallel to the nanotube axis. The periodic boundary conditions in their case results in $k_\vartheta \cdot 2bm = 2\pi\vartheta \Rightarrow k_\vartheta = \frac{2\pi}{2bm}\vartheta - \frac{2\pi}{3b} = \frac{2\pi}{2bm}\left(\vartheta - \frac{2m}{3} \right)$

This implies that the lowest subbands derived from the dispersion relation

$$\epsilon_\vartheta \left(k_x \right)^\pm = \pm at \sqrt{k_x^2 + \left(k_\vartheta \right)^2} \tag{11.20}$$

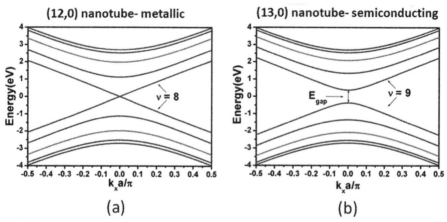

Fig. 11.4 Electronic band structure of a (12,0) metallic nanotube (A) and a (13,0) semi conducting nanotube (B) calculated using atomic P_z orbital tight binding method [Using CNTbands, Purdue University] [7]. The band structure of a metallic nanotube showing subband crossing at the "valley" point ($x=0$) and hence a non-zero density of states at the valley point.

where $k_\vartheta = \frac{2\pi}{2bm}\left(\vartheta - \frac{2m}{3}\right)$ has a linear E-K relationship only when the chiral index "m" become an integer multiple of 3, the condition for metallic nanotubes [4]. That is zig-zag nanotubes become metallic only when $m = 3 * \vartheta$. Chiral nanotubes have indices $n \neq m \neq 0$.

In Fig. 11.4, the electronic band structures of metallic and semiconducting nanotubes are shown, where the subbands cross at $k_x = 0$ for metallic nanotubes, resulting in a finite density of states (DOS) around the "K" point, while semiconducting nanotube exhibits nearly zero DOS around K.

11.5   Types of nanoribbons

Rather than rolling graphene sheets to form nanotubes, it is also possible to achieve semiconducting properties by controlling the width and edge structure of a graphene sheet. Such GNRs are classified into arm-chair edge (AGNR) and zig-zag edge ribbons (ZGNR) and both categories exhibit a width dependent band gap [8]. The dispersion relation in GNRs is different from the nearest neighbor tight binding dispersion relation in graphene due to the presence of edge states that stem from the π electron network at the edges [8,9]. But one can have reasonable agreement with ab initio band structure calculations by including an edge bond distortion and third nearest neighbor interaction in the tight binding description [10]. Using this modified TB model, the expression for tight binding band gap in AGNR is given by [10,11]

$$E_g \approx -2\pi(t_1 - 2t_3)/\sqrt{3}(m+1) \pm 6(t_3 + \Delta t_1)/(m+1) \qquad (11.21)$$

where t_1 and t_3 are first nearest neighbor and third nearest neighbor hopping energies, Δt_1 is the edge bond relaxation energy and $m = 3p$ or $3p + 1$, where p is an integer. The plus (minus) sign in Eq. (11.21) applies to $m = 3p$ ($m = 3p + 1$) family types. The width relation in AGNR is $W = (m - 1)a/2$, where $a = 0.246\,\text{nm}$ and in ZGNR is $\left(\frac{3m}{2} - 1\right)a_{cc}$, where $a_{cc} = 1.42\,\text{Å}$.

11.6 Density of states and quantum capacitance

The DOS is defined as a measure of the number of energy Eigen states per unit energy governed by the $E - k$ relationship. Considering a system in equilibrium with Fermi function $f_0(E)$ at energy E, the total number of electrons over the entire energy range can be assumed as [5,6]

$$N = \int dE\, D(E)\, f_0(E) \tag{11.22}$$

where $D(E)$ is considered as the DOS of the material at energy E. Alternately one can assume the total number of electrons as a summation of the Fermi functions at each energy $\in\left(\vec{k}\right)$, i.e.,

$$N = \sum_{\vec{k}} f_0\left(\in\left(\vec{k}\right)\right) \tag{11.23}$$

For the validity of the above two equalities, the density of states can be assumed as

$$D(E) = \sum_{\vec{k}} \delta\left(E - \in\left(\vec{k}\right)\right) \tag{11.24}$$

i.e., $N = \int dE\ D(E)\ f_0(E)\ = \sum_{\vec{k}} \int dE\ \delta\left(E - \in\left(\vec{k}\right)\right) f_0(E)\ = \sum_{\vec{k}} f_0\left(\in\left(\vec{k}\right)\right) = N$.
Here the property of delta function $\int dx\ \delta(x - a)f(x) = f(a)$ has been utilized.

For a 2-D solid with periodic boundary conditions along the x and y directions, the allowed k states in x and y are spaced by $\frac{2\pi}{L}$ and $\frac{2\pi}{W}$, where L and W are the length and width of the solid. Hence over an energy range dE, the number of available states are $\frac{dk_x}{2\pi/L} * \frac{dk_y}{2\pi/W}$

Therefore the total DOS of the system (Eq. 11.30) can be evaluated as previously shown by S. Datta in [5] as

$$D(E) = \int\int \frac{dk_x}{\frac{2\pi}{L}} * \frac{dk_y}{\frac{2\pi}{W}} \delta\left(E - \in\left(\vec{k}\right)\right) \tag{11.25}$$

Using another property of delta function, i.e., $\delta\left(E - \in\left(\vec{k}\right)\right) = \delta\left(\in\left(\vec{k}\right) - E\right)$, as discussed in [5] one can write

$$D(E) = \iint \frac{dk_x}{2\pi/L} * \frac{dk_y}{2\pi/W} \delta\left(\in\left(\vec{k}\right) - E\right) \qquad (11.26)$$

This integral in circular coordinates can be written as

$$D(E) = \frac{LW}{4\pi^2} \int_{-\pi}^{\pi} d\theta \int dk \cdot k \, \delta\left(\in\left(\vec{k}\right) - E\right) \qquad (11.27)$$

$$\Rightarrow D(E) = \frac{LW}{4\pi^2} *2\pi* \int dk \cdot k \, \delta\left(\in\left(\vec{k}\right) - E\right) \qquad (11.28)$$

But the energy eigenvalue around the valley point for graphite is

$$\in\left(\vec{k}\right) = \varepsilon \pm |h_0| = \varepsilon \pm at\sqrt{k_x^2 + \beta_y^2} = \varepsilon \pm at|k| \qquad (11.29)$$

Assuming zero site energy (i.e., $\varepsilon = 0$), $\in\left(\vec{k}\right) = \pm at|k|$ and considering only the positive value, $d\in\left(\vec{k}\right) = at*dk \Rightarrow dk = d\left[\in\left(\vec{k}\right)\right]/at$,

$$D(E) = \frac{LW}{4\pi^2} *2\pi* \int d\frac{\left[\in\left(\vec{k}\right)\right]}{at} *\in\frac{\left(\vec{k}\right)}{at} *\delta\left(\in\left(\vec{k}\right) - E\right) = \frac{A}{2\pi}\frac{|E|}{(at)^2} \qquad (11.30)$$

where A is the area of the 2-D conductor.

Therefore the density of states in a 2-D conductor (i.e., graphene) is linearly proportional to the energy E. As both the length (L) and width (W) of the conductor are large, the allowed k states are spaced very close to each other and the summation over k can be replaced by an integral as shown above (Eqs. 11.24, 11.25).

On the other hand, in the case of a 1-D material such as a CNT whose length spans over the x-direction, the distance in the y-direction is the diameter which is very small. Hence states along k_y are finitely spaced and one cannot use integration to sum up the number of states. Therefore the expression for the DOS becomes [5]

$$D(E) = \sum_{\overrightarrow{\beta_y}} \int \frac{dk_x}{2\pi/L} *\delta\left(E - \in\left(\vec{k}\right)\right) \qquad (11.31)$$

From the general dispersion relation Eq. 11.16, we have $\epsilon^2 = (at)^2(k_x^2 + \beta_y^2)$, Taking the differential with respect to k_x yields, $\epsilon d\epsilon = (at)^2 * k_x * dk_x$ and $k_x = \sqrt{\frac{\epsilon^2 - a^2 t^2 \beta_y^2}{a^2 t^2}}$.

Replacing k_x and dk_x and using the property of delta function, the expression for 1-D DOS become [6]

$$D(E) = \frac{L}{2\pi at} \sum_{\vec{\beta_y}} \frac{E}{\sqrt{E^2 - a^2 t^2 \beta_y^2}} \tag{11.32}$$

That is, corresponding to the energy values $E^2 = a^2 t^2 \beta_y^2$, the DOS exhibits a series of spikes which are typical signatures of 1-D materials called Van-Hove singularities (Fig. 11.5) [12].

For a nanotube of sufficiently large length L, the discrete energy levels in momentum space separated by a distance $\delta k = \frac{2\pi}{L}$ are close enough to form a continuum of states and the density of states can alternatively be expressed as

$$D(E) = \int \delta\left(E - \epsilon\left(\vec{k}\right)\right) dE = \int \frac{dk}{\frac{2\pi}{L}} \delta\left(E - \epsilon\left(\vec{k}\right)\right) \tag{11.33}$$

$$= \frac{L}{2\pi} \int d\epsilon \frac{dk}{d\epsilon} \delta\left(E - \epsilon\left(\vec{k}\right)\right) = \frac{L}{2\pi} \left|\frac{1}{\frac{d\epsilon}{dk}}\right|_{\epsilon = E} \tag{11.34}$$

i.e.,

$$\left|\frac{d\epsilon}{dk}\right|_{\epsilon = E} = \frac{L}{2\pi * D(E)} \tag{11.35}$$

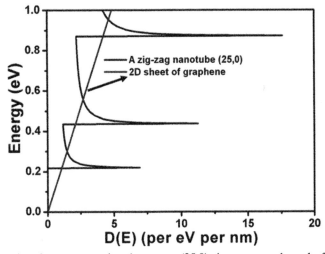

Fig. 11.5 Density of states comparison between a (25,0) zig-zag nanotube and a 2-D graphene sheet calculated using Eqs. (11.30) and (11.32) for graphene sheet and zig-zag nanotube respectively.

The energy required to add an additional electron to the nanotube at the Fermi energy is given by:

$$\delta E = \delta k \left| \frac{d\varepsilon}{dk} \right|_{\varepsilon = E_F} = \frac{1}{2\pi^* D(E_F)} \delta k \tag{11.36}$$

where $D(E_F)$ is the DOS per unit length at Fermi energy. Substituting δk yields, $\delta E = \frac{1}{D(E_F) L}$. This energy is equal to the capacitive energy $CV^2 \equiv e^2/LC$, i.e.,

$$\delta E = e^2/LC \tag{11.37}$$

And the capacitance per unit length in the case of the nanotube is given by

$$C_Q = e^2 D(E_F) \tag{11.38}$$

This term C_Q is called the intrinsic or quantum capacitance of a nanotube—representing the charge occupied in the quantized energy levels [13].

In a conventional MOSFET structure, which has been the main driver of CMOS over the last half century or so, this capacitance corresponds to that of the layer of quantized carriers confined in the inversion layer at the surface of the channel. The electrical behavior of the conventional MOSFET device is determined by the maximum potential (Φ) in the channel that controls the flow of carriers from the source into the channel via an applied gate bias (Fig. 11.6)

The potential applied at the gate terminal is dropped across the oxide and semiconductor surface to create an inversion layer of quantized electrons. A change in channel potential ($\delta\Phi$) in the device with respect to change of gate potential ($\delta\Phi_g$) can be assumed as [14–16]

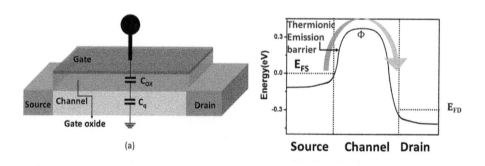

Fig. 11.6 Schematic representation of a conventional MOSFET device and the relevant capacitances (A). The conduction band profile of the device (B) shows the channel potential far from the contacts (Φ). E_{FD} represents the chemical potential of the drain terminal.

$$\delta \Phi = \delta \Phi_g \frac{C_{OX}}{C_{OX} + C_Q + C_S + C_D} \tag{11.39}$$

where C_{OX}, C_Q, C_S, C_D are the oxide, quantum, source, and drain capacitances (C_S and C_D are neglected in the following for simplicity, although C_D plays a part in short channel effects).

In a MOSFET, it is very well known that the total gate capacitance for a metal gate consists of a series combination of C_{OX} and C_Q [6]

$$C_g = \frac{C_{OX}*C_Q}{C_{OX} + C_Q} \tag{11.40}$$

The charge in the channel is described (by definition of a capacitor) as below Q_{ch},

$$C_g = \frac{\delta Q_{ch}}{\delta V_g} = q* \frac{\delta Q_{ch}}{\delta \Phi} * \frac{\delta \Phi}{\delta \Phi_g} \tag{11.41}$$

From Eq. (11.39), $\frac{\delta \Phi}{\delta \Phi_g} = \frac{C_{OX}}{C_{OX} + C_Q}$ and comparing the above two equations reveals that $q* \frac{\delta Q_{ch}}{\delta \Phi} = C_Q$. Therefore, the quantum capacitance of a conventional MOSFET of bulk material as discussed by Knoch and Appenzeller in [15,16] can be written as

$$C_q^{3D} = q \frac{\partial Q_{ch}}{\partial \Phi} \propto - q^2 t\, W\, L\, D^{3D}(E_{FS} - \Phi) \propto \sqrt{E_{FS} - \Phi} \tag{11.42}$$

In the on state, C_q^{3D} increases with the change in surface potential and finally beyond the threshold voltage (quantified by an equal amount of band bending below and above the Fermi level) reaches a stage where the surface potential hardly moves with increase in gate voltage. In this situation, $C_Q \gg C_{ox}$ because of the bulk DOS of the channel, thus limiting the scaling capacitance of the bulk MOSFET. In the device off state $C_{OX} \gg C_Q$ and $C_{OX} \gg C_D$. The relative contribution of C_Q with respect to C_{OX} determines the nature of the turn-on characteristic (called inverse subthreshold slope) which is given by the number of carriers which are thermally excited over the source/channel potential barrier. The number of electrons changes exponentially with gate voltages and N can be written as

$$N \approx N_0 \exp \left(\frac{\Phi}{k_B T} \right) \Rightarrow \log_{10} \left(\frac{N}{N_0} \right) \approx \left(\frac{\Phi}{2.3 * k_B T} \right) \tag{11.43}$$

where N_0 is the charge density corresponding to the neutral potential, i.e., the number of electrons increases by a decade with an increase in gate potential of $2.3 * k_B T = 60$ meV at room temperature as shown in Fig. 11.7A [17]. Application of a positive gate voltage lowers the overall density of states in the channel region and increases the electron density in the channel. Inverse subthreshold swing S in terms of device I–V is therefore defined by $S = \ln(10) \left(\frac{\partial I_d}{\partial V_{GS}} \frac{1}{I_d} \right)^{-1}$.

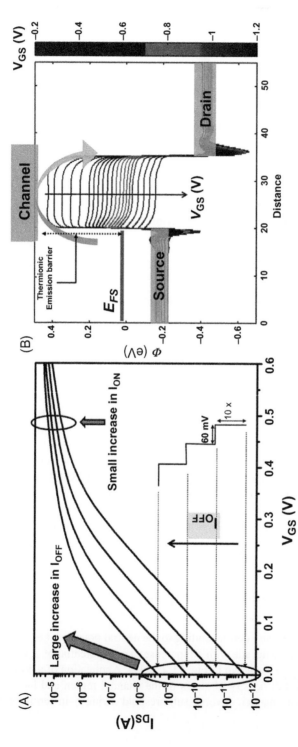

Fig. 11.7 (A) Reduction of the threshold voltage and subsequent reduction of minimum current in a MOSFET device. Each curve in the figure is left shifted by 60 mV. Figure shows a typical scenario in a conventional MOSFET where a small increase in ION by reduction of the threshold voltage of 60 mV results in a 10-fold increase in OFF current (B) Modulation of conduction band profile of an n-channel CNTFET by applied gate voltage interval of 0.05 V. As the gate voltage is increased in the negative direction, the barrier height is reduced which allows a flow of electrons from the source side and the gate voltage continues to modify the surface potential in the device on-state in contrast to bulk MOSFETs (beyond threshold)

Contrary to bulk solids, the density of states in 1-D solids such as CNTFETs, the density of states $\propto 1/\sqrt{E_{FS} - \varPhi}$. Appenzeller et al. have previously shown that the quantum capacitance per unit length in such materials is proportional to the tunneling probability of the carriers across the potential barrier in the device [16], i.e.,

$$C_q = e\frac{\partial Q_{ch}}{\partial \varPhi} \approx e\frac{\partial}{\partial \varPhi}\int_0^{\Delta\varnothing} dE\, T_{\text{WKB}} D(E)(1 - f(E - E_{FS})) \tag{11.44}$$

Hence in the Quantum capacitance limit $C_q \leq C_{\text{OX}}$ can be achieved in 1-D systems with ultrathin gate oxides and/or high-k dielectrics as shown by Jing Guo et al. in [18]. In this case the gate capacitance is dominated by C_q (rather than C_{OX}) and consequently $\frac{\delta\varPhi}{\delta\varPhi_g}$ approaches unity which means that the gate voltage continues to modify the surface potential in the device on-state (beyond threshold). In Fig. 11.7B, it is shown that one-dimensional FETs have a superior ability for electrostatic control, thereby eliminating short channel effects.

11.7 Carbon nanotube tunnel FETs (CNT-TFETs)

Since MOSFETS are limited in their switching by an $S = 60\,\text{mV/dec}$, a possible way for a faster turn-on at room temperature is by modifying the current injection mechanism at the source contact through tunneling rather than thermionic emission [19–24]. Fig. 11.8 shows the on-state and off-state band bending in a CNT-TFET. At $V_{\text{GS}} > 0\,\text{V}$, channel conduction band is below the source valence band and this makes it easier for carriers to tunnel from the source valence band to the channel conduction band. At $V_{\text{GS}} < 0\,\text{V}$, the band bending at the channel-drain results in a tunneling current between source and drain terminals. The tunneling mechanism depicted in Fig. 11.8B shows that only electrons within an energetic window of $\Delta\varnothing$ contributes to the current (I_d) as the tunneling probability is finite only within an energy window of $\Delta\varnothing$.

The inter-band tunneling probability across the contact-channel junction, T_{WKB} is given by the WKB approximation as [25]

$$T_{\text{WKB}} \approx \exp\left(-\frac{4\Lambda\sqrt{2m^*}\sqrt{E_g{}^3}}{3q\hbar\left(E_g + \Delta\varPhi\right)}\right) \tag{11.45}$$

where E_g is the band gap of the nanotube, m^* is the effective mass, Λ is the width of the interband potential barrier assumed at the source-channel interface region and $\Delta\phi$ is the energetic difference between the channel conduction band and source valence band (Fig. 11.8B).

The band-to-band tunneling (BTBT) current in a CNT-tunnel FETs can be ascertained using the NEGF or an analytic approach. The latter is given by [25,26]

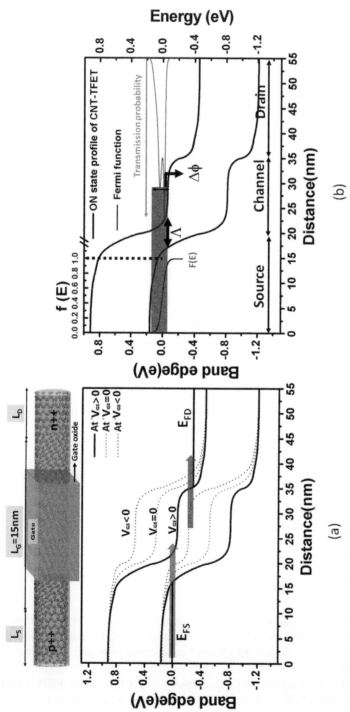

Fig. 11.8 (A) A schematic band bending in a CNT-TFET under gate voltage $V_{GS} = 0\,V$, $V_{GS} > 0\,V$ and $V_{GS} < 0\,V$. (B) BTBT at the source-channel interface where the width of the barrier is represented as Λ and the energetic difference between the channel conduction band and source valence band in the on-state is denoted as $\Delta\phi$. The Fermi distribution function plotted along with the band energy indicates that the low and high energy tails of the Fermi distribution function are cut off when the interband tunnel probability is high across the barrier. Only the *red-shaded part* of Fermi distribution is involved in device operation.

$$I_d = \frac{4e}{h} \int_0^{\Delta\phi} dE T_{\text{WKB}} (1 - f(E - E_{\text{FS}}))$$

$$= \frac{4e}{h} T_{\text{WKB}} k_B T \left(\ln \left(e^{\frac{\Delta\phi - E_{\text{FS}}}{k_B T}} + 1 \right) - \ln \left(e^{\frac{-E_{\text{FS}}}{k_B T}} + 1 \right) \right) \tag{11.46}$$

When the inter-band tunnel distance Λ is small, the tunneling probability at the interface can be considered as unity. Then, the inverse subthreshold slope for CNT-TFET in the case of 1-D CNT-TFETs can be obtained from its general form $S = \ln(10) \left(\frac{\partial I_d}{\partial V_{\text{GS}}} \frac{1}{I_d} \right)^{-1}$ and Eq. (11.46) as discussed by Knoch et al. in [26]

$$S \approx \frac{\ln(10)}{|e|} \frac{3q\hbar (\Delta\varnothing + E_g)^2}{4\Lambda \sqrt{2m^*} E_g^{3/2}} \tag{11.47}$$

In 1D-TFETs, the quantum capacitance is significantly smaller than conventional MOSFETs due to the dependence of the quantum capacitance on T_{WKB} [15,16]. For $\Lambda \ll L$, where L is the channel length, the gate capacitance $C_g = \frac{C_{\text{ox}} C_q}{C_{\text{ox}} + C_q}$. The quantum capacitance per unit length is given by [26]

$$C_q = e \frac{\partial Q_{\text{ch}}}{\partial \Phi} \approx e \frac{\partial}{\partial \Phi} \int_0^{\Delta\varnothing} dE T_{\text{WKB}} D(E)(1 - f(E - E_{\text{FS}})) \tag{11.48}$$

For devices working in the limit $C_{\text{OX}} \gg C_q$, the total capacitance is dominated by C_q which reduces the requirement of aggressive scaling of oxide thickness.

11.8   ITRS requirements-2024

Materials which can act as alternate channel for beyond CMOS device applications are expected to enable continued scaling of integrated circuits with improved energy efficiency for applications where no solutions are currently known. The ITRS roadmap targets major parameters such as saturation current, on-off ratio and device delay in the timeframe of 2024 and beyond, where carbon-based materials, can be considered viable (Table 11.1).

The 14 nm node of the ITRS roadmap is projected to operate with 0.7 V supply with a power consumption $25 \times$ higher than a V_{DD} of 0.14 V (as power consumption αV_{DD}^2). The minimum required on-off ratio of 4 orders (for HP technology) demands the subthreshold swing of 35 mV/dec for the operating voltage of $V_{\text{DD}} = 0.14$ V. A number of strategies/device architectures have been explored to realize subthreshold operation <60 mV/dec for various TFETs and a few of them are summarized in Table 11.2.

Table 11.1 ITRS-2024 targeted values for various parameters [27].

Parameter	High performance logic technology (low operating power technology)
Saturation current ($\mu A/\mu m$)	2152 (666)
On-Off ratio	$\sim 2 \times 10^4$
Intrinsic device delay (fs)	130 (280)
Supply voltage (V)	0.61 (0.46)

A comparison between the predicted and experimentally reported tunnel FETs reveals a significant gap in the ON state current and inverse subthreshold slope. In TFETs, for high ON state current, the tunneling probability at the source/channel interface should be nearly one and this require an abruptly varying doping profile (4–5 orders magnitude) between source and intrinsic channel region within a distance of few nanometers [27,28]. High trap density (D_{it}) at the channel-gate dielectric interface in experimental TFETs is detrimental to realize the steep subthreshold slope as it results in tunneling of carriers from the valence band in the p-doped source region to mid-gap trap states and a subsequent thermal emission into the conduction band that degrades S [29]. In order to realize the full potential of TFET devices, parasitic access resistance at the contact region also needs to be minimized to improve the ON current [30]. Ambipolarity in TFETs is another major issue in designing n-type and p-type TFETs. Asymmetry in the doping profile in the source/drain extensions and using a small bandgap at the source region and high bandgap at drain side (hetero junction TFETs) to minimize the BTBT at the channel/drain region are considered as suitable ways to suppress ambipolarity [31–36].

Since the first demonstration of a CNTFET by Dekker in [37], the potential of carbon-based materials for electronic applications was explored by many groups and IBM in particular (Appenzeller and Avouris et al.) [38–46]. In 2004, Ali Javey et al. demonstrated a combination of top and bottom gated, doped source/drain CNTFET with an on current up to $20\,\mu A$ and a conductance of $0.1 \times 4e^2/h$ [$4e^2/h$ is the ballistic conductance in semiconducting nanotubes] for long channel devices [47]. Eight years later, in 2012, a sub 10 nm channel Schottky barrier CNTFET of diameter 1.3 nm with subthreshold slope (S) 94 meV/dec and on current of $2410\,\mu A/\mu m$ at a $V_{DD} = 0.5\,V$ much higher than Si-based FETs was reported [43]. A short channel ($L_{ch} = 30\,nm$) gate all around n-type CNTFET also demonstrated by IBM researchers, showed an on-state current $>1000\,\mu A/\mu m$ @ $V_{DD} = 0.5\,V$ and on-off ratio $> 10^3$ [44]. More recently, ultra-short channel CNTFETs with channel length of 5 nm with on-current of $1160\,\mu A/\mu m$ [48] and steep switching CNTFETs with subthreshold swing of 40 mV/dec is demonstrated using graphene as source electrodes [49]. Fig. 11.9 summarizes the ON and OFF state performance of these best reported CNTFETs.

Table 11.2 Summary of TFET performance demonstrated using experiment and theory and its comparison with recently experimentally reported MOSFET performance [28–36].

Device Type	Device technology	Inverse Subthreshold Swing (mV/dec)	I_{ON} (μA/μm)	ON-OFF ratio	EOT (nm)	Gate length (nm)	Reference
Experimental TFETs	InGaAs/InP TFETs	93	20	6×10^5 (@ |Vgs|=1.75V, |Vds|=0.5V)	~1.3	70000	Zhou et. al,University of Notre Dame -2012
	GaSb/InAs(Sb) NW-TFETs	~235	62	>143 (@ |Vgs-V_T|=0.5V; |Vds|=0.3V)	1.3	290	Anil W. Dey, Lund University, Sweden-2013
	InAs-Si Nanowire TFET	150	2.4	10^6 (@ |V_{DD}|=1V)	~3.25	~550	Moselund et.al, IBM Research—Zurich (2012)
	Si- Tunneling Green Transistor	46	1.2	7×10^7 (@ |Vgs|=|Vds|=1V)	--	20000	Hu et. al,University of California-2010
Projected TFETs	InGaAs/AlGaAsSb heterojunction TFETs	--	400	10^4 (@ |V_{DD}|=0.3V)	0.6	100	Wang et. al, GlobalFoundries, Inc-2010
	InAs/InGaAs vertical TFETs	16	~1000	~10^6 (@ Vgs=0.6V,V_{DD}=0.4V)	0.3	20	Ganapathi et. al, University of California Berkeley-2011
Experimental MOSFETs	InGaAs-InAlAs superlattice FETs	13	4500	~4.5×10^4 (@ |V_{DD}|=0.4V)	1	40	Gnani et. al, University of Bologna-2011
	InGaAs/InP/InAlAs/InGaAs core-multishell NWFETs	75	1000	>10^6 (@ |Vgs|=|Vds|=0.5V)	2.75	200	Tomioka et.al, Hokkaido University, Japan -2012
	InGaAs Gate-all-around NW MOSFETs	91	1000	~10^4 (@ |Vgs|=|Vds|=0.5V)	1.2	20	Gu et. al,Purdue University - 2012

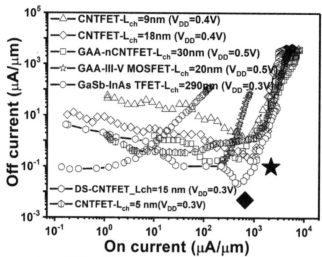

Fig. 11.9 A comparison of ON and OFF state performance of best reported short channel CNT and III–V devices [31,48–57]. The ITRS targeted values for low operating power (LOP) and high performance (HP) technologies are highlighted with filled diamond and star respectively.

The ON-OFF ratio falls short of requirements for various ITRS technologies (http://www.itrs.net/Links/2012ITRS/Home2012.htm). The main reason for the poor off-state current is the leakage current through the Schottky barrier present at the metal-CNT interface which needs to be addressed [45]. A comparison of the experimentally demonstrated performance of Schottky barrier CNTFETs with III–V MOSFET and T-FETs in Fig. 11.9 reveals that today CNTFETs exhibit better on state performance as compared to the rivaling III–V technologies; [31,48–57] however, the poor on-off ratio due to high S exhibited by the SB-CNTFETs reveals the need for short channel carbon-based FETs with steeper switching speed and minimum off-state leakage (Fig. 11.10).

Several obstacles need to be overcome before introducing CNTs into the mainstream manufacturing technology. CNTs can be metallic or semiconducting in nature and statistical analysis reveals that 33% of the nanotube chiralities are metallic. In the case of 7 nm channel devices targeted for the year 2024 by ITRS, completely eliminating metallic chiralities in devices is considered of paramount importance and a targeted impurity concentration of metallic nanotubes by year 2020 is expected to be 0.0001% (Franklin et al. (IBM)) [58]. Another important aspect is achieving high parallel packing density with a minimum pitch between the nanotubes to produce high ON state current. The targeted packing density for the year 2020 is 125 nanotubes/μm [58,59] though a packing density of 250 nanotubes/μm with a pitch of 4 nm (distance between two nanotubes) is essential to surpass silicon CMOS technology [59]. A number of methods have been explored for the purification of nanotubes such as

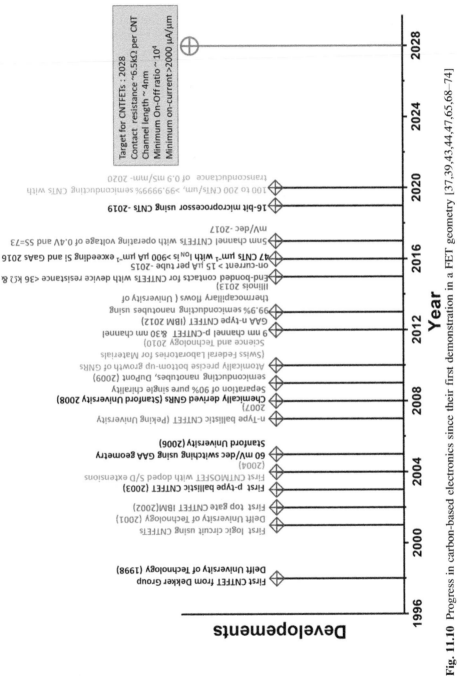

Fig. 11.10 Progress in carbon-based electronics since their first demonstration in a FET geometry [37,39,43,44,47,65,68–74]

ultra-centrifugation [60,61], chromatographic separation of semiconducting chiralities [62,63], alternating current dielectrophoresis where an opposite movement of metallic and semiconducting tubes is established along the electric field gradient [64] and DNA-assisted separation and purification [65]. The main drawback of these methods is that the resulting nanotubes are short ($\sim$1 μm) or coated with molecules/DNA so that their well-controlled parallel assembly becomes technically challenging. Perfectly aligned growth (>99.9% of SWNTs within 0.01^0 of perfect alignment) of CNTs has been realized in CVD growth on quartz substrate, but suffers from the limitation that includes metallic CNTs [66,67]. However, a recently reported CMOS compatible method in which an effective method for selective removal of metallic nanotubes has been demonstrated [68]. A long parallel array of CNTs is coated with a special organic thin film (resist) that is sensitive to thermal fluctuations. By turning off the semiconducting nanotubes by applying a suitable gate bias, current flow is achieved only through metallic nanotubes there by locally breaking down the resist around the metallic nanotubes and creating a surrounding trench. These exposed metallic nanotubes are removed by oxygen plasma treatment, giving a purity of 99.9% semiconducting nanotubes with a packing density $\sim$4 nanotubes/μm in a chip, exhibiting devices with on-off ratio $\sim 10^4$ and mobility exceeding $1000\,\mathrm{cm}^2\,\mathrm{V}^{-1}\mathrm{s}^{-1}$ [68].

However, the packing density in this approach needs to be significantly improved to meet the targeted values for 2024. A packing density of 125 nanotubes/μm is demonstrated recently with on-state current high as 1300 μA/μm and impurity concentration of metallic nanotubes as low as 0.0001% using a dimension-limited self-alignment (DLSA) procedure, however further optimization is required to improve the scaling, packing density, and switching properties [75]. Single nanotube tunnel devices using graphene as the source pad named as Dirac Source FETs (DS-FETs) have demonstrated high on-state current of up to 6.5 μA at $V_{DS} = -0.5$ V with subthreshold swing as small as 40 mV/dec and projected performance with 125 tubes/μm is expected to achieve an on state current high as 1000 μA/μm with on-off ratio > 1000 [49]. Improving contact resistance in CNTFETs is another major objective to achieve high ON state and switching performance [76,77]. CNT transistors are demonstrated with device resistance below 36 kΩ and mean resistance of 18.2 kΩ per CNT employing a channel length down to 10 nm using a CMOS compatible end-bonded and low temperature side contact methods, respectively [78,79]. An ideal contact resistance of $\sim$6.5 kΩ which is equal to the nanotube quantum resistance is the desired minimum contact resistance in nanotube devices [77,79].

CNTs exhibit a diameter dependent bandgap and hence, it is necessary to identify suitable diameter (bandgap) nanotubes which can be used to realize FETs of targeted performance proposed by ITRS for various technologies. Direct measurement of electronic band gap from spectrofluorimetric measurements is challenging as the recorded energies account not only for the electronic band gaps but also for the exciton binding energies and tunneling spectroscopic measurements underestimate the bandgaps due to screening arising from the image charge in metal substrates [80]. Hence more accurate approaches are necessary to establish a bandgap relation with diameter in nanotubes.

In this section, the performance of tunnel FETs consists of CNT diameter from 0.55 to 1.25 nm are analyzed via modeling. The TFETs IVs are evaluated using NANOTCAD ViDES in which the electron-hole concentration in the device is computed by solving the Schrödinger equation with open boundary conditions by means of the NEGF formalism [81]. An intrinsic channel length of 15 nm and doped source/drain regions of length 20 nm, an effective gate oxide thickness (EOT) of 0.5 nm and a $V_{DS} = 0.3$ V are used. A tight binding Hamiltonian with atomistic (P_z orbitals) basis has been used in the mode space approach.

Transport is calculated using the GW-band gaps of the nanotubes [82] with corresponding density of states and effective mass for various diameters are incorporated using an effective tight parameter of $\gamma = -5.2$ eV. Analyzing the tunnel FET performance of various diameter CNTs in Fig. 11.11 reveals that a low diameter nanotubes (<0.8 nm) are unsuitable for the technology due to their low ON state current (<0.1 μA per nanotube), while large diameter nanotube devices (>1.3 nm) exhibit poor ON-OFF ratios ($<10^3$) due to the high off-state leakage in those devices. A (11,0) TFET shows advantage in the case of ON state current ($\sim$2–3 μA per nanotube) while exhibiting on-off ratio $> 10^6$.

11.9 Comparison between a CNT-MOSFET and TFET

A comparison of CNTFETs using a (11,0) nanotube (diameter ≈ 0.85 nm) with bandgap ≈ 1.4 eV, an effective oxide thickness of 0.5 nm and drain bias of 0.3 V and channel length of 15 nm in the MOSFET and tunnel FET geometries are investigated in this section.

Analyzing the on-off and subthreshold performance reveals a strong device geometry dependence which stems from the fundamental differences in their operation (Figs. 11.12 and 11.13). As explained in the previous sections, for CNTMOSFETs, thermionic emission over the source-channel potential barrier limits their S value at 60 mV/dec, while a highly gate dependent tunneling probability in TFETs presents $S < 60$ mV/dec.

11.10 Carbon nanotube vs graphene nanoribbon

Despite the recent interest in GNRs, the best reported GNRFET displays an inferior performance in terms of on-off ratio ($<10^3$) and inverse subthreshold slope ($\sim$210 mV/dec) at a V_{DD} of 0.5 V [69] and their short channel performance may not be easily ascertained in the absence of ultra-narrow GNRs with atomically smooth edges [70,83]. Orientation controlled growth of GNRs is recently demonstrated via facilitating nanoribbon growth in hexagonal boron nitride trenches with on-off ratio high as 10^4 using channel length of $\sim$236 nm [84], however, further device optimization is required to improve their switching and ON current properties. Transfer characteristics of different family type GNRs in the TFET geometry are investigated using NEGF-based quantum transport simulations using NANOTCAD ViDES to identify

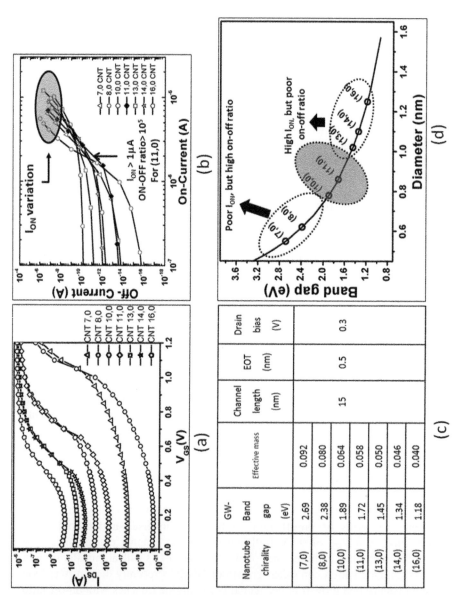

The following table appears in panel (c):

Nanotube chirality	GW-Band gap (eV)	Effective mass	Channel length (nm)	EOT (nm)	Drain bias (V)
(7,0)	2.69	0.092	15	0.5	0.3
(8,0)	2.38	0.080			
(10,0)	1.89	0.064			
(11,0)	1.72	0.058			
(13,0)	1.45	0.050			
(14,0)	1.34	0.046			
(16,0)	1.18	0.040			

Fig. 11.11 See figure legend on opposite page

(Continued)

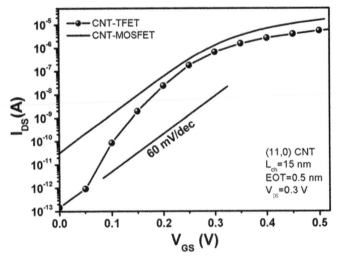

Fig. 11.12 Simulated transfer characteristics of a CNT-MOSFET, and a tunnel FET using an EOT $=0.5$ nm, $V_{DS}=0.3$ V, $L_{ch}=15$ nm for a (11,0) nanotube with GW-band gap of 1.4 eV. The gate voltage corresponding to minimum drain current (~0.1 pA in TFET) is adjusted to 0 V using a gate work function of 3.5 eV.

suitable ribbon width and family type offering high ON state current [85]. Among different types of GNRs, $3p+1$ family offers comparable ON current and ON-OFF ratio with a best identified (11,0) CNT TFET. A 16-AGNR TFET shows ON-OFF ratio $>10^6$ with an ON state current of $\sim1\,\mu$A. Narrower ribbons than considered in Fig. 11.14 from $3p+1$ family type exhibit lower ON current and hence not shown here.

In the following discussion a (11,0) nanotube with GW band gap of 1.4 eV and 16AGNR of band gap ≈ 1.4 eV is chosen for comparison. The diameter of 11,0 nanotube is ≈ 0.85 nm and width of 16AGNR is 1.87 nm. The identical band gap of the (11,0) CNT and 16AGNR in MOSFET geometries exhibit comparable performance (Fig. 11.15), but in TFET geometry (11,0) CNT exhibits better off-state and subthreshold properties as compared to 16-AGNR (Fig. 11.14). The higher on-current in CNTs

Fig. 11.11, cont'd (A) Simulated transfer characteristics of carbon nanotube TFETs with CNT diameters from 0.55 nm (7,0) to 1.25 nm (16,0) using a non-equilibrium Green function-based quantum transport simulation tool, NANOTCAD ViDES from G. Fiori et al., University of Pisa [81]. The bandgaps of the nanotubes used in the study are calculated via GW-approach as reported in [82]. (B) Comparison of ON and off current variation (at a $V_{DD}=0.3$ V) with bandgap. The effective masses and other device parameters used in the simulation are summarized in the table shown in (C). (D) an analysis shows that low diameter (<0.8 nm) nanotube TFET devices exhibit poor ON current, while large diameter (>1.25 nm) TFET devices suffer from low ON-OFF ratio.

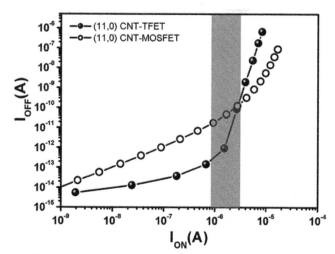

Fig. 11.13 On-Off ratio comparison in CNT-TFETs and MOSFETs. The *gray shaded part* represents the advantageous region where the devices delivers on current values between 1 and 10 μA and is beneficial for both high performance and low operating power technologies. In this region, the tunnel FETs offers maximum performance by providing off-current as low as 10^{-5} μA. Device parameters EOT = 0.5 nm, V_{DS} = 0.3 V, L_{ch} = 15 nm for a (11,0) nanotube with GW-BG ≈ 1.4 eV.

is due to the difference in quantum capacitance arising from a difference in the density of states. A 15 nm long (11,0) CNT has higher number of atoms (~1552) than a similar length 16-AGNR (~1129). It should be noted that the above comparative performance holds only for an un-optimized TFET geometry.

11.11   Summary

CNTs and GNR FETs are appealing materials for use in TFET applications due to their desirable band gap and low effective mass of the carriers however small diameter CNT devices are still limited by low on-currents and large diameter CNT devices suffer from poor on-off ratios.

Moreover, intrinsic CNTs/graphene directly connected to metal electrodes also tends to form Schottky barriers at the contact regions. Current flow through these devices is modulated by the barrier width and height using the gate electric field. Large off-state leakage in experimentally demonstrated Schottky barrier CNTFETs results in a poor ON-OFF ratio, though approximated ON state current meet the ITRS targeted values for 2024 technology (Fig. 11.16). To achieve MOSFET type or TFET characteristics requires doped source/drain regions which are still somewhat difficult to achieve and control with uniformity in carbon-based systems.

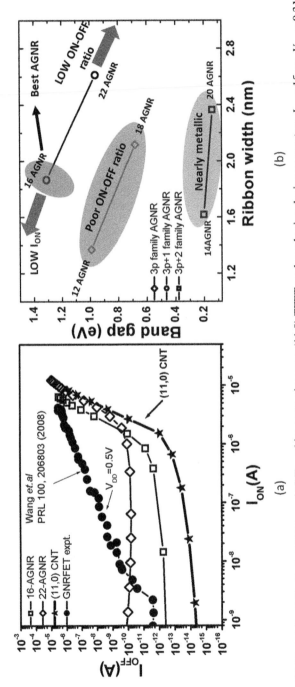

Fig. 11.14 (A) On-Off ratio in GNR tunnel FETs and its comparison to a (11,0) TFET evaluated using device parameters, $L_{\mathrm{ch}} = 15\,\mathrm{nm}$, $V_{\mathrm{DS}} = 0.3\,\mathrm{V}$, EOT = 0.5 nm, S/D doping = 0.8/nm (B) GW-band gap in GNRs vs various family types showing the ideal chirality to realize high performance TFETs is a 16-AGNR (width = 1.87 nm) with band gap ≈ 1.4 eV.

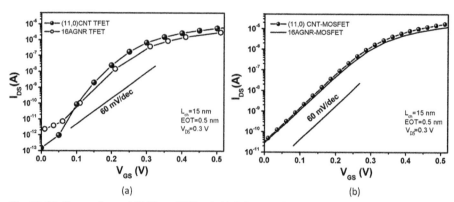

Fig. 11.15 Comparison of GNR vs CNTs. A 11,0 CNT and 16AGNR ribbon are considered in the tunnel-FET and MOSFET geometries. Device parameters EOT $= 0.5$ nm, $V_{DS} = 0.3$ V, $L_{ch} = 15$ nm. The gate voltage corresponding to minimum drain current is adjusted to 0 V using a gate work function of 3.5 eV [85].

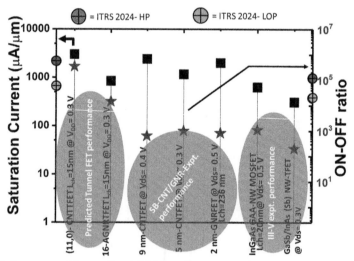

Fig. 11.16 Summary of the predicted TFET performance for CNTs and GNRs along with experimentally reported performance of Schottky barrier FETs consisting of 1.3 nm CNT and $\sim$2 nm wide GNR [43,48,50,52,69].

References

[1] B. Davari, R.H. Dennard, G.G. Shahidi, CMOS scaling for high performance and low power-the next ten years, Proc. IEEE 83 (4) (1995) 595–606.

[2] P.R. Wallace, The band theory of graphite, Phys. Rev. 71 (9) (1947) 622–634.

[3] A.H. Castro Neto, F. Guinea, N.M.R. Peres, K.S. Novoselov, A.K. Geim, The electronic properties of graphene, Rev. Mod. Phys. 81 (1) (2009) 109–162.

[4] J.-C. Charlier, X. Blase, S. Roche, Electronic and transport properties of nanotubes, Rev. Mod. Phys. 79 (2) (2007) 677–732.

[5] S. Datta, ECE 495N: Fundamentals of Nanoelectronics, 2008.
[6] S. Datta, Quantum Transport : Atom to Transistor, 2005.
[7] S. Gyungseon, Y. Youngki, J.K. Fodor, G. Jing, M. Akira, K. Diego, L. Gengchiau, K. Gerhard, L. Mark, S.A. Ibrahim, CNTbands, 2006.
[8] L. Yang, C.-H. Park, Y.-W. Son, M.L. Cohen, S.G. Louie, Quasiparticle energies and band gaps in graphene nanoribbons, Phys. Rev. Lett. 99 (18) (2007), 186801.
[9] K. Nakada, M. Fujita, G. Dresselhaus, M.S. Dresselhaus, Edge state in graphene ribbons: nanometer size effect and edge shape dependence, Phys. Rev. B 54 (24) (1996) 17954–17961.
[10] D. Gunlycke, C.T. White, Tight-binding energy dispersions of armchair-edge graphene nanostrips, Phys. Rev. B 77 (11) (2008), 115116.
[11] P. Zhao, M. Choudhury, K. Mohanram, J. Guo, Analytical theory of graphene nanoribbon transistors, in: 2008 IEEE Int. Work. Des. Test Nano Devices, Circuits Syst, 2008, pp. 3–6.
[12] L. Van Hove, The occurrence of singularities in the elastic frequency distribution of a crystal, Phys. Rev. 89 (6) (1953) 1189–1193.
[13] F. Leonard, Physics of Carbon Nanotube Devices, Elsevier Science, 2008.
[14] C.P. Auth, J.D. Plummer, Scaling theory for cylindrical, fully-depleted, surrounding-gate MOSFET's, IEEE Electron Device Lett. 18 (2) (1997) 74–76.
[15] J. Knoch, J. Appenzeller, Tunneling phenomena in carbon nanotube field-effect transistors, Phys. Status Solidi 205 (4) (2008) 679–694.
[16] J. Knoch, J. Appenzeller, Carbon nanotube field-effect transistors-the importance of being small, in: AmIware Hardware Technology Drivers of Ambient Intelligence, Springer, 2006, pp. 371–402.
[17] Y. Taur, T.H. Ning, Fundamentals of Modern VLSI Devices, Cambridge University Press, 2013.
[18] J. Guo, S. Datta, M. Lundstrom, M. Brink, P. McEuen, A. Javey, H. Dai, H. Kim, P. McIntyre, Assessment of silicon MOS and carbon nanotube FET performance limits using a general theory of ballistic transistors, in: Dig. Int. Electron Devices Meet, 2002, pp. 711–714.
[19] J.P. Leburton, J. Kolodzey, S. Briggs, Bipolar tunneling field-effect transistor: a three-terminal negative differential resistance device for high-speed applications, Appl. Phys. Lett. 52 (19) (1988) 1608–1610.
[20] W.M. Reddick, G.A.J. Amaratunga, Silicon surface tunnel transistor, Appl. Phys. Lett. 67 (4) (1995) 494–496.
[21] C. Aydin, A. Zaslavsky, S. Luryi, S. Cristoloveanu, D. Mariolle, D. Fraboulet, S. Deleonibus, Lateral interband tunneling transistor in silicon-on-insulator, Appl. Phys. Lett. 84 (10) (2004) 1780–1782.
[22] W.Y. Choi, J.Y. Song, J.D. Lee, Y.J. Park, B.-G. Park, 70-nm impact-ionization metal-oxide-semiconductor (I-MOS) devices integrated with tunneling field-effect transistors (TFETs), in: IEEE Int. Devices Meet. 2005. IEDM Tech. Dig., 2005, 2005, pp. 955–958.
[23] K.K. Bhuwalka, M. Born, M. Schindler, M. Schmidt, T. Sulima, I. Eisele, P-channel tunnel field-effect transistors down to Sub-50 nm channel lengths, Jpn. J. Appl. Phys. 45 (4B) (2006) 3106–3109.
[24] T. Nirschl, M. Weis, M. Fulde, D. Schmitt-Landsiedel, Correction to "revision of tunneling field-effect transistor in standard CMOS technologies", IEEE Electron Device Lett. 28 (4) (2007) 315.
[25] S.M. Sze, K.K. Ng, Physics of Semiconductor Devices, John wiley & sons, 2006.
[26] J. Knoch, S. Mantl, J. Appenzeller, Impact of the dimensionality on the performance of tunneling FETs: bulk versus one-dimensional devices, Solid State Electron. 51 (4) (2007) 572–578.

[27] A.M. Ionescu, K. Boucart, K.E. Moselund, V. Pott, Small Swing Switches, 2011.

[28] A.M. Ionescu, H. Riel, Tunnel field-effect transistors as energy-efficient electronic switches, Nature 479 (7373) (2011) 329–337.

[29] S. Mookerjea, D. Mohata, T. Mayer, V. Narayanan, S. Datta, Temperature-dependent I–V characteristics of a vertical $In_{0.53}Ga_{0.47}As$ tunnel FET, IEEE Electron Device Lett. 31 (6) (2010) 564–566.

[30] G. Zhou, Y. Lu, R. Li, Q. Zhang, W.S. Hwang, Q. Liu, T. Vasen, C. Chen, H. Zhu, J.-M. Kuo, S. Koswatta, T. Kosel, M. Wistey, P. Fay, A. Seabaugh, H. Xing, Vertical InGaAs/InP tunnel FETs with tunneling normal to the gate, IEEE Electron Device Lett. 32 (11) (2011) 1516–1518.

[31] L. Wang, E. Yu, Y. Taur, P. Asbeck, Design of tunneling field-effect transistors based on staggered heterojunctions for ultralow-power applications, IEEE Electron Device Lett. 31 (5) (2010) 431–433.

[32] J.-S. Jang, W.-Y. Choi, Ambipolarity factor of tunneling field-effect transistors (TFETs), J. Semicond. Technol. Sci. 11 (4) (2011) 272–277.

[33] K. Boucart, A.M. Ionescu, Double-gate tunnel FET with high-κ gate dielectric, IEEE Trans. Electron Devices 54 (7) (2007) 1725–1733.

[34] J. Trommer, A. Heinzig, U. Mühle, M. Löffler, A. Winzer, P.M. Jordan, J. Beister, T. Baldauf, M. Geidel, B. Adolphi, E. Zschech, T. Mikolajick, W.M. Weber, Enabling energy efficiency and polarity control in germanium nanowire transistors by individually gated nanojunctions, ACS Nano 11 (2) (2017) 1704–1711.

[35] S. Schneider, M. Brohmann, R. Lorenz, Y.J. Hofstetter, M. Rother, E. Sauter, M. Zharnikov, Y. Vaynzof, H.-J. Himmel, J. Zaumseil, Efficient n-doping and hole blocking in single-walled carbon nanotube transistors with 1,2,4,5-Tetrakis (tetramethylguanidino)ben-zene, ACS Nano 12 (6) (2018) 5895–5902.

[36] Q. Huang, F. Liu, J. Zhao, J. Xia, X. Liang, Ambipolarity suppression of carbon nanotube thin film transistors, Carbon N. Y. 157 (2020) 358–363.

[37] S.J. Tans, A.R.M. Verschueren, C. Dekker, Room-temperature transistor based on a single carbon nanotube, Nature 393 (6680) (1998) 49–52.

[38] R. Martel, T. Schmidt, H.R. Shea, T. Hertel, P. Avouris, Single- and multi-wall carbon nanotube field-effect transistors, Appl. Phys. Lett. 73 (17) (1998) 2447–2449.

[39] S.J. Wind, J. Appenzeller, R. Martel, V. Derycke, P. Avouris, Vertical scaling of carbon nanotube field-effect transistors using top gate electrodes, Appl. Phys. Lett. 80 (20) (2002) 3817–3819.

[40] P. Avouris, J. Appenzeller, R. Martel, S.J. Wind, Carbon nanotube electronics, Proc. IEEE 9 (11) (2003) 1772–1784.

[41] J. Appenzeller, Carbon nanotubes for high-performance electronics—progress and prospect, Proc. IEEE 96 (2) (2008) 201–211.

[42] J. Appenzeller, Y.-M. Lin, J. Knoch, Z. Chen, P. Avouris, Comparing carbon nanotube transistors—the ideal choice: a novel tunneling device design, IEEE Trans. Electron Devices 52 (12) (2005) 2568–2576.

[43] A.D. Franklin, M. Luisier, S.-J. Han, G. Tulevski, C.M. Breslin, L. Gignac, M.S. Lundstrom, W. Haensch, Sub-10 nm carbon nanotube transistor, Nano Lett. 12 (2) (2012) 758–762.

[44] A.D. Franklin, S.O. Koswatta, D. Farmer, G.S. Tulevski, J.T. Smith, H. Miyazoe, W. Haensch, Scalable and fully self-aligned n-type carbon nanotube transistors with gate-all-around, in: 2012 Int. Electron Devices Meet, 2012, pp. 4.5.1–4.5.4.

[45] J. Appenzeller, J. Knoch, V. Derycke, R. Martel, S. Wind, P. Avouris, Field-modulated carrier transport in carbon nanotube transistors, Phys. Rev. Lett. 89 (12) (2002), 126801.

[46] P. Avouris, Z. Chen, V. Perebeinos, Carbon-based electronics, Nat. Nanotechnol. 2 (10) (2007) 605–615.

[47] A. Javey, J. Guo, D.B. Farmer, Q. Wang, D. Wang, R.G. Gordon, M. Lundstrom, H. Dai, Carbon nanotube field-effect transistors with integrated Ohmic contacts and high-κ gate dielectrics, Nano Lett. 4 (3) (2004) 447–450.

[48] C. Qiu, Z. Zhang, M. Xiao, Y. Yang, D. Zhong, L.-M. Peng, Scaling carbon nanotube complementary transistors to 5-nm gate lengths, Science 355 (6322) (2017) 271–276.

[49] C. Qiu, F. Liu, L. Xu, B. Deng, M. Xiao, J. Si, L. Lin, Z. Zhang, J. Wang, H. Guo, H. Peng, L.-M. Peng, Dirac-source field-effect transistors as energy-efficient, high-performance electronic switches, Science 361 (6400) (2018) 387–392.

[50] J.J. Gu, X.W. Wang, H. Wu, J. Shao, A.T. Neal, M.J. Manfra, R.G. Gordon, P.D. Ye, 20–80nm channel length InGaAs gate-all-around nanowire MOSFETs with EOT = 1.2nm and lowest SS = 63mV/dec, in: 2012 Int. Electron Devices Meet, 2012, pp. 27.6.1–27.6.4.

[51] G. Zhou, Y. Lu, R. Li, Q. Zhang, Q. Liu, T. Vasen, H. Zhu, J.-M. Kuo, T. Kosel, M. Wistey, P. Fay, A. Seabaugh, H. Xing, InGaAs/InP tunnel FETs with a subthreshold swing of 93 mV/dec and I_{ON}/I_{OFF} ratio near 10^6, IEEE Electron Device Lett. 33 (6) (2012) 782–784.

[52] A.W. Dey, B.M. Borg, B. Ganjipour, M. Ek, K.A. Dick, E. Lind, C. Thelander, L.-E. Wernersson, High-current GaSb/InAs(Sb) nanowire tunnel field-effect transistors, IEEE Electron Device Lett. 34 (2) (2013) 211–213.

[53] K.E. Moselund, H. Schmid, C. Bessire, M.T. Bjork, H. Ghoneim, H. Riel, InAs–Si nanowire heterojunction tunnel FETs, IEEE Electron Device Lett. 33 (10) (2012) 1453–1455.

[54] C. Hu, Green transistor as a solution to the IC power crisis, in: Int. Conf. Solid-State Integr. Circuits Technol. Proceedings, ICSICT, 2008, pp. 16–19.

[55] K. Ganapathi, S. Salahuddin, Heterojunction vertical band-to-band tunneling transistors for steep subthreshold swing and high on current, IEEE Electron Device Lett. 32 (5) (2011) 689–691.

[56] E. Gnani, P. Maiorano, S. Reggiani, A. Gnudi, G. Baccarani, Performance limits of superlattice-based steep-slope nanowire FETs, in: 2011 Int. Electron Devices Meet, 2011, pp. 5.1.1–5.1.4.

[57] K. Tomioka, M. Yoshimura, T. Fukui, A III–V nanowire channel on silicon for high-performance vertical transistors, Nature 488 (7410) (2012) 189–192.

[58] A.D. Franklin, The road to carbon nanotube transistors, Nature 498 (7455) (2013) 443–444.

[59] N. Patil, J. Deng, S. Mitra, H.-S.P. Wong, Circuit-level performance benchmarking and scalability analysis of carbon nanotube transistor circuits, IEEE Trans. Nanotechnol. 8 (1) (2009) 37–45.

[60] M.S. Arnold, A.A. Green, J.F. Hulvat, S.I. Stupp, M.C. Hersam, Sorting carbon nanotubes by electronic structure using density differentiation, Nat. Nanotechnol. 1 (1) (2006) 60–65.

[61] A.A. Green, M.C. Hersam, Nearly single-chirality single-walled carbon nanotubes produced via orthogonal iterative density gradient ultracentrifugation, Adv. Mater. 23 (19) (2011) 2185–2190.

[62] M. Zheng, E.D. Semke, Enrichment of single chirality carbon nanotubes, J. Am. Chem. Soc. 129 (19) (2007) 6084–6085.

[63] H. Liu, D. Nishide, T. Tanaka, H. Kataura, Large-scale single-chirality separation of single-wall carbon nanotubes by simple gel chromatography, Nat. Commun. 2 (1) (2011) 309.

[64] R. Krupke, Separation of metallic from semiconducting single-walled carbon nanotubes, Science 301 (5631) (2003) 344–347.

[65] X. Tu, S. Manohar, A. Jagota, M. Zheng, DNA sequence motifs for structure-specific recognition and separation of carbon nanotubes, Nature 460 (7252) (2009) 250–253.

[66] N. Patil, A. Lin, E.R. Myers, K. Ryu, A. Badmaev, C. Zhou, H.-S.P. Wong, S. Mitra, Wafer-scale growth and transfer of aligned single-walled carbon nanotubes, IEEE Trans. Nanotechnol. 8 (4) (2009) 498–504.

[67] W. Zhou, C. Rutherglen, P.J. Burke, Wafer scale synthesis of dense aligned arrays of single-walled carbon nanotubes, Nano Res. 1 (2) (2008) 158–165.

[68] S.H. Jin, S.N. Dunham, J. Song, X. Xie, J. Kim, C. Lu, A. Islam, F. Du, J. Kim, J. Felts, Y. Li, F. Xiong, M.A. Wahab, M. Menon, E. Cho, K.L. Grosse, D.J. Lee, H.U. Chung, E. Pop, M.A. Alam, W.P. King, Y. Huang, J.A. Rogers, Using nanoscale thermocapillary flows to create arrays of purely semiconducting single-walled carbon nanotubes, Nat. Nanotechnol. 8 (5) (2013) 347–355.

[69] X. Wang, Y. Ouyang, X. Li, H. Wang, J. Guo, H. Dai, Room-temperature all-semiconducting sub-10-nm graphene nanoribbon field-effect transistors, Phys. Rev. Lett. 100 (20) (2008), 206803.

[70] J. Cai, P. Ruffieux, R. Jaafar, M. Bieri, T. Braun, S. Blankenburg, M. Muoth, A.P. Seitsonen, M. Saleh, X. Feng, K. Müllen, R. Fasel, Atomically precise bottom-up fabrication of graphene nanoribbons, Nature 466 (7305) (2010) 470–473.

[71] A. Javey, J. Guo, Q. Wang, M. Lundstrom, H. Dai, Ballistic carbon nanotube field-effect transistors, Nature 424 (6949) (2003) 654–657.

[72] A. Bachtold, Logic circuits with carbon nanotube transistors, Science 294 (5545) (2001) 1317–1320.

[73] Y. Lu, S. Bangsaruntip, X. Wang, L. Zhang, Y. Nishi, H. Dai, DNA functionalization of carbon nanotubes for ultrathin atomic layer deposition of high κ dielectrics for nanotube transistors with 60 mV/decade switching, J. Am. Chem. Soc. 128 (11) (2006) 3518–3519.

[74] Z. Zhang, X. Liang, S. Wang, K. Yao, Y. Hu, Y. Zhu, Q. Chen, W. Zhou, Y. Li, Y. Yao, J. Zhang, L.-M. Peng, Doping-free fabrication of carbon nanotube based ballistic CMOS devices and circuits, Nano Lett. 7 (12) (2007) 3603–3607.

[75] G.J. Brady, A.J. Way, N.S. Safron, H.T. Evensen, P. Gopalan, M.S. Arnold, Quasi-ballistic carbon nanotube array transistors with current density exceeding Si and GaAs, Sci. Adv. 2 (9) (2016), e1601240.

[76] P.M. Solomon, Contact resistance to a one-dimensional quasi-ballistic nanotube/wire, IEEE Electron Device Lett. 32 (3) (2011) 246–248.

[77] A.D. Franklin, Z. Chen, Length scaling of carbon nanotube transistors, Nat. Nanotechnol. 5 (12) (2010) 858–862.

[78] Q. Cao, S.-J. Han, J. Tersoff, A.D. Franklin, Y. Zhu, Z. Zhang, G.S. Tulevski, J. Tang, W. Haensch, End-bonded contacts for carbon nanotube transistors with low, size-independent resistance, Science 350 (6256) (2015) 68–72.

[79] G. Pitner, G. Hills, J.P. Llinas, K.-M. Persson, R. Park, J. Bokor, S. Mitra, H.-S.P. Wong, Low-temperature side contact to carbon nanotube transistors: resistance distributions down to 10 nm contact length, Nano Lett. 19 (2) (2019) 1083–1089.

[80] H. Lin, J. Lagoute, V. Repain, C. Chacon, Y. Girard, J.-S. Lauret, F. Ducastelle, A. Loiseau, S. Rousset, Many-body effects in electronic bandgaps of carbon nanotubes measured by scanning tunnelling spectroscopy, Nat. Mater. 9 (3) (2010) 235–238.

[81] G. Fiori, G. Iannaccone, G. Klimeck, A three-dimensional simulation study of the performance of carbon nanotube field-effect transistors with doped reservoirs and realistic geometry, IEEE Trans. Electron Devices 53 (8) (2006) 1782–1788.

[82] P. Umari, O. Petrenko, S. Taioli, M.M. De Souza, Communication: electronic band gaps of semiconducting zig-zag carbon nanotubes from many-body perturbation theory calculations, J. Chem. Phys. 136 (18) (2012), 181101.

[83] J. Yamaguchi, H. Hayashi, H. Jippo, A. Shiotari, M. Ohtomo, M. Sakakura, N. Hieda, N. Aratani, M. Ohfuchi, Y. Sugimoto, H. Yamada, S. Sato, Small bandgap in atomically precise 17-atom-wide armchair-edged graphene nanoribbons, Commun. Mater. 1 (1) (2020) 36.

[84] H.S. Wang, L. Chen, K. Elibol, L. He, H. Wang, C. Chen, C. Jiang, C. Li, T. Wu, C.X. Cong, T.J. Pennycook, G. Argentero, D. Zhang, K. Watanabe, T. Taniguchi, W. Wei, Q. Yuan, J.C. Meyer, X. Xie, Towards chirality control of graphene nanoribbons embedded in hexagonal boron nitride, Nat. Mater. 20 (2021) 202–207, https://doi.org/10.1038/s41563-020-00806-2.

[85] P.B. Pillai, P. Umari, M.M. De Souza, Are carbon nanotubes still a viable option for ITRS 2024? in: 2013 IEEE Int. Electron Devices Meet, 2013, pp. 32.2.1–32.2.4.

Optical spectroscopy study of two-dimensional materials

12

Miao-Ling Lin[a] and Ping-Heng Tan[a,b]
[a]State Key Laboratory of Superlattices and Microstructures, Institute of Semiconductors, Chinese Academy of Sciences, Beijing, China, [b]Center of Materials Science and Optoelectronics Engineering and CAS Center of Excellence in Topological Quantum Computation, University of Chinese Academy of Sciences, Beijing, China

12.1 Two-dimensional systems

Two-dimensional materials (2DMs) are easily exfoliable from their three-dimensional (3D) parent crystals due to the strong in-plane covalent bonds and weak interlayer van der Waals (vdW) interactions [1]. Because the physical properties in few-layer 2DMs change significantly in comparison with their bulk counterparts [2, 3], they have served as promising candidates for the applications in next-generation nanoelectronics and optoelectronics since the first discovery of a truly 2DM of graphene [4]. The two-dimensional (2D) databases consist of the graphene family (e.g., graphene [4, 5], hexagonal boron nitride [hBN] [6]), 2D dichalcogenides (e.g., MX_2, M=Mo, W, X=S, Se, Te [7]), 2D oxides (e.g., MoO_3, WO_3 [8]) and perovskite families (e.g., halide perovskite [9]), ranging from metals (e.g., $NbSe_2$ [10], TaS_2 [11]), semimetals (e.g., graphene, WTe_2 [12]), semiconductors (e.g., MoS_2 [13]) to insulators (e.g., hBN). In fact, Mounet et al. identified 5619 layered compounds from 108,423 experimentally known 3D compounds by high-throughput calculations, in which 1036 compounds are easily exfoliable and 789 compounds are potentially exfoliable [1]. According to the space group and structure similarities, the 1036 easily exfoliable compounds can be divided into 562 prototypes, among which the 10 most common ones, representing a total of 214 structures, are shown in Fig. 12.1A. Except for the FeOCl-like prototype, all these common prototypes have high-order rotational axes, especially the threefold rotational axes. The MX_2-like prototype is the most common with 64 similar structures included [14], to which many transition metal dichalcogenides (TMDs) and dihalides belong. Until now, the prototypes of CdI_2 (e.g., SnS_2 [15], etc.), MoS_2, and $FePS_3$ [16, 17] are compelling and attract much attention, while the $NdTe_3$ prototype common to many rare-earth tritellurides has not been reported.

The 2DMs with different number of layers may belong to different space groups due to the varied out-of-plane translational symmetry. Furthermore, because of the interlayer coupling, the number of layers provides a new degree of freedom for manipulating novel properties of 2DMs. For example, the semiconductor MX_2 undergoes a transition from indirect-gap semiconductor of bulk material to direct-gap

Modeling, Characterization and Production of Nanomaterials. https://doi.org/10.1016/B978-0-12-819905-3.00012-9

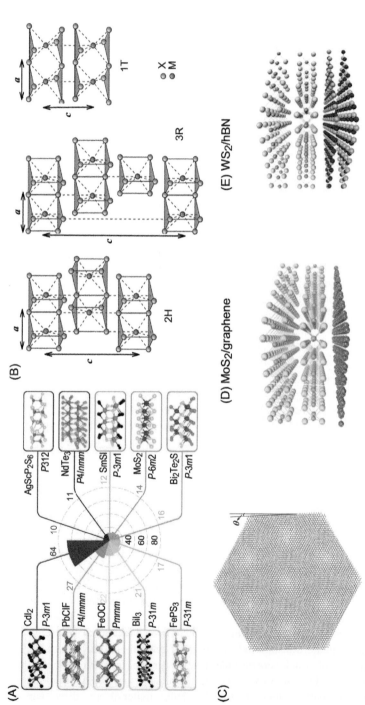

Fig. 12.1 (A) The most common 2D structural prototypes, together with the structure-type formula and the space group of the 2D systems. (B) Schematic diagrams of the three typical structural polytypes of MX_2 (M=Mo, W, X=S, Se, Te): 2H, 3R, and 1T. a and c represent the in-plane and out-of-plane lattice constants, respectively. (C) A typical schematic for twisted bilayer graphene with the twist angle of θ. The schematics of two vdWHs: (D) MoS_2/graphene and (E) WS_2/hBN.

(A) Adapted from N. Mounet, M. Gibertini, P. Schwaller, D. Campi, A. Merkys, A. Marrazzo, T. Sohier, I.E. Castelli, A. Cepellotti, G. Pizzi, N. Marzari, Two-dimensional gas of massless Dirac fermions in graphene, Nat. Nanotechnol. 13 (3) (2018) 246–252, https://doi.org/10.1038/s41565-017-0035-5; (B) adapted from Q.H. Wang, K. Kalantar-Zadeh, A. Kis, J.N. Coleman, M.S. Strano, Electronics and optoelectronics of two-dimensional transition metal dichalcogenides, Nat. Nanotechnol. 7 (11) (2012) 699–712, https://doi.org/10.1038/nnano.2012.193.

semiconductor of the monolayer [13]. In addition, for a 2DM with a given number of layers N (NL-2DM), the stacking order of the repeated rigid layer in the c-axis can be different, leading to various structural polytypes [14, 18], as one typical example of MX_2 depicted in Fig. 12.1B. There are two common polytypes, 2H (H: hexagonal) and 3R (R: rhombohedral) for the stacking of 1H-MX_2, in which the ith layer in 2H stacking is rotated by 180 degrees with respective to the $(i-1)$th layer while the adjacent layer undergoes a shift in 3R stacking. Thus, inversion symmetry exists in the even N layered (ENL) 2H-MX_2 but not in the odd N layered (ONL) 2H-MX_2. Furthermore, the metal atoms can also behave in octahedral coordination to generate the 1T stacking. The 2H-, 3R-, and 1T-stacked trilayer MX_2 belong to D_{3h}, C_{3v}, and D_{3d} symmetries, respectively, resulting in distinct novel electronic and optical properties [18, 19]. As the 2H stacking is the most common and stable stacking for MX_2, we only refer to 2H-stacked MX_2 when we talk about MX_2 later without any emphasis. In addition to the natural stacking orders for 2DMs, the twisted stacking can be naturally generated in a mechanical exfoliation process or formed by wet/dry transfer [20, 21]. Fig. 12.1C plots one typical schematic of the twisted bilayer graphene (tBLG) with a twist angle of θ, in which the moiré patterns are obviously observed. Twisted stacking can also occur between the constituents with different number of layers, forming the twisted multilayer 2DM (also denoted as $t(m + n + \cdots)$ L-2DM, $N = m + n + \cdots$). Besides, the van der Waals heterostructures (vdWHs) have recently emerged as an additional avenue to engineer new physical properties by assembling various 2DMs together in a desired fashion, since they exhibit remarkable electronic and optical properties compared with their constituents [22]. Fig. 12.1D and E shows the schematic diagrams of MoS_2/graphene and WS_2/hBN heterostructures, respectively. In twisted systems or the vdWHs formed by constituents with nearly the same lattice constants, the moiré superlattice can be generated at the interface, which can periodically modulate the behaviors of electrons [23], excitons [24], and phonons [25]. The features of moiré bands can be revealed by the transport properties [26], while the so-called moiré excitons and phonons have been probed via photoluminescence [27–29] and Raman spectroscopy [25], respectively.

The sensitivity of electronic and optical properties of 2DMs to the number of layers, stacking order, and assemblies in their hybrid systems have captured the interest of the scientific community to expand the understanding in fundamental physics behind the 2DMs and related vdWHs. Optical spectroscopy techniques, including optical contrast (OC), photoluminescence (PL) spectroscopy, Rayleigh scattering, Raman spectroscopy, and second-harmonic generation (SHG), serve as powerful techniques to investigate the electronic and optical properties through light-matter interactions. This chapter is organized as follows: Section 12.2 addresses the OC and differential reflection spectrum in 2DMs as useful tools for thickness and exciton level characterization. The typical PL spectra in semiconductor MX_2 and their valley polarization are demonstrated in detail in Section 12.3. The signatures of the moiré excitons of vdWHs in PL are also included. A simple introduction to Rayleigh scattering is given in Section 12.4. Section 12.5 provides a broad review of the Raman spectroscopy in pristine 2DMs, twisted 2DMs, and vdWHs. And finally, Section 12.6 presents the recent works on SHG in 2DMs.

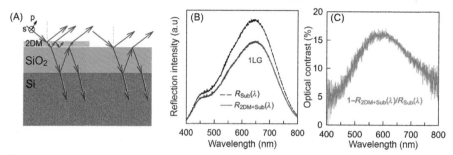

Fig. 12.2 (A) Schematic diagrams of multiple reflection and optical interference in the multilayered structures containing air, 2DM, SiO$_2$, and Si for the incident and outgoing light. s and p represent the s- and p-polarization components of the incident light, respectively. (B) The reflection spectra from bare SiO$_2$/Si substrate (R_{sub}) and from 1LG on the substrate ($R_{2DM+sub}$). (C) The corresponding OC for 1LG.

12.2   Optical contrast

By depositing a thin film onto a substrate with multilayer structure, varied multiple reflection and optical interference occur in thin films on substrate and the bare substrate, resulting in significantly different reflected light intensity of thin film from the bare substrate [30, 31], as shown in Fig. 12.2A. The relative intensity of reflection light between the thin film on substrate and the bare substrate is defined as the OC of the thin film. For physical characterization and device applications, 2DMs are usually deposited onto the SiO$_2$/Si or quartz substrates, which leads to a multilayer structure consisting of air, 2DMs, and substrates. The OC of 2DMs on substrate can be calculated by [31]

$$OC(\lambda) = 1 - R_{2DM+sub}(\lambda)/R_{sub}(\lambda), \tag{12.1}$$

where $R_{2DM+sub}(\lambda)$ and $R_{sub}(\lambda)$ are the reflection light intensity from 2DMs on substrate and that of bare substrate, respectively. Due to the multiple reflection interference and the wavelength-dependent complex refractive index of 2DMs and substrates, the reflection light intensity and thus the OC are dependent on the wavelength (λ) of the incident light, as an example of monolayer graphene (1LG) on 90 nm SiO$_2$/Si substrate shown in Fig. 12.2B and C. The setup for OC (λ) under backscattering configuration is demonstrated in detail in [31]. The OC (λ) of 2DMs on a substrate makes them be visible by the naked eye through a microscope, which can be further utilized to identify the number of layers, as discussed later.

From the perspective of theoretical calculation, OC (λ) can be quantitatively calculated by the transfer matrix based on the multiple reflection interference method [32, 33], in which the electric and magnetic components in each medium are connected by the multiplication of characteristic matrices. The detailed calculation has been reported [32]. The calculated OC (λ) is determined by the wavelength (λ),

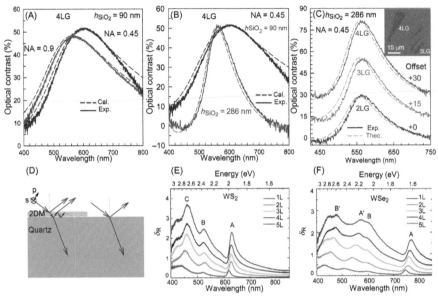

Fig. 12.3 (A) The theoretical and experimental OC (λ) of 4LG on a SiO$_2$/Si substrate ($h_{SiO2} =$ 90 nm) with objectives of NA = 0.45 and NA = 0.9. (B) The theoretical and experimental OC (λ) of 4LG on SiO$_2$/Si substrates of $h_{SiO2} =$ 90 nm and $h_{SiO2} =$ 286 nm using an objective with NA = 0.45. (C) The theoretical and experimental OC (λ) of 2LG, 3LG, and 4LG on SiO$_2$/Si substrate of $h_{SiO2} =$ 286 nm with an objective of NA = 0.45; the *inset* represents the optical images of 3LG and 4LG. (D) Schematic diagrams for the incident and outgoing light in the multilayer containing air, 2DM, and quartz. *s* and *p* represent the *s*- and *p*-polarization components of the incident light, respectively. Differential reflectance spectra of mechanically exfoliated (E) 2H-WS$_2$ and (F) 2H-WSe$_2$ with one to five layers. The peaks are labeled. (C) Adapted from Y. Lu, X.-L. Li, X. Zhang, J.-B. Wu, P.-H. Tan, Optical contrast determination of the thickness of SiO$_2$ film on Si substrate partially covered by two-dimensional crystal flakes, Sci. Bull. 60 (8) (2015) 806–811, https://doi.org/10.1007/s11434-015-0774-3; (E, F) adapted from W. Zhao, Z. Ghorannevis, L. Chu, M. Toh, C. Kloc, P.-H. Tan, G. Eda, Evolution of electronic structure in atomically thin sheets of WS$_2$ and WSe$_2$, ACS Nano 7 (1) (2013) 791–797, https://doi.org/10.1021/nn305275h.

numerical aperture (NA) of the objective, complex refractive index, and thickness of each dielectric medium. Fig. 12.3A shows the theoretical and experimental OC (λ) of one four-layer graphene (4LG) on SiO$_2$/Si substrate with 90-nm thick SiO$_2$ ($h_{SiO_2} =$ 90 nm) using an objective of NA = 0.45 and NA = 0.9. Notably, the objective of NA ≤ 0.55 is recommended for OC (λ) measurements to obtain better agreement between experimental and theoretical results [32]. In addition, when varying the thickness of SiO$_2$ covered on the Si wafer, the OC (λ) curve changes dramatically, as shown in Fig. 12.3B. Furthermore, the OC (λ) of 2DMs on the SiO$_2$/Si substrate is sensitive to the number of layers of 2DMs, which bridge the gap between OC (λ) and the determination of thickness of 2DMs. For example, as shown in Fig. 12.3C, as the thickness of *N*-layer graphene (*N*LG) increases, the OC maximum increases significantly once the NA and h_{SiO_2} are given. Therefore, once h_{SiO_2} is determined by OC [32], it is possible to distinguish the thickness of 2DMs by checking the agreements between the experimental and calculated OC (λ) based on the transfer-matrix method. The accuracy for the thickness identification of 2DM flakes is dependent on the OC (λ) difference between the flakes with adjacent number of layers. Generally, using OC (λ) can determine the number of layers of NLG up to $N = 8$.

For the effective measurement of the absorbance in semiconductor TMDs, it is common to use the transparent quartz substrate to measure the differential reflectance [34, 35], that is, the change in reflection intensity for 2DMs on quartz relative to that of the substrate, which can be defined by

$$\delta(\lambda) = R_{2DM+sub}(\lambda)/R_{sub}(\lambda) - 1. \qquad (12.2)$$

The differential reflectance of a 2DM flake relative to a transparent substrate with refractive index of n_{sub} is related to the absorbance (A) of the 2DMs by $\delta(\lambda) = \frac{4}{n_{sub}^2 - 1} A(\lambda)$ [34]. Fig. 12.3D shows the schematic diagram for the incident and outgoing light within the multilayer medium of 2DMs deposited on a quartz substrate. Fig. 12.3E and F shows the corresponding differential reflectance of WS$_2$ and WSe$_2$ with the number of layers 1–5 [34], respectively. Distinct peaks represent the great absorbance related to the excitons. It is obvious that all the exciton peaks of WS$_2$ and WSe$_2$ exhibit a gradual blueshift with the thickness decreasing. Indeed, the excitonic absorption peaks A and B correspond to the direct transitions at the K (K') point and the energy between A and B arises from the spin-orbit interaction [13]. Besides, additional peaks can be observed, in which the C peak in WS$_2$ originates from the optical transition between the density of states peaks in the valence and conduction bands [36], while the A$'$ and B$'$ peaks of WSe$_2$ arise from the splitting of the ground and excited states of A and B transitions, respectively [34]. The differential reflectance is widely utilized to determine the exciton peak positions for semiconductors and their related vdWHs, including the intralayer and interlayer excitons [29, 34, 35].

12.3 Photoluminescence spectroscopy

When a semiconductor is excited by the external energy, the electrons within the valence band jump to the conduction band, leaving a hole within the valence band. Following the excitation, the electrons and holes can spontaneously decay to the band edges and recombine again to release the energy in the form of light. This process is called luminescence, including electroluminescence and PL. PL is one form of light emission as a result of a semiconductor excited by photon energy, which is related to its unique electronic band structure, lattice structure, defects, carrier transport, and so on [13, 34, 37, 38]. For a semiconductor exhibiting a large exciton binding energy, the excited electron will be attracted to a hole by the electrostatic Coulomb force to form an exciton, and its optical properties are dominated by excitonic effects.

PL spectroscopy has been extensively studied in various material systems to reveal their optoelectronic properties. In recent years, a series of fascinating optical properties, such as strong excitonic effects and valley polarization, have been extensively reported in 2DMs and their vdWHs, which can be directly revealed by PL spectroscopy and will be demonstrated in detail later [27–29, 39–41].

Most MX_2 undergo a transition from an indirect-gap semiconductor in the bulk form to a direct-gap semiconductor in the monolayer due to the absence of weak interlayer coupling in the monolayer [13]. Furthermore, the strong dimensionality confinement and reduced dielectric screening can greatly enhance the exciton binding energy, up to hundreds of meV [42, 43]. Thus, the few-layer MX_2 tend to exhibit strong excitonic effects, which can greatly modulate their optical properties, such as PL and optical absorption. The direct bandgap in monolayer TMD can significantly enhance its PL yield. For example, the surprising bright PL from monolayer MoS_2 (1LM) can be detected in contrast to that from bilayer MoS_2, as shown in Fig. 12.4A [13]. With increasing the number of layers of MoS_2 from 1 to 6, the quantum yield decreases monotonously, as depicted in the inset of Fig. 12.4A. The corresponding normalized PL spectra of one to six layered MoS_2 by the A exciton emission are depicted in Fig. 12.4B. The 1LM exhibit a single narrow PL profile at 1.90 eV; however, the two to six layered MoS_2 flakes show multiple emission peaks, A, B, and I, corresponding to two direct-gap electronic transitions at the K (K') point and an indirect-gap transition, respectively. As the number of layers increases, the peak I shifts to smaller energy and becomes less prominent. Similar results have been revealed in other MX_2, such as WS_2, WSe_2, and $MoTe_2$ [34, 35].

Apart from the strong excitonic effects in monolayer MX_2, the inversion symmetry breaking together with spin-orbit coupling (SOC) can lead to coupled spin and valley physics in monolayer group-VI dichalcogenides [44]. This makes it possible to control the spin and valley states, shedding light on the routine toward the integration of valleytronics and spintronics in these 2DMs. Taking 1LM as an example, Fig. 12.4C plots its schematic electronic band structure, in which the conduction and valence-band edges are located at the corners (K and K' points) of the 2D Brillouin zone (BZ). The two inequivalent valleys constitute a binary index for low-energy carriers. On the one hand, due to the D_{3h} symmetry of monolayer MX_2, the inversion symmetry is broken, which results in valley-dependent optical selection rules for interband transition at the K (K') point. On the other hand, the strong SOC originating from the d orbitals of the heavy metal atoms would lead to the splitting of the valence-band edge states. In addition, the time-reversal symmetry requires that the spin splitting at different valleys must be opposite, as elucidated in Fig. 12.4C. Such strong coupling between spins and valleys can be directly detected by optical pumping with circularly polarized light [39, 41], because the interband transitions in the vicinity of the K (K') point can exclusively couple to the right-(left)-handed circularly polarized light $\sigma^+(\sigma^-)$. In principle, $\sigma^+(\sigma^-)$ excitation light can generate $\sigma^+(\sigma^-)$ PL emission. However, the PL emission under circularly polarized excitation is not perfect circularly polarized light due to the intervalley scattering process. The degree of circular polarization for PL emission is commonly defined as

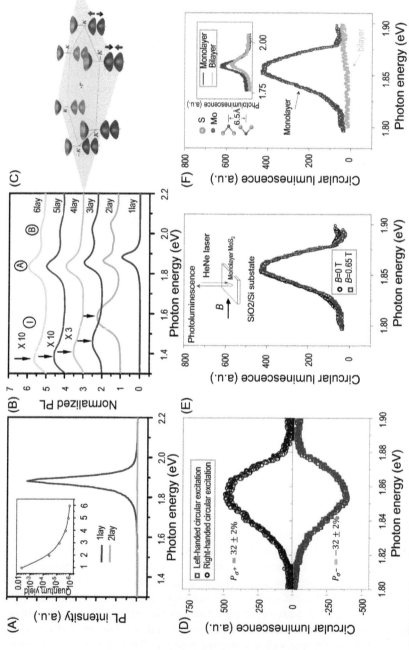

Fig. 12.4 See Figure legends on opposite page.

Continued

$$\rho = \frac{I(\sigma^+) - I(\sigma^-)}{I(\sigma^+) + I(\sigma^-)},$$ (12.3)

where $I(\sigma^\pm)$ is the left-/right-handed circular component intensity. Fig. 12.4D represents the circularly polarized PL spectra of 1LM under the σ^+ and σ^- polarized excitations of 1.96 eV (near A exciton) at 10 K. The ρ for 1LM is $\sim$0.3. The spin polarization mechanism of polarized PL can be excluded by the magnetic field-independent PL polarization shown in Fig. 12.4E. Otherwise, the spin polarization will process about the in-plane magnetic field and lead to decreased ρ. The valley polarization origin of such polarized PL can be further confirmed by a comparison of circular polarization of PL between the 1LM and bilayer MoS_2 (2LM). The preserved inversion symmetry in 2LM leads to a forbidden valley-dependent selection rule. Thus, the circular polarization of PL in 2LM is expected to be negligible, in good agreement with the experimental results, as shown in Fig. 12.4F.

In addition, abundant intriguing phenomena modified by defects, electronic doping, and lattice structure in semiconductor 2DMs have been reported recently. The neutral and charged defects in TMDs can bind the excitons to generate the so-called bound excitons [45]. Single-photon emission based on spatially localized excitons in TMD monolayers can be manipulated by electric and magnetic fields [46–48]. Electrical doping can efficiently control the neutral and charged excitons in monolayer semiconductor 2DMs [45]. All these extraordinary phenomena suggest interesting physics in 2DMs and a great potential of 2DMs in nanophotonics.

Furthermore, the optical properties and PL spectroscopy can be modulated by the band alignments of the constituents and the stacking order in vdWHs. As demonstrated in Section 12.1, the moiré superlattice can be naturally generated in twisted

---

Fig. 12.4 (A) PL spectra of monolayer and bilayer MoS_2 flakes in the range of 1.3–2.2 eV. The *inset* plots the quantum yield of 1–6L MoS_2. (B) Normalized PL spectra of 1–6L MoS_2 by the A emission peak. The spectra are scaled and offset for clarity and the scaling factors are shown. (C) Schematics of the band structure of 1LM at the K (K') point of BZ. (D) Polarization-resolved PL spectra under left-handed (σ^+) and right-handed (σ^-) circular excitations from a 1.96 eV laser at 10 K. (E) Circularly polarized components of PL spectra at zero and 0.65 T in-plane magnetic field. The *inset* shows the schematic configuration for the PL measurements. (F) Circularly polarized components of PL spectra from monolayer (*blue*) and bilayer (*green*) MoS_2. The *upper left inset* is the schematic of bilayer MoS_2 unit cell, while the *right upper inset* represents the PL spectra of monolayer (*blue*) and bilayer (*green*) MoS_2 under linearly polarized excitation of 1.96 eV at 10 K.

(A, B) Adapted from K.F. Mak, C. Lee, J. Hone, J. Shan, T.F. Heinz, Atomically thin MoS_2: a new direct-gap semiconductor, Phys. Rev. Lett. 105 (13) (2010) 136805, https://doi.org/10.1103/PhysRevLett.105.136805; (C) adapted from D. Xiao, G.-B. Liu, W. Feng, X. Xu, W. Yao, Coupled spin and valley physics in monolayers of MoS_2 and Other group-VI dichalcogenides, Phys. Rev. Lett. 108 (19) (2012) 196802, https://doi.org/10.1103/PhysRevLett.108.196802; (D–F) adapted from H. Zeng, J. Dai, W. Yao, D. Xiao, X. Cui, Valley polarization in MoS_2 monolayers by optical pumping, Nat. Nanotechnol. 7 (8) (2012) 490–493, https://doi.org/10.1038/nnano.2012.95.

systems or the vdWHs formed by constituents with nearly the same lattice constants, which periodically and laterally modulated the electronic and topographic structures [24]. For example, nanodot superstructures can be generated to confine the long-lived interlayer excitons, forming the so-called moiré excitons, as shown in Fig. 12.5A. The optical properties of moiré excitons can be constructed by moving adiabatically wave packets in the moiré potential. There are three kinds of high-symmetry sites (denoted as A, B, C sites) with threefold rotational ($\hat{C}_3$) symmetry preserved, in which the exciton wave packet (χ) is an eigenfunction of $\hat{C}_3$ rotation: $\hat{C}_3\chi_{A,s} = e^{-i\frac{2\pi}{3}s}\chi_{A,s}$, $\hat{C}_3\chi_{B,s} = e^{i\frac{2\pi}{3}s}\chi_{B,s}$, $\hat{C}_3\chi_{C,s} = \chi_{C,s}$, where s denoted the spin-valley index, $s = +(-)$ represents the exciton at the $K(-K/K')$ point with up (down) spin. Because the photons can only be convertible with excitons of the same rotational symmetry, the spin-up exciton wave packet at the A (B) site only couples to σ^+ (σ^-) polarized light while at the C site, light coupling is forbidden. The $\hat{C}_3$ symmetry also leads to energy extrema with energy variation up to $\sim$100 meV, which makes it possible to trap the exciton into a moiré potential, as shown in Fig. 12.5A. The experimental signatures of the moiré excitons have been unveiled recently [27–29]. Due to strong valley polarization of the trapped interlayer excitons, the PL peak in heterobilayers should display large circular polarization. For example, in MoSe$_2$/WSe$_2$ heterobilayers, cocircularly polarized PL can be observed in the samples with $\theta = 57$ degrees and $\theta = 20$ degrees, while cross-circularly polarized PL is observed in that of $\theta = 2$ degrees, as shown in Fig. 12.5B–D. The features of moiré excitons can be further confirmed by the magneto-PL spectroscopy, in which all PL peaks experience Zeeman splitting under a nonzero magnetic field, as shown in Fig. 12.5E–G. The energies of the σ^+ and σ^- emission shift equally but in opposite directions. The energy difference between σ^+ and σ^- under the linearly polarized excitation would increase with increasing the magnetic field value. The extracted Landé g-factor of the moiré excitons are -15.89, -15.79, and 6.72 for the MoSe$_2$/WSe$_2$ heterobilayers with $\theta = 57$ degrees, $\theta = 20$ degrees, and $\theta = 2$ degrees, respectively, as demonstrated in Fig. 12.5H–J. The g-factor is similar for all the PL peaks in heterobilayers with similar twist angles [27]. The similar g-factors for heterobilayers with $\theta = 57$ degrees and $\theta = 20$ degrees arise from the similar valley pairings of the heterobilayers with twist angles in the vicinity of 60 and 21.8 degrees. However, due to the weaker optical dipole of the umklapp recombination at 21.8 degrees than that of the direct recombination at 0 and 60 degrees, the PL emission for heterobilayers with $\theta = 20$ degrees is much weaker than of $\theta = 2$ degrees and $\theta = 57$ degrees, in line with the experimental results depicted in Fig. 12.5B–D. In addition, Tran et al. have revealed that the energy spacing between the PL peaks from interlayer exciton of MoSe$_2$/WSe$_2$ heterobilayers is twisted-angle dependent, which is consistent with the calculations of trapped excitons confined within a moiré potential [28]. Jin et al. highlighted that the moiré exciton states in WSe$_2$/WS$_2$ with a small twist angle manifest as multiple emergent peaks around the A exciton of pristine WSe$_2$ and exhibit distinct gate-dependent behaviors from that of the A exciton in monolayer WSe$_2$ and heterobilayers with a large twist angle [29]. The experimental evidence of moiré excitons presents potential platforms to manipulate the excitons and moiré optics in vdWHs for nanophotonics and quantum information applications.

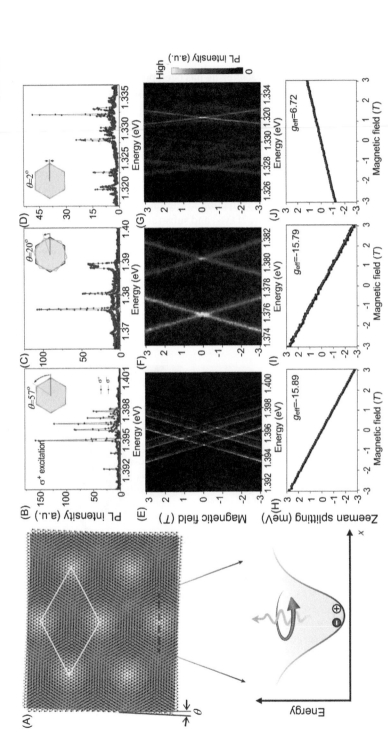

Fig. 12.5 (A) Schematic of moiré superlattice formed in a heterobilayer with θ and an exciton trapped in a moiré potential site. Helicity-resolved PL spectra of a trapped interlayer exciton of MoSe$_2$/WSe$_2$ heterobilayers with (B) $\theta = 57$ degrees, (C) $\theta = 20$ degrees, and (D) $\theta = 2$ degrees excited with σ^+ polarized light at 1.72 eV. The *insets* show their corresponding twist angles. (E–G) Total PL intensity with the change of magnetic field for the above three sample, respectively. (H–J) Zeeman splitting of the polarization-resolved PL as a function of the applied magnetic field for the above three samples, respectively.

Adapted from K.L. Seyler, P. Rivera, H. Yu, N.P. Wilson, E.L. Ray, D.G. Mandrus, J. Yan, W. Yao, X. Xu, Signatures of moiré-trapped valley excitons in MoSe$_2$/WSe$_2$ heterobilayers, Nature 567 (7746) (2019) 66–70, https://doi.org/10.1038/s41586-019-0957-1.

12.4   Rayleigh scattering

Optical spectroscopy associated with light scattering has been proved to be a fast, sensitive, and nondestructive tool in the characterization and investigation of physical properties in 2DMs [3, 49]. According to energy transfer in light-matter interactions, light scattering can be divided into elastic and inelastic, which corresponds to Rayleigh scattering and Raman/Brillouin scattering. Compared with the minority inelastic scattering photons, the number of elastically scattered photons is much larger. As the 2DMs are usually deposited on SiO_2/Si substrate, similar to the OC of 2DMs shown in Fig. 12.2A, the Rayleigh scattering intensity from the 2DMs depends on the complex refractive index of each medium, multiple reflection at the interfaces, and the interference within the multilayer containing air, 2DMs, and substrate. Therefore, the Rayleigh scattering contrast from 2DMs on substrate relative to that from substrate, which is defined by Eq. (12.1), provides an effective means to identify the number of layers of 2DMs and their dielectric constants [50, 51]. Fig. 12.6A and B shows one typical optical image of 1–3LG and 6LG deposited on the SiO_2/Si substrate ($h_{SiO2} = 300$ nm) and the corresponding Rayleigh mapping under backscattering configuration with 633 nm excitation, respectively. The Rayleigh signal intensity increases as the number of layers N increases. Similar to the OC, the theoretical Rayleigh contrast can also be

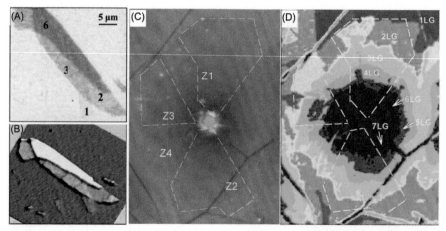

Fig. 12.6 (A) Optical image of 1–3LG and 6LG. (B) The corresponding Rayleigh mapping for the samples in (A) under 633 nm excitation. (C) Optical image of a CVD grown MLG flake. (D) The corresponding Rayleigh mapping for flake in (C) under 532 nm excitation. Stronger Rayleigh intensity is represented by *darker color*.

(A, B) Adapted from C. Casiraghi, A. Hartschuh, E. Lidorikis, H. Qian, H. Harutyunyan, T. Gokus, K.S. Novoselov, A.C. Ferrari, Rayleigh imaging of graphene and graphene layers, Nano Lett. 7 (9) (2007) 2711–2717, https://doi.org/10.1021/nl071168m; (C, D) adapted from J.-B. Wu, H. Wang, X.-L. Li, H. Peng, P.-H. Tan, Raman spectroscopic characterization of stacking configuration and interlayer coupling of twisted multilayer graphene grown by chemical vapor deposition, Carbon 110 (2016) 225–231, https://doi.org/10.1016/j.carbon.2016.09.006.

calculated based on the transfer-matrix method once the complex refractive index of each medium is known [50]. The number of layers of 2DMs can be determined by comparing the experimental Rayleigh contrast value with the theoretical one. Fig. 12.6C and D shows the optical image of a chemical vapor deposition (CVD)-grown multilayer graphene (MLG) flake and the corresponding Rayleigh contrast [51], in which the Rayleigh contrast increases with N increasing. The NLG in each region can be identified by the color in the Rayleigh contrast mapping, as shown in the image.

Although the methods to determine the number of layers of 2DMs by Rayleigh contrast and OC are similar with each other, the spatial resolution of Rayleigh contrast can be down to ~ 1 μm because laser excitation is used. As the measured Rayleigh contrast is equal to a monochromatic OC with the same excitation wavelength as the laser source for the Rayleigh contrast, one can choose an optimal wavelength at the OC maximum to achieve better signal-to-noise ratio for the Rayleigh contrast. For the number of layers identification, it would be better to choose a laser wavelength that makes Rayleigh contrast sensitive to the number of layers of 2DMs.

12.5   Raman spectroscopy

Raman scattering is one kind of inelastic scattering, in which the energy shift between the incident and scattered light corresponds to the energy of quasiparticles or elementary excitation (e.g., phonons) involved in the process of light-matter interactions. As a consequence, Raman spectroscopy is extensively utilized to explore the fundamental physics in a material, such as lattice structure and symmetry, lattice vibrations, defects, electronic band structure, interlayer coupling, electron-phonon coupling (EPC), etc., especially in 2DMs [3, 5, 7, 49, 52, 53]. Here, we focus on the Raman scattering of phonons in 2DMs, showing three typical examples of Raman spectroscopy study in 2DMs: (a) Raman spectroscopy in pristine 2DMs by taking $MoTe_2$ as a prototype [54]; (b) Raman spectroscopy of twisted 2DMs, with twisted bilayer MoS_2 (tBLM) as an example [25]; and (c) Raman spectroscopy of vdWHs by taking the hBN/WS_2 heterostructure as a model [55], which illustrates the influences of number of layers, stacking order, and interlayer/interfacial coupling on the Raman spectroscopy of 2DMs and vdWHs. The analysis here can be extended to other 2DMs and related vdWHs, paving the way to their characterizations and investigations.

12.5.1   Raman spectroscopy in pristine 2DMs

Monolayer 2DMs can be stacked layer by layer in a specific way to form multilayer 2DMs, which possess strong in-plane bonds and weak vdW interlayer coupling. Thus, the Raman spectroscopy of NL-2DMs ($N > 1$) consists of Raman modes in low- and high-frequency regions, in which the low-frequency Raman modes are induced by the relative motions of adjacent layers (so-called interlayer modes), while the high-frequency modes result from the relative motions of the atoms in a rigid layer (so-called intralayer modes). There are two kinds of interlayer modes, shear (S) and layer

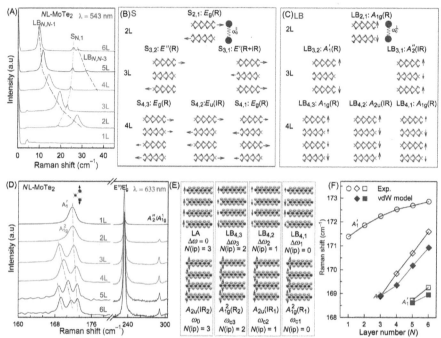

Fig. 12.7 (A) The Stokes Raman spectra of 1–6L MoTe$_2$ in the region of 0–45 cm^{-1} under the excitation at 543 nm. Symmetries, Raman activities, and normal mode displacements for the (B) S and (C) LB modes in 2–4L 2H-MoTe$_2$. R represents Raman active and IR represents infrared active. The *arrows* indicate the vibrational direction of the corresponding layers and the length represents the amplitude of the displacements. (D) The Raman spectra of 1–6L MoTe$_2$ in the range of 160–305 cm^{-1} under the excitation at 633 nm. (E) Schematic of the vdW model for Davydov splitting in 4L MoTe$_2$. N(ip) is the number of the in-phase vibration interfaces. (F) The experimental frequency (*open diamonds* and *squares*) of each Davydov entity of $A_1'(A_{1g})$ modes in 3–6L MoTe$_2$ and the corresponding calculated frequency (*solid diamonds* and *squares*) based on the experimental value of the highest-frequency component and vdW model. The experimental frequencies of the $A_1'(A_{1g})$ modes in 1–2L MoTe$_2$ are also plotted.

(A, D–F) Adapted from Q.J. Song, Q.H. Tan, X. Zhang, J.B. Wu, B.W. Sheng, Y. Wan, X.Q. Wang, L. Dai, P.H. Tan, Physical origin of Davydov splitting and resonant Raman spectroscopy of Davydov components in multilayer MoTe$_2$, Phys. Rev. B 93 (11) (2016) 115409, https://doi. org/10.1103/PhysRevB.93.115409.

breathing (LB) modes, which correspond to the interlayer vibrations parallel and perpendicular to the 2D basal plane, respectively. The S mode is also referred to as the C mode in MLG [20, 56, 57] as it provides a direct probe for the interlayer *coupling*. Other denotations, such as SM, B, and LBM are also introduced by different groups. The frequencies of S and LB modes are both dependent on the number of layers due to the interlayer characteristics, as depicted in Fig. 12.7A. For these interlayer modes, the relative vibrations between the atoms within each rigid layer can be negligible; thus, one rigid layer can be treated as a ball and only the nearest interlayer S/LB coupling

$(\alpha_0^{\parallel} / \alpha_0^{\perp})$ between adjacent balls is considered, as shown in Fig. 12.7B and C. This model is known as the linear chain model (LCM) [20, 56–58]. The frequency of the S and LB modes can be calculated by solving the following $N \times N$ (tridiagonal) dynamics matrix [3, 20, 56, 57]:

$$\omega_i^2 \mathbf{u}_i = \frac{1}{2\pi^2 c^2 \mu} \mathbf{D} \mathbf{u}_i, \tag{12.4}$$

where $\mathbf{u}_i$ is the eigenvector of the mode i with frequency ω_i, μ is the mass of single rigid layer per unit area, $c = 3.0 \times 10^{10}$ cm s^{-1} is the speed of light, and $\mathbf{D}$ is the S or LB part of the force constant matrix. Thus, the frequencies of the corresponding $S_{N, N-i}$ and $LB_{N,N-i}$ modes can be given by [3, 5, 58]

$$\omega(S_{N,N-i}) = \frac{1}{\pi c} \sqrt{\alpha_0^{\parallel}/\mu} \sin\left(\frac{i\pi}{2N}\right),$$

$$\omega(LB_{N,N-i}) = \frac{1}{\pi c} \sqrt{\alpha_0^{\perp}/\mu} \sin\left(\frac{i\pi}{2N}\right), \tag{12.5}$$

where $i = 1, 2, ..., N - 1$. There are $N - 1$ LB and $N - 1$ doubly degenerate S modes for the in-plane isotropic NL-2DMs, where N is the number of layers and i is the number of phonon branches. The $i = N - 1$ branches, that is, $S_{N,1}$ and $LB_{N,1}$ modes, correspond to the highest-frequency S and LB modes, respectively. The calculated frequencies of the S and LB modes in MoTe$_2$ as a function of N are in good agreement with the experimental ones. The corresponding ith displacement eigenvector $v_j^{(i)}$ is given by

$$v_j^{(i)} = \cos\left[\frac{i(2j - 1)\pi}{2N}\right], \tag{12.6}$$

where j labels the jth layer. For NL-2DMs with in-plane anisotropy, there are $N - 1$ LB modes, similar to that in in-plane isotropic NL-2DMs. However, as the two axes in the basal 2D plane are not equal in anisotropic NL-2DMs, for example, ReS$_2$, there are $2(N - 1)$ nondegenerate S modes, named as $S_{N,N-i}^x$ and $S_{N,N-i}^y$ modes [59].

Fig. 12.7B and C depicts the layer displacements of the S and LB modes in 2–4L MoTe$_2$. Because the ONL and ENL MoTe$_2$ belong to D_{3h} and D_{3d} symmetries, respectively, the interlayer modes in ONL and ENL MoTe$_2$ can be represented by $\frac{N-1}{2}(A_1'(R) + E''(R) + A_2''(IR) + E'(R + IR))$ and $\frac{N}{2}(A_{1g}(R) + E_g(R)) + (\frac{N}{2} - 1)(A_{2u}(IR) + E_u(IR))$, respectively. R and IR represent Raman active and infrared active, respectively. The corresponding Raman/infrared activity for $S_{N,N-i}$ and $LB_{N, N-i}$ modes in 2–4L MoTe$_2$ is indicated in Fig. 12.7B and C, which suggests that $S_{N, N-i}$ ($i = N - 1, N - 3, N - 5, ...$) and $LB_{N,N-i}$ ($i = 1, 3, 5, ...$) modes are Raman active, in line with the experimental results in Fig. 12.7A [54]. Due to the in-plane isotropy in MoTe$_2$, the $N - 1$ doubly degenerate S modes belong to E symmetry, while the

$N - 1$ LB modes belong to A symmetry. It is also the case for all in-plane isotropic MX_2. Thus, the Raman-active S modes can be detected under both parallel and cross-polarization configurations, while the Raman-active LB modes can only be observed under the parallel polarization configuration. This provides one method to distinguish the S and LB modes. The detailed assignments of the S and LB modes and demonstrations for their measurements in typical 2DMs can be found in [3].

The weak interlayer coupling can also influence the intralayer modes. Fig. 12.7D plots the Raman spectra of 1–6L $MoTe_2$ in the region of intralayer modes. Only two intralayer modes can be detected in 1L $MoTe_2$, which correspond to A_1' ($\sim$170 cm^{-1}) and E' ($\sim$233 cm^{-1}) intralayer phonons. For multilayer $MoTe_2$, the A_1' (A_{1g}) and E' (E_g) modes in ONL (ENL) $MoTe_2$ can be observed, corresponding to the A_1' and E' modes in 1L $MoTe_2$, respectively. An additional mode at $\sim$291 cm^{-1} corresponding to the Raman-inactive B_{2g} mode in 1L $MoTe_2$ can be probed due to the reduced symmetry in multilayer $MoTe_2$ [54]. One intriguing phenomenon of peak splitting is observed for A_1'-like modes in NL $MoTe_2$, in which the number and frequency of the components depend on N due to interlayer coupling. In principle, when considering the nearest interlayer coupling, the intralayer modes in $MoTe_2$ would exhibit varied frequency when the two nearest Te atoms in adjacent layers exhibit in-phase and out-of-phase vibrations. The out-of-phase vibration between the two nearest Te atoms tends to raise the frequency compared with the in-phase vibration, resulting in the so-called Davydov splitting [54]. This can be well reproduced by the vdW model, as demonstrated in detail later [54]. Assuming the frequency of the isolated entity and the uncoupled entity displaying in-plane vibration is ω_0; the frequency of coupled entity displaying out-of-phase vibration is ω_c; and the coupling frequency is $\Delta\omega$, these three frequencies are associated by the relation of $\omega_c^2 = \omega_0^2 + \Delta\omega^2$. Thus, for NL ($N > 1$) $MoTe_2$, the intralayer modes would split into N components. However, the number of components that can be detected in the Raman spectroscopy depends on their Raman activity, the splitting frequency, and the spectral resolution of the Raman system for measurements. As shown in Fig. 12.7D, the Davydov splitting is only clear for A_1'-like modes. The A_1' mode in 1L $MoTe_2$ would split into N A_1'-like modes, represented by $\frac{N+1}{2}A_1'(R)+\frac{N-1}{2}A_2''(IR)$ for ONL $MoTe_2$ and $\frac{N}{2}A_{2u}(IR)+\frac{N}{2}A_{1g}(R)$ for ENL $MoTe_2$. Thus, it is expected to observe $\frac{N+1}{2}$ and $\frac{N}{2}$ Davydov components in ONL and ENL $MoTe_2$, respectively, in good agreement with the experimental results shown in Fig. 12.7D. The atomic displacements of A_1'-like modes and LB modes for 4L $MoTe_2$ are elucidated in Fig. 12.7E, together with the number of the in-phase vibration interfaces (N(ip)) denoted. When taking detailed consideration of the frequency of the Davydov components, the coupling frequency $\Delta\omega$ can be closely related to the frequency of LB modes with the same N(ip) because the atomic displacements of the A_1'-like modes are perpendicular to the basal plane. The frequency of the $A_{2u}(IR)$ mode is that of the uncoupled entities, ω_0, and the frequencies of the other three modes (ω_{ci}, $i = 1, 2, 3$) can be calculated. This is the so-called vdW model. According to the Raman-active Davydov components with the highest frequency of 3–6L $MoTe_2$ and vdW model, the corresponding frequency of other Davydov components can be deduced (solid diamonds and squares), in line with the experimental results (open diamonds and squares), as shown in Fig. 12.7F. Notably, different Davydov components would exhibit varied resonance profiles because of their different EPC. Therefore, it is essential to choose an appropriate excitation energy to measure the Davydov components [54]. Similar Davydov splittings have been observed in other MX_2 [60, 61], such as $MoSe_2$ and WS_2. The above vdW model can be extended to reproduce the Davydov splitting associated with S and LB modes in other 2DMs [62].

12.5.2   Raman spectroscopy of twisted 2DMs

As referred to earlier, the twisted stacking at the interface of the twisted 2DMs can generate the moiré superlattice to modulate the phonon behaviors of the constituents. tBLM can be treated as a simple model system to introduce the modified Raman spectra in twisted 2DMs [25]. There exist moiré superlattices in tBLM, whose lattice constant can be represented by $L^M = a/[2(\sin(\theta/2))]$, where a is the lattice constant of the 1LM. Accordingly, the corresponding reciprocal lattices can be constructed. Fig. 12.8A shows the reciprocal lattices corresponding to the moiré superlattices of tBLM with $\theta = 10.99$ degrees, in which the large blue and orange hexagons are the Brillouin zones (BZs) of the top and bottom monolayers, respectively. Meanwhile, the red-dashed hexagons correspond to the Wigner-Seitz cells in reciprocal space related to the moiré superlattices. The basic vectors (g and g') of the moiré reciprocal superlattice can be evaluated by the relative vectors between the reciprocal lattice vectors of the top and bottom monolayers, giving an absolute value of $|g| = |g'| = 2b\sin(\theta/2)$. There are many lattice vectors of moiré reciprocal superlattices within the BZ of the monolayer constituent, because its basic vector is smaller than that of 1LM, as Γ_{j1} ($j = 1, 2, ..., 6$) labeled for the moiré reciprocal superlattice in Fig. 12.8A. With changing θ, the trajectory of g for the moiré reciprocal superlattice in the BZ of the monolayer constituent can be obtained, as the dashed line shown in Fig. 12.8B. Three representative basic vectors of the moiré reciprocal superlattice, that is, g_i ($i = 1$, 2, 3) corresponding to 30 degrees $- \theta_1$, 30 degrees $- \theta_2$, and 30 degrees are also depicted in Fig. 12.8B. Notably, the moiré superlattice with 30 degrees $- \beta$ and 30 degrees $+ \beta$ (0 degrees $< \beta <$ 30 degrees) are two commensurate partners.

In twisted bilayer 2DMs, the periodic modulation of moiré superlattices on the interlayer interactions leads to not any overall restoring force for the in-plane displacements when θ is away from 0 or 60 degrees ($\sim$5−55 degrees). Thus, the S modes locate at almost zero wave number and cannot be observed in low-frequency Raman spectroscopy. As shown in Fig. 12.8C, the $S_{2,1}$ modes are absent in the Raman spectra of tBLMs when $\theta \neq 0$ degrees and $\theta \neq 60$ degrees, further confirming that the interlayer S coupling is not enough to produce an overall restoring force between the two monolayer constituents. Similar results have been observed in twisted multilayer graphenes (tMLGs) [3, 5, 20, 63] and twisted bilayer TMDs [18, 64, 65]. In contrast, the $LB_{2,1}$ modes are present with frequencies close to those in 2H- and 3R-stacked 2LMs, which can be ascribed to the comparability between interfacial LB coupling and interlayer LB coupling with pristine 2LM. The close frequency of LB modes in tBLMs (0 degrees $< \theta <$ 60 degrees) to those in 2H-/3R-stacked 2LM in Fig. 12.8C confirms the good LB coupling between the two monolayer

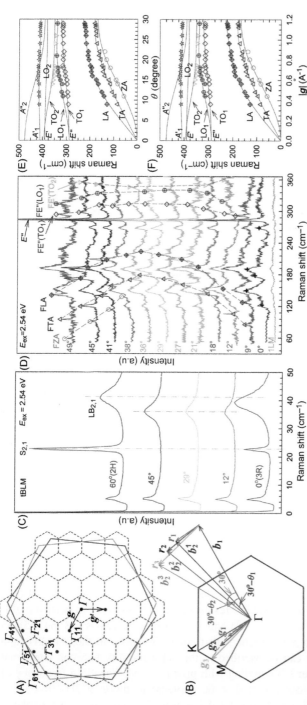

Fig. 12.8 (A) The reciprocal lattice of tBLM with $\theta = 10.99$ degrees; the *large blue* and *orange hexagons* represent the first BZ of the top and bottom MoS_2 layers, respectively. The *dashed (red) hexagons* are the Wigner-Seitz cells of the reciprocal lattices corresponding to the moiré superlattices. (B) Schematic diagram of moiré basic vectors (g_i, $i = 1, 2, 3$) corresponding to 30 degrees $- \theta_1$, 30 degrees $- \theta_2$, and 30 degrees. (C) The Stokes Raman spectra of tBLMs with different θ in the low-frequency region. (D) The Raman spectra of tBLMs with θ ranging from 9–49 degrees in the region of 50–365 cm^{-1}, with the corresponding Raman spectra of 1LM and 3R-stacked BLM ($\theta = 0$ degrees) shown together. The calculated and experimental frequencies of moiré phonons dependent on (E) θ and (F) $|g|$. The *solid gray lines* are the calculated phonon frequencies by the interpolation between those along the Γ-M and Γ-K directions in 1LM; the *scattered symbols* are the experimental data. Adapted from M.-L. Lin, Q.-H. Tan, J.-B. Wu, X.-S. Chen, J.-H. Wang, Y.-H. Pan, X. Zhang, X. Cong, J. Zhang, W. Ji, P.-A. Hu, K.-H. Liu, P.-H. Tan, Moiré phonons in twisted bilayer MoS_2, ACS Nano 12 (8) (2018) 8770–8780, https://doi.org/10.1021/acsnano.8b05006.

MoS_2 constituents, which is essential for the generation of moiré superlattices. In addition, according to the relative intensity between the $S_{2,1}$ and $LB_{2,1}$ modes [64], the tBLMs tend to exhibit 3R and 2H stacking when $\theta = 0$ degrees and $\theta = 60$ degrees, respectively.

In principle, the moiré superlattices are expected to result in phonon folding effects in tBLMs, in which the off-center phonons in the monolayer constituents linked with the corresponding reciprocal lattice vectors can be folded back to the BZ center of 1LM. Indeed, it was found that the periodic potential from moiré superlattice can modify the phonon properties of its monolayer constituents to generate the Raman modes related to moiré phonons [25]. As shown in Fig. 12.8D, a series of additional Raman modes are probed in tBLMs, with frequency monotonously changing for 0 degrees < θ < 30 degrees and 30 degrees < θ < 60 degrees, in coincidence with the case for the moiré superlattice. Based on the theory-experiment comparison from low to high frequency, the six series of θ-dependent Raman modes can be distinguished as the moiré phonons related to ZA, TA, LA, $E''(TO_1)$, $E''(LO_1)$, and $E'(TO_2)$ phonon branches in 1LM. They are the phonons *folded* back to the BZ center, denoted as FZA, FTA, FLA, $FE''(TO_1)$, $FE''(LO_1)$, and $FE'(TO_2)$, respectively. The frequencies of all the observed moiré phonon modes as a function of θ are plotted in Fig. 12.8E. The corresponding phonon branches of 1LM can be obtained according to $|g| = 2b\sin(\theta/2)$, as elucidated in Fig. 12.8F, showing good agreement with the calculated phonon branches by the interpolation between those along the Γ-M and Γ-K directions in 1LM [25]. It should be noted that the Raman modes related to moiré phonons in tBLM would be greatly enhanced when the excitation energy is close to the C exciton energy [25]. Thus, the moiré phonon modes in all tBLMs can be observed with one excitation laser.

The above analysis can be extended to other twisted bilayer systems. Indeed, the θ-dependent Raman modes in twisted bilayer graphenes, which were previously assigned to rich resonance effects [66], are actually corresponding to the moiré phonon modes [25]. Besides, the moiré superlattices are also present in vdWHs formed by two constituents with nearly the same lattice constant, and thus moiré phonon modes are expected in these vdWHs [67]. Interestingly, reconstructions of the moiré superlattice are observed in twisted bilayers with small twist angles [68–70], in which many fascinating phonon properties are discovered, such as phonon renormalization due to the ultra-strong coupling between different phonon modes [71] and localization of lattice dynamics with the presence of strain solitons [72].

Twisted bilayers are the simplest prototype of twisted 2DMs. With increase of the number of layers of twisted 2DMs, a large family of twisted 2DMs can be generated, that is, $t(m + n + \cdots)$L-2DMs ($N = m + n + \cdots$), in which an even wider range of properties is accessible. The above analysis of the moiré phonons is also applicable in these complex twisted multilayer 2DMs, which are sensitive to the twist angle and moiré superlattice. Besides, the interlayer modes and nonfolded intralayer modes in twisted multilayer 2DMs are dependent on the interfacial coupling between the constituents. For interlayer modes, the negligible interfacial shear (S) coupling leads to the localization of the S modes within its constituents, while the remarkable LB coupling gives rise to collective vibration of all the stacking layers for the LB modes [3, 18], similar to those in twisted bilayer 2DMs. The effects of the interfacial and interlayer coupling on the nonfolded intralayer modes should also be divided into two categories according to the vibration directions [73]. For nonfolded in-plane intralayer modes, the vibrations are localized within the constituents of twisted multilayer 2DMs because of the negligible interfacial S coupling at the interface. Thus, the spectral profile of nonfolded in-plane intralayer modes can be fitted by those of the corresponding modes of its constituents. In contrast, because the interfacial LB coupling at the interface is comparable to the interlayer coupling within the constituents, the number of the Davydov entities of the nonfolded out-of-plane modes depends on the number of stacking layers (i.e., N) of the whole $t(m + n + \cdots)$L-2DMs. The Davydov splitting of the intralayer modes in $t(m + n + \cdots)$L-2DMs is quantitatively reproduced by the vdW model based on the frequencies of the corresponding LB modes, as elucidated in Section 12.5.1. As the interfacial LB coupling is independent of the large twist angle, the Davydov splitting of the nonfolded out-of-plane modes is also not sensitive to the twist angle. However, the moiré superlattice with different twist angles will impose varied modulation on electron-phonon coupling. Thus, the relative Raman intensity between different Davydov entities significantly varies with the twist angle. Notably, in order to resolve all the Davydov entities of the nonfolded out-of-plane intralayer modes, high-resolution resonance Raman spectroscopy at low temperature is always necessary [73].

12.5.3 Raman spectroscopy in vdWHs

As demonstrated in Section 12.5.2, one can infer that the S modes in twisted 2DMs are confined within the constituents, while the LB modes in twisted 2DMs are from the collective vibrations of all the stacking layers. Similar phenomena are observed in vdWHs consisting of different 2DMs [3, 55, 74]. Setting the hBN/WS$_2$ formed by n-layer hBN (nL-hBN) and m-layer WS$_2$ (mLW), that is, nL-hBN/mLW as a prototype, the peak position and EPC in vdWHs are demonstrated in detail. Fig. 12.9A displays the Raman spectra of 39L-hBN/3LW excited by a laser in the range of 2.41–2.81 eV, together with the Raman spectrum of the stand-alone 3LW flake under excitation at 2.71 eV. Similar S modes can be observed in stand-alone 3LW flakes and 39L-hBN/3LW, confirming the confinement of S modes within the constituents in vdWHs. Compared with the Raman spectrum of the stand-alone 3LW flake, many new LB modes are emergent, which can be reproduced by the LCM with interlayer coupling within the constituents and interfacial coupling between the constituents considered [55]. According to the peak position of the observed LB modes, the interlayer coupling of hBN ($\alpha^\perp$(BN)) and the interfacial coupling between hBN and WS$_2$ ($\alpha^\perp$(I)) constituents can be obtained. All the observed LB modes can be well reproduced and assigned by these fitted force constants, independent of the stacking order, and twist angle due to the large mismatch of lattice constants between hBN and WS$_2$. As the LB modes in nL-hBN/mLW can be well distinguished, even n is up to tens or hundreds of layer thickness, the LB modes exhibit 3D features [55].

Interestingly, all the observed LB modes in hBN/WS$_2$ are significantly enhanced when the excitation energy approaches the C exciton energy of the corresponding stand-alone WS$_2$ flake, as demonstrated in Fig. 12.9A and B. Furthermore, the resonance profile of these LB modes is similar to the S and LB modes in the

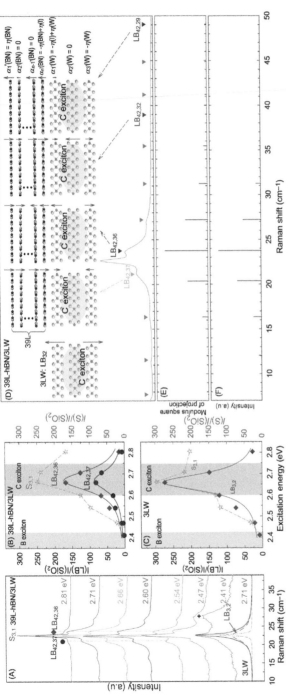

Fig. 12.9 (A) Raman spectra of 39L-hBN/3LW excited by E_{ex} ranging from 2.41 to 2.81 eV, together with the Raman spectrum of a stand-alone 3LW flake excited at 2.71 eV. The *gray dash* and *red dash-dot* lines plot the lineshape of $S_{3,1}$ and $LB_{3,2}$ modes in the stand-alone 3LW flake. The resonant profiles of (B) $LB_{42,36}$ (*red diamonds*) and $LB_{42,37}$ (*blue circles*) and $S_{3,1}$ modes (*gray stars*) in 39L-hBN/3LW, (C) $LB_{3,2}$ (*red diamonds*) and $S_{3,1}$ (*gray stars*) in 3LW flake. (D) Raman spectroscopy of 39L-hBN/3LW in the region of 5–50 cm^{-1} and the normal mode displacements (*red arrows*) of $LB_{42,37}$, $LB_{42,36}$, $LB_{42,32}$, $LB_{42,29}$ modes, together with that of $LB_{3,2}$ in a stand-alone 3LW flake. The *triangles* are the expected LB modes based on LCM. $\alpha'_i(BN)$ $(i = 1, 2, ..., n)$ and $\alpha'_j(W)(j = 1, 2, 3)$ are the polarizability derivatives of the entire layer i from the hBN constituent and layer j from the WS$_2$ constituent related to the displacement in the z-direction. (E) The modulus square of the projection from wave function of different LB modes in 39L-hBN/3LW onto that of the $LB_{3,2}$ mode in a stand-alone 3LW flake. (F) The relative intensity of LB modes in 39L-hBN/3LW according to the interlayer bond polarizability model.

Adapted from M.-L. Lin, Y. Zhou, J.-B. Wu, X. Cong, X.-L. Liu, J. Zhang, H. Li, W. Yao, P.-H. Tan, Cross-dimensional electron-phonon coupling in van der Waals heterostructures, Nat. Commun. 10 (1) (2019) 2419, https://doi.org/10.1038/s41467-019-10400-z.

corresponding stand-alone WS_2 flake, as elucidated in Fig. 12.9C. Because the hBN/WS_2 heterostructure is classified into type-I vdWH based on the band alignment of hBN and WS_2 [55], the electronic states related to the C exciton of 3LW flakes are confined within the few-layer WS_2, displaying 2D features. Therefore, the resonance profile of LB modes in hBN/WS_2 similar to that in the corresponding WS_2 flakes implies the peculiar coupling between the 2D electrons and 3D LB phonons [55]. Such cross-dimensional EPC can be understood by a microscopic picture mediated by the interfacial coupling and the interlayer bond polarizability model in vdWHs. The Raman intensity of LB modes in vdWHs is associated with the EPC strength, which is directly related to their vibration displacements (Fig. 12.9D) within the first-order approximation. As the excitation photons are directly coupled to the C exciton transition of mLW constituents, the EPC strength of a vdWH LB phonon can be approximated by the sum of the weighting factors of its normal mode displacements from all the LB modes in the corresponding mLW flake, in which the weighting factors are given by the projection between its wave function components (ψ, i.e., normal mode displacements) among the mLW constituent and those (φ_j) of the $LB_{m,m-j}$ modes ($j = 1, 2, ..., m-1$) modes in the mLW flake, that is, $p_j = |\langle\varphi_j|\psi\rangle|$. As a consequence, the Raman intensity of LB modes in vdWHs is proportional to $p^2 = \sum_j \rho_j p_j^2$, with ρ_j representing the Raman intensity of $LB_{m, m-j}$ mode in the mLW flake. The calculated relative intensity of the LB modes in 39L-hBN/3LW is shown in Fig. 12.9E, which is in good agreement with the experimental results. On the other hand, based on the interlayer bond polarizability, the Raman intensity is related to the square of the change of a system's polarizability ($\Delta\alpha^2$) [75], which is associated with the interlayer bond polarizability and bond vector, that is, $\Delta\alpha = \sum_i \alpha_i' \cdot \Delta z_i$, with α_i' and z_i representing the polarizability derivative of the entire layer i with respective to the vibrations parallel to the normal (z) and the normal displacements of layer i for one LB mode calculated by Eq. (12.6), respectively. Notably, for the LB modes, only the polarizability derivatives of the top and bottom layers in hBN and WS_2 constituents should be taken into account, as shown in Fig. 12.9E, which can be represented with the fitted parameters $\eta(BN)$, $\eta(W)$, and $\eta(I)$. $\eta(BN)$, $\eta(W)$, and $\eta(I)$ are related to the interlayer bond properties in hBN, WS_2 constituents, and that at the interface, respectively. The relative Raman intensity of LB modes in 39L-hBN/3LW can be well reproduced by $\eta(I)/\eta(W) = 0.3$ and $\eta(BN)/\eta(W) = 0.001$. The corresponding calculated Raman intensities are depicted in Fig. 12.9F [55]. Similar analyses are also applicable to the Raman spectroscopy and EPC of LB modes in other vdWHs, shedding light on manipulating the EPC of interlayer phonons in vdWHs by engineering the interface and varying the vdWH assemblies.

12.6 Second-harmonic generation

SHG is a nonlinear optical effect, where two photons with the same frequency interact with a nonlinear material to generate a new photon with twice the energy of the incident photons. The SHG intensity depends on the second-order nonlinear susceptibility ($\chi^{(2)}$) of the nonlinear material, which is zero for a material with inversion symmetry within the electric dipole approximation, similar to other even-order nonlinear optical phenomena [77, 78]. The electric-dipole-allowed SHG is known to be time invariant, denoted as i-type. The $\chi^{(2)}$ tensor related to the i-type SHG provides the signature of the crystallographic symmetry of a crystal and determines the SHG intensity, suggesting a method to identify the lattice structure and symmetry. To obtain a strong SHG, the intense pulsed laser is commonly utilized to perform the SHG measurement with careful alignment for phase matching.

As known, the ONL MX_2 does not possess an inversion center, while the ENL and bulk MX_2 exhibit inversion symmetry. Fig. 12.10A and B represents the optical image of 1–3L MoS_2 and the corresponding SHG mapping, respectively [79]. The 1LM and trilayer MoS_2 (3LM) exhibit strong SHG intensity, while the SHG is absent in 2LM, consistent with the prediction of i-type SHG. In addition, the weaker SHG intensity in 3LM than that of 1LM can be understood by the wave propagation effects [80]. As the sign of $\chi^{(2)}$ for adjacent layers alternates, the signals from adjacent layers in ONL ($N > 1$) MoS_2 would cancel each other, leading to a weaker SHG intensity than that of 1LM. On the other hand, the band gap of a material relative to the SHG energy would influence the SHG intensity. For example, the common photon energy of SHG (e.g., 3.1 eV) is higher than the bandgap of MoS_2 ($\sim$1.90 eV); thus, the SHG signal would be absorbed by MoS_2. The increased absorption of MoS_2 with increasing N will result in decreased SHG intensity. On the contrary, due to the large bandgap of hBN, the SHG photons cannot be absorbed, leading to increasing SHG intensity with respect to N [80].

The SHG intensity is also sensitive to the crystal symmetry and crystallographic orientation. As a result, the polarization-resolved SHG intensity can be utilized to distinguish the crystallographic structure and orientation [79, 80]. The generated SHG intensity of ONL MoS_2 in parallel ($I_{\parallel}$) and perpendicular ($I_{\perp}$) configurations as a function of the angle (ϕ) between the excitation polarization and the crystallographic orientation (x-direction, i.e., armchair direction shown in Fig. 12.10C) can be represented by

$$I_{\parallel} = I_0\cos^2(3(\phi + \phi_0)), \tag{12.7}$$

$$I_{\perp} = I_0\sin^2(3(\phi + \phi_0)), \tag{12.8}$$

respectively, where ϕ_0 is the initial crystallographic orientation of the MoS_2 sample. By fitting the experimental data with Eq. (12.7) or (12.8), ϕ_0 can be obtained. In addition to the SHG in MX_2, the N-dependent and strong SHG is recently found to be modulated by the stacking orders in various 2DMs, such as anisotropic ReS_2 [81], GaSe polytypes [82], etc.

Intriguingly, the layered antiferromagnetic order in material with inversion symmetric lattice can also raise the presence of SHG signals, which can be ascribed to the breakdown of the spatial-inversion symmetry and time-reversal symmetry, denoted as c-type [76, 83, 84]. The recent emergence of 2D magnets [2, 16, 17, 85] provides an ideal platform for investigating the c-type SHG effects. As known, the few-layer CrI_3 exhibits antiferromagnetic interlayer coupling and ferromagnetic coupling within a rigid layer [2, 86]. Fig. 12.10D shows the 1L CrI_3 with centrosymmetric

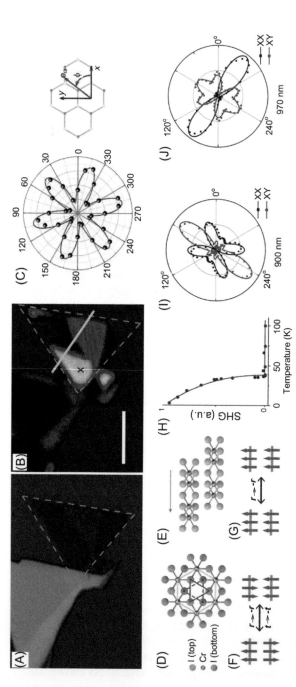

Fig. 12.10 (A) Optical image of 1–3L MoS_2. (B) SHG from the MoS_2 flake shown in (A) excited at 1.55 eV. Stronger SHG intensity is represented by *brighter color*. The scale bar is 5 μm. (C) Polar plot of SHG intensity from 1LM with varying angle between the incident light polarization and the x-direction shown in the right panel. (D) Atomic structure of 1L CrI_3, with a hexagon (*green*) formed by six Cr atoms (*gray*) and two equilateral triangles (*solid and dashed red lines*) formed by I atoms of the top (*orange*) and bottom layers. (E) The side view of 2L CrI_3 with a lateral translation. (F) Schematic diagram of interlayer antiferromagnetic states of 2L CrI_3, where spatial-inversion ($r \rightarrow -r$) and time-inversion ($t \rightarrow -t$) operations can change one state to the other. (G) Schematic diagram of interlayer ferromagnetic states of 2L CrI_3 showing a centrosymmetric spin-lattice structure. (H) SHG intensity of 2L CrI_3 with varying temperature (T). The *red solid line* is following the power law [76] $|1 - (T/T_c)^{2\beta}|$ for $T < T_c$, in which T_c is the critical temperature and β is the critical exponent. Polarization-resolved SHG intensity at 0 T excited at (I) 900 and (J) 970 nm. The excitation and detection are linearly polarized, with XX and XY indicating co- and cross-linearly polarization light, respectively.

(A–C) Adapted from L.M. Malard, T.V. Alencar, A.P.M. Barboza, K.F. Mak, A.M. de Paula, Observation of intense second harmonic generation from MoS_2 atomic crystals, Phys. Rev. B 87 (20) (2013) 201401, https://doi.org/10.1103/PhysRevB.87.201401; (D–J) adapted from Z. Sun, Y. Yi, T. Song, G. Clark, B. Huang, Y. Shan, S. Wu, D. Huang, C. Gao, Z. Chen, M. McGuire, T. Cao, D. Xiao, W.-T. Liu, W. Yao, X. Xu, S. Wu, Giant nonreciprocal second-harmonic generation from antiferromagnetic bilayer CrI_3, Nature 572 (7770) (2019) 497–501, https://doi.org/10.1038/s41586-019-1445-3.

lattice structure. By stacking two monolayers together (Fig. 12.10E), the lattice structure of 2L CrI_3 also exhibits centrosymmetry, regardless of any rigid translation between two monolayers. Within the electric dipole approximation, the SHG signal is prohibited in 2L CrI_3. However, below the critical temperature (T_c), the two monolayers of 2L CrI_3 would show antiferromagnetic interlayer coupling, as shown in Fig. 12.10F, that is, all spins in a rigid layer pointing outwards or inwards but different directions in its adjacent layer, which would break both spatial-inversion and time-reversal symmetries. Thus, it is expected to observe the c-type SHG signal in 2L CrI_3 below T_c. When applying an out-of-plane magnetic field to 2L CrI_3, it becomes ferromagnetic and the inversion symmetry restores, as demonstrated in Fig. 12.10G. The measured SHG intensity of 2L CrI_3 as a function of temperature (T) at zero magnetic field is depicted in Fig. 12.10H. The strong dependence of SHG on T is in line with the above prediction. The SHG is absent above T_c, further confirming the inversion symmetry of 2L CrI_3 [83]. In addition, the linear polarization-resolved SHG was utilized to provide information about the lattice symmetry and initial crystallographic orientation. As shown in Fig. 12.10I, the polarization dependence of 2L CrI_3 under 900 nm excitation at 0 T shows six asymmetric lobes under both co- and cross-linear polarization configurations, suggesting the broken threefold rotational symmetry in 2L CrI_3. The polarization-dependent patterns can be fitted by the c-type $\chi^{(2)}$ tensor related to C_{2h} symmetry, which implies the monoclinic stacking configuration for 2L CrI_3 [83]. The polarization-resolved SHG signal under 970 nm excitation also displaces an in-plane C_2 axis about 145 degrees with respective to the horizontal direction (Fig. 12.10J), similar to that under 900 nm excitation though details of SHG patterns vary. This further confirms the monoclinic stacking for 2L CrI_3. All these works related to SHG indicate that it is a highly sensitive probe of subtle lattice symmetry and magnetic order, opening up possibilities for the characterizations and applications of 2DMs in nonlinear optical devices.

12.7   Conclusion

In this chapter, we have addressed the recent optical spectroscopy study of 2DMs. The discussion begins with an overview of a broad portfolio of emergent 2DMs and related vdWHs, which exhibit various extraordinary physical properties modified by number of layers, stacking order, defect engineering, and even assemblies in vdWHs. The OC and differential reflection spectrum in 2DMs and vdWHs have emerged as powerful tools to identify the thickness and peak positions of excitons. We have demonstrated the strong excitonic effects and valley polarization in MX_2 via PL spectroscopy. The experimental evidence for the moiré excitons in vdWHs consisting of constituents with nearly same lattice constants is also elucidated using PL spectroscopy. Following that we give a detailed introduction of the optical spectroscopy study based on light scattering, including Rayleigh scattering and Raman scattering. Rayleigh scattering shows strong thickness dependence. Besides, Raman spectroscopy is reviewed by taking pristine 2DMs, twisted 2DMs, and vdWHs as examples, which reveal the lattice structure and symmetry, interlayer coupling, thickness-dependent Davydov splitting, electronic band structure, and EPC in the corresponding 2D systems. The optical spectroscopy related to nonlinear optical effects, that is, SHG, is also elucidated, unveiling the applications of electric dipole allowed SHG and time-noninvariant/nonreciprocal SHG in distinguishing the lattice symmetry and magnetic orders, respectively. The sensitivity of optical spectroscopy to the electronic band structure, lattice structure and symmetry, defects, doping, number of layers, and stacking order of 2DMs indicates it as a powerful tool for related physical phenomena that are otherwise challenging to probe, and also being important for the characterizations and applications of optoelectronic devices based on 2DMs.

Acknowledgments

We acknowledge support from the National Natural Science Foundation of China (Grant Nos. 12004377 and 11874350), and the CAS Key Research Program of Frontier Sciences (Grant Nos. ZDBS-LY-SLH004 and XDPB22).

References

[1] N. Mounet, M. Gibertini, P. Schwaller, D. Campi, A. Merkys, A. Marrazzo, T. Sohier, I.E. Castelli, A. Cepellotti, G. Pizzi, N. Marzari, Two-dimensional materials from high-throughput computational exfoliation of experimentally known compounds, Nat. Nanotechnol. 13 (3) (2018) 246–252, https://doi.org/10.1038/s41565-017-0035-5.

[2] B. Huang, G. Clark, E. Navarro-Moratalla, R. Klein, R.D. Cheng, K.L. Seyler, D. Zhong, E. Schmidgall, M.A. McGuire, W. Yao, D. Xiao, P. Jarillo-Herrero, X. Xu, Layer-dependent ferromagnetism in a van der Waals crystal down to the monolayer limit, Nature 546 (7657) (2017) 270–273, https://doi.org/10.1038/nature22391.

[3] P.-H. Tan, Raman Spectroscopy of Two-Dimensional Materials, Springer Nature Singapore Pte Ltd., Singapore, 2019, https://doi.org/10.1007/978-981-13-1828-3.

[4] K.S. Novoselov, A.K. Geim, S.V. Morozov, D. Jiang, Y. Zhang, S.V. Dubonos, I.V. Grigorieva, A.A. Firsov, Electric field effect in atomically thin carbon films, Science 306 (5696) (2004) 666–669, https://doi.org/10.1126/science.1102896.

[5] J.-B. Wu, M.-L. Lin, X. Cong, H.-N. Liu, P.-H. Tan, Raman spectroscopy of graphene-based materials and its applications in related devices, Chem. Soc. Rev. 47 (5) (2018) 1822–1873, https://doi.org/10.1039/c6cs00915h.

[6] G. Cassabois, P. Valvin, B. Gil, Hexagonal boron nitride is an indirect bandgap semiconductor, Nat. Photon. 10 (2016) 262–266, https://doi.org/10.1038/nphoton.2015.277.

[7] X. Zhang, X.-F. Qiao, W. Shi, J.-B. Wu, D.-S. Jiang, P.-H. Tan, Phonon and Raman scattering of two-dimensional transition metal dichalcogenides from monolayer, multilayer to bulk material, Chem. Soc. Rev. 44 (9) (2015) 2757–2785, https://doi.org/10.1039/C4CS00282B.

[8] M. Osada, T. Sasaki, Two-dimensional dielectric nanosheets: novel nanoelectronics from nanocrystal building blocks, Adv. Mater. 24 (2) (2012) 210–228, https://doi.org/10.1002/adma.201103241.

[9] E. Shi, Y. Gao, B.P. Finkenauer, Akriti, A.H. Coffey, L. Dou, Two-dimensional halide perovskite nanomaterials and heterostructures, Chem. Soc. Rev. 47 (16) (2018) 6046–6072, https://doi.org/10.1039/C7CS00886D.

[10] X. Xi, Z. Wang, W. Zhao, J.-H. Park, K.T. Law, H. Berger, L. Forró, J. Shan, K.F. Mak, Ising pairing in superconducting $NbSe_2$ atomic layers, Nat. Phys. 12 (2) (2016) 139–143,-https://doi.org/10.1038/nphys3538.

[11] B. Sipos, A.F. Kusmartseva, A. Akrap, H. Berger, L. Forró, E. Tutiš, From Mott state to superconductivity in $1T\text{-}TaS_2$, Nat. Mater. 7 (12) (2008) 960–965, https://doi.org/10.1038/nmat2318.

[12] M.N. Ali, J. Xiong, S. Flynn, J. Tao, Q.D. Gibson, L.M. Schoop, T. Liang, N. Haldolaarachchige, M. Hirschberger, N.P. Ong, R.J. Cava, Large, non-saturating magnetoresistance in WTe_2, Nature 514 (7521) (2014) 205–208, https://doi.org/10.1038/nature13763.

[13] K.F. Mak, C. Lee, J. Hone, J. Shan, T.F. Heinz, Atomically thin MoS_2: a new direct-gap semiconductor, Phys. Rev. Lett. 105 (13) (2010) 136805, https://doi.org/10.1103/PhysRevLett.105.136805.

[14] Q.H. Wang, K. Kalantar-Zadeh, A. Kis, J.N. Coleman, M.S. Strano, Electronics and optoelectronics of two-dimensional transition metal dichalcogenides, Nat. Nanotechnol. 7 (11) (2012) 699–712, https://doi.org/10.1038/nnano.2012.193.

[15] C.N. Eads, D. Bandak, M.R. Neupane, D. Nordlund, O.L.A. Monti, Anisotropic attosecond charge carrier dynamics and layer decoupling in quasi-2D layered SnS_2, Nat. Commun. 8 (2017) 1369, https://doi.org/10.1038/s41467-017-01522-3.

[16] J.-U. Lee, S. Lee, J.H. Ryoo, S. Kang, T.Y. Kim, P. Kim, C.-H. Park, J.-G. Park, H. Cheong, Ising-type magnetic ordering in atomically thin $FePS_3$, Nano Lett. 16 (12) (2016) 7433–7438, https://doi.org/10.1021/acs.nanolett.6b03052.

[17] K. Kim, S.Y. Lim, J.-U. Lee, S. Lee, T.Y. Kim, K. Park, G.S. Jeon, C.-H. Park, J.-G. Park, H. Cheong, Suppression of magnetic ordering in XXZ-type antiferromagnetic monolayer $NiPS_3$, Nat. Commun. 10 (1) (2019) 345, https://doi.org/10.1038/s41467-018-08284-6.

[18] L. Liang, J. Zhang, B.G. Sumpter, Q.-H. Tan, P.-H. Tan, V. Meunier, Low-frequency shear and layer-breathing modes in Raman scattering of two-dimensional materials, ACS Nano 11 (12) (2017) 11777–11802, https://doi.org/10.1021/acsnano.7b06551.

[19] S.N. Shirodkar, U.V. Waghmare, Emergence of ferroelectricity at a metal-semiconductor transition in a 1T monolayer of MoS_2, Phys. Rev. Lett. 112 (15) (2014) 157601, https://doi.org/10.1103/PhysRevLett.112.157601.

[20] J.-B. Wu, X. Zhang, M. Ijäs, W.-P. Han, X.-F. Qiao, X.-L. Li, D.-S. Jiang, A.C. Ferrari, P.-H. Tan, Resonant Raman spectroscopy of twisted multilayer graphene, Nat. Commun. 5 (2014) 5309.

[21] K. Liu, L. Zhang, T. Cao, C. Jin, D. Qiu, Q. Zhou, A. Zettl, P. Yang, S.G. Louie, F. Wang, Evolution of interlayer coupling in twisted molybdenum disulfide bilayers, Nat. Commun. 5 (2014) 4966, https://doi.org/10.1038/ncomms5966.

[22] K.S. Novoselov, A.H.C. Neto, Two-dimensional crystals-based heterostructures: materials with tailored properties, Phys. Scr. 2012 (146) (2012) 014006, https://doi.org/10.1088/0031-8949/2012/T146/014006.

[23] R. Bistritzer, A.H. MacDonald, Moiré bands in twisted double-layer graphene, Proc. Natl Acad. Sci. USA 108 (30) (2011) 12233–12237, https://doi.org/10.1073/pnas.1108174108.

[24] H. Yu, G.-B. Liu, J. Tang, X. Xu, W. Yao, Moiré excitons: from programmable quantum emitter arrays to spin-orbit-coupled artificial lattices, Sci. Adv. 3 (11) (2017) e1701696, https://doi.org/10.1126/sciadv.1701696.

[25] M.-L. Lin, Q.-H. Tan, J.-B. Wu, X.-S. Chen, J.-H. Wang, Y.-H. Pan, X. Zhang, X. Cong, J. Zhang, W. Ji, P.-A. Hu, K.-H. Liu, P.-H. Tan, Moiré phonons in twisted bilayer MoS_2, ACS Nano 12 (8) (2018) 8770–8780, https://doi.org/10.1021/acsnano.8b05006.

[26] Y. Cao, V. Fatemi, A. Demir, S. Fang, S.L. Tomarken, J.Y. Luo, J.D. Sanchez-Yamagishi, K. Watanabe, T. Taniguchi, E. Kaxiras, R.C. Ashoori, P. Jarillo-Herrero, Correlated

insulator behaviour at half-filling in magic-angle graphene superlattices, Nature 556 (7699) (2018) 80–84, https://doi.org/10.1038/nature26154.

[27] K.L. Seyler, P. Rivera, H. Yu, N.P. Wilson, E.L. Ray, D.G. Mandrus, J. Yan, W. Yao, X. Xu, Signatures of moiré-trapped valley excitons in $MoSe_2$/WSe_2 heterobilayers, Nature 567 (7746) (2019) 66–70, https://doi.org/10.1038/s41586-019-0957-1.

[28] K. Tran, G. Moody, F. Wu, X. Lu, J. Choi, K. Kim, A. Rai, D.A. Sanchez, J. Quan, A. Singh, J. Embley, A. Zepeda, M. Campbell, T. Autry, T. Taniguchi, K. Watanabe, N. Lu, S.K. Banerjee, K.L. Silverman, S. Kim, E. Tutuc, L. Yang, A.H. MacDonald, X. Li, Evidence for moiré excitons in van der Waals heterostructures, Nature 567 (7746) (2019) 71–75, https://doi.org/10.1038/s41586-019-0975-z.

[29] C. Jin, E.C. Regan, A. Yan, M. Iqbal Bakti Utama, D. Wang, S. Zhao, Y. Qin, S. Yang, Z. Zheng, S. Shi, K. Watanabe, T. Taniguchi, S. Tongay, A. Zettl, F. Wang, Observation of moiré excitons in WSe_2/WS_2 heterostructure superlattices, Nature 567 (7746) (2019) 76–80, https://doi.org/10.1038/s41586-019-0976-y.

[30] Z.H. Ni, H.M. Wang, J. Kasim, H.M. Fan, T. Yu, Y.H. Wu, Y.P. Feng, Z.X. Shen, Graphene thickness determination using reflection and contrast spectroscopy, Nano Lett. 7 (9) (2007) 2758–2763, https://doi.org/10.1021/nl071254m.

[31] X.-L. Li, W.-P. Han, J.-B. Wu, X.-F. Qiao, J. Zhang, P.-H. Tan, Layer-number dependent optical properties of 2D materials and their application for thickness determination, Adv. Funct. Mater. 27 (2017) 1604468, https://doi.org/10.1002/adfm.201604468.

[32] Y. Lu, X.-L. Li, X. Zhang, J.-B. Wu, P.-H. Tan, Optical contrast determination of the thickness of SiO_2 film on Si substrate partially covered by two-dimensional crystal flakes, Sci. Bull. 60 (8) (2015) 806–811, https://doi.org/10.1007/s11434-015-0774-3.

[33] X.-L. Li, X.-F. Qiao, W.-P. Han, Y. Lu, Q.-H. Tan, X.-L. Liu, P.-H. Tan, Layer number identification of intrinsic and defective multilayered graphenes up to 100 layers by the Raman mode intensity from substrates, Nanoscale 7 (17) (2015) 8135–8141, https://doi.org/10.1039/c5nr01514f.

[34] W. Zhao, Z. Ghorannevis, L. Chu, M. Toh, C. Kloc, P.-H. Tan, G. Eda, Evolution of electronic structure in atomically thin sheets of WS_2 and WSe_2, ACS Nano 7 (1) (2013) 791–797, https://doi.org/10.1021/nn305275h.

[35] C. Ruppert, O.B. Aslan, T.F. Heinz, Optical properties and band gap of single- and few-layer $MoTe_2$ crystals, Nano Lett. 14 (11) (2014) 6231–6236, https://doi.org/10.1021/nl502557g.

[36] D.Y. Qiu, F.H. da Jornada, S.G. Louie, Optical spectrum of MoS_2: many-body effects and diversity of exciton states, Phys. Rev. Lett. 111 (21) (2013) 216805, https://doi.org/10.1103/PhysRevLett.111.216805.

[37] N. Mizuochi, T. Makino, H. Kato, D. Takeuchi, M. Ogura, H. Okushi, M. Nothaft, P. Neumann, A. Gali, F. Jelezko, J. Wrachtrup, S. Yamasaki, Electrically driven single-photon source at room temperature in diamond, Nat. Photon. 6 (5) (2012) 299–303, https://doi.org/10.1038/nphoton.2012.75.

[38] W. Xu, W. Liu, J.F. Schmidt, W. Zhao, X. Lu, T. Raab, C. Diederichs, W. Gao, D.V. Seletskiy, Q. Xiong, Correlated fluorescence blinking in two-dimensional semiconductor heterostructures, Nature 541 (2017) 62–67, https://doi.org/10.1038/nature20601.

[39] K.F. Mak, K. He, J. Shan, T.F. Heinz, Control of valley polarization in monolayer MoS_2 by optical helicity, Nat. Nanotechnol. 7 (8) (2012) 494–498, https://doi.org/10.1038/nnano.2012.96.

[40] T. Cao, G. Wang, W.P. Han, H.Q. Ye, C.R. Zhu, J.R. Shi, Q. Niu, P.H. Tan, E. Wang, B.L. Liu, J. Feng, Valley-selective circular dichroism of monolayer molybdenum disulphide, Nat. Commun. 3 (2012) 887, https://doi.org/10.1038/ncomms1882.

[41] H. Zeng, J. Dai, W. Yao, D. Xiao, X. Cui, Valley polarization in MoS_2 monolayers by optical pumping, Nat. Nanotechnol. 7 (8) (2012) 490–493, https://doi.org/10.1038/nnano.2012.95.

[42] K. He, N. Kumar, L. Zhao, Z. Wang, K.F. Mak, H. Zhao, J. Shan, Tightly bound excitons in monolayer WSe$_2$, Phys. Rev. Lett. 113 (2) (2014) 026803, https://doi.org/10.1103/PhysRevLett.113.026803.

[43] Z. Ye, T. Cao, K. O'Brien, H. Zhu, X. Yin, Y. Wang, S.G. Louie, X. Zhang, Probing excitonic dark states in single-layer tungsten disulphide, Nature 513 (7517) (2014) 214–218, https://doi.org/10.1038/nature13734.

[44] D. Xiao, G.-B. Liu, W. Feng, X. Xu, W. Yao, Coupled spin and valley physics in monolayers of MoS$_2$ and Other group-VI dichalcogenides, Phys. Rev. Lett. 108 (19) (2012) 196802, https://doi.org/10.1103/PhysRevLett.108.196802.

[45] K.F. Mak, K. He, C. Lee, G.H. Lee, J. Hone, T.F. Heinz, J. Shan, Tightly bound trions in monolayer MoS$_2$, Phys. Rev. Lett. 12 (3) (2013) 207–211, https://doi.org/10.1038/nmat3505.

[46] Y.-M. He, G. Clark, J.R. Schaibley, Y. He, M.-C. Chen, Y.-J. Wei, X. Ding, Q. Zhang, W. Yao, X. Xu, C.-Y. Lu, J.-W. Pan, Single quantum emitters in monolayer semiconductors, Nat. Nanotechnol. 10 (6) (2015) 497–502, https://doi.org/10.1038/nnano.2015.75.

[47] C. Chakraborty, L. Kinnischtzke, K.M. Goodfellow, R. Beams, A.N. Vamivakas, Voltage-controlled quantum light from an atomically thin semiconductor, Nat. Nanotechnol. 10 (6) (2015) 507–511, https://doi.org/10.1038/nnano.2015.79.

[48] A. Srivastava, M. Sidler, A.V. Allain, D.S. Lembke, A. Kis, A. Imamoğlu, Optically active quantum dots in monolayer WSe$_2$, Nat. Nanotechnol. 10 (6) (2015) 491–496, https://doi.org/10.1038/nnano.2015.60.

[49] A. Jorio, R. Saito, G. Dresselhaus, M.S. Dresselhaus, Raman Spectroscopy in Graphene Related Systems, Wiley-VCH Verlag GmbH & Co. KGaA, Weinheim, 2011.

[50] C. Casiraghi, A. Hartschuh, E. Lidorikis, H. Qian, H. Harutyunyan, T. Gokus, K.S. Novoselov, A.C. Ferrari, Rayleigh imaging of graphene and graphene layers, Nano Lett. 7 (9) (2007) 2711–2717, https://doi.org/10.1021/nl071168m.

[51] J.-B. Wu, H. Wang, X.-L. Li, H. Peng, P.-H. Tan, Raman spectroscopic characterization of stacking configuration and interlayer coupling of twisted multilayer graphene grown by chemical vapor deposition, Carbon 110 (2016) 225–231, https://doi.org/10.1016/j.carbon.2016.09.006.

[52] L.M. Malard, M.A. Pimenta, G. Dresselhaus, M.S. Dresselhaus, Raman spectroscopy in graphene, Phys. Rep. 473 (5) (2009) 51–87, https://doi.org/10.1016/j.physrep.2009.02.003.

[53] A.C. Ferrari, D.M. Basko, Raman spectroscopy as a versatile tool for studying the properties of graphene, Nat. Nanotechnol. 8 (4) (2013) 235–246, https://doi.org/10.1038/nnano.2013.46.

[54] Q.J. Song, Q.H. Tan, X. Zhang, J.B. Wu, B.W. Sheng, Y. Wan, X.Q. Wang, L. Dai, P.H. Tan, Physical origin of Davydov splitting and resonant Raman spectroscopy of Davydov components in multilayer MoTe$_2$, Phys. Rev. B 93 (11) (2016) 115409, https://doi.org/10.1103/PhysRevB.93.115409.

[55] M.-L. Lin, Y. Zhou, J.-B. Wu, X. Cong, X.-L. Liu, J. Zhang, H. Li, W. Yao, P.-H. Tan, Cross-dimensional electron-phonon coupling in van der Waals heterostructures, Nat. Commun. 10 (2019) 2419, https://doi.org/10.1038/s41467-019-10400-z.

[56] P.H. Tan, W.P. Han, W.J. Zhao, Z.H. Wu, K. Chang, H. Wang, Y.F. Wang, N. Bonini, N. Marzari, N. Pugno, G. Savini, A. Lombardo, A.C. Ferrari, The shear mode of multilayer graphene, Nat. Mater. 11 (2012) 294–300, https://doi.org/10.1038/NMAT3245.

[57] J.-B. Wu, Z.-X. Hu, X. Zhang, W.-P. Han, Y. Lu, W. Shi, X.-F. Qiao, M. Ijiäs, S. Milana, W. Ji, A.C. Ferrari, P.-H. Tan, Interface coupling in twisted multilayer graphene by resonant Raman spectroscopy of layer breathing modes, ACS Nano 9 (7) (2015) 7440–7449.

[58] X. Zhang, Q.-H. Tan, J.-B. Wu, W. Shi, P.-H. Tan, Review on the Raman spectroscopy of different types of layered materials, Nanoscale 8 (12) (2016) 6435–6450, https://doi.org/10.1039/c5nr07205k.

[59] X.-F. Qiao, J.-B. Wu, L. Zhou, J. Qiao, W. Shi, T. Chen, X. Zhang, J. Zhang, W. Ji, P.-H. Tan, Polytypism and unexpected strong interlayer coupling in two-dimensional layered ReS$_2$, Nanoscale 8 (15) (2016) 8324–8332, https://doi.org/10.1039/C6NR01569G.

[60] K. Kim, J.-U. Lee, D. Nam, H. Cheong, Davydov splitting and excitonic resonance effects in Raman spectra of few-layer MoSe$_2$, ACS Nano 10 (8) (2016) 8113–8120, https://doi.org/10.1021/acsnano.6b04471.

[61] Q.-H. Tan, Y.-J. Sun, X.-L. Liu, Y. Zhao, Q. Xiong, P.-H. Tan, J. Zhang, Observation of forbidden phonons, Fano resonance and dark excitons by resonance Raman scattering in few-layer WS$_2$, 2D Mater. 4 (3) (2017) 031007, https://doi.org/10.1088/2053-1583/aa79bb.

[62] Q.-H. Tan, X. Zhang, X.-D. Luo, J. Zhang, P.-H. Tan, Layer-number dependent high-frequency vibration modes in few-layer transition metal dichalcogenides induced by inter-layer couplings, J. Semicond. 38 (3) (2017) 031006, https://doi.org/10.1088/1674-4926/38/3/031006.

[63] C. Cong, T. Yu, Enhanced ultra-low-frequency interlayer shear modes in folded graphene layers, Nat. Commun. 5 (2014) 4709, https://doi.org/10.1038/ncomms5709.

[64] S. Huang, X. Ling, L. Liang, J. Kong, H. Terrones, V. Meunier, M.S. Dresselhaus, Probing the interlayer coupling of twisted bilayer MoS$_2$ using photoluminescence spectroscopy, Nano Lett. 14 (10) (2014) 5500–5508, https://doi.org/10.1021/nl5014597.

[65] A.A. Puretzky, L. Liang, X. Li, K. Xiao, B.G. Sumpter, V. Meunier, D.B. Geohegan, Twisted MoSe$_2$ bilayers with variable local stacking and interlayer coupling revealed by low-frequency Raman spectroscopy, ACS Nano 10 (2) (2016) 2736–2744, https://doi.org/10.1021/acsnano.5b07807.

[66] V. Carozo, C.M. Almeida, E.H. Ferreira, L.G. Cançado, C.A. Achete, A. Jorio, Raman signature of graphene superlattices, Nano Lett. 11 (11) (2011) 4527–4534.

[67] P. Parzefall, J. Holler, M. Scheuck, A. Beer, K.Q. Lin, B. Peng, B. Monserrat, P. Nagler, M. Kempf, T. Korn, C. Schüller, Moiré phonons in twisted MoSe$_2$-WSe$_2$ heterobilayers and their correlation with interlayer excitons, 2D Mater. 8 (3) (2021), https://doi.org/10.1088/2053-1583/abf98e.

[68] H. Yoo, R. Engelke, S. Carr, S. Fang, K. Zhang, P. Cazeaux, S.H. Sung, R. Hovden, A.W. Tsen, T. Taniguchi, K. Watanabe, G.C. Yi, M. Kim, M. Luskin, E.B. Tadmor, E. Kaxiras, P. Kim, Atomic and electronic reconstruction at the van der Waals interface in twisted bilayer graphene, Nat. Mater. 18 (5) (2019) 448–453, https://doi.org/10.1038/s41563-019-0346-z.

[69] A. Weston, Y. Zou, V. Enaldiev, A. Summerfield, N. Clark, V. Zólyomi, A. Graham, C. Yelgel, S. Magorrian, M. Zhou, J. Zultak, D. Hopkinson, A. Barinov, T.H. Bointon, A. Kretinin, N.R. Wilson, P.H. Beton, V.I. Fal'ko, S.J. Haigh, R. Gorbachev, Atomic reconstruction in twisted bilayers of transition metal dichalcogenides, Nat. Nanotechnol. 15 (7) (2020) 592–597, https://doi.org/10.1038/s41565-020-0682-9.

[70] T.I. Andersen, G. Scuri, A. Sushko, K. De Greve, J. Sung, Y. Zhou, D.S. Wild, R.J. Gelly, H. Heo, D. Bérubé, A.Y. Joe, L.A. Jauregui, K. Watanabe, T. Taniguchi, P. Kim, H. Park, M.D. Lukin, Excitons in a reconstructed moiré potential in twisted WSe$_2$/WSe$_2$ homobilayers, Nat. Mater. 20 (4) (2021) 480–487, https://doi.org/10.1038/s41563-020-00873-5.

[71] J. Quan, L. Linhart, M.L. Lin, D. Lee, J. Zhu, C.Y. Wang, W.T. Hsu, J. Choi, J. Embley, C. Young, T. Taniguchi, K. Watanabe, C.K. Shih, K. Lai, A.H. MacDonald, P.H. Tan, F. Libisch, X. Li, Phonon renormalization in reconstructed MoS$_2$ moiré superlattices, Nat. Mater. 20 (2021) 1100–1105, https://doi.org/10.1038/s41563-021-00960-1.

[72] A.C. Gadelha, D.A.A. Ohlberg, C. Rabelo, E.G. Neto, T.L. Vasconcelos, J.L. Campos, J.S. Lemos, V. Ornelas, D. Miranda, R. Nadas, F.C. Santana, K. Watanabe, T. Taniguchi, B. van Troeye, M. Lamparski, V. Meunier, V.H. Nguyen, D. Paszko, J.C. Charlier, L.

C. Campos, L.G. Cançado, G. Medeiros-Ribeiro, A. Jorio, Localization of lattice dynamics in low-angle twisted bilayer graphene, Nature 590 (7846) (2021) 405–409, https://doi.org/10.1038/s41586-021-03252-5.

[73] Y.-C. Leng, M.-L. Lin, Y. Zhou, J.-B. Wu, D. Meng, X. Cong, H. Li, P.-H. Tan, Intrinsic effect of interfacial coupling on the high- frequency intralayer modes in twisted multilayer MoTe$_2$, Nanoscale 13 (21) (2021) 9372–9739, https://doi.org/10.1039/D1NR01309B.

[74] H. Li, J.-B. Wu, F. Ran, M.-L. Lin, X.-L. Liu, Y. Zhao, X. Lu, Q. Xiong, J. Zhang, W. Huang, H. Zhang, P.-H. Tan, Interfacial interactions in van der Waals heterostructures of MoS$_2$ and graphene, ACS Nano 11 (11) (2017) 11714–11723, https://doi.org/10.1021/acsnano.7b07015.

[75] L. Liang, A.A. Puretzky, B.G. Sumpter, V. Meunier, Interlayer bond polarizability model for stacking-dependent low-frequency Raman scattering in layered materials, Nanoscale 9 (40) (2017) 15340–15355, https://doi.org/10.1039/C7NR05839J.

[76] M. Fiebig, V.V. Pavlov, R.V. Pisarev, Second-harmonic generation as a tool for studying electronic and magnetic structures of crystals: review, J. Opt. Soc. Am. B 22 (1) (2005) 96–118, https://doi.org/10.1364/JOSAB.22.000096.

[77] R.W. Boyd (Ed.), The Nonlinear Optical Susceptibility, Nonlinear Optics, in: third ed., Academic Press, Burlington, 2008, pp. 1–67, https://doi.org/10.1016/B978-0-12-369470-6.00001-0. Chapter 1.

[78] T. Jiang, H. Liu, D. Huang, S. Zhang, Y. Li, X. Gong, Y.-R. Shen, W.-T. Liu, S. Wu, Rayleigh imaging of graphene and graphene layers, Nat. Nanotechnol. 9 (10) (2014) 825–829, https://doi.org/10.1038/nnano.2014.176.

[79] L.M. Malard, T.V. Alencar, A.P.M. Barboza, K.F. Mak, A.M. de Paula, Observation of intense second harmonic generation from MoS$_2$ atomic crystals, Phys. Rev. B 87 (20) (2013) 201401, https://doi.org/10.1103/PhysRevB.87.201401.

[80] Y. Li, Y. Rao, K.F. Mak, Y. You, S. Wang, C.R. Dean, T.F. Heinz, Probing symmetry properties of few-layer MoS$_2$ and h-BN by optical second-harmonic generation, Nano Lett. 13 (7) (2013) 3329–3333, https://doi.org/10.1021/nl401561r.

[81] Y. Song, S. Hu, M.-L. Lin, X. Gan, P.-H. Tan, J. Zhao, Extraordinary second harmonic generation in ReS$_2$ atomic crystals, ACS Photonics 5 (9) (2018) 3485–3491, https://doi.org/10.1021/acsphotonics.8b00685.

[82] W. Jie, X. Chen, D. Li, L. Xie, Y.Y. Hui, S.P. Lau, X. Cui, J. Hao, Layer-dependent nonlinear optical properties and stability of non-centrosymmetric modification in few-layer GaSe sheets, Angew. Chem. Int. Ed. 54 (4) (2015) 1185–1189, https://doi.org/10.1002/anie.201409837.

[83] Z. Sun, Y. Yi, T. Song, G. Clark, B. Huang, Y. Shan, S. Wu, D. Huang, C. Gao, Z. Chen, M. McGuire, T. Cao, D. Xiao, W.-T. Liu, W. Yao, X. Xu, S. Wu, Giant nonreciprocal second-harmonic generation from antiferromagnetic bilayer CrI$_3$, Nature 572 (7770) (2019) 497–501, https://doi.org/10.1038/s41586-019-1445-3.

[84] P. Němec, M. Fiebig, T. Kampfrath, A.V. Kimel, Antiferromagnetic opto-spintronics, Nat. Phys. 14 (3) (2018) 229–241, https://doi.org/10.1038/s41567-018-0051-x.

[85] C. Gong, L. Li, Z. Li, H. Ji, A. Stern, Y. Xia, T. Cao, W. Bao, C. Wang, Y. Wang, Z.Q. Qiu, R.J. Cava, S.G. Louie, J. Xia, X. Zhang, Discovery of intrinsic ferromagnetism in two-dimensional van der Waals crystals, Nature 546 (7657) (2017) 265–269, https://doi.org/10.1038/nature22060.

[86] P. Jiang, C. Wang, D. Chen, Z. Zhong, Z. Yuan, Z.-Y. Lu, W. Ji, Stacking tunable interlayer magnetism in bilayer CrI$_3$, Phys. Rev. B 99 (14) (2019) 144401, https://doi.org/10.1103/PhysRevB.99.144401.

Growth and characterization of graphene, silicene, SiC, and the related nanostructures and heterostructures on silicon wafer

13

Naili Yue and Yong Zhang
Department of Electrical and Computer Engineering, The University of North Carolina at Charlotte, Charlotte, NC, United States

13.1 Introduction

Following the explosive interest in the single layer of graphite, known as graphene, initiated by Geim and Novoselov in early 2000 [1], heterostructures involving graphene and other layered materials, known as Van der Waals heterostructures, have also received a great deal of attention [2]. Besides the more commonly sought after heterostructures, such as using an insulating boron nitride as the gate dielectric layer in a graphene transistor, two other less trivial ways involving graphene are discussed in this chapter. One is forming a superlattice with alternating layers of graphene and another layer material. The other is used graphene as a substrate or an interfacial layer for growing other materials that are not necessarily layered materials. For the former, a graphene/silicene superlattice was proposed in 2010 [3]; a graphene/few-layer boron nitride superlattice was demonstrated in 2012 [4]. For the latter, the idea can be traced back to the so-called Van der Waals epitaxy [5]. Examples include the early effort of growing polycrystalline Si on graphite [6], the growth of very-thin single crystalline Si on graphite or graphene on SiO_2/Si substrate and polycrystalline ZnSe on graphene/quartz substrate [7], and the recent demonstration of growth of highly oriented GaN thin-film on graphene/glass substrate and fabrication of flexible light emitting diode (LED) after the structure being lifted off from the substrate [8].

This chapter summarizes our work in two aspects: (1) the formation of graphene micro- and nanostructures on a large Si wafer by converting epitaxially deposited SiC thin-film on Si into graphene or multi-layer graphene using laser conversion; (2) deposition of the ultra-thin Si nanostructures on graphene or graphite substrates. These efforts are the necessary steps for the ultimate construction of graphene/silicene superlattices [3]. The ability to deposit at least one component on a large Si wafer is significant, because it offers the feasibility to integrate the new material systems with the existing Si-based semiconductor fabrication technologies.

Modeling, Characterization and Production of Nanomaterials. https://doi.org/10.1016/B978-0-12-819905-3.00013-0

Silicene [9], a two-dimensional (2D) silicon nanoscale material with similar structure to graphene, is also a potential candidate which could also be integrated with silicon to enhance device function and performance. Another key semiconductor material, SiC, a compound of silicon and carbon, has higher electron mobility, higher band gap, higher thermal conductivity than silicon, enables itself to be an important material for making power electronic devices, especially in high power, high frequency, and high temperature applications. Therefore, it is significantly meaningful to integrate these new materials with silicon wafer by using existing manufacturing technologies and tools. In general, there are two integration methods, one is by transferring or bonding process and the other is direct growth. In the transferring or bonding process, the new thin film materials such as graphene or silicene are fabricated on difference substrates like nickel, copper, and so forth, and then are transferred or bonded to silicon wafer. Since defects, stresses, or damage would be introduced in the new layer structures during transferring, direct growth method in which the desirable new thin films are grown on silicon wafers directly or by growing another material as an intermediate layer is used. In our work, we have investigated and demonstrated the growth methods for integrating new semiconductor materials including graphene, silicene, and silicon carbide with silicon wafers, opening possibility of introducing new materials into semiconductor manufacturing industry.

13.2   Molecular beam epitaxy (MBE) growth of 3C-SiC thin film grown on Si (111)

It was reported that graphene was formed on 3C-SiC/Si by vacuum annealing process [10]. We instead used a simple laser sublimation method to form graphene on 3C-SiC/Si.

An MBE system from SVT Associates Inc. (Eden Prairie, MN) was used to epitaxially grow thin films on different substrates in this study. Ultra-high vacuum level of over 10^{-10} Torr could be achieved by using a series of pumps in cascade including mechanical rough pump, turbomolecular pump, diffusion pump, and cryopump. Fullerene (C60) powders as carbon source and Si wafer substrate as silicon source were used to grow SiC thin films on Si substrates. At an appropriate combination of substrate temperature (T_S) and C60 evaporation temperature (T_C), C60 was evaporated from the crucible in the molecular form, decomposed into atoms when incident on the Si substrate, and then reacted with Si from the substrate to form SiC thin film on silicon wafers. On the basis of the trial experiments, the typical growth temperature ranges for forming SiC thin film were $T_S = 700$–$800°C$ and $T_C = 500$–$600°C$, where T_S and T_C were silicon substrate and carbon source temperatures, respectively. The optimal growth temperature combination was found to be $T_S = 800°C$ and $T_C = 550°C$ in terms of crystallinity and uniformity. To ensure high purity growth, the gas or vapor in the chamber was monitored with in-situ mass spectrometer, and the film surface quality was inspected in real time with in-situ reflection high-energy electron diffraction (RHEED), as shown in diffraction pattern in Fig. 13.1. Fig. 13.1A shows the diffraction pattern of clean or atomic flat substrate or thin film surface, while Fig. 13.1B

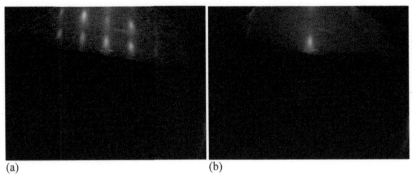

(a)   (b)

Fig. 13.1 RHEED patterns in the growth of 3C-SiC thin film on Si substrates. (A) Before growth. (B) During growth.

shows dots or concentric rings following the disappearance of the streak pattern, indicating the formation of polycrystalline structures. After 10-min growth, the samples were held at the growth temperature for 5–10 min to homogenize the epitaxial film. To reduce the residual stress in SiC film, the film was finally controlled to cool down slowly at 10°C/min to room temperature. The crystalline structure was characterized with X-ray diffraction (XRD). XRD patterns of 3C-SiC grown on both Si (111) and bare Si (111) are shown in Fig. 13.2. It can be clearly seen that newly emerging peak at $2\theta = 35.6°$ represents 3C-SiC (111) thin film grown on Si (111).

It was found that both grain size and surface roughness of 3C-SiC thin film increase with substrate temperature, and so does crystallinity. Fig. 13.3A–D show the evolution of grain size and surface roughness with substrate temperature increasing from 700°C

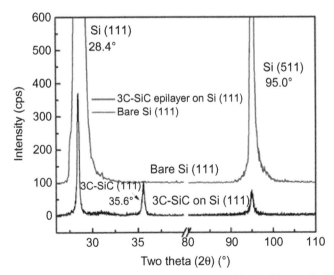

Fig. 13.2 XRD patterns of bare Si (111) substrate and 3C-SiC thin film on Si (111).

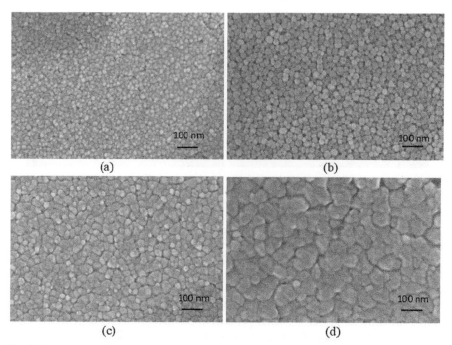

Fig. 13.3 Scanning electron microscopy (SEM) micrographs of surface morphology of 3C-SiC (111) grown in Si (111) for 5 min at $T_C = 550°C$ and different T_S temperatures: (A) 700°C; (B) 800°C; (C) 900°C; (D) 1000°C.

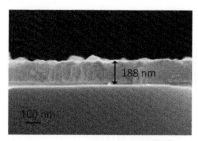

Fig. 13.4 SEM micrograph of the cross-section of a 3C-SiC (111) film on Si (111) at $T_C = 550°C$ and $T_S = 800°C$ for 5 min.

to 1000°C. Fig. 13.4 shows the cross-section image of the 3C-SiC thin film shown in Fig. 13.3B with an average thickness of about 188 nm.

Moreover, another interesting observation is that the crystallinity of the 3C-SiC (111) thin film also increases with growth temperatures, which is exhibited in the changing full width half maximum (FWHM) of the XRD peaks shown in Fig. 13.5. It can be apparently seen that FWHM of peaks at $2\theta = 35.6°$ decreases with increasing growth temperatures from 700/450 to 1000/550°C.

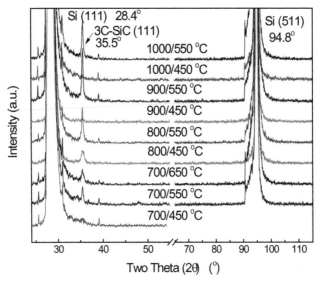

Fig. 13.5 XRD patterns of the 3C-SiC thin films grown on Si (111) substrates (Each spectrum is labeled by substrate temperature/source temperature, e.g., 700/450°C indicates that substrate and source temperatures are 700°C and 450°C, respectively).

13.3 Conversion from 3C-SiC thin film to graphene under laser illumination

Since monolayer graphene was experimentally discovered by Novoselov and Geim [11] in 2004, graphene research has evolved from the exploration and development of synthesis techniques of large size graphene [12–15] to the patterning of graphene nanostructures (GNSs) [16,17]. Conventionally, GNSs are patterned on a large size graphene film either obtained by photolithography or electron beam lithography and plasma etching or obtained by unzipping of carbon nanotube [16,17]. However, wet lithography and dry etching processes, which incur some defects or contamination to the surface-sensitive graphene, have to be involved in all patterning processes. Furthermore, the subsequent transfer process could cause some perturbation to the morphology or introduce defects to the fragile 2D graphene during transfer or handling. In contrast, laser-induced local graphene formation (LLG) is a quick and clean approach for synthesizing and patterning graphene. The LLG technique was first explored by Ohkawara [18] and further developed by Perrone [19], Lee [20], Yue [21,22], and Yannopoulos et al. [23]. Among these laser-based methods of making graphene from SiC, all other approaches focus on bulk SiC, but our work [21,22] focuses on using MBE-grown, thin film 3C-SiC on Si wafer substrates to develop the patterning of micro- and even nanoscale graphene structures. The graphene layer directly formed on the Si wafer would open the possibility of integration of graphene-based nanoscale electronic devices with existing silicon-based microelectronics.

For the laser-induced graphene approach, the fundamental principle lies in that a tightly focused laser beam of sufficiently high power density can induce local heating, which may sublimate Si atoms of SiC and leave C atoms to recrystallize into graphene layers on Si wafer. Therefore, the general idea is similar to the approach of annealing SiC in high vacuum and high temperature (typically 1200–1600°C) [24] to obtain a graphene film. This annealing technique has been shown to yield high quality graphene (e.g., with high electron mobility), in particular on the Si face of SiC where monolayer graphene can be obtained [25]. In contrast, our LLG procedure has been carried out under ambient condition (at room temperature in air), although it can also be performed in vacuum or appropriate gas environment and even under an elevated temperature. In this study, a 532-nm continuous wave (CW) Nd-YAG laser with maximum incident power of ∼30 mW was used to illuminate and then convert the surface layer of the 3C-SiC (111) grown on Si (111) into graphene layers. The 532 nm CW laser is an integrated part of a confocal scanning μ-Raman system (LabRAM HR800, Horiba Jobin Yvon). The beam size of 1–2 mm can be focused down to a diffraction limit spot size of ∼0.7 μm on the sample surface when a 100 × objective lens with numerical aperture (NA) = 0.9 is used. The structural change under laser illumination can be continuously monitored in-situ using μ-Raman. The Raman signal is collected in backscattering geometry. The Raman mapping is carried out by moving the translational stage where the sample sits, controlled by programmable software. With the addition of an external shutter synchronized with the motion control of the translation stage, the illumination site and shape can be controlled to generate different patterns, for instance, dots, lines, or areas. The Raman signal can be obtained simultaneously while the material is being illuminated or probed afterward using lower power. Micron scale patterns can be obtained by directly using the focused laser beam with a diffraction-limited spot size. Even smaller feature sizes (e.g., 100 nm) are possible, if illumination masks are used to define the illumination area or the SiC thin film is selectively deposited in desired patterns. The external and internal shutters are used to control illumination and signal acquisition times, respectively. The laser illumination setup is shown in Fig. 13.6. The phase transformation mechanism is illustrated schematically in Fig. 13.7.

Fig. 13.6 Experimental setup for SiC-to-graphene conversion under laser illumination, showing that a laser beam is focused by a microscope objective lens on a SiC thin-film grown on a Si wafer.

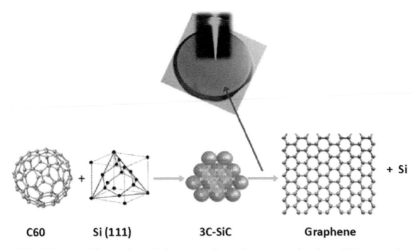

C60   **Si (111)**   **3C-SiC**   **Graphene**   **+ Si**

Fig. 13.7 Schematic illustration of phase transformation conversion from SiC to graphene.

Typically, a laser power of 20–30 mW was required to induce a conversion from SiC to graphene in the above-described setup, depending on the grain structure of the SiC thin film (to be discussed later in detail). In the subsequent Raman characterization, the laser power was reduced down to 1–2 mW to avoid any laser-heating effect during measurement. The Raman spectra of 3C-SiC thin film before and after laser illumination were displayed in Fig. 13.8A.

Three strong peaks of 1348, 1583, and 2691 cm^{-1}, which correspond to D, G, and G$'$ (2D) bands of graphene, respectively, are clearly observed to emerge after laser illumination in Fig. 13.8A. However, the spectrum of the as-grown 3C-SiC on Si (111) does not show these three peaks. A complementary observation is that the intensity of transverse optical phonon mode of 3C-SiC thin film drops significantly after laser illumination, which implies thickness of 3C-SiC decreases due to decomposition or phase change. Since fullerene (C60) was used as a carbon source to grow 3C-SiC on Si (111) and the Raman spectrum of C60 also shows D and G bands, there is a possibility that D and G peaks are from C60 after laser illumination. However, the comparison of spectra of laser-illuminated 3C-SiC and C60, as shown in Fig. 13.8B, indicates that C60 possess only two broad bands D and G, and the double resonant 2D mode only appears in the spectrum of laser-illuminated 3C-SiC. Therefore, 2D Raman mode is a more reliable confirmation of the graphene layers, and the Raman spectrum confirms the formation of graphene layers on the laser-illuminated 3C-SiC in Si (111). The comparison of the spectrum of SiC after illumination with that of C60 and that of SiC before illumination convincingly proves that graphene layers were definitely produced under laser illumination.

To resolve the number of graphene layers, transmission electron microscopy (TEM) (JEOL2100) and atomic force microscopy (AFM) (Nanoscope SPM V5r30, Veeco) were also used to image the cross-section of the sample and measure the thickness of the graphene layers, respectively. The d-spacing between graphene layers was measured and calculated to be about 3.70 Å from TEM image shown in Fig. 13.9,

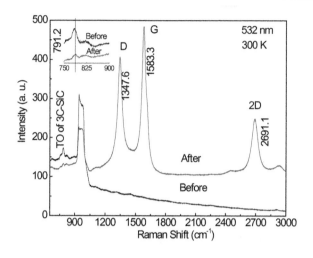

(a)

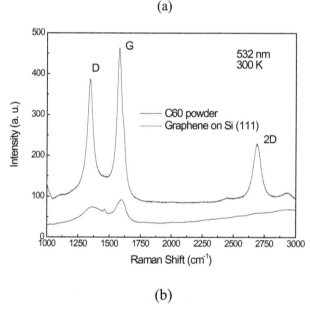

(b)

Fig. 13.8 The Raman spectra of 3C-SiC on Si (111) before and after laser illumination and of C60. (A) 3C-SiC before and after laser illumination; (B) Graphene versus C60.

but the standard interlayer spacing of graphite is $\sim$3.35 Å. Therefore, the measured spacing is larger than that of the crystalline graphite with A-B stacking or Bernal stacking ($c/2 = 3.35$ Å), but close to the theoretically predicted graphene layers separation of 3.61 Å (0 K) for the A-A stacking (turbostratic stacking) [3], implying that the stacking order of the multi-layer graphene is different from that in the crystalline graphite such as highly ordered pyrolytic graphite (HOPG) [26], and expected to have weaker

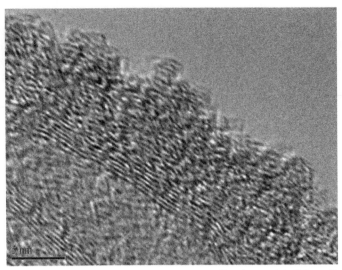

Fig. 13.9 TEM micrograph of graphene layers on laser-illuminated 3C-SiC on Si (111).

interlayer coupling. The turbostratic stacking might result from extremely fast heating and cooling rates of laser which may not give graphene layers enough time to equilibrate into the energetically more favorable stacking order. The number of graphene layers was measured and calculated to be about 8–9 layers, falling in the range of few-layer-graphene. As corroborating evidence, the tapping mode AFM image shown in Fig. 13.10 revealed a flake with thickness of 3.517 nm, which is approximately the thickness of $3.517/3.7 = 9$–10 layers in agreement with the result from the TEM image.

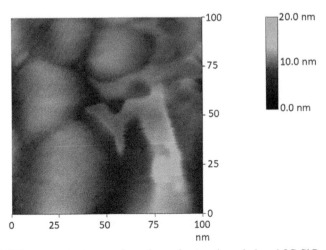

Fig. 13.10 AFM topography image of graphene sheet on laser-induced 3C-SiC on Si (111).

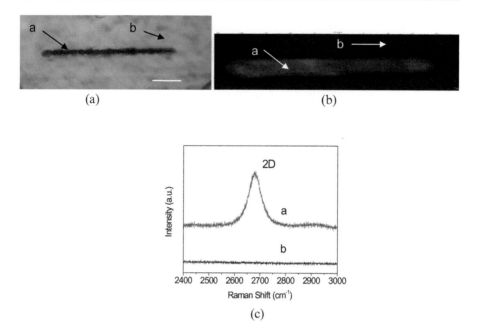

Fig. 13.11 Characterization results of laser-illuminated 3C-SiC line on Si (111). (A) Optical image (scale bar: 5 μm); (B) Raman mapping image (image size: 30 × 10 μm^2; spectral range: 2650–2750 cm^{-1}); (C) Raman spectra selected from two different points *a* and *b*.

To test the feasibility of patterning graphene structures on the SiC film with laser, we did a line scan of 20 μm long under the illumination of a 532-nm laser with a power of about 20 mW. The illuminated line appears darker than the rest area, as revealed in the optical image in Fig. 13.11A. The corresponding Raman mapping at the graphene 2D peak was performed with results shown in Fig. 13.11B and C.

The optical micrograph shown in Fig. 13.11A indicates that the line width of about 1 μm results from the laser spot size of about 0.7 μm. The comparison of Raman image shown in Fig. 13.11B with optical image shown in Fig. 13.11A indicates that the graphene layers distribute uniformly within the ribbon. The comparison between the two Raman spectra from illuminated and non-illuminated areas, as shown in Fig. 13.11C, offers an unambiguous confirmation for the conversion from SiC to graphene. These promising results provide inspiring possibility for patterning graphene-based electronic devices and electron-photon superstructures.

13.4   Programmed laser illumination to pattern graphene-based photonic structure in micro or potentially nanoscale

By using laser illumination technique, three approachescan be, in principle, applied to patterning graphene micro- or nanostructures based on the laser illumination technique: (a) Direct writing (DW), (b) Pre-pattern (PP), and (c) Illumination mask

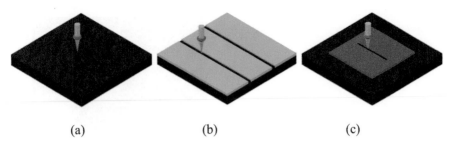

(a)   (b)   (c)

Fig. 13.12 Three approaches to pattern graphene micro- or nanostructures using laser annealing technique: (A) direct writing; (B) pre-patterning; (C) illumination mask (*black*—Si, *blue*—SiC, *light blue*—SiO$_2$, *brown*—illumination mask, and *green*—laser).

(IM), as schematically illustrated in Fig. 13.12. Here, we concentrate on evaluation of the first two methods.

The direct writing method, as schematically illustrate in Fig. 13.12A, is the simplest one of the three methods. Since the smallest feature size of graphene microstructure depends on the laser spot size determined by the NA of the microscope lens, laser wavelength, laser power, and microstructure of 3C-SiC thin film, there is a limit for the minimal feature size of the patterned microstructures. Fig. 13.13A–F show optical images (left) and corresponding Raman mapping images (right) of graphene microdiscs formed on three different 3C-SiC (111) on Si (111) samples. As the three samples have different surface morphologies and crystalline quality, different threshold laser power values are required to sublimate 3C-SiC into graphene layers. Sample A, shown in Fig. 13.13A and B, has more defects than samples B and C due to a native SiO$_2$ layer remaining on Si substrate, and thus requires the least laser power to decompose SiC into graphene. Sample B has same thickness as sample A but with less defects because native silicon oxide layer was cleaned before growth of 3C-SiC. Therefore, the SiC thin film on sample B has higher crystallinity with less defects and requires more laser power to decompose SiC into graphene layers. Sample C has the least defects and thinner than sample B; therefore, it requires more laser power than sample B because light absorption is less due to the combination of higher crystallinity with less defect and more heat dissipation through thinner SiC.

Similarly, two, four, and nine graphene dots in arrays of 1×2, 2×2, and 3×3, as well as a 10-μm long graphene ribbon were also patterned by programming laser illumination. Following the patterning of graphene dots, Raman mapping was also performed on the patterned areas. The Raman images are shown in Fig. 13.14A–D.

In contrast to the direct writing method, the pre-patterning method is a selective deposition technique in which the 3C-SiC thin films are selectively deposited on the pre-patterned areas. The pre-patterned areas are formed by using general clean room fabrication processes such as photolithography and etching. In this method, in contrast to the direct writing method, the smaller feature size of laser-induced graphene is not limited by the laser spot size. This approach is based on the finding that C$_{60}$ only reacts with the exposed Si but not SiO$_2$ [22]. However, the achievable feature size depends on the capability of photo- or e-beam lithography, dry etching, and selective deposition. To investigate the possibility of pre-patterning method, we

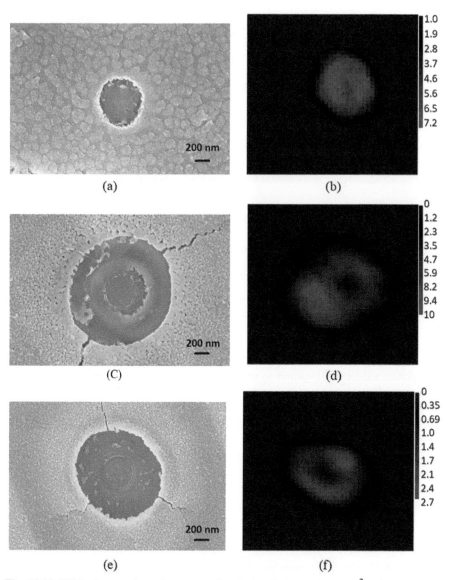

Fig. 13.13 SEM micrographs and corresponding Raman images ($4 \times 4\,\mu m^2$) of three discs illuminated with laser on three samples A, B, and C. (A) and (B) Sample A; (C) and (D) Sample B; (E) and (F) Sample C (Raman spectral range: $2650–2750\,cm^{-1}$ of the graphene 2D peak).

used a photo-lithographic method to pattern a Si (111) substrate capped with $\sim$100 nm thick SiO_2, and demonstrated the feasibility of this pre-patterning approach. Fig. 13.15A shows an optical image of an area illuminated with laser in which 3C-SiC was selectively deposited on 5 μm wide silicon strips separated by 5 μm

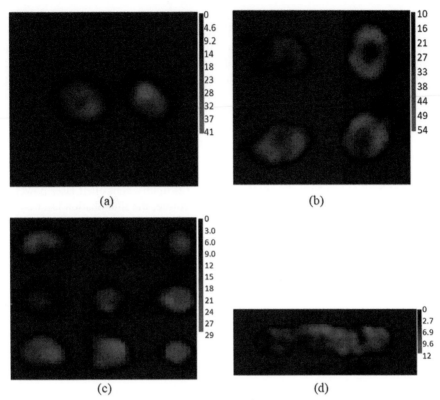

Fig. 13.14 Raman mapping images (2650–2750 cm^{-1}) of different graphene micro-disc arrays and a μ-ribbon. (A) 2 dots in (1 × 2) array (image size: 6 × 6 μm^2); (B) 4 dots in (2 × 2) array (image size: 6 × 6 μm^2); (C) 9 dots in (3 × 3) array (image size: 9 × 9 μm^2); (D) a graphene μ-ribbon (image size: 14 × 4 μm^2).

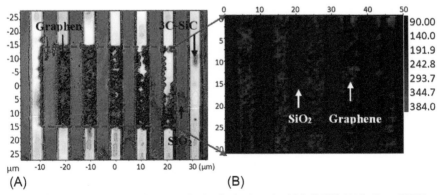

Fig. 13.15 Optical and Raman images of selectively deposited 3C-SiC/Si (111) (5 μm SiC/5 μm SiO$_2$). (A) Optical image; (B) Raman mapping image (2650–2750 cm^{-1}).

SiO_2 spacers. Since the decomposed C60 reacts with Si only and does not react with SiO_2, SiC is only formed on the exposed Si areas. Fig. 13.15B shows the corresponding Raman image of the laser-induced graphene 2D band. It can be clearly seen that only the areas with 3C-SiC formed on the exposed Si ribbons show graphene characteristics.

Following the success in fabricating above-demonstrated graphene microstructures by using the pre-patterning method, SiC ribbons in nanoscale were also pursued and selectively deposited on Si wafer successfully in order to be induced into graphene nanoribbons under laser illumination. The minimum width of the achieved SiC nano-wires is about 89 nm as shown in Fig. 13.16. However, the conversion from SiC nano-wire to graphene presents new challenges to laser energy absorption in two aspects: (1) higher laser power because some heat generated by $\sim$700 nm focused laser spot is dissipated by SiO_2 on Si wafer instead of SiC; (2) tightly controlled laser power to avoid extra sublimation of SiC nanoribbon. Therefore, the illumination laser power must be optimized. Due to the limited laser power available, we were unable to decompose 3C-SiC nanoribbons into graphene nanoribbons. However, it is possible, in principle, to convert pre-patterned SiC nanoribbons to graphene nanoribbons under the optimized laser illumination conditions.

The third technique is an illumination mask method, using a mask made of heat-resistive materials with low thermal conductivities such as metals or ceramics. The mask can be fabricated by using laser cutting or standard clean room fabrication techniques such as photo- or e-beam lithography and dry etching. Likewise, this method presents higher requirements for optimization of illumination conditions than the former two methods.

13.5 Growth of semiconductors on hetero-substrates using graphene as an interfacial layer

It has been a general interest to be able to grow a semiconductor material on a foreign substrate for obtaining either new properties of the heterojunction or simply achieving processing flexibility. Many approaches have been used for epitaxial growth of a

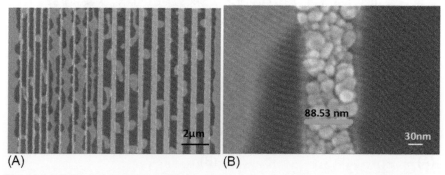

(A)   (B)

Fig. 13.16 SEM micrographs of SiC nanowires grown on Si substrate. (A) Overall view; (B) zoom-in view.

semiconductor layer on a substrate with a different lattice structure. Typically, however, a lattice mismatch will be incurred and lead to the formation of defects such as cracks and delamination. In addition, a chemical mismatch may result in undesirable electronic coupling.

Graphene, due to its unique properties, is used as an interfacial layer to grow Si and other semiconductors or crystalline materials including 2D Si and other nanostructures on any foreign substates that can withstand temperature without the strigent limitation of lattice matching typically required for epitaxial growth [7], as shown in Fig. 13.17.

Various advantages result from graphene as an interfacial layer, including high conductivity, good transparency, high atomic density, chemically inertness, high melting point, adaptability to many substrate, and easy lift-off capabilities [7].

As our first attempt to deposit silicence on graphene in 2011, thin Si films were grown on graphite and graphene on SiO_2/Si substrates in an MBE system (SVT Associates Inc.) by evaporating bulk Si powders with an e-beam evaporator. Graphite substrates of a few mm lateral size were cleaved from a large single crystal HOPG. Si was deposited in the central region of the small graphite substrate in order to obtain highly crystallized thin films with less defects. As the van der Waals bonding between the adjacent layers in graphite is very weak, the top layer of graphite can be viewed as a single layer of graphene supported by a large number of stacked graphene layers with weak bonding. Therefore, one may approximatly regard both graphite and graphene on SiO_2 as graphene substrate, although there might be some subtle differences in the charge distribution between the graphene-graphite and graphene-SiO_2, which could affect the bonding between the graphene and the epilayer. The typical growth conditions used in this study are described as follows: growth chamber base pressure was about 2×10^{-8} Torr; the graphite or graphene substrate was heated up to the growth temperature $T_g = 800$ or $850°C$ and held for 15 min; e-beam evaporator was run with acceleration voltage of 6.07 kV, emission current of 150 mA, and filament current 31 A; growth time was $t_g = 10$ or 15 min; holding time at T_g was 5 min for homogenizing the grown crystalline structure; cooling rate was controlled at 10°C/min from T_g to 500°C, then cooled down naturally to room temperature in the growth chamber. The samples were rotated during growth to improve uniformity of the grown thin films. Also, liquid nitrogen was circulated through shround in the wall of growth chamber to cool down the inner wall of the chamber to prevent the residual depositions on the wall from reevapration, avoiding the incorporation of impurities into the grown

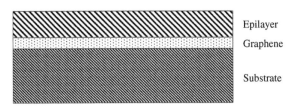

Fig. 13.17 Schematic illustration of graphene as an interfacial layer to grow epitaxial layer on silicon or other substrates.

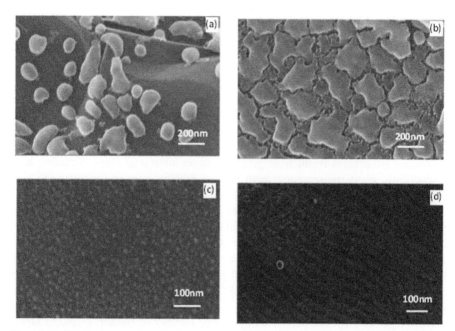

Fig. 13.18 SEM micrographs of epitaxial silicon grown on graphite and graphene on SiO_2/Si substrates. (A) and (C) from two areas on graphite; (B) and (D) from two areas on graphene on SiO_2/Si [27].

thin films. Fig. 13.18 shows the SEM images of Si films grown on graphite substrate (sample S1) and graphene on SiO2/Si substrate (sample S2) [27].

The SEM images in Fig. 13.18A and C were taken from sample S1 with $T_g/t_g = 800°C/15$ min, showing two areas of different densities of Si particles or islands under different magnifications, roughly 100 to 200 nm in size. Figs. 13.18B and D from sample S2 with $T_g/t_g = 850°C/10$ min, showing one area with very small Si particles in the order of 10 nm and a thin-film like structure of a few μm in size possibly with embedded small Si particles. The thicknesses or heights of these Si structures were measured to be in the range of 1–15 nm by AFM, as shown in the two represerative AFM images, Fig. 13.19A and B. Another sample S3 grown on a graphene/SiO_2/Si substrate was examined by TEM, which indicates that Si nanocrystals, typically a few nm in size, were observed on the surface. Fig. 13.19C is a low magnification image, showing Si layer deposited on the graphene/SiO_2 substrate. Fig. 13.19D is a high resolution image with visible Si lattice planes of a single Si nanocrystal, but the graphene layer in between Si nanocrystals and Si substrate is too thin to be seen.

The crystalline structures of the epitaxial Si nanostructures were characterized by confocal micro Raman spectroscopy using a Horiba LabRam HR800 Raman microscope with a $100\times$ lens (NA = 0.9), excited with a 532-nm Nd-YAG laser. The excitation laser power was controlled at a sufficiently low value (~1 mW) to minimize heating-induced peak shift. Fig. 13.20 shows a few represerative Raman spectra from the Si on graphite sample. Fig. 13.20A is from S1 measured on two areas: one with a Si

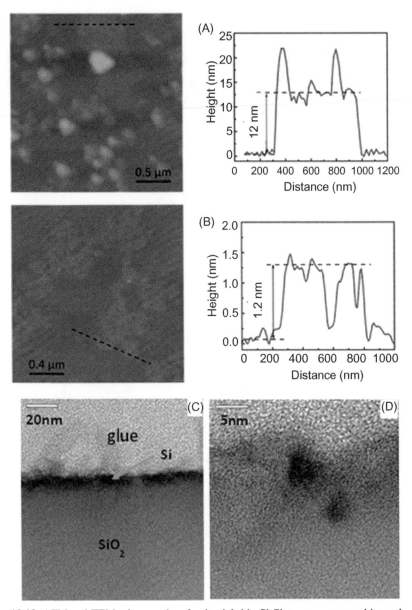

Fig. 13.19 AFM and TEM micrographs of epitaxial thin Si film grown on graphite and graphene: (A and B) AFM images from samples S1 and S2, respectively; (C and D) TEM images from sample S3.

particle and the other a uniform area, compared with a bulk Si. In contrast to the severely distorted Raman spectra reported for Si nanoparticles also grown on graphite [28], here we have oberved single crystalline Si-like Raman spectra for the epitaxial Si structures, with only a small redshift in the peak position and small broadening in linewidth.

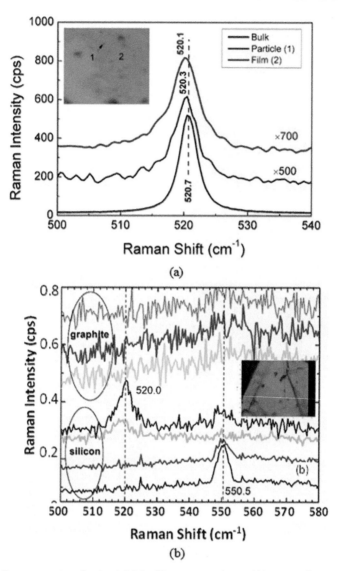

Fig. 13.20 Raman spectra of epitaxial thin silicon on graphene: (A) spectra from two sites on S1, compared with that of bulk Si; (B) spectra from multiple sites of thin Si films, compared to graphite spectra (inset: an optical image of the area).

Interestingly, the shift of the thin-film area is slightly more than the the particle that is somewhat thicker. It is obvious that despite the expected close lattice matching between graphene and Si (111), the in-plane lattice constant of Si is actually a few percent smaller. It has been documented that 2D films like monolayer MoS_2 and WS_2 usually form significant chemical bonding with the substrates on which they are grown [29–31]. Given the predicted weak but significant chemical bonding between graphene and silicene [3,32], we expect that the thin diamond-like Si strcuture could experience some tensile epitaxial strain from the graphite substrate.

The tensile strain could qualitatively explain the redshift in Raman frequency of the thinner layer is larger than that of the thicker particle. Moreover, the expected non-uniform bond lengths along the growth direction, due to the variation of the in-plane lattice constant with the thickness, might also contribute to the small Raman line broadening. In terms of Raman intensity, if we assume that Raman signal is proportional to the sample volume, based on the absorption coefficient of Si ($\alpha \sim 10^4 \mathrm{cm}^{-1}$ at 532 nm), we can offer a rough estimate for the Si film thickness to be 1.4 nm (i.e., 4–5 monolayers thick using the monolayer thickness of buck Si at 3.15 Å), which is consistent with what we measured with AFM from the thin area, such as Fig. 13.19B.

In analogy to that the phonon frequency change between diamond E_{2g} mode ($\sim 1300 \mathrm{cm}^{-1}$) and graphene E_{2g} mode ($\sim 1600 \mathrm{cm}^{-1}$), one would expect that the silicene E_{2g} phonon frequency to be roughly in proportion higher than that of bulk Si at $\sim 520 \mathrm{cm}^{-1}$. Indeed, the theoretically predicted value for free-standing silicene is $562 \mathrm{cm}^{-1}$ [33] or $575 \mathrm{cm}^{-1}$ [34]. Therefore, the spectra shown in Fig. 13.20A are likely of bulk-like Si structures. However, at certain locations that appear to have ultra-thin Si films based on the signal strength, we have instead observed a Raman mode at $550.5 \mathrm{cm}^{-1}$, as shown in Fig. 13.20B with spectra measured from multiple Si sites and graphite sites. On those Si sites, there is an anti-correlation between the 3D Si peaks near $520 \mathrm{cm}^{-1}$ and the new Si-related Raman mode near $550 \mathrm{cm}^{-1}$. We note that these Si-related spectra are distinctly different from those of graphite that do not exhibit any well defined feature in the same spectral range. This $550.5 \mathrm{cm}^{-1}$ mode is much closer to the predicted free-standing silicene mode, and the redshift from the theoretical value could be due to the presence of the tensile strain from the substrate [3,32].

In contrast to the well-known low stability in air for the silicene-like structures reported in the literature, it is unusual that the Raman spectra of these very thin Si samples remain highly stable 2–3 years after the sample were grown. One would expect that the thin Si structures had been mostly oxidized and converted into SiO_2, given the oxidation rate of 11–13 Å in 1 day [35] or about 2 nm in 1 month [36]. Interestingly enough, this amazing anti-oxidation effect of the graphene substrate was convincingly corroborated by scanning tunneling microscopy (STM) measurements done on one of the samples.

Fig. 13.21 shows the electrical characterization and STM images for three distinctly different regions on sample S1: of no Si growth (i.e., exposed graphite), of ultra-thin Si, and of relatively thick Si. These measurements were acquired using an Agilent AFM 5420 atomic force microscope with a STM nose cone and scanner. The tip was prepared by cutting the wire at a 45 degree angle prior to lowering into position. The current scans were performed in constant current mode, and STM images were obtained in constant height mode. The I-V curves were taken by bringing the tip into contact with the sample at different selected locations of interest, where the tip was held at a constant position and a voltage sweep was performed while measuring the current. The surface of the graphite substrate away from the growth region was used as one contact, and the tip was grounded. Fig. 13.21A is the current map of an

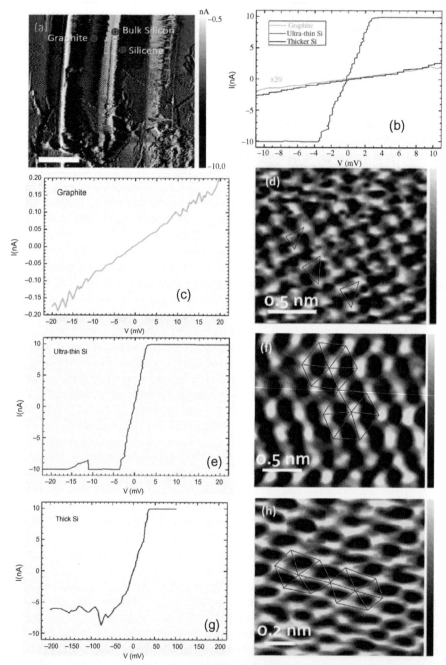

Fig. 13.21 STM images and I-V curves of epitaxial thin silicon on graphite: (A) current map over a large area containing three types of regions; (B) comparison of I-V curves of the three types of regions under low bias voltages; (C and D): I-V curve and STM image of graphite; (E and F): I-V curve and STM image of ultra-thin silicon; (G and H): I-V curve and STM image of thick silicon.

area with Si deposition, showing ribbon-like Si structures. The brownish color area is graphite, the lightest color area is the thicker Si, whereas the dark area in between is the ultra-thin, silicene-like Si, judged by their I-V characteristics and STM images. Note that the strong current contrast revealed in Fig. 13.21A is because of the very large variations in conductivity between the three regions such that, despite an attempt to measure in the constant current mode, the system was not able to maintain a constant current. Fig. 13.21B contrasts the typical I-V characteristics of the three regions under the contact mode. The graphite region is least conductive, than the thicker Si region, and the silicene-like region is most conductive, with a conductivity of up to 500 times that of the graphite region. For instance, at 3.5 mV, the current of the silicene-like region is 370 times that of the graphite region. The high conductivity of the silicene-like region could be due to the charge transfer effect from graphite to silicene [32]. Fig. 13.21C, E, and G plot the I-V curves of the three regions in an extended voltage range, respectively. Both graphite and silicene-like regions show ohmic behavior though with large difference in conductivity, whereas the thicker Si exhibits Schottky junction-type characteristic, consistent with literature reports for either graphite or graphene/bulk Si junctions [37–39]. The conductivity change with increasing film thickness is qualitatively consistent with the expectation that beyond two monolayers, the multi-layer silicene or thin-Si film becomes a semiconductor [40]. Fig. 13.21D, F, and H are the corresponding STM images obtained under ambient condition from the three regions, respectively. They show distinctively different patterns. The pattern of Fig. 13.21D resembles that expected for graphite, a triangular lattice [41], although it is highly distorted, and the bright-spot separation of 2.40 ± 0.43 Å is in good agreement with the lattice constant of graphite at 2.46 Å [3]. Despite the topmost layer of graphite is a graphene layer with a hexagonal structure, the STM image should instead be a triangular lattice, due to the interference of the underneath layer [41,42]. The patterns in Fig. 13.21F and H for the Si areas are more regular. They both are triangular lattices, but the spacings are quite different from each other and from that of the graphite region. In Fig. 13.21F for the silicene-like Si, the pattern is consistent with what is expected for the Si version of graphite [41], the bright-spot separation is 3.53 ± 0.19 Å, somewhat smaller than the silicene lattice constant (about 3.8 Å). The structure revealed by Fig. 13.21H for the thicker silicon region shows a bright-spot spacing of 1.93 ± 0.22 Å, which does not match any of the known reconstructed Si surfaces [43]. Nevertheless, it is a total surprise that one could observe the Si (111) by STM in air after the long air exposure of the sample (grown in December 2011 and measured in August 2013). The exact underlying structures corresponding to these STM images remain to be confirmed through other means, but the differentiations between them confirm that they exhibit distinctively different material properties.

A freshly cleaved Si (111) surface will undergo surface reconstruction if being kept in high vacuum, otherwise will be oxidized into a SiO_2 capping layer. Either case, the surface modification is to remove the dangling bonds or minimize the surface energy. Besides SiO_2, hydrogen atoms are often used to passivate the dangling bonds in Si. These processes apply to a thick bulk Si. When the layer is sufficiently thin and electronically coupled to a substrate, charge transfer across the heterostructure interface may drastically change the picture. If a very thin Si slab remains in its idealistic sp^3 bonding, it will have one dangling bond on the top layer and one on the bottom layer. There are at least two ways to mitigate the dangling bonds: (1) If the slab is only one monolayer thick, partially collapsing the buckled (111) monolayer will allow the upper and lower dangling bonds to form a partial π bond, yielding the so-called silicene that is in-principle structurally stable, although remains chemically unstable (because the weak partial π bond is susceptible to chemical reaction). In contrast, a fully collapsed diamond (111) monolayer forms a much stronger π bond, namely graphene, thus chemically much more stable. (2) Accepting charge from the substrate to passivate the dangling bonds, which has been shown possible theoretically for a silicene/graphene superlattice [32]. Charge transfer-induced passivation has been demonstrated to yield stable inorganic–organic hybrid superlattices with two mono-layer thick II-VI slabs in reality [44]. It requires more precise growth control and structure characterization to achieve and confirm the feasibility of growing a single layer silicene. However, a self-passivated ultra-thin Si film or multi-layer silicene could potentially be more useful for practical applications than monolayer silicene, because it retains the basic properties of the bulk Si, most importantly the bandgap [40], whereas silicene is metallic.

13.6 Conclusions

In contrast to the limitation that fabrication of graphene-based devices involves a series of processes including deposition, transfer, and photolithography, which inherently incur defects or damage to graphene surface structure. The direct growth of graphene on silicon wafer by using a SiC epilayer as an intermediate layer reported hereby opens a route to fabricate graphene-based devices directly on a large silicon wafer. The selective growth of SiC on silicon wafer and subsequent laser illumination can pattern graphene layer in any size and shape on silicon wafer, enabling the integration of graphene-based electronic devices with the existing electronic devices fabricated on silicon wafers. On the other hand, the direct growth of silicon nanocrystals such as silicene on graphene were also investigated and confirmed. This proves the possibility of fabricating silicon nanodevices on graphene. The ability to grow a single-crystalline thin Si film on graphene substrate opens up new avenues for future generation Si electronic devices. It is highly desirable to fabricate Si-based devices on a flexible substrate either directly or through post growth transfer. Before a large and uniform graphene substrate is available, it might be challenging to grow a large and continuous thin Si film. However, it may not be necessary after all if the goal is to make nanoscale Si devices, because a large film is only needed for the traditional top-down approach. Our work suggests the possibility to selectively deposit high quality nanoscale Si structures: silicene, a-few-layer silicene, and Si nanocrystals, using a template of nanoscale graphene structures. For instance, one possible way to obtain such a template could be firstly growing an array of SiC nanostructures on a large Si wafer then converting them into GNSs with a laser beam [22], followed by the growth of Si nanostructures on the GNSs [27]. Moreover, the direct growth of SiC on silicon wafer also enables the fabrication of SiC-based devices on silicon wafer.

Therefore, the explorations discussed above suggest the possible avenues to integrate graphene-, silicon nanostructure-, and/or SiC-based devices on silicon wafers, which offer novel approaches for integrating new materials with silicon wafers for manufacturing high-performance devices.

Acknowledgment

The authors acknowledge Army Research Office (ARO), Charlotte Research Institute, Northrop Grumman Corp, and Bissell Distinguished Professorship from UNCC for financial support.

References

[1] A.K. Geim, K.S. Novoselov, The rise of graphene, Nat. Mater. 6 (2007) 183.

[2] A.K. Geim, I.V. Grigorieva, Van der Waals heterostructures, Nature 499 (2013) 419.

[3] Y. Zhang, R. Tsu, Binding graphene sheets together using silicon-graphene/silicon superlattice, Nanoscale Res. Lett. 5 (2010) 805.

[4] S.J. Haigh, A. Gholinia, R. Jalil, R. Romani, L. Britnell, D.C. Elias, K.S. Novoselov, L.A. Ponomarenko, A.K. Geim, R. Gorbachev, Cross-sectional imaging of individual layers and buried interfaces of graphene-based heterostructures and superlattices, Nat. Mater. 11 (2012) 764–767.

[5] A. Koma, K. Sunouchi, T. Miyajima, Summary abstract: fabrication of ultrathin heterostructures with van der Waals epitaxy, J. Vac. Sci. Technol. B 3 (1985) 724.

[6] T. Kunze, S. Hauttmann, J. Seekamp, J. Muller, Recrystallized and epitaxially thickenedpoly- silicon layers on graphite substrates, in: Conference Record of the Twenty Sixth IEEE Photovoltaic Specialists Conference, 1997, 1997, p. 735.

[7] Y. Zhang, R. Tsu, N. Yue, Growth of Semiconductors on Hetero-Substrates using Graphene as an Interfacial Layer, 2017. US Patent No.: 2017/0194437A1.

[8] F. Ren, B. Liu, Z. Chen, Y. Yin, J. Sun, S. Zhang, B. Jiang, B. Liu, Z. Liu, J. Wang, M. Liang, G. Yuan, J. Yan, T. Wei, X. Yi, J. Wang, Y. Zhang, J. Li, P. Gao, Z. Liu, Van der Waals epitaxy of nearly single- crystalline nitride films on amorphous graphene-glass wafer, Sci. Adv. 7 (2021) 1–7. eabf5011.

[9] P. Vogt, G. Le Lay, Silicene: Prediction, Synthesis, Application, Springer, 2018.

[10] B. Gupta, M. Notarianni, N. Mishra, M. Shafiei, F. Iacopi, N. Motta, Evolution of epitaxial graphene layers on 3C-SiC/Si (111) as a function of annealing temperature in UHV, Carbon 68 (2014) 563.

[11] K.S. Novoselov, A.K. Geim, S.V. Morozov, D. Jiang, Y. Zhang, I.V. Dubonos, I.V. Grigorieva, A.A. Firsov, Electric field effect in atomically thin carbon films, Science 306 (2004) 666.

[12] J. Hass, W.A. de Heer, E.H. Conrad, The growth and morphology of epitaxial multilayer graphene, J. Phys. Condens. Matter 20 (2008), 323202.

[13] K.S. Kim, Y. Zhao, H. Jang, S.Y. Lee, J.M. Kim, K.S. Kim, J.H. Ahn, P. Kim, J.Y. Choi, B. H. Hong, Large-scale pattern growth of graphene films for stretchable transparent electrodes, Nature (London) 457 (2009) 706–710.

[14] Y.S. Li, W.W. Cai, J.H. An, S.Y. Kim, J.H. Nah, D.G. Yang, R. Piner, A. Velamakanni, I. H. Jung, E. Tutuc, S.K. Banerjee, L. Colombo, R.S. Ruoff, Large-area synthesis of high-quality and uniform graphene films on copper foils, Science 324 (2009) 1312–1314.

[15] K.S. Novoselov, A.K. Geim, S.V. Morozov, D. Jiang, M.I. Katsnelson, I.V. Grigorieva, S. V. Dubonos, A.A. Firsov, Two-dimensional gas of massless Dirac fermions in graphene, Nature (London) 438 (2005) 197.

[16] L. Jiao, L. Zhang, X. Wang, G. Diankov, H. Dai, Narrow graphene nanoribbons from carbon nanotubes, Science 458 (2009) 877–880.

[17] D.V. Kosynkin, A.L. Higginbotham, A. Sinitskii, J.R. Lomeda, A. Dimiev, B.K. Price, J. M. Tour, Longitudinal unzipping of carbon nanotubes to form graphene nanoribbons, Nature 458 (2009) 872–876.

[18] Y. Ohkawara, T. Shinada, Y. Fukada, S. Ohshio, H. Saitoh, H. Hiraga, Synthesis of graphite using laser decomposition of SiC, J. Mater. Sci. 30 (2003) 2447–2453.

[19] D. Perrone, G. Maccioni, A. Chiolerio, C. Martinez de Marigorta, M. Naretto, P. Pandolfi, P. Martino, C. Ricciardi, A. Chiodoni, E. Celasco, L. Scaltrito, S. Ferrero, Study on the possibility of graphene growth on 4H-silicon carbide surfaces via laser processing, in: Proceedings of the Fifth International WLT-Conference on Laser Manufacturing, Munich, June, 2009.

[20] S. Lee, M.F. Toney, W. Ko, J.C. Randel, H.J. Jung, K. Munakata, J. Lu, T.H. Geballe, M. R. Beasley, R. Sinclair, H.C. Manoharan, A. Salleo, Laser-synthesized epitaxial graphene, ACS Nano 4 (2010) 7524–7530.

[21] N. Yue, Y. Zhang, R. Tsu, Selective formation of graphene on a Si wafer, in: Mater. Res. Soc. Symp. Proc, vol. 1407, 2012.

[22] N. Yue, Y. Zhang, R. Tsu, Ambient condition laser writing of graphene structures on polycrystalline SiC thin film deposited on Si wafer, Appl. Phys. Lett. 102 (2013), 071912.

[23] S.N. Yannopoulos, A. Siokou, N.K. Nasikas, V. Dracopoulos, F. Ravani, G.N. Papatheodorou, CO2-laser-induced growth of epitaxial graphene on 6H-SiC (0001), Adv. Funct. Mater. 22 (2012) 113–120.

[24] S. Shivaraman, R.A. Barton, X. Yu, J. Alden, L. Herman, M.V.S. Chandrashekhar, J. Park, P.L. McEuen, J.M. Parpia, H.F. Craighead, M.G. Spencer, Free-standing epitaxial graphene, Nano Lett. 9 (9) (2009) 3100–3105.

[25] W.A. de Heer, C. Berger, M. Ruan, M. Sprinkle, X. Li, Y. Hu, B. Zhang, J. Hankinson, E. Conrad, Large area and structured epitaxial graphene produced by confinement controlled sublimation of silicon carbide, Proc. Natl. Acad. Sci. U. S. A. 108 (41) (2011) 16900–16905.

[26] A.C. Ferrari, Raman spectroscopy of graphene and graphite: disorder, electron-phonon coupling, doping and nonadiabatic effects, Solid State Commun. 143 (2007) 47–57.

[27] N. Yue, J. Myers, L. Su, W. Wang, F. Liu, R. Tsu, Y. Zhuang, Y. Zhang, Growth of oxidation-resistive silicene-like thin flakes and Si nanostructures on graphene, J. Semicond. 40 (2019), 062001.

[28] L. Wang, H. Tu, S. Zhu, X. Chen, J. Du, Dispersed Si nanoparticles with narrow photoluminescence peak prepared by laser ablated deposition, Chin. J. Nonferrous Met. 20 (2010) 724.

[29] L. Su, Y. Zhang, Y. Yu, L. Cao, Dependence of coupling of quasi 2-D MoS$_2$ with substrates on substrate types, probed by temperature dependent Raman scattering, Nanoscale 9 (2014) 4920.

[30] L. Su, Y. Yu, L. Cao, Y. Zhang, Effects of substrate type and material-substrate bonding on high-temperature behavior of monolayer WS$_2$, Nano Res. 8 (2015) 2686.

[31] L. Su, Y. Yu, L. Cao, Y. Zhang, In situ monitoring of the thermal-anneal effect in a monolayer of MoS2, Phys. Rev. Appl. 7 (2017), 034009.

[32] S. Yu, X. Li, S. Wu, Y. Wen, S. Zhou, S. Zhu, Novel electronic structures of superlattice composed of graphene and silicene, Mater. Res. Bull. 50 (2014) 268.

[33] J.A. Yan, R. Stein, D.M. Schaefer, Electron-phonon coupling in two-dimensional silicene and germanene, Phys. Rev. B 88 (2013), 121403.

[34] E. Scalise, M. Houssa, G. Pourtois, B. van den Broek, V. Afanas'ev, A. Stesmans, 2D vibrational properties of silicene and germanene, Nano Res. 6 (2013) 19–28.

[35] S. Raider, R. Flitsch, M. Palmer, Oxide growth on etched silicon in air at room temperature, J. Electrochem. Soc. 122 (1975) 413.

[36] J. Ryckman, R. Reed, R. Weller, D. Fleetwood, S. Weiss, Enhanced room temperature oxidation in silicon and porous silicon under 10 keV irradiation, J. Appl. Phys. 108 (2010), 113528.

[37] S. Riazimehr, D. Schneider, C. Yim, S. Kataria, V. Passi, A. Bablich, G. Duesberg, M. Lemme, EUROSOI-ULIS 2015: Joint International EUROSOI Workshop and International Conference on Ultimate Integration on Silicon, 2015, p. 77.

[38] D. Sinha, J. Lee, Ideal graphene/silicon Schottky junction diodes, Nano Lett. 14 (2014) 4660.

[39] S. Tongay, T. Schumann, A. Hebard, Graphite based Schottky diode formed on Si, GaAs, and 4H-SiC substrate, Phys. Rev. Appl. 95 (2009), 222103.

[40] Z. Guo, Y. Zhang, H.J. Xiang, X. Gong, A. Oshiyama, Structural evolution and optoelectronic application of multilayers silicene, Phys. B 92 (2015), 201413.

[41] H. Mizes, S. Park, W.A. Harrison, Multiple-tip interpretation of anomalous scanning-tunneling-microscopy images of layered materials, Phys. Rev. B 36 (1987) 4491 (R).

[42] S. Hembacher, F. Gilessibl, J. Mannhart, C. Quate, Revealing the hidden atom in graphite by low-temperature atomic force microscopy, Proc. Natl. Acad. Sci. U. S. A. 100 (22) (2003) 12539–12542.

[43] H. Neddermeyer, Scanning tunneling microscopy of semiconductor surface, Rep. Prog. Phys. 59 (1996) 701.

[44] Y. Zhang, G. Dalpian, B. Fluegel, S. Wei, A. Mascarenhas, X. Huang, L. Wang, Novel approach to tuning the physical properties of organic-inorganic hybrid semiconductors, Phys. Rev. Lett. 96 (2006), 026405.

Raman spectroscopy and molecular dynamics simulation studies of graphitic nanomaterials

Daniel Casimir[a], Raul Garcia-Sanchez[a], Olasunbo Farinre[a], Lia Phillips[b], and Prabhakar Misra[a]

[a]Laser Spectroscopy Laboratory, Department of Physics & Astronomy, Howard University, Washington, DC, United States, [b]Department of Physics & Astronomy, Appalachian State University, Boone, NC, United States

14.1 Overview

Many new crystalline forms of carbon have been discovered experimentally in the past few decades. These crystalline graphitic materials or allotropes include exotic materials, such as C60 Buckyballs, carbon nanotubes (CNTs), and graphene, to name a few. These newer forms of carbon have significantly different properties compared to the more common forms of C in graphite and diamond [1]. As a rough guide of the dimensions involved, carbon nanotubes (CNTs) are about 10,000 times thinner than a strand of hair, while graphene is approximately 300,000 times thinner than a single sheet of paper. Also, all three of the low-dimensional carbon allotropes mentioned above, Buckyballs, CNTs, and graphene, share a common sp^2 covalently bonded, hexagonal arrangement of C-atoms.

14.1.1 CNT synthesis

The three primary modes of producing CNTs are through Chemical Vapor Deposition (CVD), The Arc Discharge Method (AD) and Laser Ablation (LA). While a multitude of varieties of each of these techniques now exist, which can lead to more subcategories of nanotube manufacturing methods, the above broad categorization serves to highlight the similarities shared by essentially all types of CNT synthesis. The majority of industrially practical methods of CNT production basically involve the thermal decomposition of a carbon containing source material in a sealed chamber that is heated to very high temperatures. Metal catalysts present in the chamber serve as nucleation sites for nanotube growth, along with empirically determined pressure and temperature conditions that result in the desired nanotube type and diameter distribution.

Modeling, Characterization and Production of Nanomaterials. https://doi.org/10.1016/B978-0-12-819905-3.00014-2

(a) Chemical Vapor Deposition (CVD): In CVD CNT production, CH_4, CO, or some other carbon containing gas, is decomposed by metal catalysts in a high temperature chamber at pressures at or below 1 atm. The carbon atoms dissociate from the gaseous material that is flowing into the chamber and nucleate on the metal catalysts, and subsequently selfassemble into cylindrical CNTs. The resultant CNT material is then removed from the furnace after being allowed to cool, in order to prevent damage to the tubes due to oxidation.

(b) Arc Discharge (AD): In this technique, a graphite electrode with a gap of a few mm is used to produce carbon plasma. Again, the process takes place in a sealed chamber that is cooled at or below 1 atm pressure, where the vapor then condenses on metal catalyst particles as the plasma travels towards the cathode. This technique is usually applied in the production of C60, whereby various graphitic materials may be present in the resultant material, necessitating post-production purification and separation of CNTs.

(c) Laser Ablation (LA): This method refers to the use of a laser in the vaporization of graphite contained in a highly heated chamber. The graphitic target material is usually mixed with catalyst metals, resulting in nanotube formation, as the vapors of the ablated material condense on the cooled surface of the chamber.

In all the above-mentioned nanotube production techniques, the exact growth mechanism of these tubular nanostructures is still unclear. Therefore, at present, production of specific chirality or diameter nanotubes is extremely challenging, leaving one to rely on the use of optimized arrangements and densities of catalyst particles to produce statistically correlated distributions of nanotube diameters.

14.1.2 Graphene and graphene nanoplatelets

Graphene is an allotrope of carbon, which consists of covalently bonded atoms arranged in a planar hexagonal structure depicted in Fig. 14.1. This 2-dimensional material is considered as the conceptual base material for all other carbon allotropes, where single planes can be rolled into spheres, cylinders, or stacked one atop the other to form 0-D Buckeyballs, 1-D CNTs, or graphite, respectively. Graphene was experimentally isolated and identified in 2004 and has arguably rapidly become the most

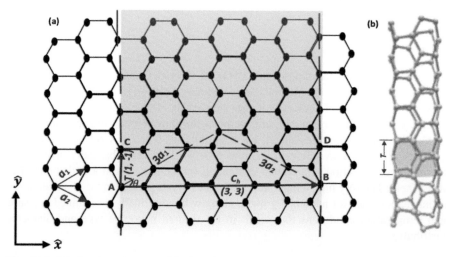

Fig. 14.1 (A) Graphene hexagonal lattice; (B) Resulting single-walled carbon nanotube.

researched topic in materials science, as evidenced by the Nobel Prize recognition awarded to its discoverers—Andre Geim and Konstantin Novoselov—six years later in 2010. This one-atom thick material has provided avenues in the study of low-dimensional physics, where its unique band structure has been and is still being used as a testing ground of quantum relativistic effects in materials science in addition to other "basic Physics" [2].

In carbon, a group IV element, the four valence electrons in the 2s and 2p atomic orbitals of neighboring atoms mix with each other, or "hybridize" into sp^2 orbitals, resulting in very strong s-bonded electrons confined along the planar graphene surface and the $2p_z$ electrons that form weaker p bonds through electron clouds that are distributed perpendicular to the graphene plane. A few of the remarkable properties of graphene follow. Owing to its 2-D structure, the behavior of charge carriers in graphene is governed more closely by the relativistic 3-D Dirac equation, instead of the Schrodinger equation. Hence, π electrons in graphene mimic relativistic massless Dirac fermions, with speeds just two orders of magnitude less than the speed of light. Two other interesting effects that have been demonstrated are the appearance of new Quantum Hall effects and the anomalous minimum of the conductivity of graphene in zero-field. In contrast to its remarkable properties, some of which were mentioned above, the first successful technique used in the isolation of single graphene layers is the simple micro-mechanical cleavage of graphite layers through the repeated use of scotch tape.

A pair of research studies has reported the mass fabrication of defect-free graphene using two notable processes: chemical vapor deposition on copper [3] and liquid phase exfoliation of graphite [4]. However, some factors such as high sales costs and low fabrication rates hinder the mass production of high-quality graphene [5]. Such limitations make graphene nanoplatelets (GnPs) more attractive for industrial applications because they can be produced commercially and defect-free at low cost as compared to graphene and carbon nanotubes. Commercially available GnPs consist of monolayer graphene, few layer graphene (FLG) and nanostructured graphite as shown in Fig. 14.2, and they typically range in thickness: 0.34–100 nm [6]. GnPs are widely used in technological applications due to their light weight, high mechanical rigidity, high electrical conductivity and planar structure [5]. Numerous research studies have shown that GnPs can be covalently and non-covalently functionalized at the edges and added to polymers to form nanocomposites with the goal of improving the mechanical and thermal properties of the nanocomposites [7,8]. The functionalization of GnPs removes their inert nature and improves chemical bonding with bulk materials and dispersion in a variety of solvents and polymers. The geometry of graphite with functional groups shown in Fig. 14.3 was built using the Avogadro visualization software [9].

14.2 Literature review on CNTs

Before specifically discussing thermal expansion solely, this section presents a more general overview of the study of the CNTs' mechanical properties, including a range of properties, such as elastic constants, Poisson's ratios, and Young's Moduli of CNTs

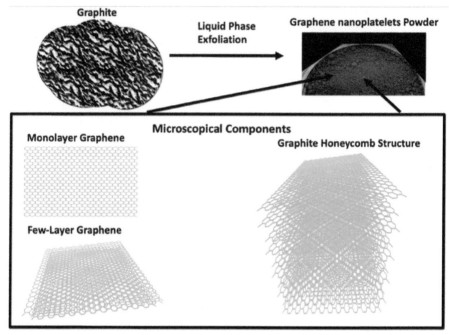

Fig. 14.2 Model showing the manufacture of GnPs from natural graphite through liquid phase exfoliation.

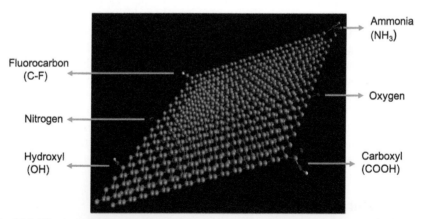

Fig. 14.3 Diagram showing functional groups (ammonia, fluorocarbon, nitrogen, hydroxyl, oxygen, and carboxyl) attached to the edges of two layers of graphene. The interlayer distance (d) is 3.35Å. Carbon atom is *gray*, oxygen atom is *red*, hydrogen atom is *silver*, nitrogen atom is *navy blue* and fluorine atom is *light blue*.

of the single-walled and multi-walled variety, and also bundles or "ropes" of these nano-objects. The dependence of these structural properties on chirality and diameter were discussed in some of the earliest such studies, which relied heavily on continuum elasticity theory. Ruoff and Lorents [10] early on realized that "The thermal conductivity and thermal expansion of CNTs are also fundamentally interesting and technologically important properties." They were also one of the few initial investigators to exploit the structural similarities between graphite/graphene in their studies. Regarding the tensile strength of nanotubes, for example, Ruoff and Lorents [10] obtained the stiffness constant of CNTs in the axial dimension, as shown in Eq. (14.1), by scaling the intra-planar Young's modulus of graphite with an estimate of the cross-sectional area of a CNT; earlier estimates of the latter dimension of the inter-planar spacing in graphitic layers was approximately 0.34 nm. The use of this estimate as the "thickness" of a nanotube has been, and still proves problematic, especially with regards to the wide scatter in the reported values of the structural properties mentioned previously.

$$K = \frac{E_0\left(r_o^2 - r_i^2\right)}{r_o^2} \tag{14.1}$$

The tensile stiffness value computed by Ruoff and Lorents [10] was ~800 GPa, nearly three-quarters of the ideal in plane graphite value of 1060 GPa. They also showed that the behavior of the thermal expansion of CNTs differ significantly from that of graphite and carbon fibers, noting that the radial expansion coefficient for not only single-walled CNTs, but also multi-walled types, would be similar to the in-plane axial values. Their topologically based reasoning was that although multi-walled tubes have multiple tubes with separations and Van der Waals interactions similar to graphite, the tubes making up the MWCNT wrap around onto themselves, in which case the Van der Waals forces between individual tubes does not play a critical role in radial thermal expansion.

Tersoff and Ruoff [11] examined ordered hexagonal arrays of CNTs modeled as cylinders in their calculations, resulting again in some of the earlier predictions of the unique unexpected mechanical properties of CNTs. Their early modeling for example predicted the energetically favorable flattening of the sides of nanotubes against each other in nanotube bundles for tubes with diameters of 0.25 nm or greater due to Van der Waals interactions. Tubes with 0.1 nm diameter or smaller were shown to behave essentially as rigid cylinders. The authors in Tersoff and Ruoff [11] also note the unexpected strange behavior of the rigidity of nanotube bundles. The expectation is for rigidity to decrease as the diameter of the tubes in the bundle increase, since materials soften with lower density, and density in the case of nanotube bundles varies inversely with diameter. Tersoff and Ruoff's study, however, predicted nanotube bundle material initially stiffens, then softens, and eventually settles at a compressibility value that does not depend on any additional decrease of the density.

Robertson et al. [12] used empirical potentials in their study of the mechanical properties of nanotubes, where they established the inverse squared relationship between the strain energy and nanotube radius. The same relationship between radius and strain energy is also derivable from continuum mechanics, thus showing the applicability of continuum elasticity theory to the study of CNTs early on.

In Robertson et al.'s, study interactions were modeled with the many-body potentials developed by Tersoff and Ruoff [11] and Brenner et al. [13], which were parameterized by the lattice constant, binding energy, and elastic constants of various carbon lattices. Robertson et al. [12] also verified their findings with ab initio calculations.

Sankar and Kumar [14] studied the mechanical, transport, and electronic properties of metallic (Armchair (4,4)) and semi-conducting (Zigzag (4,0)) SWCNTs. The Young's modulus for the SWCNTs was determined using Density Functional Theory (DFT) within the local density approximation (LDA) and the non-equilibrium green function (NEGF) approach. Strain was applied axially to the constructed CNT segment of Armchair and Zigzag, both consisting of a total number of 48 atoms, in order to calculate the Young's modulus using the ab initio atomistic approach. Two different methods were employed in determining the Young's modulus of the SWCNTs within the first principles calculations: a graph of stress (Force/Area) as a function of applied strain (%) was plotted, the equation is shown in Eq. (14.2) and a graph of change in energy of the system with increasing strain was plotted, the equation is shown in Eq. (14.3).

$$Y = \frac{Force/Area}{\Delta L/L_0} \tag{14.2}$$

where $\frac{Force}{\Delta L/L_0}$ is the slope of the graph and the area is πR^2

$$Y = \frac{1}{L_0} \left(\frac{d^2E}{d\varepsilon^2} \right) \tag{14.3}$$

where $\left(\frac{d^2E}{d\varepsilon^2} \right)$ is the slope of the graph and L_0 is the initial length of the SWCNTs.

Sankar and Kumar [14] found the Young's modulus for metallic SWCNTs (4,4) to be 2.73 TPa and 4.17 TPa using the Force-Strain and Energy-Strain methods, respectively, while Young's modulus for the semi-conducting SWCNTs (4,0) was found to be 1.8 TPa and 2.15 TPa using the Force-Strain and Energy-Strain methods, respectively.

Rao et al. [15] investigated the mechanical properties of Zigzag, Armchair and Chiral SWCNTs, using molecular dynamics-based finite elemental analysis and considering the effect of the C—C Van der Waals bond (non-bonded atomic interactions) on the mechanical properties of the SWCNTs. The potential functions used in describing the interaction between the carbon atoms were the Lennard-Jones 6–12 interaction potential function and the Morse potential function. It was reported that the inclusion of the C—C Van der Waals bond (non-bonded interactions) significantly reduced the mechanical properties (tensile strength and Young's modulus) of the SWCNTs and thereby making them important for further study. The average values of the longitudinal tensile strength of Armchair, Zigzag and Chiral SWCNTs were reported to be 119.73 GPa, 102.27 GPa and 116.09 GPa, respectively, while the average values of the longitudinal Young's modulus of Armchair, Zigzag and Chiral SWCNTs were reported to be 1.26 TPa, 1.26 TPa, and 1.18 TPa, respectively. Rao et al. [15] reported that the maximum tensile strength of Armchair SWCNTs was greater than that of the Zigzag and Chiral SWCNTs and increased with the increase in diameter of the SWCNT.

Experimental attempts at measuring the Young's modulus of CNTs also occurred concurrently with the above modeling/atomistic studies. One experimental method adopted an initial nanotube investigation with the use of Transmission Electron Microscopy (TEM) imaging of the thermally induced vibrations of the free end of CNTs in bulk roped samples. Krishnan et al. [16] employed this technique to measure the Young's modulus of 27 single-walled CNTs, reporting an average value of 1.25 TPa. Krishnan's study was one of the early ones using this technique to obtain a CNT Young's modulus close to the in-plane C11 elastic constant of graphite, but more importantly it added more support to the initial claims of the remarkable mechanical properties of SWCNTs. Krishnan performed a least-squares analysis on the TEM images of the nanotubes in order to determine the various dimensions of the sample tubes, which also involved analyzing the "blur" of free vibrating end of the tubes to determine the vibrational amplitude. The Young's moduli were then calculated by basing the behavior of the nanotubes on that of homogeneous cylindrical shells of a suitably chosen thickness.

X-Ray diffraction of bundled SWCNT samples was also another experimental method in the determination of nanotube mechanical properties [17,18], after applying XRD to CNTs produced by the direct current arc discharge method, the authors reported that the triangular lattice constant of SWCNT bundles L0, shrank over an entire temperature range of 290–1600 K. Values of the lattice constant were obtained from the (10) reflection Bragg angle at 2θ approximately equal to $6°$. A coefficient of linear thermal expansion perpendicular to the tube bundle axis over the temperature ranges of 290–330 K and 1300–1550 K were determined to be (11.52×10^{-6}) and $(2.48 \times 10{-6})$ K^{-1}, respectively.

Atomic Force Microscopy (AFM), which is probably one of the more costly methods, is also used in investigating nanotube mechanical properties [19]. Salvetat et al. [20] used an AFM probe to apply lateral forces on bundles of single wall nanotubes or isolated multi-walled tubes whose sections were laying across 0.4 µm diameter holes on a porous membrane. Based on the nanotube deflection, Salvetat et al. [20] concluded with Young's moduli values of 0.87 TPa and 0.755 TPa, respectively, for unpurified nanotube samples and samples that underwent thermal treatment. Yang and Li [21] also used the AFM (tapping and contact mode) to investigate the mechanical properties of SWCNTs on a quartz substrate. The effective radial modulus (E_{radial}) of the SWCNT was calculated using equation Eq. (14.4).

$$E_{radial} = \frac{F/A}{\delta/D} \tag{14.4}$$

where F is the force acting normal on the SWCNT, A is the area of the normal force, δ is the radial displacement of the SWCNT, and D is the initial diameter of the SWCNT. It was reported that the effective radial modulus decreases from 57 GPa to 9 GPa with the increase in diameter of the SWCNT from 0.92 nm to 1.91 nm following the relationship E_{radial} $R_{SWCNT}^{-\alpha}$.

Bozovic et al. [22] investigated the effect of applying mechanical strain on the electronic properties of SWCNTs on a SiO_2 substrate using Scanned Gate Microscopy (SGM). Mechanical strain of about 2%–30% was applied by using AFM (tapping mode) to manipulate and elongate the nanotubes. The average strain was measured by comparing the lengths of the SWCNT segments before and after applying strain. It was reported that when the applied strain exceeded a value of about 5%–10%, a drastic increase in electron backscattering was observed, which induces an elongation in the nanotubes of about 7% over a length of 520 nm.

In addition, Salvetat et al.'s study also demonstrated the ease with which nanotubes can slide pass one another, revealing a significantly low shear modulus for bundles of SWCNTs, which was not surprising to the authors considering how easily graphite is cleaved, using scotch tape for example in the mechanical cleavage technique used in the successful isolation of graphene.

While being stretched at high temperatures, $\sim 10^3 K$, nanotubes were also discovered to display "a superplastic behavior." Both single and multi-walled nanotubes, while being stretched axially at extreme temperatures using two AFM tips connected at both ends, suffered fractional increases in length of almost 300%, while concurrently undergoing a more than 10-fold decrease in their diameter. This extremely elastic response of nanotubes at high temperatures was in stark contrast to the brittle behavior they displayed in the same type of experiments at room temperature, with multi-walled and single wall nanotube bundles fracturing at 3%–12% and 3.1%–5.3% strains, respectively. Some possible explanations of this "super plastic" behavior have been proposed over time, with one example being the 5/7/7/5 defect mechanism [23]. It refers to the formation of defects consisting of two pentagons and two heptagons in the hexagonal CNT lattice through induced Stone-Wales rotations of certain bonds. This defect formation is an endothermic process when no stress is applied to the CNT and becomes exothermic after a specific amount of applied stress. The resulting 5/7/7/5 defect then traverses the length of the nanotube as two pairs of pentagon and heptagon, appearing as a distorted kink in the nanotube. The overview in this section shows some of the exciting mechanical properties of CNTs, together with the promise of significant advances that are yet to be made in understanding the behavior of these objects under mechanical stress at high temperatures and pressures.

14.3   Literature review on graphene and graphene nanoplatelets (GnPs)

This section gives a brief overview of earlier research studies that have investigated the thermal, structural and vibrational properties of graphene and GnPs using various experimental tools and Molecular Dynamics (MD) simulations. Yoon et al. [24] reported the temperature dependence of the G-band frequencies of monolayer or single-layer graphene (SLG), bilayer graphene (BLG), and graphite on SiO_2 in the temperature range 4.2–475 K using Raman spectroscopy. The thermal expansion coefficient (TEC) was found to be $-(8.0 \pm 0.7) \times 10^{-6} \, K^{-1}$ and strongly dependent on temperature. It was observed that in both SLG and BLG, there is a blueshift in the

G-band frequency when the temperature is cooled down from room temperature and a redshift in the G-band frequency when the temperature rises from room temperature to 400 K. However, at temperatures greater than 400 K in SLG and BLG, Yoon et al. [24] reported blueshifts in the G-band frequency. They explained that the redshift which occurred at temperatures 200–400 K is due to a mismatch between the TEC's of the substrate (SiO_2) and graphene which leads to a strain on graphene. Also, at temperatures greater than 400 K, it was explained that the blueshift of the G-band frequency is as a result of slips which occurred in SLG and MLG on the surface of the substrate. In the case of graphite on SiO_2, the G-band frequency shits to lower frequencies as the temperature rises in the range 4.2–475 K because no significant amount of strain was observed in the experiment to cause the graphite sample to slip.

Jarosinski et al. [25] reported the thermal conductivity of GnPs epoxy resin composites as a function of temperature in the range 2–300 K. They also studied the effect of adding different concentrations of GnPs at 0.1 wt%, 0.5 wt%, 1.0 wt%, 2.0 wt%, and 4.0 wt% in the composite preparation on the thermal conductivity, with the goal of maintaining the electrical insulating properties of the nanocomposites. The electrical insulating properties of the nanocomposites were achieved by using a smaller GnPs particle size of 4.5 μm and a concentration ≤4.0 wt%. The electrical resistivity of the nanocomposites was measured at room temperature using a four-point probe and an in-sandwich configuration with a Faraday cage, while the thermal conductivity was measured using the Hot Disc Thermal Constant Analyzer (TPS 500) technique. Results from Jarosinski et al. [25] showed that the thermal conductivity of the nanocomposites increases as the concentration of GnPs (wt%) added to epoxy resin increases. The nanocomposite with 4.0 wt% concentration of GnPs was shown to yield 132% increase in thermal conductivity. Also, they reported that at low temperatures, the thermal conductivity shows a T^2 dependence, as a result of the quadratic dispersion of out-of-plane phonon modes [26]; while at high temperatures (over 20 K) the thermal conductivity shows a linear dependence as a result of the Umklapp process, which occurs in phonon-phonon scattering [27].

Dai et al. [7] investigated the effect of covalently functionalized GnPs (3-aminopropyltriethoxysilane (KH550)) and non-covalently functionalized GnPs (Triton X-100) on the thermal and mechanical properties of boron phenolic resin (BPR). The GnPs were functionalized to ensure the homogenous dispersion of GnPs in the polymer matrix. Fourier Transform Infrared (FTIR) spectroscopy, Raman spectroscopy and Scanning Electron Microscopy (SEM) were used to confirm the successful functionalization of the GnPs. The thermal properties of the composites were investigated using Thermogravimetric Analysis (TGA) under a nitrogen atmosphere, while the mechanical properties were investigated using flexural tests. Their results showed that the addition of functionalized GnPs to BPR increased the flexural strength due to a strong adhesion between BPR and the surface of the functionalized GnPs. Also, it was reported that the addition of the functionalized GnPs to BPR showed higher thermal conductivity (0.28 W/mK for KH550-GnPs and 0.27 W/mK for Triton X-100-GnPs) compared to the addition of pristine GnPs to BPR (0.25 W/mK) due to a stronger interfacial interaction between the functionalized GnPs and BPR.

Molecular Dynamics (MD) simulations have been attempted by numerous researchers for investigating the thermal properties of graphene and GnPs. Koukaras et al. [28] reported the temperature dependence (in the range 60–1500 K) of the Raman active phonon mode (Γ-E_{2g}), corresponding to the G-band in the Raman spectra, using different interatomic potentials: Reparameterized and Optimized Tersoff [29], AIREBO [30], LCBOP [31], incorporated within the MD simulation. MD simulations were performed using the Large-scale Atomic/Molecular Massively Parallel Simulator (LAMMPS) software package [32]. Koukaras et al. [28] reported that the results from temperature dependence of the Γ-E_{2g} vibrational mode of graphene using the Optimized Tersoff potential [29] shows a linear temperature dependence with a small numeric discrepancy in the temperature coefficient ($\chi_G = -0.05\,\text{cm}^{-1}/\text{K}$) when compared with the experimental data. The AIREBO [30] and LCBOP [31] potentials failed at reproducing correctly the temperature dependence of the Γ-E_{2g} vibrational mode of graphene reported by experimental observations.

Zou et al. [33] also investigated the vibrational and thermal properties of graphene using MD simulations. They calculated the phonon dispersion curves using different interatomic potentials: Tersoff [34,35], Reparametrized and Optimized Tersoff [29], AIREBO [30], and REBO [13]). Results from Zou et al. [33] showed the G peak of the graphene Projected Density Of States (PDOS) using the Reparametrized and Optimized Tersoff potential [29] to be at a frequency of 46 THz at room temperature. This value was said to have agreed well with experimental results, unlike the Tersoff [34,35], AIREBO [30] and REBO [13] potentials that produce G peak values that do not agree well with experimental data. Also, Zou et al. [33] calculated the thermal conductivity of graphene using the Green-Kubo method. The calculated values of the thermal conductivity of graphene using the Reparametrized and Optimized Tersoff [29], Tersoff [34,35] and REBO [13] potentials were reported to be 1192 W/mK, 560 W/mK, and 275 W/mK, respectively. Zou et al. [33] therefore concluded that the above potentials failed to accurately model the thermal conductivity of graphene because the calculated values did not agree well with experimental data.

14.4   Methodology

In this section, we will discuss the characterization tools employed to study the morphological, vibrational, structural, and thermal properties of purified single-walled CNTs and GnPs. In addition, this section covers the study of the thermal properties of purified single-walled CNTs using MD simulations within the LAMMPS software.

14.4.1   Raman spectroscopy of SWCNTs

Raman spectral data were collected on the CNT samples under thermal loading, from room temperature up to 200°C, in steps of 10°C. The sample analyzed was CNT powder from Unidym Carbon Nanotubes. The SWCNTs were produced by the High-Pressure Carbon Monoxide Process (HiPco), followed by purification. According to the manufacturer, the diameter range of the nanotubes was from 0.8 to 1.2 nm,

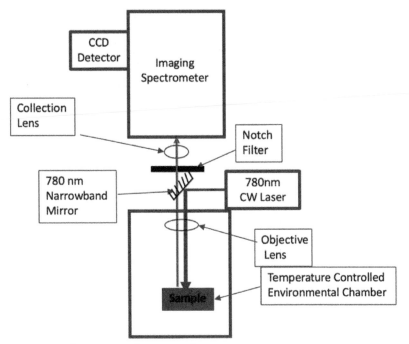

Fig. 14.4 Raman Spectroscopy Instrument Setup.

and the lengths varied in the range $\sim$100–1000 nm. After thermo gravimetric analysis, the manufacturers also claimed to have less than 15% residual Fe catalyst by weight [36].

The instrument used to carry out the Raman spectroscopy of our samples is a Thermo-Scientific DXR Smart Raman Spectrometer, with a 780 nm wavelength laser and $M^2 \leq 1.5$ beam quality. A laser power setting of 6 mW was used for all spectra collections.

Fig. 14.4 shows the internal setup and schematic of the instrument. The laser beam is focused on the sample via lenses and the scattering that results from the laser striking the sample is sent back into the CCD detector and a Raman signature is obtained. The spectrometer resolution using a 25 μm slit is $\sim$1.93 cm^{-1}. The temperature controlled environmental chamber used the Ventacon Heated Cell, which houses the sample on its surface (for solid samples such as plates) or within it (as is the case for nanopowder and CNTs) and can heat the sample from room temperature up to 200°C.

The experimental method carried out the collection of spectra at a constant laser power, with the sample heated externally with a heat cell up to 200°C. All data were collected only for the purified batch of SWCNT, with much more data collection needed. Initial analysis of the peak-shift variation of the various Raman features of the externally heated SWCNT are shown below in Figs. 14.16–14.20, and the tabulated data in Table 14.1 agree well with the results reported by Raravikar [37] and Gregan [38].

Table 14.1 Temperature slopes of the various Raman peaks of bundled single-walled carbon nanotubes, measured and compared to two other references.

Temperature slope	Results (Ours) $\frac{\Delta\omega}{\Delta T}$ (cm^{-1}/°C) 780 nm excitation Wavelength	Reference [37] $\frac{\Delta\omega}{\Delta T}$ (cm^{-1}/K) 514 nm excitation Wavelength	Reference [38] $\frac{\Delta\omega}{\Delta T}$ (cm^{-1}/K) 514.5 nm excitation Wavelength
G$^+$ Peak	−0.0187	−0.0189	−0.016
G$^-$ Peak	−0.0204	−0.0238	−0.016
G$'$ Peak	−0.0264	–	–
"Defect" D Peak	−0.0182	–	–
RBM $\sim$267 cm^{-1}	–	–	

SEM analysis was also utilized in order to identify candidate nanotube models to be used in the MD simulations. The temperature-dependent shift of the peak locations (with corrected and uncorrected intensities), peak widths, and other spectral variations obtained from the data served as the means for experimental validation of the atomistic/MD simulation results.

14.4.2 MD simulations of SWCNTs

MD simulation is one of the primary atomistic/simulation techniques used in modeling the thermal/structural properties of SWCNTs. In this classical technique, the trajectory of a model of the nano-system is generated according to Newton's equations of motion, and equilibrium and transport properties are subsequently computed from averages over the time-evolution of the model system [39]. In MD simulations of any complexity, the algorithm of the simulations is usually some variation of a series of sequential steps [40].

The energetics of the model nano-system in MD simulations is handled by empirical inter-atomic potentials. Also dependent on the inter-atomic potentials are the calculated structural/thermal properties of interest in this chapter, such as elastic constants, thermal expansion, etc. The form and parameters of these potentials are derived from many sources, such as quantum-mechanical calculations and experimental data, and with the empirical potential undergoing continual refinement. The potential used in the present study is the Adaptive Intermolecular Reactive Empirical Bond Order Potential (AIREBO), developed by Stuart et al. [30] specifically for modeling chemical reactions and interactions of hydrocarbons. It is one of the popular variants of many-body potentials that employ Tersoff's original bond-order concept [41], which refers to the modification of the strength of bonded interactions based on the local environment of the bonded interaction being calculated.

The Raman spectral analysis will help determine which type of model nanotubes to use in the MD simulations, where expansion coefficients and other elastic constants for individual representative nanotube models will be calculated based on the AIREBO interatomic potential.

The REBO term of the potential refers to the Reactive Empirical Bond Order potential formulated by Brenner et al. [13]. This part of the AIREBO potential is switched on for short-ranged (less than ~2 Angstrom separation) interactions between the C—C, H—H, and C—H species. It is responsible for covalently bonded interactions, such as those between C-atoms in graphene, nanotubes, etc. A variety of chemical influences on the bond strength between two neighbor atoms, such as coordination numbers, conjugation effects, bond angles, etc. are accounted for in the term b_{ij} in Eq. (14.5).

$$E^{REBO} = V_{ij}^R\left(r_{ij}\right) + b_{ij}V_{ij}^A\left(r_{ij}\right) \tag{14.5}$$

$$E = \frac{1}{2}\sum_i \sum_{i \neq j}\left[E_{ij}^{REBO} + E_{ij}^{LJ} + \sum_{k \neq i}\sum_{l \neq i,j,k} E_{kijl}^{TORSION}\right] \tag{14.6}$$

The Lennard-Jones (L-J) contribution to the AIREBO potential, denoted as E^{LJ} in Eq. [14.6], considers Van der Waals and dispersion interactions based on the 6–12 potential. The third contribution to the AIREBO potential, denoted as $E_{kijl}^{Torsion}$, calculates the energy of dihedral angles present in the modeled system. This contribution was an added improvement to the AIREBO potential's predecessor, the REBO potential, which did not consider the energetics of 4-body torsional interactions about individual bonds [30]. The rule governing the switching on and off of the long-range Lennard-Jones interactions between two atoms are what give this potential the "reactive" nature embodied in its name. The decision to include L-J interactions depends on the non-bonded separation between two atoms, the strength of the bonds between bonded atoms, and the network of intermediate bonds between two interacting atoms. For example, atoms commonly experience L-J repulsion at close separations only if they are not 1–4 neighbor atoms between two consecutive bonds, or if there is a negligible chance of the two atoms bonding themselves.

These rules governing the rules for switching on the long range L-J interactions makes the AIREBO potential better suited to take into account the chemical nature of the modeled system, and are handled by the universal switching function shown in Eq. (14.7), which varies continuously between 0 and 1, where $\Theta(t)$ is a Heaviside step function.

$$\Theta(t) = \begin{cases} 0, t < 0 \\ 1, t \geq 0 \end{cases} \tag{14.7}$$

$$S(t) = \theta(-t) + \theta(t)\theta(1 - t)\left[1 - t^2(3 - 2t)\right] \tag{14.8}$$

Table 14.2 Average intensity ratios (I_D/I_G) of the D and G bands with experimental uncertainties for pristine and functionalized graphene nanoplatelets.

Samples	Band D (cm^{-1})	Band G (cm^{-1})	Band 2D (cm^{-1})	I_D/I_G
Pristine-GnPs	1353.90 ± 3.30	1581.30 ± 1.10	2722.10 ± 2.50	0.25 ± 0.06
GnPs-Nitrogen	1354.80 ± 1.00	1580.60 ± 0.90	2719.60 ± 1.00	0.36 ± 0.03
GnPs-7 wt% Carboxyl	1353.90 ± 2.00	1581.40 ± 1.70	2721.10 ± 3.70	0.33 ± 0.03
GnPs-35 wt% Carboxyl	1349.53 ± 5.41	1581.39 ± 1.75	2707.50 ± 10.75	0.93 ± 0.04
GnPs-Ammonia	1355.20 ± 2.20	1582.10 ± 1.40	2721.80 ± 2.30	0.36 ± 0.02
GnPs-Fluorocarbon	1356.00 ± 1.40	1582.30 ± 0.70	2722.70 ± 3.30	0.29 ± 0.02
GnPs-Argon	1356.30 ± 2.20	1582.80 ± 0.20	2723.20 ± 1.40	0.34 ± 0.01
GnPs-Oxygen	1357.20 ± 0.60	1583.20 ± 0.30	2723.80 ± 1.70	0.32 ± 0.01

Some initial data on an MD simulation of a (10,2) chirality SWCNT, the dominant chiral species present in the purified nanotube sample, was based on the Raman spectra presented earlier. The vibrations of a 200 Å long SWCNT is based on the AIREBO potential discussed previously. A non-bonded Lennard-Jones cut-off of 3 Å was used in addition to the other default parameterized constants in Table 14.2 [30]. Non-periodic boundaries were used in all 3 dimensions, of the simulation box with x, y, and z dimensions of $100.0 \times 100.0 \times 240.0$ Å. A 0.5 fs time-step, and 100,000 total time-steps, were used. Also, simple velocity rescaling was used to achieve a mean temperature of 300.0 K over the course of the run. The sampling geometry used to describe the length variation of the nanotube was based on the summation of the center of mass between consecutive cylindrical disk sections of the model nanotube as shown in Fig. 14.5. The behavior of the axial dimension in this case of an initial room temperature MD simulation exhibits the anomalous thermal expansion behavior also observed in other MD studies.

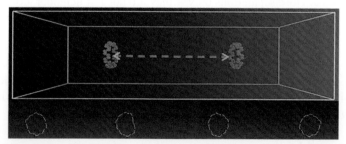

Fig. 14.5 The axial sampling length L along the undeformed axis of the armchair (10, 10) SWCNT used at constant temperature molecular dynamics (MD) simulation (*above*). The time evolution of the radial cross-section of the (10, 10) SWCNT during the constant temperature (NVT) MD simulation.

14.4.3 Characterization of GnPs

This section discusses different experimental tools which were used to study the various properties of pristine and functionalized GnPs. The pristine GnPs used for this research were supplied by US Research Nanomaterials, Inc., with a planar size of 4–12 µm, planar thickness of 2–8 nm, and each nanoplatelet is composed of 3–6 layers of graphene sheets. The functionalized GnPs were supplied by Graphene Supermarket, with a planar size of 0.3–5 µm and planar thickness < 50 nm. The carboxyl (COOH) functionalized GnPs with 35 wt% functionality was supplied by Cheap Tubes, Inc., with a planar size of 1–2 µm, planar thickness of 3–10 nm, and each nanoplatelet is composed of approximately 4 layers. The Renishaw inVia Raman microscope with a 514 nm wavelength laser, laser beam quality of 0.65 mm and laser maximum output power of 50 mW, was used to carry out the Raman spectroscopy of the samples. The laser was focused on the GnP samples through a ×20 microscope objective lens. The laser exposure time was set to 10 s, with 10 accumulations to enhance the signal-to-noise ratio, and a laser power of 100%, were used to collect Raman spectra of the GnP samples.

Fig. 14.6 shows the internal schematic for the Renishaw inVia Raman microscope. It consists of a microscope which is used to direct light on the sample and collect the scattered light interacting with the sample. The scattered light then passes through the Rayleigh filter, which filters all except a tiny fraction of the Raman scattered light, and the collimated Raman light passes through a diffraction grating, which in turn splits

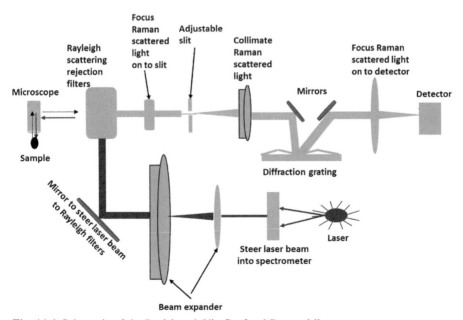

Fig. 14.6 Schematic of the Renishaw inVia Confocal Raman Microscope.

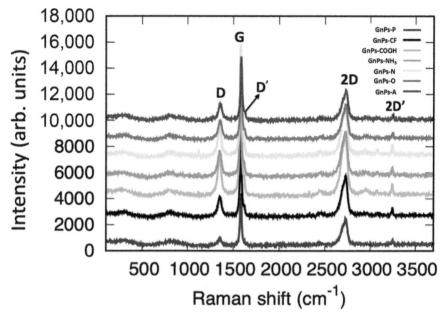

Fig. 14.7 Raman spectra of pristine and functionalized graphene nanoplatelets.

the Raman scattered light into its component wavelengths [42,43]. The setup also consists of a CCD camera for the final detection of the Raman spectrum, which is subsequently displayed on a monitor.

The Raman spectra of pristine and functionalized GnPs with ammonia, argon, carboxyl, nitrogen, oxygen, and fluorocarbon are shown in Fig. 14.7, and the D, G, and 2D bands are clearly observed. The G band is as a result of the presence of in-plane sp^2 bonded carbon atoms [7] and corresponds to the E_{2g} irreducible representation of longitudinal optical (LO) and transverse optical (TO) phonon modes at the gamma (Γ) point of the Brillouin zone. The D band is due to the in-plane breathing modes of the sp^2 bonded carbon atoms and corresponds to the A_{1g} irreducible representation of the transverse optical (TO) phonon modes at the K point of the Brillouin zone [44], while the 2D band is the second-order of the D band. The D-band intensity of the functionalized GnPs is higher compared to the D-band intensity of the pristine GnPs, which is expected because the D peak is activated by the presence of defects in graphene. However, no shifts in the frequencies of the 2D and G bands were observed, which can be explained as a result of the low percentage of functionalization in the samples. The presence of a weak D' peak is also observed in the Raman spectra of the functionalized nanoplatelet samples, which is related to near-K point phonons in the first Brillouin zone [45]. However, the D' peak disappears when the percentage of carboxyl in the GnPs is increased from 7% to 35%, as shown in Fig. 14.8.

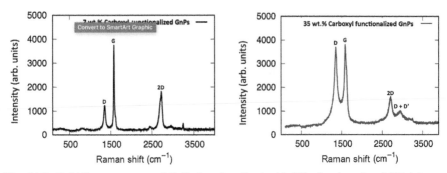

Fig. 14.8 *(Left)* Raman spectra of GnPs functionalized with 7% of carboxyl and *(Right)* Raman spectra of GnPs functionalized with 35% of carboxyl showing the D, G, 2D, and D+D′ peaks.

Fig. 14.8 shows that the D-band intensity drastically increases in the Raman spectra of GnPs functionalized with 35 wt% of carboxyl as a result of the increase in the percentage of carboxyl in the GnPs. The 2D peak of GnPs functionalized with 35 wt% of carboxyl shifts to a lower frequency of 2705.8 cm^{-1}, when compared to the 2D peak of GnPs functionalized with 7 wt% of carboxyl of frequency 2723.4 cm^{-1}, and a reduction of the intensity of the 2D peak is also observed in the Raman spectrum of GnPs functionalized with 35 wt% of carboxyl.

Table 14.2 shows the average positions of the characteristic D, G, 2D peaks, and the ratio of the intensities of the D and G peaks (I_D/I_G) of pristine and functionalized GnPs. The experiment was conducted five times for each of the GnPs samples to determine the experimental uncertainties. The intensity ratios (I_D/I_G) of the D and G peaks correspond to the level of defects in the GnP samples. A considerable increase in the intensity ratio (I_D/I_G) of the D and G bands of the functionalized GnPs compared to the pristine GnPs shows a break in the symmetry of the honeycomb structure of graphene as a result of functionalization. Functionalization of graphene changes the hybridization of the carbon atom from sp^2 to sp^3. However, the GnP samples functionalized with 35 wt% of carboxyl is shown to have the highest D to G ratio (I_D/I_G) as a result of the increase in functionalization.

The SEM images provide key insight into the morphology of nanoparticles at the nanometer and micrometer levels. The Phenom Pure SEM with magnification ranging between 20 × to 65,000 × was used to record the SEM images of the pristine and functionalized GnPs samples, as shown in Fig. 14.9. The SEM images of pristine and functionalized GnPs show generally the platelet-shaped GnPs stacked randomly on each other, whereas at 35 wt% concentration of carboxyl in the GnPs, a distinct change is observed in the platelet-shape to form large agglomerates.

The 3D views of the SEM images of pristine and functionalized GnPs were generated using the Gwyddion Software [46] and are shown in Fig. 14.9, which

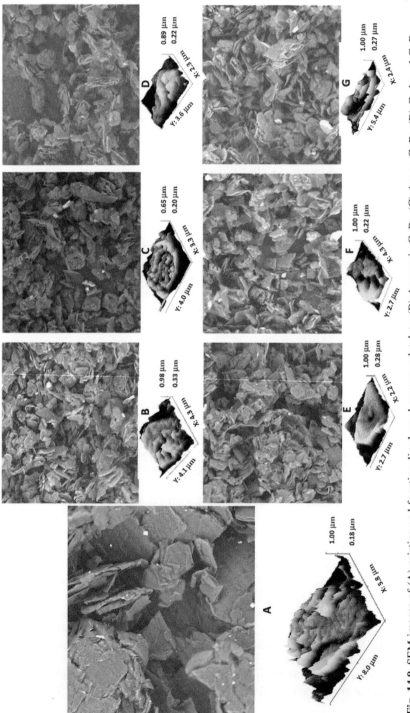

Fig. 14.9 SEM images of (A) pristine and functionalized graphene nanoplatelets: (B) Ammonia-GnPs, (C) Argon-GnPs, (D) Carboxyl-GnPs, (E) Fluorocarbon-GnPs, (F) Nitrogen-GnPs, (G) Oxygen-GnPs, and (H) 35wt.% carboxyl-GnPs showing the platelet-like shape of the individual nanoplatelets stacked randomly.

Table 14.3 Average x, y, z measurements of aggregates of pristine and functionalized graphene nanoplatelets.

Graphene nanoplatelets (GNPs) samples	X Average (μm)	Y Average (μm)	Z Average (μm)
GNPs	5.2	6.4	0.79
GNPs (Ammonia)	3.6	3.5	0.70
GNPs (Argon)	3.3	3.4	0.65
GNPs (Carboxyl)	3.1	4.5	0.70
GNPs (Fluorocarbon)	3.0	2.9	0.69
GNPs (Nitrogen)	3.9	2.8	0.76
GNPs (Oxygen)	2.9	3.8	0.65
GNPs (35 wt% Carboxyl)	10.0	10.2	0.73

illustrates the aggregates of sub-micron platelets within the GnPs samples with diameters ranging from 2.8 μm to 11.8 μm, depending on the particle size specified by the manufacturer and the number of platelets stacked in the aggregates. The average of five aggregates cropped from different spots within the SEM image is taken to obtain a rough estimate of the dimensions in the x, y, and z directions for the pristine and functionalized GnP aggregates and these values are summarized in Table 14.3.

The FTIR spectra of the pristine and functionalized GnPs were collected using the Perkin Elmer Frontier FT-IR/NIR Spectrometer equipped with the universal attenuated total reflection (UATR) polarization accessory. The UATR uses a DiComp crystal, which is composed of a diamond ATR and ZnSe focusing element in direct contact with the diamond crystal. The ATR spectroscopy works based on the principle of total internal reflection. A beam of radiation passes through the focusing element (ZnSe) and the radiation undergoes total internal reflection in the diamond crystal; an evanescent wave is produced, which extends a few microns beyond the diamond crystal and decays rapidly.

A pressure arm shown in Fig. 14.10 is used to maintain a close contact between the crystal surface and the sample during ATR measurements. The sample absorbs part of the radiation which then produces the absorption spectrum on the screen. Fig. 14.11 shows the FT-IR data of the pristine and functionalized GnPs. The wavenumbers of the FT-IR spectra were recorded in the range $4000 \, \text{cm}^{-1}$–$600 \, \text{cm}^{-1}$.

Surface analysis was carried out using FTIR on pristine and particularly functionalized GnPs to identify the functional groups present in the samples as displayed in Fig. 14.11. The peaks at 3735, 2685, 1459, 1292, and $872 \, \text{cm}^{-1}$ in the FTIR spectrum of pristine GnPs are assigned to free hydroxy (OH), C—H stretching, O—C—O, C—N and C—H bending vibrations, respectively, which were introduced during the exfoliation process of graphite. The sp^2 C=C stretch in the aromatic regions of the carbon rings also appears at frequency $1670 \, \text{cm}^{-1}$ in the pristine GnPs spectrum. Also, the diamond peaks appearing in spectra of all the GnPs samples

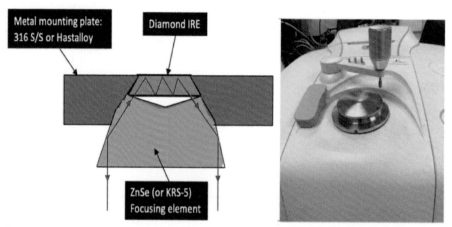

Fig. 14.10 *(Left)* Schematic of a Universal-ATR top plate; *(Right)* Perkin Elmer Frontier FT-IR/NIR Spectrometer with UATR accessory.

shown in Fig. 14.11 originate from the diamond ATR crystal in the FTIR spectrometer which was utilized. The FTIR spectra of GnPs-Carboxyl, GnPs-Argon, GnPs-Ammonia, GnPs-Nitrogen, GnPs-Oxygen, and GnPs-Fluorocarbon show relatively weak peaks compared to spectra of GnPs-35wt.% carboxyl due to the low concentration of functional groups present. The FTIR spectra of GnPs-35wt.% carboxyl display sharps peaks at frequencies 3749, 2658, 1687, 1844, 1539, and 992 cm^{-1} assigned to free O—H, C—H stretching, symmetric, and asymmetric C=O stretching, C=C and C—O vibrations, respectively. Also, a redshift in frequency of the C—O symmetric stretching and C=C vibrations is observed in the spectra of GnPs-35wt.% carboxyl because of an increase in the mass of carboxyl molecule introduced; recall that the frequency of vibration is inversely proportional to the mass of a vibrating molecule.

The X-ray diffraction spectra of the GnPs samples were collected using the Thermo Scientific™ ARL EQUINOX 100 X-Ray diffractometer (XRD), with a beam size of approximately 5 mm × 300 μm and a spinning stage for powder samples. The spectra were taken using a wavelength of 1.54056 Å and the diffraction patterns were recorded from 0° to 120° (Fig. 14.12).

Fig. 14.13 shows the diffraction peaks (002), (100), (004), and (110) in the XRD spectra of pristine and functionalized GnPs. The interlayer spacings (d_{002}) of pristine and functionalized GnPs are calculated using the Bragg formula shown in Eq. (14.9), while the crystallite sizes (out-of-plane (D_c) and in-plane (D_a)) are calculated using the Scherrer equation shown in Eq. (14.10) (Table 14.4).

$$2d_{002}\, sin\theta = \lambda \qquad\qquad (14.9)$$

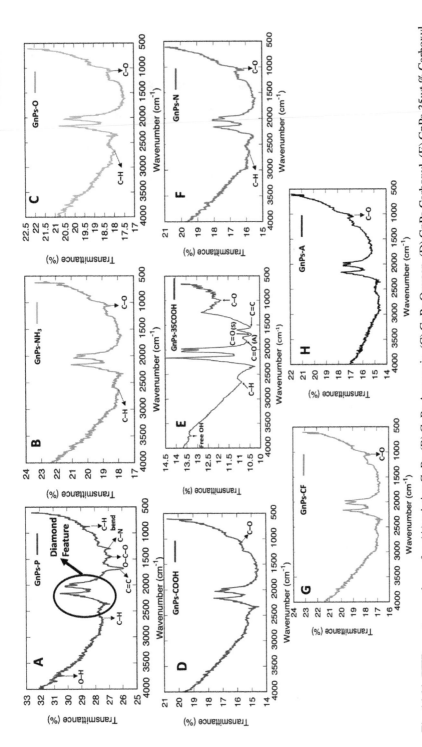

Fig. 14.11 FTIR spectra are shown for: (A) pristine GnPs, (B) GnPs-Ammonia, (C) GnPs-Oxygen, (D) GnPs-Carboxyl, (E) GnPs-35wt.% Carboxyl, (F) GnPs-Nitrogen, (G) GnPs-Fluorocarbon, and (H) GnPs-Argon. The diamond ATR crystal contributes peaks near the center of the horizontal axis range and are identified in (A).

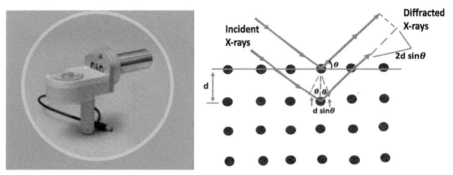

Fig. 14.12 *(Left)* Spinning stage for powder samples of Thermo Scientific ARL EQUINOX 100 X-Ray diffractometer adapted from the ThermoFisher Scientific Manual; *(Right)* Bragg's Law showing Diffraction of X-rays from Crystal Lattice.

$$D_c = \frac{0.89\lambda}{FWHM\ (002)\ (2\theta)\ x\ COS\theta} \quad \text{and}$$

$$D_a = \frac{0.89\lambda}{FWHM\ (100)\ (2\theta)\ x\ COS\theta} \quad (14.10)$$

where the constant 0.89 is the Scherrer constant, FWHM (002) and FWHM (100) are the full-width-half-maximum of the diffraction peaks (002) and (100), respectively, and θ is the Bragg angle corresponding to the diffraction peaks: (002) and (100) [47].

The diffraction peak (002) found at approximately 26^0 for the pristine and functionalized GnPs corresponds to an interlayer distance in the range 0.335–0.338 nm, which agrees well with the interlayer distance for graphite [48]. The (002) peak gives important information about the orientation of the aromatic carbon ring in the hexagonal lattice and its strong peak signifies that it has a preferred orientation in the (002) plane [47]. The (004) peak corresponds to the reflection from graphitic planes [49].

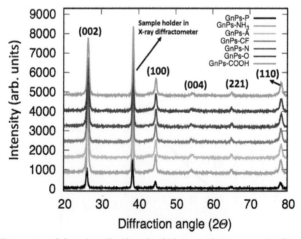

Fig. 14.13 XRD spectra of functionalized and pristine graphene nanoplatelets.

Table 14.4 Out-of-plane crystallite size (D_c), interlayer distance (d) and in-plane crystallite size (D_a) of pristine and functionalized graphene nanoplatelets.

Samples	$(2\theta)^0$ (002)	FWHM (002) $(2\theta)^0$	$(2\theta)^0$ (100)	FWHM (100) $(2\theta)^0$	d (nm)	D_c (nm)	D_a (nm)
Pristine-GnPs	26.50	0.4734	44.56	1.0986	0.3360	17.05	7.73
GnPs-Ammonia	26.54	0.8367	44.64	3.2530	0.3356	9.65	2.61
GnPs-Argon	26.71	0.5326	44.63	4.0596	0.3350	15.15	2.09
GnPs-Carboxyl	26.54	0.6214	44.63	4.3000	0.3356	13.04	1.98
GnPs-Fluorocarbon	26.54	0.6214	44.63	3.6255	0.3356	13.04	2.34
GnPs-Nitrogen	26.37	0.5030	44.58	3.7426	0.3376	16.00	2.27
GnPs-Oxygen	26.36	0.6806	44.66	4.9943	0.3376	11.83	1.7

14.5   Temperature-dependent Raman spectra, bulk vs. individual SWCNT

The three references, Huong et al. [50], Huang et al. [51], and Iliev et al. [52], were some of the earliest to observe and investigate the possible use of the temperature-induced frequency shift of the Raman bands of CNTs and other allotropes. In these initial and subsequent studies, such as that of Gregan [38], the extent of aggregation of the nanotube samples, and effect of inter-tube interactions on temperature studies were also recognized; and in that study on SWCNTs Gregan collected Raman spectra at increasingly greater temperatures using both highly bundled SWCNT powdered samples, and also samples dispersed in 1,2 dichloroethane (DCE) at concentrations both below and above the dispersion limit, thus ensuring adequate de-bundling of the SWCNTs. Further information on the use of discontinuities in the absorption spectra of SWCNT samples dispersed in DCE and other organic solvents to determine the level of aggregation of nanotube samples is available in Gregan [38].

Gregan's temperature-dependent Raman study on dispersed de-bundled single walled nanotubes elucidated two differences in the temperature-induced shifts of peak positions compared to bulk bundled samples, supporting the conclusion that not only temperature, but the level of bundling and inter-tube interactions play a role in the peak positions in the Raman spectra of SWCNTs. 514.5 nm laser excitation was used to collect spectra in the temperature range 83–343 K. The nanotubes used in our study and Gregan's were produced through the HiPco technique, with Gregan's samples being cooled and heated on a Linkham THMS600 stage. The first effect of de-bundling on the temperature-induced Raman peak shift observed by Gregan [38] was the frequencies being greater in magnitude compared to the bundled bulk samples. This result was not surprising, since the upshift in the Raman bands of SWCNTs after aggregating into bundled "ropes" had been known for some time now [38]. More importantly, Gregan [38] showed a significant change in the temperature coefficient itself of de-bundled SWCNTs compared to the case of bundled tubes, up to one order of magnitude.

Iliev et al. [52] settled an important issue regarding Raman spectroscopy applied to macroscopic bundled SWCNT samples outlined below. The peaks from the Raman spectra of bundled nanotubes are dominated by the diameters that are resonant with the excitation wavelength [53]. Because of this resonant enhancement, it is important in the analysis of Raman spectra of bulk samples to know if the bands observed are homogeneous peaks primarily from the resonant tubes or are complicated in-homogeneously broadened peaks composed of contributions from several different tubes with very similar diameters [52]. Iliev et al. [52] concluded that the latter is the case based again on the temperature variation of band components, only this time it was the temperature variation of the peak width, not peak location that was used. Nanotube samples used in [52] were produced by the laser arc method, and spectra were collected with 632.8 nm He—Ne laser excitation at temperatures from 5 to 500 K.

$$\Delta(T) = \gamma + \Gamma_0 \left(1 + \frac{2}{e^x - 1}\right) \qquad (14.11)$$

In Iliev et al. [52], the authors show that the application of Eq. (14.11), which expresses the variation of the phonon line width Δ with temperature due to phonon scattering, did not reproduce the experimental linewidth temperature variation based on the assumption of a homogeneous band. The temperature behavior of the linewidth was able to be modeled only after multiple simulations of various nearly similar diameter tubes.

Huong et al. [50] in one of the earliest known temperature-dependent Raman studies of SWCNT were able to identify a significant structural property of CNTs, namely, hemispherical end caps that contained pentagons in the sp_2 skeleton of carbon fullerenes. The Raman peak at $1470\,cm^{-1}$, observed in C60 "Buckyballs", exhibited resistance to increased temperatures in SWCNT samples compared to the latter fullerene type. Huong et al. [50] concluded that the resilient behavior of the pentagon containing end caps on nanotubes was due to the constraint of their being attached to the cylindrical walls, while the spherical C60 fullerenes would more easily "explode" under laser-induced heating due to thermal expansion. CNTs produced by the arc-discharge method and 514.5 nm wavelength excitation from an Ar ion laser were used in this Micro-Raman study. Like subsequent similar studies, the increase in frequency observed in CNT samples under thermal loading was attributed to lengthening of C—C distances.

A third and final study of the temperature dependence of the Raman bands of SWCNT discussed in this section is that of Huang et al. [51]. The authors subjected single-wall CNTs, Highly Ordered Pyrolytic Graphite (HOPG), and "Active-Carbon", a mixture of amorphous carbon and crystalline graphite to increasing power levels (0.2–4.0 mW) of a 632.8 nm He—Ne laser, and again demonstrated the linear frequency downshift of the Raman bands with temperature in all samples except for the HOPG. Huang et al. [51] also noted the need for caution when comparing the temperature behavior of the Raman peaks of CNTs, especially for the still not fully understood dispersive D "Defect" band and its overtones. Huang et al. [51] suggested that the defects present in the amorphous carbon and nanotube samples allowed for the flexibility necessary for bond elongation in these samples, and that such defects were not sufficiently present in crystalline HOPG to allow for the resulting temperature-induced peak shift in the latter.

14.5.1 Linear frequency downshift with increasing temperature

Two separate Raman spectral collections were recorded every 10°C in the range 30–200°C, with an exposure time of 10.00 s averaged over 3 exposures for each collection. Spectra were also collected after allowing enough time for a maximum temperature fluctuation of ±0.1°C as displayed by the heat cell's electronic display.

Numerous studies have shown that the effect of increased temperature is to redshift the different mode frequencies of the Raman peaks of carbon allotropes, including SWCNT [42,54]. Such highly visible temperature shifts of the Raman bands of CNTs are the focus of this section and serve as one of the many examples of the earlier stated usefulness of Raman spectroscopy in probing the structural/mechanical properties of CNTs. According to Dresselhaus and Eklund [54] the cause of this temperature-induced peak shift can be decomposed into an exclusively thermal effect, and a volume effect due to thermal expansion of the lattice, represented by the first and second terms respectively in Eq. (14.12).

$$\Delta\omega = \left(\frac{\partial\omega}{\partial T}\right)_V \Delta T + \left(\frac{\partial\omega}{\partial V}\right)_T \left(\frac{\partial V}{\partial T}\right)_P \Delta T \tag{14.12}$$

Such studies, however, have also demonstrated the prediction that this frequency shift, $\Delta\omega$, is dominated by the temperature effect, with a negligible volume-related effect term, based on actual measurements. In terms of empirical polynomials, such as Eq. (14.13), where ω_0 is the fitted frequency at $0\,\mathrm{K}$, and a_1 and a_2 represent first- and second-order temperature coefficients, the previously mentioned "temperature effect" dominance expresses itself as a negligible second order term $a_2 \approx 0$, with the temperature variation in fact being linear [38], which is also the case for us.

$$\omega = \omega_0 - a_1 T - a_2 T^2 \tag{14.13}$$

The temperature coefficients obtained have magnitudes comparable to literature values as shown in Table 14.1. The peak locations in the ω vs Temperature plots, Figs. 14.14–14.18, illustrate the linear frequency downshift slope, as temperature increases, utilizing the average of the values from both separate Raman collections.

In Figs. 14.14–14.18, we were able to reproduce the same linear red-shift with increasing temperature for the major Raman peaks in bundled samples of SWCNTs,

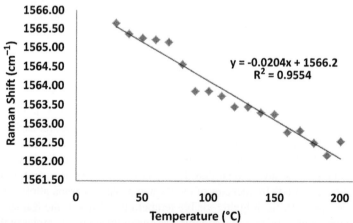

Fig. 14.14 Linear frequency downshift with temperature of the G$^-$ Raman peak of purified bundled SWCNT.

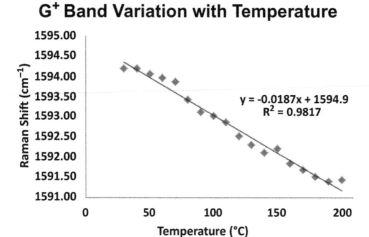

Fig. 14.15 Linear frequency downshift with temperature of the G^+ Raman peak of purified bundled SWCNT.

namely, the G^+, G^-, D, G' and Radial Breathing Mode (RBM) peaks. The linear plot for the peak location versus temperature for the G' plot has the most scatter compared to the other Raman bands, as evidenced by its R^2 value. It should be noted that this is the only second-order Raman band included in Figs. 14.14–14.18, and while the cause of the large variation in the redshift trend is unknown, this feature is a highly dispersive band that is very sensitive to electronic and photonic perturbations [55].

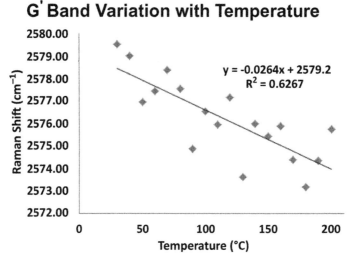

Fig. 14.16 Linear frequency downshift with temperature of the G' Raman peak of purified bundled SWCNT.

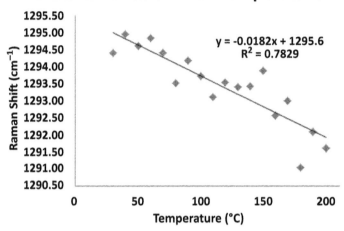

Fig. 14.17 Linear frequency downshift with temperature of the D Raman peak of purified bundled SWCNT.

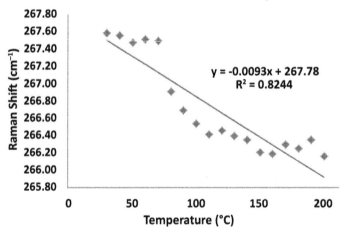

Fig. 14.18 Linear frequency downshift with temperature of the RBM Raman peak of purified bundled SWCNT.

14.6   Temperature-dependent Raman spectra of GnPs

In this section, we present results from the temperature dependence of the G-band of pristine GnPs. The pristine GnPs sample was compressed in a brass sample holder and placed in an aluminum disk cell which contains apertures for heating up the cell. A variable variac transformer for voltage supply was connected to the setup through

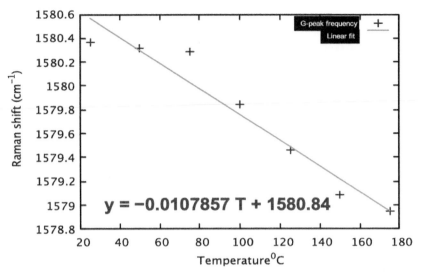

Fig. 14.19 Temperature dependence of G-peak frequency of pristine graphene nanoplatelets.

one of the apertures and a thermocouple placed in another aperture to read the temperature of the disk cell. The samples were heated from room temperature (25°C) to 175°C in steps of 25°C by increasing the voltage. Our results show that the frequency of the G-band decreases linearly as the temperature is increased from 25°C to 175°C, resulting in a negative temperature coefficient [56]. This result agrees well with earlier research studies [24,56] on graphene and graphite. The temperature coefficient of the G-band corresponds to the slope of the graph in Fig. 14.19 and is expressed in Eq. (14.14).

$$\chi_G = \frac{\Delta \omega}{\Delta T} = -0.0107857 \text{ cm}^{-1}/°C \qquad (14.14)$$

where $\Delta \omega$ and ΔT are the changes in frequency and temperature of the G band, respectively.

The frequency downshift of the G band can be attributed to the direct coupling of the phonon modes and as a result of thermal expansion induced by volume change [56].

14.7 Application of molecular dynamics (MD) simulation to SWCNT structural analysis

The examination of the thermo-mechanical properties of CNTs is deeply linked with computational nanoscience [57]. This broad field, which includes numerical modeling and MD simulation, provides an integral avenue in the research of the extremely reduced time, length, and energy scale processes occurring with these objects.

Much information on the mechanical properties is derivable from the inter-atomic potentials used to model them. The atomic level stress tensor, for example, which is obtained by the method of finite strain, involves the first derivative of the potential, whereas the elastic constants and bulk moduli are based on the second derivative. The usual course of action to help meet the challenge of limited computing resources is referred to as a "multi-scale" technique [39], where data from quantum calculations and or continuum mechanics acts as input to MD simulations or vice versa. The results of WenXing et al.'s [58] MD analysis of the Young's Modulus demonstrates this and was also chosen because of Wenxing's use of the earlier variant of the AIREBO potential. Based on MD simulations using Verlet time integrations with a 1 fs time step, Wenxing et al. calculated Young's Moduli of 1026.176 and 929.8 ± 11.5 GPa for graphite and SWCNTs, respectively. Both values are close to experimental values and to those from quantum calculations.

Based on the initial Raman analysis, SEM visualization, and manufacturer data on representative nanotubes, the most dominant resonant tube based on Raman spectra at 780 nm wavelengths were used as the initial model nanotubes in the MD simulations. Large portions of the script used to perform the MD simulations at increasing temperatures have been developed. The sampling geometry and other geometrical parameters of the nanotube models have been incorporated into the simulation script and used to compute thermal expansion data.

Some initial data on an MD simulation of a (10, 2) chirality SWCNT, the dominant chiral species present in the purified nanotube sample, are presented here based on the Raman spectra mentioned previously. The vibrations of a 200 Å long SWCNT is based on the AIREBO potential discussed earlier. The model nanotube for this chirality consists of a total of 2232 atoms. A non-bonded Lennard-Jones cut-off of 3 Å was used in addition to the other default parameterized constants of the AIREBO potential [30]. Non-periodic boundaries were used in all 3 dimensions of the simulation box with x, y, and z dimensions of $100.0 \times 100.0 \times 240.0$ Å, respectively. A 0.5 fs time-step, and 100,000 total time-steps were used. Verlet integration was used in addition to simple velocity rescaling in order to achieve a mean temperature of 300.0 K over the course of the run. The sampling geometry used to describe the length variation of the nanotube was based on the summation of the center of mass between consecutive cylindrical disk sections of the model nanotube. The behavior of the axial dimension in this case of an initial room temperature MD simulation exhibits the anomalous thermal expansion behavior also observed in other MD studies. The MD simulations of nanotube models are currently in the equilibration phase. The MD simulations were done using the Large-scale Atomic/Molecular Massively Parallel Simulator (LAMMPS) program [66]. They were also run remotely on the computing clusters at the National Energy Research Scientific Computing Center (NERSC), a division of the Lawrence Berkley National Laboratory, in order to meet the very large computational and time demands of the simulations. The sampling phase of the calculations where structural data were collected followed the MD simulations.

As indicated previously, the value of the volume thermal expansion coefficient used in the fitting of the experimental thermal Raman peak shift in Yang et al. [59] came from the expansion coefficient of the triangular lattice constant of SWCNT ropes, as determined by X-ray diffraction [60]. An often used method in the determination of the structural/thermo-mechanical properties of CNTs has been the use of continuum elasticity theory results on the behavior of an equivalent hollowed cylinder or cylindrical shell in the reduction of MD simulation data [61]. However, although macroscopic continuum theory applied to SWCNTs happened to be surprisingly relevant and useful, as noted by Harik and Gates [62] and many others during the early days of nanotube structural analysis, extreme caution is still required in approaches based on macroscopic continuum mechanics, since "… its relevance for a covalent bonded system of only a few atoms in diameter is far from obvious [62]." In fact, the determination of the cross-sectional thickness of CNTs remains an active research issue [61]. In Batra and Gupta [61], the structural properties of the "Equivalent Continuum Structure" represented by isotropic shells were used to elucidate these properties by equating the first Rayleigh mode and RBM frequencies of the two structures. Another technique based on the axial, twisting, and first Love vibrational modes of an elastic shell and its corresponding model SWCNT was also used. Batra and Gupta [63] were then able to derive the Young's modulus, Poisson's ratio and an expression for the thickness of a SWCNT as a function of radius and the bond-length in the equilibrated SWCNT model used in the MD simulations. The key result of Batra and Gupta [63] was the SWCNT thickness calculated as 0.88 Angstroms when the nanotube radius was $\sim$3.68 Angstroms, and this wall thickness value increased to and remained at approximately 1.37 Angstroms when the diameter was at or greater than 15 Angstroms, both values being much smaller than the inter-planar spacing between graphitic planes, $\sim$3.4 Angstroms, often used as a first approximation.

Based on the Raman analysis, SEM visualization, and manufacturer data, representative nanotube types with reasonable probabilities of being present in the samples used in the simulations have been modeled. Large portions of the script used to perform the MD simulations at increasing temperatures have now been developed. The sampling geometry and other geometrical parameters of the nanotube models have been incorporated into the simulation script and used to compute the thermal expansion data and validated by experiment.

14.8 Support vector machine algorithms and machine learning

Support Vector Machines, otherwise known as SVMs, implement simple algorithms to analyze and separate data using a hyperplane located in N-dimensional space, where N is the number of parameters. By adding parameters, the data are differentiated and transformed between planes. The transformations that data undergo are referred to as kernels. Applying kernels allows users to discriminate even extreme cases where data might overlap. SVM uses these extreme values of the dataset to draw the decision surface. The values that the hyperplane's margin divides on are referred to as the support vectors [64]. One of the many positive attributes of SVM is the fact that a huge dataset is not necessary to begin the computation. However, result accuracy improves with larger datasets. In addition, the MATLAB software has a beginning

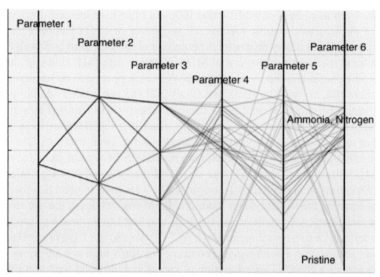

Fig. 14.20 Network diagram of SVM learning using 6 parameters to create a 6-dimensional quadratically separable dataset. As the data move throughout the dimensions, one can see that the pristine and functionalized GnPs become highly separated. Conversely, for the ammonia and nitrogen functionalized nanoplatelets, slight overlapping is still noticed. However, in this overlapping, a hyperplane surface was created, separating the ammonia and nitrogen functionalized GnPs with complete success.

classification learner toolbox available for people to use [64]; it facilitates the uploading of a dataset in the form of a CSV file into the computer program. We prepared the dataset from the Raman spectral results of the ammonia (NH_3) and nitrogen (N_2) functionalized GnPs. Twelve trials were performed for each of the functionalized GnPs, while five trials were collected for the pristine GnPs for improved results. For each trial, the D, G, and 2D-band peak wavelengths were documented, as well as all possible peak-to-peak ratios. The peak wavelengths used in the modeling and analysis have been summarized in Section 14.4.3. In total, there were 6 parameters: the three peak wavelengths and their comparative ratios. The ratios were taken into consideration as the number of defects of the graphene should be altered with the functional groups present. In turn, these parameters created a 6-dimensional space where the data could be separated. For visualization purposes, Fig. 14.20 depicts the layers, or dimensions, that the data filters through to eventually be quadratically separated in the end.

Although Fig. 14.20 is complex, it holds great promise, as it was successful in correctly identifying which functional groups are present in given samples; that is, the functional groups present—or lack thereof—were separated into their respective categories without the computer mistaking one sample for another. With a working model, it would now be possible to add on more functional groups, one at a time sequentially, and expand the model even further with a computer algorithm that would enable differentiation between all 6 functional groups associated with the GnPs investigated, and eventually be able to characterize those that have made contact with harmful poisonous compounds. Looking ahead, large datasets must be prepared in order to verify the results and build more efficient and versatile models and then be able to develop practical graphitic toxic gas sensors for field deployment.

14.9 Conclusions

We have attempted to capture in this chapter the varied array of techniques used in both experimental and computational methods, and the inevitable mixing of these techniques, in the study of mechanical properties and thermal characteristics of CNTs and GnPs. The parameters and attributes of our own classical MD simulations, most importantly basing the intra-nanotube energetics on the robust (AIREBO) many body potential, were discussed in Sections 14.4.2 and 14.7 The Raman spectra of actual SWCNT samples collected under increasing temperature, and our reproduction of the linear downshift of frequency with temperature, is a key part of Section 14.5. Also, initial MD simulations based on the AIREBO potential have already provided promising results, by reproducing the anomalous axial contraction of SWCNTs at this stage of the investigation. In addition, Section 14.4.3 reports a detailed characterization of pristine and functionalized pristine GnPs using X-ray diffraction (XRD), Raman spectroscopy, SEM, and FT-IR techniques. Results from the Raman spectroscopy data show a change in the hybridization of the carbon atoms from sp^2 to sp^3 upon functionalization at the edges of the nanoplatelets through analysis of the D, D', G and 2D peaks. Our FT-IR data support this claim and have shown that the concentration (wt%) and type of functionalization present largely affects the electronic and vibrational properties of the nanoplatelets. The temperature dependence of the G band of pristine GnPs has been discussed in Section 14.6 and the results show a linear relationship between the G-band frequency and temperature (shift to lower frequencies), which is attributed to the direct coupling of the phonon modes and thermal expansion. Other characterization tools, such as Scanning Tunneling Microscopy (STM), X-ray Photoelectron Spectroscopy (XPS), and Electron Energy Loss Spectroscopy (EELS) are required to obtain information about the surface topography and chemical composition on the surface and at the edge sites of the functionalized nanoplatelets. These characterization tools are also necessary in determining the configuration of the edges (armchair and zigzag) of the nanoplatelets. The edge sites play a crucial role in determining the chemical reactivity of the carbon atoms [65], which in turn affects their electronic and vibrational properties. We are also using machine learning algorithms to discriminate between the various functionalized GnPs studied. Theoretical studies using MD simulation and DFT are important to further support experimental studies in investigating the electronic and vibrational properties of the functionalized GnPs and would be beneficial for the fabrication of certain types of devices, such as toxic gas sensors.

Much progress remains to be made in examining and understanding the mechanical properties of SWCNTs and responses to stresses and temperature gradients. Some of the extreme elastic properties of SWCNTs were also discussed and exhibited through SEM imaging. Also, such data on the mechanical properties of actual CNT samples are extremely important for comparison with computational studies, which is becoming the preferred method of study of this subject in lieu of the more expensive ab initio and Scanning Probe Method (SPM) techniques.

Acknowledgments

Financial support from the National Science Foundation (REU Site in Physics at Howard University, NSF Award# PHY-1659224 and Excellence in Research NSF Award# DMR-2101121) is gratefully acknowledged, as well as allocation support from the Extreme Science and Engineering Discovery Environment (XSEDE Grant Nos. TG-DMR190126 and TG-PHY210066). We also would like to acknowledge the assistance of the technical staff and the use of the X-ray diffraction and Raman instrumentation at the Thermo Fisher Scientific facilities in Lanham, MD, for some of the GnP spectral measurements reported here.

References

[1] H.S.P. Wong, D. Akinwande, Carbon Nanotube and Graphene Device Physics, Cambridge University Press, New York, 2011.

[2] A.K. Geim, K.S. Novoselov, The rise of graphene, Nat. Mater. 6 (2007) 183–191.

[3] X. Li, W. Cai, J. An, S. Kim, J. Nah, D. Yang, R. Piner, A. Velamakanni, I. Jung, E. Tutuc, S.K. Banerjee, L. Colombo, R.S. Ruoff, Large-area synthesis of high-quality and uniform graphene films on copper foils, Science 324 (5932) (2009) 1312–1314.

[4] Y. Hernandez, V. Nicolosi, M. Lotya, F.M. Blighe, Z. Sun, S. De, I. McGovern, B. Holland, M. Byrne, Y.K. Gun'Ko, et al., High-yield production of graphene by liquid-phase exfoliation of graphite, Nat. Nanotechnol. 3 (9) (2008) 563.

[5] P. Cataldi, A. Athanassiou, I. Bayer, Graphene nanoplatelets-based advanced materials and recent progress in sustainable applications, Appl. Sci. 8 (9) (2018) 1438.

[6] B.Z. Jang, A. Zhamu, Processing of nanographene platelets (NGPs) and NGP nanocomposites: a review, J. Mater. Sci. 43 (15) (2008) 5092–5101.

[7] J. Dai, C. Peng, F. Wang, G. Zhang, Z. Huang, Effects of functionalized graphene nanoplatelets on the morphology and properties of phenolic resins, J. Nanomater. 2016 (2016) 1–8.

[8] M.Y. Shen, T.Y. Chang, T.H. Hsieh, Y.L. Li, C.L. Chiang, H. Yang, M.C. Yip, Mechanical properties and tensile fatigue of graphene nanoplatelets reinforced polymer nanocomposites, J. Nanomater. 2013 (2013) 1.

[9] M.D. Hanwell, D.E. Curtis, D.C. Lonie, T. Vandermeersch, E. Zurek, G.R. Hutchison, Avogadro: an advanced semantic chemical editor, visualization, and analysis platform, J. Chem. 4 (1) (2012) 17.

[10] R.S. Ruoff, D.C. Lorents, Mechanical and thermal properties of carbon nanotubes, Carbon 33 (7) (1995) 925–930.

[11] J. Tersoff, R.S. Ruoff, Structural properties of a carbon-nanotube crystal, Phys. Rev. Lett. 73 (5) (1994) 676–679.

[12] D.H. Robertson, D.W. Brenner, J.W. Mintmire, Energetics of nanoscale graphitic tubules, Phys. Rev. B Condens. Matter. 45 (21) (1992) 12592–12595.

[13] D.W. Brenner, et al., A second-generation reactive empirical bond order (REBO) potential expression for hydrocarbons, J. Phys. Condens. Matter 14 (2002) 783–802.

[14] P.G. Sankar, K.U. Kumar, Mechanical and electrical properties of single walled carbon nanotubes: a computational study, Eur. J. Sci. Res. 60 (3) (2011) 342–358.

[15] P.S. Rao, S. Anandatheertha, G.N. Naik, S. Gopalakrishnan, Estimation of mechanical properties of single wall carbon nanotubes using molecular mechanics approach, Sadhana 40 (4) (2015) 1301–1311.

[16] A. Krishnan, et al., Young's Modulus of single-walled carbon nanotubes, Phys. Rev. B Condens. Matter. 58 (20) (1998) 14013–14019.

[17] Y. Maniwa, et al., Thermal expansion of single-walled (SWNT) bundles: X-ray diffraction studies, Phys. Rev. B Condens. Matter. 64 (2001) 1–3.

[18] Y. Yosida, High-temperature shrinkage of single-walled carbon nanotube, J. Appl. Phys. 87 (7) (2000) 3338–3341.

[19] T. Yamamoto, K. Watanabe, E.R. Hernandez, A. Jorio, et al., Mechanical properties, thermal stability and heat transport in carbon nanotubes, in: Carbon Nanotubes: Advanced Topics in the Synthesis, Structure, Properties and Applications, Springer, 2008, pp. 165–194.

[20] J.P. Salvetat, G.A.D. Briggs, J.M. Bonard, R.R. Basca, A.J. Kulik, Elastic and shear moduli of singlewalled carbon nanotube ropes, Phys. Rev. Lett. 82 (1999) 944.

[21] Y.H. Yang, W.Z. Li, Radial elasticity of single-walled carbon nanotube measured by atomic force microscopy, Appl. Phys. Lett. 98 (4) (2011), 041901.

[22] D. Bozovic, M. Bockrath, J.H. Hafner, C.M. Lieber, H. Park, M. Tinkham, Plastic deformations in mechanically strained single-walled carbon nanotubes, Phys. Rev. B 67 (3) (2003), 033407.

[23] M.B. Nardelli, B.I. Yakobson, J. Bernhole, Mechanism of strain release in carbon nanotubes, Phys. Rev. B 57 (1998) R4277.

[24] D. Yoon, Y.W. Son, H. Cheong, Negative thermal expansion coefficient of graphene measured by Raman spectroscopy, Nano Lett. 11 (8) (2011) 3227–3231.

[25] L. Jarosinski, A. Rybak, K. Gaska, G. Kmita, R. Porebska, C. Kapusta, Enhanced thermal conductivity of graphene nanoplatelets epoxy composites, Mater. Sci.-Pol. 35 (2) (2017) 382–389.

[26] S. Choi, J. Kim, Thermal conductivity of epoxy composites with a binary-particle system of aluminum oxide and aluminum nitride fillers, Compos. Part B Eng. 51 (2013) 140–147.

[27] K.M. Shahil, A.A. Balandin, Graphene–multilayer graphene nanocomposites as highly efficient thermal interface materials, Nano Lett. 12 (2) (2012) 861–867.

[28] E.N. Koukaras, G. Kalosakas, C. Galiotis, K. Papagelis, Phonon properties of graphene derived from molecular dynamics simulations, Sci. Rep. 5 (2015) 12923.

[29] L. Lindsay, D.A. Broido, Optimized Tersoff and Brenner empirical potential parameters for lattice dynamics and phonon thermal transport in carbon nanotubes and graphene, Phys. Rev. B 81 (20) (2010), 205441.

[30] S.J. Stuart, A.B. Tutein, J.A. Harrison, A reactive potential for hydrocarbons with intermolecular interactions, J. Chem. Phys. 112 (2000) 6472.

[31] J.H. Los, A. Fasolino, Intrinsic long-range bond-order potential for carbon: performance in Monte Carlo simulations of graphitization, Phys. Rev. B 68 (2) (2003), 024107.

[32] J.M. Haile, Molecular Dynamics Simulation: Elementary Methods, Wiley, New York, 1997.

[33] J.H. Zou, Z.Q. Ye, B.Y. Cao, Phonon thermal properties of graphene from molecular dynamics using different potentials, J. Chem. Phys. 145 (13) (2016), 134705.

[34] J. Tersoff, Modeling solid-state chemistry: interatomic potentials for multicomponent systems, Phys. Rev. B 39 (8) (1989) 5566.

[35] J. Tersoff, Erratum: modeling solid-state chemistry: interatomic potentials for multicomponent systems, Phys. Rev. B 41 (5) (1990) 3248.

[36] P. Misra, D. Casimir, R. Garcia-Sanchez, Thermal expansion properties of single-walled carbon nanotubes, by Raman spectroscopy at 780 nm wavelength, in: Annual International Conference on Opto Electronics, Photonics & Applied Physics (OPAP), 2013, pp. 52–55.

[37] N.R. Raravikar, P. Keblinski, A.M. Rao, M.S. Dresselhaus, Temperature dependence of radial breathing mode Raman frequency of single-walled carbon nanotubes, Phys. Rev. B Condens. Matter. 66 (2002) 1–9.

[38] E. Gregan, The Use of Raman Spectroscopy in the Characterization of Single Walled Carbon Nanotubes, Doctoral Dissertation, Dublin Institute of Technology, School of Physics, 2009.

[39] H. Rafii-Tabar, Computational Physics of Carbon Nanotubes, Cambridge University Press, Cambridge, 2008.

[40] D. Frenkel, B. Smith, Understanding Molecular Simulation: From Algorithms to Applications, Academic Press, New York, 2002.

[41] C.Q. Sun, Size dependence of nanostructures: Impact of bond order deficiency, Prog. Solid State Chem. 35 (2007) 1–159.

[42] D. Casimir, I. Ahmed, R. Garcia-Sanchez, P. Misra, F. Diaz, Raman spectroscopy of graphitic nanomaterials, in: Raman Spectroscopy, IntechOpen, 2017.

[43] D. Casimir, H. Alghamdi, I.Y. Ahmed, R. Garcia-Sanchez, P. Misra, Raman spectroscopy of graphene, graphite and graphene nanoplatelets, in: 2D Materials, IntechOpen, 2019.

[44] D. Sfyris, G.I. Sfyris, C. Galiotis, Stress Interpretation of Graphene E-2g and A-1g Vibrational Modes: Theoretical Analysis, 2017. arXiv preprint arXiv:1706.04465.

[45] R. Beams, L.G. Cançado, L. Novotny, Raman characterization of defects and dopants in graphene, J. Phys. Condens. Matter 27 (8) (2015), 083002.

[46] D. Nečas, P. Klapetek, Gwyddion: an open-source software for SPM data analysis, Centr. Eur. J.Phys. 10 (1) (2012) 181–188.

[47] T. Qiu, J.G. Yang, X.J. Bai, Y.L. Wang, The preparation of synthetic graphite materials with hierarchical pores from lignite by one-step impregnation and their characterization as dye absorbents, RSC Adv. 9 (22) (2019) 12737–12746.

[48] M. Mohr, J. Maultzsch, E. Dobardžić, S. Reich, I. Milošević, M. Damnjanović, A. Bosak, M. Krisch, C. Thomsen, Phonon dispersion of graphite by inelastic x-ray scattering, Phys. Rev. B 76 (3) (2007), 035439.

[49] V. García, M.R. Gude, A. Ureña, Effect of graphene nanoplatelets features on cure kinetics of benzoxazine composites, in: 20th International Conference on Composite Materials, Copenhagen, Denmark, 2015.

[50] P.V. Huong, R. Cavagnot, P.M. Ajayan, D. Stephan, Temperature dependent vibrational spectra of carbon nanotubes, Phys. Rev. B Condens. Matter. 51 (15) (1995) 10048–10051.

[51] F. Huang, K.T. Yue, P. Tan, S.L. Zhang, Z. Shi, X. Zhou, Z. Gu, Temperature dependence of the Raman spectra of carbon nanotubes, J. Appl. Phys. 84 (7) (1998) 4022–4024.

[52] M.N. Iliev, A.P. Litvinchuk, S. Arepalli, P. Nikolaev, C.D. Scott, Fine structure of the low-frequency Raman phonon bands of single-wall carbon nanotubes, Chem. Phys. Lett. 316 (2000) 217–221.

[53] R. Saito, G. Dresselhaus, M.S. Dresselhaus, Physical Properties of Carbon Nanotubes, Imperial College Press, London, 1998.

[54] M.S. Dresselhaus, P.C. Eklund, Phonons in carbon nanotubes, Adv. Phys. 49 (6) (2000) 705–814.

[55] A. Jorio, M.S. Dresselhaus, R. Saito, G. Dresselhaus, Raman Spectroscopy in Graphene Related Systems, Wiley-VCH Verlag GmbH & Co. KGaA, Weinheim, 2011.

[56] I. Calizo, A.A. Balandin, W. Bao, F. Miao, C.N. Lau, Temperature dependence of the Raman spectra of graphene and graphene multilayers, Nano Lett. 7 (9) (2007) 2645–2649.

[57] H. Rafii-Tabar, Computational modelling of thermomechanical and transport properties of carbon nanotubes, Phys. Rep. 390 (2004) 235–452.

[58] B. WenXing, Z. ChangChun, C. WanZhao, Simulation of Young's modulus of single-walled carbon nanotubes by molecular dynamics, Physica B 352 (2004) 156–163.

[59] X.X. Yang, Z.F. Zhou, Y. Wang, J.W. Li, N.G. Guo, W.T. Zheng, J.Z. Peng, C.Q. Sun, Raman spectroscopic determination of the length, energy, Debye temperature, and compressibility of the C-C bond in Carbon allotropes, Chem. Phys. Lett. 575 (2013) 86–90.

[60] B. Ingham, M.F. Toney, X-ray diffraction for characterizing metallic films, in: Metallic Films for Electronic, Optical and Magnetic Applications, Woodhead Publishing, 2014, pp. 3–38.

[61] R.C. Batra, S.S. Gupta, Wall thickness and radial breathing modes of single-walled carbon nanotubes, J. Appl. Mech. 75 (2008) 1–6.

[62] V.M. Harik, T.S. Gates, Applicability of the continuum-shell theories to the mechanics of carbon nanotubes, in: NASA/CR-2002-211460 ICASE Report No. 2002-7, 2002.

[63] R.C. Batra, S.S. Gupta, Wall thickness and elastic moduli of single-walled carbon nanotubes from frequencies of axial, torsional and inextensional modes of vibration, Comput. Mater. Sci. 47 (2010) 1049–1059.

[64] S. Patel, Chapter 2: SVM (support vector machine)-theory, in: Machine Learning, 2017, p. vol. 101.

[65] A. Bellunato, H. Arjmandi Tash, Y. Cesa, G.F. Schneider, Chemistry at the edge of graphene, ChemPhysChem 17 (6) (2016) 785–801.

[66] S. Plimpton, et al., LAMMPS—a flexible simulation tool for particle-based materials modeling at the atomic, meso, and continuum scales, Comput. Phys. Commun. 271 (2022) 108171, https://doi.org/10.1016/j.cpc.2021.108171.

Nanoalloys and catalytic applications

Shan Wang, Shiyao Shan, Dominic Caracciolo, Aolin Lu, Richard Robinson, Guojun Shang, and Chuan-Jian Zhong
Department of Chemistry, State University of New York at Binghamton, Binghamton, NY, United States

15.1 Introduction

Design, synthesis, and characterization of metal nanoparticles, especially alloy nanoparticles ranging from sub-nanometer to a few nanometers, are important in heterogeneous catalysis [1–4]. It is the nanoscale size range over which metal nanoparticles undergo a transition from metallic to atomic properties, leading to unique electronic and catalytic properties different from their bulk counterparts. Significant advances have been made in harnessing the nanoscale catalytic properties in the energy and environmental fronts [2–7]. However, challenges remain, especially in the preparation and characterization of active, robust, low-cost nanocatalysts with controllable sizes, shapes, compositions, and structures. One promising approach involves alloying. Alloy is a mixture of two or more metallic species, which can exist either in a complete solid solution state exhibiting a single-phase characteristics or in a partial or phase-segregated solid solution state with two or more phases. Nanoalloy (NA) differs from bulk alloy in several significant aspects in terms of mixing patterns and geometric shapes [8]. Considering both binary and ternary NAs, there are different types of mixing patterns, two of which include completely phase segregated NAs, where the different phases share either an extended mixed interface or a very limited number of heterometal bonds, and mixed NAs with chemically ordered or random structures. The degree of segregation, mixing, and atomic ordering depend on a number of factors [8], including relative strengths of homoatomic versus heteroatomic bonds, surface energies of the component element, differences in atomic sizes, charge transfer between the different atomic species, and strength of binding to surface ligands or support materials. The observed atomic arrangement, and geometric shape as well, for a particular NA depends critically on the balance of these factors and on the preparation and usage conditions. In addition, the surface structures of NAs could be very complex due to the enrichment of certain elements in the core or shell. In the case of metal dissolution (dealloying) in the presence of acidic electrolytes, often being referred as Pt-skin structure formation [9–14], the details of the noble metal skin could be influenced strongly by the structural types of the NAs, the understanding of which in terms of structural evolution, noble metal skin, or d-band structure has attracted increasing interests in catalysis and electrocatalysis.

Modeling, Characterization and Production of Nanomaterials. https://doi.org/10.1016/B978-0-12-819905-3.00015-4

Supported metal nanoparticles from traditional preparative methods have been well demonstrated for various catalytic reactions [5,15,16]. New approaches to the synthesis of metal nanoparticles capped in molecular monolayers, dendrimers, and polymers have been rapidly emerging. While some of the catalysts exploit the functional groups at the capping shell of the nanoparticles [17,18], most explore the surface sites on the metal nanoparticles either after removing the capping layers [15,16,19] or through open channels of the capping layers [20–22]. In addition, the nanoscale facet is an important factor in catalysis. For noble metal (e.g., Pt and Pd) alloyed with other transition metals (e.g., Ni, Cu, Co, etc.), a volcano curve has been observed for certain metal ratios (e.g., 1:1, 1:3, or 3:1) in binary nanocatalysts [3,23–28]. Moreover, the strong metal-support interaction is shown to play a key role in the catalytic reactions [29–31]. In addition to various nanostructural design parameters [32–34], the understanding of how the activation of oxygen on alloy catalysts plays an important role in different catalytic reactions remains elusive.

Insights into the NAs in catalysis applications have also been gained through theoretical studies. In a study of ternary PtVFe particle [35] in terms of molecular chemisorption configurations, the O_2 bond is found to be weakened the most with the Yeager configuration, followed by the Griffith and the Pauling configurations. The charge transfer and the adsorption energy are dependent on the metal and adsorption site. The modeled PtVFe clusters happened to be most effective in transferring electrons to oxygen species, thus weakening the O_2 bonds when the oxygen atoms bind to non-Pt atoms, providing useful hints for the design of trimetallic NAs. Similarities have also been found for O_2 adsorption and dissociation on $Pt_4V_2Fe_7$. Another important fundamental aspect of the ternary NA's electrocatalytic activity and stability is related to the oxophilicity for different metals [36,37]. It is hypothesized that the removal of O-containing intermediates (—O, —OH, —OOH) resulting from the O=O breaking on the Pt-bridge sites, that recovers the catalytically active sites, is possibly more effective by alloying Pt with two different metal components (M and M') than with a single metal component. There are several recent studies supporting this hypothesis. Extended X-ray absorption fine structure (EXAFS) and X-ray photoelectron spectroscopy (XPS) characterizations of Pt-trimetallic alloy electrocatalysts for oxygen reduction reaction (ORR) prepared by electrodeposition [38] showed that the electronic states such as the d-band centers, geometric factors, and Pt-Pt distance change with the composition. Fe and Co might play bifunctional roles: adsorption and desorption of oxygen species, respectively, due to their different electronic properties. Also, recent density functional theory (DFT) calculation of oxygen and hydroxyl adsorption energies showed that there is an approximately linear trend in oxygen reduction activity vs O or OH binding energy for a large number of different transition and noble metals [37]. Another DFT study of Pt_3Ni surface doped with a third transition metal (V, Fe, Co, Ir, etc.) found negative Pt surface segregation energies whereas doping with Pd, Ag, and Au was found to completely suppress the Pt segregation [39]. There are a number of relevant critical issues, including how the initial structure of an as-prepared NA catalyst impacts the structural evolution (e.g., Pt-skin or d-band shift) and how the structural integrity can be sustained by the alloying components and processing strategies.

NA catalysts and selected examples of catalytic oxidation reactions in gas phase and in electrolytes [40–46] are discussed in this chapter. One important emphasis is placed on understanding atomic-scale chemical and structural ordering and coordination in correlation with the catalytic or electrocatalytic properties, which can be obtained based on synchrotron X-ray characterizations, such as high-energy X-ray diffraction coupled to atomic pair distribution functions (HE-XRD/PDFs) and X-ray absorption fine structure (XAFS).

15.2   Synthesis of NAs and preparation of catalysts

15.2.1   Synthesis of NAs

The synthesis of molecularly capped metal nanoparticles as building blocks for engineering the nanoscale catalytic materials takes advantage of diverse attributes, including monodispersity, processability, solubility, stability, and self-assembly capability in terms of size, shape, composition, and surface properties. Indeed, the preparation of nanoparticles capped in monolayers, polymers, or dendrimers is rapidly emerging, demonstrating remarkable parallels to catalytic activities for supported nanoparticles [15,17,19,47,48]. One important approach involves core-shell-type synthesis [49–52]. The core-shell-type nanostructure can be broadly defined as core and shell of different matters in close interaction, including inorganic/organic and inorganic/inorganic combinations [19,53–58]. The synthesis of metal nanoparticles in the presence of organic capping agents to form encapsulated metal nanoparticles has demonstrated promises for preparing nanocatalysts with controllable size, shape, composition, and surface properties [19,52,54–58].

The use of capping agents in the synthesis is perhaps one of the most effective strategies to control the size, shape, and composition of metal nanoparticles [58–64]. There are increasing examples demonstrating that the metal nanoparticles synthesized by this strategy display intriguing catalytic properties for a number of chemical reactions. Among many methods of synthesis, wet chemical synthesis is very effective in producing monodispersed nanoparticles [16,49,54,65]. The two-phase synthesis reported first by Brust et al. [48] is now widely used for fabricating stable and soluble gold and many other metal nanoparticles of a few nm core sizes. Further modifications of the synthetic protocols or processing of the produced nanoparticles have been shown to produce a variety of chemically tunable nanoparticles in terms of size, shape, and surface properties [66]. Perhaps one of the most important attributes of these nanoparticles is the reactivity of both the metal nanoparticles and the capping molecules that allows reengineering or assembly of nanoparticles. The capping molecules can be reconstituted to impart the desired functional properties, which was demonstrated in a place-exchange reaction first by Murray's group [67]. For example, the synthesis of AuPt nanoparticles involves transfer of $AuCl_4^-$ and $PtCl_6^-$ from aqueous to organic solution by a phase transfer reagent followed by reduction in the presence of thiols and/or amines, producing thiolate (or amine)-capped gold-platinum (AuPt) nanoparticles of a few nm sizes [48,60]. Briefly, $AuCl_4^-$ and $PtCl_6^{2-}$ were first transferred from aqueous solution into toluene solution using a phase-transfer reagent (tetraoctylammonium bromide). Thiols (e.g., decanethiol, DT) or amine compounds (e.g., oleylamine, OAM) were added to the organic solution as capping agents. An excess of aqueous $NaBH_4$ was slowly added for the reduction reaction. The resulting DT/OAM-encapsulated AuPt nanoparticles in toluene were collected by removing the solvent and cleaned using ethanol. The control of the bimetallic composition in the entire bimetallic composition range and the particle size with high monodispersity (<0.5 nm) was achieved by manipulating the precursor feeding ratio [60]. The as-synthesized Au_mPt_{100-m} nanoparticles with different compositions are capped with thiol/amine monolayer shells. Varying the feeding ratio of the metal precursors to capping agents used in the synthesis controls the bimetallic composition and sizes of the nanoparticles. Bimetallic nanoparticles in which the nanocrystal core consists of one metal core and another metal shell (core@shell), such as Au@Pt and Pt@Au, could also be synthesized by seeded growth method [68]. Similar approach was used for synthesizing other binary nanoparticles, such as AuAg [69], core-shell AuAg [70], and Cu nanoparticles [71]. Similar strategies in terms of molecular encapsulation have also been used for the synthesis of other binary and ternary nanoparticles under controlled reaction conditions, such as AuM (M = Cu, Ag, etc.), PdM, PtM, and PtMM′ (M, M′ = Ni, Co, Fe, V, Cu, etc.) [39–45,69,70,72]. The synthesis of these NA particles is in contrast to traditional approaches in preparing supported nanoparticle catalysts, which involve coprecipitation, deposition-precipitation, ion-exchange, impregnation, successive reduction, calcination, etc. [16,73–79], and often have limited ability to control the size and composition due to the propensity of aggregation.

15.2.2   Preparation of catalysts

A key strategy in our approaches to the preparation of nanocatalysts involves molecularly mediated synthesis of the nanoparticles and post-synthesis thermochemical processing under controlled temperatures and atmospheres (Fig. 15.1). The removal of the organic encapsulation from metal nanoparticles is also an important step in the catalyst preparation. Among different strategies [1], thermochemical processing strategy has demonstrated not only effective needs in removing the encapsulation but also in refining the nanostructural parameters (Fig. 15.1). The combination of the molecular encapsulation-based synthesis and thermochemical processing strategies thus involves a sequence of steps for the preparation of NA catalysts: (1) chemical synthesis of the metal nanocrystal cores capped with ligands, (2) assembly of the encapsulated nanoparticles on supporting materials (e.g., carbon powders, TiO_2, or SiO_2) [23,31,41,80], and (3) thermal treatment of the supported nanoparticles [81–83]. In the molecularly engineered synthesis and thermochemically controlled processing of NA catalysts, the understanding of whether alloying or phase-segregation in multimetallic nanoparticles is different from bulk scale materials and how the catalytic activity and stability are influenced by size, composition, morphology, support, and phases have been an important focus of the recent research activities.

The size and composition of the nanoparticles produced by thermochemical processing are controllable, as demonstrated for a series of binary and ternary nanoparticles supported

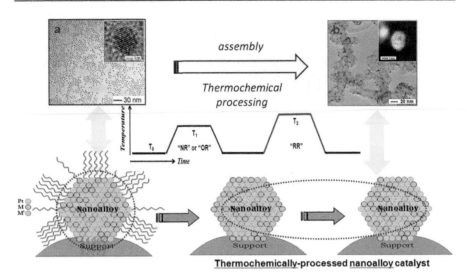

Thermochemically-processed nanoalloy catalyst

Fig. 15.1 (A) schematic illustration of molecularly engineered synthesis and thermochemical processing of NA catalysts under non-reactive (N_2) (NR), oxidative (O_2) reaction (OR), and reductive (H_2) reaction (RR) atmosphere. Example TEM/high-resolution transmission electron microscopy (HR-TEM) images showing as-synthesized $Pt_{39}Ni_{22}Co_{39}$ nanoparticles (3.4 ± 0.4 nm), thermochemically treated carbon-supported $Pt_{39}Ni_{22}Co_{39}$ nanoparticles (4.4 ± 0.5 nm).
Figure reprinted with permission from S. Shan, J. Luo, N. Kang, J. Wu, W. Zhao, H. Cronk, Y. Zhao, Z. Skeete, J. Li, P. Joseph, S. Yan, C.J. Zhong, Metallic nanoparticles for catalysis applications, in: Modeling, Characterization, and Production of Nanomaterials, Elsevier, 2015, pp. 253–288.

on carbon [41,45,60]. One example is shown in Fig. 15.1 for PtNiCo nanoparticles in as-synthesized and carbon-supported forms [41,60,84]. The as-synthesized $Pt_{39}Ni_{22}Co_{39}$ nanoparticles are highly monodispersed with a good crystallinity. The subtle increase of the particle sizes for the thermochemically treated carbon-supported $Pt_{39}Ni_{22}Co_{39}$ nanoparticles was due to thermal sintering of the nanoparticles. AuPt and PdCu NAs [40,72,85,86] are two examples of binary NAs. These nanoparticles display crystalline characteristics and mostly uniform distributions of the two metal components. The AuPt nanoparticles are fully alloyed with uniform bimetallic distribution in the nanoparticles (Fig. 15.2A). The observation of the indicated lattice fringes of 0.235 nm indicates that the carbon-supported nanoparticles are highly crystalline with a subtle difference in interatomic distance between catalysts treated under H_2 at 400 and 800 °C. Carbon-supported AuPt nanoparticles are shown to exhibit single-phase alloying characteristics, and also a controllable degree of partial alloying or phase segregation depending on the thermochemical processing conditions [81,85,86]. The fully alloyed AuPt nanoparticles have a uniform distribution of the two metals across the nanoparticles. The bimetallic nanoparticles are not only crystalline with highly faceted surfaces but also have a uniform distribution of Au and Pt. Similar nanocrystalline features and relatively uniform bimetallic distribution have also

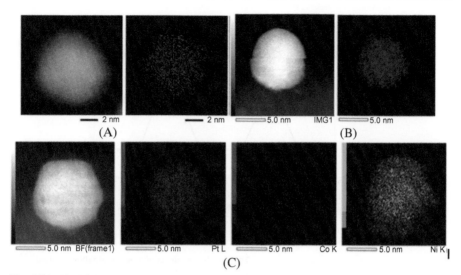

Fig. 15.2 (A) High-resolution transmission electron microscopy (HR-TEM) and energy dispersive spectroscopy (EDS) composition mapping for $Au_{22}Pt_{78}/C$ (*red*: Pt, *blue*: Au), (B) $Pd_{21}Cu_{79}/C$ (*red*: Cu, *blue*: Pd), and (C) $Pt_{39}Ni_{22}Co_{39}$ nanoparticles (*blue*: Pt, *red*: Co, *green*: Ni).

Panel A: Figure reprinted with permission from S. Shan, J. Luo, N. Kang, J. Wu, W. Zhao, H. Cronk, Y. Zhao, Z. Skeete, J. Li, P. Joseph, S. Yan, C.J. Zhong, Metallic nanoparticles for catalysis applications, in: Modeling, Characterization, and Production of Nanomaterials, Elsevier, 2015, pp. 253–288; Panel B: Figure reprinted with permission from S. Shan, J. Luo, N. Kang, J. Wu, W. Zhao, H. Cronk, Y. Zhao, Z. Skeete, J. Li, P. Joseph, S. Yan, C.J. Zhong, Metallic nanoparticles for catalysis applications, in: Modeling, Characterization, and Production of Nanomaterials, Elsevier, 2015, pp. 253–288; Panel C: Figure reprinted with permission from S. Shan, J. Luo, L. Yang, C.-J. Zhong, Technology, nanoalloy catalysts: structural and catalytic properties, Catal. Sci. Technol. 4 (10) (2014) 3570–3588.

been observed for PdCu nanoparticles (Fig. 15.2B) [72], and PtNiCo nanoparticles (Fig. 15.2C) [87] revealing that the bimetallic or the trimetallic distribution is basically a uniform alloy across the entire nanoparticle.

    As will be discussed in the latter sections, the catalytic or electrocatalytic properties of the NA catalysts prepared by the combination of molecular encapsulation-based synthesis and thermochemical processing strategies depend on the detailed synthesis and thermochemical processing parameters, showing synergistic correlation with the atomic-scale structural evolution of the NAs.

15.3   Structural characterizations

The atomic-scale chemical and structural ordering of NAs plays an important role in determining the catalytic properties of NAs. In addition to TEM, HRTEM, and EDX imaging techniques, there are many analytical techniques that have been used to

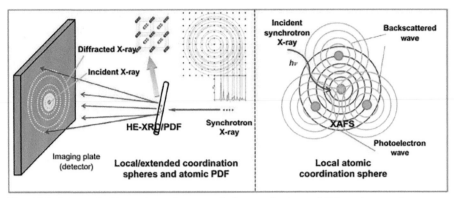

Fig. 15.3 Schematic illustrations of structural characterizations of NAs using synchrotron X-ray based techniques: average structure sensitive HE-XRD and local structure sensitive XAFS.

Figure reprinted with permission from S. Shan, J. Luo, N. Kang, J. Wu, W. Zhao, H. Cronk, Y. Zhao, Z. Skeete, J. Li, P. Joseph, S. Yan, C.J. Zhong, Metallic nanoparticles for catalysis applications, in: Modeling, Characterization, and Production of Nanomaterials, Elsevier, 2015, pp. 253–288.

characterize the NA structures. In particular, synchrotron X-ray techniques, including HE-XRD/PDFs and XAFS, have been demonstrated to be powerful for gaining insights into both extended and local atomic structures of NAs (Fig. 15.3.). The HE-XRD/PDF study can be coupled with 3D simulations for determining the overall atomic-scale structure across the NA, including details of the chemical ordering. XANES/EXAFS has been demonstrated as a powerful technique being used to determine the near neighbor metal–metal coordination, metal oxide content, electronic properties, and charge transfer phenomena in the studied alloys, including the effect of support providing valuable information for assessing O_2 activation on the NA binding sites in terms of local structural configurations. EXAFS provides structural information on the coordination environment around each component for assessing the alloy formation in terms of coordination number (CN) of M or M′ around Pt. Metal oxide contents can be estimated from XANES spectra using a linear combination fitting. In addition, the Pt L-edge absorption edge peak (or white line) is sensitive to the particle size, shape, and support type [88,89]. The understanding of how both local and extended atomic ordering parameters, including structural/chemical alloying degree, interatomic distances, CNs, and oxophilicity, correlate with the NA catalyst's activity and stability is important for the design of NA catalysts. Using high-energy X-rays (80–1000 keV, about one order of magnitude higher than conventional X-rays), the unique advantages of HE-XRD include high penetration into materials, no radiation damage of the sample, and many other qualities enabling the characterization of the nanoscale materials and detail structures. Both ex situ and in situ HE-XRD coupled to atomic PDF analysis could be used to obtain fundamental insight into atomic-scale structure of nanoparticles in various gas atmospheres (Fig. 15.3.). The PDF technique has emerged recently as a powerful and unique tool for the characterization and structure refinement of crystalline materials with the intrinsic disorder [90]. The strength of the technique stems from the fact that it takes into account the total diffraction, that is Bragg as well as diffuse scattering, and so it can be applied to both crystalline [91] and amorphous materials [92]. In contrast, other techniques used to study polycrystalline samples (e.g., Rietveld) [93] use only the Bragg scattering portion of the pattern, and therefore any information originating from the disorder present is lost. A reactor-type cell [94] can be used in the in-situ experiments. The XRD diffraction data can be reduced to the so-called structure factors, $S(q)$, and then Fourier transformed into atomic PDFs $G(r)$. The wave vector q is defined as $q = 4\pi\sin(\theta)/\lambda$, where θ is half of the scattering angle and λ is the wavelength of the X-rays used. The atomic PDFs $G(r)$ are experimental quantities that oscillate around zero and show positive peaks at real space distance; r, where the local atomic density $\rho(r)$ exceeds the average one ρ_o. The PDF $G(r)$ can be expressed by the equation: $G(r) = 4\pi r \rho_o \rho(r)/\rho_o - 1$. High-energy XRD and atomic PDFs have already proven to be very efficient in studying the atomic-scale structure of nanosized materials [95,96].

XAFS (Fig. 15.3.) is another spectroscopic technique that uses X-rays to probe the physical and chemical structure of matter at an atomic scale. XAFS is element-specific, in that X-rays are chosen to be at and above the binding energy of a particular core electronic level of a particular atomic species. Unlike regular diffraction-based techniques for studying the atomic structure of matter, XAFS does not require a crystalline sample. XAFS provides information on how X-rays are absorbed by an atom at energies near and above the core-level binding energies. Because XAFS spectra are very sensitive to the formal oxidation state, coordination chemistry, and the distances, CN, and species of the atoms immediately surrounding the selected element, it provides a relatively simple way to determine the chemical state and local atomic structure for a selected atomic species. Since XAFS is an atomic probe, all atoms have core level electrons, and importantly, crystallinity is not required for XAFS measurements, making it one of the few structural probes available for non-crystalline and highly disordered materials, including solutions. Because X-rays are fairly penetrating in the matter, XAFS is not inherently surface-sensitive, though special measurement techniques can be applied to enhance its surface sensitivity [97]. As discussed in the next two subsections, these two techniques provide information for short-range interatomic bonding and coordination structure about a specific atom as well as more extended structural/chemical ordering across the entire nanoparticle, regardless of whether the particle is crystalline or amorphous [98,99].

15.3.1 Nanophase and atomic coordination structures

NAs of binary PtM and ternary PtMM' (M/M' = Co, Fe, V, Ni, etc.) compositions have been studied to understand the structure–activity correlation [41,45]. Consider PtVCo NAs [41] as an example to illustrate the structural evolution by thermochemical processing treatments of oxidation reaction in O_2 (OR) followed by reduction reaction in H_2 (RR) at different temperatures (Fig. 15.4). There is a clear structural evolution for the catalyst treated in "OR"–"RR" treatment atmosphere [41]. A hypothetical face-centered cubic (fcc) lattice inferred from the increase of the major Bragg peaks

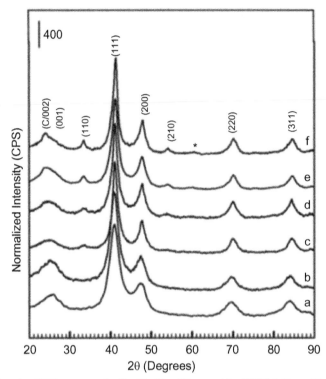

Fig. 15.4 Regular XRD patterns for $Pt_{45}V_{18}Co_{37}/C$ treated by "OR" followed by "RR" treatment at different temperatures: 400 °C (A), 500 °C (B), 600 °C (C), 700 °C (D), 800 °C (E), and 926 °C (F).

Figure reprinted with permission from B.N. Wanjala, B. Fang, S. Shan, V. Petkov, P. Zhu, R. Loukrakpam, Y. Chen, J. Luo, J. Yin, L. Yang, Design of ternary nanoalloy catalysts: effect of nanoscale alloying and structural perfection on electrocatalytic enhancement, Chem. Mater. 24 (22) (2012) 4283–4293.

showed a gradual decrease (Fig. 15.4). The appearance of the low-intensity peaks for $T > 600$ °C suggests that the particle's structure rearranges away from the fcc type at higher temperatures. This rearrangement transforms the fcc-type structure of the as-prepared particles into particles with a face-centered tetragonal (fct)-type structure [100], bearing some resemblance to phase transitions observed in binary alloy particles such as PtCo and PtV [101–103].

Traditionally, regular in-house XRD is used for determining the detailed structures of nanoparticles, which, however, has limitations largely because of the fact that small-sized nanoparticles do not exhibit extended crystalline structures. In contrast, HE-XRD/PDF technique can provide insights into the atomic-scale details of the phase state and structures, especially for small-sized nanoparticles. One example is to study the phase state of PtVCo/C NAs [41]. To reveal the nature of the atomic ordering in $Pt_{45}V_{18}Co_{37}/C$ and its temperature-driven evolution in more details, we conducted HE-XRD experiments coupled to atomic PDF analysis that is very well suited for nanosized particles [103,104]. HE-XRD patterns for $Pt_{45}V_{18}Co_{37}/C$ catalysts treated by "OR" followed by "RR" at 400, 600, and 926 °C are shown in Fig. 15.5A. Atomic PDFs extracted from the HE-XRD patterns are shown in Fig. 15.5B together with that for pure Pt particles used here as a standard. The PDFs exhibit a series of well-defined peaks reflecting the presence of a sequence of well-defined coordination spheres in the ternary catalysts. In the PDF for pure Pt particles, the first peak is positioned at 2.76 Å which is very close to the Pt–Pt distance in the respective bulk metal. The first peak in the experimental PDFs for PtVCo/C catalysts is positioned at 2.70 Å indicating a shortening of the metal–metal distances in the alloy particles due to the smaller size of the Co/V atoms as compared to Pt atoms. All features in the PDF for pure Pt can be very well fit with a structure model featuring an fcc-type structure with a parameter of 3.92 Å. The results revealed a random alloy fcc-type or chemically ordered alloy fct-type structure for 400 °C treated catalyst. The chemically ordered fct lattice model, however, becomes increasingly a better match to the experimental PDFs for the particles annealed at >400 °C. Indeed only the fct-type model can explain well the fine features of the experimental PDF data (see the encircled region) that are characteristic benchmarks of a chemically ordered structure where Pt, Co, and V atoms are not randomly distributed within the nanoparticles but assume preferential ordering with respect to each other. The PDF refined fct lattice parameters change from $a = 2.710$ Å and $c = 3.776$ Å for the sample annealed at 400 °C, to $a = 2.700$ Å and $c = 3.773$ Å at 600 °C and then to $a = 2.698$ Å and $c = 3.770$ Å at 926 °C. If Pt and $Pt_{45}V_{18}Co_{37}$ samples are compared based on a hypothetical fcc model, the respective lattice parameters change from $a = 3.82$ Å (400 °C) to $a = 3.809$ Å (600 °C) and then to $a = 3.803$ Å (926 °C). The synchrotron-based PDF studies unambiguously show that with increasing the temperature, the catalyst undergoes a phase transition from a random alloy to a chemically ordered alloy-type structure, which is accompanied by shrinking of the distances between the metal atoms.

EXAFS data have also shown that the formation of oxygenated metal species in the NAs is highly dependent on the thermal treatment condition. The atomic M-M coordination in $Pt_{45}V_{18}Co_{37}/C$ catalysts was analyzed by EXAFS on selected samples annealed at 400 and 800 °C [41]. The understanding of the structural changes between these two temperatures provides information for assessing how the change in activity with annealing temperature is related to the structural change of the catalysts. The data can be analyzed by the fitting of the Pt EXAFS data using *PtCo* and *PtV* models. Based on analysis of the structural parameters from fitting Pt L3 edge EXAFS data using *PtCo* model and the Co K edge data using *CoCo* model for $Pt_{45}V_{18}Co_{37}/C$ catalyst, the results clearly show a significant presence of Co/V neighbors around Pt and Pt neighbors around Co, which is a direct evidence of the formation of PtVCo alloy in the samples studied here. Also, the EXAFS data indicate a higher probability of Pt bonding with transition metal atoms for the catalyst treated at 800 °C as compared to the catalyst treated at 400 °C. For example, the CN of Pt-Co/V atomic pairs changes from 2.3 for the 400 °C treated sample to 3.1 for the 800 °C sample, while the corresponding change in CN of Pt–Pt is insignificant (from 5.7 to 5.4). Although the increase in particle size may lead to an increase of total CN, the more significant

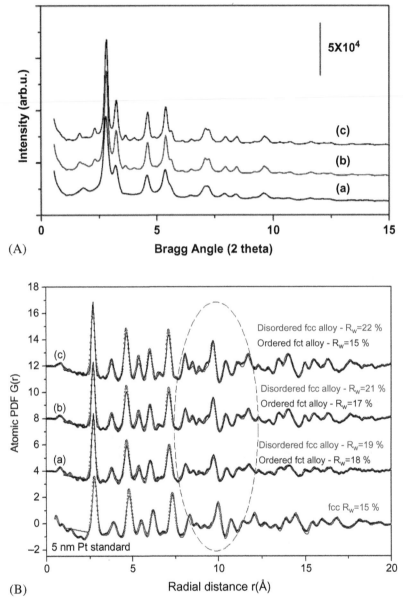

Fig. 15.5 (A) HE-XRD patterns for $Pt_{45}V_{18}Co_{37}/C$ catalysts treated by "OR" followed by "RR" treatment as a function of treatment temperature: 400 °C (a), 600 °C (b), and 926 °C (c) and (B) atomic PDFs extracted from the HE-XRD diffraction patterns for $Pt_{45}V_{18}Co_{37}/C$ catalysts at "RR" annealing temperature: (a) 400 °C, (b) 600 °C, and (c) 926 °C. The PDF for 5 nm Pt particle is shown for comparison.

Figure reprinted with permission from B.N. Wanjala, B. Fang, S. Shan, V. Petkov, P. Zhu, R. Loukrakpam, Y. Chen, J. Luo, J. Yin, L. Yang, Design of ternary nanoalloy catalysts: effect of nanoscale alloying and structural perfection on electrocatalytic enhancement, Chem. Mater. 24 (22) (2012) 4283–4293.

increase in CN of Pt-Co/V than that of Pt–Pt suggests better alloying between Pt and Co/V. The EXAFS results also indicate a clear reduction of the oxide content (e.g., reduction in the metal-oxygen CNs) for the 800 °C treated sample.

HE-XRD/PDFs and EXAFS techniques have also been used under in situ/operando conditions for characterizing structural changes of the catalysts upon thermochemical treatments [41,45]. A comparison of the structural evolutions of PtVCo and PtNiCo NAs under both oxidative (O_2) and reductive (H_2) atmosphere [45] reveals differences in the changes of the lattice parameters in the oxidation–reduction cycles between the two systems (Fig. 15.6). One intriguing finding is that the changes in the lattice parameter of the PtVCo catalyst are only half of those observed for the PtNiCo system. In other words, PtVCo maintains a more stable lattice parameter, that is, a more stable atomic-scale structure, than PtNiCo system in the catalytic reaction process. The PtVCo/C NA's interatomic distances increase when the catalyst is treated in the oxidizing atmosphere, and shrink uniformly when the catalyst is treated in the reducing atmosphere. The NA particles remain "expanded" when cooled back to room temperature (RT) in oxidizing atmosphere and "shrunk" if cooled back to RT in reducing atmosphere. The expansion of the interatomic distances upon oxidation is consistent with the enhanced interaction of oxygen with the particle's surface that is favored by the oxophilicity of the base transition metal. The overall shrinking of the interatomic distances in the NA particles treated in a reducing atmosphere is consistent with freeing their surface from the oxygen species. Interestingly, ternary nanoparticles show a

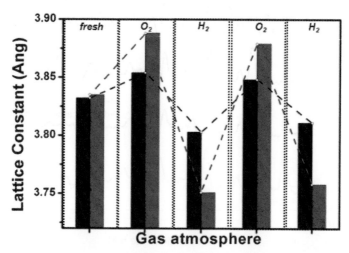

Fig. 15.6 A comparison of the fcc-lattice parameters for $Pt_{45}V_{18}Co_{37}$ (*open circles black*) and $Pt_{25}Ni_{16}Co_{39}$ (*black squares red*) in the thermochemical treatment cycles in O_2 (at 240 °C) and H_2 (400 °C) atmosphere, as denoted on the horizontal axis, and then cooled down to room temperature.
Figure reprinted with permission from S. Shan, J. Luo, N. Kang, J. Wu, W. Zhao, H. Cronk, Y. Zhao, Z. Skeete, J. Li, P. Joseph, S. Yan, C.J. Zhong, Metallic nanoparticles for catalysis applications, in: Modeling, Characterization, and Production of Nanomaterials, Elsevier, 2015, pp. 253–288.

great degree of tunability in lattice parameters. For PtVCo NA, there is a clear lattice expansion upon oxidation ($\delta a/a = 0.9\%$) and lattice shrinking upon reduction ($\delta a/a = -1.0\%$). In comparison, the lattice expansion upon oxidation ($\delta a/a = 1.4\%$) and shrinking upon reduction ($\delta a/a = -3.6\%$) are somewhat larger for that of PtNiCo NA (Fig. 15.6). The introduction of a third transition metal ($M' = V$, Ni, Ir, etc.) into PtCo NAs minimizes or suppresses changes in the lattice parameters of the nanoparticle during the catalytic processes, consistent with the order of oxophilicity of these metals.

The characterization of AuPt nanoparticles [40] provides another example demonstrating the power of HE-XRD/PDFs in gaining atomic-scale insights into the NA structure. AuPt NAs less than 10 nm in size are analyzed using element-specific resonant HE-XRD coupled with atomic PDFs and computer simulations (Fig. 15.7). The bond lengths in the alloys differing in 0.1 Å leads to extra structural distortion as compared to pure Pt and Au particles. In the NAs, Au–Au and Pt–Pt bond lengths differing by 0.1 Å are revealed. The results demonstrate the presence of an extra structural distortion for AuPt nanoparticles as compared to pure Pt and Au nanoparticles. Pt–Pt and Pt–Au pairs were identified in the first coordination shell of Pt atoms, providing atomic-scale insights into the alloy characteristic for the AuPt nanoparticles. Reverse Monte Carlo (RMC) simulations refined models are also used to aid the analysis of the alloy structures. Pure Au and Pt nanoparticles (NPs) are used to produce model configurations for Au_mPt_{100-m} NPs where Au and Pt atoms show various patterns of chemical order–disorder effects. All peaks in the partial Au–Au PDFs for

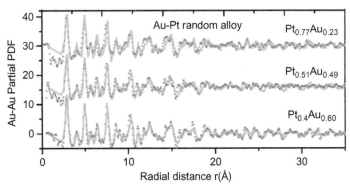

Fig. 15.7 Experimental (*symbols*) and model (*cyan line*) Au–Au partial PDFs for 5.1 nm Au_mPt_{100-m} particles annealed at 400 °C. The datasets are shifted by a constant factor for clarity. Au_core/Pt_shell model Au–Au PDFs show first peaks that are too strong in intensity than the peaks in the respective experimental Au–Au PDFs, and also decay to zero much faster than the experimental data. Peaks in the Au–Au PDFs for Pt_core/Au_shell model show characteristic step-down loss in intensity at short interatomic distances that is inconsistent with the behavior of the experimental data. Au–Pt random alloy model Au–Au PDFs agrees with the experimental data in very good detail.

Figure reprinted with permission from V. Petkov, B.N. Wanjala, R. Loukrakpam, J. Luo, L. Yang, C.-J. Zhong, S. Shastri, Pt–Au alloying at the nanoscale. Nano Lett. 12 (8) (2012) 4289–4299.

Au_mPt_{100-m} NPs are with largely diminished areas corresponding to reduced Au–Au CNs that are characteristic of random-type Pt–Au alloys. The alloying of Pt and Au occurs not only within a wide range of Pt–Au concentrations but is also stable in NPs of different sizes. Importantly, the alloying characteristics of the less than 5-nm sized AuPt particles were found to remain unchanged for a long period of time under ambient conditions, demonstrating structural stability of the NA.

The formation of NAs from mixed single-component nanoparticles during thermochemical evolution processes have also been studied by synchrotron X-ray techniques. One example involves the formation of AuCu NA from Au and Cu nanoparticles [42]. Fig. 15.8 shows the behavior of the wide-angle XRD patterns during uniform heating from 25 to 225 °C of a binary solution of 2-nm Au and 0.5-nm Cu cast on a planar Si substrate. The temperature evolution of the binary sample is compared with the evolution of unary Cu and unary Au nanoparticles (Fig. 15.8B). Due to the smaller size and lower melting point, the Cu particles start to aggregate immediately after heating above 60 °C. In sharp contrast, the Au nanoparticles restructure significantly only above 140 °C. In terms of the "half-way" temperature for the increase of the integrated intensity, the temperature for Cu + Au (~115 °C) falls in between those for Au (~160 °C) and Cu (~90 °C). Since this temperature is relatively low compared to the melting temperature of 2 nm Au nanoparticles, the restructuring process likely involves solid-phase sintering and recrystallization. The physical mixture exhibits an intermediate behavior, but it is clear that the particles restructure at considerably lower temperatures than pure Au nanoparticles, indicating a significant effect of the Cu nanoparticles.

15.3.2 Computational modeling

The nanoscale alloying or phase segregation depends strongly on the composition, support, and thermal treatment temperature [104,105]. For supported Au_mPt_{100-m} nanoparticles, fundamental insights have been gained for assessing the alloying or phase-segregation states. Theoretical studies revealed a negative heat of formation for AuPt nanoparticles smaller than ~6 nm, which constituted as a thermodynamic basis for favoring the formation of alloyed phase [106]. The other involves the theoretical finding of a possible phase segregation of AuPt particles as a result of the combination of larger surface free energy of Pt ($2.48 J/m^2$) than Au ($1.50 J/m^2$) [106] and the lower melting point of smaller-sized nanoparticles than larger-sized particles ($\Delta T \propto 1/r_{(particle\ radius)}$) [107]. The nanoscale phase of Au_mPt_{100-nm} nanoparticles is a very important parameter for the exploration of the nanoscale bimetallic functional properties. The lattice constants of the bimetallic nanoparticles were found not only to scale linearly with Pt%, in contrast to the bulk AuPt counterparts which display a miscibility gap at 20–90% Au, but also to be smaller than those of the bulk counterparts [105,108]. This finding was the first example demonstrating that the nanoscale AuPt nanoparticles could exhibit single-phase character and reduced interatomic distances in the entire bimetallic composition range in contrast to the bulk counterparts [105,108]. In addition to the latest HE-XRD/PDFs evidence [40], results from theoretical modeling and DFT computation for small AuPt nanoparticles (~1.2 nm) have

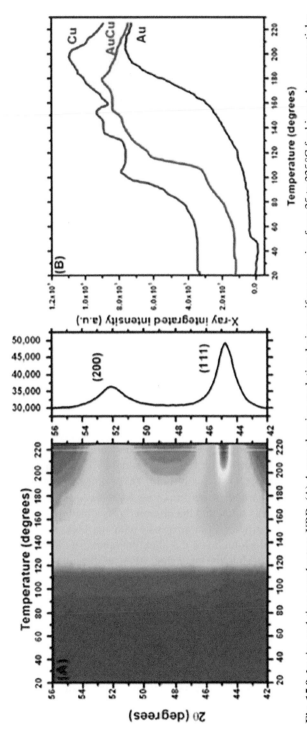

Fig. 15.8 In-situ real-time synchrotron XRD. (A) A map showing evolution during uniform ramping from 25 to 225°C for binary Au nanoparticles (2 nm) and Cu nanoclusters with a Cu:Au molar ratio of 8.5 cast on a Si surface, and a snapshot of XRD pattern at 220°C and (B) comparison of temperature evolution of (111)–intensity for Au, Cu, and binary (1:1) Au and Cu (X-ray energy: 6.9 keV).

Figure reprinted with permission from J. Yin, S. Shan, L. Yang, D. Mott, O. Malis, V. Petkov, F. Cai, M. Shan Ng, J. Luo, B.H. Chen, Gold–copper nanoparticles: nanostructural evolution and bifunctional catalytic sites, Chem. Mater. 24 (24) (2012) 4662–4674.

also revealed a difference in the lattice constant—composition relationship between the nanoscale and the bulk systems in agreement with the experimental data [85,86]. In the theoretical modeling of realistic particle sizes, a negative heat is revealed for the formation for AuPt particles smaller than ∼6 nm [106], thermodynamically favoring the formation of alloyed phase [106]. The atomic distributions of Au and Pt are also shown to change in distance from the center of mass in the nanoparticle depending on temperature and support [109]. The complex phase properties are also supported by in situ time-resolved X-ray diffraction study of phase change kinetics at high temperature [68]. The nanoscale phase evolution has a significant impact on the surface alloying or segregation [104,108,110], as supported by Fourier transform infrared (FTIR) probing of CO adsorption as a function of the bimetallic composition [108] and XPS analysis of the relative surface composition as a function of the treatment temperature [41,104].

DFT computation of small AuPt nanoparticles (∼1.2 nm) has provided some insights into the phase properties [86]. Seventeen 55-atom AuPt nanoparticles (∼1.2 nm), 6 $Au_{24}Pt_{76}$ (13 Au atoms and 42 Pt atoms), 5 $Au_{49}Pt_{51}$ (27 Au atoms and 28 Pt atoms), and 6 $Au_{76}Pt_{24}$ (42 Au atoms and 13 Pt atoms) were studied using DFT calculations. These calculations were carried out using the spin-polarized DFT method that is implemented in Vienna Ab-initio Simulation Package [106,111,112]. The structures and the corresponding stability of these nanoparticles are shown in Fig. 15.9A.

A complete Pt-core/Au-shell structure is found to be the most stable among all isomers. The least stable isomer has the least number of surface Au atoms. When temperature increases, the alloy nanoparticles evolve in such a way that the outer shell Pt atoms migrate into the inner shell. This is further supported by molecular dynamics simulations [113], where upon heating above 600 K such Pt migration takes place in an alloy nanoparticle of ∼3.6 nm. The computational results further indicate that thermal treatments of alloy AuPt nanoparticles should be below certain temperatures to prevent the loss of outershell Pt atoms. Fig. 15.9B shows the DFT-calculated lattice constants, which are indeed smaller than that of bulk. This finding is in agreement with our earlier finding of the nanoscale phase properties for the alloyed AuPt/C catalysts [105]. The computation results supported the observed overall lattice shrinking of the metal and bimetallic nanoparticles in comparison with the bulk counterparts, and showed the possibility of an evolution of the alloy nanoparticles in such a way that the outer shell Pt atoms migrate into the inner shell when temperature increases, which is qualitatively consistent with the XRD analysis results. There are many recent examples on DFT computational modeling of AuPt nanoparticle structures in terms of composition and structures [114]. For example [114], the study of AuPt NAs with 55 atoms with an intermediate composition range by a procedure combining global optimization searches within an atomistic potential model with DFT relaxation of the lowest-energy isomers identified two structural motifs in close competition: the Mackay icosahedron and an asymmetric capped decahedron.

The ability to tune the atomic-level nanophase structure of alloy nanoparticles is important for the design and preparation of active and stable catalysts. For PdCu alloy nanoparticles, the structure can be tuned by varying the thermochemical treatment

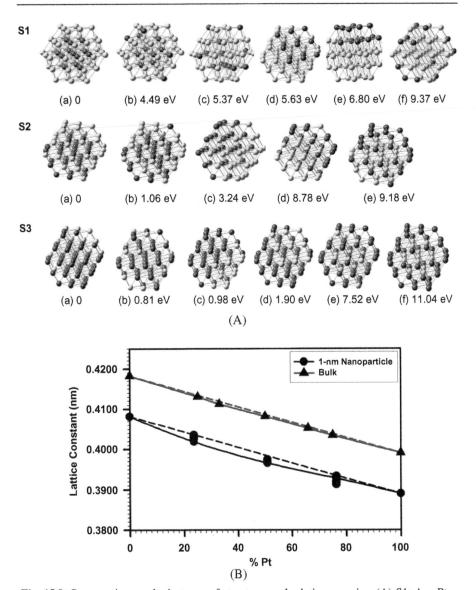

Fig. 15.9 Computation results in terms of structures and relative energies: (A) S1: $Au_{76}Pt_{24}$ nanoparticles. A positive number means that the corresponding structure is less stable than structure (a). S2: $Au_{49}Pt_{51}$ nanoparticles. A positive number means that the corresponding structure is less stable than structure (a). S3: $Au_{24}Pt_{76}$ nanoparticles. A positive number means that the corresponding structure is less stable than structure (a). and (B) DFT-calculated lattice constant—composition relationship for the nanoscale and bulk bimetallic systems.
Figure reprinted with permission from B.N. Wanjala, J. Luo, R. Loukrakpam, B. Fang, D. Mott, P.N. Njoki, M. Engelhard, H.R. Naslund, J.K. Wu, L. Wang, Nanoscale alloying, phase-segregation, and core–shell evolution of gold–platinum nanoparticles and their electrocatalytic effect on oxygen reduction reaction, Chem. Mater. 22 (14) (2010) 4282–4294.

condition [115]. As shown in Fig. 15.10B, phase structure transformation between fcc, body-centered cubic (bcc), and fcc/bcc mixture can be achieved by calcination temperature. Using HE-XRD/PDF technique, the experimental data shows that the 100°C/H_2-treated $Pd_{50}Cu_{50}$ NPs yield a pure fcc phase structure, whereas the 400°C/H_2-treated $Pd_{50}Cu_{50}$ NPs exhibits a mixed structure of bcc alloy core and fcc alloy shell (Fig. 15.10B). The $Pd_{50}Cu_{50}$ NPs with a pure fcc structure is shown to exhibit an enhanced ORR catalytic activity, but a decreased durability than

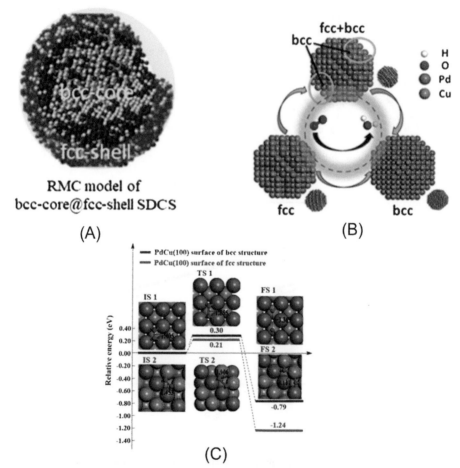

Fig. 15.10 (A) RMC model of the bcc-core@fcc-shell structure. (B) A schematic illustration of phase evolution in terms of fcc and bcc structures manipulated by thermochemical treatment temperature. (C) DFT calculation of O—O bond cleavage in O_2 molecule on fcc- and bcc-structured PdCu alloy models.

Figure reprinted with permission from Z.-P. Wu, S. Shan, Z.-H. Xie, N. Kang, K. Park, E. Hopkins, S. Yan, A. Sharma, J. Luo, J. Wang, Revealing the role of phase structures of bimetallic nanocatalysts in the oxygen reduction reaction, ACS Catal. 8 (12) (2018) 11302–11313.

$Pd_{50}Cu_{50}$ NPs with a bcc-core@fcc-shell structure. Further, the DFT calculation of the reaction mechanism (Fig. 15.10C) shows that the O—O bond cleavage in O_2 molecule on fcc-phase model exhibits a lower reaction barrier than bcc-phase model. This result is in agreement with the experimental results.

15.4   Applications in heterogeneous catalysis

15.4.1   Catalytic oxidation of carbon monoxide and propane in the gas phase

The study of NAs in gas-phase reactions allows probing how the catalytic activity and stability of the NAs can be correlated with the detailed structure by design. Under gas-phase catalytic conditions, one important reaction involves water gas shift reaction [116], which is important to hydrogen energy production. A key challenge is to remove the trace of CO from hydrogen fuel which is a poison for some catalysts used downstream. The typical process of CO removal is catalytic oxidation of CO at low temperatures:

$$2CO(g) + O_2(g) \rightarrow 2CO_2(g) \tag{15.1}$$

The catalytic CO oxidation has been an important subject of interest [41,45,98], which is often used as a probe molecule to NA catalysts [3,29,41,45] for the understanding of catalytic synergy in relation to atomic surface structure, phase properties, and NA-support interactions. For CO oxidation over PtNiCo NAs on different supports [41], both supports and thermochemical treatment conditions were found to affect substantially the overall atomic-scale structure of the NA particles and the surface oxygenated Ni/Co species. It is the combination of the particle's atomic-scale structural/chemical ordering and the surface chemistry that determines the NA's catalytic activity. As illustrated in Fig. 15.11, there are two types of catalytic surface sites for

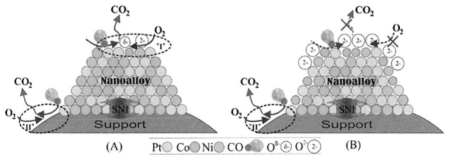

Fig. 15.11 Illustration of catalytic CO reaction over supported nanoalloy catalyst in (A) reduced nanoalloy or (B) oxidized nanoalloy state.
Figure reprinted with permission from L. Yang, S. Shan, R. Loukrakpam, V. Petkov, C. J. Zhong, Role of support—nanoalloy interactions in the atomic-scale structural and chemical ordering for tuning catalytic sites, J. Am. Chem. Soc. 134 (36) (2012) 15048—15060.

supported NAs. The NA particles alone can provide Type-I sites (Co-Pt-Ni sites) involving Pt atoms for activating CO and base transition metal atoms (Ni/Co) for assisting the oxygen activation. Supports like titania can provide catalyst-support perimeter sites (Type-II sites), which are inactive in the case of silica supports and hardly available in the case of carbon supports. The catalytic properties of Type-I sites depend on whether the NA is in a reduced or oxidized state and the support-NA interaction, both of which are significantly influenced by the atomic-scale ordering and phase structures of the NA. The surface of the NA involves oxygen storage and release capacity where the transition base metals Ni/Co provide sites for the activated O^{2-} and $O^{\delta-}$ species. While both reduced and oxidized surfaces may involve Type-II sites for O_2 activation, the Type-I site for O_2 activation exists only on reduced NA surface (A), and does not exist on the oxidized NA surface where the surface is completely blocked by the oxide species (B).

The study of CO oxidation over PtCoM (M = Ni,V) NAs provided some insights into the active surface sites (Fig. 15.12) [41]. In Fig. 15.12A, the catalytic CO reaction rates of the ternary PtNiCo/C catalyst are compared with those of the binary PtNi/C and PtCo/C in an as-prepared state. For the as-prepared catalyst, which is likely partially oxidized at the surface by air, PtNiCo has an activity higher than PtCo and PtNi. PtCo's activity is slightly higher than Pt, whereas PtNi shows almost no improvement in activity. Similar results were also obtained for the H_2-reduced catalysts. Upon reactivation in-situ by H_2, which reduces the particles back to alloy state, the activity for the ternary catalyst shows an increase by a factor of 2–3. PtCo shows the largest increase in activity (by a factor of 10–20), though being still lower than the ternary catalyst, especially at the high reaction temperature. In contrast, PtNi shows practically no change in activity, similar to Pt.

The comparison of the catalytic CO oxidation over PtVCo NA with those over different binary and ternary NAs also led to an intriguing finding [45]. The catalytic activity of PtVCo/C catalyst is compared with those for PtNiCo/C and PtIrCo/C catalysts (Fig. 15.12B). It is evident that the PtVCo/C exhibits the highest activity, in the order PtVCo ≫ PtNiCo > PtIrCo ≥ PtCo. At RT, the activity of PtVCo/C is greater than that of PtNiCo/C by a factor of 10–20. In comparison with PtCo/C, the addition of V into the binary system leads to an increase in the activity by a factor of 10^2–10^3, whereas the addition of Ni into the binary system leads to an increase in the activity by a factor of 10–20. The addition of Ir into the binary system showed the smallest increase in the catalytic activity. The catalytic activity order appears to be consistent with the order of oxophilicity of the third transition metal (M') in PtCoM' (M' = Ni, V, and Ir), which is in the order of V > Ni > Ir. The introduction of M' with oxophilicity larger (i.e., V) than Co results in a greater enhancement to the activity than introducing a metal (i.e., Ni) with oxophilicity close to Co (Fig. 15.12B). The understanding of this type of Pt-based NA properties in terms of M or M' (Fig. 15.12B) is useful for creating a bifunctional or multifunctional synergy for effective maneuvering surface CO or oxygenated species over the M/M' sites through structural or compositional effect.

Interesting insights have also been gained by detailed analysis of the in situ HE-XRD/PDF results of PtVCo and PtNiCo NA catalysts in terms of the self-tunable oxophilicity and structural integrity. For $Pt_{45}V_{18}Co_{37}$/C NAs, the fits to the atomic

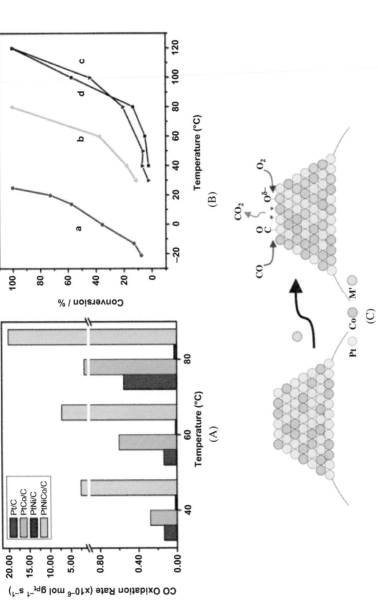

Fig. 15.12 (A) Comparisons of Pt mass-specific CO oxidation rates of $Pt_{39}Ni_{22}Co_{39}/C$ with those of $Pt_{64}Ni_{36}/C$, $Pt_{45}Co_{55}/C$, and Pt/C catalysts in the as-prepared state*. The Pt mass-specific CO oxidation rate is normalized against the metal loading on carbon and the Pt composition in the nanoparticles. (*The as-prepared catalyst underwent "OR" "RR" or "NR" "RR" treatment by exposure to ambient air for an extensive period of time before catalytic measurement.) (B) Comparison of CO conversion% over $Pt_{45}V_{18}Co_{37}/C$ (*red*, a), $Pt_{39}Ni_{22}Co_{39}/C$ (*green*, b), and $Pt_{23}Ir_{29}Co_{48}/C$ (*blue*, c) with $Pt_{55}Co_{45}/C$ (*black*, d) catalysts. (C) A schematic illustration of the introduction of a third transition metal (M') into PtCo nanoalloy forming PtCoM' nanoalloy for formation/removal of —O or —CO species in CO oxidation. Panel A: Figure reprinted with permission from L. Yang, S. Shan, R. Loukrakpam, V. Petkov, C.J. Zhong, Role of support–nanoalloy interactions in the atomic-scale structural and chemical ordering for tuning catalytic sites, J. Am. Chem. Soc. 134 (36) (2012) 15048–15060; Panels B and C: Figure reprinted with permission from S. Shan, V. Petkov, L. Yang, D. Mott, B.N. Wanjala, F. Cai, B.H. Chen, J. Luo, C.-J. Zhong, Oxophilicity and structural integrity in maneuvering surface oxygenated species on nanoalloys for CO oxidation, ACS Catal. 3 (12) (2013) 3075–3085.

PDFs reveal (1) little change in lattice parameter for the $CO+O_2$ mixed gas, as evidenced by 3.842 Å at RT, 3.845 Å at 80 °C, and 3.844 Å when the sample is cooled back down to RT; and (2) a considerable shrinking of the lattice constant from 3.850 Å at RT to 3.848 Å at 400 °C, and to 3.831 Å when the sample is cooled back to RT under the reducing gas (H_2). A similar in situ experiment was performed for $Pt_{25}Ni_{16}Co_{59}$/C catalyst. Little structural change was seen when the catalyst was subjected to a gas mixture of CO and O_2 at 80 °C. In particular, the change in the lattice constant was relatively small when the temperature was changed from RT (3.885 Å) to 80 °C (3.887 Å), and then back to RT (3.886 Å). However, upon exposing to H_2 gas atmosphere at 400 °C, the change was noticeable, as evidenced by the difference of the respective lattice constants: from 3.861 Å at RT to 3.856 Å at 100 °C, then to 3.851 Å at 200 °C, 3.846 Å at 300 °C, 3.828 Å at 400 °C, and to 3.851 Å when the sample is cooled to RT. Under CO reaction conditions (80 °C, $CO+O_2$), the change in the lattice parameter is very small (<0.1%). Upon reductive treatment following the catalytic reaction, there is a small lattice parameter shrinking (−0.5%). It may be conjectured that the insignificant change of the lattice parameter under the catalytic reaction condition is indicative of a surface self-regulation of the oxygenated species by the second and third transition metals (V and Co) in the PtVCo NA. The insignificant change in the lattice parameter may be one of the important factors for the significant enhancement of the catalytic activities for CO oxidation. The introduction of a third transition metal (M′ = V, Ni, Ir, etc.) into PtCo NAs minimizes or suppresses changes in the lattice parameters of the nanoparticle during the catalytic processes, leading to an activity enhancement in the order of V > Ni > Ir, consistent with the order of oxophilicity of these metals. There is a significant contrast of the lattice expansion and shrinking between M′ = V and Ni in the NAs under the oxidation/reductive and catalytic reaction conditions. The lattice expansion and shrinking under CO oxidation conditions are very small for PtVCo and PtNiCo catalysts.

CO oxidation reaction over PdNi NA catalyst is another example demonstrating a remarkable optimal catalytic activity at Pd:Ni ratio of ∼50:50, which also exhibits remarkable tunability in terms of phase state, bimetallic composition, and atomic-scale structure [117]. The catalytic synergy and structure tunability are achieved by thermochemical treatment under controlled conditions. Using HE-XRD/PDF technique, insights into both local and extended structures across the entire nanoparticle can be obtained. In Fig. 15.13A, the partial interatomic distances and total first CNs involving Pd atoms at the surface of PdNi NPs and the CO oxidation activity are plotted as a function of the bimetallic composition. The results show a strong correlation between the atomic-scale structure and catalytic activity. A maximum catalytic activity is revealed at Pd:Ni ratio of ∼50:50, at which there are also maximum interatomic separation and CN for the surface Pd atoms in the PdNi NPs. In Fig. 15.13B, the catalytic synergy for CO adsorption and oxidation over PdNi NPs with Pd/Ni ratio of ∼50:50 is depicted by two different sites "A" and "B" in terms of high- and low-coordinated surface atomic sites. Site "A" may be considered more catalytically active than site "B" for two reasons: (1) for site "A" atoms, the first neighbor distance is elongated to a greater extent which is associated with a d-band upshift and an increase in oxygen binding energy, as known from DFT calculation [118]. (2) For site

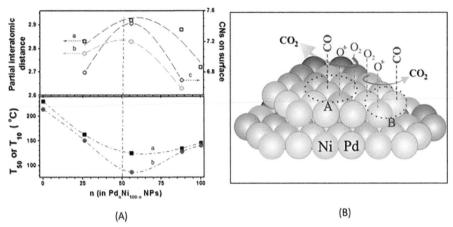

(A)   (B)

Fig. 15.13 (A) Comparison of partial Pd–Pd (*black*, a) and Pd–Ni (*green*, b) interatomic distances and the average total CNs (*red*, c) for Pd atoms at the surface of Pd_nNi_{100-n} NPs (*top panel*) with their T_{50} (*black*, a) and T_{10} (*red*, b) values for CO oxidation reaction (*bottom panel*). (B) Illustration of the structural synergy for CO adsorption and oxidation over PdNi NPs with Pd/Ni ratio of ~50:50 in terms of high ("A" in a "terrace-like" environment) and low ("B" in a "step/corner-like" environment) coordinated surface atomic sites (Pd atoms in *gray* and Ni atoms in *green*). *The larger arrow size* illustrates the higher catalytic activity for site "A" than for site "B."
Figure reprinted with permission from S. Shan, V. Petkov, L. Yang, J. Luo, P. Joseph, D. Mayzel, B. Prasai, L. Wang, M. Engelhard, C.-J. Zhong, Atomic-structural synergy for catalytic CO oxidation over palladium–nickel nanoalloys, J. Am. Chem. Soc. 136 (19) (2014) 7140–7151.

"A" atoms, the average first CNs are relatively higher which could lead to a lower propensity of the poisoning of the surface active sites.

Hydrocarbon (e.g., propane) oxidation, which plays an important role in the remediation of hydrocarbon pollutants and the production of sustainable energy, is another example of gas-phase catalytic reactions. The total catalytic oxidation of propane (C_3H_8) is shown by:

$$C_3H_8 \ (g) + 5O_2 \ (g) \rightarrow 3CO_2 \ (g) + 4H_2O \ (l) \qquad (15.2)$$

Propane oxidation over PtNiCo NAs is shown to involve an intriguing catalytic synergy with surface oxygenation forming a Pt-NiOCoO surface layer and disordered ternary alloy core [119]. Fig. 15.14A shows an aberration-corrected high-angle annular dark-field scanning transmission electron microscopy (HAADF-STEM) image of as-synthesized PtNiCo NA nanoparticles with an average size of 4.9 (±0.6) nm and a lattice spacing of 0.188 nm characteristic of (111) crystal plane. Fig. 15.14A also shows HRTEM image of PtNiCo supported on Al_2O_3 after a certain degree of thermochemical oxidation. The image reveals an oxygenated Pt-NiOCoO surface layer and disordered ternary alloy core with a lattice spacing of 0.204 nm at edges and

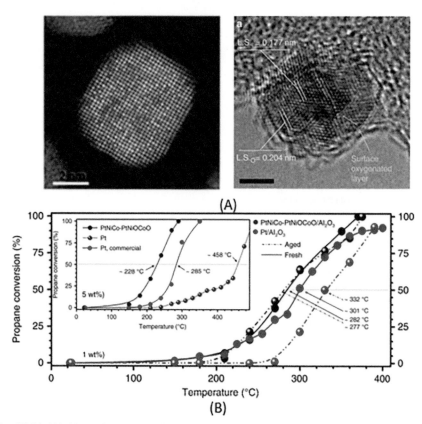

Fig. 15.14 (A) Aberration-corrected HAADF-STEM images (*left*) for as-synthesized $Pt_nNi_mCo_{100-n-m}$ ($n = 42$, $m = 39$) and high-resolution transmission electron microscopy (HR-TEM) (*right*) of PtNiCo-PtNiOCoO/Al_2O_3 ($n = 42$, $m = 39$, 1.0 wt%), scale bar = 2 nm. (B) Plot of propane conversion over the ternary catalyst derived from PtNiCo-PtNiOCoO/Al_2O_3 ($n = 42$, $m = 39$, 1.0 wt%, *black*) and Pt/Al_2O_3 (1.0 wt%, *red*): freshly prepared (*solid curve*), and hydrothermally aged under 10% CO_2 + 10% H_2O + N_2 at 800 °C for 16 h (*dash curve*). Values of T_{50} (i.e., the temperature at which 50% conversion is achieved) are indicated in the plots.

Figure reprinted with permission from S. Shan, J. Li, Y. Maswadeh, C. O'Brien, H. Kareem, D.T. Tran, I.C. Lee, Z.-P. Wu, S. Wang, S. Yan, Surface oxygenation of multicomponent nanoparticles toward active and stable oxidation catalysts, Nat. Commun. 11 (1) (2020) 1–9.

0.177 nm in the center. The surface oxygenated layer is ~1.4 nm in thickness which accounts for 6–7 atomic layers. The combination of X-ray absorption fine structure (EXAFS) and HE-XRD/PDF analyses show that the oxygenated PtNiCo catalyst features a long-range disordered alloy character in the core with different degrees of metal oxygenation. The results show a high level of oxygenation for Ni and Co (N (Ni–O) ~4.7, N(Co–O) ~2.8) and a low level of oxygenation for Pt (N(Pt–O) ~ 1.1), demonstrating the ability to harness the surface oxygenation of different metals. Fig. 15.14B shows that the PtNiCo-PtNiOCoO/Al_2O_3 catalyst exhibits higher catalytic activity and stability than the commercial Pt/Al_2O_3. The high catalytic activity is evidenced by $T_{50} \sim 282\,°C$ being lower than commercial $Pt/Al_2O_3(T_{50} \sim 301\,°C)$ with the same total metal loading. The $PtNiCo-PtNiOCoO/Al_2O_3$ catalyst features superior stability, showing no indication of deactivation after $800\,°C$ hydrothermal aging. For commercial Pt/Al_2O_3 catalyst, there is an obvious decay of catalytic activity which exhibits $\sim30\,°C$ increase of T_{50} after the $800\,°C$ hydrothermal aging.

Further, in situ/operando time-resolved HE-XRD/PDF and diffuse reflectance infrared Fourier transform spectroscopy characterizations of propane oxidation over $PtNiCo-PtNiOCoO/Al_2O_3$ catalyst reveal a largely irregular oscillatory kinetics associated with the dynamic lattice expansion/shrinking, ordering/disordering, and formation of surface-oxygenated sites and intermediates [119]. The catalytic synergy is responsible for the high activity and stability of $PtNiCo-PtNiOCoO/Al_2O_3$ catalyst in comparison with Pt/Al_2O_3 catalyst. Further refinements of the surface oxygenation may lead to a paradigm shift in the design of active and stable catalysts for hydrocarbon oxidation reactions.

15.4.2 Electrocatalytic oxygen evolution and oxygen reduction reaction in electrolytes

The development of water splitting cells for hydrogen production from renewable sources and fuel cells for effective conversion of hydrogen to electricity has become a global drive toward a sustainable power package of the future. For water electrolysis, the overall reaction can be divided into two half-cell reactions: hydrogen evolution reaction (HER) and oxygen evolution reaction (OER). OER is the key half-reaction in water-splitting reaction. This reaction occurs at the anode and involves a four-electron transfer process which requires a remarkably high overpotential compared to HER. OER is known to be the major bottleneck in improving the overall efficiency of electrochemical water splitting. Therefore, it is imperative to seek highly efficient OER catalysts that can effectively reduce the kinetic limitation. AuCo NAs are shown to exhibit an enhanced electrocatalytic activity for OER by controlling the morphology and composition of the catalyst, which is critical for achieving the desired electrocatalytic properties [120]. AuCo nanoparticles feature a uniform size distribution with a core-shell structure (Fig. 15.15A). The catalyst showed a composition dependence of the activity, displaying a maximum OER activity for Au:Co ratio of 2:3, as shown in Fig. 15.15C. The AuCo nanoparticles are partially alloyed with segregated phases of fcc Au, hcp Co, and fcc Co, as detected by XRD. The partially phase-segregated surface sites are shown to exhibit a bifunctional synergy where Co acts as an active center in a higher valent state, whereas the surface Au serves as a strong electron sink promoting various steps of OER (Fig. 15.15B).

For proton-exchange membrane fuel cell (PEMFC), reactions include the ORR at the cathode and the hydrogen oxidation reaction (HOR) at the anode. In a hydrogen fuel cell, the reaction kinetics for the cathodic ORR is much more sluggish than that for the anodic HOR. Therefore, the ORR electrocatalysts are considered as the most critical component for PEM fuel cell performance. The composition-structure–activity synergy in Pt/Pd-based alloy electrocatalysts is one of the most

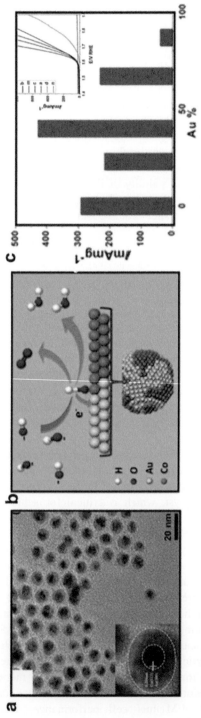

Fig. 15.15 (A) TEM images of AuCo core-shell nanoparticles (*inset*: high-magnification TEM showing the lattice fringes corresponding to fcc Au and fcc Co), (B) schematic illustration of OER on the partially phase segregated AuCo nanoparticles, (C) comparison of catalytic activities at 1.67 V in 0.1 M KOH (*inset*: polarization curves of (a) $Au_{23}Co_{77}/C$, (b) $Au_{40}Co_{60}/C$, (c) $Au_{71}Co_{29}/C$, (d) $Au_{95}Co_5/C$, (m) CoO_x/C, (n) Au/C in 0.1 M KOH). Figure reprinted with permission from A. Lu, D.-L. Peng, F. Chang, Z. Skeete, S. Shan, A. Sharma, J. Luo, C.-J. Zhong, Interfaces, composition-and structure-tunable gold–cobalt nanoparticles and electrocatalytic synergy for oxygen evolution reaction, ACS Appl. Mater. Interfaces 8 (31) (2016) 20082–20091.

important topics for the design and fabrication of efficient catalysts. In situ HE-XRD/ PDFs technique has been used to study the composition-structure-activity synergy of electrocatalysts in PEMFC [121]. This relationship displays a dynamic nature during potential cycling. One example involves the correlation between ORR mass activity and bimetallic compositions for PdNi NA electrocatalysts before, during, and after potential cycles. The composition of the alloy NPs changes due to Ni leaching during potential cycling. A maximum ORR mass activity is found at ~30 at.% Pd for fresh catalysts and ~70 at.% Pd for cycled catalysts after extensive potential cycles. The maximum ORR mass activity after a certain number of potential cycles is likely linked to a PdNi alloy NP with a composition near ~50%. The analysis of the Ni-leaching process in terms of atomic-scale structure evolution sheds further light on the activity-composition-structure correlation. It reveals an oscillatory behavior of interatomic distances, which reflects a dynamic reconstruction of PdNi nanoparticles under ORR and fuel cell operating conditions. PdNi nanoparticles are shown to maintain an alloy characteristic despite the reconstruction. There are two possible scenarios for understanding the atomic structure changes accompanying the leaching of Ni from Pt–Ni alloy catalysts during electrochemical potential cycling under ORR condition (Fig. 15.16A). One scenario involves realloying to reach another NA state with a Pd-enriched composition. The other envisages the formation of PdNi core with Pd-rich shell. Further understanding of this type of interatomic distance fluctuation in the catalyst in correlation with the base metal leaching and realloying mechanisms under the electrocatalytic operation condition may have important implications for the design and preparation of catalysts with controlled activity and stability.

In comparison with Pt-based NP electrocatalysts, Pt-based alloy nanowires (NWs) constitute another family of promising electrocatalysts. The NWs exhibit unique anisotropic nature, effective mass and charge transfer, and lower tendencies of metal dissolution and Ostwald ripening/aggregation. One example involves the study of PtFe NWs with different bimetallic compositions as fuel cell electrocatalysts [122]. Among PtFe NWs with different compositions, the NWs with an initial composition of ~24% Pt are shown to exhibit the highest mass activity and durability for ORR and under fuel cell operating conditions. The catalysts are studied using ex-situ and in situ HE-XRD/PDF and 3D modeling, revealing a dynamic structure evolution process under electrochemical and PEMFC operating conditions. Both $Pt_{24}Fe_{76}$ and $Pt_{42}Fe_{58}$ NWs show a mixed-phase structure consisting fcc and bcc types, whereas $Pt_{71}Fe_{29}$ NWs feature a long-range correlation/ordered fcc phase. Under ORR or fuel cell operating conditions, $Pt_{24}Fe_{76}$, and $Pt_{42}Fe_{58}$ NWs both experience rapid leaching of Fe metal. It reaches a final Pt content of about 70% within the first few potential cycles and then a slight composition fluctuation during further 40,000 cycles. Fig. 15.16B and C shows compositions and lattice parameters change during the potential cycling. The lattice parameters exhibit a dramatic increase in the first six potential cycles and then keep fluctuating slightly in the further extended ~1500 potential cycles. $Pt_{71}Fe_{29}$ NWs show a steady composition throughout the entire potential cycling process. It is evident that the PtFe NWs transform into a PtFe alloy core and Pt-rich shell structure. The core-shell structure undergoes a slow dynamic reconstruction upon the further ~1500 potential cycles. This type of core-sheath structure evolution is originated from the initial quick dealloying and the subsequent realloying processes, leading to high

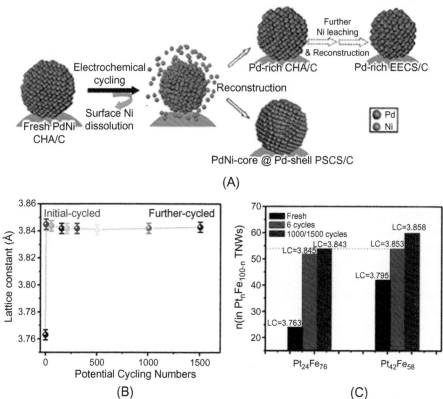

Fig. 15.16 (A) A schematic illustration of the dynamic evolution process for PdNi/C NPs under fuel cell operating conditions, (B) a plot of lattice constant of $Pt_{24}Fe_{76}$ NWs vs potential cycling number, and (C) compositions and lattice parameters change during fuel cell operation. Panel A: Figure reprinted with permission from Z.P. Wu, S. Shan, S.Q. Zang, C.J. Zhong, Dynamic core−shell and alloy structures of multimetallic nanomaterials and their catalytic synergies, Acc. Chem. Res. 53 (12) (2020) 2913–2924; Panels B and C: Figure reprinted with permission from Z. Kong, Y. Maswadeh, J.A. Vargas, S. Shan, Z.-P. Wu, H. Kareem, A.C. Leff, D.T. Tran, F. Chang, S. Yan, N. Sanghyun, X. Zhao, J.M. Lee, J. Luo, S. Shastri, G. Yu, P. Valeri, C.-J. Zhong, Origin of high activity and durability of twisty nanowire alloy catalysts under oxygen reduction and fuel cell operating conditions, J. Am. Chem. Soc. 142 (3) (2019) 1287–1299.

durability. This finding has implications for guiding the design of high-performance NA electrocatalysts for applications in PEMFCs.

15.5 Summary and future perspectives

Alloying Pt or Pd with a second (M) and/or third (M′) transition metal (PtM/M′) via molecularly encapsulated nanoparticle synthesis and thermochemical processing of the nanostructure is a promising pathway for the design and preparation of active, stable, and low-cost catalysts for a wide range of catalytic or electrocatalytic reactions. The NA catalysts exhibit enhanced catalytic activities for several reactions related to sustainable energy production and conversion, including catalytic oxidation of carbon monoxide in a gas phase and electrocatalytic ORR in a fuel cell reaction condition. In addition to the usual size, shape, and composition phenomena, the enhanced catalytic or electrocatalytic activities are linked to the unique nanoscale phenomena in terms of atomic-scale alloying, interatomic distances, metal coordination structures, structural/ chemical ordering, and phase states that operate synergistically in activating oxygen and maneuvering surface oxygenated species. These structural details provide important information for the design of catalysts with high activity at significantly reduced use of noble metals (low cost), which is a sustainable pathway in green energy exploration. There are still technical issues yet to be addressed in the area of understanding how the structural synergy can be harnessed for the stability of base metals in the NA catalysts under harsh catalytic or electrocatalytic operation conditions. In addition to further investigations of the refinement of the synthesis and processing strategies, there is a clear need for in-situ experiments for understanding the structural evolution processes such as de-alloying process in the electrolyte, poisoning of catalytic sites, and atomic-scale rearrangements leading to changes in size, shape, or surface energy by enrichment of certain elements in the core or shell. To determine the structural changes in these processes, studies using various in situ X-ray or spectroscopic techniques [123], part of our ongoing investigation, are expected to generate some fresh insights for advancing our ability in achieving high stability of the NA catalysts by design. This ability will also be useful for creating new opportunities to explore the NA catalysts for a wide range of sustainable energy production, conversion, and storage, including oxygen evolution reaction in lithium-air batteries [124,125], where the catalytic synergy of the NA plays an important role in the catalytic activity and stability.

Acknowledgments

The authors express their gratitude to former and current members of the Zhong Research Group and collaborators who have made contributions to the work described in this chapter. The research work was supported by the National Science Foundation (CHE 2102482, CHE 1566283, IIP 1640669) and the Department of Energy, Basic Energy Sciences (DE-SC0006877).

References

[1] Y. Grin, J. Reedijk, K. Poeppelmeier, Comprehensive Inorganic Chemistry II, Elsevier, Oxford, 2013.
[2] J. Wu, H. Yang, Platinum-based oxygen reduction electrocatalysts, Acc. Chem. Res. 46 (8) (2013) 1848–1857.
[3] A.K. Singh, Q. Xu, Synergistic catalysis over bimetallic alloy nanoparticles, ChemCatChem 5 (3) (2013) 652–676.
[4] S. Guo, S. Zhang, S. Sun, Tuning nanoparticle catalysis for the oxygen reduction reaction, Angew. Chem. Int. Ed. 52 (33) (2013) 8526–8544.

[5] N.V. Long, Y. Yang, C.M. Thi, N. Van Minh, Y. Cao, M. Nogami, The development of mixture, alloy, and core-shell nanocatalysts with nanomaterial supports for energy conversion in low-temperature fuel cells, Nano Energy 2 (5) (2013) 636–676.

[6] Y.X. Xie, N. Dimitrov, Ultralow Pt loading nanoporous Au-Cu-Pt thin film as highly active and durable catalyst for formic acid oxidation, Appl. Catal. B Environ. 263 (2020) 118366.

[7] Y.X. Xie, C. Li, S. Razek, J.Y. Fang, N. Dimitrov, Synthesis of nanoporous Au−Cu−Pt alloy as a superior catalyst for the methanol oxidation reaction, ChemElectroChem 7 (2) (2020) 569–580.

[8] R. Ferrando, J. Jellinek, R.L. Johnston, Nanoalloys: from theory to applications of alloy clusters and nanoparticles, Chem. Rev. 108 (3) (2008) 845–910.

[9] H.A. Gasteiger, S.S. Kocha, B. Sompalli, F.T. Wagner, Activity benchmarks and requirements for Pt, Pt-alloy, and non-Pt oxygen reduction catalysts for PEMFCs, Appl. Catal. B Environ. 56 (1–2) (2005) 9–35.

[10] A. Brouzgou, S. Song, P. Tsiakaras, Low and non-platinum electrocatalysts for PEMFCs: current status, challenges and prospects, Appl. Catal. B Environ. 127 (2012) 371–388.

[11] C. Wang, N.M. Markovic, V.R. Stamenkovic, Advanced platinum alloy electrocatalysts for the oxygen reduction reaction, ACS Catal. 2 (5) (2012) 891–898.

[12] M.K. Jeon, C.H. Lee, G.I. Park, K.H. Kang, Combinatorial search for oxygen reduction reaction electrocatalysts: a review, J. Power Sources 216 (2012) 400–408.

[13] M. Watanabe, D.A. Tryk, M. Wakisaka, H. Yano, H. Uchida, Overview of recent developments in oxygen reduction electrocatalysis, Electrochim. Acta 84 (2012) 187–201.

[14] C.-H. Cui, S.-H. Yu, Engineering interface and surface of noble metal nanoparticle nanotubes toward enhanced catalytic activity for fuel cell applications, Acc. Chem. Res. 46 (7) (2013) 1427–1437.

[15] J.D. Aiken III, R.G. Finke, A review of modern transition-metal nanoclusters: their synthesis, characterization, and applications in catalysis, J. Mol. Catal. A Chem. 145 (1–2) (1999) 1–44.

[16] C. Sorensen, K. Klabunde, Nanoscale Materials in Chemistry, John Wiley & Sons, Inc., New York, NJ, 2001, pp. 169–221.

[17] H. Li, Y.-Y. Luk, M. Mrksich, Catalytic asymmetric dihydroxylation by gold colloids functionalized with self-assembled monolayers, Langmuir 15 (15) (1999) 4957–4959.

[18] R.S. Ingram, R.W. Murray, Electroactive three-dimensional monolayers: anthraquinone ω-functionalized alkanethiolate-stabilized gold clusters, Langmuir 14 (15) (1998) 4115–4121.

[19] C.-J. Zhong, M.M. Maye, Core–shell assembled nanoparticles as catalysts, Adv. Mater. 13 (19) (2001) 1507–1511.

[20] H. Qian, M. Zhu, Z. Wu, R. Jin, Quantum sized gold nanoclusters with atomic precision, Acc. Chem. Res. 45 (9) (2012) 1470–1479.

[21] R. Philip, P. Chantharasupawong, H. Qian, R. Jin, J. Thomas, Evolution of nonlinear optical properties: from gold atomic clusters to plasmonic nanocrystals, Nano Lett. 12 (9) (2012) 4661–4667.

[22] D.E. Cliffel, F.P. Zamborini, S.M. Gross, R.W. Murray, Mercaptoammonium-monolayer-protected, water-soluble gold, silver, and palladium clusters, Langmuir 16 (25) (2000) 9699–9702.

[23] L. Liu, X. Gu, Y. Cao, X. Yao, L. Zhang, C. Tang, F. Gao, L. Dong, Crystal-plane effects on the catalytic properties of Au/TiO2, ACS Catal. 3 (12) (2013) 2768–2775.

[24] K. Kusada, H. Kobayashi, R. Ikeda, Y. Kubota, M. Takata, S. Toh, T. Yamamoto, S. Matsumura, N. Sumi, K. Sato, Solid solution alloy nanoparticles of immiscible Pd

and Ru elements neighboring on Rh: changeover of the thermodynamic behavior for hydrogen storage and enhanced CO-oxidizing ability, J. Am. Chem. Soc. 136 (5) (2014) 1864–1871.

[25] M. Liao, Q. Hu, J. Zheng, Y. Li, H. Zhou, C.-J. Zhong, B.H. Chen, Pd decorated Fe/C nanocatalyst for formic acid electrooxidation, Electrochim. Acta 111 (2013) 504–509.

[26] P.N. Duchesne, G. Chen, N. Zheng, P. Zhang, Local structure, electronic behavior, and electrocatalytic reactivity of CO-reduced platinum–iron oxide nanoparticles, J. Phys. Chem. C 117 (49) (2013) 26324–26333.

[27] Y.X. Xie, C. Li, E. Castillo, J.Y. Fang, N. Dimitrov, Nanoporous Pd-Cu thin films as highly active and durable catalysts for oxygen reduction in alkaline media, Electrochim. Acta 385 (2021) 138306.

[28] Y.X. Xie, Y. Yang, D. Muller, H. Abruña, N. Dimitrov, J.Y. Fang, Enhanced ORR kinetics on Au-doped Pt–Cu porous films in alkaline media, ACS Catal. 10 (17) (2020) 9967–9976.

[29] J.C. Bauer, D.R. Mullins, Y. Oyola, S.H. Overbury, S. Dai, Structure activity relationships of silica supported AuCu and AuCuPd alloy catalysts for the oxidation of CO, Catal. Lett. 143 (9) (2013) 926–935.

[30] I. Geukens, D.E. De Vos, Organic transformations on metal nanoparticles: controlling activity, stability, and recyclability by support and solvent interactions, Langmuir 29 (10) (2013) 3170–3178.

[31] X.Y. Liu, A. Wang, T. Zhang, C.-Y. Mou, Catalysis by gold: new insights into the support effect, Nano Today 8 (4) (2013) 403–416.

[32] S. Khanal, G. Casillas, N. Bhattarai, J. Velázquez-Salazar, U. Santiago, A. Ponce, S. Mejía-Rosales, M. José-Yacamán, CuS2-passivated Au-core, Au3Cu-shell nanoparticles analyzed by atomistic-resolution Cs-corrected STEM, Langmuir 29 (29) (2013) 9231–9239.

[33] L. Leppert, R. Albuquerque, A. Foster, S. Kummel, Interplay of electronic structure and atomic mobility in nanoalloys of Au and Pt, J. Phys. Chem. C 117 (33) (2013) 17268–17273.

[34] J.M. Ting, T.S. Navale, F.S. Bates, T.M. Reineke, Precise compositional control and systematic preparation of multimonomeric statistical copolymers, ACS Macro Lett. 2 (9) (2013) 770–774.

[35] L. Wang, J.I. Williams, T. Lin, C.-J. Zhong, Spontaneous reduction of O2 on PtVFe nanocatalysts, Catal. Today 165 (1) (2011) 150–159.

[36] J.L. Fernández, D.A. Walsh, A.J. Bard, Thermodynamic guidelines for the design of bimetallic catalysts for oxygen electroreduction and rapid screening by scanning electrochemical microscopy. M–Co (M: Pd, Ag, Au), J. Am. Chem. Soc. 127 (1) (2005) 357–365.

[37] J.K. Nørskov, J. Rossmeisl, A. Logadottir, L. Lindqvist, J.R. Kitchin, T. Bligaard, H. Jonsson, Origin of the overpotential for oxygen reduction at a fuel-cell cathode, J. Phys. Chem. B 108 (46) (2004) 17886–17892.

[38] S.J. Hwang, S.J. Yoo, S. Jang, T.-H. Lim, S.A. Hong, S.-K. Kim, Ternary Pt–Fe–Co alloy electrocatalysts prepared by electrodeposition: elucidating the roles of Fe and Co in the oxygen reduction reaction, J. Phys. Chem. C 115 (5) (2011) 2483–2488.

[39] Y. Zhang, Z. Duan, C. Xiao, G. Wang, Density functional theory calculation of platinum surface segregation energy in Pt3Ni (111) surface doped with a third transition metal, Surf. Sci. 605 (15–16) (2011) 1577–1582.

[40] V. Petkov, B.N. Wanjala, R. Loukrakpam, J. Luo, L. Yang, C.-J. Zhong, S. Shastri, Pt–Au alloying at the nanoscale, Nano Lett. 12 (8) (2012) 4289–4299.

[41] B.N. Wanjala, B. Fang, S. Shan, V. Petkov, P. Zhu, R. Loukrakpam, Y. Chen, J. Luo, J. Yin, L. Yang, Design of ternary nanoalloy catalysts: effect of nanoscale alloying and structural perfection on electrocatalytic enhancement, Chem. Mater. 24 (22) (2012) 4283–4293.

[42] J. Yin, S. Shan, L. Yang, D. Mott, O. Malis, V. Petkov, F. Cai, M. Shan Ng, J. Luo, B.H. Chen, Gold–copper nanoparticles: nanostructural evolution and bifunctional catalytic sites, Chem. Mater. 24 (24) (2012) 4662–4674.

[43] R. Loukrakpam, S. Shan, V. Petkov, L. Yang, J. Luo, C.-J. Zhong, Atomic ordering enhanced electrocatalytic activity of nanoalloys for oxygen reduction reaction, J. Phys. Chem. C 117 (40) (2013) 20715–20721.

[44] V. Petkov, Y. Ren, S. Shan, J. Luo, C.-J. Zhong, A distinct atomic structure–catalytic activity relationship in 3–10 nm supported Au particles, Nanoscale 6 (1) (2014) 532–538.

[45] S. Shan, V. Petkov, L. Yang, D. Mott, B.N. Wanjala, F. Cai, B.H. Chen, J. Luo, C.-J. Zhong, Oxophilicity and structural integrity in maneuvering surface oxygenated species on nanoalloys for CO oxidation, ACS Catal. 3 (12) (2013) 3075–3085.

[46] X. Peng, M.C. Schlamp, A.V. Kadavanich, A.P. Alivisatos, Epitaxial growth of highly luminescent CdSe/CdS core/shell nanocrystals with photostability and electronic accessibility, J. Am. Chem. Soc. 119 (30) (1997) 7019–7029.

[47] T.H. Galow, U. Drechsler, J.A. Hanson, V.M. Rotello, Highly reactive heterogeneous Heck and hydrogenation catalysts constructed through 'bottom-up' nanoparticle self-assembly, Chem. Commun. 10 (2002) 1076–1077.

[48] M. Brust, M. Walker, D. Bethell, D.J. Schiffrin, R. Whyman, Synthesis of thiol-derivatised gold nanoparticles in a two-phase liquid–liquid system, J. Chem. Soc. Chem. Commun. 7 (1994) 801–802.

[49] A.C. Templeton, W.P. Wuelfing, R.W. Murray, Monolayer-protected cluster molecules, Acc. Chem. Res. 33 (1) (2000) 27–36.

[50] G. Schmid, V. Maihack, F. Lantermann, S. Peschel, Ligand-stabilized metal clusters and colloids: properties and applications, J. Chem. Soc. Dalton Trans. 5 (1996) 589–595.

[51] T. Schmidt, M. Noeske, H.A. Gasteiger, R. Behm, P. Britz, H. Bönnemann, PtRu alloy colloids as precursors for fuel cell catalysts: a combined XPS, AFM, HRTEM, and RDE study, J. Electrochem. Soc. 145 (3) (1998) 925.

[52] R.L. Whetten, J.T. Khoury, M.M. Alvarez, S. Murthy, I. Vezmar, Z. Wang, P.W. Stephens, C.L. Cleveland, W. Luedtke, U. Landman, Nanocrystal gold molecules, Adv. Mater. 8 (5) (1996) 428–433.

[53] M.A. El-Sayed, Some interesting properties of metals confined in time and nanometer space of different shapes, Acc. Chem. Res. 34 (4) (2001) 257–264.

[54] F. Caruso, Nanoengineering of particle surfaces, Adv. Mater. 13 (1) (2001) 11–22.

[55] J.J. Storhoff, C.A. Mirkin, Programmed materials synthesis with DNA, Chem. Rev. 99 (7) (1999) 1849–1862.

[56] S. Sun, C.B. Murray, D. Weller, L. Folks, A. Moser, Monodisperse FePt nanoparticles and ferromagnetic FePt nanocrystal superlattices, Science 287 (5460) (2000) 1989–1992.

[57] M. Chen, D.E. Nikles, Synthesis, self-assembly, and magnetic properties of Fe x Co y Pt100-x-y nanoparticles, Nano Lett. 2 (3) (2002) 211–214.

[58] C. Wang, H. Daimon, T. Onodera, T. Koda, S. Sun, A general approach to the size-and shape-controlled synthesis of platinum nanoparticles and their catalytic reduction of oxygen, Angew. Chem. 120 (19) (2008) 3644–3647.

[59] Y. Kang, C.B. Murray, Synthesis and electrocatalytic properties of cubic Mn–Pt nanocrystals (nanocubes), J. Am. Chem. Soc. 132 (22) (2010) 7568–7569.

[60] C.-J. Zhong, J. Luo, B. Fang, B.N. Wanjala, P.N. Njoki, R. Loukrakpam, J. Yin, Nano-structured catalysts in fuel cells, Nanotechnology 21 (6) (2010), 062001.

[61] Y. Niu, R.M. Crooks, Preparation of dendrimer-encapsulated metal nanoparticles using organic solvents, Chem. Mater. 15 (18) (2003) 3463–3467.

[62] R. Narayanan, M.A. El-Sayed, Effect of catalysis on the stability of metallic nanoparticles: Suzuki reaction catalyzed by PVP-palladium nanoparticles, J. Am. Chem. Soc. 125 (27) (2003) 8340–8347.

[63] J. Zhang, K. Sasaki, E. Sutter, R. Adzic, Stabilization of platinum oxygen-reduction electrocatalysts using gold clusters, Science 315 (5809) (2007) 220–222.

[64] B. Lim, M. Jiang, P.H. Camargo, E.C. Cho, J. Tao, X. Lu, Y. Zhu, Y. Xia, Pd-Pt bimetallic nanodendrites with high activity for oxygen reduction, Science 324 (5932) (2009) 1302–1305.

[65] S. Sun, E.E. Fullerton, D. Weller, C. Murray, Compositionally controlled FePt nanoparticle materials, IEEE Trans. Magn. 37 (4) (2001) 1239–1243.

[66] M.J. Hostetler, C.-J. Zhong, B.K. Yen, J. Anderegg, S.M. Gross, N.D. Evans, M. Porter, R.W. Murray, Stable, monolayer-protected metal alloy clusters, J. Am. Chem. Soc. 120 (36) (1998) 9396–9397.

[67] M.J. Hostetler, A.C. Templeton, R.W. Murray, Dynamics of place-exchange reactions on monolayer-protected gold cluster molecules, Langmuir 15 (11) (1999) 3782–3789.

[68] J. Luo, L. Wang, D. Mott, P.N. Njoki, Y. Lin, T. He, Z. Xu, B.N. Wanjana, I.I.S. Lim, C.J. Zhong, Core/shell nanoparticles as electrocatalysts for fuel cell reactions, Adv. Mater. 20 (22) (2008) 4342–4347.

[69] N.N. Kariuki, J. Luo, L. Han, M.M. Maye, L. Moussa, M. Patterson, Y. Lin, M.H. Engelhard, C.J. Zhong, Nanoparticle-structured ligand framework as electrode interfaces, Electroanalysis 16 (1-2) (2004) 120–126.

[70] S. Nishimura, A.T.N. Dao, D. Mott, K. Ebitani, S. Maenosono, X-ray absorption near-edge structure and X-ray photoelectron spectroscopy studies of interfacial charge transfer in gold–silver–gold double-shell nanoparticles, J. Phys. Chem. C 116 (7) (2012) 4511–4516.

[71] D. Mott, J. Galkowski, L. Wang, J. Luo, C.-J. Zhong, Synthesis of size-controlled and shaped copper nanoparticles, Langmuir 23 (10) (2007) 5740–5745.

[72] J. Yin, S. Shan, M.S. Ng, L. Yang, D. Mott, W. Fang, N. Kang, J. Luo, C.-J. Zhong, Catalytic and electrocatalytic oxidation of ethanol over palladium-based nanoalloy catalysts, Langmuir 29 (29) (2013) 9249–9258.

[73] U. Paulus, A. Wokaun, G. Scherer, T. Schmidt, V. Stamenkovic, V. Radmilovic, N. Markovic, P. Ross, Oxygen reduction on carbon-supported Pt–Ni and Pt–Co alloy catalysts, J. Phys. Chem. B 106 (16) (2002) 4181–4191.

[74] E. Antolini, Formation of carbon-supported PtM alloys for low temperature fuel cells: a review, Mater. Chem. Phys. 78 (3) (2003) 563–573.

[75] H. Yang, W. Vogel, C. Lamy, N. Alonso-Vante, Structure and electrocatalytic activity of carbon-supported Pt–Ni alloy nanoparticles toward the oxygen reduction reaction, J. Phys. Chem. B 108 (30) (2004) 11024–11034.

[76] D.L. Fedlheim, C.A. Foss, Metal Nanoparticles: Synthesis, Characterization, and Applications, CRC Press, 2001.

[77] P. Waszczuk, G.-Q. Lu, A. Wieckowski, C. Lu, C. Rice, R. Masel, UHV and electrochemical studies of CO and methanol adsorbed at platinum/ruthenium surfaces, and reference to fuel cell catalysis, Electrochim. Acta 47 (22 – 23) (2002) 3637–3652.

[78] T. Schmidt, H. Gasteiger, R. Behm, Methanol electrooxidation on a colloidal PtRu-alloy fuel-cell catalyst, Electrochem. Commun. 1 (1) (1999) 1–4.

[79] R. Raja, T. Khimyak, J.M. Thomas, S. Hermans, B.F. Johnson, Single-step, highly active, and highly selective nanoparticle catalysts for the hydrogenation of key organic compounds, Angew. Chem. Int. Ed. 40 (24) (2001) 4638–4642.

[80] M. Haruta, S. Tsubota, T. Kobayashi, H. Kageyama, M.J. Genet, B. Delmon, Low-temperature oxidation of CO over gold supported on TiO2, α-Fe2O3, and Co3O4, J. Catal. 144 (1) (1993) 175–192.

[81] J. Luo, P.N. Njoki, Y. Lin, L. Wang, C.J. Zhong, Activity-composition correlation of AuPt alloy nanoparticle catalysts in electrocatalytic reduction of oxygen, Electrochem. Commun. 8 (4) (2006) 581–587.

[82] J. Luo, N. Kariuki, L. Han, L. Wang, C.-J. Zhong, T. He, Preparation and characterization of carbon-supported PtVFe electrocatalysts, Electrochim. Acta 51 (23) (2006) 4821–4827.

[83] L. Han, W. Wu, F.L. Kirk, J. Luo, M.M. Maye, N.N. Kariuki, Y. Lin, C. Wang, C.-J. Zhong, A direct route toward assembly of nanoparticle–carbon nanotube composite materials, Langmuir 20 (14) (2004) 6019–6025.

[84] B.N. Wanjala, B. Fang, R. Loukrakpam, Y. Chen, M. Engelhard, J. Luo, J. Yin, L. Yang, S. Shan, C.-J. Zhong, Role of metal coordination structures in enhancement of electrocatalytic activity of ternary nanoalloys for oxygen reduction reaction, ACS Catal. 2 (5) (2012) 795–806.

[85] B.N. Wanjala, J. Luo, B. Fang, D. Mott, C.-J. Zhong, Gold-platinum nanoparticles: alloying and phase segregation, J. Mater. Chem. 21 (12) (2011) 4012–4020.

[86] B.N. Wanjala, J. Luo, R. Loukrakpam, B. Fang, D. Mott, P.N. Njoki, M. Engelhard, H.R. Naslund, J.K. Wu, L. Wang, Nanoscale alloying, phase-segregation, and core–shell evolution of gold–platinum nanoparticles and their electrocatalytic effect on oxygen reduction reaction, Chem. Mater. 22 (14) (2010) 4282–4294.

[87] S. Shan, J. Luo, L. Yang, C.-J. Zhong, Technology, nanoalloy catalysts: structural and catalytic properties, Catal. Sci. Technol. 4 (10) (2014) 3570–3588.

[88] F. Behafarid, L. Ono, S. Mostafa, J. Croy, G. Shafai, S. Hong, T. Rahman, S.R. Bare, B.R. Cuenya, Electronic properties and charge transfer phenomena in Pt nanoparticles on γ-Al2O3: size, shape, support, and adsorbate effects, Phys. Chem. Chem. Phys. 14 (33) (2012) 11766–11779.

[89] D. Bazin, D. Sayers, J. Rehr, C. Mottet, Numerical simulation of the platinum LIII edge white line relative to nanometer scale clusters, J. Phys. Chem. B 101 (27) (1997) 5332–5336.

[90] V. Petkov, S.J. Billinge, J. Heising, M.G. Kanatzidis, Application of atomic pair distribution function analysis to materials with intrinsic disorder. Three-dimensional structure of exfoliated-restacked WS 2: not just a random turbostratic assembly of layers, J. Am. Chem. Soc. 122 (47) (2000) 11571–11576.

[91] V. Petkov, I.-K. Jeong, J. Chung, M. Thorpe, S. Kycia, S.J. Billinge, High real-space resolution measurement of the local structure of Ga $1-x$ In x as using X-ray diffraction, Phys. Rev. Lett. 83 (20) (1999) 4089.

[92] Y. Waseda, The structure of non-crystalline materials, in: Liquids and Amorphous Solids, McGraw Hill Higher Education, 1980.

[93] H.M. Rietveld, A profile refinement method for nuclear and magnetic structures, J. Appl. Crystallogr. 2 (2) (1969) 65–71.

[94] S.M. Oxford, P.L. Lee, P.J. Chupas, K.W. Chapman, M.C. Kung, H.H. Kung, Study of supported PtCu and PdAu bimetallic nanoparticles using in-situ X-ray tools, J. Phys. Chem. C 114 (40) (2010) 17085–17091.

[95] T. Egami, S.J. Billinge, Underneath the Bragg Peaks: Structural Analysis of Complex Materials, Elsevier, 2003.

[96] V. Petkov, Nanostructure by high-energy X-ray diffraction, Mater. Today 11 (11) (2008) 28–38.

[97] M. Newville, The IFEFFIT Reference Guide. Consortium for Advanced Radiation Sources, University of Chicago, Chicago, USA, 2004.

[98] J. Evans, J.T. Gauntlett, J.F.W. Mosselmans, Characterisation of oxide-supported alkene conversion catalysts using X-ray absorption spectroscopy, Faraday Discuss. Chem. Soc. 89 (1990) 107–117.

[99] C. Antoniak, Extended X-ray absorption fine structure of bimetallic nanoparticles, Beilstein J. Nanotechnol. 2 (1) (2011) 237–251.

[100] M. Watanabe, K. Tsurumi, T. Mizukami, T. Nakamura, P. Stonehart, Activity and stability of ordered and disordered Co-Pt alloys for phosphoric acid fuel cells, J. Electrochem. Soc. 141 (10) (1994) 2659.

[101] S. Koh, J. Leisch, M.F. Toney, P. Strasser, Structure-activity-stability relationships of Pt–Co alloy electrocatalysts in gas-diffusion electrode layers, J. Phys. Chem. C 111 (9) (2007) 3744–3752.

[102] S. Koh, C. Yu, P. Mani, R. Srivastava, P. Strasser, Activity of ordered and disordered Pt-Co alloy phases for the electroreduction of oxygen in catalysts with multiple coexisting phases, J. Power Sources 172 (1) (2007) 50–56.

[103] V. Petkov, 3D Structure of Nanosized Catalysts by High-Energy X-Ray Diffraction, Taylor & Francis, 2009.

[104] B.N. Wanjala, R. Loukrakpam, J. Luo, P.N. Njoki, D. Mott, C.-J. Zhong, M. Shao, L. Protsailo, T. Kawamura, Thermal treatment of PtNiCo electrocatalysts: effects of nanoscale strain and structure on the activity and stability for the oxygen reduction reaction, J. Phys. Chem. C 114 (41) (2010) 17580–17590.

[105] J. Luo, M.M. Maye, V. Petkov, N.N. Kariuki, L. Wang, P. Njoki, D. Mott, Y. Lin, C.-J. Zhong, Phase properties of carbon-supported gold–platinum nanoparticles with different bimetallic compositions, Chem. Mater. 17 (12) (2005) 3086–3091.

[106] S. Xiao, W. Hu, W. Luo, Y. Wu, X. Li, H. Deng, Size effect on alloying ability and phase stability of immiscible bimetallic nanoparticles, Eur. Phys. J. B 54 (4) (2006) 479–484.

[107] M.M. Maye, Y. Lou, C.-J. Zhong, Core–shell gold nanoparticle assembly as novel electrocatalyst of CO oxidation, Langmuir 16 (19) (2000) 7520–7523.

[108] D. Mott, J. Luo, P.N. Njoki, Y. Lin, L. Wang, C.-J. Zhong, Synergistic activity of gold-platinum alloy nanoparticle catalysts, Catal. Today 122 (3–4) (2007) 378–385.

[109] B.H. Morrow, A. Striolo, Supported bimetallic Pt-Au nanoparticles: structural features predicted by molecular dynamics simulations, Phys. Rev. B 81 (15) (2010), 155437.

[110] O. Malis, M. Radu, D. Mott, B. Wanjala, J. Luo, C. Zhong, An in situ real-time X-ray diffraction study of phase segregation in Au–Pt nanoparticles, Nanotechnology 20 (24) (2009), 245708.

[111] G. Kresse, J. Hafner, Ab initio molecular dynamics for liquid metals, Phys. Rev. B 47 (1) (1993) 558.

[112] G. Kresse, J. Furthmüller, Efficient iterative schemes for ab initio total-energy calculations using a plane-wave basis set, Phys. Rev. B 54 (16) (1996) 11169.

[113] G. Kresse, J. Furthmüller, Efficiency of ab-initio total energy calculations for metals and semiconductors using a plane-wave basis set, Comput. Mater. Sci. 6 (1) (1996) 15–50.

[114] D. Bochicchio, F. Negro, R. Ferrando, Competition between structural motifs in gold–platinum nanoalloys, Comput. Theor. Chem. 1021 (2013) 177–182.

[115] Z.-P. Wu, S. Shan, Z.-H. Xie, N. Kang, K. Park, E. Hopkins, S. Yan, A. Sharma, J. Luo, J. Wang, Revealing the role of phase structures of bimetallic nanocatalysts in the oxygen reduction reaction, ACS Catal. 8 (12) (2018) 11302–11313.

[116] S.D. Senanayake, D. Stacchiola, J.A. Rodriguez, Unique properties of ceria nanoparticles supported on metals: novel inverse ceria/copper catalysts for CO oxidation and the water-gas shift reaction, Acc. Chem. Res. 46 (8) (2013) 1702–1711.

[117] S. Shan, V. Petkov, L. Yang, J. Luo, P. Joseph, D. Mayzel, B. Prasai, L. Wang, M. Engelhard, C.-J. Zhong, Atomic-structural synergy for catalytic CO oxidation over palladium–nickel nanoalloys, J. Am. Chem. Soc. 136 (19) (2014) 7140–7151.

[118] J. Miller, A. Kropf, Y. Zha, J. Regalbuto, L. Delannoy, C. Louis, E. Bus, J.A. van Bokhoven, The effect of gold particle size on AuAu bond length and reactivity toward oxygen in supported catalysts, J. Catal. 240 (2) (2006) 222–234.

[119] S. Shan, J. Li, Y. Maswadeh, C. O'Brien, H. Kareem, D.T. Tran, I.C. Lee, Z.-P. Wu, S. Wang, S. Yan, Surface oxygenation of multicomponent nanoparticles toward active and stable oxidation catalysts, Nat. Commun. 11 (1) (2020) 1–9.

[120] A. Lu, D.-L. Peng, F. Chang, Z. Skeete, S. Shan, A. Sharma, J. Luo, C.-J. Zhong, Inter-faces, composition- and structure-tunable gold–cobalt nanoparticles and electrocatalytic synergy for oxygen evolution reaction, ACS Appl. Mater. Interfaces 8 (31) (2016) 20082–20091.

[121] J. Wu, S. Shan, V. Petkov, B. Prasai, H. Cronk, P. Joseph, J. Luo, C.J. Zhong, Composition-structure-activity relationships for palladiumalloyed nanocatalysts in oxy-gen reduction reaction: an ex-situ/in-situ high energy X-ray diffraction study, ACS Catal. 5 (9) (2015) 5317–5327.

[122] Z. Kong, Y. Maswadeh, J.A. Vargas, S. Shan, Z.-P. Wu, H. Kareem, A.C. Leff, D.T. Tran, F. Chang, S. Yan, N. Sanghyun, X. Zhao, J.M. Lee, J. Luo, S. Shastri, G. Yu, P. Valeri, C.-J. Zhong, Origin of high activity and durability of twisty nanowire alloy catalysts under oxygen reduction and fuel cell operating conditions, J. Am. Chem. Soc. 142 (3) (2019) 1287–1299.

[123] J. Timoshenko, B. Cuenya, *In situ/operando* electrocatalyst characterization by X-ray absorption spectroscopy, Chem. Rev. 121 (2) (2021) 882–961.

[124] Y.C. Lu, E. Crumlin, G. Veith, J. Harding, S.H. Yang, *In situ* ambient pressure X-ray photoelectron spectroscopy studies of lithium-oxygen redox reactions, Sci. Rep. 2 (2012) 715.

[125] C.L. Dong, L. Vayssieres, In situ/operando X-ray spectroscopies for advanced investiga-tion of energy materials, Chem. Eur. J. 24 (69) (2018) 18356–18373.

Lower dimensional nontoxic perovskites: Structures, optoelectronic properties, and applications

16

Nasir Ali, Xiaoyu Wang, and Huizhen Wu
Zhejiang Province Key Laboratory of Quantum Technology and Devices, Department of Physics, State Key Laboratory for Silicon Materials, Zhejiang University, Hangzhou, People's Republic of China

16.1   Introduction

The appealing properties of the organometal halide perovskites (PVKs) in numerous optoelectronic applications grasped greater scientific attention. Along with facile solution processibility, the structural flexibility of PVKs made them unique among semiconducting materials [1–4]. PVK materials appear to have a defect tolerant nature, high absorption coefficient (α), adjustable energy bandgap (E_{BG}), long carrier lifetime, small exciton bonding, structural flexibility, and exceptional photo-harvesting and radiating properties [5,6]. These properties enabled PVK solar cells (SCs) to yield power conversion efficiency (PCE) beyond 25.5%, standing as a strong contender to the Si-based SCs [7]. Nonetheless, PVK materials encountered two main issues, that is, environmental instability and the inclusion of a bioaccumulative lead (Pb), hindering their commercialization. To address these issues, numerous methods have been adopted; for instance, lower-dimensional PVKs (i.e., 2D, 1D, and 0D) were introduced, which not only endowed environmental stability to the PVKs but also eliminated the toxic Pb from their framework. Lower dimensional (i.e., 0D and 1D) PVKs mostly contain Bi, which is not only far less toxic than the conventional Pb but also exhibited better environmental durability [8]. Besides Bi-based lower-dimensional PVKs, Pb-based layered 2D-PVKs also demonstrated better environmental stability and improved performance in optoelectronic devices, for example, in photovoltaics. The following sections of this chapter discuss the various lower-dimensional PVKs, including Pb-based layered 2D PVKs and Bi-based 0D, 1D, and 2D PVKs. In addition, the optoelectronic properties and various electronic and optoelectronic applications of Bi-based PVKs are discussed.

Modeling, Characterization and Production of Nanomaterials. https://doi.org/10.1016/B978-0-12-819905-3.00016-6

16.2   2D layered PVKs

Layered 2D PVKs are one of the most widely studied counterparts of the organometal halide PVKs. Their general formula is $(RNH_3)_2A_{n-1}Pb_nX_{3n+1}$, where $n =$ the number of octahedra in the 2D layer, R = longer chained organic cation, A = organic or inorganic cation (e.g., Cs^+, FA^+, and MA^+), and Xishalogens (e.g., Cl^-, Br^-, and I^-) [9,10]. Upon subsequent annealing of the PVK precursor solution, strong ionic bonding compels the free-standing halogen octahedra ($[PbI_6]^{4-}$) to form a laterally extended 2D network. The organic long-chain cations (NH^{3+}) attached themselves with the halogens inside the octahedral voids, sandwiching the octahedral 2D layers, see Fig. 16.1A [11]. Van-der-Waals forces are responsible for attaching the long tails (R-groups), isolating the whole structure from either side. These long R-group chains act as dangling bonds, restricting the growth of the PVK layer to lateral directions, thus giving it a 2D structure [12]. The quantum and dielectric confinement effects enable the formation of a series of quantum wells in the 2D PVKs, where "barriers" and "wells" are, respectively, formed by the organic and inorganic layers, as given in Fig. 16.1B. Excitons restrained inside such quantum wells have higher binding energies, thus making them highly stable and empowering them with inimitable optical properties [13]. Interestingly, the barrier height, along with the quantum well's width and depth can be adjusted by altering the organic chain lengths, "n" values, and halogens, respectively [14]. Remarkably, the longer organic chains possess comparatively greater hydrophobicity than the smaller organic cations (i.e., MA^+ and FA^+); thus protecting the PVK active layer from moisture ingress [15]. Although, 2D PVKs offered better environmental stability, still, the localization of charge in the 2D layer and their wide E_{BG} render them poor PCEs [16–21]. However, 2D PVKs found their applications in other optoelectronic devices, for example, photodetectors, light-emitting diodes, and field-effect transistors [22]. In addition, akin to 3D PVKs,

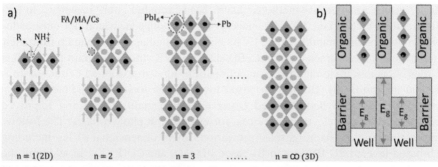

Fig. 16.1 (A) Schematic of the 2D PVKs with varying octahedral layers (thickness), at $n = \infty$ 2D PVK become 3D PVK. (B) The formation of natural quantum well formed by the longer organic chains and PbI_6 octahedra.

Reproduced with permission N. Ali, et al., An overview of the decompositions in organo-metal halide perovskites and shielding with 2-dimensional perovskites, Renew. Sust. Energ. Rev. 109 (2019) 160–186.

2D PVKs also contained toxic Pb as a central part of their octahedral network, thus making them objectionable. Therefore, a replacement in the PVK structure was needed to substitute the toxic-Pb, hence several other lower-dimensional PVKs were introduced. Among these, Bi-based PVKs were widely studied and demonstrated for many optoelectronic devices. The following section will discuss about various Bi-based lower-dimensional PVKs, their optoelectronic properties, and applications.

16.3   Bi-based lower-dimensional PVKs

Bi is superior to all Pb substitutes in terms of similar electronegativity, electronic ordering, and nearly the same ionic size as Pb. Bi exhibits some distinctive properties than Pb, for example, less toxicity, remarkable air stability, improved optophysical properties, and exceptional defect tolerance [23]. Similar to Pb, the ns^2 electronic configuration of Bi-based PVKs confine the defects to shallow states by setting antibonding interaction at VBM (valence band maximum); therefore, Bi is capable of retaining its optoelectronic properties [24]. The existence of different oxidation states makes it difficult for Bi to achieve charge neutrality in the 3D ABX_3PVK framework [25,26]. Therefore, to attain this neutrality, a hybrid valence approach is applied, where Pb (which is a divalent cation) is exchanged by a pair of trivalent and monovalent cations to get a collective divalent state [27]. Bi-based PVKs have distorted MX_6 octahedra lattices, hence, like typical Pb-based PVKs, single corner-sharing octahedra is not possible [28]. Instead, they build up a network of the corner- and edge-sharing octahedral, which are classified as dimer units (0D), chain-like motifs (1D), and other 2D structures [29].

16.3.1   Bi-based 0D-PVKs

Such PVKs are known as 0D PVKs because they do not share their octahedral network in any direction. To visualize the structure of 0D Bi-based PVKs, we need to have an idea about how bi-octahedra is formed? In 0D structure, Bi is surrounded by six halogen atoms, acting as the center of the octahedron $[BiX_6]$, as depicted in Fig. 16.2A. This octahedron attaches itself to another octahedron through its three halogen atoms (X), which forms bi-octahedra $[Bi_2X_9]$, see Fig. 16.2B. Accordingly, Bi occupies 66.67% of new octahedral sites and 33.3% of metal sites remain unoccupied. The halogens, which are shared by octahedron to form bi-octahedron, are named as "bridging halides," whereas other halides (encircled in Fig. 16.2B) are known as "terminal halides"; consequently, the hexagonal close-packed structure is formed, as visualized in Fig. 16.2C [30,31]. This hexagonal structure consists of an anion and cation combination (i.e., AX_3) that are further stacked in hexagonal layers (i.e., AX_{12}), having A as a common atom, as shown in Fig. 16.2D. In this way, $[Bi_2X_9]^{3-}$ complex is obtained, which forms a face-shared bi-nuclear octahedral, where A-cations fill up the empty spaces at the terminal halide side that becomes the reason to restrain further attachment [32–34]. The general formula for 0D PVK is $A_3Bi_2X_9$; where a prototype $(MA)_3Bi_2I_9(MA = CH_3NH_3)$ has been widely studied for different optoelectronic applications [8].

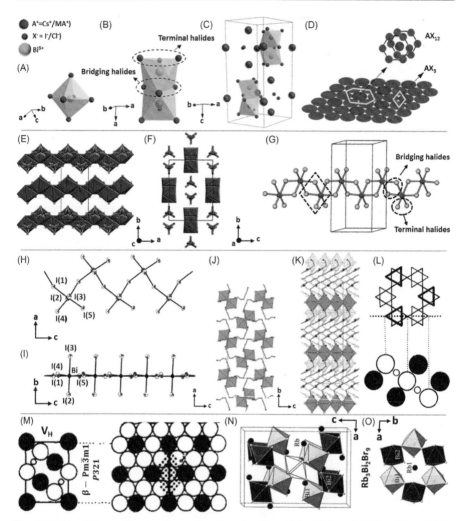

Fig. 16.2 Crystallographic structures of various families of the Bi-based PVKs: (A) and (B) are BiX_6 octahedron and Bi_2X_9 bi-octahedra. (C) Crystal structure of the 0D $A_3Bi_2X_9$ perovskite. (D) Visualization of the hexagonal closed packed structure formed by the combination of AX_3. (E) and (F) The crystal structure of $LiBiI_4 \cdot 5H_2O$, viewed along "c" and "a" axis, respectively. (G) A fragment of 1D $(BiI4)^-$ chain. (H) and (I) The BiI_5^{2-} chain structure for $(H_2AETH)BiI_5$: top and side views. (J) Crystal structure of $(H_2DAH)BiI_5$: viewed down b-axis, that is, parallel to the BiI_5^{2-} chains. (K) Crystal structure of $(H_2AETH)BiI_5$, viewed down the a-axis. *Dashed lines* depict the unit cell outline. (L) BX_6 grouping with "c" type stacking of layers where the octahedra are connected through vertices. The $A_3B_2X_9$ crystal structure type is depicted by the atom arrangement in a plane $\perp$ to the AX_3 layers. (M) 2D corrugated crystal structure ($A_3B_2X_9$) presented by the atomic arrangement in a plane $\perp$ to AX_3 layers. Atoms are represented by large filled circles, B atoms by small open circles, and X atoms by large open circles. (N) Projection view of the 2D corrugated structure ($Rb_3Bi_2Br_9$). (O) Represents that Rb_1 atoms are coordinated by the characteristic rings of Bi_1Br_6 and Bi_2Br_6 octahedra.

(Continued)

16.3.2 Bi-based 1D-PVKs

In such a configuration of Bi-based PVKs, BiX_6 octahedra are combined such that they conform to the 1D structure. Based on their connectivity, a 1D structure can further be categorized into (i) edge-shared and (ii) corner-shared octahedra or zigzag symmetry [35].

16.3.2.1 Edge-sharing octahedra

To comprehend the 1D configuration of edge-sharing octahedra in Bi-based PVK, the edge of octahedron should be focused, which is indeed a Bi-trihalide (BiX_3), where Bi is centered between three halogens, as shown in Fig. 16.2G. 1D chains of $[BiX_4]^-$ are formed when the opposite edges (i.e., BiX_3) of the octahedra (BiX_6) are shared in such a way that a total of four halogen atoms (known as bridging halides) bridge together, as shown in Fig. 16.2E–G. This anionic arrangement extends like a chain along with two free halogens, known as terminal halides. The distance between Bi and terminal halides is shorter than the distance between Bi and bridging halides, which validates the edge (BiX_3) sharing. In addition, the distance between two Bi atoms in the adjoining octahedra is greater than 4.4 Å, hence, ruling out the possibility of Bi—Bi bonding and eliminating the Bi-octahedra formation, which abstains them to attach further [36]. Here, a 1D monoclinic compound (i.e., $LiBiI_4 \cdot 5H_2O$) with space group $C2/c$ is taken as an example to elaborate edge-shared 1D PVK structure, as shown in Fig. 16.2E–G.

16.3.2.2 Corner-sharing octahedra (zigzag chained structure)

In this structure, the deformed octahedra (BiX_6) conform to a 1D network via attaching the octahedral corners employing cis-halogen bridges, hence, attain a zigzag chained configuration, see Fig. 16.2H. The A-site cations acquire the space between these chains, keeping them isolated [37,38]. Thus, the A-site cations govern the chain formation and give it a 1D symmetry. The BiX_6 octahedra obtain a zigzag symmetry by utilizing three distinct bonding pairs: (i) bridging, (ii) terminal trans to bridging, and (iii) terminal cis to bridging. Unlike the common case, herein the bond lengths

---

Fig. 16.2, Cont'd (A–G) Reproduced with permission N.A. Yelovik, et al., Iodobismuthates containing one-dimensional BiI4–anions as prospective light-harvesting materials: synthesis, crystal and electronic structure, and optical properties. Inorg. Chem. 55(9) (2016) 4132–4140. (H–K) Reproduced with permission D.B. Mitzi, P. Brock, Structure and optical properties of several organic — inorganic hybrids containing corner-sharing chains of bismuth iodide octahedra, Inorg. Chem. 40(9) (2001) 2096–2104. (L and M) Reproduced with permission B. Chabot, E. Parthe, Cs3Sb2I9 and Cs3Bi2I9 with the hexagonal Cs3Cr2Cl9 structure type, Acta Crystallogr. Sect. B: Struct. Crystallogr. Cryst. Chem. 34(2) (1978) 645–648. (N and O) Reproduced with permission J.H. Chang, T. Doert, M. Ruck, Structural variety of defect perovskite variants M3E2X9 (M = Rb, Tl, E = Bi, Sb, X = Br, I), Z. Anorg. Allg. Chem. 642(13) (2016) 736–748.

are different, that is, bridging > cis > trans bridging bond. The longest of them is the bridging bond (Bi—I_1), which acts as the backbone of the inorganic chain, whereas Bi—I_1, Bi—I_2 and Bi—I_3, Bi—I_4 are, respectively, cis and trans to long bridging bonds [39], as given in Fig. 16.2H and I. In Fig. 16.2J and K, (H_2AETH)BiI_5 is an example of the 1D zigzag structure with an orthorhombic crystal system.

16.3.3 Bi-based 2D-PVKs

By interchanging A-site cation with a smaller one (e.g., Rb^+, K^+, and NH^{4+}) and replacing the halogens (e.g., I with Cl and Br) turns the 0D Bi-based PVK ($A_3Bi_2I_9$) into a corrugated layered defect 2D PVKs [40]. Take the same aforementioned 0D Bi-based PVK structure under consideration, where AX_3 layers acquire a closed pack hexagonal structure (Fig. 16.2D). Such AX_3 layers can be stacked in three different modes: (i) $(h)_6$, (ii) $(hcc)_2$, and (iii) $(c)_3$, where the double octahedral (Bi_2X_9) separation, resulted by $(h)_6$ stacking mode can be observed [41]. The $(c)_3$ stacking mode enables the bonding between single halogens, whereas $(hcc)_2$ mode enables the above two networking. The crystal structure of 2D Bi-based PVKs is analogous to that of $(c)_3$ mode, where AX_3's stack as cubic layers, forming a complex polyhedral symmetry, as shown in Fig. 16.2L. Large ionic radii difference between the A and X site elements restrain their close packing and one is unable to observe AX_{12} layers [31]. Moreover, the BiX_6 octahedra are also deformed, adopting a structure with three closer halogens (known as bridging halides) and the rest of the three atoms are farther away from Bi, as given in Fig. 16.2B. Such bridging (i.e., BiX_3) is set inside AX_3 layers in such a way that all bridging atoms are shared with three other octahedra, forming corrugated Bi_2X_9 layers, as shown in Fig. 16.2M. Moreover, these corrugated layers run all over the structure and form a 2D structure [40,42,43], see Fig. 16.2O. In Fig. 16.2N, $Rb_3Bi_2I_9$ with a monoclinic structure (space group Pc) is taken as an example of a 2D corrugated structure.

16.4   Optoelectronic properties of Bi-based PVKs

Efficient optoelectronic devices demand desirable optoelectronic characteristics of PVK materials. These characteristics primarily comprise photoabsorption, the nature and width of the E_{BG}, carrier mobility, photoemission and photoresponsivity, exciton binding energy, defect density, and lifetimes of the charge carriers. The succeeding section gives a brief introduction to the key optoelectronic properties concerning Bi.

16.4.1   Photoabsorption

To efficiently harvest sunlight, a single junction PV device should have an E_{BG} within the optimal range of the Shockley-Queisser (SQ) efficiency-limit curve [44,45], Fig. 16.3A. The E_{BG}s of the Pb-based ABX_3PVKs are near the optimal values of the SQ-limit (i.e., ~1.34 eV). Conversely, Bi-based PVKs have relatively larger E_{BG}'s. However, preliminary studies have shown that $MA_3Bi_2I_9$ and

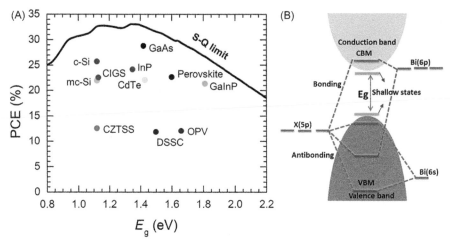

Fig. 16.3 (A) Maximum PCE (Shockley-Queisser limit) for a solar-cell operated under AM 1.5G illumination at 298.15 K, as a function of the E_{BG}. (B) Electronic band structure of the Bi-based perovskites and formation of shallow states, indicating their defect tolerance nature. (A) Reproduced with permission N.-G. Park, H. Segawa, Research direction toward theoretical efficiency in perovskite solar cells, ACS Photonics 5(8) (2018) 2970–2977. (B) Reproduced with permission S. Attique, et al., A potential checkmate to lead: bismuth in organometal halide perovskites, structure, properties, and applications, Adv. Sci. 7(13) (2020) 1903143.

$Cs_3Bi_2I_9$ display similar photoabsorption to that of typical Pb-based PVKs [46]. The 0D, 1D, and 2D categories of Bi-based PVKs distinguish them from the ordinary 3DABX$_3$ type PVKs. Since the BiX$_6$ octahedra in the Bi-based PVKs are responsible for building the valence and conduction band (CB) electronic states, the structural dimensionality has a strong influence on the optoelectronic properties of Bi-based PVKs [47]. To elaborate on all these Bi-based PVKs is beyond the scope of this discussion; herein, the discussion will be limited to the prototype Bi-based 0D $A_3Bi_2X_9$ PVKs. Nonetheless, some other important Bi-based PVKs will also be discussed for their optoelectronic properties.

16.4.1.1 Optical bandgap (E_{BG})

Bi-based PVKs possess an electronic structure similar to that of Pb-based PVKs. Their CB has a partial contribution of the p-orbitals of Bi and completely filled VB have mixed s-orbitals of Bi and p-orbitals of halides, forming an E_{BG} between CBM and VBM as depicted in Fig. 16.3B. Along with 0D PVKs, other lower-dimensional Bi-based PVKs exhibit an intense excitonic peak near 500 nm, screening their actual E_{BG}; thus making it challenging to accurately determine their true E_{BG} [48]. Therefore, a discrepancy in the E_{BG} values of the reported Bi-based 0D PVKs is always found; for instance, the E_{BG} for $Cs_3Bi_2I_9$ and $MA_3Bi_2I_3$ are reported in the range of 1.9 to 2.8 eV and 1.75 to 2.94 eV, respectively [8,49]. The E_{BG}'s for the reported lower-dimensional Bi-based PVKs are listed in Fig. 16.4. The average E_{BG} value of

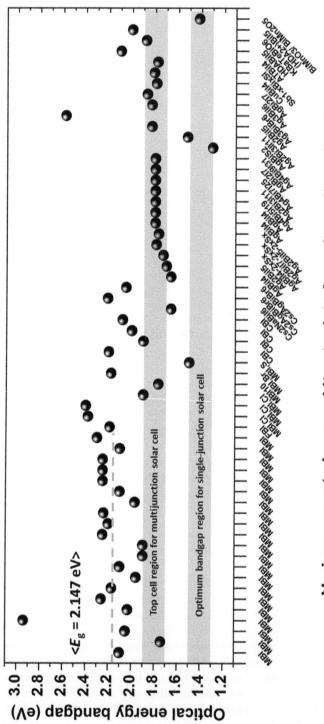

Various reported perovskites to date for solar-cell applications

Fig. 16.4 Summary of the overall reported E_{BG}'s of Bi-based PVKs utilized for solar cells. The *horizontal gray line* in the *pink-shaded portion* represents the average E_{BG} line of the most extensively studied PVK compound, that is, $MA_3Bi_2I_9$ for SCs applications. The *blue and red horizontal shaded portions* indicate the optimum E_{BG} regions of the single and multi-junction (top cell) SCs. Most of the reported E_{BG}'s are in the optimal bandgap region of the multi-junction SC (top cell).
Reproduced with permission S. Attique, et al., A potential checkmate to lead: bismuth in organometal halide perovskites, structure, properties, and applications, Adv. Sci. 7(13) (2020) 1903143.

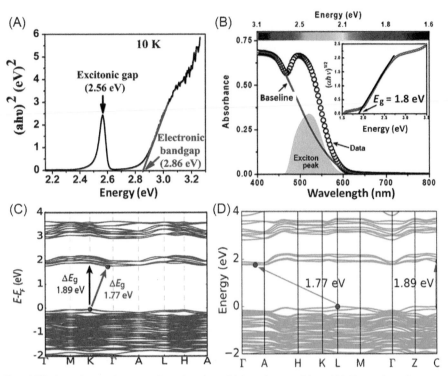

Fig. 16.5 Optoelectronic properties of various Bi-based perovskites: (A) Tauc plot of $Cs_3Bi_2I_9$ nanocrystals at 10 K. (B) Absorption spectra of the $MA_3Bi_2I_9$ film. After the exciton peak was extracted, an indirect bandgap of 1.8 eV was obtained. (C) Band structure of $MA_3Bi_2I_9$ calculated by DFT: 1.77 eV (indirect bandgap, *red arrow*) and 1.89 eV (direct bandgap, *black arrow*). (D) Electronic band structure of $Cs_3Bi_2I_9$ (based on PBE + SOC) along with high symmetry points in the reciprocal space, exhibiting an indirect bandgap (1.77 eV) between the L-point in the VB and along the $\Gamma \rightarrow A$ direction in the CB. The direct bandgap of 1.89 eV occurs at the C-point.

(A) Reproduced with permission J. Pal, et al., Synthesis and optical properties of colloidal M3Bi2I9 (M = Cs, Rb) perovskite nanocrystals, J. Phys. Chem. C 122(19) (2018) 10643–10649. (B and C) Reproduced with permission X. Chen, et al., Atmospheric pressure chemical vapor deposition of methylammonium bismuth iodide thin films, J. Mater. Chem. A 5(47) (2017) 24728–24739. (D) Reproduced with permission X. Huang, et al., Band gap insensitivity to large chemical pressures in ternary bismuth iodides for photovoltaic applications, J. Phys. Chem. C 120(51) (2016) 28924–28932.

2.147 eV is determined (from Fig. 16.4) for $MA_3Bi_2I_3$ PVK. Phase-pure Bi-based PVKs demonstrate a significantly larger E_{BG} than the SQ limit of single-junction PV devices, although reside mostly in the optimum range of the top-cell in the multi-junction tandem SCs, as visualized in Fig. 16.4.

The presence of the excitonic peak associated with 0D PVKs ($Cs_3Bi_2I_9$) was observed at 2.56 eV, where its E_{BG} was 2.86 eV [48], see Fig. 16.5A. Such excitonic peaks were also confirmed from the temperature-dependent absorption spectra, measured at 296–10 K. Identical results were shown by the bulk $Cs_3Bi_2I_9$ PVKs as well as by their nanocrystals (NCs). It was also realized that these excitonic peaks were not caused by the change in structure and size of the crystals but ascribed to the dimer-like structures [50]. In addition, different E_{BG}s for the thin films and single-crystal of the same PVK material have been observed. For instance, different E_{BG}s of 2.26, 2.0, and 1.96 eV have been reported for thin film, polycrystalline powder, and single crystal of the same $MA_3Bi_2I_9$ PVK [51]. Such a variation in the E_{BG} was accredited to the crystal orientation, crystallinity, and defect sites in PVKs, as a single crystal usually exhibits high crystallinity and lesser defect sites, resulting in a smaller E_{BG}. It was further justified by a blueshift in the wavelength, when a single crystal was grounded and its deep red color turned into an orange blue. This effect became even more prominent in the case of thin films, where the crystal orientations were further disrupted, which leads to the formation of defect sites [51].

16.4.1.2   Nature of the bandgap

Other than the E_{BG} of the photoabsorber, its nature is also crucial. The majority of the energy is lost due to nonradiative recombinations in the semiconductors having indirect bandgap, whereas direct bandgap materials do not show any significant energy loss, as they undergo radiative transitions [52]. In this regard, 0D $MA_3Bi_2I_9$ PVK exhibits a direct bandgap while high-energy direct and low-energy indirect bandgaps have been reported for the $Cs_3Bi_2I_9$ [52–56]. Nonetheless, $MA_3Bi_2I_9$ has also exhibited an indirect bandgap, Fig. 16.5B–D [54,57]. Contrary to 0D $MA_3Bi_2I_9$ and $Cs_3Bi_2I_9$, the 2D $Rb_3Bi_2I_9$ PVK NCs displayed two noticeable sharp absorption peaks at 2.28 and 2.65 eV, respectively [48]. Likewise, $Rb_3Bi_2I_9$ and other 2D PVKs, for example, $Cs_3Bi_2Br_9$ and hybrid Ruddlesden-Popper have shown the same absorption properties [58–60]. However, neither bulk $Rb_3Bi_2I_9$ nor its NCs showed any observable photoluminescence (PL), which might be indicative of the existence of intrinsic defect states or nonradiative recombinations in the bandgap [58].

16.4.1.3   Absorption coefficient (α)

The photoabsorption coefficient (α) is an important parameter of a photoabsorber to be considered for optoelectronic applications. Concerning this, Bi-based PVKs have shown ideal light absorption properties both in visible and infrared regions with α as high as $10^5\,cm^{-1}$ for a 450 nm thin film [46]. These excellent absorption characteristics are attributed to the absorption between VB states of halogen 5p and CB states of Bi 6p orbitals of the Bi-based PVKs. In addition, the better absorption properties of Bi-based PVKs could also be attributed to the optical transition between Bi-6s states near the VBM and Bi-6p CB states [61].

16.4.2 Photoluminescence

PL is a spontaneous light emission from a semiconducting material (when optically excited), which directly analyzes the optical properties of the surfaces and interfaces [62]. The occurrence of the PL peak and absorption edge at the same wavelength confirms the absence of defects in the PVK material; however, a material with defects exhibits a mismatch between the PL peak and absorption edge. This deviation between the absorption edge and PL peak is called Stoke's shift, which may be due to either extrinsic or intrinsic mechanisms in the PVK material [63]. A broadened PL peak is common in Bi-based PVKs, which ultimately indicates the presence of defects [64]. In Bi-based PVKs, BiX_6 octahedra require to keep their charge balance, which renders them to structural deterioration [65]. Consequently, polarons are created that are similar to free electrons and can create optically active states between the conduction and VB, acting as emission centers and are known as "self-trapping states" [65,66]. Accordingly, these excited electrons undergo nonradiative transitions prior to reaching the corresponding energy levels. Such self-trapping results in low-energy emissions, which explains the origin of the red-shifted wavelength and larger Stoke's shift, also observed theoretically [67].

The PL of 0D $A_3Bi_2X_9$ single-crystals has also been measured experimentally both at room and lower temperatures [48,58,68], as shown in Fig. 16.6. $Cs_3Bi_2I_9$ typically exhibited a broader but weak PL emission, ranging from 1.85 to 2.03 eV with a peak at 1.93 eV, as shown in Fig. 16.6B. The broadening in the PL was ascribed to octahedral or local lattice distortions [58]. However, in contrast to this, $Rb_3Bi_2I_9$ showed no room temperature PL. Moreover, the PL properties of $Cs_3Bi_2Br_9$ PVK were also studied, which were sensitive to the phase transition, caused by a decrease in temperature.

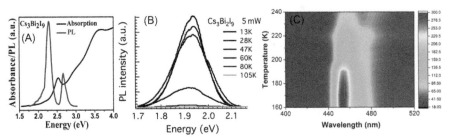

Fig. 16.6 (A) Optical absorption and PL spectra of the CBI NCs. (B) PL spectra of $Cs_3Bi_2I_9$ measured as a function of temperature. (C) Pseudocolor map of the temperature-dependent PL spectra of $Cs_3Bi_2Br_9$ single-crystal.
(A) Reproduced with permission J. Pal, et al., Synthesis and optical properties of colloidal M3Bi2I9 (M = Cs, Rb) perovskite nanocrystals, J. Phys. Chem. C 122(19) (2018) 10643–10649. (B) Reproduced with permission K.M. McCall, et al., Strong electron–phonon coupling and self-trapped excitons in the defect halide perovskites A3M2I9 (A = Cs, Rb; M = Bi, Sb), Chem. Mater. 29(9) (2017) 4129–4145. (C) Reproduced with permission M. Leng, et al., All-inorganic bismuth-based perovskite quantum dots with bright blue photoluminescence and excellent stability, Adv. Funct. Mater. 28(1) (2018) 1704446.

Therefore, the PL mechanism of $Cs_3Bi_2Br_9$ under the temperature range (160–240 K) was studied [68,69]. As shown in Fig. 16.6C, a redshift (453–457 nm) along with declined PL intensity was detected and accredited to the nonradiative recombination, as the temperature was raised to higher than 200 K. Other than single-crystals, the PL characteristics of the quantum dots and thin films have also remained under consideration, where inorganic PVKs ($Cs_3Bi_2X_9$, X = I, Br, or Cl) QDs showed high photoluminescence quantum yield than conventional Pb-based PVKs. Besides exceptional PL characteristics, such quantum dots demonstrated outstanding environmental stability, ascribed to the reduction in the defects caused by the smaller size effect [68,70–72]. In addition to a common broad single PL peak, two narrow and distinguished PL peaks at 2.66 and 2.28 eV, respectively, were also reported for $Cs_3Bi_2I_9NCs$ [48], see Fig. 16.6A. By altering the crystal sizes and their morphologies, it was concluded that such peaks were not originated due to different sizes and morphologies, rather 2.6 eV and 2.28 eV were accredited to the de-excitation and the excitonic emission, respectively.

16.4.3 Defect tolerance

By defect tolerance is the capability of semiconducting material to retain its optical characteristics, irrespective of the presence of defects. The formation of defects is unavoidable during the course of material processing. Deep-level defects within the E_{BG} mostly affect the optoelectronic properties of material, while defects outside the bandgap seldom affect the optoelectronic properties of the semiconducting material. Consequently, those deep-state defects cause nonradiative recombinations or act as trap centers by themselves and hindering the charge transport [73].

The existence of active ns^2 lone pairs in Bi-based PVKs enables the creation of antibonding interactions at VBM, which ultimately helps to restrain the defects to shallow states; thus making them tolerant to defects [74]. To understand this, focus is on the electronic configuration of Bi-halide, in which the X-5p halogen orbital is overlapped by the Bi-6p orbital, creating VB deeper states and the bottom of VBM. Likewise, the X-5p and Bi-6s hybridize to create the minima of VB (bonding) and maxima of VB (antibonding). The middle of the VB is formed by the interaction of one X-5p orbital with Bi-6s and Bi-6p orbitals [75]. This way, an E_{BG} is established between VBM and CBM, as elaborated in Fig. 16.3B. This band structure tends to have defects closer to the band edge, which induces shallow defect states between CBM and VBM. Such shallow states make the charge transfer much easier and therefore, increases their mobility [28,74,76].

In addition, it is also reported that $MAPbI_3$ and $Cs_3Bi_2I_9$ have a similar foundation of VBM and different CBM from Pb-based PVKs [77]. Theoretical calculations showed that $MA_3Bi_2I_9$ has a better defect tolerance ability than that of $Cs_3Bi_2I_9$ [77]. Similarly, it has been reported that due to a weak coupling of Bi-6s and Br-6p orbitals in $Cs_2AgBiBr_6$ double PVKs, they contribute lesser to the VBM. As a result, Ag acts as a shallow acceptor, which could form p-type conduction, thus exhibit poor defect tolerance [78]. Moreover, to make this material more defect tolerant, Bi-rich/Br-poor conditions are required.

16.4.4 Carrier lifetime (τ)

A prolonged carrier lifetime is a prerequisite for enhanced carrier extraction, which subsequently enables improved PV performance [79,80]. An energetic laser pulse excites the PVK material and creates an electron-hole pair (exciton) that tends to recombine via Coulombic force. Time-resolved PL (TRPL) measures the time duration of the excited electron-holepair until they recombine; for longer they are separated, an extended lifetime is expected. A longer charge carrier lifetime facilitates efficient charge extraction and charge transport, making the material more favorable for PV applications [81].

The reported lifetime for Pb-free alternative PVKs is ranging from <0.1 ns to ≈ 10 ns [82]. Although, exceptionally, $Cs_2AgBiBr_6$ and $Cs_2AgBiCl_6$ displayed a faster exponential decay trailed by a slower decay tail [83–85]. Hence, bi- or even tri-exponential functions were applied to fit these TRPL curves. Crystallographic defects explain the initial faster decay, whereas the longest decay tails were ascribed to the fundamental exciton lifetime, as shown in Fig. 16.7A and B [86]. The plausible reason for the longer carrier lifetime might be indicative of an indirect energy bandgap and defect tolerance [78,82]. Likewise, it was previously realized that the decay lifetime was considerably different, that is, 145 and 54 ns, respectively, for single-crystal and powder of $Cs_2AgBiBr_6$. It was because a powder sample had a greater number of traps than that of a thin film. Conversely, longer carrier lifetimes of 668 and 657 ns were reported and were not considerably different for the two samples and accredited to the fundamental recombinations, as shown in Fig. 16.7A [83]. Besides double PVKs, the lifetime of the pure and DCl-doped 0D $Cs_3Bi_2I_9$ have also been studied, which are given in Fig. 16.7B and C, respectively. In comparison with the pure $Cs_3Bi_2I_9$ (1.8 ns), DCl-doped $Cs_3Bi_2I_9$ exhibited a longer PL lifetime of 15.5 ns, which was attributed to the improved crystallinity and reduced defects sites, initiated by DCl doping.

16.4.5 Exciton binding energy

Another important property of PVK material is the exciton binding energy, which governs optoelectronic properties, and excitons with longer diffusion length and increased lifetime are attributed to the smaller binding energy. Unfortunately, Bi-based PVKs have greater exciton binding energies (i.e., few 100 meV) due to their lower dimensionality and structure isolation (e.g., CBI). Nevertheless, Bi-based PVKs demonstrated better optical properties, comparable to those of Pb-based 3D PVKs [58]. For instance, a 0D PVK ($Cs_3Bi_2I_9$) reported by Machulin et al. [50] showed a higher exciton binding energy (279 meV). It was proposed that the discrepancy in the E_{BG} and crystal color could have been caused by the higher exciton binding energy. In addition, a higher exciton binding energy was observed for ionic alkali halide compounds that induced a larger Stoke's shift between absorption and emission spectra [58,87,88]. The appearance of the excitonic peak in the absorption spectra (even at room temperatures) was attributed to such a higher exciton binding energy [89]. Furthermore, 0D PVK ($MA_3Bi_2I_9$) showed exciton with yet a higher binding energy (i.e., 300 meV) [58].

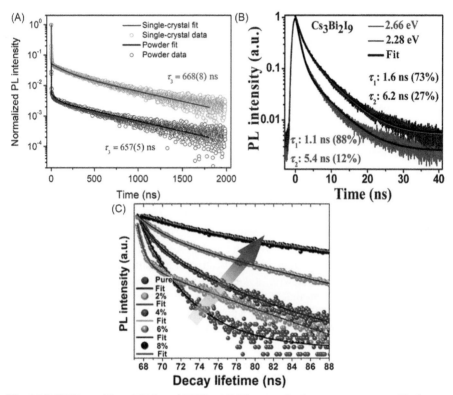

Fig. 16.7 TRPL profiles of Bi-based PVKs: (A) Time-resolved room-temperature PL decay profiles of the powder and single-crystal samples of $Cs_2Ag^IBi^{III}Br_6$. (B) PL decay profiles of colloidal $Cs_3Bi_2I_9$ NCs measured at two different emission wavelengths (i.e., 466 and 544 nm). (C) TRPL curves of pure and dodecylammonium chloride (DCl)-doped $Cs_3Bi_2I_9$. The arrow is a guide to the eye, indicating the increase in PL lifetime.
(A) Reproduced with permission A.H. Slavney, et al., A bismuth-halide double perovskite with long carrier recombination lifetime for photovoltaic applications, J. Am. Chem. Soc. 138(7) (2016) 2138–2141. (B and C) Reproduced with permission J. Pal, et al., Synthesis and optical properties of colloidal M3Bi2I9 (M = Cs, Rb) perovskite nanocrystals, J. Phys. Chem. C, 122(19) (2018) 10643–10649; Reproduced with permission N. Ali, et al., The effect of dodecylammonium chloride on the film morphology, crystallinity, and performance of lead-free Bi-based solution-processed photovoltaics devices, Sol. Energy 207 (2020) 1356–1363.

16.4.6 Charge carrier mobility (μ)

Charge carriers' mobility (μ) is another important parameter for optoelectronic applications, which states how rapidly the charge carriers move through the metals or semiconductors under the applied electric field. Based on films with and without electrodes, numerous approaches are used to determine μ [90]. Here, the discussion will be narrowed to the space charge limited current (SCLC) method that is commonly used to determine the μ of the Bi-based lower-dimensional PVKs. In this method, IV characteristics obey Lampert's theory of current in solids. As shown in Fig. 16.8A–C [51],

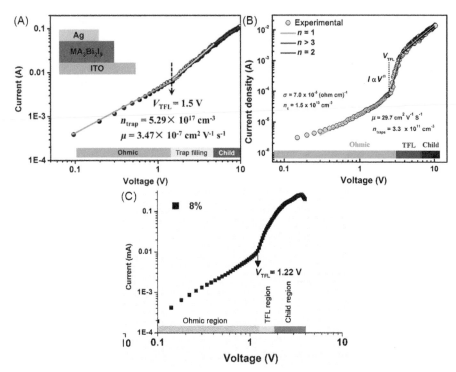

Fig. 16.8 SCLC results of the 0D Bi-based PVKs with three regions, that is, Ohmic, trap-filled, Child regions: (A) and (B) are SCLC curves of ITO/MBI/Ag and ITO/MBI/Au (electrons only) devices plotted under double-logarithmic scales measured in positive sweep-modes. (C) The SCLC results of the electron-only device of $Cs_3Bi_2I_9$.
(A and B) Reproduced with permission M. Abulikemu, et al., Optoelectronic and photovoltaic properties of the air-stable organohalide semiconductor $(CH_3 NH_3)_3 Bi_2I_9$, J. Mater. Chem. A 4(32) (2016) 12504–12515; C. Ran, et al., Construction of compact methylammonium bismuth iodide film promoting lead-free inverted planar heterojunction organohalide solar cells with open-circuit voltage over 0.8 V, J. Phys. Chem. Lett. 8(2) (2017) 394–400. (C) Reproduced with permission N. Ali, et al., The effect of dodecylammonium chloride on the film morphology, crystallinity, and performance of lead-free Bi-based solution-processed photovoltaics devices, Sol. Energy 207 (2020) 1356–1363.

its plot is classified into three regions, that is, (i) Small voltage region, linear region, following I α V relation. (ii) Higher biasing conduction carried out by traps, following I α $V^{n>3}$ and labeled as trap-filled limit region. This region can be used for the estimation of trap density in the material. (iii) The Child regime, followed by Mott Gurney relation, i.e., I α $V^{n=2}$, is used to evaluate μ, with

$$\mu = 8JL^3/9\varepsilon\varepsilon_0 V^2, \tag{16.1}$$

where, μ =mobility, J =current density, L =film thickness, ε = material's dielectric constant, ε_0 =free space permittivity, and V =applied voltage. By SCLS analyses, the

μ of 29.7 cm^2 V^{-1} s^{-1} was determined for MBI PVK, which was equivalent to the Pb-analogs, showing its suitability for PV applications, see Fig. 16.8A–C [51,91]. High-quality thin films are essential to ensure better carrier mobility, that is, a coarse film containing pinholes has a lower μ. For instance, a considerably lower μ of 3.47×10^{-7} cm^2 V^{-1} s^{-1} for MA$_3$Bi$_2$I$_9$ was reported, which was ascribed to the poor film morphology, see Fig. 16.8A [92]. Besides the μ values, trap densities of 5.2×10^{17} to 3.3×10^{11} cm^{-3} and 1.1×10^{16} cm^{-3} have also been calculated from SCLC results for 0D MA$_3$Bi$_2$I$_9$ and Cs$_3$Bi$_2$I$_9$ PVKs thin films, respectively.

16.5   Optoelectronic applications of Bi-based PVKs

Owing to their remarkable optoelectronic properties and better ambient stability, Bi-based lower-dimensional PVKs are utilized in numerous optoelectronic applications, including photovoltaics, photodetectors, memristors, and capacitors. Therefore, lower-dimensional Bi-based PVKs provided a versatile structural framework for novel nontoxic PVKs where various A-, B-, and X-site materials can be exploited for several applications. Since MA and Cs persisted to be an important part of the Pb-based PVKs SCs, most of the lower-dimensional Bi-based PVKs also contain MA and Cs as their A-site cations. Therefore, in the following section, some important initial works on 0D Bi-based PSCs (with MA and Csat A-site cations) will be discussed. Moreover, the overall advancement in lower-dimensional Bi-based PSCs will be summarized.

16.5.1   PV applications of 0D Bi-based PVKs

Undeniably, the prototype Pb-based ABX$_3$PVKs are popular for their unprecedently increased efficiency in SCs [93]. As Bi-based PVKs have comparable optoelectronic properties with that of Pb analogs, similar PV characteristics are expected from them. Therefore, after the pioneering effort of Park et al. [46], Bi-based PVKs are widely employed for PV applications [75,94,95]. Three kinds of thin films composed of MA$_3$Bi$_2$I$_9$, Cs$_3$Bi$_2$I$_9$, and MA$_3$Bi$_2$I$_9$Cl$_x$ were prepared in the ambient environment via solution processability and showed long-term environmental stability [46]. While changing the A-site cations (MA or Cs), almost similar changes in optical properties have been observed in Bi-based PVKs as perceived for Pb-based PVKs, see Fig. 16.9A–D. All these PVKs possessed a high α of $\sim 2 \times 10^5$ cm^{-1}, a wide range of E_{BG} (2.1–2.4 eV), and exciton binding energies (70–300 meV). Even though exhibited higher α, analogous to conventional Pb-based PVKs, still they had higher E_{BG}s and exciton binding energies than Pb-based PVKs. Consequently, the above three PVKs were utilized in SCs, where Cs$_3$Bi$_2$I$_9$-based devices demonstrated the highest PCE of 1.09%, see Fig. 16.9E and F.

Similarly, Lyu et al. [96] presented the first single-crystal as well as solution-processed thin films of MA$_3$Bi$_2$I$_9$ and demonstrated them for PV applications. The PV device yielded a marginally improved PCE of 0.192% having V_{OC} 0.51 V, J_{SC} 1.16 mA cm^{-2}, and FF of 46%. Likewise, a single crystal of A$_3$Bi$_2$I$_9$ (A = MA and Cs)

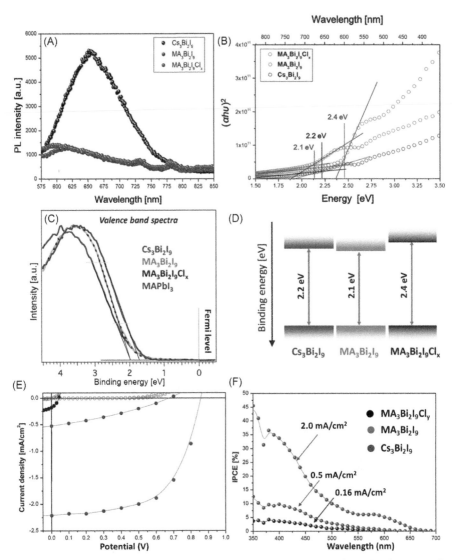

Fig. 16.9 (A) Tauc plots for E_{BG} calculations. (B) PL spectra. (C) Valance band spectra of Bi-based PVKs. (D) Comparison of their binding energies with the typical Pb-based PVK MAPbI$_3$. (E) J-V characteristics: inset is the device parametric metrics. (F) The IPCE curve for the three different Bi-based PVKs.

Reproduced with permission B.-W. Park, et al., Bismuth based hybrid perovskites A3Bi2I9 (A: Methylammonium or cesium) for solar cell application, Adv. Mater. 27(43) (2015) 6806–6813.

with a millimeter size was synthesized through a hydrothermal process [97]. Single crystals showed an E_{BG} of 1.9 eV and their spin-coated thin films exhibited a homogenous surface morphology, reduced traps, remarkable μ, and long-term stability. Finally, MA$_3$Bi$_2$I$_9$-based PSC gave a PCE of ~0.2% along with V_{OC} 0.53 V, J_{SC}

0.65 mA cm^{-2}, and FF of 65%. Correspondingly, environmentally stable MA$_3$Bi$_2$I$_9$ single-crystal, powder, and thin films were prepared whose optoelectronic properties were sensitive to the crystallinity of the compounds [51]. For example, different E_{BG}, that is, 1.96 and 2.26 eV, were observed for single crystal and thin-film, respectively. In addition, a remarkable photocurrent generation (in the visible wavelength range), longer exciton lifetime, and greater mobilities were measured using transient absorption spectroscopy and SCLC measurements. A underperformed SC device yielded a PCE of 0.11% along with V_{OC} 0.72 V, J_{SC} 0.49 mA cm^{-2}, and FF 31.8%, which was ascribed to the poor surface features of the thin film. Afterward, Oz et al. [98] synthesized and studied 0D MA$_3$Bi$_2$I$_9$ PVK for their structural properties. The synthesized PVK compound showed a relatively higher E_{BG} (2.9 eV) with a PL peak at 1.65 eV. A heterojunction SC made of this material produced PCE of 0.1%, with V_{OC} 0.66 V, J_{SC} 0.22 mA cm^{-2}, and FF 0.49%.

In the above discussed initial works, MA$_3$Bi$_2$I$_9$ SCs yielded considerably lower PCE (0.12% and 0.1%, respectively) than that of Cs$_3$Bi$_2$I$_9$ (with slightly higher PCE). There were many reasons for the lower PCE's, for example, surface roughness and poor crystallinity of thin films, deeper defect states and wider bandgaps, higher exciton binding energies, and other materials around the PVK active layer. Thus, the performance of the SC device made of the MA$_3$Bi$_2$I$_9$ PVK as an active layer was studied under the influence of the electron-transporting layer [99]. Using a one-step solution methodology, a polycrystalline film was fabricated and three basic types of device structures, that is, planar, on brookite meso-TiO$_2$, and anatase meso-TiO$_2$ were probed systematically. It was revealed that the film morphology was mostly dependent on the substrate chosen and the architecture of the respective device. Films grown on the planar substrate were nonuniform, where those on anatase meso-TiO$_2$ layers grew better due to enhanced nucleation. In the case of brookite-TiO$_2$, the interparticle necking hindered the MA$_3$B$_2$I$_9$ percolation in the pores; hence, the nucleation and uniform growth of the PVK film ultimately stopped. The anatase meso-TiO$_2$SC device performed much better (e.g., J_{SC} 0.8 mA cm^{-2}) than that of the SC fabricated on the brookite meso-TiO$_2$ layer. SCs showed a better PCE of about 0.2% with an outstanding stability of more than 10 weeks in the ambient conditions. Likewise, Zhang et al. [100] utilized a one-step solution process to fabricate MA$_3$B$_2$I$_9$ PVK and systematically studied the process of film formation and its morphology on meso-TiO$_2$/ITO surface. It was comprehended that the substrate structure, as well as the precursor concentration, play a vital role in the fabrication of a highly crystalline film. Subsequently, the PCE of the device was effectively enhanced to 0.42% with V_{OC} 0.66 V, J_{SC} 0.91 mA cm^{-2}, and FF 65%. Recently, it was realized that the longer organic chain cations (i.e., DCl) can also improve the film morphology and crystallinity of the Cs$_3$Bi$_2$I$_9$ film [101]. It was also observed that the defect concentration in the DCl-doped Cs$_3$Bi$_2$I$_9$ films was reduced, which consequently led to a higher PCE of 1.308% (the highest among solution-processed Cs$_3$Bi$_2$I$_9$-based devices), see Fig. 16.10A–C.

With a small amendment in the stoichiometry of Cs$_3$Bi$_2$I$_9$, a new PVK compound, that is, Cs$_3$Bi$_2$I$_{10}$, was introduced [102]. This new PVK exhibited a smaller and extended absorption spectrum and narrow E_{BG} (1.77 eV). Unlike Cs$_3$Bi$_2$I$_9$, the new PVK compound (Cs$_3$Bi$_2$I$_{10}$) had a layered structure and altered crystal orientation.

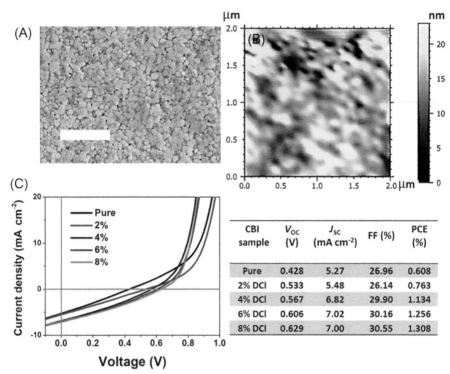

Fig. 16.10 (A) SEM image of the morphology of a DCl-doped $Cs_3Bi_2I_9$ thin film. (B) Corresponding AFM image. (C) J-V characteristics of the solar cell device of DCl-doped $Cs_3Bi_2I_9$ and device parameters at different doping concentrations. Reproduced with permission N. Ali, et al., The effect of dodecylammonium chloride on the film morphology, crystallinity, and performance of lead-free Bi-based solution-processed photovoltaics devices, Sol. Energy 207 (2020) 1356–1363.

Its broad photoabsorption spectrum enabled an enhanced photocurrent up to 700 nm, indicative of increased optical absorption and photocurrent in SC. PCE of 0.4% with V_{OC} of 0.31 V, J_{SC} of 3.40 mA cm^{-2}, and FF of 38% were recorded for the device.

Inspired by the above preliminary works, a large number of Bi-based lower-dimensional PVKs have been synthesized via different routes and demonstrated for their PV applications [8,101,103–106]. Along with better operational stability, SCs exhibited improved PV performance. The overall progress of these SCs in terms of their PV characteristics as well as device stability is given in Fig. 16.11.

16.5.2 Other applications

The remarkable photoabsorption characteristics of Bi-based PVKs enable them to be utilized for PV applications; nevertheless, they also found their applications in other optoelectronic and electronic devices; for example, photodetectors, capacitors, and memory storage devices, as summarized in the succeeding section.

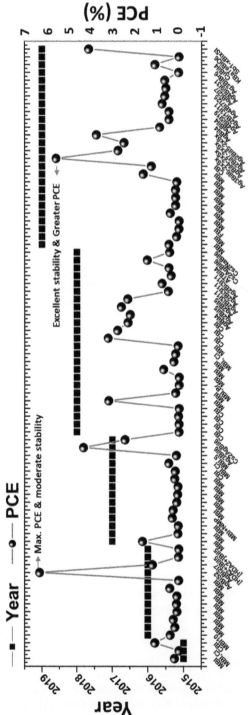

Bi-based Pb-free perovskite compounds

Fig. 16.11 The overall summary of Bi-based perovskites solar cell along with the compound names and recorded PCE's. The better performance devices based on the power conversion efficiency and stability are labeled. The data were updated (by adding references [101, 103–106]) from the Web of Science, searching the keywords "Bismuth + Perovskites + Solar cells."
Reproduced with permission S. Attique, et al., A potential checkmate to lead: bismuth in organometal halide perovskites, structure, properties, and applications, Adv. Sci. 7(13) (2020) 1903143.

16.5.2.1 Photodetectors

Photodetectors instantly convert the absorbed photons into electric current and are vital in communications, computation, remote sensing, biological and chemical sensing, and imaging. The greater α of Bi-based PVKs enables efficient light absorption in a few hundred nanometer-thin films. To ensure fast photo-response, the interelectrode distance should be very small. Recently, Bi-based PVKs have been demonstrated for photodetection purposes [107–113]. The first photodetector based on $CsBi_3I_{10}$ was put forward by Tong et al. [108] in 2017, which was very sensitive to 650 nm wavelength (red light), and exhibited a higher on/off ratio of 10^5, and a faster response time, that is, $\tau_{rise} = 0.33$ ms and $\tau_{decay} = 0.38$ ms. In comparison with Pb-based photodetectors, the $CsBi_3I_{10}$-based device displayed a specific detectivity and photoresponsivity of about 21.8 A W^{-1} and 1.93×10^{13} Jones, respectively. The device also yielded a high EQE of $4.13 \times 10^3\%$ and excellent ambient stability for more than 3 months. Likewise, highly oriented TMHDBiBr$_5$PVK SCs (volume $= 32 \times 24 \times 12$ mm^3) were utilized to fabricate an efficient photodetector with an on/off ratio of $\sim 10^3$ and a faster response speed, with the rise and decay times of 8.9 ms and 10.2 ms, respectively [109]. In addition, the photodetector demonstrated a 100 mA W^{-1} photoresponsivity, which remained constant in the 365–700 nm wavelength range, see Fig. 16.12A–F. Such improved photodetection properties were ascribed to the improved crystallinity, longer carrier lifetime, and better charge transport in the PVK active layer. Besides, $Cs_3Bi_2I_9$ nanoplates were also utilized to fabricate a visible light detector through simple solution processing on the ITO substrate, which could also be deposited on any other kind of substrate [110]. It was the first $Cs_3Bi_2I_9$-based photodetector, exhibiting 33.1 mA W^{-1} photoresponsivity under the illumination of a laser beam with a wavelength of 450 nm. Such devices exhibited a remarkable photoresponse, sixfolds higher than the solution-processed MAPbI$_3$ nanowires. In addition to their operational stability, the detectivity of the photodetecting device has reached as high as 10^{10} Jones. The device also demonstrated a rapid response time, that is, τ_{rise} and τ_{decay} of 10.2 ms and 37.2 ms, respectively.

Recently, another photodetector based on the vertically oriented single crystals of 0D PVK-like compound $((PD)_2Bi_2I_{10}\cdot2H_2O)$ [111]. The photodetecting device of such PVK comparatively underperformed, that is, showing a low photocurrent of 194 mA, on/off ratio of 2.1, and a 1.14 mA W^{-1} photoresponsivity. Moreover, the photodetector showed a lower specific detectivity (1.9×10^6 Jones) and only 0.4% EQY. Likewise, several other underperformed photodetectors were constructed using 1D [$(C_6H_{13}N)_2BiI_5$, (TMP) BiX$_5$ (X = I, Br, or Cl)] and 2D [$(TMP)_{1.5}(Bi_2I_7Cl_2)$] PVK [112,113].

16.5.2.2 Memristors and capacitors

To exploit lower-dimensional Bi-based PVKs for resistive memory applications, $Cs_3Bi_2I_9$ and $MA_3Bi_2I_9$ nanosheets were utilized to fabricate novel flexible memristors through the process of dissolution and recrystallization [114]. The prepared ultrathin memory devices exhibited a typical resistive-switching at a smaller operating voltage of ~ 0.3 V, exhibited a high on/off ratio ($\sim 10^3$), and protracted retention of data ($> 10^4$ s). The devices also showed outstanding ambient durability ($Cs_3Bi_2I_9$ only), flexibility, reproducibility, and endurability, see Fig. 16.13A–F.

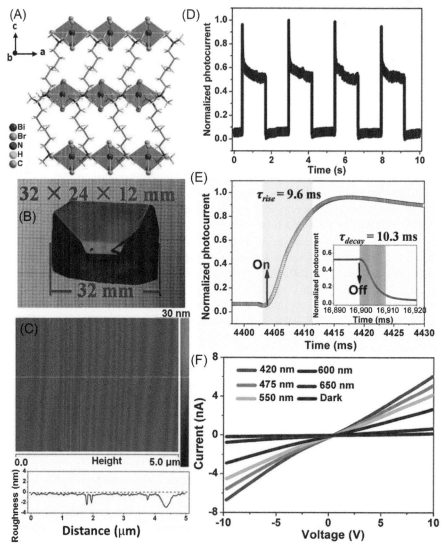

Fig. 16.12 Single crystals and photodetecting device characteristics of Bi-based perovskites: (A) and (B) Crystal structure, structural unit, and an optical image of the as-grown single-crystal of (TMHD)BiBr$_5$. (C) AFM image showing the surface roughness profile of the inch-size (TMHD)BiBr$_5$ single-crystal. (D) Attenuation-less time-dependent photocurrent characteristics of the single-crystal photodetector under chopped light irradiation measured for several cycles. (E) Temporal photocurrent response; the highlighted regions are showing a rise and decay times 9.6 ms and 10.3 ms, respectively. (F) IV characteristics of the (TMHD)BiBr$_5$ single-crystal photodetector measured in dark as well as under light illuminations with various wavelengths. Reproduced with permission C. Ji, et al., Inch-size single crystal of a lead-free organic–inorganic hybrid perovskite for high-performance photodetector, Adv. Funct. Mater. 28(14) (2018) 1705467.

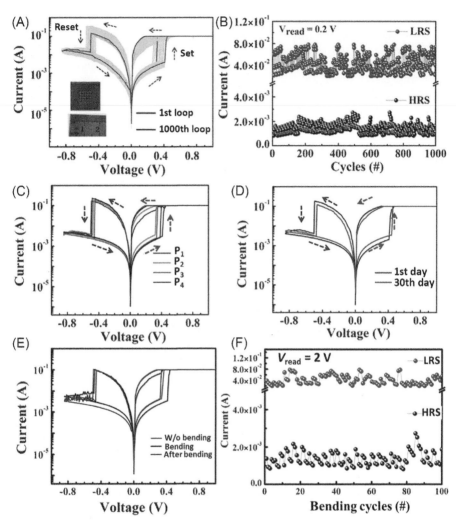

Fig. 16.13 Resistive switching performance and stability of the Bi-based resistive memory devices: (A) Repeatability analysis of the bipolar memristor monitored by testing the IV characteristics over 1000 cycles. (B) Endurance of the memristor over 1000 cycles, where the high-resistive state (HRS) and low resistive state (LRS) values were read at 0.2 V. (C) IV characteristics of the same memristor monitored at four different spots. (D) Stability analysis of the device over 30 days. (E) IV characteristics without and during bending stress (bending radius = 0.9 cm). (F) Endurance test of the memristor for 100 bending cycles, where HRS and LRS values were read at 0.2 V.

Reproduced with permission Y. Hu, et al., Ultrathin Cs3Bi2I9 nanosheets as an electronic memory material for flexible memristors, Adv. Mater. Interfaces 4(14) (2017) 1700131.

However, the environmental instability of $MA_3Bi_2I_9$ nanosheets made them unfavorable for the manufacturing of memristors. Conversely, in another study, $MA_3Bi_2I_9$-based memristors, fabricated via solution-processing exhibited a speed of 100 ns, along with consistent retention of $\sim 10^4$ s, and endurance of about 300 cycles [115]. The device exhibited a multi-level storage capacity with four resistive states and also displayed better environmental stability for 5 months.

These PVKs were also employed for the preparation of ultrafast capacitors. For example, Pious et al. [33] introduced a double-layered $MA_3Bi_2I_9$-based electrochemical capacitor. The device was stable enough up to 84.4% and displaying 10^4 cycles of charge/discharge with a capacitance of about 5.5 mF/cm^2. Moreover, spectroscopy results have shown that the $MA_3Bi_2I_9$ layer provides additional surface area for electrolytes.

16.6   Conclusion

Organo-lead halide PVKs have gained overnight prominence in photovoltaics and other optoelectronic applications; however, confronted with poor material and chemical stability in the ambient air and the toxicity of Pb. The issue of environmental instability was partially addressed using device encapsulation, interfacial and compositional engineering, the employment of additional hydrophobic and heat resistive layers, and via lowering the dimension of PVKs. Among these strategies, the 2D PVKs not only offered environmental stability but also retained mostly the performance of the device. Still, 2D PVKs have Pb as the central component, toxic to human health; therefore, other lower-dimensional PVKs were introduced. Among the lower-dimensional PVKs, Bi-based PVKs (with comparable optoelectronic properties) successfully reduced the toxicity and offered superior ambient stability to the conventional PVK material. Based on their structure and dimensionality, Bi-based PVKs were divided into several classes. Among those, Bi-based 0D PVKs were widely used for many optoelectronic applications, most commonly for SCs, which exhibited excellent durability and lesser toxicity than Pb. However, they showed only satisfactory performance, much inferior to the conventional Pb-based PVKs. Nonetheless, Bi-based devices, other than SCs, demonstrated comparable, or even better performance than Pb-based ones. Lower-dimensional PVKs (including Bi-based PVKs) are in the initial phase of material development; therefore, to further improve their performance, a copious amount of multi-disciplinary effort is required in terms of material processing and engineering, device architecture optimization, and introduction of new lower-dimensional PVKs with similar optoelectronic properties to Pb.

References

[1] S. Gamliel, L. Etgar, Organo-metal perovskite based solar cells: sensitized versus planar architecture, RSC Adv. 4 (55) (2014) 29012–29021.
[2] Z. Song, et al., Impact of processing temperature and composition on the formation of methylammonium lead iodide perovskites, Chem. Mater. 27 (13) (2015) 4612–4619.

 [3] S. Aharon, L. Etgar, Two dimensional organometal halide perovskite nanorods with tunable optical properties, Nano Lett. 16 (5) (2016) 3230–3235.
 [4] A. Kojima, et al., Organometal halide perovskites as visible-light sensitizers for photovoltaic cells, J. Am. Chem. Soc. 131 (17) (2009) 6050–6051.
 [5] P. Zhu, et al., Direct conversion of perovskite thin films into nanowires with kinetic control for flexible optoelectronic devices, Nano Lett. 16 (2) (2016) 871–876.
 [6] F. Zhang, et al., Comparative studies of optoelectrical properties of prominent PV materials: halide perovskite, CdTe, and GaAs, Mater. Today 36 (2020) 18–29.
 [7] K. Yadaiah, et al., Synthesis and characterization of $(CdSe)(1-x)(ZnS)(x)$ mixed polycrystalline semiconductors by co-precipitation method, Int. J. Mod. Phys. B 16 (19) (2002) 2885–2899.
 [8] S. Attique, et al., A potential checkmate to lead: bismuth in organometal halide perovskites, structure, properties, and applications, Adv. Sci. 7 (13) (2020) 1903143.
 [9] Z. Cheng, J. Lin, Layered organic–inorganic hybrid perovskites: structure, optical properties, film preparation, patterning and templating engineering, CrystEngComm 12 (10) (2010) 2646–2662.
[10] C. Huo, et al., Two-dimensional metal halide perovskites: theory, synthesis, and optoelectronics, Small Methods 1 (3) (2017) 1600018.
[11] H. Huang, et al., Colloidal lead halide perovskite nanocrystals: synthesis, optical properties and applications, NPG Asia Mater. 8 (2016) e328, https://doi.org/10.1038/am.2016.167.
[12] D.B. Mitzi, Solution-processed inorganic semiconductors, J. Mater. Chem. 14 (15) (2004) 2355–2365.
[13] T. Ishihara, J. Takahashi, T. Goto, Optical properties due to electronic transitions in two-dimensional semiconductors $(CnH2n+1NH3)2PbI4$, Phys. Rev. B 42 (17) (1990) 9.
[14] K. Wang, et al., Quasi-two-dimensional halide perovskite single crystal photodetector, ACS Nano 12 (5) (2018) 4919–4929.
[15] T.M. Koh, et al., Enhancing moisture tolerance in efficient hybrid 3D/2D perovskite photovoltaics, J. Mater. Chem. A 6 (5) (2018) 2122–2128.
[16] J.A. Sichert, et al., Quantum size effect in organometal halide perovskite nanoplatelets, Nano Lett. 15 (10) (2015) 6521–6527.
[17] I.C. Smith, et al., A layered hybrid perovskite solar-cell absorber with enhanced moisture stability, Angew. Chem. 126 (42) (2014) 11414–11417.
[18] D.H. Cao, et al., 2D homologous perovskites as light-absorbing materials for solar cell applications, J. Am. Chem. Soc. 137 (24) (2015) 7843–7850.
[19] M. Safdari, et al., Layered 2D alkyldiammonium lead iodide perovskites: synthesis, characterization, and use in solar cells, J. Mater. Chem. A 4 (40) (2016) 15638–15646.
[20] F. Zhang, et al., Advances in two-dimensional organic–inorganic hybrid perovskites, Energy Environ. Sci. 13 (4) (2020) 1154–1186.
[21] N. Ali, et al., An overview of the decompositions in organo-metal halide perovskites and shielding with 2-dimensional perovskites, Renew. Sust. Energ. Rev. 109 (2019) 160–186.
[22] W. Honglei, et al., Two-dimensional perovskites and their applications on optoelectronic devices, Prog. Chem. 29 (8) (2017) 859.
[23] R. Mohan, Green bismuth, Nat. Chem. 2 (4) (2010) 336.
[24] Z. Ran, et al., Bismuth and antimony-based oxyhalides and chalcohalides as potential optoelectronic materials, npj Comput. Mater. 4 (1) (2018) 14.
[25] Z. Wang, et al., Stability of perovskite solar cells: a prospective on the substitution of the A cation and X anion, Angew. Chem. Int. Ed. 56 (5) (2017) 1190–1212.

[26] B. Yang, et al., Lead-free, air-stable all-inorganic cesium bismuth halide perovskite nanocrystals, Angew. Chem. Int. Ed. 56 (41) (2017) 12471–12475.
[27] C. Zhang, et al., Design of a novel and highly stable lead-free Cs 2 NaBiI 6 double perovskite for photovoltaic application, Sustain. Energy Fuels 2 (11) (2018) 2419–2428.
[28] N.C. Miller, M. Bernechea, Research update: bismuth based materials for photovoltaics, APL Mater. 6 (8) (2018), 084503.
[29] S.E. Creutz, et al., Structural diversity in cesium bismuth halide nanocrystals, Chem. Mater. 31 (13) (2019) 4685–4697.
[30] O. Lindqvist, et al., Crystal structure of Caesium bismuth iodide, Cs3Bi2I9, Acta Chem. Scand. 22 (1968) 2943–2952.
[31] A.J. Lehner, et al., Crystal and electronic structures of complex bismuth iodides A 3Bi2I9 (A = K, Rb, Cs) related to perovskite: aiding the rational design of photovoltaics, Chem. Mater. 27 (20) (2015) 7137–7148.
[32] M.-M. Yao, et al., General synthesis of Lead-free metal halide perovskite colloidal nanocrystals in 1-Dodecanol, Inorg. Chem. (2019).
[33] J.K. Pious, et al., Zero-dimensional methylammonium bismuth iodide-based Lead-free perovskite capacitor, ACS Omega 2 (9) (2017) 5798–5802.
[34] R. Jakubas, J. Zaleski, L. Sobczyk, Phase transitions in (CH3NH3) 3Bi2I9 (MAIB), Ferroelectrics 108 (1) (1990) 109–114.
[35] S.F. Hoefler, G. Trimmel, T. Rath, Progress on lead-free metal halide perovskites for photovoltaic applications: a review, Monatsh. Chem. 148 (5) (2017) 795–826.
[36] N.A. Yelovik, et al., Iodobismuthates containing one-dimensional BiI4–anions as prospective light-harvesting materials: synthesis, crystal and electronic structure, and optical properties, Inorg. Chem. 55 (9) (2016) 4132–4140.
[37] D.B. Mitzi, P. Brock, Structure and optical properties of several organic – inorganic hybrids containing corner-sharing chains of bismuth iodide octahedra, Inorg. Chem. 40 (9) (2001) 2096–2104.
[38] D.M. Fabian, S. Ardo, Hybrid organic–inorganic solar cells based on bismuth iodide and 1, 6-hexanediammonium dication, J. Mater. Chem. A 4 (18) (2016) 6837–6841.
[39] G.A. Mousdis, et al., Preparation, structures and optical properties of [H3N (CH2) 6NH3] BiX5 (X= I, Cl) and [H3N (CH2) 6NH3] SbX5 (X= I, Br), Z. Naturforsch. B 53 (8) (1998) 927–932.
[40] V. Sidey, et al., Crystal growth and X-ray structure determination of Rb3Bi2I9, J. Alloys Compd. 296 (1–2) (2000) 53–58.
[41] B. Chabot, E. Parthe, Cs3Sb2I9 and Cs3Bi2I9 with the hexagonal Cs3Cr2Cl9 structure type, Acta Crystallogr. Sect. B: Struct. Crystallogr. Cryst. Chem. 34 (2) (1978) 645–648.
[42] J.H. Chang, T. Doert, M. Ruck, Structural variety of defect perovskite variants M3E2X9 (M= Rb, Tl, E= Bi, Sb, X= Br, I), Z. Anorg. Allg. Chem. 642 (13) (2016) 736–748.
[43] H. Jagodzinski, Der Symmetrieeinfluss auf den allgemeinen Lösungsansatz eindimensionaler Fehlordnungs-probleme, Acta Crystallogr. 7 (1) (1954) 17–25.
[44] N.-G. Park, H. Segawa, Research direction toward theoretical efficiency in perovskite solar cells, ACS Photonics 5 (8) (2018) 2970–2977.
[45] F. Wang, et al., Defects engineering for high-performance perovskite solar cells, npj Flex. Electron. 2 (1) (2018) 22.
[46] B.-W. Park, et al., Bismuth based hybrid perovskites A3Bi2I9 (A: Methylammonium or cesium) for solar cell application, Adv. Mater. 27 (43) (2015) 6806–6813.
[47] B. Saparov, D.B. Mitzi, Organic–inorganic perovskites: structural versatility for functional materials design, Chem. Rev. 116 (7) (2016) 4558–4596.

[48] J. Pal, et al., Synthesis and optical properties of colloidal M3Bi2I9 (M = Cs, Rb) perovskite nanocrystals, J. Phys. Chem. C 122 (19) (2018) 10643–10649.

[49] Z. Xiao, et al., Bandgap optimization of perovskite semiconductors for photovoltaic applications, Chem. Eur. J. 24 (10) (2018) 2305–2316.

[50] V. Machulin, et al., Effect of temperature variation on shift and broadening of the exciton band in Cs 3 Bi 2 I 9 layered crystals, Low Temp. Phys. 30 (12) (2004) 964–967.

[51] M. Abulikemu, et al., Optoelectronic and photovoltaic properties of the air-stable organohalide semiconductor (CH 3 NH 3) 3 Bi 2 I 9, J. Mater. Chem. A 4 (32) (2016) 12504–12515.

[52] Y. Zhang, et al., Direct-indirect nature of the bandgap in lead-free perovskite nanocrystals, J. Phys. Chem. Lett. 8 (14) (2017) 3173–3177.

[53] X. Huang, et al., Band gap insensitivity to large chemical pressures in ternary bismuth iodides for photovoltaic applications, J. Phys. Chem. C 120 (51) (2016) 28924–28932.

[54] M. Pazoki, et al., Bismuth iodide perovskite materials for solar cell applications: electronic structure, optical transitions, and directional charge transport, J. Phys. Chem. C 120 (51) (2016) 29039–29046.

[55] Z. Zhang, et al., High-quality (CH3NH3)3Bi2I9 film-based solar cells: pushing efficiency up to 1.64%, J. Phys. Chem. Lett. 8 (17) (2017) 4300–4307.

[56] A. Kulkarni, M. Ikegami, T. Miyasaka, Morphology evolution of non-toxic MA3Bi2I9 based lead free perovskite solar cells, Mater. Sci. (2017). https://www.semanticscholar. org/paper/Morphology-Evolution-of-Non-toxic-MA3Bi2I9-Based-Kulkarni-Ikegami/ 40e888edc6edb44e03b82c41d74a8130c66f86b4.

[57] X. Chen, et al., Atmospheric pressure chemical vapor deposition of methylammonium bismuth iodide thin films, J. Mater. Chem. A 5 (47) (2017) 24728–24739.

[58] K.M. McCall, et al., Strong electron–phonon coupling and self-trapped excitons in the defect halide perovskites A3M2I9 (A= Cs, Rb; M= Bi, Sb), Chem. Mater. 29 (9) (2017) 4129–4145.

[59] K.K. Bass, et al., Vibronic structure in room temperature photoluminescence of the halide perovskite Cs3Bi2Br9, Inorg. Chem. 56 (1) (2016) 42–45.

[60] C.C. Stoumpos, et al., Ruddlesden–Popper hybrid lead iodide perovskite 2D homologous semiconductors, Chem. Mater. 28 (8) (2016) 2852–2867.

[61] I. Benabdallah, et al., Lead-free perovskite based bismuth for solar cells absorbers, J. Alloys Compd. 773 (2019) 796–801.

[62] T.H. Gfroerer, Photoluminescence in analysis of surfaces and interfaces, in: R.A. Meyers, G.E. McGuire (Eds.), Encyclopedia of Analytical Chemistry, 2006, https:// doi.org/10.1002/9780470027318.a2510.

[63] Z. Gan, et al., External stokes shift of perovskite nanocrystals enlarged by photon recycling, Appl. Phys. Lett. 114 (1) (2019), 011906.

[64] N.H. Linh, et al., Alkali metal-substituted bismuth-based perovskite compounds: a DFT study, J. Sci.: Adv. Mater. Devices (2019).

[65] Y. Zhou, et al., Ultrabroad near-infrared photoluminescence from bismuth doped CsPbI 3: polaronic defects vs. bismuth active centers, J. Mater. Chem. C 4 (12) (2016) 2295–2301.

[66] P. Sun, et al., Prediction of the role of bismuth dopants in organic-inorganic lead halide perovskites on photoelectric properties and photovoltaic performance, J. Phys. Chem. C 123 (20) (2019) 12684–12693.

[67] E. Mosconi, et al., First-principles modeling of bismuth doping in the MAPbI3 perovskite, J. Phys. Chem. C 122 (25) (2018) 14107–14112.

[68] M. Leng, et al., All-inorganic bismuth-based perovskite quantum dots with bright blue photoluminescence and excellent stability, Adv. Funct. Mater. 28 (1) (2018) 1704446.

[69] Y.N. Ivanov, et al., Phase transitions of Cs3Sb2I9, Cs3Bi2I9 , and Cs3Bi2Br9 crystals, Inorg. Mater. 37 (6) (2001) 623–627.

[70] M. Sykora, et al., Effect of air exposure on surface properties, electronic structure, and carrier relaxation in PbSe nanocrystals, ACS Nano 4 (4) (2010) 2021–2034.

[71] M.V. Kovalenko, et al., Prospects of Nanoscience With Nanocrystals, ACS Publications, 2015.

[72] M.A. Boles, et al., The surface science of nanocrystals, Nat. Mater. 15 (2) (2016) 141.

[73] A. Maiti, et al., Band-edges of bismuth-based ternary halide perovskites (A3Bi2I9) through scanning tunneling spectroscopy Vis-à-Vis impact of defects in limiting the performance of solar cells, Sol. Energy Mater. Sol. Cells 200 (2019), 109941.

[74] R.E. Brandt, et al., Identifying defect-tolerant semiconductors with high minority-carrier lifetimes: beyond hybrid lead halide perovskites, MRS Commun. 5 (2) (2015) 265–275.

[75] L.C. Lee, et al., Research update: bismuth-based perovskite-inspired photovoltaic materials, APL Mater. 6 (8) (2018), 084502.

[76] R.E. Brandt, et al., Investigation of bismuth triiodide (BiI3) for photovoltaic applications, J. Phys. Chem. Lett. 6 (21) (2015) 4297–4302.

[77] B. Ghosh, et al., Poor photovoltaic performance of Cs3Bi2I9: an insight through first-principles calculations, J. Phys. Chem. C 121 (32) (2017) 17062–17067.

[78] T. Ou, et al., Visible light response, electrical transport, and amorphization in compressed organolead iodine perovskites, Nanoscale 8 (22) (2016) 11426–11431.

[79] N.-G. Park, Organometal perovskite light absorbers toward a 20% efficiency low-cost solid-state mesoscopic solar cell, J. Phys. Chem. Lett. 4 (15) (2013) 2423–2429.

[80] A.D. Jodlowski, et al., Large guanidinium cation mixed with methylammonium in lead iodide perovskites for 19% efficient solar cells, Nat. Energy 2 (12) (2017) 972–979.

[81] W. Metzger, et al., Time-Resolved Photoluminescence and Photovoltaics, National Renewable Energy Lab (NREL), Golden, CO, 2005.

[82] R.L. Hoye, et al., Fundamental carrier lifetime exceeding 1 μs in Cs2AgBiBr6 double perovskite, Adv. Mater. Interfaces 5 (15) (2018) 1800464.

[83] A.H. Slavney, et al., A bismuth-halide double perovskite with long carrier recombination lifetime for photovoltaic applications, J. Am. Chem. Soc. 138 (7) (2016) 2138–2141.

[84] E. Greul, et al., Highly stable, phase pure Cs 2 AgBiBr 6 double perovskite thin films for optoelectronic applications, J. Mater. Chem. A 5 (37) (2017) 19972–19981.

[85] G. Volonakis, et al., Lead-free halide double perovskites via heterovalent substitution of noble metals, J. Phys. Chem. Lett. 7 (7) (2016) 1254–1259.

[86] L. Song, et al., Efficient inorganic perovskite light-emitting diodes with polyethylene glycol passivated ultrathin CsPbBr3 films, J. Phys. Chem. Lett. 8 (17) (2017) 4148–4154.

[87] R. Knox, Introduction to exciton physics, in: Collective Excitations in Solids, Springer, 1983, pp. 183–245.

[88] C. Kittel, Introduction to Solid State Physics, John Wiley & Sons. Inc, New York, 2005.

[89] T. Kawai, et al., Optical absorption in band-edge region of (CH 3 NH 3) 3 B i 2 I 9 single crystals, J. Phys. Soc. Jpn. 65 (5) (1996) 1464–1468.

[90] J. Peng, et al., Insights into charge carrier dynamics in organo-metal halide perovskites: from neat films to solar cells, Chem. Soc. Rev. 46 (19) (2017) 5714–5729.

[91] N. Ali, et al., Enhanced stability in cesium assisted hybrid 2D/3D-perovskite thin films and solar cells prepared in ambient humidity, Sol. Energy 189 (2019) 325–332.

[92] C. Ran, et al., Construction of compact methylammonium bismuth iodide film promoting lead-free inverted planar heterojunction organohalide solar cells with open-circuit voltage over 0.8 V, J. Phys. Chem. Lett. 8 (2) (2017) 394–400.

[93] A.K. Jena, A. Kulkarni, T. Miyasaka, Halide perovskite photovoltaics: background, status, and future prospects, Chem. Rev. 119 (5) (2019) 3036–3103.

[94] O. Stroyuk, Lead-free hybrid perovskites for photovoltaics, Beilstein J. Nanotechnol. 9 (1) (2018) 2209–2235.

[95] H. Fu, Review of lead-free halide perovskites as light-absorbers for photovoltaic applications: from materials to solar cells, Sol. Energy Mater. Sol. Cells 193 (2019) 107–132.

[96] M.Q. Lyu, et al., Organic-inorganic bismuth (III)-based material: a lead-free, air-stable and solution-processable light-absorber beyond organolead perovskites, Nano Res. 9 (3) (2016) 692–702.

[97] Z. Ma, et al., Air-stable layered bismuth-based perovskite-like materials: structures and semiconductor properties, Phys. B Condens. Matter 526 (2017) 136–142.

[98] S. Öz, et al., Zero-dimensional (CH3NH3)3Bi2I9 perovskite for optoelectronic applications, Sol. Energy Mater. Sol. Cells 158 (2016) 195–201.

[99] T. Singh, et al., Effect of electron transporting layer on bismuth-based Lead-free perovskite (CH3NH3)3 Bi2I9 for photovoltaic applications, ACS Appl. Mater. Interfaces 8 (23) (2016) 14542–14547.

[100] X. Zhang, et al., Active-layer evolution and efficiency improvement of $(CH_3NH_3)_3Bi_2I_9$-based solar cell on TiO_2-deposited ITO substrate, Nano Res. 9 (10) (2016) 2921–2930.

[101] N. Ali, et al., The effect of dodecylammonium chloride on the film morphology, crystallinity, and performance of lead-free Bi-based solution-processed photovoltaics devices, Sol. Energy 207 (2020) 1356–1363.

[102] M.B. Johansson, H. Zhu, E.M.J. Johansson, Extended photo-conversion Spectrum in low-toxic bismuth halide perovskite solar cells, J. Phys. Chem. Lett. 7 (17) (2016) 3467–3471.

[103] W. Hu, et al., Bulk heterojunction gifts bismuth-based lead-free perovskite solar cells with record efficiency, Nano Energy 68 (2020), 104362.

[104] S. Sanders, et al., Chemical vapor deposition of organic-inorganic bismuth-based perovskite films for solar cell application, Sci. Rep. 9 (1) (2019) 9774.

[105] C. Momblona, et al., Co-evaporation as an optimal technique towards compact methylammonium bismuth iodide layers, Sci. Rep. 10 (1) (2020) 10640.

[106] R. Waykar, et al., Environmentally stable lead-free cesium bismuth iodide (Cs3Bi2I9) perovskite: synthesis to solar cell application, J. Phys. Chem. Solids 146 (2020), 109608.

[107] W. Zhang, et al., Triiodide-induced band-edge reconstruction of a lead-free perovskite-derivative hybrid for strong light absorption, Chem. Mater. 30 (12) (2018) 4081–4088.

[108] X.-W. Tong, et al., High-performance red-light photodetector based on Lead-free bismuth halide perovskite film, ACS Appl. Mater. Interfaces 9 (22) (2017) 18977–18985.

[109] C. Ji, et al., Inch-size single crystal of a lead-free organic–inorganic hybrid perovskite for high-performance photodetector, Adv. Funct. Mater. 28 (14) (2018) 1705467.

[110] Z. Qi, et al., Highly stable lead-free Cs3Bi2I9 perovskite nanoplates for photodetection applications, Nano Res. 12 (8) (2019) 1894–1899.

[111] J.K. Pious, et al., Zero-dimensional Lead-free hybrid perovskite-like material with a quantum-well structure, Chem. Mater. 31 (6) (2019) 1941–1945.

[112] W. Zhang, et al., (C6H13N)2BiI5: a one-dimensional Lead-free perovskite-derivative photoconductive light absorber, Inorg. Chem. 57 (8) (2018) 4239–4243.

[113] X. Shai, et al., Efficient planar perovskite solar cells using halide Sr-substituted Pb perovskite, Nano Energy 36 (2017) 213–222.
[114] Y. Hu, et al., Ultrathin Cs3Bi2I9 nanosheets as an electronic memory material for flexible memristors, Adv. Mater. Interfaces 4 (14) (2017) 1700131.
[115] B. Hwang, J.-S. Lee, Lead-free, air-stable hybrid organic–inorganic perovskite resistive switching memory with ultrafast switching and multilevel data storage, Nanoscale 10 (18) (2018) 8578–8584.

Transmission electron microscopy (TEM) studies of functional nanomaterials

Weilie Zhou[a], Y.H. Ikuhara[b], Zhi Zheng[a], K. Wang[a], B. Cao[a], and J. Chen[a]
[a]University of New Orleans, New Orleans, LA, United States, [b]Japan Fine Ceramics Center, Nagoya, Japan

17.1 Introduction

Ever since the discovery of the transmission electron microscope (TEM), it has played an essential role in revealing the crystal structure and surface and chemical composition of materials [1]. These capabilities become increasingly crucial in characterizing nanomaterials toward an atomic-level control of structure and composition. Using nanowires as examples, in addition to the bright field (BF) imaging of nanowire morphology, TEM allows for the determination of crystallinity and growth direction of nanowires by analyzing the select-area electron diffraction (SAED) patterns, assignment of surface polarizations by studying the convergent beam electron diffraction (CBED), and identification of planar defects with atomic resolution via high-resolution transmission microscopy (HRTEM) imaging. Dark-field (DF) imaging can be used to visualize crystalline defects such as twinning and secondary phase precipitates and determine the interfaces in nanoheterostructures. As a well-established compositional analysis technique, energy-dispersive X-ray spectroscopy (EDS) is capable of identifying and quantifying nanowire composition with heavy elements. A more chemically sensitive spectroscopic technique, electron energy loss spectroscopy (EELS), can provide better identification and qualification of low atomic number elements, such as oxygen and carbon. Beyond these conventional TEM techniques, Z-contrast or high-angle annular dark-field (HAADF) images obtained in scanning transmission electron microscope (STEM) mode can offer good composition contrast, for instance, of a nanowire superlattice with elements comprising of differentiable atomic numbers or noble-metal nanoparticles decorated nanowires [2]. Despite the limitations in imaging elements with low atomic numbers, TEM enables us to map compositional variations on the nanoscale along or across the growth direction of nanowires when combined with EDS or EELS. However, because of the effects of spherical aberration (Cs) in the electromagnetic lens, further achieving atomic resolution imaging and analysis in a conventional STEM is limited by the probe size and electron-beam current density. By applying the Cs-correction technology, one can expect a higher signal-to-noise ratio and a reduced probe broadening effect [3]. Consequently, performance for the analysis of structure, composition,

Modeling, Characterization and Production of Nanomaterials. https://doi.org/10.1016/B978-0-12-819905-3.00017-8

and physical properties on the atomic scale can be significantly improved. For instance, newly developed annular bright-field (ABF) imaging technique in a *Cs*-corrected STEM equipment allows for the direct visualization of oxygen and nitrogen atom columns, opening new routes for the fine characterization of oxide and nitride nanostructures [4]. Moreover, electron tomography in STEM mode can retrieve 3D structure information, such as shape and morphology, of nanostructures by multiple-angle imaging and tomographic reconstruction [5]. In addition, in situ TEM techniques provide opportunities to probe the nucleation and growth, transformation, measurement, and function of nanomaterials in real time [6].

The goal of this chapter is to illustrate the capabilities of TEM techniques in nanostructures with a number of nanomaterials in different dimensions. We will start with some examples of TEM study in 0D nanomaterials in the second section of this chapter. After that, we will focus on 1D nanomaterials, in which the third section will focus on using TEM to determine the growth mechanism using ZnO and SnO_2 (ZnO: Sn) as examples; the TEM characterization of 1D heterostructure will be exemplified in the fourth section. Recent advances of TEM studies on 2D materials will be presented in the fifth section. In addition, the recent progress of using advanced TEM characterizations in practical nanodevice applications will be reviewed in the sixth section. Finally, a brief summary of TEM in nanomaterials characterization will be presented.

17.2   TEM characterizations of 0D nanostructures

0D nanostructures, that is, nanoparticles, is the most commonly studied and fundamental component in the nanotechnology. It differs from bulk materials due to its extremally small size, generally in the range of 1–100 nm. Owing to their unique size-dependent electronic, optical, magnetic, and catalytic properties, 0D nanostructures have been utilized in many applications such as biological labeling, light emitting diodes, and photovoltaics. One famous example with commercialized applications is the semiconducting quantum dots with emission wavelengths that can be adjusted through their sizes, which are used in commercial TV branded as quantum dot light emitting diodes. In both fundamental research and practical applications of nanotechnology, understanding the structure, composition, and physical properties of nanomaterial itself is of great importance.

17.2.1   TEM characterization of In_2O_3 semiconductor nanoparticles

Chemical sensing is one of the most promising applications for nanomaterials, showing extremely high sensitivity resulting from the large surface-to-volume ratio. In our previous works, we have demonstrated the superior sensitivities of metal oxide nanoparticles (NPs) to different target gases at room temperature and found that surface modification is the most crucial factor to improve the sensitivity [7,8].

Although significant progress has been made in the preparation of nanoparticles, the fine control of the size and shape, especially for nanoparticles smaller than 10 nm, remains a challenge. Thus, we first synthesized highly crystalline, variable-shaped In_2O_3 nanoparticles by the heat treatment of different indium-organic precursors in oleyl alcohol solutions [9]. The morphological features and composition of the resulting nanoparticles can be tuned upon changing the metal precursor and the atmosphere in which the reactions are conducted. As shown in Fig. 17.1A, the reaction of In(acac)$_3$ with oleyl alcohol at 320 °C under open air produces nanoparticles with a well-developed morphology. The cubic-shape particles have an average edge length of 7.0 nm (relative standard deviation (RSD) of 9.4%) and aspect ratios ranging from 1.0 to 1.2. The HRTEM image of a single 7-nm size cube (inset of Fig. 17.1A) shows crossed lattice fringes with fringe separations of 2.92 Å, which can be assigned to the (222) planes of cubic phase In$_2$O$_3$. However, when using In(ac)$_3$ precursor, particles with a less uniform morphology were obtained, as shown in Fig. 17.1B. The as-prepared nanoparticles have less regular shape with an average diameter of 9.2 nm with a RSD of 13%. Nanocrystals with a cubic morphology were also obtained when the In(O^{i}Pr)$_3$ precursor was heated in the presence of oleyl alcohol as shown in Fig. 17.1C. The resulting In$_2$O$_3$ nanocubes have round corners and an average edge length of 19.9 nm (RSD of 11% and aspect ratios of 1.0–1.15).

Different noble metals, such as Au and Pt nanocrystals, were then sputtered on the surface of metal oxide nanoparticles and served as surface modification in order to improve the sensing performance, the sizes of these noble metal nanocrystals were tuned by the power rate and duration of sputtering [10]. As shown in Fig. 17.2, the larger nanoparticles of In$_2$O$_3$ and the relatively small nanoparticles of Pt nanocrystals can be clearly identified. When the nominal thickness of Pt is controlled to be 0.1, 0.2, and 0.4 nm, the actual size of the Pt nanocrystals observed is about sub-1, 1, and 2 nm by HAADF-STEM observation, respectively. Importantly, the HAADF-STEM image of Fig. 17.2G provides detailed information of the atomic structure of both In$_2$O$_3$ nanoparticle and Au nanocrystal, in which we can clearly see the {111} facets of Au nanocrystal as highlighted. The adsorption energy of CO gas on Au(111) is the lowest compared to other metals, leading to the highest binding affinity and sticking coefficient, which results in the good CO gas sensing.

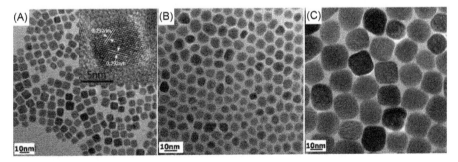

Fig. 17.1 TEM images of In$_2$O$_3$ nanoparticles obtained using different precursors with oleyl alcohol: (A) In(acac)$_3$, (B) In(ac)$_3$, and (C) In(O^{i}Pr)$_3$.

Adapted with permission from D. Caruntu, K. Yao, Z. Zhang, T. Austin, W. Zhou, C.J. O'Connor, One-step synthesis of nearly monodisperse, variable-shaped In$_2$O$_3$ nanocrystals in long chain alcohol solutions. J. Phys. Chem. C 114 (2010) 4875–4886. https://doi.org/10.1021/jp911427b. Copyright © 2010 American Chemical Society.

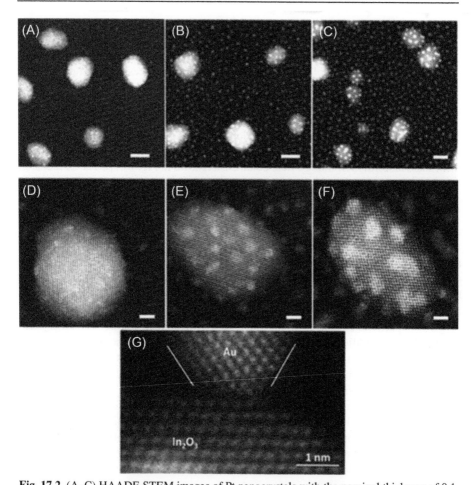

Fig. 17.2 (A–C) HAADF-STEM images of Pt nanocrystals with the nominal thickness of 0.1, 0.2, and 0.4 nm in low magnification. The scale bar is 10 nm. (D–F) HAADF-STEM images of Pt nanocrystals with the nominal thickness of 0.1, 0.2, and 0.4 nm in high magnification. The scale bar in (A–F) is 2 nm. (G) A HAADF-STEM image of a Au nanocrystal on the surface of an In_2O_3 nanoparticle, where {111} facets are highlighted by *white lines*.
Adapted with permission from K. Yao, D. Caruntu, S. Wozny, R. Huang, Y.H. Ikuhara, B. Cao, C.J. O'Connor, W. Zhou, Towards one key to one lock: catalyst modified indium oxide nanoparticle thin film sensor array for selective gas detection. J. Mater. Chem. 22 (2012) 7308. https://doi.org/10.1039/c2jm15179k. Copyright © 2012 Royal Society of Chemistry.

17.2.2 Direct observation of yolk-Shell transforming to gold single atoms and clusters

Understanding the growth mechanism and dynamic behavior under external stimuli of nanomaterials is critical in order to design and engineer nanomaterials for various applications. To this end, in-situ TEM has become increasingly important for nanomaterials characterization. With the rapid progress of in situ TEM in the recent decades, scientists can now have a powerful tool to characterize the dynamic changes in size, shape, interface structure, electronic state, and chemical composition under various external conditions, such as heat, stress, optical excitation, and magnetic or electric fields, for nanomaterials [11].

One of the examples is the in-situ observation of yolk-shell transforming to gold single atoms and clusters [12]. Fig. 17.3A shows the HAADF-STEM image of the fresh Au@Ni$_2$P yolk-shell structure. The Au yolk exhibits brighter contrast due to a higher atomic number than the Ni$_2$P shell. The chemical identity of the yolk and shell is further proved by EDS mapping in Fig. 17.3B–E. Fig. 17.3F is the magnified high-resolution Z-contrast image of Au@Ni$_2$P. The simultaneously acquired BF STEM image is shown in Fig. 17.3G. The fast Fourier transform (FFT) of the shell and Au yolk indicate that the Au(111) planes are epitaxial with the Ni$_2$P(301) crystal plane, as shown in Fig. 17.3G. In addition, we can clearly observe that the Au yolk is polycrystalline rather than single-crystalline. The yellow line points out a grain boundary that separates the two parts of Au with different crystal orientations. Another atomic-scale BF STEM image is shown in Fig. 17.3H, which clearly shows the epitaxial relationship of Au [200]//Ni$_2$P [021].

To clearly visualize the structural changes, the in-situ TEM was performed at elevated temperatures to probe the dynamical structural changes in real time. As shown in Fig. 17.4A, at room temperature, the Au yolk shows a bright color due to a heavier atomic number and presents itself as the yolk in the Ni$_2$P shell. The size of the original vacant hole in Ni$_2$P is around 16 nm. At 350 °C, the Au atoms began to diffuse into the Ni$_2$P and a lot of bright clusters and single atoms appear in the Ni$_2$P shell as circled in red in Fig. 17.4B. At high temperature (350 °C or higher), the gold yolk became unstable, and Au atoms began to diffuse into Ni$_2$P. These Au atoms quickly diffuse to cover the inner surface of Ni$_2$P due to a much larger surface diffusivity than bulk diffusivity. With a longer time at elevated temperature, these Au atoms began to diffuse into the Ni$_2$P, forming Au single atoms and bright clusters inside the Ni$_2$P shell and on its surface, as labeled in red in Fig. 17.4B. With further heating to 500 °C for 5 min, the yolk disappeared, and the contrast of the whole Ni$_2$P became evenly bright for the whole particle (Fig. 17.4C). With high-resolution STEM in Fig. 17.4D, the crystalline lattice spacing was observed as 2.89 Å, which corresponds to the (110) surface of Ni$_2$P. However, the diffraction spots in the FFT of Fig. 17.4E exhibited elongation and splitting, indicating lattice distortion by diffused Au atoms in the Ni$_2$P lattice. The diameter of the inner hole of the nanoparticles changed from 16 nm to about 11 nm after the whole heating process. In comparison, it is found that the shell on the bottom left region expanded from 8.5 nm to about 13.4 nm. The hole shrank, but the shell thickness increased significantly due to the dissolution of Au in Ni$_2$P. In conclusion, initiated by annealing the heterogeneous Au@Ni$_2$P catalyst at elevated temperatures, gold atoms can diffuse through the Ni$_2$P matrix and form single atoms and tiny clusters. With controlled temperature and time, the gold atoms can diffuse into the Ni$_2$P shell and form gold single atoms well dispersed in the Ni$_2$P matrix and on surfaces. With further heating at elevated temperature up to 500 °C, all gold atoms can dissolve in Ni$_2$P and form a uniform solid-solution phase.

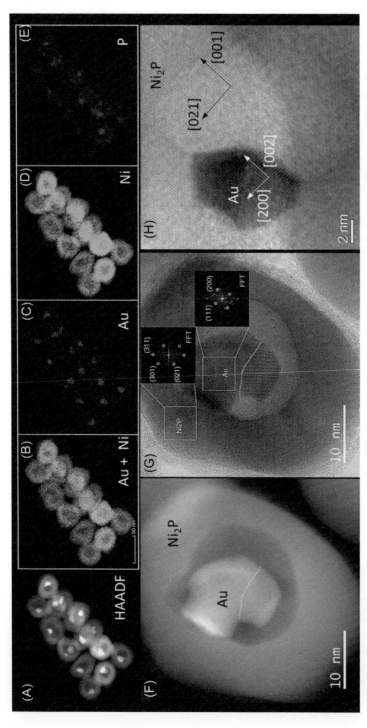

Fig. 17.3 (A) HAADF-STEM image of the fresh Au@Ni$_2$P yolk–shells; (B) overlaid Au and Ni EDS map; (C) Au map; (D) Ni map; (E) P map; (F) high-resolution HAADF image; and (G) high-resolution bright field STEM and FFT analysis of one Au@Ni$_2$P yolk-shell; and (H) atomic-scale analysis of the orientation relationship between the yolk and shell of another particle.

Adapted with permission from C. Cai, S. Han, Q. Wang, M. Gu, Direct observation of yolk–shell transforming to gold single atoms and clusters with superior oxygen evolution reaction efficiency. ACS Nano 13 (2019) 8865–8871. https://doi.org/10.1021/acsnano.9b02135. Copyright © 2019 American Chemical Society.

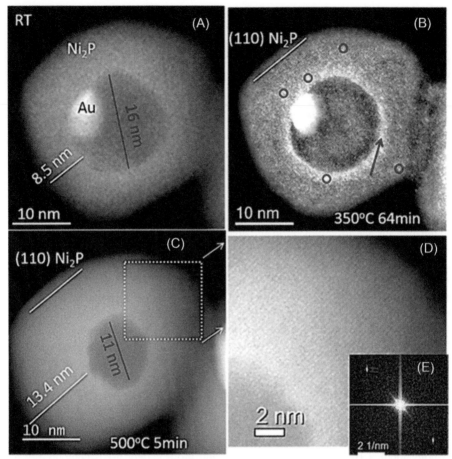

Fig. 17.4 (A) Fresh sample, (B) after heating at 350 °C for 64 min, and (C) after additional heating at 500 °C for 5 min; (D) atomic-scale image and FFT of the structure after the Au dissolved in the Ni_2P lattice at the square region in panel (C).
Adapted with permission from C. Cai, S. Han, Q. Wang, M. Gu, Direct observation of yolk–shell transforming to gold single atoms and clusters with superior oxygen evolution reaction efficiency. ACS Nano 13 (2019) 8865–8871. https://doi.org/10.1021/acsnano. 9b02135. Copyright © 2019 American Chemical Society.

17.3 1D nanowire growth characterization

1D nanostructures are commonly grown by vapor-liquid-solid (VLS) [13], vapor-solid [14], template-assisted electrochemical deposition [15], and solution growth [16]. The VLS mechanism, where metallic nanoparticles catalyze the anisotropic growth, offers more controllability in nanowire diameters, locations, and density and has attracted special interest [17]. However, the shape and the composition uniformity of the nano-structure are strongly associated with the growth conditions and the intrinsic crystallographic features, such as polarity and planar defects. Note that incorporating a metal particle in the nanowire generally degrades the nanowire's physical properties [18]. In this manner, the polarity-driven or defect-induced growth becomes more intriguing in the case where the 1D nanostructure growth is governed by a self-catalyzed mechanism (i.e., without catalysts).

17.3.1   Polarity determination in 1D ZnO nanostructures

Most of the II-VI and III-V compound semiconductors adopt either hexagonal wurtzite (WZ) or cubic zinc blende (ZB) structures. The lack of a center of symmetry in both crystal structures, that is, the crystal polarity, may induce spontaneous polarization in WZ or piezoelectric polarization in ZB, which strongly impacts the impurity incorporation, etch rate, defect formation, and surface roughness of the crystals. This impact becomes more significant when the size reduces to nanoscale, partially owing to the high surface-to-volume ratio and quantum confinement effects. In fact, the crystal polarity influences the spontaneous nucleation and growth of nanoarchitectures, and it is strongly correlated with the optoelectronic and transport properties of the nanowire devices [19]. Therefore, polarity determination in nanowires is essential for understanding the growth mechanism. With a low spatial resolution, the polarity of surfaces or layers can be determined by various methods including chemical or thermal etching, resonant X-ray diffraction (XRD), and high-energy ion channeling, hemispherically scanned X-ray photoelectron diffraction. In contrast, as a microscopic technique, TEM allows for investigating the polarity even with unit cell resolution, defining itself as a better choice for determining the polarity of nanowires. Various TEM-based methods are used to determine the polarity of layers or surfaces, including CBED, spectrum imaging by EELS and EDS, and HAADF and ABF performed in STEM mode. The relative merits and demerits in their capability of polarity determination depend highly on the crystals and the corresponding morphologies. Among them, CBED is a well-established and more widely used methodology to determine the polarity. Fundamentally, it uses the breakdown of Friedel's law caused by the electron dynamical diffraction to determine the polarity. CBED patterns vary depending upon different factors, such as incident direction and absorption of the electrons, sample thickness, accelerating voltage, and the observation temperature [20]. Here, we use 1D ZnO nanoneedles as an example to demonstrate the polarity determination through CBED. The growth mechanism will also be discussed based on the CBED results.

The ZnO nanoneedles were synthesized on a silicon substrate by chemical vapor deposition [21]. Briefly, mixed powders of zinc oxide, manganese oxide, and graphite were used as the precursor. The entire system was naturally cooled down to room temperature after the reaction. This slow cooling was believed to be crucial to achieve nanoneedle morphology. A low-magnification TEM image, as shown in Fig. 17.5A, revealed that the collected sample exhibited a needle-like morphology with two opposite tips. Fig. 17.5B illustrates the CBED patterns taken from the areas marked by left cycle in Fig. 17.5A . The corresponding simulated CBED patterns were shown in Fig. 17.5C. Similar operation can be performed on the right marked area in

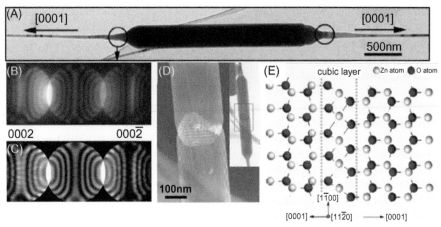

Fig. 17.5 (A) TEM image showing a double tip nanoneedle. (B) CBED patterns gained from the zone marked by the left cycle in (A). (C) The corresponding simulated CBED pattern with the CBED in (B). (D) Magnified dark-field image of the zone marked by a *rectangle* in the inserted TEM bright field image. (E) The schematic illustration of the stacking sequence of atoms and the growth direction of the perfect ZnO nanoneedles with a thin pseudo-cubic layer.

Fig. 17.5A. The nanoneedle is a c-axis-grown WZ structure described as a number of alternating planes composed of four-fold tetrahedral-coordinated O^{2-} and Zn^{2+} ions, stacked alternately along the c-axis. The (0001) and (000-1) surfaces are terminated with Zn and O, respectively, giving rise to positively and negatively charged polar surfaces. The intensity distributions in the (0002) and the (000-2) disks are significantly different, which is due to the noncentral symmetric structure of ZnO. Experimental and simulated CBED patterns are in excellent agreement, indicating that the two tips of nanoneedles grow out of the positively charged (0001)-Zn surfaces.

A close investigation of the double tip nanoneedles by TEM reveals that some of the double-tip nanoneedles have a kink in the middle as shown in the rectangular region of the insert in Fig. 17.5D, implying the possibility that two single-tip nanoneedles fused together and formed a double-tip nanoneedle during the growth. The corresponding enlarged DF image is presented in Fig. 17.5D, and the projection of interface is clearly seen, suggesting that the two single-tip nanoneedles were fused together to form a double-tip nanoneedle. Possible models revealing the growth of the single-tip and double-tip nanoneedles are given in Fig. 17.5E. Initially, the system reaches a critical value, and ZnO nanorods were formed by the effect of a temperature gradient and airflow. When the temperature started to descend during the cooling period, the nanorods still grew, but thinner than the original rod, forming sharp tips until the temperature reached a critical value, which was compared with synthesis condition of the nanorods formed under a fast cooling speed. The tips formed along the [0001] direction by the effect of the molecular polarity, resulting in the formation of many single-tip nanoneedles. Meanwhile, some single-tip nanoneedles encountered and aligned, growing into double-tip nanoneedles with kinks in the middle.

However, CBED is not very straightforward because it requires image simulations associated with an accurate measure of defocus and sample thickness. Moreover, a certain volume of crystals is needed to obtain accurate CBED patterns. This limits its capability of polarity determination in ultrathin nanostructures. In addition, its pattern is very sensitive to the defects in the illuminated area, possibly causing misleading interpretations when high-density stacking faults exist in the nanostructure. HAADF-STEM, also called Z-contrast imaging, provides a direct and robust imaging mode for determining the location of heavy atoms and thus can be used to assign the polarity of nanostructures composed by heavy elements. Contrary to CBED, no image simulation is involved in this technique, and more importantly, it is insensitive to sample thickness and defocus. However, it is limited in visualizing the light atoms, such as, Li, O, and N, which imposes difficulties in determining the polarity of nitride and oxide nanostructures. Fortunately, the recently established ABF-STEM technique enables the direct imaging of both lighter and heavier atomic columns simultaneously [4], providing a reliable and efficient approach in assigning polarities in compound semiconductor nanostructures that consist of lighter elements [19].

It should be noted that TEM is a microscopic method, providing information on individual nanowires but not the whole population. The polarity determination in nanostructures is more accurate when one combines TEM with other macroscopic techniques, for instance, resonant XRD.

17.3.2 Stacking fault-induced growth of ultrathin ZnO nanobelts

As one type of planar defects, stacking faults are formed by a change in the stacking sequence of atomic planes. They are commonly observed in nanowire and nanobelts and play important roles in determining the physical properties of the nanostructures. Theoretically, an appropriate number of stacking faults or nanoscale twins in metal nanowires may give rise to ultrahigh yield strengths and flow stresses in metals. However, the presence of stacking faults in functional semiconducting nanostructures generally results in reduced carrier mobility and thus degraded performance in corresponding nanodevices. Therefore, great effort has been dedicated to achieving stacking-fault-free nanostructures, such as III-V nanowires by molecular beam epitaxy [22,23]. From a growth perspective, the formation of stacking faults in nanostructures can stabilize the high-energy surfaces and influence a preferred growth direction. HRTEM imaging, along with image simulation, is in high demand to probe the nature of stacking faults, allowing for an increased understanding of the nanowire growth.

Ultrathin ZnO nanobelts were synthesized via a vapor transport technique, and no catalyst was introduced [24]. The as-synthesized ZnO nanobelt has a width of ~5.5 nm, evident in the HRTEM image taken along the $\langle 2\text{-}1\text{-}10 \rangle$ zone axis (Fig. 17.6A). The inset shows the corresponding SAED pattern, suggesting that the thin ZnO nanobelts grow along the $\langle 01\text{-}10 \rangle$ direction, with top/bottom $\{2\text{-}1\text{-}10\}$ surfaces and $\{0001\}$ side surfaces. Interestingly, the thin ZnO nanobelt exhibits stacking fault contrast along its $\{0002\}$ basal plane, as denoted by white arrows in Fig. 17.6A. An enlarged HRTEM image of the $\{0002\}$ stacking-fault plane is shown in Fig. 17.6B.

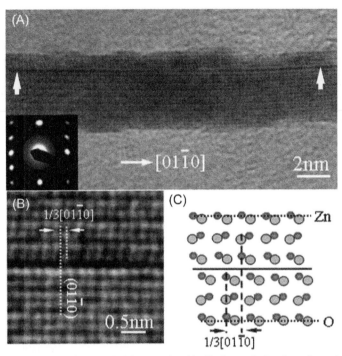

Fig. 17.6 (A) Representative HREM image of a thin ZnO nanobelt taken along the ⟨2-1-10⟩ zone axis. Inset in the left bottom corner is the corresponding SAED pattern. The *white arrows* denote the {0002} stacking fault. (B) Enlarged HREM image of the {0002} stacking fault plane. (C) Schematic drawing of the projection of the atomic structure along the ⟨2-1-10⟩ direction. The *solid line* denotes the stacking fault position.

The {01-10} lattice planes on both sides of the {0002} stacking fault share one-third ⟨01-10⟩ with each other, which is the projection of one-sixth ⟨2-203⟩ Frank partial dislocation on the {0002} basal plane in ZnO. It is believed that the terminating partial dislocations glide out of the thin nanobelts, leaving the {0002} stacking-fault plane penetrating through the entire length of the thin nanobelts.

In the case of the ionic compound ZnO, in order to maintain nearest-neighbor bonding across the stacking-fault plane, intrinsic and extrinsic stacking faults arrive at the same structure of the stacking-fault plane by a simple shear, as shown in the schematics of Fig. 17.6C. The stacking sequence is $A\alpha B\beta A\alpha B\beta A\alpha | B\gamma C\alpha B\gamma C\alpha B\gamma$, where the Roman and Greek letters denote Zn and O atoms in the {0002} double planes, respectively. It is apparent that the electronic environment across the stacking-fault plane is dramatically disturbed, which might be the reason for the formation of the dark contrast along the {0002} stacking-fault plane.

Crystallographically, the {0002} stacking-faults plane would result in a step with a translation of one-third ⟨01-10⟩ at each side of the {01-10} growth fronts. The steps might provide preferential deposition sites for the reactants Zn^{2+} and O^{2-} ions during the growth process of the thin ZnO nanobelt, which could result in fast axial growth of the thin ZnO nanobelt. Consequently, the stacking fault induced much faster axial growth along $\langle 01\text{-}10 \rangle$ than the lateral $\{0002\}$ growth, under the assumption that the reactant supplies are constant. In fact, constant supplies of the reactants Zn^{2+} and O^{2-} ions were reasonable since all of the synthesis parameters were carefully maintained without variations during the synthesis process. Thus, it is believed that the $\{0002\}$ stacking faults caused the formation of the thin ZnO nanobelt in our experiment.

17.3.3   Structure analysis of a superlattice nanowire by TEM: A case of SnO_2 $(ZnO:Sn)_n$ nanowire

A superlattice is a periodic (strict) or aperiodic (broad) structure that consists of alternate layers of two (or more) different materials or two phases of a single material. It can help maintain the stoichiometry of the materials and manipulate the carrier concentration down to the intrinsic level, making it a very versatile and powerful candidate in electronic and photonic applications. In the last decade, 1D superlattice nanomaterials have attracted significant attention for their potential quantum confinement effects on both circumferential and axial directions. Generally, superlattice nanowires can be prepared via two approaches: artificial growth and natural self-assembly. The former method applies to group III-V and IV materials, such as GaP/GaAs [25,26], InAs/InP [27], and Si/SiGe [28], by the alternate introduction of vapor phase reactants that react with the same catalyst. While the latter method mostly works for a specific group of homologous compounds $InMO_3$ $(ZnO)_n$ (M = In, Ga, Fe, Al, etc.), superlattice structures will form naturally when the compositional elements reach a specific ratio.

Compared with superlattice nanowires of group III-V and IV materials, the structure of homologous compounds $InMO_3(ZnO)_n$ is more complicated, consisting of slabs of WZ M-doped ZnO $(M/Zn_nO_{n+1}^+)$ separated by octahedrally coordinated InO_2^- (In–O) layers [29]. Because two segments have the same amount of positive and negative charges, the nanowire itself is charge-balanced. The In ions in the In–O layer occupy the octahedral sites, while the M and Zn ions in the M/Z-O layer take the tetrahedral or trigonal bipyramidal ones. These In–O layers induce a polarity inversion in the two adjacent ZnO layers forming an inverse domain boundary [30]. Zigzag-shaped contrast polarity inversion boundaries would form between two In–O layers in some $InMO_3(ZnO)_n$ nanowires with a larger n $(n > 6)$ [31,32] to relax the strain energy. Although $InMO_3(ZnO)_n$ has been extensively studied in the forms of bulk and thin film structures, it was not until 2004 that $In_2O_3/(ZnO)_n$ superlattice nanowires were synthesized with diverse compositions $(n = 5, 7, 18)$ [33], followed by the synthesis of $In_2O_3(ZnO)_4$ and $In_2O_3(ZnO)_5$ nanowires, doped with 6%–8% Sn [34]. In addition to $In_2O_3(ZnO)_n$ nanowires, $InGaO_3(ZnO)_n$ [35,36], $InAlO_3(ZnO)_n$ [37], and $InFeO_3(ZnO)_n$ [31] superlattice nanowires were successively reported later on.

A new superlattice nanowire was found when we tried to substitute In with Sn during the growth, which has never been observed in bulk form. It has similar features of layered structures, but the distribution of Sn atoms among the nanowires is not clear.

Fig. 17.7A is a typical TEM image of the new nanowire mentioned above that shows modulated contrast along its entire length with no obvious amorphous shell attached. The inset is the corresponding SAED pattern taken from the [100] zone axis. Notice that, in addition to the main diffraction spots of the normal ZnO phase, a series of streaks can be seen along the [001] direction, which is the typical sign of planar defects. The EDS measurement (Fig. 17.7B) indicates the presence of Sn in the nanowire, and the average Sn content in the nanowire, defined as the atomic ratio of Sn/(Zn +Sn), is identified to be about 23.8%. Fig. 17.7C is a high-resolution TEM image

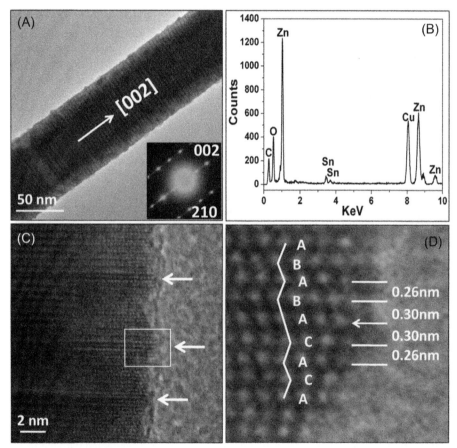

Fig. 17.7 TEM image (A) of the $SnO_2(ZnO:Sn)_n$ nanowire with [002] growth direction. Inset is the [100] electron diffraction pattern. (B) EDS shows the presence of Sn. (C and D) HRTEM images reveal that modulated contrast comes from those individual atomic planes which separate the nanowire into slabs.

taken at the edge of a nanowire. It turns out that the black contrast under low magnification comes from many individual atomic layers as illustrated by white arrows, separating the body of nanowires into a series of slabs. Evidently, these isolated layers are also the origin of the streaks in the diffraction pattern. The highlighted area in the white square is enlarged in Fig. 17.7D. It is found that the atomic layers' stacking sequence switches from *BABA* to *ACAC* over the marked layer.

Moreover, the interplanar distances within the slab are measured as 0.26 nm, corresponding to the characteristic *d*-spacing of the ZnO (002) plane, while the distance between the arrow-marked layer and its neighboring layers is 0.30 nm, which is larger than 0.26 nm. It is known that the radius of the ion is correlated with its coordination number (CN). Sn^{4+} ions have two possible radii, 69 pm for CN = 4 and 83 pm for CN = 6 [38]. Because the Zn^{2+} ions have a radius of 74 pm, the bigger Sn^{4+} ion (83 pm) with octahedral coordination should result in a thicker atomic layer. Thus, it is conceivable that these isolated atomic layers are composed of an octahedral Sn–O plane. However, it is challenging to discriminate Sn atoms from Zn atoms under conventional TEM investigation.

Z-contrast imaging under the STEM-HAADF mode is very sensitive to atomic weight, making it effective for discriminating Sn and Zn atoms. The intensity of the atomic image is approximately proportional to the mean square of the atomic number of the constituent atoms (Z^2). The position of Sn atomic columns should be able to be unambiguously determined because Sn ($Z = 50$) is much heavier than Zn ($Z = 30$). Thus, an advanced *Cs*-corrected JEM-2100F scanning TEM was applied to further study the fine structure of this nanowire. Fig. 17.8A is a Z-contrast image of a typical superlattice nanowire, clearly showing the presence of Sn atomic layers (bright lines). Sn columns are well resolved with obviously higher contrast compared with Zn columns, and individual planes of Sn atoms are clearly separated by ZnO slabs. The numbers of Zn–O layers inside each slab are in the range of 11–18, and the average is 12. The stacking sequence of Zn–O slabs exhibits no periodicity, and thus as a broad definition, it can be considered as an aperiodic superlattice nanowire. If we assume that there are no Sn atoms in ZnO slabs, then the ratio of Sn/(Sn + Zn) should be about 1:13, which is significantly smaller than the Sn/(Sn + Zn) ratio of 23.8% measured by EDS. Therefore, a certain amount of Sn should exist in ZnO slabs, which was confirmed further by nanoscale EDS (not shown here). Fig. 17.8B shows a HAADF image of the nanowire with a higher magnification, from which zigzag-shaped contrast has been clearly observed. Because it is not energetically favorable for Sn atoms in ZnO slabs to locate near the SnO_2 octahedral layers, the wave-like contrast caused by doped Sn atoms does not get touched with the Sn–O layers, which is different from $InMO_3(ZnO)_n$ systems where the top of the zigzag contrast is connected with the In–O layer [39].

Although HAADF imaging can distinguish Sn and Zn atoms, it does not offer any information about the O atom distribution, which is also crucial to determine the coordination of Zn atoms in the slab and Sn atoms in the interface layer. The structure model of the superlattice oxide has been well established in previous works [29], and the distribution of In and Zn/M atoms has been confirmed with Z-contrast imaging. Direct visualization of O atoms in the structure has not been provided due to the

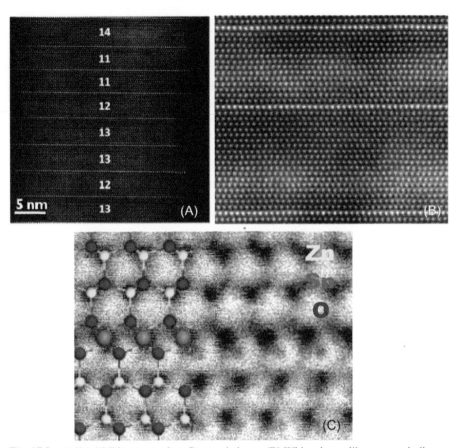

Fig. 17.8 (A) HAADF image resolves Sn atomic layer. (B) White zigzag-like contrast indicates a certain amount Sn atom is doped into ZnO slabs. (C) ABF image shows position of O column, presenting the detailed configuration of O around Zn (tetrahedral) and Sn (octahedral). Adapted with permission from B. Cao, T. Shi, S. Zheng, Y.H. Ikuhara, W. Zhou, D. Wood, M. Al-Jassim, Y. Yan, New polytypoid SnO2(ZnO:Sn)m nanowire: characterization and calculation of its electronic structure. J. Phys. Chem. C 116 (2012) 5009–5013. https://doi.org/10.1021/jp211135r. Copyright © 2012 American Chemical Society.

limitation of HAADF in detecting light elements. Fig. 17.8C shows the ABF image collected with detection angles between 7 and 33 mrad, in which O atoms (gray spots) are visualized around Sn (bigger dark spots) and Zn atoms (smaller dark spots) simultaneously. Each Sn atom is surrounded by four O atoms, which is actually a 2D projection of an Sn-centered octahedron (6 O atoms at the vertex). Analogously, each Zn atom is surrounded by three O atoms, providing a 2D projection of the Zn-centered tetrahedron (4 O atoms at the vertices). The proposed model (yellow, Zn; green, Sn; red, O) is overlaid on the left part of the ABF image, showing a perfect match.

17.4   TEM analysis of 1D nanoheterostructure

Based on the single-component 1D nanostructure, one can fabricate simple nanodevices, such as resistors, transistors, and optical waveguides, to extract fundamental physical parameters of the nanostructures or attain basic device functionality in applications [40]. Nonetheless, most semiconductor devices require junctions to manipulate the transport of electrons, phonons, and photons for exploring broader applications, which inspires numerous works on growth, assembly, and property studies of 1D nanoheterostructures. Here, 1D heterostructures can be defined as structures that consist of two or more geometrically 1D components. These junctions, including homojunctions, type-I and type-II heterojunctions, and Schottky junctions, in nanoheterostructures are generally present at the interfaces between different 1D components. These interfaces can allow or block the carrier transport, tolerate the strains and defects, and create new band structures, eventually determining the material's properties and device performance. From a material perspective, acquiring structure and composition information across the interfaces is the primary interest in characterizing 1D nanoheterostructures. On the other hand, surface modification of a single-component 1D nanostructure is an efficient way to enhance the device properties, as well as providing an ideal platform for investigating the interaction between materials with different dimensionalities. For example, loading nanocrystals (0D) on wide-band gap oxide nanowires (1D) facilitates carrier injection for photovoltaics [41]. The size distribution and faceting of the nanocrystals, along with the interface quality between the nanocrystals and nanowire, can significantly impact the final device behavior.

TEM has played an indispensable role in the structure analysis of 1D nanoheterostructures, such as identifying the interfaces, mapping the strain and composition, and imaging the extended defects and dopants. To simplify the discussion, here we divide 1D nanoheterostructures into three categories: axially heterostructured nanowire, radial core-shell nanowire, and surface-decorated 1D nanostructures. For each of these categories, we will present one or a few examples to elucidate the TEM analysis.

17.4.1   Axially heterostructured nanowires

Compared to radial heterojunction growth, synthesis of axial heterojunction is relatively challenging mainly due to more stringent requirements of crystallographic relationships and reactant atmospheres during growth. Thus, understanding crystallographic relationships between two components is quite crucial to predict or design a new axial heterojunction growth. Here is an example of growing a new ternary/binary oxide axial nanowire heterojunction based on the detailed structure analysis between Zn_2GeO_4 nanowire and a small segment of ZnO residue formed during the growth of Zn_2GeO_4 nanowire array. Controllable growth of ZnO with various lengths onto the Zn_2GeO_4 nanowire was realized based on the proposed epitaxial growth relationship and growth mechanism.

The ternary Zn_2GeO_4 nanowires were synthesized using a typical chemical vapor deposition method with a mixture of ZnO, GeO_2, and graphite powder as the precursor [42]. During the conventional TEM examination of as-grown ternary Zn_2GeO_4 nanowires, it is interesting to notice that a tiny unknown segment around 20 nm (Fig. 17.9A) is always observed between the Zn_2GeO_4 nanowire and Au catalyst, which is seldom seen in other VLS-grown ternary nanowires. The high-resolution TEM image of the interface area is displayed in Fig. 17.9B, revealing that the interplanar d-spacings on two sides are 0.41 and 0.28 nm, corresponding to Zn_2GeO_4 (300) plane and ZnO (100) plane, respectively. Generally, SAED at a large tilting range is a good way for phase identification. However, for a nanoparticle around 20 nm, it is very hard to obtain a diffraction pattern. Therefore, FFTs are performed at locations of C, D, and E, as shown in Fig. 17.9C–E, respectively. The FFT pattern of the unknown segment has the same typical hexagonal pattern as ZnO in the [001] zone axis. So this mysterious section could very likely be ZnO. The FFT pattern collected at the interface is a superimposed pattern from both Zn_2GeO_4 and ZnO phases, suggesting a crystallographic relationship of Zn_2GeO_4[010]//ZnO[001].

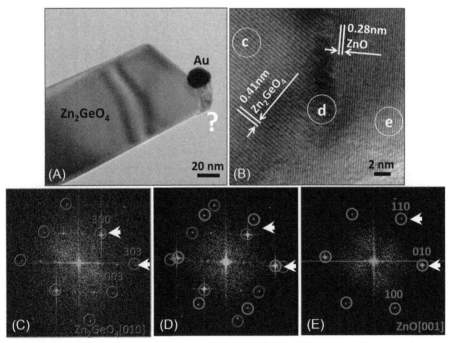

Fig. 17.9 (A and B) Low-magnification and high-resolution TEM image of a junction area. (c–e) FFT diffraction patterns of marked locations (C, D, and E) in (B), respectively, indicating Zn_2GeO_4-[010]//ZnO-[001].

Adapted with permission from B. Cao, J. Chen, R. Huang, Y.H. Ikuhara, T. Hirayama, W. Zhou, Axial growth of Zn_2GeO_4/ZnO nanowire heterojunction using chemical vapor deposition. J. Cryst. Growth 316 (2011) 46–50. https://doi.org/10.1016/J.JCRYSGRO.2010.12.060.

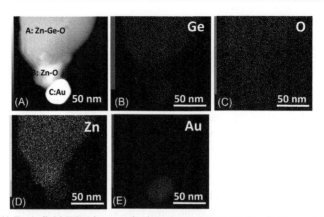

Fig. 17.10 (A) Dark-field TEM image of a junction area. (B–E) Ge, O, Zn, and Au elemental mapping, showing unknown section is composed of Zn and O.
Adapted with permission from B. Cao, J. Chen, R. Huang, Y.H. Ikuhara, T. Hirayama, W. Zhou, Axial growth of Zn_2GeO_4/ZnO nanowire heterojunction using chemical vapor deposition. J. Cryst. Growth 316 (2011) 46–50. https://doi.org/10.1016/J.JCRYSGRO.2010.12.060. Copyright © 2011 American Chemical Society.

Only one FFT pattern is insufficient for confirming the ZnO phase, but compositional information will help validate the presumption of ZnO. The EDS mappings, using a probe size of 0.1 nm, were conducted to further study the junction part, as seen in Fig. 17.10B–E, from which we can see that the region of unknown particles is majorly composed of Zn and O. Based on both the structural and compositional information, it is concluded that this particle residue is the ZnO phase. Thus, at the tip of Zn_2GeO_4 nanowire, a Zn_2GeO_4/ZnO heterojunction is formed. Because Zn_2GeO_4 and ZnO have absolutely different crystal structures, this heterojunction strongly suggests possible epitaxial growth relationships between these two phases. However, due to the small size of the ZnO section, the FFT pattern achieved is limited to a small range. Based on the rotation angle between two FFT patterns from Zn_2GeO_4 and ZnO phases, an interface model is built using Mac Tempas to simulate the superimposed electron diffraction pattern (EDP), as presented in Fig. 17.11. This gives us a bigger view of the overlapped FFT of the two phases. The FFT pattern in Fig. 17.11 is only within the range of the blue circle, beyond which more orientation relationships are indicated by green circles and dashed lines. It is found that diffraction spots of Zn_2GeO_4 $(3m, 0, 0)$ and $(0,0,3n)$ are almost aligned with diffraction spots of ZnO $(-h,h,0)$ and $(k,k,0)$, respectively, indicating orientation relationship of Zn_2GeO_4 (300)//ZnO(-110) and Zn_2GeO_4 (003)//ZnO (110), where m, n, h, and k are integers. Moreover, the diffraction spots of Zn_2GeO_4 (9, 0, 0) and (0, 0, 6) are found nearly overlapped with those of ZnO(-2,2,0) and (1,1,0), respectively. This implies good lattice matching along with these two sets of planes with small mismatches of 2.7% and 2.3%; it also confirms the epitaxial growth of the heterojunction.

Based on the close relationships between these two phases, possible epitaxial growth may lead to a longer heterojunction if more Zn vapor is provided. To grow

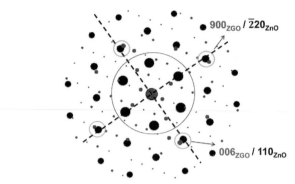

Fig. 17.11 Simulated superimposed diffraction patterns from Zn_2GeO_4/ZnO interface imply more orientation relationships beyond experimental FFT pattern (within the *blue circle*).

a longer ZnO segment, a second synthesis was conducted by extending growth time from 1 to 2 h without changing other experimental conditions. It was observed that the ZnO segment grew much longer (500 nm), forming the typical Zn_2GeO_4/ZnO heterojunction as shown in Fig. 17.12A and B (enlarged). The SAED patterns in Fig. 17.12C–E were recorded from the ZnO junction area and Zn_2GeO_4 region (denoted by C, D, and E in Fig. 17.12B, respectively). Two sets of diffraction patterns, highlighted by rectangles and rhombohedra in Fig. 17.12D, were observed at the junction region D, which correspond to the [001] zone axis of ZnO (C) and [010] zone axis of Zn_2GeO_4 (E) in Fig. 17.12C and E, respectively, indicating that Zn_2GeO_4 [010] is parallel to ZnO [001]. Fig. 17.12D matches the simulated diffraction pattern in Fig. 17.11 very well.

From the above experimental results, we proposed a possible VLS growth mechanism. First, the Zn and Ge vapors from precursors dissolve into the melted gold catalyst downstream and form an Au-Zn-Ge droplet. With increasing the content of Zn and Ge, a Zn_2GeO_4 nanowire grows out from the oversaturated Au-Zn-Ge droplet with the assistance of residual oxygen in the furnace tube. Due to the higher proportion of ZnO in the source material (Zn:Ge > 2:1), the GeO_2 source runs out first, resulting in the ending of the Ge vapor supply during the growth of ternary Zn_2GeO_4 nanowire. While excess ZnO source material continues to provide Zn vapor downstream, Au–Zn oversaturated droplets form, and the ZnO nanowires start to nucleate. By changing growth time, the length of ZnO segment can be manipulated along axial direction of the Zn_2GeO_4 nanowire.

17.4.2   Coaxial core-shell nanowires

Compared to axially heterostructured nanowires, coaxial core-shell nanowires, namely radial heterostructured nanowires, promise more exotic applications because of their unique geometry-induced properties. First, the shell can act as a sidewall passivation layer to tune the emission of the core nanowire. This is a well-known strategy for improving the luminescence for core-shell QDs [43], and it also holds true for the 1D quantum core-shell nanowire system. Second, alignment of energy states at the

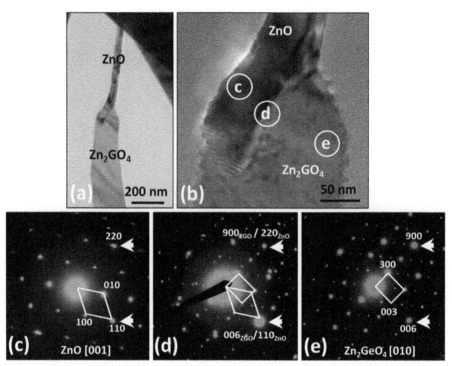

Fig. 17.12 (A) Typical morphology of a Zn_2GeO_4/ZnO axial heterojunction in the second synthesis. (B) Enlarged heterojunction area where the ZnO section, interface region, and Zn_2GeO_4 section is denoted as (C), (D), and (E), respectively. (C) SAED pattern of ZnO [001] (region C). (D) Superimposed SAED pattern from interface region D, where ZGO denotes Zn_2GeO_4. (E) SAED pattern of Zn_2GeO_4 [010] (region E).

Adapted with permission from B. Cao, J. Chen, R. Huang, Y.H. Ikuhara, T. Hirayama, W. Zhou, Axial growth of Zn_2GeO_4/ZnO nanowire heterojunction using chemical vapor deposition. J. Cryst. Growth 316 (2011) 46–50. https://doi.org/10.1016/J.JCRYSGRO.2010.12.060. Copyright © 2011 Elsevier.

interface between the two components (core and shell) generally facilitates an efficient charge separation by isolating charge carriers from the impurities and defects on both surfaces and interfaces. This, along with the extended light absorption profile through type-II interfacial transition, makes the core-shell nanowires useful for solar-energy applications such as solar cells and hydrogen generation via photoelectrochemical water splitting and thermoelectrics [44]. In addition, a nanowire surrounded by a radial shell can function as a wrap-gate structure, which provides better controllability over the conducting channel [45]. Finally, this unique coaxial geometry allows for a significant relaxation of lattice mismatching strains that would otherwise lead to incoherent growth in the planar structure [46].

Despite the possibility for growing coaxial nanowires by a one-step technique [47], one can usually fabricate core-shell nanowires by depositing a material (shell) on the template nanowire (core). Multiple depositions in a modulated growth condition may even produce more complex radial heterostructures [48]. To pursue a high-performance device, of particular importance in the fabrication process is achieving an abrupt interface between the core and shell. Nevertheless, interfacial diffusion and incoherent growth can block the way. The former can result in the occurrence of a transition layer that has to be minimized by modulating the growth conditions. The latter, mostly arising from lattice mismatching between the two materials, can also be related to the side facets of the template nanowire, and it likely induces an inhomogeneous shell thickness.

The lattice mismatch strain in planar heterostructures results in a critical thickness and composition, thus limiting the exploration of high-performance planar devices. The nanowire geometry provides new strain relaxation mechanisms that facilitate heteroepitaxy at unusually high lattice mismatch between dissimilar materials, enabling strain-induced band structure engineering [49].

The ZnO/ZnSe core-shell nanowires were synthesized by using a two-step technique, combining chemical vapor deposition of ZnO and pulsed laser deposition (PLD) of ZnSe [50]. Fig. 17.13A shows a low magnification TEM image of a ZnO/ZnSe core/shell nanowire with a rough external surface, but a sharp interface between the core and shell. The ZnSe shell grew directly in the radial direction from the surface of ZnO nanowire with a thickness of $\sim$5 to 8 nm. A HRTEM image taken from the interface region between the ZnO core and ZnSe shell (specified in the rectangular area in Fig. 17.13A), as shown in Fig. 17.13B, reveals that ZnO and ZnSe are of WZ and ZB crystal structures, respectively. At the interface, an epitaxial growth of ZnSe from ZnO core is observed. The interface is smooth, and no transitional layer is found in between. The axis of the ZnO nanowire is identified to be the WZ c-axis. The epitaxial growth relationship of the WZ ZnO core and ZB ZnSe shell has been identified as [0001]ZnO//[001] ZnSe and (2-1-10)ZnO//(011)ZnSe. Inset in Fig. 17.13B is the FFT patterns of ZnO core and ZnSe shell, indexed as WZ and ZB structure with zone axis of [2-1-10] and [011], respectively, which also confirms the above epitaxial growth relationship. Defects were also observed in the interface along the c-axis of ZnO due to the large lattice mismatch (8.8%) along the nanowire axis between $c = 0.521$ nm for ZnO and $a = 0.567$ nm for ZnSe. Note that the lattice mismatch along the nanowire axis would be much larger ($\sim$25%) if ZnO and ZnSe were both in the WZ phase. The spatial distributions of the atomic composition across the ZnO/ZnSe core/shell nanowire were obtained by a nanoprobe EDS line-scan analysis (marked by a line in Fig. 17.13A), showing that the ZnO nanowire was homogeneously coated, as shown in Fig. 17.13C. It is also found that epitaxial growth only occurred during high-temperature deposition, and no epitaxial growth was found at room temperature deposition, indicating that the epitaxial growth between ZnO and ZnSe demands favorable thermodynamics as well as kinetic conditions.

17.5 TEM analysis of 2D nanostructures

Since the discovery of graphene, we have witnessed the rapid development and enormous advancement of a new material family—2D materials. One of the unique features of 2D materials is that they are single or few atomic layers thin in 1D, thus offer them dramatically different properties, even compared to other nanomaterials. With

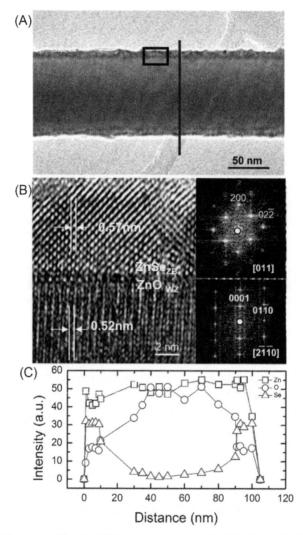

Fig. 17.13 (A) Low-magnification TEM micrograph of a ZnO/ZnSe core/shell nanowire. A thin layer of ZnSe was coated on the ZnO nanowire. (B) High-resolution TEM image of the interface of the core/shell heterostructure, enlarged from the *rectangular area* in (A), showing the epitaxial growth relationship of the ZnO wurtzite core and ZnSe zinc blende shell. *Inset* shows fast Fourier transfer (FFT) patterns of rectangular areas in (B). (C) EDS nanoprobe linescan on elements Zn, Se, and O, across the ZnO/ZnSe core/shell nanowire, indicated by a line shown in (A).

Adapted with permission from K. Wang, J. Chen, W. Zhou, Y. Zhang, Y. Yan, J. Pern, A. Mascarenhas, Direct growth of highly mismatched type II ZnO/ZnSe core/shell nanowire arrays on transparent conductingoxide substrates for solar cell applications. Adv. Mater. 20 (2008) 3248–3253. https://doi.org/10.1002/adma.200800145. Copyright © 2008 WILEY-VCH Verlag GmbH & Co. KGaA, Weinheim.

the intensive study, the 2D nanostructures have been expanded to a large number of materials, from graphene to its elemental analogues silicene, germanene, stanene, phosphorene, arsenene, and antimonene; from group IVA–VIA compounds, such as SnO, SnS_2, and $SnSe_2$, to transition metal dichalcogenides (TMDC), such as MoS_2, WS_2, $MoSe_2$, WSe_2, $MoTe_2$ [51,52]. Nevertheless, to study their properties, TEM is still an irreplaceable tool.

17.5.1 Atomic resolution imaging of graphene

In 2004, single-layer graphene was first discovered by Andre Geim and Kostya Novoselov at The University of Manchester and soon became the superstar of the material world [53]. The TEM has demonstrated remarkable flexibility in character-izations of graphene at the atomic level. Here, we will show some examples of TEM studies on graphene. For a more comprehensive review, the interested reader is referred to Huang et al. [54] and Robertson and Warner [55] for more details.

It has been revealed that defects in graphene can contribute strongly to the physical properties, such as electrical conductivity, mechanical strength, chemical reactivity, and thermal conductivity. Therefore, characterizing the structure of defects in graphene is practically interested in understanding and controlling its large-scale properties. Robertson et al. used focused electron beam to create defects in graphene and studied their behavior by TEM [56]. As shown in Fig. 17.14A, the pristine graphene sheet has perfect hexagonal structures. Fig. 17.14B was taken directly after irradiation, indicating a single divacancy being created. The TEM images of three pristine regions after 30 s exposures (Fig. 17.14C–E) show different behaviors of defect evolution. In the case of Fig. 17.14C, the divacancy structures observed have two adjacent atoms removed from the lattice. For Fig. 17.14D, the divacancy has undergone a single Stone-Wales transformation. In Fig. 17.14E, two divacancy defects are joined, forming an extended defect along the armchair lattice direction.

Besides imaging the crystal structure, TEM can also be used to investigate the elec-tronic properties of graphene in the nanoscale directly [57]. Fig. 17.15A shows a typ-ical ADF image of the edge region of a single graphene layer. The possible carbon atom positions derived from the local intensity maxima are marked by yellow circles in Fig. 17.15B. Fig. 17.15D shows three characteristic carbon K-edge fine structures extracted using sequential EELS with probe-scanning, in which their corresponding atomic positions are shown in Fig. 17.15C. All of the three spectra exhibit the typical features of sp^2 coordinated carbon atoms, for example, the sharp π^* peak around 286 eV and the exciton peak of σ^* at 292 eV. By comparison of the EELS spectrum obtained from carbon atoms with different positions, it is clear that the edge atom located at the border of the hexagonal network with two-coordination (spectrum in blue) has an extra peak around 282.6 ± 0.2 eV (labeled D in Fig. 17.1D), with the π^* peak having reduced intensity. Similarly, the edge atom which is single-bonded to its neighbor (spectrum in red) also shows weaker π^* peak and broadened σ^* peak, with the extra peak occurs at a different energy position of 283.6 ± 0.2 eV (labeled S in Fig. 17.1D).

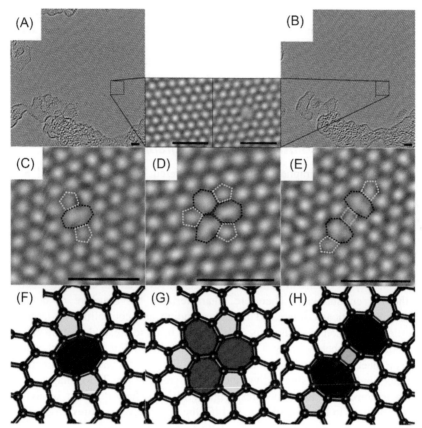

Fig. 17.14 (A) TEM image of a pristine graphene sheet before 30 s exposure to a focused electron beam. (B) Divacancy formed in exposure area directly after irradiation. *Insets* show magnified images of highlighted areas in (A) and (B). TEM images of three different 30 s exposures, resulting in (C) a divacancy, (D) a divacancy having undergone a single Stone-Wales bond rotation, and (E) two linked divacancies along the armchair direction. (F–H) Atomic models corresponding to the TEM images in (C–E), respectively. The color scheme (*dashed line annotations and model filling color*) denotes the number of carbons in the respective carbon ring, such that *green, yellow, blue, and dark blue* represent 4-, 5-, 7-, and 8-membered carbon rings, respectively. Scale bars represent 1 nm.

Adapted with permission from A.W. Robertson, C.S. Allen, Y.A. Wu, K. He, J. Olivier, J. Neethling, A.I. Kirkland, J.H. Warner, Spatial control of defect creation in graphene at the nanoscale. Nat. Commun. 3 (2012) 1144. https://doi.org/10.1038/ncomms2141. Copyright © 2012 Macmillan Publishers Limited.

17.5.2 TEM characterization of 2D TMDC materials and heterostructures

Unlike graphene, which lacks an electronic band gap, TMDCs are semiconductor materials. The typical type of TMDC is MX_2, where M is a transition metal atom (such as Mo or W) and X is a chalcogen atom (such as S, Se, or Te). Because of the

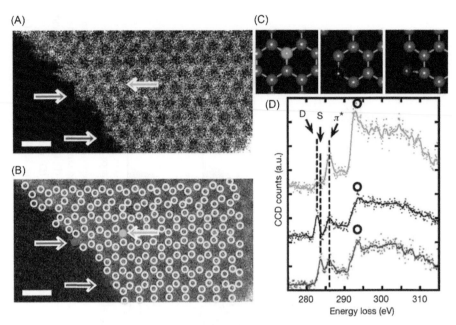

Fig. 17.15 (A) ADF image of single graphene layer at the edge region. No image-processing has been done. Atomic positions are marked by *circles* in a smoothed image (B). Scale bars, 0.5 nm. (D), ELNES of carbon K (1s) spectra taken at the *color-coded atoms* indicated in (B). *Green, blue, and red spectra* correspond to the normal sp^2 carbon atom, a double-coordinated atom and a single-coordinated atom, respectively. These different states of atomic coordination are marked by *colored arrows* in (A) and (B) and illustrated in (C) CCD, charge-coupled device. Adapted with permission from K. Suenaga, M. Koshino, Atom-by-atom spectroscopy at graphene edge. Nature 468 (2010) 1088–1090. https://doi.org/10.1038/nature09664. Copyright © 2010 Macmillan Publishers Limited.

combination of atomic-scale thickness, direct band gap, strong spin-orbit coupling, and favorable electronic and mechanical properties, 2D TMDCs have received tremendous attention recently and are promising in various applications, such as electronics, spintronics, optoelectronics, and energy harvesting [52]. Here, we will discuss some of the recent studies of 2D TMDC using TEM.

As we know, band structure engineering is of great importance to tune the properties of semiconductor materials, including TMDC. One of the effective strategies to precisely control the band structure for desired applications is by element doping. In this regard, it is vital to study the atomic structure and chemical composition in order to control the synthesis process better. As an example, monolayer $MoSe_2$ and $Mo_{1-x}W_xSe_2$ were synthesized by a low-pressure CVD method [58]. As shown in the atomic resolution ADF image of Fig. 17.16A, a honeycomb-structured $MoSe_2$ monolayer was obtained. The simulated ADF image was also given in Fig. 17.16B for comparison. The lower row in Fig. 17.16C shows the intensity line profile along the green line. Because the image intensity depends on the total atomic number of the atom occupying each site,

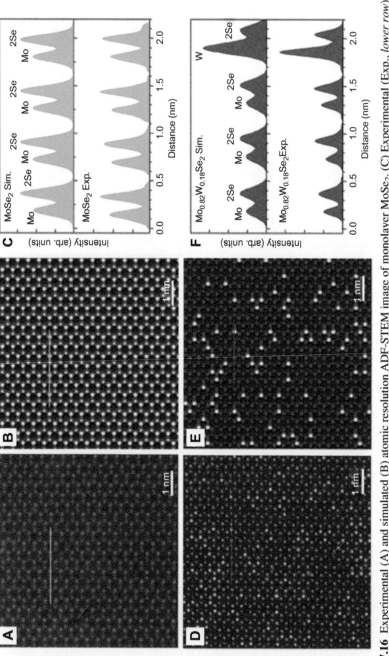

Fig. 17.16 Experimental (A) and simulated (B) atomic resolution ADF-STEM image of monolayer MoSe₂. (C) Experimental (Exp., *lower row*) and simulated (Sim., *upper row*) intensity line profiles along the solid green line in (A) and (B). Experimental (D) and simulated (E) atomic resolution ADF-STEM image of monolayer Mo₀.₈₂W₀.₁₈Se₂. The brightest spots correspond to sites where W substitutes Mo. (F) Experimental (*lower row*) and simulated (*upper row*) intensity line profiles along solid blue line in the (D) and (E). The *red arrows* in (A) and (D) indicate Se vacancies. Adapted with permission from X. Li, M.-W. Lin, L. Basile, S.M. Hus, A.A. Puretzky, J. Lee, Y.-C. Kuo, L.-Y. Chang, K. Wang, J.C. Idrobo, A.-P. Li, C.-H. Chen, C.M. Rouleau, D.B. Geohegan, K. Xiao, Isoelectronic tungsten doping in monolayer MoSe 2 for carrier type modulation. Adv. Mater. 28 (2016) 8240–8247. https://doi.org/10.1002/adma.201601991. Copyright © 2016 WILEY-VCH Verlag GmbH & Co. KGaA, Weinheim.

it is revealed that the brighter sites correspond to the atomic columns. Fig. 17.16D and E show the experimental and simulated ADF images for a $Mo_{0.82}W_{0.18}Se_2$ monolayer, which exhibit a similar honeycomb atomic structure with a $MoSe_2$ monolayer. In contrast, some of the Mo sites show brighter contrast than others, as indicated by the intensity line profile in Fig. 17.16F. Those sites correspond to doped W atoms. Based on the ADF images, it is confirmed that the W atoms are substituted for Mo atoms in the honeycomb lattice and no interstitial atoms can be found.

17.6   TEM characterization of 1D nanostructures for nanodevice applications

With all the results we described before, it is no doubt that TEM is a powerful and indispensable tool in nanomaterial research. One of the ultimate goals for material research is to apply the newly discovered materials to certain applications. To do that, we must have a thorough understanding of the fundamental physical and chemical properties of the materials and their relations to the material's structures. The detailed information can provide us useful guidelines for further material design and optimization, and in turn, improve the final device performance. In this section, we will give some examples of how TEM techniques can help in the study of representative nanodevice applications.

17.6.1   Structural characterization of polycrystalline SnO_2-coated ZnO nanowires with Pd decoration for chemical sensors

Polycrystalline SnO_2-coated ZnO nanowires with Pd decoration have been proven as high-performance nanomaterials that can detect low concentrations of industrial gases (i.e., H_2S, NO_2, and NH_3) at room temperature. A polycrystalline coating is preferred for gas sensor applications. First, the polycrystalline film has a larger surface area to capture gas molecules. Second, the energy barriers between the nanocrystals can be tuned by surface gas adsorption, which induces device conductivity changes [59]. In this study, the Pd decoration was used to further enhance the sensitivity by helping the dissociation of oxygen molecules during the transduction process [60].

A typical fabrication process for coating SnO_2 nanocrystals and Pd-nanoparticles on ZnO nanowires is shown in Fig. 17.17A. The ZnO nanowire arrays were first grown by a hydrothermal method [61] or chemical vapor deposition [50]. Physical vapor deposition methods, such as PLD or RF sputtering, were used to deposit the SnO_2 shell layer on the ZnO nanowire surfaces. In the last step, well-controlled DC sputtering was employed to apply isolated Pd nanoparticles on the SnO_2 surface. The electrical measurements [62] confirmed that the ZnO nanowires play the role of structural templates, and the electrical current mostly flowed through the polycrystalline SnO_2 shell layers. To understand the structures of the SnO_2 coating, a basic TEM technique was used to separate the SnO_2 phase from the ZnO nanowire. In Fig. 17.17C, a high-contrast BF image was formed using the direct electron beam that was selected by the objective aperture. Although we can

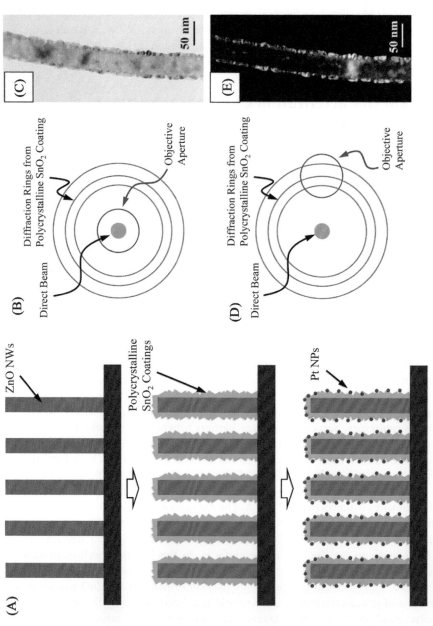

Fig. 17.17 (A) Schematic diagrams of nanowire cross sections showing the growth processes of polycrystalline Pd/SnO₂ nanowire using a ZnO-nanowire template. (C) and (E) Conventional BF and DF images of a Pd/SnO₂-coated ZnO nanowire. The diagrams (B) and (D) show the selection of electron beam for forming BF and DF images, respectively. Diffraction from single-crystal ZnO is not shown in the diagrams.

identify the particulate nature of the surface coating in the BF image, we cannot fully separate the shell layer coating from the ZnO core. To highlight the SnO_2 phase, a DF image (Fig. 17.17E) was formed using a portion of the Bragg diffracted beam. In such a DF image, only the SnO_2 phase showed the bright contrast, which clearly highlighted the polycrystalline SnO_2 coating on the nanowire. Although DF imaging is a very simple method, it is very effective in characterizing this type of core/shell structures.

However, it is difficult to identify the Pd decorations using conventional imaging techniques in TEM because of the much smaller size of Pd nanoparticles ($\sim$2–3 nm). The HRTEM image can provide the lattice fringes of the whole structure, including the single-crystal core and polycrystalline shell, as shown in Fig. 17.18A, but it is difficult to separate Pd from SnO_2 nanoparticles. Fortunately, Pd has a much higher atomic number compared to Sn or Zn, which can be highlighted by Z-contrast imaging. Z-contrast images can be formed by HAADF detector under the STEM mode. Using the forward-scattered electron beam at high angle ($>$3 degrees) can avoid the signal of Bragg scattering from crystalline specimens and maintain the contrast signal associated with the atomic number [20]. One important advantage of the HAADF image is that, compared with the phase contrast image, the contrast is generally unaffected by small changes in objective lens defocusing and specimen thickness [20]. As shown in Fig. 17.18B, the HAADF-STEM image of the nanowire has high brightness of Pd nanoparticles. These nanoparticles were homogeneously dispersed on the surface of the nanowire with a size of about 2–3 nm and interparticle distance of 2 nm. Because the Z-contrast image resolution is close to the probe diameter, the atomic resolution of HAADF-STEM image is possible in TEM with a field-emission gun with a probe size of $<$0.3 Å or a Cs-corrected lens with a sub-Angstrom probe. The inset of Fig. 17.18B is a HAADF-STEM image of a single Pd nanoparticle attached on SnO_2 surface collected with Cs-corrected TEM. It clearly showed the bright lattice fringes of Pd, which can be easily distinguished from the background SnO_2 lattice. This method is extremely useful for the identification of noble metal catalysts in metal oxides for gas-sensing applications because the structures and distributions of the catalysts determine the performance of the sensing layers. By taking advantage of the sub-Angstrom probe in Cs-corrected TEM with the nanoprobe line scan using EDS (Fig. 17.18C and D), it can be confirmed that the SnO_2 nanocrystals are coated on the surface of ZnO nanowire and the Pd nanocrystals were deposited on the outmost layer of the nanowire.

17.6.2 Structural characterization of nanowire-based quantum-dot-sensitized solar cells (QDSSC)

Nanowire-based QDSSCs may work as the platform for low-cost high-efficiency solar energy conversion. There are multiple advantages to directly attach QDs on the nanowire surface for sunlight absorption and conversion. First, the energy band gap of QDs can be tuned by adjusting the QD size to match the sunlight spectrum for efficient light absorption. Second, multi-exciton generation [63] in QDs may circumvent the Shockley–Queisser limit of conventional p-n junction-based solar cells. Third, the

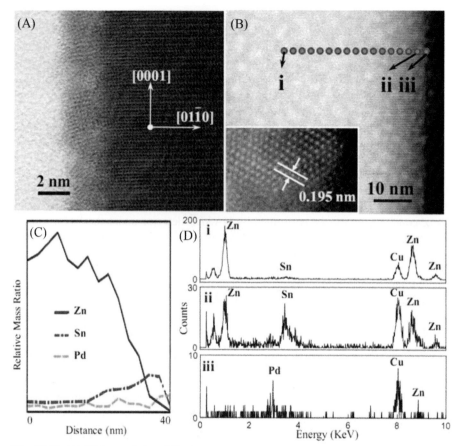

Fig. 17.18 (A) High-resolution TEM image showing the ZnO nanowire with polycrystalline SnO$_2$ and Pd-nanocrystal coating. (B) HAADF Z-contrast image taken under a Cs-corrected field-emission TEM highlighting the Pd-nanocrystal distribution. *Inset* shows the lattice image of a Pd nanocrystal taken by HAADF-STEM through [110] zone. (C) Composition distribution obtained by nanoprobe XEDS linescan across the nanowire. The sampling points are indicated in (B). (D) Three EDS spectra obtained at the (i) ZnO core, (ii) SnO$_2$ layer, and (iii) outermost Pd coating.

Adapted with permission from J. Chen, K. Wang, R. Huang, T. Saito, Y.H. Ikuhara, T. Hirayama, W. Zhou, Facile route to polycrystalline Pd/SnO$_2$ nanowires using ZnO-nanowire templates for gas-sensing applications. IEEE Trans. Nanotechnol. 9 (2010) 634–639. https://doi.org/10.1109/TNANO.2010.2052629. Copyright © 2010 IEEE.

single-crystalline nanowires provide direct fast electron transport channels for efficient charge collection. The performance of QDSSCs highly relies on the structural configuration of the QDs and the metal oxide nanowires. The structure of the contacts between the QDs and the metal oxide is one of the important factors for efficient electron transfer from QDs to metal oxides [64]. The surface condition of the QDs also determines the hole-harvesting process between the QDs and electrolyte [65].

We use solar cells based on CdSe QD-sensitized Zn_2SnO_4 nanowires as an example to explain how the detailed structural characterization with a TEM can help our understanding of the relation between the structure and solar cell performance. The QDSSC photoanodes were prepared with a two-step method [66]. The single-crystal Zn_2SnO_4 nanowires were first grown by conventional CVD through a VLS mechanism and transferred onto transparent conducting metal oxide using a printing method [67]. The CdSe QDs were directly grown on the Zn_2SnO_4 nanowire surface by PLD. Fig. 17.19A–H show a series of TEM images of CdSe QDs-coated Zn_2SnO_4 nanowires prepared under different laser fluences. These low-magnification

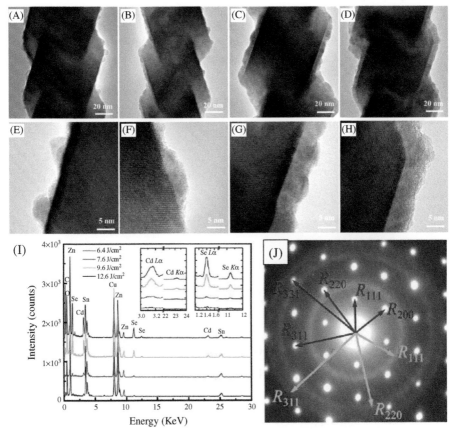

Fig. 17.19 Low- and high-magnification TEM images of CdSe QDs-coated Zn_2SnO_4 nanowires coated using PLD with different laser energies: (A and E) 6.4 J/cm²; (B and F) 7.6 J/cm²; (C and G) 9.6 J/cm²; (D and H) 12.6 J/cm². (I) XEDS spectra of Zn_2SnO_4 nanowires after PLD coating of CdSe QDs with different laser energy densities. The *insets* are enlarged peaks for Cd and Se. The spectra are shifted for clarification. (J) SAED patterns of Zn_2SnO_4 nanowire coated with CdSe QDs along $[011]_{ZTO}$ zone axis.

Adapted with permission from Q. Dai, J. Chen, L. Lu, J. Tang, W. Wang, Pulsed laser deposition of CdSe quantum dots on Zn_2SnO_4 nanowires and their photovoltaic applications, Nano Lett. 12 (2012) 4187–4193, https://doi.org/10.1021/nl301761w. Copyright © 2012 American Chemical Society.

images indicated that the CdSe QDs sparsely distributed on the nanowire surfaces when a low laser fluence was used but started forming dense QD layers when the laser fluence increased. Interestingly, the QDs have a similar size ($\sim$5 nm) that is independent of the fluence. The deposition of CdSe QDs on different substrates also showed that the size of QDs is self-limited [68] and independent of the substrate's material. The XEDS measurements (Fig. 17.19I) further confirmed that the amount of CdSe QDs coating increased as the laser fluence increased. However, despite the thicker coating of CdSe QDs for higher absorption of sunlight, the QDSSCs coated with higher laser fluences had lower efficiency [66], which may be explained by the band gap change and the lower hole harvesting for inner layers of QDs as QDs stack on each other [69]. The SAED shown in Fig. 17.19J also indicated that the QDs were randomly attached onto the Zn_2SnO_4 nanowires without specific orientation. To further understand the interfaces of QD-to-nanowire and QD-to-electrolyte, HRTEM image (Fig. 17.20) was taken at the Zn_2SnO_4 surface. An amorphous shell layer was clearly

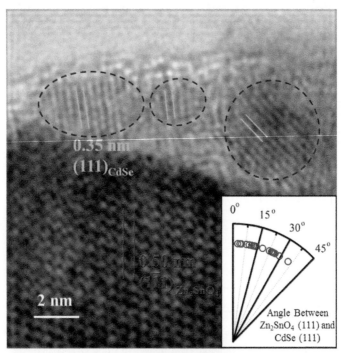

Fig. 17.20 A typical HRTEM image of CdSe QDs coated on Zn_2SnO_4 nanowires by pulsed laser deposition observed along Zn_2SnO_4 [011] zone axis. The (111) planes of each CdSe QDs and (1-11) planes of Zn_2SnO_4 were labeled by *green and red color*, respectively. The *inset* shows the angles between CdSe (111) planes and (1-11) planes were measured from HRTEM images. Twenty-four CdSe QDs were examined.

Adapted with permission from Q. Dai, J. Chen, L. Lu, J. Tang, W. Wang, Pulsed laser deposition of CdSe quantum dots on Zn_2SnO_4 nanowires and their photovoltaic applications, Nano Lett. 12 (2012) 4187–4193, https://doi.org/10.1021/nl301761w. Copyright © 2012 American Chemical Society.

shown on the surface of CdSe QDs. The role of this amorphous layer in the electron transfer from QDs to metal oxide or hole harvesting from QDs to electrolyte is still unclear, although a number of studies [64,65,70] have been conducted to investigate the effects of surface ligands or inorganic coatings. However, this observation on the surface structures provides a valuable reference for future characterization of solar cell performance. In addition, the measurement of the angles between the lattice fringes of CdSe and Zn_2SnO_4 confirmed that there is no epitaxial relation between the QDs and the nanowire.

17.6.3   Structural characterization of nanowire-based piezo-phototronics photodetector

Piezo-phototronic effect, a three-way coupling among semiconducting, piezoelectricity, and photon excitation, has attracted much attention since its first application in a ZnO-based UV photodetector and has widely been applied to enhance the performance of optoelectronic devices, such as solar cells, light emitting diodes, electrochemical processes, and photodetectors, through tuning and controlling charge carrier generation, separation, transport, and/or recombination. One of the strategies to improve the performance of piezo-phototronics effect is to fabricate a heterojunction structure by forming a staggered type II band alignment where the valence and conduction bands of one semiconductor are lower (or higher) than those of another semiconductor, enabling efficient separation and transportation of photogenerated electrons/holes through spatial confinement. One of the major requirements for the high-performance device is to fabricate abrupt interface to archive high-quality heterojunction. Therefore, atomic TEM analysis is proven to be essential in understanding the fundamental and proving guidelines for material synthesis and device fabrication.

The unique geometric feature of the core-shell nanowires offers great flexibility to release the strain at epitaxial interfaces. Nonetheless, the extended defects, such as dislocation and stacking faults, in the core-shell nanowire, could still negatively impact the electronic and optical properties. CdSe and ZnTe stand out as an optimal combination in the II-VI semiconductor materials for constructing a high-quality heteroepitaxial junction because of their small lattice mismatch, similar thermal expansion coefficients, and well-established doping capability [71]. Fig. 17.21A shows a low-magnification TEM image of several CdSe/ZnTe nanowires. The morphology of coated nanowires presented some nanoparticles on the surface. The presence of these nanoparticles could be attributed to the emission of microscopic particulates from solid targets when irradiated by a laser beam. One representative lower magnification TEM image of a single-coated nanowire in Fig. 17.21B revealed that the nanowire exhibited core-shell geometry if we neglect the microscopic particulates on the surface. The interface of the core and shell can be easily distinguished by the phase contrast, as highlighted by the dash lines. The core had a diameter of 80 nm, and the thickness of the shell was around 20 nm. To obtain detailed information about the interface, a high-resolution TEM image was recorded at the interface, shown in

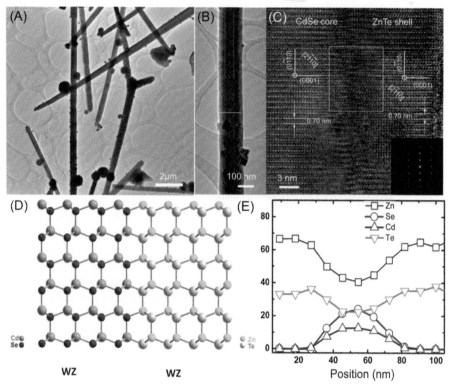

Fig. 17.21 (A) Low-magnification TEM image of CdSe/ZnTe core-shell nanowires. (B) A representative TEM image of an individual CdSe/ZnTe core-shell nanowire, clearly showing the smooth surface and core-shell geometry. (C) HRTEM images of a typical core-shell interface manifesting the epitaxial growth relationship of CdSe$_{WZ}$ and ZnTe$_{WZ}$; *Inset* shows the FFT of the area marked as a *dashed frame*. (D) Ball and stick model of the relaxed CdSe (WZ)/ ZnTe (WZ) structure. (E) Nanoprobe EDS scanning along the *line* marked in (B), suggesting the conformal shelling and absence of interfacial diffusion.

Fig. 17.21C. Most of the lattice fringes of the core and shell along the interface of the nanowire were well matched and continuous, despite a few dislocations. Further structural analysis revealed that both the core and shell were crystallized in WZ structures, and an epitaxial growth relationship could be assigned as $(0001)_{CdSe}//(0001)_{ZnTe}$ and $(0\bar{1}10)_{CdSe}//(0\bar{1}10)_{ZnTe}$. The FFT of the dashed rectangular area across the interface shows only one set of diffraction patterns, further confirming the epitaxial growth between the core and shell. An atomistic model of a WZ/WZ heterostructure was illustrated in Fig. 17.21D. The Cd and Se atoms have been marked in green and red for WZ CdSe, respectively, while Zn and Te atoms have been marked in cyan and gray, respectively, for WZ ZnTe. Owing to the same WZ crystal structure and close lattice parameters, nearly identical atom arrangements were demonstrated. A special concern in fabricating core-shell nanowires is the interfacial diffusion between the core and shell. To clarify this in CdSe/ZnTe core-shell nanowires, we employed nanoprobe

EDS line scanning along the white line marked in Fig. 17.21A. The element dispersive distribution data, as shown in Fig. 17.21E, verified that there was minimal interfacial diffusion occurring between the core and shell in this synthesis condition. In addition, this result also further confirmed the conformal coating of ZnTe on CdSe nanowires. We later used such system for a broadband photodetection through piezo-phototronic effect with broad spectral detection, high sensitivity, and fast response [72].

The ZnO/ZnS core-shell nanowire arrays were synthesized using a two-step technique [73]. Fig. 17.22A is a representative low-magnification TEM image for a ZnO/ZnS core-shell nanowire, showing sharp interfaces and a shell layer of ~12 nm. The HRTEM image taken from the area labeled as b in Fig. 17.22A reveals the detailed interface structure between the core and shell. Despite the presence of lattice distortion and stacking faults, both the core and shell exhibit lattice fringes and can be further

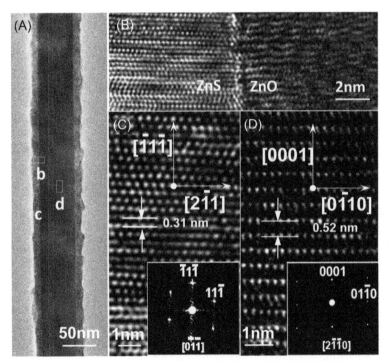

Fig. 17.22 (A) Low-magnification TEM micrograph of a ZnO/ZnS core/shell nanowire, showing a thin layer of ZnS coated on the ZnO nanowire. (B) High-resolution TEM image of the interface of the core/shell heterostructure, enlarged from the rectangular area in (a), showing the epitaxial growth relationship of ZnO wurtzite core and ZnS zinc blende shell. (C and D) Atomic resolution images of the core and shell areas taken from the *rectangular areas* in (C), respectively. The *insets* in (C and D) represent the corresponding fast Fourier transfer patterns. Adapted with permission from K. Wang, J.J. Chen, Z.M. Zeng, J. Tarr, W.L. Zhou, Y. Zhang, Y. F. Yan, C.S. Jiang, J. Pern, A. Mascarenhas, Synthesis and photovoltaic effect of vertically aligned ZnO/ZnS core/shell nanowire arrays. Appl. Phys. Lett. 96 (2010) 123105. https://doi.org/10.1063/1.3367706. Copyright © 2010 AIP Publishing LLC.

identified as WZ and zinc blende (ZB) structures, respectively, thus indicating that the ZB-ZnS shell layer was grown over the WZ-ZnO core. The magnified HRTEM image of the core and shell taken from rectangles c and d in Fig. 17.22A and their corresponding FFT patterns are also shown in Fig. 17.22C and D. The marked interplanar d spacings of 0.31 and 0.52 nm correspond, respectively, to the (-11-1) lattice plane of ZB ZnS with the [011] zone axis and the (0001) lattice plane of WZ ZnO with the [2-1-10] zone axis. Thus, the core and shell are determined as epitaxial growth with the growth relationship of $[0001]_{ZnO}//[-11-1]_{ZnS}$ and $[01-10]_{ZnO}//[21-1]_{ZnS}$, in contrast to that for the previously studied ZnO/ZnSe system, where $[0001]_{ZnO}//[001]_{ZnSe}$ was observed.

The epitaxial growth of WZ ZnS and ZnO has been reported in many nanoheterostructures in the planar form even though a large lattice mismatch (~20%) exists [74,75]. In our case, although the lattice mismatch between the WZ ZnO and ZB ZnS is still fairly large (~19%), it appears that various factors, including perhaps the geometry of the "substrate" (the ZnO nanowire), size of the core, growth temperature, and nonthermal equilibrium condition, play roles in successfully growing such an epitaxial core/shell heterostructure. However, due to the large lattice mismatch, structure defects, such as stacking faults and lattice distortion as shown in Fig. 17.22B, inevitably exist in order to release strain energy when the shell layer grows thicker. Further shell growth results in roughening and grain formation that provides additional strain relief. This could explain the rough external surface of the ZnS shell. Such a system was successfully applied in a piezo-phototronic effect enhanced UV/visible photodetector [76]. Because of the spatially indirect type-II transition facilitated by the abrupt interface between the ZnO core and ZnS shell, photons with energies significantly smaller (2.2 eV) than the band gap of ZnO (3.2 eV) and ZnS (3.7 eV) can be detected, which demonstrates a prototype UV/visible photodetector based on the truly wide band gap semiconductors.

17.6.4 Structural characterization of nanowire-based energy storage devices

One-dimensional materials have been wildly studied as promising candidates for energy storage devices, such as supercapacitors and batteries. Nanowire-based materials exhibit excellent electrochemical performance owing to their unique properties. First, nanostructured electrodes enlarge the surface area noticeably compared to bulk electrodes, therefore provide more ion adsorption sites and charge transfer active sites. Second, both electronic and ionic conductivity can be enhanced by precisely tailoring the pore structure, resulting in the improved specific capacitance and rate capability. Third, nanostructured engineering of electrode materials can also modify the phase change during charge/discharge, avoid undesirable side reactions, and improve the mechanical stability, therefore leading to higher cycling stability. Finally, nanostructured materials offer additional properties for advanced device configuration otherwise difficult to achieve for bulk or thin film materials, such as lightweight, flexibility, and transparency. As we mentioned before, one advantage of modern

TEM is its capability to collect the structural and chemical information of electrode materials at nanometer and even at the atomic scale, by combining it with electron diffraction, EELS, and EDS. This unique feature of TEM and STEM has been wildly used in energy-related materials [77–79]. We will show some examples of our recent works in this section to illuminate the importance of TEM and STEM study in energy-related materials.

The first example is CoP nanowire arrays synthesized by a two-step process [80]. The $Co(CO_3)_{0.5}(OH)\cdot0.11H_2O$ precursor was first synthesized by a hydrothermal method, followed by low-temperature phosphidation to convert it into CoP. As shown in the TEM image of Fig. 17.23A, CoP nanowire shows a rough surface with the width of 50–100 nm. The inset HRTEM lattice image shows the lattice fringes with an interplane spacing of 0.247 nm, representing the (111) planes of CoP. Correspondingly, the diffraction rings in Fig. 17.23B reveal that the CoP nanowire exhibits a polycrystalline feature with indexed diffraction rings of (011), (111), (112), (211), (202), and

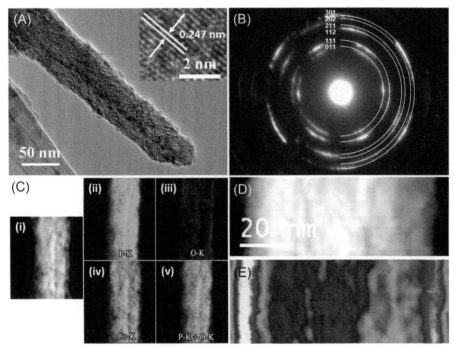

Fig. 17.23 (A) TEM image of a single CoP nanowire and the inset showing the HRTEM lattice image of the (111) plane. (B) Indexed SAED pattern of the CoP nanowire. (C) HAADF-STEM and corresponding EDS elemental maps imaged by P, O, Co, and P+O. (D and E) HAADF-STEM image and corresponding EELS maps using the energy range of 6.0–7.0 eV for CoP nanowire, respectively.

Adapted with permission from Z. Zheng, M. Retana, X. Hu, R. Luna, Y.H. Ikuhara, W. Zhou, Three-dimensional cobalt phosphide nanowire arrays as negative electrode material for flexible solid-state asymmetric supercapacitors. ACS Appl. Mater. Interfaces 9 (2017) 16986–16994. https://doi.org/10.1021/acsami.7b01109. Copyright © 2017 American Chemical Society.

(301) planes of the orthorhombic CoP phase ($a = 5.077$ Å, $b = 3.281$ Å, $c = 5.587$ Å). Fig. 17.23C shows the high-angle annular dark-field scanning TEM (HAADF STEM) image and the corresponding EDS elemental mappings (P, O, Co, and P+O) of a single CoP nanowire, further confirming the homogeneous distribution of P and Co elements along the nanowire. The combined mapping image of P and O elements (Fig. 17.23C-(v)) reveals a thin layer of oxygen on the surface of the CoP nanowire, indicating a formation of a thin oxidation surface layer. Furthermore, a low-loss EELS mapping (Fig. 17.23D and E) using an energy range of 6.0–7.0 eV indicates the occurrence of Co^{2+} on the surface. The Co^{2+} can be assigned to the interband transition of the extended O 2p states to the conduction band (O 2p to Co 3d (eg)) in cobalt oxide [81], which confirmed that the cobalt oxide layer with a thickness of about 8 nm was formed and concentrated on the surface of the CoP nanowire. The formation of the surface oxidation layer is caused by the reaction of CoP and O_2 in ambient air due to the metallic nature of CoP. Based on the detailed TEM and STEM characterization and later the electrochemical measurement, we found that the surface oxidation layer on the CoP nanowire increased during the cycling process by the possible irreversible electrochemical oxidation reaction, thus prevent ions from penetrating into the CoP core in the nanowire and causing the reduction of capacitance. However, we are able to suspend the irreversible electrochemical oxidation reaction by replacing the aqueous electrolyte with solid-state electrolyte, in turn dramatically improve the device stability [80].

The second example of using advanced TEM and STEM characterization in supercapacitor study is PEDOT coated FeP nanorod arrays [82]. The FeP nanorod arrays were synthesized using a similar method as CoP. The PEDOT (poly (3,4-ethylenedioxythiophene)) was coated on FeP nanorod arrays using an in situ polymerization method [83]. As shown in Fig. 17.24A, a layer of shell has been successfully observed on the surface of the FeP nanorod. The EDP in Fig. 17.24B reveals the polycrystalline nature of the FeP nanorod, which can be indexed as the orthorhombic FeP phase. The HRTEM, as shown in Fig. 17.24C, matches well with the simulation results of FeP from the [1$\bar{1}$0] zone axis. EELS displays the useful information of the characteristic signals of elements included, as shown in Fig. 17.24D and E. The P, Fe, and O signals with the phosphorus $L_{2,3}$-edge, iron $L_{2,3}$-edge, and oxygen K-edge at 148, 713, and 540 eV, respectively, can be detected via EELS on the FeP/PEDOT nanorod [78,82]. A clear and intense pre-peak is observed before the oxygen K-edge major peak at 540 eV, which is closely related to the existence of amorphous FeO in the nanorod. In addition, a carbon K-edge at 288 eV from PEDOT appears as shown in Fig. 17.24D. The FeP/PEDOT core/shell nanorod structure was further analyzed using a combination of STEM and EDS imaging, which has been proven to be a powerful means of characterizing compositional distribution in the core/shell structure [84,85]. Fig. 17.24F shows the existence of Fe, P, O, C, and S using EDS and the distribution of these elements was obtained via STEM-EDS maps as shown in Fig. 17.24G. The EDS maps of Fe, P, and O have almost the same elemental distribution qualitatively, confirming that the core is FeP. However, the C and S originated from PEDOT are widely distributed on the FeP nanorods. A superimposed display of Fe and S (Fe+S) maps

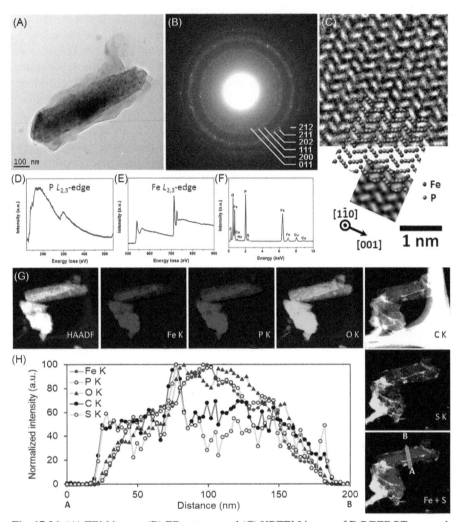

Fig. 17.24 (A) TEM image, (B) ED pattern, and (C) HRTEM image of FeP/PEDOT nanorod, (D and E) EELS spectrum and (F) EDS spectrum of FeP/PEDOT nanorods, and (G) HAADF-STEM image and the corresponding EDS elemental mapping images of Fe, P, O, C, S, and Fe+S, respectively. (H) EDS line profiles extracted from each map along with the *white arrow* (from A to B; 200nm in distance) represented on the Fe+S map. Each profile is normalized at maximum intensity.

Adapted with permission from J. Luo, Z. Zheng, A. Kumamoto, W.I. Unah, S. Yan, Y.H. Ikuhara, X. Xiang, X. Zu, W. Zhou, PEDOT coated iron phosphide nanorod arrays as high-performance supercapacitor negative electrodes. Chem. Commun. 54 (2018) 794–797. https://doi.org/10.1039/C7CC09163J. Copyright © 2018 The Royal Society of Chemistry.

shows distinctly different distributions between them. Fig. 17.24H shows the line profiles of EDS maps along the white arrow (from A to B) in the overlapped map with Fe and S. The normalized intensity profiles of C and S for the PEDOT region with the thickness of about 20 nm are higher than those of Fe, P, and O for the FeP nanorod, which indicates that FeP exists as the core and PEDOT as the shell. As mentioned before, metal phosphides generally suffer from the inferior cycling stability resulting from the physicochemical and structural deformation during the long-time redox reaction. In this study, the highly conductive PEDOT layer could benefit the charge transfer in the electrode/electrolyte interface and improve the electrode conductivity, as well as serve as a protection layer to prevent the FeP nanorod arrays from structure deformation. The electrochemical performance and cycling stability of FeP can be both greatly enhanced.

17.7   Concluding remarks

TEM investigations accompany almost every publication regarding nanostructure growth and play an increasingly important role in studying the structure–property relationship. In this chapter, we have demonstrated the capabilities of TEM on the structure analysis of nanostructures with imaging, diffraction, and spectroscopy. Besides the conventional TEM techniques, such as BF, DF imaging, and SAED, which can help us study the size, morphology, and crystal structure, modern TEM has become more and more important in nanomaterial characterization giving its ability to identify and quantify the chemical and electronic structure down to atomic scale. Taking ZnO nanoneedles as an example, we have shown the polarity determination by CBED and proposed a rational growth mode. With the help of HAADF-STEM, we now have the capabilities to investigate the growth mechanism, interface structure, and composition mapping in complex nanoheterostructures. Atomic resolution EDS and EELS provide not only the element distribution but also rich information of characteristic signals such as oxidation states. The in situ TEM techniques, although only briefly mentioned in this chapter, provide the opportunity to investigate the nanowire growth and response to external stimuli in a real-time manner, facilitating the basic understanding of the growth mechanism and the surface energetics of nanostructures.

In conclusion, the full spectrum of the TEM characterization has already shown significant impact in the field of materials science and nanotechnology and will continue to benefit the scientific and industrial communities.

Acknowledgments

The authors acknowledge the collaboration with colleagues who have been cited in the references. This work was supported by the DARPA Grant No. HR0011-09-1-0047, research grants from Louisiana Board of Regents Contract Nos. LEQSF(2007-12)-ENH-PKSFI-PRS-04, LEQSF(2008-11)-RD-B-10, and LEQSF(2011-13)-RD-B-08, and American Chemical Society Petroleum Research Fund under PRF No 48796-DN110.

References

[1] J.W. Edington, Practical Electron Microscopy in Materials Science, N.V. Philips, 1976.

[2] S.J. Pennycook, P.D. Nellist, Scanning Transmission Electron Microscopy, Springer New York, New York, NY, 2011, https://doi.org/10.1007/978-1-4419-7200-2.

[3] Y. Kotaka, T. Yamazaki, Y. Kataoka, Atomic-resolution imaging and analysis with C s-corrected scanning transmission electron microscopy, Fujitsu Sci. Tech. J. 46 (2010) 249–256.

[4] S.D. Findlay, S. Azuma, N. Shibata, E. Okunishi, Y. Ikuhara, Direct oxygen imaging within a ceramic interface, with some observations upon the dark contrast at the grain boundary, Ultramicroscopy 111 (2011) 285–289, https://doi.org/10.1016/J.ULTRAMIC.2010.12.022.

[5] Y. Ding, F. Zhang, Z.L. Wang, Deriving the three-dimensional structure of ZnO nano-wires/nanobelts by scanning transmission electron microscope tomography, Nano Res. 6 (2013) 253–262, https://doi.org/10.1007/s12274-013-0301-2.

[6] N. Petkov, In situ real-time TEM reveals growth, transformation and function in one-dimensional nanoscale materials: from a nanotechnology perspective, ISRN Nanotechnol. 2013 (2013) 1–21, https://doi.org/10.1155/2013/893060.

[7] K. Yao, D. Caruntu, B. Cao, C.J. O'Connor, W. Zhou, Investigation of gas-sensing performance of SnO_2 nanoparticles with different morphologies, IEEE Trans. Nanotechnol. 9 (2010) 630–633, https://doi.org/10.1109/TNANO.2010.2047728.

[8] K. Yao, D. Caruntu, Z. Zeng, J. Chen, C.J. O'Connor, W. Zhou, Parts per billion-level H2S detection at room temperature based on self-assembled In_2O_3 nanoparticles, J. Phys. Chem. C 113 (2009) 14812–14817, https://doi.org/10.1021/jp905189f.

[9] D. Caruntu, K. Yao, Z. Zhang, T. Austin, W. Zhou, C.J. O'Connor, One-step synthesis of nearly monodisperse, variable-shaped In_2O_3 nanocrystals in long chain alcohol solutions, J. Phys. Chem. C 114 (2010) 4875–4886, https://doi.org/10.1021/jp911427b.

[10] K. Yao, D. Caruntu, S. Wozny, R. Huang, Y.H. Ikuhara, B. Cao, C.J. O'Connor, W. Zhou, Towards one key to one lock: catalyst modified indium oxide nanoparticle thin film sensor array for selective gas detection, J. Mater. Chem. 22 (2012) 7308, https://doi.org/10.1039/c2jm15179k.

[11] H. Zheng, Y.S. Meng, Y. Zhu, Frontiers of in situ electron microscopy, MRS Bull. 40 (2015) 12–18, https://doi.org/10.1557/mrs.2014.305.

[12] C. Cai, S. Han, Q. Wang, M. Gu, Direct observation of yolk–shell transforming to gold single atoms and clusters with superior oxygen evolution reaction efficiency, ACS Nano 13 (2019) 8865–8871, https://doi.org/10.1021/acsnano.9b02135.

[13] R.S. Wagner, W.C. Ellis, Vapor-liquid-solid mechanism of single crystal growth, Appl. Phys. Lett. 4 (1964) 89–90, https://doi.org/10.1063/1.1753975.

[14] Y.-J. Hsu, S.-Y. Lu, Vapor–solid growth of Sn nanowires: growth mechanism and super-conductivity, J. Phys. Chem. B 109 (2005) 4398–4403, https://doi.org/10.1021/jp046354k.

[15] Y. Zhao, M. Chen, T. Xu, W. Liu, X. Liu, Preparation of porous anodic oxide film on aluminum and its application in synthesis of 1-dimensional nanomaterials, Chin. J. Chem. Phys. 17 (2004) 369–374.

[16] J. Ye, L. Qi, Solution-phase synthesis of one-dimensional semiconductor nanostructures, J. Mater. Sci. Technol. 24 (2008) 529–540.

[17] S.A. Fortuna, X. Li, Metal-catalyzed semiconductor nanowires: a review on the control of growth directions, Semicond. Sci. Technol. 25 (2010), https://doi.org/10.1088/0268-1242/25/2/024005, 024005.

[18] A. Fasoli, A. Colli, F. Martelli, S. Pisana, P.H. Tan, A.C. Ferrari, Photoluminescence of CdSe nanowires grown with and without metal catalyst, Nano Res. 4 (2011) 343–359, https://doi.org/10.1007/s12274-010-0089-2.

[19] M. de la Mata, C. Magen, J. Gazquez, M.I.B. Utama, M. Heiss, S. Lopatin, F. Furtmayr, C. J. Fernández-Rojas, B. Peng, J.R. Morante, R. Rurali, M. Eickhoff, A. Fontcuberta i Morral, Q. Xiong, J. Arbiol, Polarity assignment in ZnTe, GaAs, ZnO, and GaN-AlN nanowires from direct dumbbell analysis, Nano Lett. 12 (2012) 2579–2586, https://doi.org/10.1021/nl300840q.

[20] D.B. Williams, C.B. Carter, Transmission Electron Microscopy: A Textbook for Materials Science, second ed., Springer US, Boston, MA, 2009.

[21] Y. Li, Q.F. Xing, Y. Yan, W.L. Zhou, A large quantity synthesis of ZnO nanoneedles and their polarity determination, J. Nanosci. Nanotechnol. 10 (2010) 2023–2027, https://doi.org/10.1166/jnn.2010.2129.

[22] H. Huang, X. Ren, X. Ye, J. Guo, Q. Wang, Y. Yang, S. Cai, Y. Huang, Growth of stacking-faults-free zinc blende GaAs nanowires on Si substrate by using AlGaAs/GaAs buffer layers, Nano Lett. 10 (2010) 64–68, https://doi.org/10.1021/nl902842g.

[23] H. Shtrikman, R. Popovitz-Biro, A. Kretinin, M. Heiblum, Stacking-faults-free zinc blende GaAs nanowires, Nano Lett. 9 (2009) 215–219, https://doi.org/10.1021/nl8027872.

[24] K. Wang, J. Chen, Y. Chen, W. Zhou, A large quantity synthesis of ultra-thin ZnO nanobelts induced from stacking faults, J. Nanosci. Nanotechnol. 10 (2010) 2257–2260, https://doi.org/10.1166/jnn.2010.2128.

[25] M.T. Borgström, M.A. Verheijen, G. Immink, T. de Smet, E.P.A.M. Bakkers, Interface study on heterostructured GaP–GaAs nanowires, Nanotechnology 17 (2006) 4010–4013, https://doi.org/10.1088/0957-4484/17/16/002.

[26] M.S. Gudiksen, L.J. Lauhon, J. Wang, D.C. Smith, C.M. Lieber, Growth of nanowire superlattice structures for nanoscale photonics and electronics, Nature 415 (2002) 617–620, https://doi.org/10.1038/415617a.

[27] M.T. Björk, B.J. Ohlsson, T. Sass, A.I. Persson, C. Thelander, M.H. Magnusson, K. Deppert, L.R. Wallenberg, L. Samuelson, One-dimensional steeplechase for electrons realized, Nano Lett. 2 (2002) 87–89, https://doi.org/10.1021/nl010099n.

[28] Y. Wu, R. Fan, P. Yang, Block-by-block growth of single-crystalline Si/SiGe superlattice nanowires, Nano Lett. 2 (2002) 83–86, https://doi.org/10.1021/nl0156888.

[29] J.L.F. Da Silva, Y. Yan, S.-H. Wei, Rules of structure formation for the homologous InMO3 (ZnO)n compounds, Phys. Rev. Lett. 100 (2008), https://doi.org/10.1103/PhysRevLett.100.255501, 255501.

[30] Y. Yan, S.J. Pennycook, J. Dai, R.P.H. Chang, A. Wang, T.J. Marks, Polytypoid structures in annealed In2O3–ZnO films, Appl. Phys. Lett. 73 (1998) 2585–2587, https://doi.org/10.1063/1.122513.

[31] D.L. Huang, L.L. Wu, X.T. Zhang, Size-dependent InAlO3(ZnO) m nanowires with a perfect superlattice structure, J. Phys. Chem. C 114 (2010) 11783–11786, https://doi.org/10.1021/jp102964p.

[32] N. Wang, Y.H. Yang, J. Chen, N. Xu, G. Yang, One-dimensional Zn-doped In2O3–SnO2 superlattice nanostructures, J. Phys. Chem. C 114 (2010) 2909–2912, https://doi.org/10.1021/jp910802z.

[33] J. Jie, G. Wang, X. Han, J.G. Hou, Synthesis and characterization of ZnO:In nanowires with superlattice structure, J. Phys. Chem. B 108 (2004) 17027–17031, https://doi.org/10.1021/jp0484783.

[34] C.W. Na, S.Y. Bae, J. Park, Short-period superlattice structure of Sn-doped In2O3(ZnO)4 and In2O 3(ZnO)5 nanowires, J. Phys. Chem. B 109 (2005) 12785–12790, https://doi.org/10.1021/jp0442246.

[35] S.C. Andrews, M.A. Fardy, M.C. Moore, S. Aloni, M. Zhang, V. Radmilovic, P. Yang, Atomic-level control of the thermoelectric properties in polytypoid nanowires, Chem. Sci. 2 (2011) 706, https://doi.org/10.1039/c0sc00537a.

[36] D.P. Li, G.Z. Wang, X.H. Han, J.S. Jie, S.T. Lee, Synthesis and characterization of in-doped ZnO planar superlattice nanoribbons, J. Phys. Chem. C 113 (2009) 5417–5421, https://doi.org/10.1021/jp810158u.

[37] D.P. Li, G.Z. Wang, Q.H. Yang, X. Xie, Synthesis and photoluminescence of $InGaO_3$ $(ZnO)_m$ nanowires with perfect superlattice structure, J. Phys. Chem. C 113 (2009) 21512–21515, https://doi.org/10.1021/jp906381h.

[38] R.D. Shannon, Revised effective ionic radii and systematic studies of interatomic distances in halides and chalcogenides, Acta Crystallogr. Sect. A 32 (1976) 751–767, https://doi.org/10.1107/S0567739476001551.

[39] B. Cao, T. Shi, S. Zheng, Y.H. Ikuhara, W. Zhou, D. Wood, M. Al-Jassim, Y. Yan, New polytypoid $SnO_2(ZnO:Sn)_m$ nanowire: characterization and calculation of its electronic structure, J. Phys. Chem. C 116 (2012) 5009–5013, https://doi.org/10.1021/jp211135r.

[40] J.-P. Colinge, C.-W. Lee, A. Afzalian, N.D. Akhavan, R. Yan, I. Ferain, P. Razavi, B. O'Neill, A. Blake, M. White, A.-M. Kelleher, B. McCarthy, R. Murphy, Nanowire transistors without junctions, Nat. Nanotechnol. 5 (2010) 225–229, https://doi.org/10.1038/nnano.2010.15.

[41] K.S. Leschkies, R. Divakar, J. Basu, E. Enache-Pommer, J.E. Boercker, C.B. Carter, U.R. Kortshagen, D.J. Norris, E.S. Aydil, Photosensitization of ZnO nanowires with CdSe quantum dots for photovoltaic devices, Nano Lett. 7 (2007) 1793–1798, https://doi.org/10.1021/nl070430o.

[42] B. Cao, J. Chen, R. Huang, Y.H. Ikuhara, T. Hirayama, W. Zhou, Axial growth of Zn2GeO4/ZnO nanowire heterojunction using chemical vapor deposition, J. Cryst. Growth 316 (2011) 46–50, https://doi.org/10.1016/J.JCRYSGRO.2010.12.060.

[43] P. Reiss, M. Protière, L. Li, Core/shell semiconductor nanocrystals, Small 5 (2009) 154–168, https://doi.org/10.1002/smll.200800841.

[44] Y. Zhang, Wang, A. Mascarenhas, "Quantum coaxial cables" for solar energy harvesting, Nano Lett. 7 (2007) 1264–1269, https://doi.org/10.1021/nl070174f.

[45] J. Xiang, W. Lu, Y. Hu, Y. Wu, H. Yan, C.M. Lieber, Ge/Si nanowire heterostructures as high-performance field-effect transistors, Nature 441 (2006) 489–493, https://doi.org/10.1038/nature04796.

[46] H. Wang, M. Upmanyu, C.V. Ciobanu, Morphology of epitaxial core–shell nanowires, Nano Lett. 8 (2008) 4305–4311, https://doi.org/10.1021/nl8020973.

[47] D. Moore, J.R. Morber, R.L. Snyder, Z.L. Wang, Growth of ultralong ZnS/SiO2 core–shell nanowires by volume and surface diffusion VLS process, J. Phys. Chem. C 112 (2008) 2895–2903, https://doi.org/10.1021/jp709903b.

[48] L.J. Lauhon, M.S. Gudiksen, D. Wang, C.M. Lieber, Epitaxial core–shell and core–multishell nanowire heterostructures, Nature 420 (2002) 57–61, https://doi.org/10.1038/nature01141.

[49] J.K. Hyun, S. Zhang, L.J. Lauhon, Nanowire heterostructures, Annu. Rev. Mater. Res. 43 (2013) 451–479, https://doi.org/10.1146/annurev-matsci-071312-121659.

[50] K. Wang, J. Chen, W. Zhou, Y. Zhang, Y. Yan, J. Pern, A. Mascarenhas, Direct growth of highly mismatched type II ZnO/ZnSe core/shell nanowire arrays on transparent conducting oxide substrates for solar cell applications, Adv. Mater. 20 (2008) 3248–3253, https://doi.org/10.1002/adma.200800145.

[51] D. Geng, H.Y. Yang, Recent advances in growth of novel 2D materials: beyond graphene and transition metal dichalcogenides, Adv. Mater. 30 (2018) 1800865, https://doi.org/10.1002/adma.201800865.

[52] S. Manzeli, D. Ovchinnikov, D. Pasquier, O.V. Yazyev, A. Kis, 2D transition metal dichalcogenides, Nat. Rev. Mater. 2 (2017) 17033, https://doi.org/10.1038/natrevmats.2017.33.

[53] K.S. Novoselov, A.K. Geim, S.V. Morozov, D. Jiang, Y. Zhang, S.V. Dubonos, I.V. Grigorieva, A.A. Firsov, Electric field in atomically thin carbon films, Science 306 (2004) 666–669, https://doi.org/10.1126/science.1102896.

[54] P.Y. Huang, J.C. Meyer, D.A. Muller, From atoms to grains: transmission electron microscopy of graphene, MRS Bull. 37 (2012) 1214–1221, https://doi.org/10.1557/mrs.2012.183.

[55] A.W. Robertson, J.H. Warner, Atomic resolution imaging of graphene by transmission electron microscopy, Nanoscale 5 (2013) 4079, https://doi.org/10.1039/c3nr00934c.

[56] A.W. Robertson, C.S. Allen, Y.A. Wu, K. He, J. Olivier, J. Neethling, A.I. Kirkland, J.H. Warner, Spatial control of defect creation in graphene at the nanoscale, Nat. Commun. 3 (2012) 1144, https://doi.org/10.1038/ncomms2141.

[57] K. Suenaga, M. Koshino, Atom-by-atom spectroscopy at graphene edge, Nature 468 (2010) 1088–1090, https://doi.org/10.1038/nature09664.

[58] X. Li, M.-W. Lin, L. Basile, S.M. Hus, A.A. Puretzky, J. Lee, Y.-C. Kuo, L.-Y. Chang, K. Wang, J.C. Idrobo, A.-P. Li, C.-H. Chen, C.M. Rouleau, D.B. Geohegan, K. Xiao, Isoelectronic tungsten doping in monolayer $MoSe_2$ for carrier type modulation, Adv. Mater. 28 (2016) 8240–8247, https://doi.org/10.1002/adma.201601991.

[59] Y. Shimizu, M. Egashira, Basic aspects and challenges of semiconductor gas sensors, MRS Bull. 24 (1999) 18–24, https://doi.org/10.1557/S0883769400052465.

[60] A. Kolmakov, D.O. Klenov, Y. Lilach, S. Stemmer, M. Moskovits, Enhanced gas sensing by individual SnO2 nanowires and nanobelts functionalized with Pd catalyst particles, Nano Lett. 5 (2005) 667–673, https://doi.org/10.1021/nl050082v.

[61] L. Vayssieres, Growth of arrayed nanorods and nanowires of ZnO from aqueous solutions, Adv. Mater. 15 (2003) 464–466, https://doi.org/10.1002/adma.200390108.

[62] J. Chen, K. Wang, R. Huang, T. Saito, Y.H. Ikuhara, T. Hirayama, W. Zhou, Facile route to polycrystalline Pd/SnO_2 nanowires using ZnO-nanowire templates for gas-sensing applications, IEEE Trans. Nanotechnol. 9 (2010) 634–639, https://doi.org/10.1109/TNANO.2010.2052629.

[63] J.B. Sambur, T. Novet, B.A. Parkinson, Multiple exciton collection in a sensitized photovoltaic system, Science 330 (2010) 63–66, https://doi.org/10.1126/SCIENCE.1191462.

[64] I. Robel, V. Subramanian, M. Kuno, P.V. Kamat, Quantum dot solar cells. Harvesting light energy with CdSe nanocrystals molecularly linked to mesoscopic TiO2 films, J. Am. Chem. Soc. 128 (2006) 2385–2393, https://doi.org/10.1021/ja056494n.

[65] Q. Shen, J. Kobayashi, L.J. Diguna, T. Toyoda, Effect of ZnS coating on the photovoltaic properties of CdSe quantum dot-sensitized solar cells, J. Appl. Phys. 103 (2008), https://doi.org/10.1063/1.2903059, 084304.

[66] Q. Dai, J. Chen, L. Lu, J. Tang, W. Wang, Pulsed laser deposition of CdSe quantum dots on Zn_2SnO_4 nanowires and their photovoltaic applications, Nano Lett. 12 (2012) 4187–4193, https://doi.org/10.1021/nl301761w.

[67] J. Chen, L. Lu, W. Wang, Zn_2SnO_4 nanowires as photoanode for dye-sensitized solar cells and the improvement on open-circuit voltage, J. Phys. Chem. C 116 (2012) 10841–10847, https://doi.org/10.1021/jp301770n.

[68] T. Donnelly, S. Krishnamurthy, K. Carney, N. McEvoy, J.G. Lunney, Pulsed laser deposition of nanoparticle films of Au, Appl. Surf. Sci. 254 (2007) 1303–1306, https://doi.org/10.1016/J.APSUSC.2007.09.033.

[69] N. Guijarro, T. Lana-Villarreal, I. Mora-Seró, J. Bisquert, R. Gómez, CdSe quantum dot-sensitized TiO_2 electrodes: effect of quantum dot coverage and mode of attachment, J. Phys. Chem. C 113 (2009) 4208–4214, https://doi.org/10.1021/jp808091d.

[70] J.B. Sambur, S.C. Riha, D. Choi, B.A. Parkinson, Influence of surface chemistry on the binding and electronic coupling of CdSe quantum dots to single crystal TiO_2 surfaces, Langmuir 26 (2010) 4839–4847, https://doi.org/10.1021/la903618x.

[71] M.C. Phillips, E.T. Yu, Y. Rajakarunanayake, J.O. McCaldin, D.A. Collins, T.C. McGill, Characterization of CdSe/ZnTe heterojunctions, J. Cryst. Growth 111 (1991) 820–822, https://doi.org/10.1016/0022-0248(91)91089-S.

[72] S.C. Rai, K. Wang, J. Chen, J.K. Marmon, M. Bhatt, S. Wozny, Y. Zhang, W. Zhou, Enhanced broad band photodetection through piezo-phototronic effect in CdSe/ZnTe Core/Shell nanowire array, Adv. Electron. Mater. 1 (2015) 1400050, https://doi.org/10.1002/aelm.201400050.

[73] K. Wang, J.J. Chen, Z.M. Zeng, J. Tarr, W.L. Zhou, Y. Zhang, Y.F. Yan, C.S. Jiang, J. Pern, A. Mascarenhas, Synthesis and photovoltaic effect of vertically aligned ZnO/ZnS core/shell nanowire arrays, Appl. Phys. Lett. 96 (2010), https://doi.org/10.1063/1.3367706, 123105.

[74] M.-Y. Lu, J. Song, M.-P. Lu, C.-Y. Lee, L.-J. Chen, Z.L. Wang, ZnO − ZnS heterojunction and ZnS nanowire arrays for electricity generation, ACS Nano 3 (2009) 357–362, https://doi.org/10.1021/nn800804r.

[75] X. Wu, P. Jiang, Y. Ding, W. Cai, S. Xie, Z.L. Wang, Mismatch strain induced formation of ZnO/ZnS heterostructured rings, Adv. Mater. 19 (2007) 2319–2323, https://doi.org/10.1002/adma.200602698.

[76] S.C. Rai, K. Wang, Y. Ding, J.K. Marmon, M. Bhatt, Y. Zhang, W. Zhou, Z.L. Wang, Piezo-phototronic effect enhanced UV/visible photodetector based on fully wide band gap type-II ZnO/ZnS core/shell nanowire array, ACS Nano 9 (2015) 6419–6427, https://doi.org/10.1021/acsnano.5b02081.

[77] R. Huang, T. Hitosugi, S.D. Findlay, C.A.J. Fisher, Y.H. Ikuhara, H. Moriwake, H. Oki, Y. Ikuhara, Real-time direct observation of Li in $LiCoO_2$ cathode material, Appl. Phys. Lett. 98 (2011), https://doi.org/10.1063/1.3551538, 051913.

[78] Y.H. Ikuhara, X. Gao, C.A.J. Fisher, A. Kuwabara, H. Moriwake, K. Kohama, H. Iba, Y. Ikuhara, Atomic level changes during capacity fade in highly oriented thin films of cathode material $LiCoPO_4$, J. Mater. Chem. A 5 (2017) 9329–9338, https://doi.org/10.1039/C6TA10084H.

[79] Y.H. Ikuhara, X. Gao, R. Huang, C.A.J. Fisher, A. Kuwabara, H. Moriwake, K. Kohama, Epitaxial growth of $LiMn_2O_4$ thin films by chemical solution deposition method for multilayer lithiumion batteries, J. Phys. Chem. C 118 (2014) 19540–19547.

[80] Z. Zheng, M. Retana, X. Hu, R. Luna, Y.H. Ikuhara, W. Zhou, Three-dimensional cobalt phosphide nanowire arrays as negative electrode material for flexible solid-state asymmetric supercapacitors, ACS Appl. Mater. Interfaces 9 (2017) 16986–16994, https://doi.org/10.1021/acsami.7b01109.

[81] H.A.E. Hagelin-Weaver, G.B. Hoflund, D.M. Minahan, G.N. Salaita, Electron energy loss spectroscopic investigation of Co metal, CoO, and Co3O4 before and after Ar+ bombardment, Appl. Surf. Sci. 235 (2004) 420–448, https://doi.org/10.1016/j.apsusc.2004.02.062.

[82] J. Luo, Z. Zheng, A. Kumamoto, W.I. Unah, S. Yan, Y.H. Ikuhara, X. Xiang, X. Zu, W. Zhou, PEDOT coated iron phosphide nanorod arrays as high-performance supercapacitor negative electrodes, Chem. Commun. 54 (2018) 794–797, https://doi.org/10.1039/C7CC09163J.

[83] J. Han, Y. Dou, J. Zhao, M. Wei, D.G. Evans, X. Duan, Flexible CoAl LDH@PEDOT core/shell nanoplatelet array for high-performance energy storage, Small 9 (2013) 98–106, https://doi.org/10.1002/smll.201201336.

[84] K. Cui, A. Kumamoto, R. Xiang, H. An, B. Wang, T. Inoue, S. Chiashi, Y. Ikuhara, S. Maruyama, Synthesis of subnanometer-diameter vertically aligned single-walled carbon nanotubes with copper-anchored cobalt catalysts, Nanoscale 8 (2016) 1608–1617, https://doi.org/10.1039/C5NR06007A.

[85] K. Matsui, H. Yoshida, Y. Ikuhara, Nanocrystalline, ultra-degradation-resistant zirconia: its grain boundary nanostructure and nanochemistry, Sci. Rep. 4 (2015) 4758, https://doi.org/10.1038/srep04758.

SMM studies on high-frequency electrical properties of nanostructured materials

18

Y. Zhuang, J. Myers, Z. Ji, and K. Vishal
Department of Electrical Engineering, Wright State University, Dayton, OH, United States

18.1 Introduction

Characterization of nanostructured materials has been consistently experiencing technological revolution in the past a few decades. One groundbreaking technology in materials characterization is the non-invasive and non-destructive near-field scanning microwave microscopy (SMM) [1]. First, this technology demonstrates a nanometer scale spatial resolution which is well beyond the Abbe diffraction limit: at radio frequency (RF) and microwaves (MW), the resolution is about $10^4 \sim 10^7$ smaller than the propagation wavelength (λ). Second, the great versatility of SMM allows it to be widely applied in materials-related research, chemical sensing, electrochemical devices, life science in-situ characterization, etc. Third, due to the penetration of the electromagnetic waves (EMWs) in RF/MW frequency range (300 MHz $\sim$ 300 GHz), SMM can uniquely perform detection below the surface. This feature will promote SMM to play a significant role in blocking hardware obfuscation and in building trusted electronics. The following applications highlight the breadth and depth of this technology.

18.1.1 High speed electronics

Integrated circuits (ICs) have been playing a significant role in the semiconductor industry by integrating billions of transistors into a small chip aiming for high performance. The demands for the fifth generation (5G) communication drive the IC technology toward higher frequencies: (1) Sub-6 GHz 5G to mid and low frequency bands under 6 GHz; (2) mmWave 5G to higher frequency radio bands ranging from 24 to 40 GHz. Despite the ICs design and fabrication, materials involved in ICs are required to work properly at RF/MW frequencies. However, the electromagnetic properties of the materials exhibit frequency-dependency. It is well known that the dielectric loss due to the relaxation at the resonant frequencies of the polarization causes declination of the materials' permittivity vs frequency [2]. Permeability of the magnetic materials is governed by the ferromagnetic resonance frequency (FMR) [3]. Furthermore, except a few layers above the substrate are single crystalline, a number of polycrystalline layers are also employed either aiming for fabrication completion such as

Modeling, Characterization and Production of Nanomaterials. https://doi.org/10.1016/B978-0-12-819905-3.00018-X

insulation, metallic ground, electrical shield, and interconnects, or targeting for functional enhancements like high-k materials for gate insulation and high performance capacitive components, magnetic materials to improve integrated inductive components, and multiferroic materials for multi-functionality. Besides the structural-property correlation of the polycrystalline materials, the grain boundaries exhibit frequency-dependent mixed electromagnetic properties. As the device dimension approaches to or smaller than the grain size, the impact of the grain boundary on device performance is rising and varies from grain to grain. However, the underlying physics of the grain boundaries is still not fully understood at RF/MW frequencies, in part because of its tiny size which generally in the range of sub-micron, in part because of its structural complexity, and in part because of the lack of characterization tool capable of performing electromagnetic properties at RF/MW frequency with nanometer spatial resolution. Therefore, in-situ characterization of materials at RF/MW becomes indispensable to predict device performance with high integration density.

18.1.2 Testing in versatile environment

Scanning probe microscopy (SPM) provides an excellent platform capable of performing characterization in versatile environment including living cells, electrochemical devices, liquids, toxic, etc. Most of them rely on intimate physical contacts between SPM tip and the specimen to probe the corresponding physical–chemical interactions. This, however, imposes an insurmountable challenge for quantitative measurements conducted at low frequencies. The excessively large and uncontrollable contact and spreading resistances dominate the probe-sample interactions, dwarfing the response from all the other physical–chemical interactions. This problem can be circumvented by combining the use of near-field created around the SPM tip and developing tools capable of characterization at RF/MW frequencies. The contact and spreading resistances will be shorted at the high frequencies by the increased capacitive coupling between the tip and sample, which makes the signals from the interested physical–chemical interactions outstand.

In the environment-sensitive characterization, non-invasive and non-destructive methods are crucial in detection of natural physical–chemical interactions [4]. The penetration ability of EMWs at RF/MW frequencies and the heat created by absorbing the EMWs energy can lead to damages to the samples, specifically for biomaterials. At RF/MW frequencies, the energy of MW photons is in the range of 10^{-6} electron volts, which is far below the strength of chemical bonds in matter. The tiny size of the SMM tip not only restricts the power conducted to it but also localizes the field enhancement near to the tip. As a result, the impact of EMWs radiation at RF/MW frequencies is negligible, making SMM a superior non-invasive and non-destructive testing method.

18.1.3 Hardware obfuscation

In the past 2 decades, semiconductor industry has been experiencing rapid globalization of its supply chain. The continuous increase of the complexity makes it impossible to fabricate the all ICs in a single company or nation, which comes to the concern of security. Among all the others, hardware obfuscation induced at the untrusted foundries presents a rising threat for its resistance to today's most available detection techniques by modifying the doping configuration at nanometer scale underneath the surface [5,6]. At RF/MW frequency range, the penetration depth of the EMWs can reach a few microns to millimeters depending on the materials' electrical conductivity and permittivity. The nanometer spatial resolution and the in-depth analysis capability might render SMM to be potentially useful in trusted electronics to detect hardware Trojans at the physical layer.

The chapter is organized in five sections including introduction (I), methodology (II), modeling of SMM (III), examples of SMM in investigations of nanostructured materials (IV), and summary (V). After introducing technical advances of SMM comparing to other SPMs in Section 18.2—methodology, physical model of SMM at RF/MW frequency has been established by using finite element method in Section 18.3. Section 18.4 discusses a number of examples of using SMM in investigations of nanostructured materials. At the end, a summary of SMM is given in Section 18.5.

18.2 Methodology

Non-invasive and non-destructive characterization methods are essential in obtaining natural response from nanostructured materials. In the past, optical microscope is the dominant technical approach in the investigation of materials at micron size. However, further improvement of its spatial resolution to nanometer scale imposes a formidable challenge due to the Abbe diffraction limit. This becomes even more unfavorably on the technology operated at frequencies lower than the visible light, like MW imaging (MIG) with wavelength in the range from a few millimeters to meters. Electron microscopy has thus been extensively explored and becomes the major technique to probe the materials' properties of having sub-nanometer spatial resolution due to its ultra-short wavelength (de Broglie wavelength). However, despite the incapability of in-situ measurement, the overdue time consumption required for specimen preparation for transmission electron microscopy and its induced fetal damages to materials structure present a big technical barrier in the investigation of the structural-property correlation. Field-emission scanning electron microscopy can reach nanometer spatial resolution and does not need any special specimen preparation, but other than surface topographical property, it can provide only limited additional physical–chemical properties information.

SPM characterizes the interaction between a physical probe and samples' surface based on a large variety of physical–chemical principles. The first of its type is the scanning tunneling microscopy (STM) by measuring the tunneling current between the conductive probe and the specimen. This Nobel Prize-winning technique was founded by Gerd Binnig and Heinrich Rohrer in 1981. Due to the ultra-sensitivity, STM demonstrates capability of in situ imaging with atomic resolution. Later on, there are versatile SPM techniques originated from STM but built on different physical–chemical principles, such as atomic force microscopy (AFM), scanning probe electrochemistry, magnetic resonance force microscopy, near-field scanning optical microscopy, scanning spreading resistance microscopy, scanning thermal microscopy, and scanning thermo-ionic microscopy. Most of these SPM techniques are developed for non-destructive measurements and are capable to access to the real environment of the samples natural exist. However, the measurements are mostly conducted at surfaces and are in general incapable of depth-dependent detection.

MIG has been widely used in military and law enforcement such as remote sensing, security, and surveillance in the past. This is largely because of the fact that EMWs in this frequency range can propagate through optical opaque objects. Similarly, limited by the Abbe diffraction limit as in the optical microscopy, the spatial resolution of MIG is about half of the propagation wavelength. The first MIG was reported by Bryant and Gunn [7], showing a millimeter spatial resolution at 450 MHz. The milestone work in breaking the diffraction limit is the demonstration of a spatial resolution close to one-sixtieth of the propagation wavelength in a near-field instrument [8]. The principle behind is that in the near-field approximation, as the MW signals travel along a transmission line, the electromagnetic fields will be significantly enhanced along the narrow signal line and surrounded within a distance approximately to the dimension of the signal line (Fig. 18.1). The concentrated electromagnetic field dominates the interaction between the MW field and sample, thus promoting the spatial resolution approximately to the dimension of the signal line. In today's technology, the diameter of the signal line can be scaled down to 50 nm or below, making sub-100 nm resolution feasible. This equivalents to a better than $\lambda/10^6$ spatial resolution at 1 GHz.

The first demo of near-field SMM was demonstrated in 1989 [9] using a coaxial tip to achieve 30 ps waveform at a spatial resolution set by the tip radius. Since then, tremendous efforts have been done to develop near-field SMM. In general, SMM can be operated in the reflection mode and transmission mode. The reflection mode has been broadly used in characterizing surface inhomogeneity of the electrical properties [10–13], with spatial resolution up to atomic level [14]. The drawback is the limitation of the detection depth. Very recently, transmitted mode SMM has been developed to circumvent this challenge and enables in-depth analysis [15].

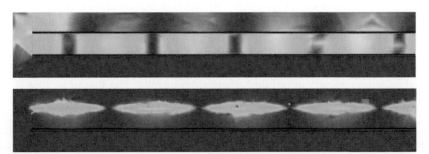

Fig. 18.1 Transmission line electric field propagation for a transmission line with an equal size signal and ground (*top*), and a transmission line with a much smaller signal than ground (*bottom*).

Practically SMM can be operated in contact mode and non-contact mode. Contact mode is performed by landing the SMM tip intimately on the surface of the specimen. This mode assures a high spatial resolution which is directly proportional to the tip size. In the configuration of non-contact mode, there is a gap between the SMM tip and the specimen surface whose resolution depends on the size of the SMM tip as well as the tip-sample separation distance. Although the existence of the tip-sample gap deteriorates the spatial resolution in general, the release of the demand for intimate contact significantly improves its flexibility by allowing measurements in versatile environments, including liquid, cells, active sample, soft-materials, and so on, in which other characterization techniques have difficulties to get access.

As used in remote sensing, security, and surveillance, MW can penetrate through optically opaque materials. This feature allows near-field SMM obtain in-depth information and probe materials' physical–chemical properties under the surface. In the frequency range 1–20 GHz within which most of SMM measurements are conducted, the penetration depth (or skin depth) can be up to a few-hundreds micron depending on the conductivity and permittivity of the specimen. Although the SMM tip-sample interaction is mediated by evanescent fields which drops sharply over a distance beyond the tip radius of 30–100 nm, MW can penetrate far-deep in the sample, exemplified by MW image recorded of metallic lines embedded at 2.3 μm below the surface [16,17]. In addition to the in-depth analysis, SMM also releases the needs of intimate contact which are required by most of SPM measurements. Poor physical contact results in exceedingly large contact resistance. This is an inevitable effect among all the SPM measurements performed at low frequencies. At MW frequencies, the tip-sample is conducted via the coupled capacitance which behaves as an electrical-short at high frequencies, guaranteeing a good electrical contact between the tip and sample.

18.3   Modeling of SMM system

There has been a continuous effort in developing quantitative analysis of correlation between SMM imaging contrast and materials properties [18–23]. Attempts of quantitative investigations via sophisticate calibration of the SMM systems have also been proposed [24,25]. To investigate the experimental results, a number of EM models have been reported, which in general discuss the tip-sample interaction by using either finite element analysis or electrodynamics-based analytical solution under quasi-static condition [26–28], and the complex reflection coefficient (Γ_{in}) by transmission line theory [27]. A few works reporting the image contrast in SMM measurements reveals strong frequency dependency, and has been found turned over at different harmonic resonant frequencies [10,13,29]. Without identifying the resonant frequencies, investigation of SMM imaging may lead to substantial uncertainties, or even contradictory conclusions. Comprehensive understanding of the frequency-dependent SMM image contrast requires modeling of the surrounding RF/MW circuit and electrodynamic modeling of SMM tip-sample interactions.

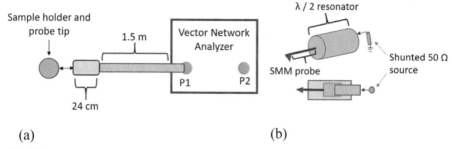

(a)   (b)

Fig. 18.2 (A) Block diagram of VNA, SMM sample holder, and cables. A 1.5-m coaxial cable is connected to one port of PNA, and the other end of the cable is connecting to the 24 cm coaxial cable. (B) EMPro model of SMM sample holder and probe tip including a 3-cm long TrLR with its first λ resonance at around 2.0 GHz, and a 50Ω shunt resistance to enhance measurement sensitivity. The tip was a cone with a base diameter of 100 μm and an end diameter of 100 nm.

18.3.1 Modeling of SMM RF/MW circuit

In our SMM system, the probe is connected to the performance network analyzer (PNA) via a 1.5-m coaxial cable connected to a 24-cm coaxial cable as shown in Fig. 18.2A. Prior to the measurements, one-port calibration has been performed at one end of the 24 cm coaxial cable, which will largely suppress the coupling interferences from these two cables. Between the tip and the end of 24 cm coaxial cable, there is a transmission line resonator (TrLR). The length of the TrLR is about 3.0 cm and is designed to have $\lambda/2$ resonance in the vicinity of 2.0 GHz. By varying the operating frequency, the TrLR resonates at a series of frequencies when the length of the TrLR equals to integer multiplies of half-($\lambda/2$).

As the impacts from the 1.5 m and 24 cm long cables have been calibrated out, the equivalent circuits of the surrounding RF/MW circuit including the tip, TrLR, and shunt resistor can be simplified as shown in Fig. 18.3A and B.

Following the transmission line theory, the complex input reflection coefficient Γ_{in} of the surrounding RF/MW circuit can be computed by

$$\Gamma = -\frac{y_L + j\tan(\beta l)}{(2 + y_L) + j(1 + 2y_L)\tan\beta l} \tag{18.1}$$

$$y_L = g_L + jb_L \tag{18.2}$$

Fig. 18.3 The equivalent transmission line models of the SMM system are shown in (A) with arbitrary length l and (B) $l = \lambda/2$.

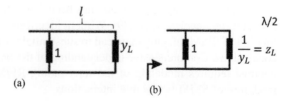

(a)   (b)

where z_0 is the characteristic impedance of the transmission line, β the wave number, and l the length of TrLR. The normalized complex load admittance y_L reflects the tip-sample interaction, where the normalized conductance g_L represents the conductive loss and the normalized susceptance b_L represents the capacitive coupling at the end of the tip. The correlations of the g_L and b_L with the samples' electromagnetic properties can be computed from the electrodynamic modeling of SMM tip-sample interactions in Section 18.3.2. It should be pointed out that Eq. (18.2) does not include materials' magnetic property. On one hand, SMM measurements are usually conducted at frequency in a few GHz, where the magnetic permeability for most magnetic materials drop dramatically due to ferromagnetic (FM) resonant effect. On the other hand, the inductance enhancement caused by materials' magnetic permeability is proportional to the dimension of SMM tip which is in the range of a few tens of nanometer. This further weakens the magnetic impacts on the SMM imaging contrast. For simplicity, Section 18.3.2 focuses discussions on modeling including samples' electrical properties only.

Due to frequency dependency (Eq. 18.1), investigation of the input reflection coefficient Γ_{in} at frequencies offset from the resonances becomes very challenging, and in many cases, these calculations are insignificant because of the slow variation of the electromagnetic properties in RF/MW frequency range. It should be mentioned that the above argument becomes invalid for materials experiencing dielectric characteristic absorption or FM resonance for which the permittivity or magnetic permeability changes abruptly with frequencies. Without losing generality, investigations of materials' electromagnetic properties are performed onset of $\lambda/2$ resonances of the TrLR. Practically, characterization performed at TrLR resonances can also lead to significant improvement of the measurement sensitivity due to the enhanced electromagnetic field. At TrLR resonances, the Γ_{in} can be simplified from Eq. (18.1) to:

$$\Gamma_{in-\lambda/2} = -\frac{y_L}{2 + y_L} \tag{18.3}$$

18.3.2 Electrodynamic modeling of SMM tip-sample interactions

It is known that the guided electromagnetic field in a transmission line propagates between the signal line and the electrical ground. The intensity distribution of the electromagnetic field is in general not even but more concentrated to the one with smaller physical size. As the size of the SMM tip (10–50 nm) is much smaller than the dimension of the electrical ground (~mm), the electromagnetic field tends to concentrate along the SMM tip down to the samples. Such an assumption is valid for most currently existing SMM systems. From modeling of the tip-sample interaction using finite element theory, the g_L and b_L in Eq. (18.2) can be determined.

Materials generally exhibit anisotropic electrical properties, that is, the conductivity and permittivity along the vertical direction (normal to the surface) are different from their in-plane counterparts (parallel to the surface). This has been widely observed in low dimensional thin film such as two-dimensional sheets. Such anisotropy properties will practically result in significant impacts for measurements performed at DC or low frequencies, like contact resistance, etc. At RF and MW frequencies, this problem can be circumvented due to the tip-sample capacitive interaction. In general, the capacitive effect is inversely proportional to film thickness and is proportional to frequency. Due to the superficial penetration depth at RF and MW frequencies, the capacitive effect (i.e., admittance) is enhanced along the vertical direction, making the tip-sample interaction as a "short circuit." This leaves the in-plane conductivity and permittivity of the sample materials the major impact on the SMM measurements. By neglecting the vertical anisotropy, low dimensional materials can be modeled as three-dimensional materials by taking the corresponding conductivity and permittivity.

Simulations of Γ_{in} at frequencies $f = 2.1$ GHz and 17.9 GHz (1st and 8th $\lambda/2$ resonant frequency of the resonator) with varied conductivity (from 1 to 10^6 S/m) and the permittivity (from 1 to 10^5) are shown in Fig. 18.4. The slight offset of the first resonant frequency from 2.0 GHz is obtained by tuning the length of the TrLR to match

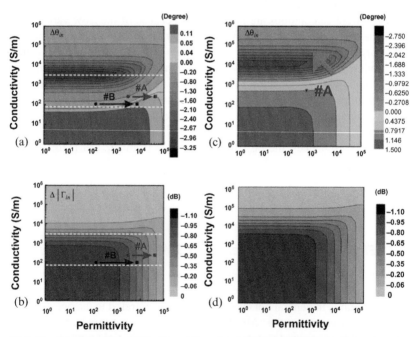

Fig. 18.4 Simulated the phase shifts (A and C: and amplitude modifications (B and D) at $f = 2.1$ and 17.9 GHz of the Γ_{in} vs the electrical conductivity (σ) and the relative permittivity (ε_r) regarding to the pristine graphene ($\sigma = 10^6$ S/m and $\varepsilon_r = 10^5$), respectively. The *solid circles* and squares in (A) and (B) are used to schematically represent the measurements carried out at $f = 2.1$ and 17.9 GHz, respectively. The greater capacitive effect at higher frequencies is modeled by increase of the permittivity. The shifts from the *solid circles to the solid squares* demonstrate the phase shifts in (A) and the magnitude modifications in (B) as frequency increases. The *dashed lines* mark the region where σ is in the range of 80–3000 S/m.

the experimental observation in Section 18.4.2. In addition, the simulations were performed by taking the tip diameter of 50 nm. Practically, the tip size is in the range between 20 and 50 nm. The simulated results were found very sensitive to the tip cross-sectional profile, which is however difficult to obtain for each individual tip and varies significantly from tip to tip. Thus, the choice of 50 nm tip diameter of the tip in the simulations, without losing generality, is close to the "worst"-scenario in demonstrating the relative change of the reflection coefficient. Further reducing the tip diameter will enhance the relative change of the reflection coefficient. The relative phase shifts ($\Delta\theta_{in}$) and the amplitude difference $\Delta|\Gamma_{in}|$ are counted from the phase and magnitude calculated at $\sigma = 10^6$ S/m and $\varepsilon_r = 10^5$. The up-limit of conductivity σ is set to coincide with the reported conductivity of pristine graphene monolayer [30], while the limit $\varepsilon_r = 10^5$ is restricted by our numerical simulation software.

From the mapping of the $\Delta|\Gamma_{in}|$ (Fig. 18.4B and D) calculated at $f = 2.1$ and 17.9 GHz, respectively, the contrast turns monotonically into more gray for smaller permittivity and conductivity. This manifests that a decrease of either the samples' conductivity or the permittivity will inevitably lead to a negative shift (NES) of $\Delta|\Gamma_{in}|$. Besides, the contour levels of the contrast shift toward up-right as frequency increases, indicating a higher gradient of $\Delta|\Gamma_{in}|$ vs the conductivity and permittivity. This means that a higher sensitivity of SMM measurements can be obtained at higher frequency, that is, more rapid variation of the amplitude contrast at higher frequency for the same changes of the σ and ε. On the other hand, the contrast of the phase variation vs samples conductivity and permittivity shows a complex manner (Fig. 18.4A and C). The mapping of the phase shifts ($\Delta\theta_{in}$) can be divided into regions "gray"—NES and "green"—positive shift (POS). This induces ambiguousness in investigation of the materials properties from SMM's phase diagrams. In the POS region, the contrast shows a monotonic variation vs the samples' permittivity and conductivity, that is, POS for lower σ and ε. This has been observed in both Fig. 18.4A and C, computed at $f = 2.1$ and 17.9 GHz, respectively. As frequency increases, the border between the NES and POS regions shift upwards, showing the possibility of phase contrast reversion, that is, at low frequency $\Delta\theta_{in}$ is in the NES region, while in higher frequency, $\Delta\theta_{in}$ shifts to POS region. This phenomenon has been found in a very good agreement with our experimental observations [31]. In the region of NES, the mapping of $\Delta\theta_{in}$ presents an "Onion" structure (ONS). The maximum of the negative-shift of $\Delta\theta_{in}$ regarding to the reference ($\sigma = 10^6$ S/m and $\varepsilon_r = 10^5$) is located at the center of ONS. Evaluating the gradient of the $\Delta|\Gamma_{in}|$ and $\Delta\theta_{in}$ vs the conductivity, it turns out the SMM imaging contrast exhibits the highest sensitivity at the center of the ONS, that is, the greatest density of the contour levels, which coincides to the observations in Ref. [30]. Since the center of the ONS can be modified by changing the measurement frequency (Fig. 18.4A and C), SMM measurements should be optimized to the proper frequency regarding to the specimen's conductivity. The contrast of $\Delta\theta_{in}$ in NES shows a radical pattern vs variation of σ and ε: (1) the phase shifts in NES exhibit monotonic dependency on the materials' permittivity as the conductivity remains the constant. The increase of permittivity leads to a phase shift less negative; (2) the phase shifts experience a maximum NES at conductivity at around 3000 S/m (marked by the upper yellow dashed line in Fig. 18.4A) when $\varepsilon < 1000$. In addition, the contour levels in this region are aligned parallel to the horizontal axis, referring to the predominant contribution by the conductivity. Permittivity starts to play as an important role when ε is above 1000. In summary, the phase shifts in SMM measurements depend on a number of factors including the variation of materials' electromagnetic properties, the frequency at which the SMM measurements carried out, the σ and ε of the reference samples.

Usually the electromagnetic properties σ and ε do not show strong frequency dependency (except dielectric characteristic absorption or FM resonance), so SMM contrast variation of the magnitude and phase at different frequencies can be analyzed by computation at a fixed frequency. Due to the large separation from the physical ground in SMM configuration, the load can be, in general, considered as an open circuit consisting of a shunt resistor and a shunt capacitor to describe the conductive and capacitive effects which are related to the conductivity and permittivity of the sample, respectively. The impacts generated by the capacitive effects on Γ_{in} are indeed determined by an admittance (ωC), where ω is the angular frequency and C is the sample induced capacitance. Hence, the increase of frequency will have the same impacts on the capacitive effects as increase of materials' permittivity (i.e., capacitance) as long as the admittance keeps consistent. Based upon the above arguments, the capacitive effects at higher frequency can be interpreted by performing simulations at lower frequencies while taking a larger value of permittivity into account.

18.4 Examples of SMM in investigations of nanostructured materials

18.4.1 Probing conductivity of grain boundary in nanostructured ferrites

Metallic FM thin films provide a unique opportunity for integrated RF/MW magnetic devices like antennas, filters, inductors, and so on, in the GHz range [32–34]. Despite continuous efforts to develop the FM materials in the past 3 decades [35–37], developing IC-compatible magnetic materials with a high permeability, high FMR, and low losses still impose a formidable challenge. Bulk MW ferrite materials have been widely used in the early RF/MW devices. However, the high processing temperature, Snoek's frequency limit, and the need for external biasing magnetic fields impede their integration into the standard complementary metal oxide semiconductor technology.

Low-temperature spin-sprayed ferrite films (Fe_3O_4) with a high self-biased magnetic anisotropy field have been reported showing a FMR frequency > 5 GHz [38]. Such films hold great potentials for RF/MW devices and find immediate applications in patch antennas and bandpass filters [39–43]. The ferrite specimen consisted of three layers: Fe_3O_4 (1.2 μm)/Photoresist (60 nm)/Fe_3O_4 (1.2 μm) [10]. AFM surface morphological image of the ferrite multilayer shows pebble-stone-shaped particles with size ranging from a few hundred nanometers to a couple of microns (Fig. 18.5).

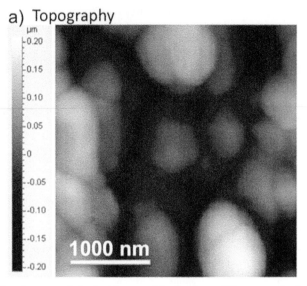

Fig. 18.5 AFM surface morphological image of the ferrite multilayer acquired simultaneously with the SMM images. The SMM system used in the measurements has different channels for surface morphology and the SMM images by recording reflection coefficient at RF/MW frequencies (i.e., laser diode for surface morphology, PNA for SMM), ensuring decoupling between the surface morphology and the SMM image.

The SMM system renders simultaneously acquired information of the amplitude and phase of s_{11}, and the surface topography. Practically, identification of the $\lambda/2$ resonances play a crucial role in investigating the SMM imaging contrast. However, calibration to the end of the SMM tip imposes a formidable technical challenge due to its tiny size and lack of calibration standards. The uncalibrated 24 cm coaxial cable (Fig. 18.2A), the TrLR, tip cantilever, and the SMM tip cause uncertainty in determination of the $\lambda/2$ resonances (Fig. 18.6). In the experiments, prior to the measurements, the SMM system was calibrated to the end of the 1.5 m coaxial cable (Fig. 18.2A). This eliminated the resonant frequencies that arise due to the long cable. The $\lambda/2$ resonant frequencies were then determined by raising the SMM tip of about 40 μm above the sample surface and then performing a frequency sweep (Fig. 18.6). As the tip is hanging in the air and the physical ground in the SMM system is 3 mm away from the SMM tip, the load can be, in general, considered as an open circuit with perturbed resistance and capacitance contributed by the sample. Thus, the following assumptions become valid: $g_L \sim 0$, $|b_L| \ll 1$, due to open circuit. And Eq. (18.3) can be simplified to

$$\Gamma_{in-\lambda/2} = \frac{e^{j\theta_{\lambda/2}}}{\sqrt{4 + b_L^2}}; \theta_{\lambda/2} = \tan^{-1}\left(\frac{2}{b_L}\right) \sim \pm\frac{\pi}{2} \tag{18.4}$$

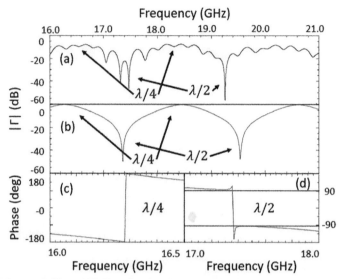

Fig. 18.6 Measured (A) and EMPro simulated (B) magnitude of the complex reflection coefficient. The phase angles of the complex reflection coefficient are calculated at $\lambda/4$ (C) and $\lambda/2$ (D). The phase angel of $\lambda/4$ resonance approaches to ±180 degree and the phase angle of $\lambda/2$ approaches to ±90 degree.

Clearly, the above equations show that after shunting a normalized resistor in Fig. 18.3A, the phase of Γ_{in} approaches to $\pm\frac{\pi}{2}$ at $\lambda/2$ resonance. This can be used as the criterion in experiments to identify the $\lambda/2$ resonances. Afterwards, the tip was lowered onto the sample and raster scanned across.

SMM measurement results of the s_{11} parameter recorded at half-wavelength resonant frequencies are shown in Fig. 18.7. It has been observed that the grain boundaries are light while the grains are dark in amplitude (Fig. 18.7A). In the phase

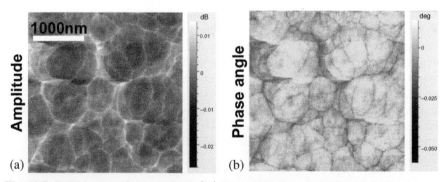

Fig. 18.7 SMM images of the amplitude $|\Gamma|$ (A) and phase angle θ (B) of the reflection coefficient recorded at $\theta = -93.4°$ and $f=2.4868$ GHz. The length of the transmission line is designed to be $\lambda/2$ at 2.5 GHz.

measurement, grain boundary turns into darker while the grains reverse to brighter (Fig. 18.7B). These observations can be explained by the electrodynamic modeling of SMM tip-sample interactions in Section 18.3.2. Reported in Ref. [44], the conductivity and permittivity of the ferrite films synthesized by spin coating are in the range of 40.0 s/m $< \sigma <$ 167.0 S/m, and $\varepsilon <$ 300.0. Compared to Fig. 18.4A, the contour levels of the phase shifts in this area are parallel to the horizontal axis, indicating the minor impacts of permittivity on the phase shifts. By omitting permittivity-induced contribution, the contrast of the SMM amplitude and phase shift will be dominant by the change of conductivity. The contrast of the amplitude and phase of the grains and their boundaries shown in Fig. 18.7A and B evidence that the conductivity of the grain boundary is higher than the grain.

This observation is consistent with the findings of Visoly-Fisher et al. [45], who has discovered a conductive path for electrons along the grain boundary, which proves to be beneficial for the performance of polycrystalline solar cells. According to the grain-boundary space charge model, the mixed ionic and electronic boundary contains a grain boundary core and two adjacent space charge regions [46–48]. Accumulation of charges in the space charge region has been revealed by a number of groups [49–51]. The accumulated charges, we believe, play an important role in the enhancement of the local grain boundary conductivity, leading to the contrast in the SMM image. It is worth mentioning that such enhancement only occurs along the grain boundary. Perpendicular to the grain boundary, the two adjacent space charge regions can be considered as double Schottky barrier [46], blocking current flow from grain to grain, and imposing high resistance cross the grain boundary.

18.4.2 Micro-patterned graphene

Graphene has attracted tremendous interests in RF and MW electronic devices including graphene-based field effect transistors [52–60], resonators [61,62], switches [63], mixers [64–66], and transmission lines [67–69], because of its unique electrical properties, which are high carrier mobility [70–72], high saturation velocity [73], large critical current density [64], and exceedingly large permittivity at RF/MW [65]. A graphene-transistor with a self-aligned gate has been demonstrated with a cut-off frequency above 300 GHz [54,55]. The demand for large-scale integration makes the top-down lithographic process the prevailing technology in fabrication of graphene-based devices at micro- and nano-meter scale. Due to the feature of single atomic layer, the electrical properties of graphene monolayer are very sensitive to structural defects which are inevitably induced during the devices' process. The conductivity of micro-patterned graphene monolayer has been found to be exponentially decreased as the etching-induced defects density increases at low frequencies ($<$1000 Hz) [66,74]. To date, understanding of the structure–property correlation of graphene monolayer is still missing. Investigation of etching-induced surface impedance modulation in graphene monolayer has been reported using SMM. Monolayer graphene was grown on a 100-μm thick Cu foil (Graphene Supermarket) by chemical vapor deposition (CVD). The transferred CVDG shows a sheet resistance of around 1200 Ω/□ measured by Hall measurements. The conductivity σ of graphene

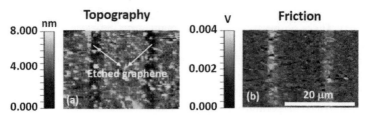

Fig. 18.8 Surface topography (A) and friction (B) of etched graphene grating. The etched graphene shows a reduced step height in the surface morphology (A), while the unetched graphene (i.e., pristine graphene) strips presented a smoother surface than the etched graphene characterized by using the lateral force microscope (B).

monolayer is then estimated around 10^6 S/m by taking the thickness of 1.0 nm [75]. Later, the graphene monolayer was transferred onto a Si (550 μm)/SiO$_2$ (300 nm) substrate and structured into grating (Fig. 18.8). Two samples (#A and #B) were fabricated with varied etching time of 10 s for #A and 30 s for #B. AFM surface morphology images of #A show a reduced step height and a rougher surface of the etched graphene.

SMM images of #A were recorded at a series of half-wavelength ($\lambda/2$) harmonic resonances (Fig. 18.9): first harmonic resonance ($f=2.1$ GHz) (Fig. 18.9A and B);

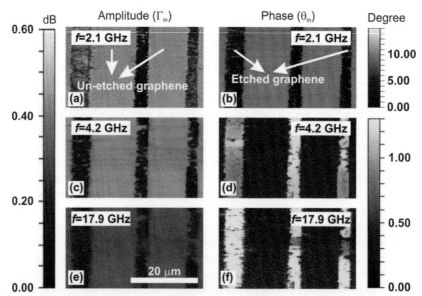

Fig. 18.9 SMM images of sample #A measured at various half-wavelength harmonic resonances: amplitude $|\Gamma_{in}|$ (A) and phase angle θ_{in} (B) at first harmonic resonance ($f=2.1$ GHz); amplitude (C) and phase (D) at second harmonic resonance ($f=4.2$ GHz); amplitude (E) and phase (F) at ninth harmonic resonance ($f=17.9$ GHz). The imaging contrast of phase at $f=2.1$ GHz (B) is reversed from (D) and (F) recorded at $f=4.2$ and 17.9 GHz, respectively.

second harmonic resonance ($f=4.2\,$GHz) (Fig. 18.9C and D); and ninth harmonic resonance ($f=17.9\,$GHz) (Fig. 18.9E and F). As the Γ_{in} is a complex number at RF/MW frequencies, both its amplitude ($|\Gamma_{in}|$) (Fig. 18.9A, C, and E) and phase angle (θ_{in}) (Fig. 18.9B, D, and F) were recorded. The imaging contrast is formed by the difference of amplitude between the etched graphene (Γ_{in-A}^{et}) and the unetched graphene (Γ_{in-A}^{unet}) and the phase angle between the etched graphene (θ_{in-A}^{et}) and the unetched graphene (θ_{in-A}^{unet}). Prior to the measurements, the PNA was calibrated to identify the half-wavelength resonances by getting rid of the unnecessary interferences from the surrounding cable connections. The uncertainty of resonant frequency determination has been improved to $\pm0.1\,$GHz. At $\lambda/2$ harmonic resonances, the unetched graphene exhibits a greater amplitude $|\Gamma_{in}|$ over the etched graphene strips at all three frequencies. Imaging contrast reversion has been observed in the SMM phase images as frequency increased from 2.1 to 4.2 GHz and 17.9 GHz (Fig. 18.9B, D, and F): $\theta_{in-A}^{unet} > \theta_{in-A}^{et}$ at $f=2.1\,$GHz, $\theta_{in-A}^{unet} < \theta_{in-A}^{et}$ at $f=4.2$ and 17.9 GHz. The smaller magnitudes of $|\Gamma_{in}|$ measured at different frequencies evidence the lower conductivity and permittivity of the etched graphene from the pristine graphene due to the monotonous change of $|\Gamma_{in}|$ vs conductivity and permittivity as shown in Fig. 18.4A and C. The variation of the SMM phase imaging contrast can be explained by the transition marked in Fig. 18.4A, because the enhanced capacitive effects at higher frequencies can be interpreted by simulations at lower frequencies while taking into account a larger value of permittivity. Thus, the greater capacitive effects at higher frequencies will favorite in moving $\triangle\theta_{in}$ toward a larger permittivity, resulting in shifting the position on Fig. 18.4A from the negative (NES) to the positive region (POS). Further investigations of the impact of the etching-induced defects on the electrical properties of graphene monolayer were performed on sample with longer etching time.

Further investigation of the impact of the etching-induced defects on the electrical properties of graphene monolayer was performed on sample #B with longer etching time. The SMM measurements show similar imaging contrast between the unetched and etched graphene as those for sample #A at all harmonic resonances. Figs. 18.10 and 18.11 compared the $\triangle|\Gamma_{in}|$ and $\triangle\theta_{in}$ of samples #A and #B measured at frequencies $f=2.1$ and 17.9 GHz, respectively. Results are summarized in Table 18.1. At $f=2.1\,$GHz, #B shows a 0.4 dB reduction of $\triangle|\Gamma_{in}|$, and $\sim$3.0 degree increase (less negative) of $\triangle\theta_{in}$ over #A. As marked in Fig. 18.4B, the reduction of $\triangle|\Gamma_{in}|$ indicates the conductivity and permittivity of the etched graphene in #B are further reduced due to the longer etching time. To achieve the increase of $\triangle\theta_{in}$ (less negative) from #A to #B, the conductivity of the etched graphene in #A and #B should be in the range between 80 and 3000 S/m (Fig. 18.4A). Because only in this region (marked as dashed lines in Fig. 18.4A), the $\triangle\theta_{in}$ can be less negative as the conductivity and permittivity become smaller.

When frequency is increased to $f=17.9\,$GHz, #B shows a 0.5 dB reduction of $\triangle|\Gamma_{in}|$, and $\sim$4.5 degree increase (more positive) of $\triangle\theta_{in}$ over #A. The observations are in very good agreement with the numerical simulations: (i) the lowered $\triangle|\Gamma_{in}|$ of #B coincides with the simulations by taking a further reduced conductivity and permittivity of etched graphene in #B into account (Fig. 18.4B); (ii) the measured positive values of $\triangle\theta_{in}$ for both #A and #B can be interpreted by the greater capacitive effects at higher frequencies, as schematically shown on Fig. 18.4A; and (iii) the observation

Fig. 18.10 Comparison of the magnitude modification $\Delta|\Gamma_{in}|$ (A), and the phase shift $\Delta\theta_{in}$ (B) of the complex input reflection coefficient Γ_{in} of samples #A and #B at $f=2.1\,\text{GHz}$.

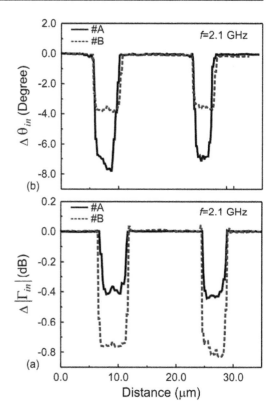

of a larger positive value of $\Delta\theta_{in}$ in #B than in #A is also demonstrated by the simulations (Fig. 18.4A). By correlating all the SMM imaging contrast of the $\Delta|\Gamma_{in}|$ and $\Delta\theta_{in}$, #B shows lower conductivity and permittivity of the etched graphene than #A, which may be caused by the augmentation of the induced structural disorders due to elongated etching time. In addition, the conductivity of the etched graphene of #A and #B should be in the range between 80 and 3000 S/m, which is more than three orders of magnitude reduction compared to the pristine graphene.

The SMM measurements show similar imaging contrast between the unetched and etched graphene as those for sample with longer etching time at all harmonic resonances. Oxygen plasma etching-induced defects in graphene monolayer has been found to cause considerable surface impedance modification, leading to significant enhancement of the SMM imaging contrast between the unetched and etched graphene layer. Increase of the etching time and consequently the higher defect density results in significant enhancement of the amplitude and phase of the input reflection coefficient, that is, $|\Gamma_{in}|$ and θ_{in}. This coincides with the observations in Raman spectroscopy measurements. The reversion of the SMM imaging contrast recorded at low and high frequencies has been observed, demonstrating that the SMM imaging contrast is controlled by both the conductivity and permittivity. It has been found that the etching-induced defects result in more than 3 orders of magnitude reduction in conductivity compared to the pristine graphene.

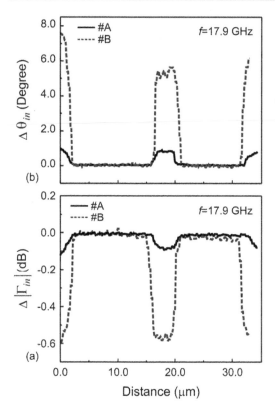

Fig. 18.11 Comparison of the magnitude modification $\Delta|\Gamma_{in}|$ (A), and the phase shift $\Delta\theta_{in}$ (B) of the complex input reflection coefficient Γ_{in} of samples #A and #B at $f=17.9\,\text{GHz}$.

Table 18.1 Comparison of $\Delta|\Gamma_{in}|$ and $\Delta\theta_{in}$ between #A and #B measured at $f=2.1$ and 17.9 GHz.

Sample	$f=2.1\,\text{GHz}$		$f=17.9\,\text{GHz}$	
	$\Delta|\Gamma_{in}|$ (dB)	$\Delta\theta_{in}$ (degree)	$\Delta|\Gamma_{in}|$ (dB)	$\Delta\theta_{in}$ (degree)
A	−0.4	−6.8	−0.1	0.8
B	−0.8	−3.7	−0.6	5.3

18.5   Summary

Near-field SMM has proven to be a unique technique by implementing RF/MW microelectronics in nanometer scale analysis. The non-invasive and non-destructive in situ MIG technique can find applications in the investigation of broad range of materials in versatile environments. In addition to semiconductor materials, which has been studied most in the past using SMM, its breadth may extend to soft materials, biomaterials, living cells, and so on. The carrier frequency in SMM is usually between

1 and 20 GHz, corresponding to a few pico-seconds. This allows SMM be employed in characterization of radioactive materials, highly reactive, and toxic materials.

Much of the work done in the past is to perform qualitative investigations of the SMM images due to the lack of precise calibration technique at RF/MW frequencies. The general calibration in MW engineering relies on fabrication of calibration standards, which however can not be applied directly to SMM. Due to the tiny size, the SMM tip will inevitably obtain slight damage during the scanning in contact-mode measurement. This will lead to the SMM system drift away from the calibration. Developing of advanced calibration standards, SMM tip and the corresponding calibration methods are highly demanded. For non-contact mode measurements, calibration method for quantitative analysis does not exist due to the radiation effect at the tip of the SMM probe. Developing of reliable calibration technique presents a high risk challenge in the future for both contact- and non-contact mode characterization in versatile environments.

Combining the use of near-field approach and high carrier frequency makes it viable for SMM to probe the structural information and physical–chemical properties under the surface. A few micron meters penetration depth at RF/MW frequencies and the nanometer spatial resolution are crucial in detection of hardware Trojans added by malicious foundry. This opens opportunity for SMM to be the major instrument in developing technology-based solutions for built-in trust measures, provided protection of semiconductor IP at various levels of abstraction and manufacturing of trusted electronics are becoming the fast growth field in the next decades.

References

[1] S. Berweger, T.M. Wallis, A. Kabos, Nanoelectronic characterization, IEEE Microw. Mag. (2020) 36.
[2] K.C. Kao, Dielectric Phenomena in Solids, Elsevier Academic Press, London, 2004.
[3] A.G. Gurevich, G.A. Melkov, Magnetization Oscillations and Waves, vol. 24–25, CRC, Boca Raton, FL, 1996.
[4] A. Tselev, Near-field microwave microscopy: subsurface imaging for in situ characterization, IEEE Microw. Mag. 21 (10) (Oct. 2020) 72–86, https://doi.org/10.1109/MMM.2020.3008241.
[5] G.T. Becker, F. Regazzoni, C. Paar, W.P. Burleson, Stealthy Dopant-Level Hardware Trojans, 2013, pp. 197–214.
[6] T. Sugawara, et al., Reversing stealthy dopant-level circuits, J. Cryptogr. Eng. 5 (2) (2015) 85–94, https://doi.org/10.1007/s13389-015-0102-5.
[7] C.A. Bryant, J.B. Gunn, Noncontact technique for the local measurement of semiconductor resistivity, Rev. Sci. Instrum. 36 (11) (1965) 1614–1617, https://doi.org/10.1063/1.1719404.
[8] E.A. Ash, G. Nicholls, Super-resolution aperture scanning microscope, Nature 237 (5357) (1972) 510–512, https://doi.org/10.1038/237510a0.
[9] M. Fee, S. Chu, T.W. Hänsch, Scanning electromagnetic transmission line microscope with sub-wavelength resolution, Opt. Commun. 69 (3–4) (1989) 219–224, https://doi.org/10.1016/0030-4018(89)90103-X.

[10] Y. Xing, J. Myers, O. Obi, N.X. Sun, Y. Zhuang, Scanning microwave microscopy char-acterization of spin-spray-deposited ferrite/nonmagnetic films, J. Electron. Mater. 41 (3) (2012) 530–534, https://doi.org/10.1007/s11664-011-1874-8.

[11] V.V. Talanov, A. Scherz, A.R. Schwartz, Noncontact electrical metrology of Cu/low-k interconnect for semiconductor production wafers, Appl. Phys. Lett. 88 (26) (2006) 262901, https://doi.org/10.1063/1.2216898.

[12] A. Tselev, et al., Mesoscopic metal–insulator transition at ferroelastic domain walls in VO2, ACS Nano 4 (8) (2010) 4412–4419, https://doi.org/10.1021/nn1004364.

[13] J. Myers, T. Nicodemus, Y. Zhuang, T. Watanabe, N. Matsushita, M. Yamaguchi, Char-acterization of grain boundary conductivity of spin-sprayed ferrites using scanning micro-wave microscope, J. Appl. Phys. 115 (17) (2014) 17A506, https://doi.org/10.1063/1.4860941.

[14] J. Lee, C.J. Long, H. Yang, X.-D. Xiang, I. Takeuchi, Atomic resolution imaging at 2.5 GHz using near-field microwave microscopy, Appl. Phys. Lett. 97 (18) (2010) 183111, https://doi.org/10.1063/1.3514243.

[15] A.O. Oladipo, et al., Analysis of a transmission mode scanning microwave microscope for subsurface imaging at the nanoscale, Appl. Phys. Lett. 105 (13) (2014) 133112, https://doi.org/10.1063/1.4897278.

[16] C. Plassard, et al., Detection of defects buried in metallic samples by scanning microwave microscopy, Phys. Rev. B Condens. Matter Mater. Phys. 83 (12) (2011) 2–5, https://doi.org/10.1103/PhysRevB.83.121409.

[17] L. You, J.-J. Ahn, Y.S. Obeng, J.J. Kopanski, Subsurface imaging of metal lines embedded in a dielectric with a scanning microwave microscope, J. Phys. D Appl. Phys. 49 (4) (2016) 045502, https://doi.org/10.1088/0022-3727/49/4/045502.

[18] C. Gao, X.-D. Xiang, Quantitative microwave near-field microscopy of dielectric proper-ties, Rev. Sci. Instrum. 69 (11) (1998) 3846–3851, https://doi.org/10.1063/1.1149189.

[19] D.E. Steinhauer, et al., Quantitative imaging of dielectric permittivity and tunability with a near-field scanning microwave microscope, Rev. Sci. Instrum. 71 (7) (2000) 2751–2758, https://doi.org/10.1063/1.1150687.

[20] J.H. Lee, S. Hyun, K. Char, Quantitative analysis of scanning microwave microscopy on dielectric thin film by finite element calculation, Rev. Sci. Instrum. 72 (2) (2001) 1425, https://doi.org/10.1063/1.1342032.

[21] C. Gao, B. Hu, P. Zhang, M. Huang, W. Liu, I. Takeuchi, Quantitative microwave eva-nescent microscopy of dielectric thin films using a recursive image charge approach, Appl. Phys. Lett. 84 (2004) 4647, https://doi.org/10.1063/1.1759389.

[22] Z.Y. Wang, M.A. Kelly, Z.X. Shen, L. Shao, W.K. Chu, H. Edwards (Eds.), Quantitative measurement of sheet resistance by evanescent microwave probe, Appl. Phys. Lett 86 (2005), https://doi.org/10.1063/1.1891296, 153118.

[23] A.N. Reznik, E.V. Demidov, Quantitative determination of sheet resistance of semicon-ducting films by microwave near-field probing, J. Appl. Phys. 113 (9) (2013) 094501, https://doi.org/10.1063/1.4794003.

[24] H.P. Huber, et al., Calibrated nanoscale capacitance measurements using a scanning microwave microscope, Rev. Sci. Instrum. 81 (11) (2010) 113701, https://doi.org/10.1063/1.3491926.

[25] G. Gramse, M. Kasper, L. Fumagalli, G. Gomila, P. Hinterdorfer, F. Kienberger, Cali-brated complex impedance and permittivity measurements with scanning microwave microscopy, Nanotechnology 25 (14) (2014) 145703, https://doi.org/10.1088/0957-4484/25/14/145703.

[26] C. Balusek, B. Friedman, D. Luna, B. Oetiker, A. Babajanyan, K. Lee, A three-dimensional finite element model of near-field scanning microwave microscopy, J. Appl. Phys. 112 (8) (2012) 084318, https://doi.org/10.1063/1.4759253.

[27] K. Lai, W. Kundhikanjana, M. Kelly, Z.X. Shen, Modeling and characterization of a cantilever-based near-field scanning microwave impedance microscope, Rev. Sci. Instrum. 79 (6) (2008) 063703, https://doi.org/10.1063/1.2949109.

[28] A.N. Reznik, Electromagnetic model for near-field microwave microscope with atomic resolution: determination of tunnel junction impedance, Appl. Phys. Lett. 105 (8) (2014) 083512, https://doi.org/10.1063/1.4894369.

[29] Y. Xing, J. Myers, O. Obi, N.X. Sun, Y. Zhuang, Excessive grain boundary conductivity of spin-spray deposited ferrite/non-magnetic multilayer, J. Appl. Phys. 111 (7) (2012) 07A512, https://doi.org/10.1063/1.3676242.

[30] J. Myers, S. Mou, K.-H. Chen, Y. Zhuang, Scanning microwave microscope imaging of micro-patterned monolayer graphene grown by chemical vapor deposition, Appl. Phys. Lett. 108 (5) (2016) 053101, https://doi.org/10.1063/1.4940991.

[31] Z. Ji, et al., Microwave imaging of etching-induced surface impedance modulation of graphene monolayer, J. Vac. Sci. Technol. A 36 (5) (2018) 05G508, https://doi.org/10.1116/1.5035417.

[32] Y. Zhuang, M. Vroubel, B. Rejaei, J.N. Burghartz, No title, in: Techn. Dig. IEDM2002, vol. 475, 2011.

[33] M. Yamaguchi, T. Kuribara, K.-I. Arai, Two-port type ferromagnetic RF integrated inductor, in: 2002 IEEE MTT-S International Microwave Symposium Digest (Cat. No. 02CH37278), 2002, pp. 197–200, https://doi.org/10.1109/MWSYM.2002.1011592.

[34] V. Korenivski, R.B. van Dover, Design of high frequency inductors based on magnetic films, IEEE Trans. Magn. 34 (4) (1998) 1375–1377, https://doi.org/10.1109/20.706553.

[35] P. Wang, H. Zhang, R. Divan, A. Hoffmann, Tailoring high-frequency properties of permalloy films by submicrometer patterning, IEEE Trans. Magn. 45 (1) (2009) 71–74, https://doi.org/10.1109/TMAG.2008.2005330.

[36] T.J. Klemmer, K.A. Ellis, L.H. Chen, B. van Dover, S. Jin, Ultrahigh frequency permeability of sputtered Fe–Co–B thin films, J. Appl. Phys. 87 (2) (2000) 830–833, https://doi.org/10.1063/1.371949.

[37] S.X. Wang, N.X. Sun, M. Yamaguchi, S. Yabukami, Properties of a new soft magnetic material, Nature 407 (6801) (2000) 150–151, https://doi.org/10.1038/35025142.

[38] K. Kondo, S. Yoshida, H. Ono, M. Abe, Spin sprayed Ni(–Zn)–Co ferrite films with natural resonance frequency exceeding 3GHz, J. Appl. Phys. 101 (9) (2007) 09M502, https://doi.org/10.1063/1.2710465.

[39] G.-M. Yang, et al., Electronically tunable miniaturized antennas on magnetoelectric substrates with enhanced performance, IEEE Trans. Magn. 44 (11) (2008) 3091–3094, https://doi.org/10.1109/TMAG.2008.2003062.

[40] G.-M. Yang, et al., Tunable miniaturized patch antennas with self-biased multilayer magnetic films, IEEE Trans. Antennas Propag. 57 (7) (2009) 2190–2193, https://doi.org/10.1109/TAP.2009.2021972.

[41] G.-M. Yang, et al., Planar annular ring antennas with multilayer self-biased NiCo-ferrite films loading, IEEE Trans. Antennas Propag. 58 (3) (2010) 648–655, https://doi.org/10.1109/TAP.2009.2039295.

[42] G.-M. Yang, et al., Miniaturized antennas and planar bandpass filters with self-biased NiCo-ferrite films, IEEE Trans. Magn. 45 (10) (2009) 4191–4194, https://doi.org/10.1109/TMAG.2009.2023996.

[43] G.M. Yang, et al., Planar circular loop antennas with self-biased magnetic film loading, Electron. Lett. 44 (5) (2008) 332, https://doi.org/10.1049/el:20080200.

[44] J. Myers, T. Nicodemus, Y. Zhuang, T. Watanabe, N. Matsushita, M. Yamaguchi, Characterization of grain boundary conductivity of spin-sprayed ferrites using scanning microwave microscope, J. Appl. Phys. 115 (2014) 17A506, https://doi.org/10.1063/1.4860941.

[45] I. Visoly-Fisher, S.R. Cohen, K. Gartsman, A. Ruzin, D. Cahen, Understanding the beneficial role of grain boundaries in polycrystalline solar cells from single-grain-boundary scanning probe microscopy, Adv. Funct. Mater. 16 (5) (2006) 649–660, https://doi.org/10.1002/adfm.200500396.

[46] X. GUO, R. WASER, Electrical properties of the grain boundaries of oxygen ion conductors: acceptor-doped zirconia and ceria, Prog. Mater. Sci. 51 (2) (2006) 151–210, https://doi.org/10.1016/j.pmatsci.2005.07.001.

[47] H.L. Tuller, J. Maier, Solid State Ionics, vol. 131, 2000, p. 1.

[48] S. Kim, J. Maier, On the conductivity mechanism of nanocrystalline ceria, J. Electrochem. Soc. 149 (10) (2002) J73, https://doi.org/10.1149/1.1507597.

[49] S.N. Flengas, T.H. Estell, The electrical properties of solid oxide electrolytes, Chem. Rev. 70 (1970) 339.

[50] S.S. Lion, W.L. Worrell, Electrical properties of novel mixed-conducting oxides, Appl. Phys. A Solid Surf. 49 (1) (1989) 25–31, https://doi.org/10.1007/BF00615461.

[51] F. Capel, C. Moure, P. Durán, A.R. González-Elipe, A. Caballero, Titanium local environment and electrical conductivity of TiO2-doped stabilized tetragonal zirconia, J. Mater. Sci. (2000) 345–352.

[52] Y. Wu, et al., High-frequency, scaled graphene transistors on diamond-like carbon, Nature 472 (7341) (2011) 74–78, https://doi.org/10.1038/nature09979.

[53] K.L.S. Inanc Meric, N. Baklitskaya, P. Kim, RF performance of top-gated, zero-bandgap graphene field-effect transistors, in: 2008 IEEE Int. Electron. Devices Meet, 2008.

[54] Y. Wu, et al., State-of-the-art graphene high-frequency electronics, Nano Lett. 12 (6) (2012) 3062–3067, https://doi.org/10.1021/nl300904k.

[55] L. Liao, et al., High-speed graphene transistors with a self-aligned nanowire gate, Nature 467 (7313) (2010) 305–308, https://doi.org/10.1038/nature09405.

[56] Y.-M. Lin, C. Hsin-Ying, K.A. Jenkins, D.B. Farmer, P. Avouris, A. Valdes-Garcia, Dual-gate graphene FETs with f_{T} of 50 GHz, IEEE Electron. Device Lett. 31 (1) (2010) 68–70, https://doi.org/10.1109/LED.2009.2034876.

[57] Y.-M. Lin, C. Dimitrakopoulos, K.A. Jenkins, D.B. Farmer, H.-Y. Chiu, A. Grill, P.H. Avouris, 100-GHz transistors from wafer-scale epitaxial graphene, Science 327 (5966) (2010) 662, doi:10.1126/science.1184289.

[58] Y.-M. Lin, K.A. Jenkins, A. Valdes-Garcia, J.P. Small, D.B. Farmer, P. Avouris, Operation of graphene transistors at gigahertz frequencies, Nano Lett. 9 (1) (2009) 422–426, https://doi.org/10.1021/nl803316h.

[59] J.S. Moon, et al., Epitaxial-graphene RF field-effect transistors on Si-face 6H-SiC substrates, IEEE Electron. Device Lett. 30 (6) (2009) 650–652, https://doi.org/10.1109/LED.2009.2020699.

[60] J. Lee, et al., RF performance of pre-patterned locally-embeddedback-gate graphene device, in: Proc. IEDM Tech. Dig., San Francisco, CA, USA, 2010, pp. 23.5.1–23.5.4.

[61] J.S. Bunch, et al., Electromechanical resonators from graphene sheets, Science 315 (5811) (2007) 490, https://doi.org/10.1126/science.1136836.

[62] Y. Xu, et al., Radio frequency electrical transduction of graphene mechanical resonators, Appl. Phys. Lett. 97 (24) (2010) 243111, https://doi.org/10.1063/1.3528341.

[63] K.M. Milaninia, M.A. Baldo, A. Reina, J. Kong, All graphene electromechanical switch fabricated by chemical vapor deposition, Appl. Phys. Lett. 95 (18) (2009) 183105, https://doi.org/10.1063/1.3259415.

[64] R. Murali, Y. Yang, K. Brenner, T. Beck, J.D. Meindl, Breakdown current density of graphene nanoribbons, Appl. Phys. Lett. 94 (24) (2009) 243114, https://doi.org/10.1063/1.3147183.

[65] Y. Wu, et al., Microwave transmission properties of chemical vapor deposition graphene, Appl. Phys. Lett. 101 (5) (2012) 053110, https://doi.org/10.1063/1.4737424.

[66] K. Kim, H.J. Park, B.-C. Woo, K.J. Kim, G.T. Kim, W.S. Yun, Electric property evolution of structurally defected multilayer graphene, Nano Lett. 8 (10) (2008) 3092–3096, https://doi.org/10.1021/nl8010337.

[67] D.-Y. Jeon, et al., Radio-frequency electrical characteristics of single layer graphene, Jpn. J. Appl. Phys. 48 (9) (2009) 091601, https://doi.org/10.1143/JJAP.48.091601.

[68] H.-J. Lee, E. Kim, J.-G. Yook, J. Jung, Intrinsic characteristics of transmission line of graphenes at microwave frequencies, Appl. Phys. Lett. 100 (22) (2012) 223102, https://doi.org/10.1063/1.4722585.

[69] H.-J. Lee, et al., Radio-frequency characteristics of graphene monolayer via nitric acid doping, Carbon NY 78 (2014) 532–539, https://doi.org/10.1016/j.carbon.2014.07.037.

[70] K.S. Novoselov, et al., Electric field effect in atomically thin carbon films, Science 306 (5696) (2004) 666–669, https://doi.org/10.1126/science.1102896.

[71] E.V. Castro, et al., Limits on charge carrier mobility in suspended graphene due to flexural phonons, Phys. Rev. Lett. 105 (26) (2010) 266601, https://doi.org/10.1103/PhysRevLett.105.266601.

[72] A.S. Mayorov, et al., Micrometer-scale ballistic transport in encapsulated graphene at room temperature, Nano Lett. 11 (6) (2011) 2396–2399, https://doi.org/10.1021/nl200758b.

[73] I. Meric, M.Y. Han, A.F. Young, B. Ozyilmaz, P. Kim, K.L. Shepard, Current saturation in zero-bandgap, top-gated graphene field-effect transistors, Nat. Nanotechnol. 3 (11) (2008) 654–659, https://doi.org/10.1038/nnano.2008.268.

[74] D.C. Kim, D.-Y. Jeon, H.-J. Chung, Y. Woo, J.K. Shin, S. Seo, The structural and electrical evolution of graphene by oxygen plasma-induced disorder, Nanotechnology 20 (37) (2009) 375703, https://doi.org/10.1088/0957-4484/20/37/375703.

[75] P. Nemes-Incze, Z. Osváth, K. Kamarás, L.P. Bíró, Anomalies in thickness measurements of graphene and few layer graphite crystals by tapping mode atomic force microscopy, Carbon NY 46 (11) (2008) 1435–1442, https://doi.org/10.1016/j.carbon.2008.06.022.

Thermal analysis of nanoparticles: Methods, kinetics, and recent advances ☆

19

Elisabeth Mansfield[a] and Mark Banash[b]
[a]Applied Chemicals and Materials Division, National Institute of Standards and Technology, Boulder, CO, United States, [b]Neotericon LLC, Bedford, NH, United States

19.1   Introduction

Nanoparticles continue to remain an important and diverse class of materials with applications in all major areas of the economy, including manufacturing, medicine and energy, with new products constantly finding their way to market. The Project on Emerging Nanotechnologies reported over 1600 nano-containing consumer products on the market as of October 2013, as compared to 54 products in 2005 when the first edition of this book was published [1]. Many of today's nanoparticle containing product databases are now retired or obsolete, although the Dutch Nanodatabase (www.nanodb.dk) seems to be the most up-to-date reporting over 5000 nanomaterial-containing products available in Europe [2]. The main classes of nanoparticles utilized in products include carbon-based (nanotubes and fullerenes), metallic (including gold, silver, iron and copper), metal oxide (iron oxide, ZnO, TiO_2, CeO_2, SiO_2 among others), and quantum dots [3]. A quick review of the literature shows there are endless possibilities of nanoparticle coatings that can be applied to the surfaces of these particles in order to improve solubility, improve biocompatibility, target interactions with other molecules, or prevent interactions with the nanoparticle's environment [4]. In fact, most nanoparticles used in commercial products have employed surface modifications such as an oxide layer or organic functional groups. More complex nanoparticles, specifically those targeted for drug delivery, continue to grow more advanced in their composition and ability to perform a given task [5]. Analytical characterization of the materials, including the core, coatings, and any other factors that may impact nanoparticle use, are essential to understanding their potential in a commercial product, evaluating the effects of their release into the environment, or developing them for biological or medical applications.

Nanoparticle characterization is not a trivial task and becomes more difficult as we advance our technology [4]. The methods available for bulk-scale analysis of materials are not necessarily applicable to nanoparticles. Because of these and sample preparation challenges, complementary analytical techniques are often used to confirm the

---

☆ Contribution of NIST, an agency of the US government; not subject to copyright in the United States.

Modeling, Characterization and Production of Nanomaterials. https://doi.org/10.1016/B978-0-12-819905-3.00019-1

values of specific measurements [6]. Some of the most common methods for analysis include microscopy (scanning electron microscopy, transmission electron microscopy), spectroscopy (Raman, UV/visible, Fourier-transform infrared spectroscopy (FTIR), fluorescence), elemental characterization (neutron activation analysis, inductively coupled plasmon techniques) and particle sizing methods (dynamic light scattering, field flow fractionation) [3].

Nanoparticle characterization techniques and measured properties.

Technique	Properties
Scanning electron microscopy	Particle sizes, morphology, degree of agglomeration/aggregation/networking
Transmission electron microscopy	Particle internal structure, especially crystallinity
Raman spectroscopy	Vibrational properties, sizes for very small particles, order/disorder
UV/visible spectroscopy	Electronic structure
Fourier transform infrared spectroscopy	Surface chemistry, especially chemical groups
Magnetic coercivity	Magnetic properties, presence of magnetic domains
Light scattering	Particle sizes/shapes, esp./for larger particles (>100 nm)

In general, microscopy methods provide a means for visualizing the sample's size, shape, and morphology. Spectroscopy and elemental characterization are often used to identify the nanoparticles' elemental and structural characteristics, and particle sizing methods yield information about size and shape. Many standard protocols have been adapted, when necessary, to address the unique challenges of small particle size and limited sample volumes that come with nanoparticle analysis. The above table is not a complete list of techniques or their measurable results but is an introduction to the analytical data most commonly reported.

Thermal analysis methods, which measure the properties of the sample because of change in temperature or heat flow, are often used to provide quick information on nanoparticles in the manufacturing or laboratory setting. A thermal investigation can provide information about particle and coating composition, crystallinity, and formation kinetics. Thermogravimetric analysis (TGA), calorimetry, and differential scanning calorimetry are the most common thermal analysis methods utilized for nanoparticle work and are discussed here in more detail. The application of these methods can be direct in some cases using the same procedures developed for bulk-scale materials. However, with small quantities of material, it may be necessary to concentrate and dry a sample to ensure the minimum sample volume for the analytical technique is met (typically ~10 mg). Future applications of thermal analysis methods for nanoparticle samples are discussed at the end of the chapter.

19.2 Thermal analysis methods

19.2.1 Thermogravimetric analysis (TGA)

Thermogravimetric analysis (TGA) describes the process in which the sample mass is monitored as it is heated at a controlled rate in a controlled atmosphere. The sample is introduced onto a balance within a furnace and heated using a specific temperature program. The atmosphere can be reactive (e.g., oxygen, air), inert (e.g., argon, nitrogen) or vacuum. The major data outputs are thermograms, either the % weight loss with time or temperature. First and second derivatives of the % weight loss vs temperature thermogram are heavily utilized for data analysis.

As a result of the heating, the sample mass will either increase or decrease as a result of changes in the material. Some of the most commonly observed transitions include desorption of an adsorbed species such as water, oxidation of metal species, decomposition via pyrolysis, and even combustion in the case of carbonaceous materials. Further information can be obtained by coupling the TGA to an evolved-gas analysis technique, such as FTIR, mass spectrometry (MS), or gas chromatography mass spectrometry (GC-MS). A key TGA measurement is the residual mass (Mres) of the sample at a given temperature, usually given as a percentage of the original sample mass:

$$Mres\ (\%) = \frac{M_{r,T}}{M_s}\ 100\%$$

where $Mres$ (%) is the residual mass expressed as a percentage, $M_{r,T}$ is the mass remaining at a specific temperature (T), and M_s is the initial mass of the sample. At the end of an oxidizing sample run, the residual mass is often referred to as the ash content. Other important TGA metrics are the temperatures in which events occur. Temperatures of importance can be determined by differentiating the thermogram with respect to temperature or time and identifying the peaks. This can show features that may not be easily discerned or quantified in the thermogram (e.g., shoulders). The onset temperature at which major transitions occur may also be quantified. Step-by-step methods for using TGA to quantitate the amount of surface coating on a nanoparticle have been published [7].

19.2.2 Differential scanning calorimetry (DSC)

Calorimetry is the measurement of the evolution or absorption of heat during chemical reactions, phase transitions, and other physical changes. Thermodynamic properties such as changes in enthalpy or entropy can be obtained. In the case where a chemical reaction occurs, kinetic data such as reaction rate and activation energy can be determined. Many types of calorimetry can be applied to nanoparticle samples, but Differential Scanning Calorimetry (DSC) is perhaps the most widely used. In DSC, the difference in heat flow between two cells, one containing the sample and the other, a reference material, is determined. The sample and reference cells are heated independently such that the temperature difference is monitored and maintained at zero; that is $\Delta T = T_S - T_R = 0$. As a material undergoes a phase transition, the amount of heat needed to maintain a constant sample temperature will vary, depending on whether the transition is endothermic or exothermic. The difference in energy needed to match the sample temperature to the reference is the amount of excess heat released or absorbed as a result of sample transitions. In the most basic form, the information from DSC can be described by

$$\frac{dH}{dt} = mC_p \frac{dT}{dt}$$

where dH/dt is the DSC heat flow signal, m is the sample mass, C_p is the sample heat capacity, and dT/dt is the heating rate [8]. Physical changes in the material, such as melting, crystallization, and glass transitions, can all be observed using DSC. As differential thermal analysis (DTA) and TGA both involves direct temperature measurement, they can be combined in the same instrument and their data can thus be recorded simultaneously (see Fig. 19.1).

Other calorimetry methods can determine other thermodynamic parameters of nanoparticle systems. Isothermal titration calorimetry (ITC) can be used to measure enthalpy changes and interactions between nanoparticles and materials binding to the surface, including binding affinities and stoichiometry. ITC is well suited for measurements in biological and/or nanoparticle solutions for drug delivery because it allows for solution-based measurements [9].

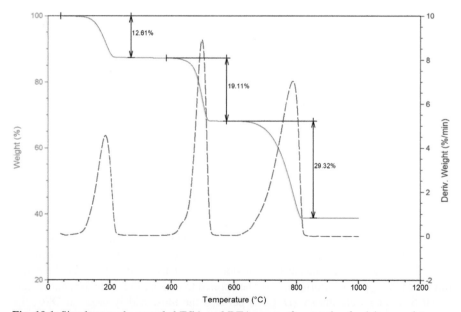

Fig. 19.1 Simultaneously recorded TGA and DTA scans of a sample of calcium oxalate.

19.3   Thermal analysis of nanoparticle purity and composition

As with any material, nanoparticle purity and composition are vital to know from both an applications perspective and as key output variables for any process control-based quality plan. The wide variety of production methods hint at the challenges involved. For example, some nanoparticles are made through solution-based processes in which nanoparticles form as the precipitates of a chemical reaction. Here, by-products can be trapped inside or adsorbed onto the particles. For drug-delivery applications, a common method involves sol–gel chemistry where some of the suspension solution can become trapped within the nanoparticles [10]. TGA can be used to determine how much of a certain component is present in the final mixture as a mass percentage. One challenge is that often, prior to analysis, nanoparticles must be isolated from the encapsulating solution through centrifugation or other means of separation. The chemical composition is determined by heating the nanoparticle sample and comparing the oxidation temperature transitions to those of the pure components [11]. The mass loss at each temperature is used to quantitatively determine chemical constituents.

TGA and DSC have also been used to validate protein encapsulation in nanoparticles by measuring the composition and heat flow results from the sample containing encapsulated proteins versus the pure nanoparticle samples. The difference in the thermograms provides evidence of protein encapsulation and can give an estimate of the mass of protein per nanoparticle [12]. TGA and DSC are bulk-scale techniques, with even their relatively small sample sizes of a few milligrams still containing very large numbers of nanoparticles. Therefore, in order to relate the total mass loss determined by their techniques to the mass loss experienced by the individual nanoparticles, information about the size, and morphology of the nanoparticles is required. Such data includes the average size of the nanoparticle as measured with light scattering or electron microscopy. Similarly, the mass of nanoparticles loaded into other systems, such as liposomal membranes, can be evaluated using similar TGA methods [13].

DSC can be used to evaluate crystallization behaviors and interactions of drugs with the nanoparticle-based delivery systems [14]. DSC has been used, for example, to evaluate interactions between indomethacin, a lipophilic drug, and solid lipid nanoparticles prepared for pharmaceutical drug delivery. Kinetic studies in the DSC were used to evaluate whether the indomethacin was associated with the lipid system and if so, how it was migrating through the bulk of the nanoparticle [15]. Microemulsion synthetic methods also benefit from calorimetric determination of the nanoparticles and synthesis matrix to garner information about how changes in the synthetic process may impact nanoparticle composition [16]. From measurements such as these, the synthesis of the nanomaterials could be tuned to obtain particles of desired sizes.

By comparison to standards, the purity of synthesized nanomaterials can be measured by DSC and TGA. TGA can provide composition information for nonmetal-based nanoparticles. One example of this is the application of TGA for the analysis of carbon nanotubes (CNTs) [3,17]. Besides the nanotubes themselves, the soot samples typically consist of amorphous carbons, adsorbed hydrocarbons that are by-products of the synthesis conditions, other types of structured carbon, and metal catalyst particles. By heating the CNT powder in an oxidizing environment, CNTs will decompose at a given temperature characteristic of the nanomaterial, with individual transitions observable for each component. The observed oxidation temperatures are around 200°C for amorphous carbons, $\sim$ 400°C for single-wall CNTs, $\sim$ 600°C for multi-wall CNTs, and anything over 650°C is attributed to metal catalyst and its oxidation products [18]. Desorption of associated hydrocarbons typically occurs at temperatures below 200°C. The use of TGA in the case of CNTs, can provide a measure of purity for CNT material, as measured by the percentage of the sample that degrades in the temperature range of interest. Furthermore, TGA is one of the fastest methods to measure relative percentages of single-wall versus multi-walled CNTs in a sample, but there are many challenges in obtaining accurate ratios, as many of the components tend to have overlapping temperatures [18]. There is significant work toward understanding peak fitting to truly understand composition [19] and this remains an area of development for carbon nanomaterials.

Organic materials such as polymer-based particles or lipid nanoparticles can be measured with thermal methods. For example, TGA can determine purity, and DSC can measure component interactions [10]. Nanoparticles made from polyelectrolytes with opposite charges form more complex mixtures than just the physical mixture of the two components, which yields a distinct broad exothermic peak in the DSC [10]. When the two components are mixed but do not interact significantly, two distinct exothermic peaks (one for alginate (negatively charged), one for chitosan (positively charged)) are observable. These typically occur at the same temperatures as those for the pure components. Mixtures that interact with each other usually have a different thermal signature, often with one broad peak that defines the mixture. By utilizing TGA to evaluate the composition, one can distinguish if particles are a mixture of the two components, or a more complex, thermally stable system. This technique is also commonly used to demonstrate miscible and immiscible polymer blends by the same technique.

Chemical composition is also important for determining what is attached to the core of a nanoparticle. Simultaneous thermal analysis (STA) can be used to validate the attached coating, as can TGA and DSC individually. In some systems, composition is theoretically known but needs to be validated after a synthesis. In a [Ni(en)$_3$] (NO$_3$)$_2$ nanoparticle population [20] two major transitions were observed in the thermogram which validated the theoretical calculated weights of each ligand (en, or ethylenediamine, and NO$_3$) from the complex, indicating the nanoparticle population was the expected composition. The DTA curve showed that these mass losses consisted of an endothermic event (release of ethylenediamine) followed by exothermic decomposition of the complex, giving additional information about the kinetics of the reactions to form the nanoparticles [20].

Other nanoparticle coatings, such as adsorbed non-covalent polymer or protein coatings, are particularly important to study, as these can be naturally occurring and lead to dramatic changes in how the nanoparticle interacts with the environment. Adsorbed coatings are often added to protect nanoparticles from the environment or improve solubility in the given environment. Adventitious coatings may develop as a nanoparticle travels through a natural environment. Measuring these adsorbed coatings is complex, as many factors associated with the nanoparticle (surface properties such as size, charge) and properties of the adsorbed moiety (charge, hydrophobicity) govern these interactions. Some examples include protein coronas, which form as a result of exposure to protein systems, and adsorbed natural organic matter in water samples. Quantitating the number of associated proteins (either different types or quantities of a single protein) on the nanoparticle surface under biological conditions is of importance as more and more nanomaterials get introduced into biological systems for therapeutic or inadvertent means [21]. Isothermal titration calorimetry has been used to measure stoichiometric affinity, and enthalpy of protein-nanoparticle interactions [22].

Mass loss due to decomposition of nanoparticles at specific temperatures can be used to determine (a) dehydration of the nanoparticle or associated protein, (b) decomposition of the surface ligands/associated proteins, and (c) decomposition of the nanoparticle. The nanoparticle concentration, protein concentration and surface area are used to calculate the number of proteins per nanoparticle. In some cases, it may be necessary to correct the mass measurements for the loss of water to obtain the mass loss due only to surface coatings [23]. The corrected mass measurements can then be used to estimate functional group density and coverage on the nanoparticle surface. Surface coverage can be estimated through simple surface packing models [22]. Nanoparticle coating thermal analysis can further be complicated by changes in the nanoparticles prompted by the removal of the coating. Secondary analysis, such as mass spectrometry or FTIR, can be used to distinguish between physi- and chemisorbed materials and the oxidation or reduction of the inner material and this has been successful demonstrated for fatty acid coated magnetite nanoparticles [24].

19.4   Evaluation of nanoparticle-containing composites

Nanocomposites include materials in which nanoparticles are added to a matrix material to tailor the optical, mechanical, magnetic, electrical, or thermal properties. Thermal analysis methods are often used to examine the differences between the matrix versus the matrix with incorporated nanoparticles. Changes in the thermal performance of composites with the addition of nanoparticles can be monitored using DSC and TGA. Shifts to higher decomposition temperatures, as measured by TGA, indicate increased thermal stability of the composite material [25]. The measurement of glass transition temperatures and melting points with DSC can be used to measure differences in degree of crystallinity and heat of fusion [25–27]. Mass loss profiles via TGA can also be used, in conjunction with mass spectrometry, to determine what is being released from the composite at a given time (water, decomposition products), which can provide further information of the composite composition [28]. Using the information obtained from TGA and DSC, any enhancements or negative changes from the original composite structure can be evaluated. Not only does thermal analysis give information about the bulk properties that may affect the processability of the composites, but it also can give information about how the particles are integrated into the composite network.

19.5 Monitoring kinetics of thermal transitions

Thermal analytic techniques can be used to monitor kinetic behavior of nanoparticle systems. One example is using DSC to determine the cure kinetics of nanocomposites, as the cure can be affected by nanoparticle composition and density within the composite. DSC can also measure glass transition temperatures, which can be used to predict nanoparticle mobility within the matrix [27]. The cure kinetics can be measured by monitoring the evolution of heat and measuring the activation energy. Curing is typically exothermic, so the degree of cure is defined as the heat released during a cure, divided by the total heat of reaction, with the cure itself identified as a maximum in the measured total heat. The rate of cure is calculated as:

$$\frac{d\alpha}{dt} = \frac{1}{H_g} \frac{dH}{dt}$$

where the $d\alpha/dt$ is the reaction rate, H_g is the total heat of the reaction, and dH/dt is the change in the heat evolved during the cure at a certain time [29].

Calorimetry provides a means for the measurement of many types of nanoparticle kinetics. Direct measurement via a micromechanical calorimeter of particle energies opened up the field for nanoparticle energetic measurements via thermal analysis [30]. When combined with size classification, it is possible to obtain the size dependent enthalpies of formation, an invaluable metric as it is a direct probe of how surface-to-volume ratio affects cohesive energy density [31]. The technique can be further extended to allow for nanoparticle energetics in catalyst sintering applications to be measured, which is extremely useful as many of these nanoparticles are used to make other nanomaterials such as CNTs [32]. A related measurement, the enthalpy of adsorption, can also be used to predict energetics of nanoparticle oxide formation [33]. Oxide formation is perhaps the most important mechanism in terms of catalyst deactivation, so knowledge of the thermodynamics is important in developing methods of mitigation and prevention. A complete study of oxidation of nickel nanoparticles was completed in 2008 [34] in which multiple theories were examined to represent the oxidation kinetics. In the end, it was found that nanoparticle oxidization occurred in a diffusion dominated way for this particular nickel nanoparticle system and the work is a great reference for others interested in oxidation kinetics in metallic nanoparticle systems.

Nanoparticle energetic state and the growth process as a result of reactant concentrations are measurable through calorimetric means. Experimental enthalpies are obtained by integrating the dynamic calorimeter signal over time. Many different microemulsion synthesized nanoparticle systems have been studied using this method.

Rate equations can also be determined from the work if the time constants used are greater than that of the calorimeter [35]. Rates of reactions for kinetic processes, such as thermal dehydration of nanoparticle populations have been measured by constructing Arrhenius plots from DSC thermograms [36]. The Arrhenius plots can be used to determine the energy of activation for the hydration process. In one case, the higher activation energy indicated a chemical reaction rather than diffusion-controlled reaction (hematite nanoparticles [36]) and there was a highly ordered transition state in the dehydration process. Other aspects, such as reactivity of nanoparticles have been studied using non-isothermal TGA measurements [37].

19.6   Innovations in thermal analysis for nanoparticles

There is room for improvement for today's thermoanalytical techniques to become more appropriate for nanoparticle systems. Smaller sample sizes, ability to increase sensitivity, and monitoring nanomaterial-matrix interactions are of importance. Recent developments in thermoanalytical methods have made nanoparticle analysis on small scales possible.

Nanocalorimetry is a microchip-based system capable of measuring samples on the order of nL in volume or micrograms to nanograms in mass [38] and has a resolution below 5 nW when measuring heats of reactions [39]. The small sample volumes allow for the measurement of as-produced samples and their interactions with the environment, as well as interactions between nanomaterials and cells, which is of importance in the area of nanomedicine [38]. As with any thermoanalytical technique, accurate temperature, power, and mass measurements are needed. The measurements of these small-scale interactions are difficult as the compensation for the heat loss in these systems is not as controlled as in large scale instrumentation. Overall, the application of nanocalorimetry has led to advances in understanding binding reactions between biological systems and nanoparticles, crystallization and melting behaviors of particles, and size-dependent thermodynamics and kinetics [40]. Further development to push toward femtowatt measurements [41], improved uncertainty analysis and the push for aqueous environmental analysis will yield the next phase of nanocalorimeters [38].

TGA has also been advanced significantly to provide decomposition information on nanotechnology-relevant scales. In the late 1970s, the use of more sensitive balances for TGA, specifically utilizing quartz crystal microbalances, was suggested as a potential improvement. Work at the National Institute of Standards and Technology led to the development of a microscale-TGA instrument which requires 1000-fold less sample than traditional instrumentation [18] and overcomes previous challenges associated with use of quartz crystal microbalances. TGA has been limited by sensitivity of the microbalance, with commercial instruments requiring at least 1 mg of sample. As nanoparticle applications require tighter controls on nanoparticle compositions, often the yield from synthesis is far below the 1 mg size limit for traditional instruments. The microscale-TGA method replaces the microbalances in traditional instruments with piezoelectric quartz crystal microbalances to allow for analysis on sample masses of 1–10 µg and sensitivities of mass detection on the order of ng

[42]. After a correction for thermal changes in quartz at elevated temperatures (which previously limited use of quartz crystal microbalances), decomposition data have been obtained for a number of materials [43]. CNT decomposition can be monitored using microscale-TGA, as well as quantitative measurement of coatings on gold and silica nanoparticles [18,21,44], all of which could be compared to conventional TGA or other analytical results. This new TGA instrumentation will enable measurements of limited nanoparticle populations, such as those that may be obtained from environmental or biological samples without complex sample preparation or tedious sample analysis. Further development of the microscale-TGA measurements for nanomaterial analysis would include the linking of microscale-TGA to secondary detection instrumentation (Raman, mass spectrometry) to yield more information about the decomposition products, improving sensitivity and commercialization of the instrumentation [45].

19.7 Conclusions

Thermoanalytical methods are now widely used for the study of nanomaterials and products containing nanoparticles. Thermal analysis has many advantages that complement other nanoparticle analysis techniques. In most cases, preparation of the nanoparticle sample is minimal (i.e., concentration of the sample) and no additional modifications (i.e., fluorescent labeling, etc.) are needed to complete the analysis. A wide range of techniques can be used including TGA, DSC, and other calorimetric systems. Commercial instrumentation is available and interpretation of results is often easy enough that thermal analysis can be applied in the manufacturing setting to compare batch-to-batch reproducibility. Nanoparticle purity and composition can be assessed quickly on the laboratory scale with minimal sample preparation and results can be compared to other analytical methods to provide further certainty of the composition. Interactions that may influence nanoparticle coatings can also be determined using thermal methods, including those in which the nanoparticle may be exposed to environmental systems such as water or biological media. Kinetics of nanoparticle systems is increasingly important as more complex synthetic methods are used. Nanoparticle systems, such as nanocomposites, can be characterized for composition and kinetics of reactions as related to nanoparticle content.

References

[1] Project on Emerging Nanotechnologies, 2013. http://www.nanotechproject.org/.
[2] S.F. Hansen, O.F.H. Hansen, M.B. Nielsen, Advances and challenges towards consumerization of nanomaterials, Nat. Nanotechnol. 15 (2020) 964–965.
[3] G.L. Hornyak, H.F. Tibbals, J. Dutta, J.J. Moore, Introduction to Nanoscience & Nanotechnology, CRC Press, Boca Raton, FL, 2009.
[4] E.K. Richman, J.E. Hutchison, The nanomaterial characterization bottleneck, ACS Nano 3 (2009) 2441–2446.
[5] M.J. Mitchell, M.M. Billingsley, R.M. Haley, M.E. Wechsler, N.A. Peppas, R. Langer, Engineering precision nanoparticles for drug delivery, Natl. Rev. 20 (2021) 101–124.

[6] J.E. Decker, A.R. Hight Walker, K. Bosnick, C.A. Clifford, L. Dai, J.A. Fagan, S.A. Hooker, Z. Jakubek, C. Kingston, J. Makar, M. Elisabeth, M.T. Postek, B. Simard, S. Ralph, S. Wise, A.E. Vladar, L. Yang, R. Zeisler, Sample preparation protocols for realization of reproducible characterization of single-walled carbon nanotubes, Metrologia 46 (2009) 682–692.

[7] A.A. Dongargaonkar, J.D. Clogston, Quantitation of surface coatings on nanoparticles using thermogravimetric analysis, in: S.E. McNeil (Ed.), Methods in Molecular Biology, Springer Science+Business Media LLC, 2018.

[8] M.J. O'Neill, Measurement of specific heat functions by differential scanning calorimetry, Anal. Chem. 38 (1966) 1331–1336.

[9] K. Bouchemal, New challenges for pharmaceutical formulations and drug delivery systems characterization using isothermal titration calorimetry, Drug Discov. Today 13 (2008) 960–972.

[10] B. Sarmento, D. Ferreira, F. Veiga, A. Ribeiro, Characterization of insulin-loaded alginate nanoparticles produced by ionotropic pre-gelation through DSC and FTIR studies, Carbohydr. Polym. 66 (2006) 1–7.

[11] Y. Dong, S.-S. Feng, Poly(D,L-lactide-co-glycolide)/montmorillonite nanoparticles for oral delivery of anticancer drug, Biomaterials 26 (2005) 6068–6076.

[12] D. Ma, M. Li, A.J. Patil, S. Mann, Fabrication of protein/silica core-shell nanoparticles by microemulsion-based molecular wrapping, Adv. Mater. 16 (2004) 1838–1841.

[13] E. Amstad, J. Kohlbrecher, E. Muller, T. Schweizer, M. Textor, E. Reimhult, Triggered release from liposomes through magnetic actuation of iron oxide nanoparticle containing membranes, Nano Lett. 11 (2011) 1664–1670.

[14] B. Siekmann, K. Westesen, Thermoanalysis of the recrystallization process of melt-homogenized glyceride nanoparticles, Colloids Surf. B Biointerfaces 3 (1994) 159–175.

[15] F. Castelli, C. Puglia, M.G. Sarpietro, L. Rizza, F. Bonina, Characterization of indomethacin-loaded lipid nanoparticles by differential scanning calorimetry, Int. J. Pharm. 304 (2005) 231–238.

[16] P. Fini, M.L. Curri, M. Castagnolo, F. Ciampi, A. Agostiano, Calorimetric study of CdS nanoparticle formation in w/o microemulsions, Mater. Sci. Eng. C 23 (2003) 1077–1081.

[17] A.G. Bannov, M.V. Popov, P.B. Kurmashov, Thermal analysis of carbon nanomaterials: advantage and problems of interpretations, J. Therm. Anal. Calorim. 142 (2020) 349–370.

[18] E. Mansfield, A. Kar, S.A. Hooker, Applications of TGA in quality control of SWCNTs, Anal. Bioanal. Chem. 396 (2010) 1071–1077.

[19] M.A. Banash, Identification of carbon nanomaterials by deconvolution of thermogravimetric analysis signals, Thermochim. Acta 700 (2021), https://doi.org/10.1016/j.tca.2021.178928, 178928.

[20] S. Farhadi, Z. Roostaei-Zaniyani, Preparation and characterization of NiO nanoparticles from thermal decomposition of the $[Ni(en)_3](NO_3)_2$ complex: a facile and low-temperature route, Polyhedron 30 (2011) 971–975.

[21] K.B. Sebby, E. Mansfield, The stability and surface coverage of polymer stabilized gold nanoparticles, eCells Mater. 20 (2010) 234.

[22] T. Cedervall, I. Lynch, S. Lindman, T. Berggard, E. Thulin, H. Nilsson, K.A. Dawson, S. Linse, Understanding the nanoparticle-protein corona using methods to quantify exchange rates and affinities of proteins for nanoparticles, Proc. Natl. Acad. Sci. 104 (2007) 2050–2055.

[23] B. Wang, P. Wu, R.A. Yokel, E.A. Grulke, Influence of surface charge on lysozyme adsorption to ceria nanoparticles, Appl. Surf. Sci. 258 (2012) 5332–5341.

[24] M. Rudolph, J. Erler, U.A. Peuker, A TGA-FTIR perspective of fatty acid adsorbed on magnetite nanoparticles—decomposition steps and magnetite reductions, Colloids Surf. A Physicochem. Eng. Asp. 297 (2012) 16–23.

[25] M.M. Hasan, Y. Zhou, H. Mahfuz, S. Jeelani, Effect of SiO_2 nanoparticle on thermal and tensile behavior of nylon-6, Mater. Sci. Eng. A 429 (2006) 181–188.

[26] P. Rittigstein, J.M. Torkelson, Polymer-nanoparticle interfacial interactions in polymer nanocomposites: confinement effects on glass transition temperature and suppression of physical aging, J. Polym. Sci. B 44 (2006) 2935–2943.

[27] D. Shah, P. Maiti, D.D. Jiang, C.A. Batt, E.P. Giannelis, Effect of nanoparticle mobility on toughness of polymer nanocomposites, Adv. Mater. 17 (2005) 525–528.

[28] V. Kanniah, B. Wang, Y. Yang, E.A. Grulke, Graphite functionalization for dispersion in a two-phase lubricant oligomer mixture, J. Appl. Polym. Sci. 125 (2012) 165–174.

[29] P. Rosso, Y. Lin, Epoxy/silica nanocomposites: nanoparticle-induced cure kinetics and microstructure, Macromol. Rapid Commun. 28 (2007) 121–126.

[30] T. Bachels, R. Schäfer, H.-J. Güntherodt, Dependence of formation energies of tin nanoclusters on their size and shape, Phys. Rev. Lett. 84 (2000) 4890–4893.

[31] D. Xie, M.P. Wang, W.H. Qi, A simplified model to calculate the surface-to-volume atomic ratio dependent cohesive energy of nanocrystals, J. Phys. Condens. Matter 16 (2004) L401–L405.

[32] C.T. Campbell, S.C. Parker, D.E. Starr, The effect of size-dependent nanoparticle energetics on catalyst sintering, Science 298 (2002) 811–814.

[33] A. Navrotsky, Energetics of nanoparticle oxides: interplay between surface energy and polymorphism, Geochem. Trans. 4 (2003) 34–37.

[34] P. Song, W. Dongsheng, Z.X. Guo, T. Korakianitis, Oxidation investigation of nickel nanoparticles, Phys. Chem. Chem. Phys. 2008 (2008) 5057–5065.

[35] F. Aliotta, V. Arcoleo, S. Buccoleri, G. La Manna, V. Turco Liveri, Calorimetric investigation on the formation of gold nanoparticles in water/AOT/n-heptane microemulsions, Thermochim. Acta 265 (1995) 15–23.

[36] A.S. Al-Kady, M. Gaber, M.M. Hussein, E.-Z.M. Ebeid, Structural and fluorescence quenching characterization of hematite nanoparticles, Spectrochim. Acta A 83 (2011) 398–405.

[37] K. Park, D. Lee, A. Rai, D. Mukherjee, M.R. Zachariah, Size-resolved kinetic measurements of aluminum nanoparticle oxidation with single particle mass spectrometry, J. Phys. Chem. B 109 (2005) 7290–7299.

[38] F. Yi, D.A. La Van, Nanoscale thermal analysis for nanomedicine by nanocalorimetry, WIREs Nanomed. Nanobiotechnol. 4 (2012) 31–41.

[39] W. Lee, W. Fon, B.W. Axelrod, M.L. Roukes, High-sensitivity microfluidic calorimeters for biological and chemical applications, Proc. Natl. Acad. Sci. 106 (2009) 15225–15230.

[40] F. Li, D.A. LaVan, Nanocalorimetry: exploring materials faster and smaller, Appl. Phys. Rev. 6 (2019), 031302.

[41] B.G. Burke, D.A. LaVan, Laser heating and detection of bilayer microcantilevers for non-contact thermodynamic measurements, Appl. Phys. Lett. 102 (2013), 021916.

[42] W.L. Johnson, E. Mansfield, Thermogravimetric analysis with a heated quartz crystal microbalance, in: Frequency Control Symposium (FCS), 2012 IEEE International, 2012, pp. 1–5.

[43] E. Mansfield, A. Kar, T.P. Quinn, S.A. Hooker, Quartz crystal microbalances for microscale thermogravimetric analysis, Anal. Chem. 82 (2010) 9977–9982.

[44] E. Mansfield, K.M. Tyner, C.M. Poling, J.L. Blacklock, Determination of nanoparticle surface coatings and nanoparticle purity using microscale thermogravimetric analysis, Anal. Chem. 86 (2014) 1478–1484.

[45] E. Mansfield, T.P. Quinn, Microscale thermogravimetric device analyzes nanoparticle purity and coatings, in: SPIE Newsroom, 2011.

Impedance humidity sensors based on metal oxide semiconductors: Characteristics and mechanism

Hongyan Zhang and Shuguo Yu
School of Physical Science and Technology, Xinjiang University, Urumqi, People's Republic of China

20.1 Introduction

Humidity sensor is one kind of sensor that can realize humidity measurement by transforming physical or chemical information generated by reactions between the sensing material and target water molecules to create optical, electrical, or sound signals that can be effectively measured. Humidity sensors play an important role in the fields of applied pharmaceuticals, agriculture, industrial manufacturing, weather forecast, smart home, material processing, food packaging, and so on [1–4]. In addition, humidity detection becomes more interested in medical research in recent years. For example, a sensor may monitor the characteristics of humidity exhaled from human breath and provide basic information about physiological conditions related to anxiety, apnea, epilepsy, and early heart attack. Furthermore, monitoring skin's moisture level in various situations also helps to establish physiological conditions related to growth, repairing, and aging [5–7]. To meet the requirements of humidity sensors in different fields, low processing costs, high sensitivity, long-term stability, and rapid response/recovery time are still the main directions of improving humidity sensors [8–12]. Up to now, people have mainly concentrated on the development of high-performance humidity sensors based on sensitive materials, such as metal oxide semiconductors (MOS) [13,14], polymers [15], and inorganic/organic composite materials [16]. Among these materials, MOS (Fig. 20.1) have been widely used in manufacture of humidity sensors because of their excellent surface-to-volume ratio, strong adsorption capability, high surface reactivity, and catalytic efficiency [17,18].

Researchers use precise control of the grain size distribution, local doping, grain boundaries, and surface states of MOS materials to obtain better sensitive materials and meet the requirements of humidity sensors in different fields [19,20]. Feng et al. used a chemical vapor deposition and successive ion layer adsorption reaction method to grow ZnO and CdS on silicon nanoporous pillar array (Si-NPA). The humidity response of CdS/ZnO/Si-NPA exceeds more than 201,530% in relative humidity (RH) from 11% to 95%, the response/recovery time is 110/32 s, and the

Modeling, Characterization and Production of Nanomaterials. https://doi.org/10.1016/B978-0-12-819905-3.00020-8

Fig. 20.1 Metal oxide semiconductor materials commonly used in the preparation of humidity sensors.

maximum hysteresis error is 2.67% [21]. Gong et al. used a self-assembly method to synthesize ultra-thin two-dimensional TiO_2 nanosheets with abundant surface oxygen vacancy defects and large specific surface areas [22]. This humidity sensor has a resistance change of more than 4 orders of magnitude from 11% RH to 95% RH, the response/recovery time is 3/50 s, and the maximum hysteresis error is 4.6%. Wang et al. synthesized an urchin-like CuO nanostructure modified by reduced graphene oxide (rGO) using microwave hydrothermal method. The resistance changes more than 5 orders of magnitude in a range from 11% RH to 95% RH, response/recovery time is 2/17 s [23]. Kuang et al. used a Au-catalyzed vapor-liquid-solid (VLS) growth process to synthesize high-yield SnO_2 nanowires (NWs) with a large number of oxygen vacancies [24], making the humidity sensor change approximately 3 orders of magnitude in the range of 11% RH to 85% RH; response and recovery time are also greatly shortened. From the results of the above-mentioned MOS sensors, large specific surface area, abundant surface oxygen vacancies, and active sites are the key factors to improve performance of humidity sensors.

Various types of humidity sensors based on MOS sensitive materials have been prepared, such as impedance [9–12], capacitance [25], quartz crystal microbalance (QCM) [26], surface-acoustic wave (SAW) [9–12], field effect transistors (FETs) [27], fiber optic [28], and microwave sensors. Impedance humidity sensor has attracted much attention due to its high sensitivity, simple structure, convenient preparation, low cost, and compatibility with integrated circuit technology, which has become the main type of commercial humidity sensors. In this chapter, we introduce the MOS humidity sensor in terms of measurement system, detection parameters, and sensing mechanism.

20.2   Humidity sensing measurement standard

20.2.1   Humidity measuring device

Fig. 20.2 shows a typical test equipment and the process of impedance humidity sensing. A Zennium workstation (Zahner, Germany) is used to monitor and record the resistance change. RH sensing measurement is carried out in a closed chamber using a two-electrode configuration. The supersaturated salt solutions of LiCl, $MgCl_2$,

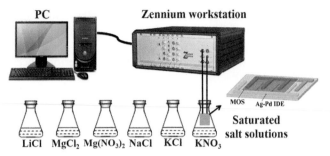

Fig. 20.2 The schematic of the MOS humidity measurement system.

Mg(NO$_3$)$_2$, NaCl, KCl, and KNO$_3$ have been formed under 11%, 33%, 54%, 75%, 85%, and 95% RH in a closed room. MOS materials is mixed with deionized water to form a paste, sprayed on Ag-Pd interdigital electrode (IDE) surface, and dried at 60°C in oven for 1 h. Then, the Ag-Pd IDE is placed in different RH environments to test the humidity characteristics.

20.2.2 Characteristics of humidity sensors

Quantifiable technical parameters of a humidity sensor include humidity measurement range, sensitivity, linearity, response/recovery time, hysteresis, and stability, which are often used to measure the performance of humidity sensors to meet the specific requirements in different application scenarios. The following is a brief introduction of characteristic parameters for the humidity sensor.

20.2.2.1 Humidity range

Humidity is the percentage of water in the environment. RH is defined as the ratio of water vapor pressure P_1 in the air to saturated water vapor pressure P_2 at the same temperature in percentage, namely RH$=(P_1/P_2) \times 100\%$. Moreover, humidity range is usually divided into three stages as low humidity (<30% RH), medium humidity (30%–70% RH), and high humidity ('70% RH) [29].

20.2.2.2 Sensitivity (response)

Sensitivity is an important factor of static characteristics of the sensor. It is defined as the ratio of the increment of output to one of corresponding inputs that causes the increment. Furthermore, sensitivity of a humidity sensor is the most important characteristic standard to describe its performance to ensure that the sensor can be used to accurately detect humidity in environment. The sensitivity(S) of humidity sensor is represented by the formula below,

$$S(\Omega/\%\,RH) = \frac{X_l - X_h}{\Delta RH}$$

where X_l is the value of humidity characteristic parameter in low humidity environment, X_h is the value of humidity characteristic parameter in high-humidity environment, and ΔRH is the change in RH [30,31].

Furthermore, response (R) is also used to measure the performance of a humidity sensor, the formula is:

$$R = \frac{X_l - X_h}{X_h} \times 100\%$$

where X_l or X_h is the value of humidity characteristic parameter in low- or high-humidity environment. Moreover, for humidity sensors of capacitance, QCM, SAW, or FET types, it is only necessary to replace the humidity characteristic parameters when sensitivity is to be calculated.

20.2.2.3 Linearity

Linearity is another important factor to describe static characteristics of a sensor. It refers to the degree to which the actual relation curve between the output and the input of the sensor deviates from the fitted line, where it is assumed that the measured input is in a stable state. It is defined as the ratio of the maximum deviation value between the actual characteristic curve and the fitting line within the full range to the output value of the full range. In humidity detection, linearity refers to linear relationship of humidity characteristic curve at a specific operating frequency. Better linearity means better performance of a humidity sensor [32]. For example, Fig. 20.3 shows the impedance of a ZnO-SiO$_2$ humidity sensor versus RH at different frequencies [33]. The humidity sensor maintains a good linear relationship when the test frequency is 100 Hz.

20.2.2.4 Response and recovery time

Response time is defined as the time required for the humidity sensing characteristic quantity to change from a low humidity stable state to a high-humidity stable state when humidity increases [34]. Moreover, recovery time is defined as the time required for the humidity sensing characteristic of humidity sensor to change from a high-humidity stable state to a low humidity stable state when humidity decreases. In general, humidity sensing characteristics will be slightly delayed from the change of environmental RH being tested, so the time required for the humidity sensing characteristic to reach 90% of stable value is defined as response/recovery time [35–40].

20.2.2.5 Humidity hysteresis

Hysteresis means that the input and the output characteristic curves of a sensor do not coincide when input quantity changes from small to large (positive test) and input quantity changes from large to small (negative test). For the same value of input signal, the positive and the negative output signals of the same sensor are not exactly the same, which is called hysteresis difference.

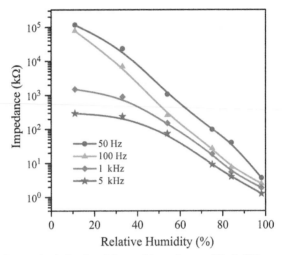

Fig. 20.3 Dependence of relative humidity and impedance of ZnO-SiO$_2$ at different frequencies.
Adapted with permission from V.K. Tomer, S. Duhan, A.K. Sharma, R. Malik, S.P. Nehra, S. Devi, One pot synthesis of mesoporous ZnO-SiO$_2$ nanocomposite as high performance humidity sensor, Colloids Surf. A Physicochem. Eng. Asp. 483 (2015) 121–128, https://doi.org/10.1016/j.colsurfa.2015.07.046. Copyright © 2015 American Colloids and Surfaces A: Physicochemical and Engineering Aspects.

Since the change of the humidity sensing characteristic lags behind the change of environmental humidity, it will inevitably lead to the curve of adsorption process and the curve of desorption process not completely coinciding, and there is a hysteresis error occurred during this time. The curve of adsorption process and desorption process will form a non-closed loop, which is called hysteresis loop. The hysteresis loop shows the degree of non-coincidence between adsorption process of humidity sensor and desorption process, which further shows that these two processes are not completely reversible. Reducing the hysteresis error of a humidity sensor can help to quickly detect different humidity environments. The following is the calculation formula for the hysteresis error,

$$\gamma H = \pm \frac{\Delta H_{max}}{2 F_{FS}}$$

where ΔH_{max} is the difference between characteristic quantities of adsorption and desorption process and F_{FS} is the output of full scale [36–41].

20.2.2.6 Stability

The stability of humidity sensor means that the humidity sensing performance does not change with time [42,43]. To meet the actual needs in production, it is necessary to develop a humidity sensor that can work stably for a long time. There are two main reasons for the poor stability of a humidity sensor. First, the performance of humidity sensitive material itself deteriorates. Second, the sensing film of the device is uneven, deformed, or separated from the substrate during the production of the sensor.

20.3   Humidity sensors based on MOS materials

The key of a humidity sensor is the sensitive material, which determines the performance of the humidity sensor, especially its sensitivity and stability. MOS meet the requirements of high-performance humidity sensitive materials because of their high-specific surface area and direct path of charge transport. High-specific surface area means that there exists a large number of active and adsorption sites on the surface of the material, and the direct route of charge transport has a higher electron mobility. Furthermore, structure, specific surface area, active site, hydrophilicity, and coating thickness of the sensitive material are important factors affecting the performance of MOS humidity sensors.

There are many types of MOS materials with humidity-sensitive characteristics, including ZnO, CuO, TiO_2, SnO_2, Al_2O_3, and In_2O_3. ZnO is a typical n-type MOS material with a direct broad band gap of 3.37 eV and a large exciton binding energy of 60 meV at room temperature [44,45]. Moreover, ZnO has surface hydrophilic functional groups, high-specific surface area and stability, which are suitable for humidity sensors. In this section, ZnO is taken as an example to illustrate how a high-performance humidity sensor can be obtained.

20.3.1   ZnO humidity sensor

Alkaline conditions are favorable for the synthesis of ZnO with rich oxygen vacancies, hydrophilic functional groups, high-specific surface area, and electron mobility, which are important factors to determine the performance of ZnO humidity sensors [36–40]. Fig. 20.4A shows the impedance change of a ZnO humidity sensor in a range of 11% to 95% RH when pH of the precursor solution is 9, 10, and 11, labeled as ZnO-1, ZnO-2, and ZnO-3, respectively. The responses of ZnO-1, ZnO-2, and ZnO-3 humidity sensors can be calculated to be 954,696%, 2,515,127%, or 517%. The impedance change of ZnO-2 humidity sensor is more than 4 orders of magnitude with good linearity. In Fig. 20.4B and C, ZnO-2 shows a small hysteresis with a maximum hysteresis error of 0.9% and the response/recovery time is 31/14 s. Fig. 20.4D shows the stability of ZnO-2 sensor from 11% to 95% RH. It can be found that the performance of ZnO-2 has long-term stability from 33% to 95% RH except for slight fluctuations in impedance at 11% RH.

Fig. 20.5A and B show that ZnO-2 with rough surface and evenly distributed ZnO particles about 300 nm in size. The rough surface makes this device have more surface active sites. In Fig. 20.5C, Fourier transform-infrared (FTIR) spectra show absorption peaks within 3100 to 3600 cm^{-1} which corresponds to O-H tensile vibration modes of surface adsorption (-OH) [46]. It can be observed that the absorption peak of O-H stretching vibration of ZnO slightly shifts toward low wave number with the increase

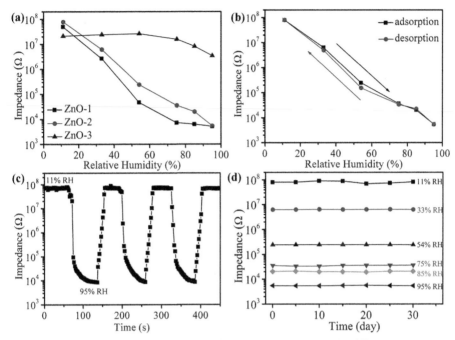

Fig. 20.4 (A) Impedance response curves of a ZnO humidity sensor in different humidity environments; (B) hysteresis characteristics; (C) response/recovery time; and (D) stability test. Adapted with permission from S. Yu, H. Zhang, J. Zhang, Z. Li, Effects of pH on high-performance ZnO resistive humidity sensors using one-step synthesis, Sensors 19(23) (2019) 5267, https://doi.org/10.3390/s19235267. Copyright © 2019 Sensors.

of pH of the precursor, which is mainly due to the fact that a large amount of hydroxyl groups (-OH) are generated. Hydroxyl group (-OH) as a hydrophilic functional group can adsorb more water molecules to enhance adsorption process of the sensor. In Figs. 20.5D–F, the content of oxygen vacancies defects (O_2) on the surface of ZnO also gradually increases with the increase of pH of the precursor solution. Strong electrostatic field near oxygen vacancy defect causes rapid decomposition of water molecules to produce a large amount of conductive H_3O^+ [22]. The increase of oxygen vacancy defects indicates that water molecules adsorbed on the surface of the material will be rapidly decomposed, thereby shortening the response time of the sensor. In conclusion, the synthesis of ZnO under alkaline conditions increases surface oxygen vacancies, active sites, and the concentration of hydroxyl groups (-OH) in ZnO, which enhances the performance of the ZnO humidity sensor.

Although ZnO can be used as humidity materials with high performance, there still exist some deficiencies' such as long response/recovery time and poor linearity and selectivity, which has become an obstacle to the miniaturization and the integration development of the humidity sensor system.

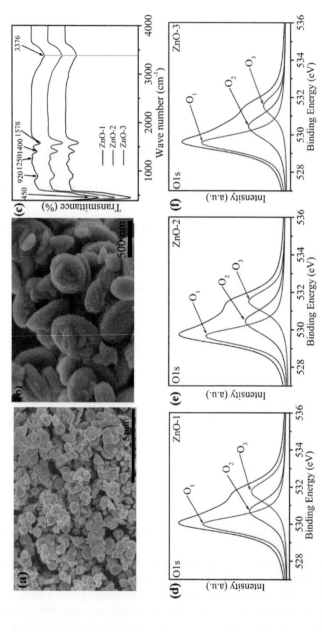

Fig. 20.5 (A) and (B) SEM of ZnO particles at pH = 10; (C) FTIR; and (D–F) XPS spectra of O 1s of ZnO at pH = 9, 10, and 11, respectively. Adapted with permission from S. Yu, H. Zhang, J. Zhang, Z. Li, Effects of pH on high-performance ZnO resistive humidity sensors using one-step synthesis, Sensors 19(23) (2019) 5267, https://doi.org/10.3390/s19235267. Copyright © 2019 Sensors.

20.3.2 Mental doped ZnO humidity sensor

Doping is an effective method to improve chemical composition, energy band structure, and photoelectric properties of MOS materials. Metals have good electrical conductivity, and metal doping can increase the carrier concentration and reduce the influence of the interaction between orbitals of doped atoms on the scattering of the carriers. In addition, metal doping can improve surface morphology and microscopic states of ZnO, thereby improving the performance of the humidity sensor.

20.3.2.1 Mg-doped ZnO humidity sensor

There are many types of metal element ions, such as Cu^{2+}, Cd^{2+}, Al^{3+}, Fe^{3+}, and Mg^{2+}, with radii similar than that of Zn^{2+}. Morphology and surface microscopic states of ZnO will be changed and performance of humidity sensors based on ZnO will be improved when doped with those metal elements.

For example, a high-performance humidity sensor was designed with Mg-doped black ZnO [47]. It can be observed from Fig. 20.6A that the impedance change of Mg-doped humidity sensor is about 4 orders of magnitude in the range of 11% to 95% RH when the molar ratio of Mg^{2+} ions is 1.5 mol%. In Fig. 20.6B, by changing the test frequency to test the performance of the sensor under different humidity levels, the humidity sensor exhibits the best linear relationship when working frequency is 100 Hz. Furthermore, the Mg^{2+}-doped ZnO humidity sensor shows a small hysteresis with a maximum hysteresis error of 4.1% as in Fig. 20.6C. Fig. 20.6D shows response/recovery time of Mg-doped ZnO (1.5 mol%) from 11% RH to 95% RH to be 24/12 s, and it has good repeatability during 4 consecutive adsorption and desorption cycles.

Compared with ZnO, the particle sizes of Mg-doped ZnO microspheres become smaller and more uniform, which increases the surface roughness and specific surface area of the material as shown in Fig. 20.6E and F. Fig. 20.6G and H show X-ray photoelectron spectroscopy (XPS) spectra for ZnO and Mg-doped ZnO. Compared with ZnO, the proportion of oxygen vacancies on the surface of Mg-doped ZnO is greatly increased from 14.19% to 45.15%, which generates high density of H_3O^+ and promotes the conduction process.

Results show that the metal Mg-doped ZnO improves sensitivity, linearity, hysteresis, and response/recovery characteristics over the ZnO humidity sensors, which is due to better surface morphology and more oxygen vacancy defects of Mg-doped ZnO.

20.3.2.2 Rare earth element doped ZnO humidity sensor

Rare earth metals such as europium (Eu), cerium (Ce), erbium (Er), and dysprosium (Dy) have excellent electrical conductivity, electrochemical performance, highly efficient catalytic performance, and high oxygen ion mobility. As dopants, rare earth metals have great impacts on morphology and surface oxygen vacancy defects of ZnO, making such a sensor to achieve high sensitivity and rapid response.

In Fig. 20.7, the sol-gel method was used to prepare Eu-doped ZnO resistive humidity sensors [36–40]. Fig. 20.7A shows the impedance change of Eu-doped

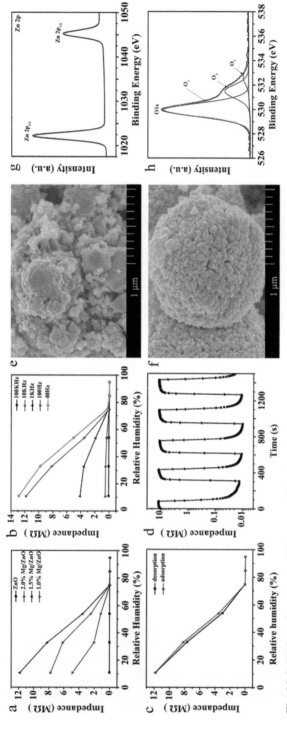

Fig. 20.6 (A) Impedance versus RH curves of Mg-doped ZnO humidity sensors; (B) at difference frequency; (C) humidity hysteresis; (D) response and recovery characteristic curve of Mg-doped ZnO microspheres (1.5 mol%); SEM images of (E) ZnO and (F) Mg-doped ZnO (1.5 mol%); XPS spectra of O 1s of (G) ZnO and (H) Mg-doped ZnO (1.5 mol%).

Adapted with permission from C. Lin, H. Zhang, J. Zhang, C. Chen, Enhancement of the humidity sensing performance in mg-doped hexagonal ZnO microspheres at room temperature, Sensors 19(3) (2019) 519, https://doi.org/10.3390/s19030519. Copyright © 2019 Sensors.

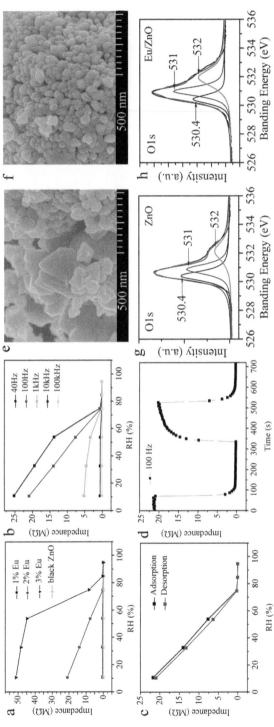

Fig. 20.7 (A) Impedance versus RH curve of Eu-doped ZnO; (B) Frequency selection, (C) humidity hysteresis characteristics, and (D) response and recovery properties of Eu-doped ZnO (2 mol%); (E, F) SEM images and (G, H) XPS of O 1s of ZnO and Eu-doped ZnO (2 mol%). Adapted with permission from S. Yu, H. Zhang, C. Lin, M. Bian, The enhancement of humidity sensing performance based on Eu-doped ZnO, Curr. Appl. Phys. 19(2) (2019) 82–88, https://doi.org/10.1016/j.cap.2018.11.015. Copyright © 2019 Current Applied Physics.

ZnO with different molar ratios in the range from 11% to 95% RH. The response of the humidity sensor gradually increases with the increase of Eu^{3+} content, and Eu-doped ZnO (2 mol%) shows better linearity. To determine the best measurement frequency, the impedance changes of the sensor from 40 Hz to 100 kHz were tested as shown in Fig. 20.7B. The sensitivity and the linearity are still higher when test frequency is 100 Hz, so 100 Hz is the optimal operation frequency. Fig. 20.7C shows the humidity hysteresis of Eu-doped ZnO. The characteristic of adsorption process and desorption process of Eu-doped ZnO humidity sensor almost completely overlap, indicating a very small hysteresis or high reversibility. Fig. 20.7D shows the response/recovery time of Eu-doped ZnO (2 mol%) humidity sensor to be 5/19 s, respectively.

Fig. 20.7E and F show that doping with Eu^{3+} ions will make the particle size of ZnO smaller and the distribution more uniform, which makes more water molecules adsorbed on the surface of Eu-doped ZnO. Furthermore, Eu^{3+} ion doping increases the concentration of oxygen vacancy defects on the surface of ZnO as shown both in Fig. 20.7G and H, which leads to rapid decomposition of water molecules, thereby increasing the response speed of the ZnO sensor.

Another example is the Er-doped ZnO impedance humidity sensor. The curves of impedance of Er-doped ZnO sensors with different doping concentrations [9–12] can be shown in Fig. 20.8A. The Er-doped ZnO (3%) sensor exhibits 3 orders of magnitude impedance variation when RH increases from 11% to 95%. By changing the frequency to test the performance of the Er-doped ZnO humidity sensor under different humidity levels, it is found that the linearity of the sensor is the best when the operating frequency is 100 Hz as in Fig. 20.8B. Fig. 20.8C and D show hysteresis characteristics and response/recovery time of the humidity sensor, respectively. Er-doped ZnO exhibits a smaller humidity hysteresis and the response/recovery times are 32.3/39.6 s. It can be concluded that sensitivity, hysteresis, and response/recovery time of the ZnO humidity sensor have been improved by Er doping.

It can be seen from Figs. 20.9A–D that compared with the ZnO sensor, the morphology of Er-doped ZnO has been significantly improved after Er being doped. The high activity and large specific surface area of the humidity sensor provides more adsorption sites for water molecules. XPS spectra show that Er doping greatly increases the number of oxygen vacancies on surface of ZnO as in Fig. 20.9E and F. More surface oxygen vacancies promote dissociation process of water molecules on the surface of Er-doped ZnO.

The above two examples illustrate that rare earth metal doped MOS improves surface microscopic states and surface oxygen vacancies, which can enhance performance of humidity sensors based on MOS.

20.3.3   Noble metal-modified ZnO humidity sensor

Noble metals (Hg, Pt, Ag, and Au) as modifiers often attract more attention due to local plasmon resonance effects and good conductivity. Electrons are transferred from MOS surface to such noble metals, which broadens the electron depletion layer on the MOS surface and results in the formation of a noble metal-modified ZnO Schottky contact [48] to enhance the sensitivity of humidity sensors. Moreover, it has a strong

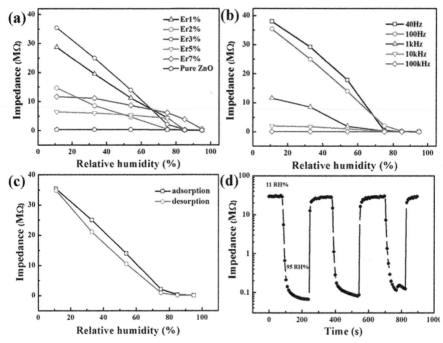

Fig. 20.8 (A) Impedance versus RH curves of undoped and doped ZnO with different Er contents. (B) Frequency dependence, (C) humidity hysteresis characteristics, and (D) response and recovery behaviors of the 3% Er:ZnO humidity sensor.
Adapted with permission from M. Zhang, H. Zhang, L. Li, K. Tuokedaerhan, Z. Jia, Er-enhanced humidity sensing performance in black ZnO-based sensor, J. Alloys Compd. 744 (2018) 364–369, https://doi.org/10.1016/j.jallcom.2018.02.109. Copyright © 2018 Journal of Alloys and Compounds.

ability to split water molecules when the size of the noble metal reaches a nanometer level. Furthermore, noble metal-modified MOS materials will cause surface defects, surface electron transfer, and microstructure of the materials and exhibit unique physical/chemical properties, which are all helpful to improve performance of humidity sensors.

Among many noble metals, gold (Au) is a popular modifier for its unique electrical storage capability and good electrical conductivity [49–51]. Modification of gold nanoparticles (AuNPs) can enhance the humidity sensing properties of MOS materials. According to previous reports, Au-modified ZnO (Au/ZnO) nanosheets synthesized by a hydrothermal method as a sensitive material show excellent humidity characteristics [36–40]. It was found that the impedance change of AuNPs-modified ZnO was more than 5 orders of magnitude in a range from 11% RH to 95% RH as in Fig. 20.10A. It can be seen from Fig. 20.10B, the performance of Au/ZnO is the best when the operating frequency is 100 Hz. Fig. 20.10C and D show the maximum hysteresis error of the Au/ZnO humidity sensor is reduced from 5.4% to 3.0%, and

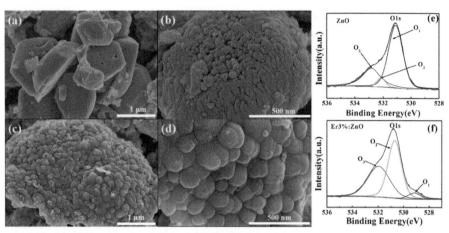

Fig. 20.9 SEM images of (A, B) ZnO and (C, D) 3% Er:ZnO under low and high resolutions; (E, F) XPS spectra of O 1s of undoped and doped ZnO.
Adapted with permission from M. Zhang, H. Zhang, L. Li, K. Tuokedaerhan, Z. Jia, Er-enhanced humidity sensing performance in black ZnO-based sensor, J. Alloys Compd. 744 (2018) 364–369, https://doi.org/10.1016/j.jallcom.2018.02.109. Copyright © 2018 Journal of Alloys and Compounds.

response and recovery time is reduced from 36/44 s to 16/28 s. Fig. 20.10E shows that this kind of sensor has good stability and can be used effectively for a long time.

As shown in Fig. 20.10F, the synthesized Au/ZnO exhibits a sheet-like structure with a diameter of 500 to 700 nm, and the distribution is relatively loose. Such a structure can provide more adsorption sites for water molecules. XPS spectra show that Au nanoparticle modification increases the proportion of oxygen vacancy defects on the surface of ZnO from 31% to 35% as shown in Fig. 20.10G and H. The excellent performance of the Au/ZnO humidity sensor is mainly due to the introduction of Au nanoparticles, which increases both the active sites and the number of oxygen vacancies on the surface of ZnO to capture more water molecules and to accelerate their decomposition. In addition, the activity of Au nanoparticles is relatively strong, which increases the decomposition rate of water molecules to a certain extent.

Another example is the effects of AuNPs-modified ZnO (sample 1) and chloroauric acid ($HAuCl_4$)-modified ZnO (sample 2) on performance of humidity sensing [20]. Fig. 20.11A shows the changes of impedance of ZnO and the two Au-modified ZnO humidity sensors from 11% RH to 95% RH at an operating frequency of 100 Hz. It can be found that the impedance changes of Au-modified ZnO (sample 1 and sample 2) have achieved more than 4 orders of magnitude and $HAuCl_4$-modified ZnO (sample 2) exhibits better linearity, which means that Au nanoparticles greatly improve the sensitivity of the ZnO humidity sensors. As shown in Fig. 20.11B and C, the $HAuCl_4$-modified ZnO humidity sensor exhibits small hysteresis error (2.35%) and response/recovery times (15/6 s).

Fig. 20.12A and B show the diameters of ZnO and AuNPs-modified ZnO (sample 1) nanodiscs from 1 to 1.15 μm. In Fig. 20.12C, the sizes of $HAuCl_4$-modified ZnO (sample 2) particles are reduced to the range of 80 to 180 nm after the addition of

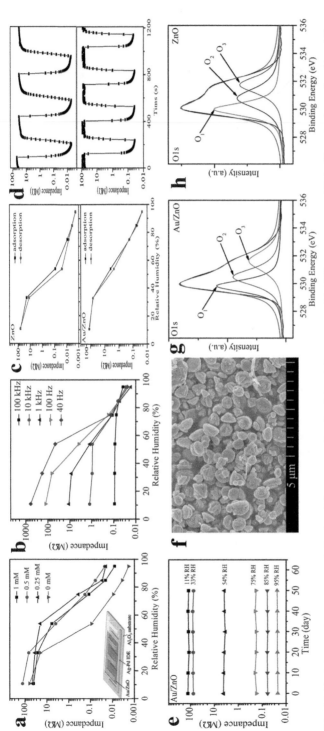

Fig. 20.10 (A) Impedance versus relative humidity curves of Au/ZnO humidity sensors with different amounts of HAuCl₄ and schematic of humidity sensing substrate (inset). (B) Relationship of impedance and relative humidity based on Au/ZnO at various frequencies; (C–E) hysteresis, response/recovery characteristics, and stability of humidity sensor based on ZnO and Au/ZnO (0.5 mM HAuCl₄); (F) SEM image of Au/ZnO; (G, H) XPS spectrum of O 1s of ZnO and Au/ZnO.

Adapted with permission from S. Yu, H. Zhang, C. Chen, C. Lin, Investigation of humidity sensor based on Au modified ZnO nanosheets via hydrothermal method and first principle, Sensors Actuators B Chem. 287 (2019) 526–534, https://doi.org/10.1016/j.snb.2019.02.089. Copyright © 2019 Sensors and Actuators B: Chemical.

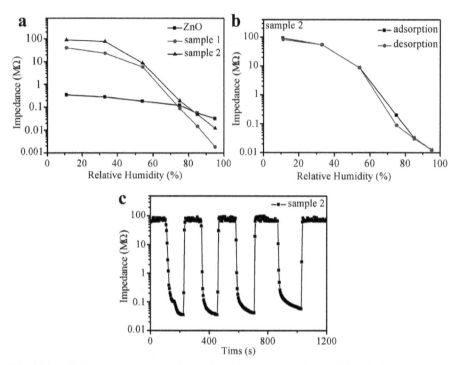

Fig. 20.11 (A) Impedance versus relative humidity curves of Au-modified ZnO humidity sensors; (B) hysteresis; and (C) response/recovery characteristics of Au-modified ZnO humidity sensors.

Adapted with permission from H. Zhang, S. Yu, C. Chen, J. Zhang, J. Liu, P. Li, Effects on structure, surface oxygen defects and humidity performance of Au modified ZnO via hydrothermal method, Appl. Surf. Sci. 486 (2019) 482–489, https://doi.org/10.1016/j.apsusc.2019.04.266. Copyright © 2019 Applied Surface Science.

$HAuCl_4$, which will provide more adsorption sites to increase sensitivity of the sensors. Compared with ZnO as shown in Fig. 20.12D and E, Au modification makes the number of oxygen vacancies on the surface of $HAuCl_4$-modified ZnO increase, and the stronger electrostatic field near oxygen vacancies accelerates decomposition of water molecules, which shortens response time of the sensor. The above studies further illustrate that the precious metal modification can enhance surface oxygen vacancies, active sites, and electron transfer of the MOS materials, improving the humidity sensing performance of MOS sensors.

20.3.4   ZnO two-dimensional (2D) material composite humidity sensor

Composite material is a new material which is an optimized combination of different material components made by advanced material preparation technologies. Combination of MOS and different materials can increase surface hydrophilic functional

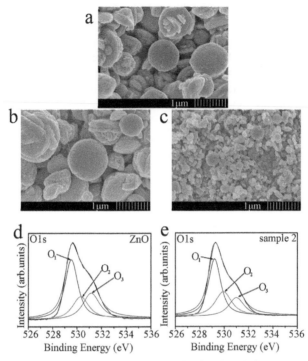

Fig. 20.12 SEM image of (A) ZnO, (B) AuNPs-modified ZnO and (C) HAuCl$_4$-modified ZnO. XPS spectrum of O 1s of (D) ZnO and (E) HAuCl$_4$-modified ZnO.
Adapted with permission from H. Zhang, S. Yu, C. Chen, J. Zhang, J. Liu, P. Li, Effects on structure, surface oxygen defects and humidity performance of Au modified ZnO via hydrothermal method, Appl. Surf. Sci. 486 (2019) 482–489, https://doi.org/10.1016/j.apsusc. 2019.04.266. Copyright © 2019 Applied Surface Science.

groups and surface defect states of the MOS composite material, which is beneficial to capture a large number of water molecules. Since the discovery of graphene materials in 2004, various 2D materials with single layer or few layers have immediately attracted attention in sensor fields due to the large specific surface area, excellent hygroscopicity, high carrier mobility, and electrical conductivity. The 2D materials can overcome all the shortcomings of small specific surface area, high resistivity, and poor hygroscopicity of some MOS materials, which opens up a new stage for the development of high-performance humidity sensors based on MOS materials. In this part, we introduce humidity sensors based on ZnO and 2D materials to reveal the application of 2D materials in humidity detection.

20.3.4.1   g-C$_3$N$_4$/ZnO humidity sensor

As a typical 2D nanomaterial, g-C$_3$N$_4$ has been used in high-sensitivity humidity sensors due to its large surface area ratio and inherent defects to improve proton conductivity inside water layer. However, high-humidity hysteresis and long response/recovery time in high-humidity environment limit its application in the field of humidity sensing. Recent studies have shown that the compound of MOS materials and g-C₃N₄ is an effective way to improve the performance of MOS humidity sensor.

Zhang et al. [52] designed an impedance humidity sensor based on the g-C₃N₄/ZnO nanocomposite. Fig. 20.13A shows a comparative study of the responses of ZnO, g-C₃N₄, and g-C₃N₄/ZnO in different humidity levels from 11% RH to 95% RH. Compared with ZnO and g-C₃N₄, the humidity sensing response of g-C₃N₄/ZnO is greatly improved. The response of g-C₃N₄/ZnO is 10,500 which is 58.6 times that of g-C₃N₄ nanoflakes. Fig. 20.13B shows response and recovery characteristic test of g-C₃N₄/ZnO from 11% RH to 95% RH. In order to avoid the influence of external environment, the relative humidity environment change time is controlled within 1 s. The humidity sensor shows a quick response to the change of RH, and response time and recovery time were 22 s and 5 s, respectively. g-C₃N₄/ZnO humidity sensor can quickly adsorb and lose water molecules during the continuous adsorption (11%–95% RH) and desorption (95%–11% RH) processes. g-C₃N₄/ZnO humidity sensor shows a faster response and recovery speed, mainly due to the special mutual embedded structure of g-C₃N₄/ZnO which makes a large number of water molecules to be quickly adsorbed on its surface, and a large number of oxygen vacancy defects on surface

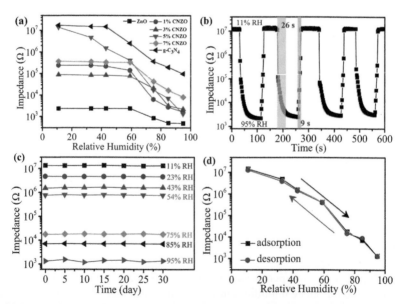

Fig. 20.13 (A) Comparative response values of composite and g-C₃N₄ nanoflake based sensor device; (B) response/recovery characteristics, (C) stability and (D) hysteresis test of the g-C₃N₄/ZnO film.

of g-C$_3$N$_4$/ZnO make water molecules to be quickly decomposed into H$_3$O$^+$ to participate in conduction, thereby improving the response speed of the humidity sensor. To ensure stability, g-C$_3$N$_4$/ZnO humidity sensor was repeatedly tested for 30 days. It can be seen from Fig. 20.13C that the response is quite stable during this period and humidity hysteresis of the sensor is very small (Fig. 20.13D).

Transmission electron microscope (TEM) image of g-C$_3$N$_4$/ZnO are shown in Figs. 20.14A–D. In Fig. 20.14A and B, ZnO is in the shape of nanoparticle with a diameter of about 70 nm, and g-C$_3$N$_4$ exhibits a thin porous nanosheet structure. Fig. 20.14C and D shows HRTEM of g-C$_3$N$_4$/ZnO, where significant lattice fringes of ZnO and g-C$_3$N$_4$ can be found. The lattice spacing of ZnO is 0.26 nm corresponding to the (002) crystal surface of ZnO, and the lattice spacing of g-C$_3$N$_4$ is 0.32 nm corresponding to its (002) crystal surface. The intercalation structure will produce a large specific surface area, which will enhance the response of g-C$_3$N$_4$/ZnO composite sensing layer. From the results of XPS spectra in Fig. 20.14E and F, it is obvious that there are more oxygen vacancies on the surface of g-C$_3$N$_4$/ZnO. Surface oxygen vacancies can be used as active sites for the dissociation of water molecules, so that a large number of water molecules are decomposed to form conductive ions, which help to improve conductivity of the sensor. Under the combined effects of the specific surface area and oxygen vacancies, g-C$_3$N$_4$/ZnO exhibits excellent humidity sensing performance. g-C$_3$N$_4$/ZnO composite material not only overcomes the high resistance of ZnO but also solves the high-humidity hysteresis and long response/recovery time of g-C$_3$N$_4$ in high-humidity environments, making g-C$_3$N$_4$/ZnO possible for applications in humidity sensing.

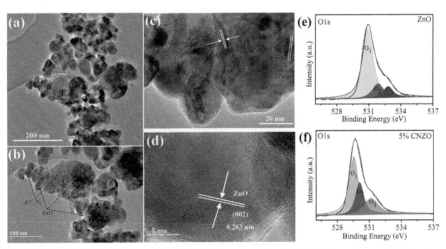

Fig. 20.14 (A–D) TEM image of g-C$_3$N$_4$/ZnO composite. (E, F) XPS spectrum of O 1s of g-C$_3$N$_4$/ZnO.

Adapted with permission from S. Yu, C. Chen, H. Zhang, J. Zhang, J. Liu, Design of high sensitivity graphite carbon nitride/zinc oxide humidity sensor for breath detection, Sensors Actuators B Chem. 332 (2021) 129536, https://doi.org/10.1016/j.snb.2021.129536. Copyright © 2021 Sensors and Actuators B: Chemical.

20.3.4.2   rGO/ZnO/Cu humidity sensor

rGO is a 2D material with a large specific surface area, high chemical stability, and carrier mobility. Unfortunately, it has poor sensitivity to RH due to its low content of oxygen-containing functional groups on edges of the rGO substrate. Considered that rGO is a high-conductivity material, it can transfer electrons between it and other sensing materials. Therefore, some researchers used rGO and MOS materials to design a humidity sensor with high performance. Kuntal et al. [53] successfully prepared rGO/ZnO/Cu nanocomposite by one-step method and used it in humidity detection. Fig. 20.15A shows the changes in resistance and sensing response of rGO/ZnO/Cu from 11% RH to 97% RH at room temperature. The sensor shows a decrease in

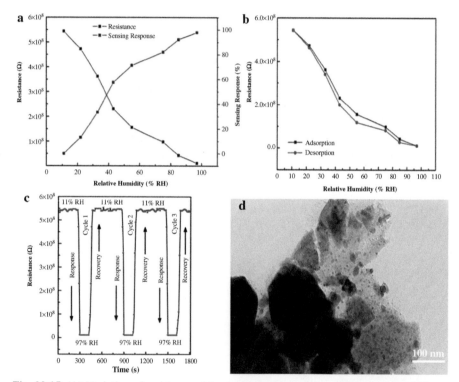

Fig. 20.15 (A) Variation of resistance and sensing response of the rGO/ZnO nanorod/Cu nanocomposite as a function of RH; (B) humidity hysteresis characteristic curve, (C) and the recycling/repeatability with the same response and recovery time; (D) TEM images of the nanocomposite.

Adapted with permission from D. Kuntal, S. Chaudhary, A.B.V. Kiran Kumar, R. Megha, CH. V.V. Ramana, Y. T. Ravi Kiran, S. Thomas, D. Kim, rGO/ZnO nanorods/Cu based nanocomposite having flower shaped morphology: AC conductivity and humidity sensing response studiesat room temperature, J. Mater. Sci. Mater. Electron. 30 (2019) 15544–15552, https://doi.org/10.1007/s10854-019-01931-8. Copyright © 2019 Journal of Materials Science: Materials in Electronics.

impedance and an increase in sensing response due to adsorption of water molecules on the surface of rGO/ZnO/Cu, and the maximum sensing response is 97.79%. The humidity sensor exhibits a smaller hysteresis and a relatively short response and recovery time (19/42 s) in Fig. 20.15B and C.

Better performance of rGO/ZnO/Cu humidity sensor is mainly due to its large surface area to volume ratio (Fig. 20.15D), high vacancy density, and hydrophilic functional group on the surface of rGO, which makes water molecules more easily to be adsorbed on the material surface and thus improve response of sensor. Furthermore, special structure of rGO/ZnO/Cu allows the charge to be transferred from water to ZnO nanorods, thereby increasing conductivity.

It is believed that ZnO-based humidity sensors still have more room of developments and prospects. Defects of surface oxygen vacancies, surface active sites, and hydrophobicity of ZnO are continuously increased through different methods, which provide promising directions for the manufacture of high-performance commercial ZnO-based humidity sensors.

The performance of the most impedance humidity sensors based on ZnO and its composite materials in past years are summarized in Table 20.1.

Table 20.1 Comparison of the performance of reported impedance ZnO humidity sensors.

Materials	Measurement range (% RH)	Response/ sensitivity	Hysteresis	T_{Res}/ T_{Rec} (s)	Ref.
Cd/ZnO	Air–95	–	–	3/5	[54]
ZnO nanowire	12–97	4 orders of magnitude	–	3/10	[55]
ZnO-SiO$_2$	11–95	5 orders of magnitude	2%	50/100	[56]
ZnO NSs	12–96	220	~5%	600/3	[57]
ZnO-SiO$_2$	11–98	5 orders of magnitude	1.2%	15/16	[33]
ZnWO$_4$-ZnO	5–98	3416	–	50/100	[58]
PVA-modified ZnO	40–90	12.2 KΩ/0.1% RH	1%–2%	40/60	[59]
MoS$_2$@ZnOQDs	11–95	4 orders of magnitude	–	1/20	[60]
ZnO/V$_2$O$_5$	35–95	16.88 MΩ/% RH	6%	–	[61]
Er3%:ZnO	11–95	Over 3 orders of magnitude	–	32.3/39.6	[9–12]
Eu-doped ZnO	11–95	3 orders of magnitude	–	5/19	[36]
ZnO nanowires	1.73–82.13	1189.93	–	202/57.4	[1,2]
ZnO/MoS$_2$	35–85	301 times	–	138/166	[52]
ZnO nanosheets	11–97	86%	–	25/5	[62]
ZnO	11–95	2,515,127%	0.9%	31/15	[36–40]

20.4 Humidity sensing mechanism

To obtain a humidity sensor with high sensitivity, good reproducibility, low cost, easy to be manufactured, and long-term stability, studying the humidity sensing mechanism of the sensors has important guiding significance for us to improve the performance and to develop new materials and devices. In this part, we introduce the sensing mechanism for the impedance humidity sensor in three aspects.

20.4.1   Water molecule adsorption mechanism

Over the past years, researchers have proposed different mechanisms for impedance humidity sensors based on MOS materials, and the theory of electron-proton conductivity is generally recognized by most researchers. The impedance value of sensitive material will change when water molecules are adsorbed onto the surface and grain boundaries of MOS materials.

Fig. 20.16 shows a multi-layer structure of MOS surface to adsorb water molecules. At low humidity level, water molecules in the environment will first form a layer of chemical adsorption on surface and grain boundaries of MOS. Due to the surface collision or self-ionization of water molecules, decomposition of protons (H^+) and hydroxide ions (OH^-) are induced ($H_2O \Leftrightarrow H^+ + OH^-$) [63,64]. The OH^- formed by decomposition of water molecules will adsorb cations on the surface and grain boundaries of MOS materials to reduce electron affinity of the cations. Electrons originally captured by the cation are released, and electron accumulation will occur on the surface and grain boundaries of the material, thereby lowering potential barrier and making depletion layer disappeared, resulting in an increase in dielectric constant and bulk conductivity. In addition, chemically adsorbed OH^- can also ionize out H^+, which results in conduction near OH^- [65]. Since the ionization activation energy of OH^- is relatively large, the amount of ionized H^+ is small. In this process, electronic conduction is dominant and ions conduction is minor.

At medium humidity level, chemical adsorption process is basically completed with the increase of humidity. Water molecules are adsorbed on two OH^- of chemical adsorption layer by a double hydrogen bond to form the first physical adsorption layer on the surface and interface of the material. The physically adsorbed water molecules

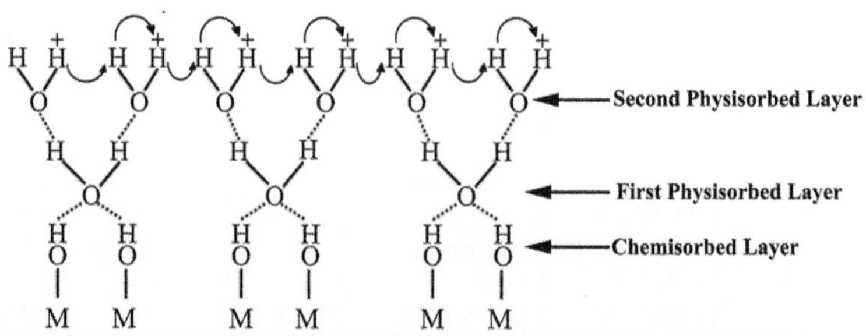

Fig. 20.16 Multilayer structure of adsorbing water molecules on the surface of MOS.

are dissociated into H_3O^+ under the action of an electrostatic field ($H_2O + H_3O^+ \Leftrightarrow H_3O^+ + H_2O$) due to the local charge density and electrostatic field strength around chemically adsorbed OH^- and H_3O^+ diffuses conduction between chemically adsorbed OH^- [9–12,66]. Furthermore, there is an additional layer formed on the first physical adsorption layer with water molecules further increasing. At this time, water molecules are bonded to the upper layer of water molecules with a single hydrogen bond, and electric charge is transferred between adjacent water molecules through the hydrogen bond.

At high humidity level, a large number of water molecules are adsorbed on surface of the material to increase physically adsorbed water molecules to form a continuous water layer. The individually combined water molecules become mobile and can form a continuous dipole and electrolyte layer between electrodes, resulting in an increase in dielectric constant and bulk conductivity [67]. In this process, the main reason for electrical conductivity is that physically adsorbed water ionizes a large amount of H_3O^+ to transfer conductivity. In addition, there may be a large amount of soluble ions in the grain boundaries, which also plays a certain role in conducting electricity [36].

20.4.2   Complex impedance and equivalent circuit (EC) mechanism

Complex impedance-ECs are the common method to analyze humidity sensing mechanism of sensors. It can help us to understand the causes of a series of problems of a humidity sensor such as hysteresis, drift, stability, anti-interference ability, and high resistance of the sensor, so as to provide theoretical guidance for the selection of the humidity-sensitive material and the optimization of the device structure. Complex impedance spectroscopy (CIS) and EC are introduced to explain the influence of different humidity levels on the material interface and the interface states of the humidity sensing film and electrode. In this part, impedance represents the conduction process and capacitance represents the polarization process. Commonly used parameters are C_f, R_f, Z_w, R_{ct}, and CPE (constant phase element), which are as follows:

(1) CPE represents the "non-Debye capacitance" between the interfaces of sensitive materials.
(2) R_{ct} represents the charge transfer resistance.
(3) Z_w represents the impedance of the diffusion element caused by water molecules adsorbed between the material and electrode interface.
(4) R_f represents the impedance caused by water molecules entered the film.
(5) C_f represents the capacitance between the material and electrode interface.

As a method to analyze the humidity sensing mechanism, CIS and EC can be used to intuitively understand dielectric loss characteristics of impedance humidity sensors and to analyze the mechanism for other types of humidity sensors. The difference of impedance spectra illustrates the difference of conductivity and polarization of sensing film under different RH values. Ze et al. and Yu et al. used CIS and corresponding EC to analyze the conduction process of MoS_2-modified ZnO and Au-modified ZnO humidity sensors [36–40,60]. The analysis of CIS and EC is usually divided into low RH, medium RH, and high RH.

At low RH, a small amount of water molecules is adsorbed on the surface of the sensitive materials and the electron layer enhances the charge transfer effect, resulting in a rapid drop in impedance of the humidity sensor. Physically, the adsorbed water molecules cannot move freely because of the interaction of surface hydrogen bonds. With the increase of RH, CIS gradually becomes a semicircular in Figs. 20.17A and 20.18A, and the corresponding EC is composed of a film capacitor and an inherent resistance is parallel in Figs. 20.17C and 20.18C. The inherent resistance and the capacitance of the material are mainly caused by proton jumping between hydroxyl groups (OH^-). The double hydrogen bond restricts free movement of protons which can only jump and conduct electricity between the hydroxyl groups (OH^-), making the ion conduction difficult. At this stage, a small amount of water molecules is adsorbed on the surface of sensitive materials through the double hydrogen bonds, and there is no a continuous water layer formed.

At medium RH, multiple layers of physically adsorbed water molecules are gradually formed between the sensitive material and electrode, resulting in polarization and diffusion of water molecules. As a result, the contribution of sensitive materials to conductivity becomes smaller and physically synthesized water molecules are dominant. CIS shows that the semicircle gradually becomes smaller, and there is a short line appeared at the end of the semicircle in low frequency range in Figs. 20.17B and 20.18B. The short line represents Warburg impedance (Z_w) in Figs. 20.17D and 20.18D, which is originated from the diffusion process of ions or carriers between the sensing membrane and the electrode interface. The corresponding EC can be

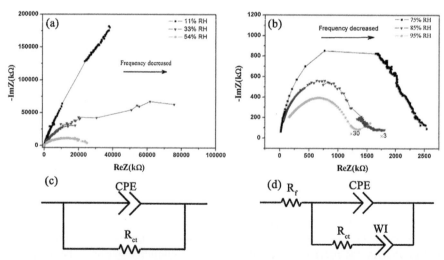

Fig. 20.17 (A–D) Complex impedance curves of a MoS_2@ZnO QDs humidity sensor measured at 11%, 33%, 54%, 75%, 85%, and 95% RH in a frequency range from 10Hz to 100kHz. Adapted with permission from L. Ze, G. Yueqiu, L. Xujun, Z. Yong, MoS_2-modified ZnO quantum dots nanocomposite: synthesis and ultrafast humidity response, Appl. Surf. Sci. 399 (2017) 330–336, https://doi.org/10.1016/j.apsusc.2016.12.034. Copyright © 2017 Applied Surface Science.

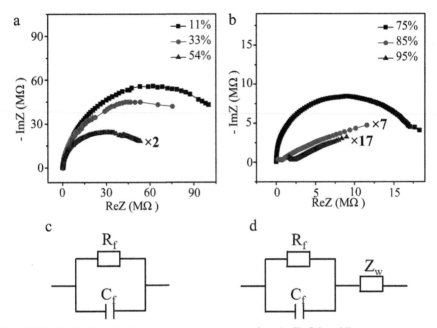

Fig. 20.18 (A, B) Complex impedance spectroscopy of an Au/ZnO humidity sensor measured at different RH from 11% to 95%; (C, D) corresponding equivalent circuits of Au/ZnO under different RH levels.

Adapted with permission from S. Yu, H. Zhang, C. Chen, C. Lin, Investigation of humidity sensor based on Au modified ZnO nanosheets via hydrothermal method and first principle, Sensors Actuators B Chem. 287 (2019) 526–534, https://doi.org/10.1016/j.snb.2019.02.089. Copyright © 2019 Sensors and Actuators B: Chemical.

connected in series with impedance Z_w. With the increase of RH, the characteristic of impedance Warburg gradually increases, and there are a large number of water molecules physically adsorbed under the electrostatic field to be ionized, forming hydronium ions (H_3O^+) as the main charge carriers. Ion conduction plays a leading role in this stage.

At high RH, water molecules in multi-physical layers gradually form a liquid film with the further increase of RH. In this process, Z_w plays a leading role. Proton hopping transport can easily go through the Grotthuss chain reaction ($H_2O + H_3O^+ \rightarrow H_3O^+ + H_2O$) in the multiple continuous water layers [9–12], resulting in an increase in proton hopping transport.

20.4.3 Density functional theory (DFT) modeling

Based on quantum mechanics, first-principles or ab initio modeling simplifies a multi-atomic composition system into a multi-particle system composed of nuclei and electrons. The first-principles calculation process does not contain any adjustable parameter to minimize the empirical treatment of the problem. First-principles computational simulation has been widely used in the field of materials science, especially when the properties of materials are expected to be traced back to the electronic level, which has important scientific research value and practical role in revealing the adsorption mechanism of gas-sensitive materials and developing related material designs. First-principles DFT can be used to calculate electronic properties of the composite material and water molecules during the adsorption process under the condition of zero adsorption, and the mechanism of humidity sensor can also be obtained.

Taken Au/ZnO humidity sensor as an example, DFT was used to study its humidity sensing mechanism [36–40]. The calculations in Fig. 20.19 are obtained using Dmol3 code based on the DFT theory. A ZnO supercell includes 48 Zn atoms and 47 O atoms,

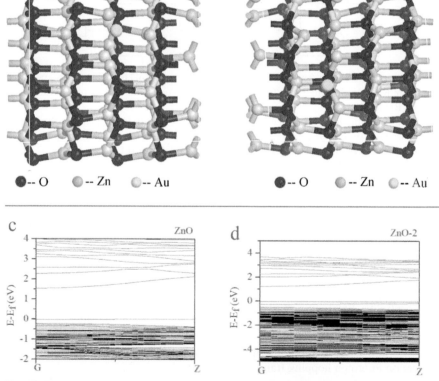

Fig. 20.19 Au adsorption on (A) oxygen-vacancy (ZnO-1) and (B) other sites on the surface of ZnO supercell (ZnO-2); energy band of (C) pure ZnO and (D) ZnO supercell (ZnO-2). Adapted with permission from S. Yu, H. Zhang, C. Chen, C. Lin, Investigation of humidity sensor based on Au modified ZnO nanosheets via hydrothermal method and first principle, Sensors Actuators B Chem. 287 (2019) 526–534, https://doi.org/10.1016/j.snb.2019.02.089. Copyright © 2019 Sensors and Actuators B: Chemical.

and the removed O atom indicates an O vacancy on the surface of the ZnO supercell. Au atom absorption on the ZnO supercell can be considered as either adsorbed Au on O-vacancy site (ZnO-1) or on the surface of the ZnO supercell (ZnO-2), which are shown in Fig. 20.19A and B. Calculation results show that when Au atoms are adsorbed on the surface of ZnO without occupying oxygen vacancies, there will be more oxygen vacancy defects available. Oxygen vacancies can accelerate decomposition of water and produce more H_3O^+ when water molecules are adsorbed on the surface of Au/ZnO, which increases conductivity of ZnO.

Fig. 20.19C and D show that the band gap of ZnO is 1.54 eV and the band gap of Au/ZnO can be reduced to 1.28 eV. Although the calculated band gap sizes are somewhat lower than the experimental results, the change trend of the band gap is consistent with experiments. A smaller band gap means that Au/ZnO is more sensitive to water molecules, which can absorb more water molecules and improve the response of the sensor. Furthermore, Au nanoparticles do not occupy oxygen vacancies. A large number of water molecules are decomposed into conductive ions under the action of oxygen vacancies, so that the sensor achieves the goal of rapid response. In addition to analyzing the performance of the sensor from the perspective of energy bands and surface oxygen vacancies, it is also an effective method to analyze the mechanism of the humidity sensor by calculating adsorption energy between water molecules and materials.

20.5   Conclusions

Humidity sensors, as multi-disciplinary integrated application products of materials, chemistry, physics, and informatics, have been developed rapidly with the development of science and technology. Strong stability, high sensitivity, low hysteresis, and good linearity enable humidity sensors to meet different needs of various fields in industries. In this chapter, we mainly introduce the detection parameters of resistance MOS humidity sensor, humidity sensing mechanism, and the ways to improve the performance of MOS. By taking the ZnO humidity sensor as an example, it is found that a large surface specific area, a large density of surface oxygen vacancies, and good hydrophobicity are very important factors for the preparation of sensitive materials with excellent humidity sensing properties. With the further expansion of material synthesis and detection methods, humidity sensors will have a broader development prospect.

Acknowledgments

This work was supported by the National Natural Science Foundation of China (No. 62064011, 61665011) and the Xinjiang Science and Technology Project (No.2016D01C057). The authors acknowledge the collaboration with colleagues who have been cited in the references.

References

[1] S.Y. Park, J.E. Lee, Y.H. Kim, J.J. Kim, Y.-S. Shim, S.Y. Kim, et al., Room temperature humidity sensors based on rGO/MoS$_2$ hybrid composites synthesized by hydrothermal method, Sensors Actuators B Chem. 258 (2018) 775–782, https://doi.org/10.1016/j.snb.2017.11.176.

[2] S. Park, D. Lee, B. Kwak, H.-S. Lee, S. Lee, B. Yoo, Synthesis of self-bridged ZnO nanowires and their humidity sensing properties, Sensors Actuators B Chem. 268 (9) (2018) 293–298, https://doi.org/10.1016/j.snb.2018.04.118.

[3] J. Qian, Z. Peng, Z. Shen, Z. Zhao, G. Zhang, X. Fu, Positive impedance humidity sensors via single-component materials, Sci. Rep. 6 (2016) 25574, https://doi.org/10.1038/srep25574.

[4] S. Yu, H. Zhang, C. Chen, J. Zhang, P. Li, Preparation and mechanism investigation of highly sensitive humidity sensor based on two-dimensional porous gold/graphite carbon nitride nanoflake, Sensors Actuators B Chem. 307 (2020), https://doi.org/10.1016/j.snb.2020.127679, 127679.

[5] C. Ashley, D. Burton, Y.B. Sverrisdottir, M. Sander, D.K. Mckenzie, V.G. Macefield, Firing probability and mean firing rates of human muscle vasoconstrictor neurones are elevated during chronic asphyxia, J. Physiol. 588 (4) (2010) 701–712, https://doi.org/10.1113/jphysiol.2009.185348.

[6] R.J. Goldberg, P.G. Steg, I. Sadiq, C.B. Granger, E.A. Jackson, A. Budaj, et al., Extent of, and factors associated with, delay to hospital presentation in patients with acute coronary disease (the GRACE registry), Am. J. Cardiol. 89 (7) (2002) 791–796, https://doi.org/10.1016/S0002-9149(02)02186-0.

[7] T.Q. Trung, T.D. Le, S. Ramasundaram, N.-E. Lee, Transparent, stretchable, and rapid-response humidity sensor for body-attachable wearable electronics, Nano Res. 10 (006) (2017) 2021–2033, https://doi.org/10.1007/s12274-016-1389-y.

[8] A.S. Ismail, M.H. Mamat, M.M. Yusoff, M.F. Malek, A.S. Zoolfakar, R.A. Rani, et al., Enhanced humidity sensing performance using Sn-doped ZnO nanorod Array/SnO$_2$ nanowire heteronetwork fabricated via two-step solution immersion, Mater. Lett. 210 (2018) 258–262, https://doi.org/10.1016/j.matlet.2017.09.040.

[9] H. Zhang, L. Xiang, Y. Yang, M. Xiao, J. Han, L. Ding, et al., High-performance carbon nanotube complementary electronics and integrated sensor systems on ultrathin plastic foil, ACS Nano 12 (3) (2018) 2773–2779, https://doi.org/10.1021/acsnano.7b09145.

[10] D. Zhang, X. Zong, Z. Wu, et al., Ultrahigh-performance impedance humidity sensor based on layer-by-layer self-assembled tin disulfide/titanium dioxide nanohybrid film, Sens. Actuators B 266 (2018) 52–62, https://doi.org/10.1016/j.snb.2018.03.007.

[11] M. Zhang, H. Zhang, L. Li, K. Tuokedaerhan, Z. Jia, Er-enhanced humidity sensing performance in black ZnO-based sensor, J. Alloys Compd. 744 (2018) 364–369, https://doi.org/10.1016/j.jallcom.2018.02.109.

[12] D. Zhang, Y. Cao, P. Li, et al., Humidity-sensing performance of layer-by-layer self-assembled tungsten disulfide/tin dioxide nanocomposite, Sensors Actuators B Chem. 265 (2018) 526–538, https://doi.org/10.1016/j.snb.2018.03.043.

[13] J.J. Steele, M.T. Taschuk, M.J. Brett, Response time of nanostructured relative humidity sensors, Sens. Actuators B 140 (2) (2009) 610–615, https://doi.org/10.1016/j.snb.2009.05.016.

[14] Z. Ying, C. Yu, Y. Zhang, C. Xin, C. Feng, L. Chen, et al., A novel humidity sensor based on NaTaO$_3$ nanocrystalline, Sens. Actuators B 174 (2012) 485–489, https://doi.org/10.1016/j.snb.2012.08.050.

[15] Y. Sakai, M. Matsuguchi, T. Hurukawa, Humidity sensor using cross-linked poly(chloromethyl styrene), Sens. Actuators B 66 (1–3) (2000) 135–138, https://doi.org/10.1016/S0925-4005(00)00313-0.

[16] C.W. Lee, H.S. Park, J.G. Kim, B.K. Choi, S.W. Joo, M.S. Gong, Polymeric humidity sensor using organic/inorganic hybrid polyelectrolytes, Sens. Actuators B 109 (2) (2005) 315–322, https://doi.org/10.1016/j.snb.2004.12.063.

[17] D. Nunes, A. Pimentel, A. Goncalves, S. Pereira, R. Branquinho, P. Barquinha, et al., Metal oxide nanostructures for sensor applications, Semicond. Sci. Technol. 34 (4) (2019) 043001–043060, https://doi.org/10.1088/1361-6641/ab011e.

[18] Y. Zhang, J. Chung, J. Lee, J. Myoung, S. Lim, Synthesis of ZnO nanospheres with uniform nanopores by a hydrothermal process, J. Phys. Chem. Solids 72 (12) (2011) 1548–1553, https://doi.org/10.1016/j.jpcs.2011.09.016.

[19] A. Dey, Semiconductor metal oxide gas sensors: a review, Mater. Sci. Eng. B 229 (2018) 206–217, https://doi.org/10.1016/j.mseb.2017.12.036.

[20] H. Zhang, S. Yu, C. Chen, J. Zhang, J. Liu, P. Li, Effects on structure, surface oxygen defects and humidity performance of Au modified ZnO via hydrothermal method, Appl. Surf. Sci. 486 (2019) 482–489, https://doi.org/10.1016/j.apsusc.2019.04.266.

[21] M.H. Feng, W.C. Wang, X.J. Li, Capacitive humidity sensing properties of CdS/ZnO sesame-seed-candy structure grown on silicon nanoporous pillar array, J. Alloys Compd. 698 (2017) 94–98, https://doi.org/10.1016/j.jallcom.2016.11.370.

[22] M. Gong, Y. Li, Y. Guo, X. Lv, X. Dou, 2D TiO_2 nanosheets for ultrasensitive humidity sensing application benefited by abundant surface oxygen vacancy defects, Sensors Actuators B Chem. 262 (2018) 350–358, https://doi.org/10.1016/j.snb.2018.01.187.

[23] Z. Wang, Y. Xiao, X. Cui, P. Cheng, B. Wang, Y. Gao, et al., Humidity-sensing properties of Urchinlike CuO nanostructures modified by reduced graphene oxide, ACS Appl. Mater. Interfaces 6 (6) (2014) 3888–3895, https://doi.org/10.1021/am404858z.

[24] Q. Kuang, C. Lao, Z.L. Wang, Z. Xie, L. Zheng, High-sensitivity humidity sensor based on a single SnO_2 nanowire, J. Am. Chem. Soc. 129 (19) (2007) 6070–6071, https://doi.org/10.1021/ja070788m.

[25] W. Wang, A.V. Virkar, A conductimetric humidity sensor based on proton conducting perovskite oxides, Sens. Actuators B 98 (2/3) (2004) 282–290, https://doi.org/10.1016/j.snb.2003.10.035.

[26] J. Du, M. Yao, L. Wenying, et al., Novel QCM humidity sensors using stacked black phosphorus nanosheets as sensing film, Sens. Actuators B 244 (2017) 259–264, https://doi.org/10.1016/j.snb.2017.01.010.

[27] S. Wang, G. Xie, Y. Su, et al., Reduced graphene oxide-polyethylene oxide composite films for humidity sensing via quartz crystal microbalance, Sens. Actuators B 255 (2018) 2203–2210, https://doi.org/10.1016/j.snb.2017.09.028.

[28] P. Kronenberg, P.K. Rastogi, P. Giaccari, H.G. Limberger, Relative humidity sensor with optical fiber Bragg gratings, Opt. Lett. 27 (16) (2002) 1385–1387, https://doi.org/10.1364/OL.27.001385.

[29] S. Wang, Z. Chen, A. Umar, Y. Wang, T. Tian, Y. Shang, et al., Supramolecularly modified graphene for ultrafast-responsive and highly stable humidity sensor, J. Phys. Chem. C 119 (2015) 28640–28647, https://doi.org/10.1021/acs.jpcc.5b08771.

[30] J.-W. Han, B. Kim, J. Li, M. Meyyappan, Carbon nanotube based humidity sensor on cellulose paper, J. Phys. Chem. C 116 (41) (2012) 22094–22097, https://doi.org/10.1021/jp3080223.

[31] W. Meng, S. Wu, X. Wang, D. Zhang, High-sensitivity resistive humidity sensor based on graphitic carbon nitride nanosheets and its application, Sensors Actuators B Chem. 315 (2020), https://doi.org/10.1016/j.snb.2020.128058, 128058.

[32] P. Zhu, Y. Liu, Z. Fang, Y. Kuang, Y. Zhang, C. Peng, et al., Flexible and highly sensitive humidity sensor based on cellulose nanofibers and carbon nanotube composite film, Langmuir 35 (14) (2019) 4834–4842, https://doi.org/10.1021/acs.langmuir.8b04259.

[33] V.K. Tomer, S. Duhan, A.K. Sharma, R. Malik, S.P. Nehra, S. Devi, One pot synthesis of mesoporous ZnO-SiO$_2$ nanocomposite as high performance humidity sensor, Colloids Surf. A Physicochem. Eng. Asp. 483 (2015) 121–128, https://doi.org/10.1016/j.colsurfa.2015.07.046.

[34] J. Wu, Y.-M. Sun, Z. Wu, X. Li, N. Wang, K. Tao, et al., Carbon nanocoil-based fast-response and flexible humidity sensor for multifunctional applications, ACS Appl. Mater. Interfaces 11 (4) (2019) 4242–4251, https://doi.org/10.1021/acsami.8b18599.

[35] M.M. Arafat, B. Dinan, S.A. Akbar, A.S. Haseeb, Gas sensors based on one dimensional nanostructured metal-oxides: a review, Sensors 12 (6) (2012) 7207–7258, https://doi.org/10.1002/chin.201350224.

[36] X. Yu, X. Chen, X. Ding, X. Chen, X. Yu, X. Zhao, High-sensitivity and low-hysteresis humidity sensor based on hydrothermally reduced graphene oxide/nanodiamond, Sensors Actuators B Chem. 283 (2019) 761–768, https://doi.org/10.1016/j.snb.2018.12.057.

[37] H. Yu, C. Wang, F.-Y. Meng, J.-G. Liang, N.-Y. Kim, Design and analysis of ultrafast and high-sensitivity microwave transduction humidity sensor based on belt-shaped MoO$_3$ nanomaterial, Sensors Actuators B Chem. 304 (2019), https://doi.org/10.1016/j.snb.2019.127138, 127138.

[38] S. Yu, H. Zhang, J. Zhang, Z. Li, Effects of pH on high-performance ZnO resistive humidity sensors using one-step synthesis, Sensors 19 (23) (2019) 5267, https://doi.org/10.3390/s19235267.

[39] S. Yu, H. Zhang, C. Lin, M. Bian, The enhancement of humidity sensing performance based on Eu-doped ZnO, Curr. Appl. Phys. 19 (2) (2019) 82–88, https://doi.org/10.1016/j.cap.2018.11.015.

[40] S. Yu, H. Zhang, C. Chen, C. Lin, Investigation of humidity sensor based on Au modified ZnO nanosheets via hydrothermal method and first principle, Sensors Actuators B Chem. 287 (2019) 526–534, https://doi.org/10.1016/j.snb.2019.02.089.

[41] R. Douani, N. Lamrani, M.H. Oughanem, M. Saidi, B. Boudart, Improvement of humidity sensing performance of BiFeO$_3$ nanoparticles-based sensor by the addition of carbon fibers, Sensors Actuators A Phys. 307 (2020), https://doi.org/10.1016/j.sna.2020.111981, 111981.

[42] K.N. Chappanda, A. Chaix, S.G. Surya, B.A. Moosa, N.M. Khashab, K.N. Salama, Trianglamine hydrochloride crystals for a highly sensitive and selective humidity sensor, Sensors Actuators B Chem. 294 (2019) 40–47, https://doi.org/10.1016/j.snb.2019.05.008.

[43] Y.-G. Han, Relative humidity sensors based on microfiber knot resonators-a review, Sensors 19 (23) (2019) 5196, https://doi.org/10.3390/s19235196.

[44] A. Chelouche, T. Touam, M. Tazerout, F. Boudjouan, D. Djouadi, A. Doghmane, Low cerium doping investigation on structural and photoluminescence properties of sol-gel ZnO thin films, J. Lumin. 181 (2017) 448–454, https://doi.org/10.1016/j.jlumin.2016.09.061.

[45] S.S. Shinde, P.S. Shinde, Y.W. Oh, et al., Structural, optoelectronic, luminescence and thermal properties of Ga-doped zinc oxide thin films, Appl. Surf. Sci. 258 (2012) 9969–9976, https://doi.org/10.1016/j.apsusc.2012.06.058.

[46] L.V. Trandafilović, D.J. Jovanović, X. Zhang, S. Ptasińska, M.D. Dramićanin, Enhanced photocatalytic degradation of methylene blue and methyl orange by ZnO: Eu nanoparticles, Appl. Catal. B Environ. 203 (2017) 740–752, https://doi.org/10.1016/j.apcatb.2016.10.063.

[47] C. Lin, H. Zhang, J. Zhang, C. Chen, Enhancement of the humidity sensing performance in mg-doped hexagonal ZnO microspheres at room temperature, Sensors 19 (3) (2019) 519, https://doi.org/10.3390/s19030519.

[48] K. Khojier, F. Teimoori, S. Zolghadr, M.B. Pashazanousi, Fast and low concentration detection of liquefied petroleum gas by au-activated ZnO sensor, Mater. Res. Bull. 108 (2018) 96–100, https://doi.org/10.1016/j.materresbull.2018.08.044.

[49] Z. Kang, X. Yan, Y. Wang, Y. Zhao, Z. Bai, Y. Liu, et al., Self-powered photo-electrochemical biosensing platform based on Au NPs@ZnO nanorods array, Nano Res. 9 (2) (2015) 344–352, https://doi.org/10.1007/s12274-015-0913-9.

[50] Z. Kang, X. Yan, L. Zhao, Q. Liao, K. Zhao, H. Du, et al., Gold nanoparticle/ZnO nanorod hybrids for enhanced reactive oxygen species generation and photodynamic therapy, Nano Res. 8 (6) (2015) 2004–2014, https://doi.org/10.1007/s12274-015-0712-3.

[51] Y. Li, T. Lv, F.-X. Zhao, Q. Wang, X.-X. Lian, Y.-L. Zou, Enhanced acetone-sensing performance of Au/ZnO hybrids synthesized using a solution combustion method, Electron. Mater. Lett. 11 (5) (2015) 890–895, https://doi.org/10.1007/s13391-015-5146-2.

[52] S. Yu, C. Chen, H. Zhang, et al., Design of high sensitivity graphite carbon nitride/zinc oxide humidity sensor for breath detection, Sensors Actuators B Chem. 332 (2021) 129536, https://doi.org/10.1016/j.snb.2021.129536.

[53] D. Kuntal, S. Chaudhary, A.B.V. Kiran Kumar, R. Megha, C.H.V.V. Ramana, Y.T. Ravi Kiran, S. Thomas, D. Kim, rGO/ZnO nanorods/cu based nanocomposite having flower shaped morphology: AC conductivity and humidity sensing response studiesat room temperature, J. Mater. Sci. Mater. Electron. 30 (2019) 15544–15552, https://doi.org/10.1007/s10854-019-01931-8.

[54] Q. Wan, Q.H. Li, Y.J. Chen, T.H. Wang, X.L. He, X.G. Gao, et al., Positive temperature coefficient resistance and humidity sensing properties of cd-doped ZnO nanowires, Appl. Phys. Lett. 84 (16) (2004) 3085–3087, https://doi.org/10.1063/1.1707225.

[55] Y. Zhang, K. Yu, D. Jiang, Z. Zhu, H. Geng, L. Luo, Zinc oxide nanorod and nanowire for humidity sensor, Appl. Surf. Sci. 242 (1–2) (2005) 212–217, https://doi.org/10.1016/j.apsusc.2004.08.013.

[56] Q. Yuan, N. Li, J. Tu, X. Li, R. Wang, T. Zhang, et al., Preparation and humidity sensitive property of mesoporous ZnO-SiO$_2$ composite, Sens. Actuators B 149 (2) (2010) 413–419, https://doi.org/10.1016/j.snb.2010.06.036.

[57] F.-S. Tsai, S.-J. Wang, Enhanced sensing performance of relative humidity sensors using laterally grown ZnO nanosheets, Sensors Actuators B Chem. 193 (2014) 280–287, https://doi.org/10.1016/j.snb.2013.11.069.

[58] M.V. Arularasu, R. Sundaram, Synthesis and characterization of nanocrystalline ZnWO$_4$-ZnO composites and their humidity sensing performance, Sens. Bio-Sens. Res. 11 (2016) 20–25, https://doi.org/10.1016/j.sbsr.2016.08.006.

[59] S. Patil, N. Ramgir, S. Mukherji, V.R. Rao, PVA modified ZnO nanowire based micro-sensors platform for relative humidity and soil moisture measurement, Sensors Actuators B Chem. 253 (2017) 1071–1078, https://doi.org/10.1016/j.snb.2017.07.053.

[60] L. Ze, G. Yueqiu, L. Xujun, Z. Yong, MoS$_2$-modified ZnO quantum dots nanocomposite: synthesis and ultrafast humidity response, Appl. Surf. Sci. 399 (2017) 330–336, https://doi.org/10.1016/j.apsusc.2016.12.034.

[61] N.K. Pandey, Application of V$_2$O$_5$-ZnO nanocomposite for humidity sensing studies, Int. J. Mater. Sci. Appl 6 (2017) 119, https://doi.org/10.11648/j.ijmsa.20170603.12.

[62] S.P. Gupta, A.S. Pawbake, B.R. Sathe, D.J. Late, P.S. Walke, Superior humidity sensor and photodetector of mesoporous ZnO nanosheets at room temperature, Sensors Actuators B Chem. 293 (8) (2019) 83–92, https://doi.org/10.1016/j.snb.2019.04.086.

[63] K. Wang, X. Qian, L. Zhang, Y. Li, H. Liu, Inorganic–organic p-n heterojunction nanotree arrays for a high-sensitivity diode humidity sensor, Appl. Mater. Interfaces 5 (12) (2013) 5825–5831, https://doi.org/10.1021/am4014677.

[64] Y. Xiong, W. Xu, D. Ding, W. Lu, L. Zhu, Z. Zhu, et al., Ultra-sensitive NH_3 sensor based on flower-shaped SnS_2 nanostructures with sub-ppm detection ability, J. Hazard. Mater. 341 (2018) 159–167, https://doi.org/10.1016/j.jhazmat.2017.07.060.

[65] Z. Chen, C. Lu, Humidity sensors: a review of materials and mechanisms, Sens. Lett. 3 (4) (2005) 274–295, https://doi.org/10.1166/sl.2005.045.

[66] M.A. Dwiputra, F. Fadhila, C. Imawan, V. Fauzia, The enhanced performance of capacitive-type humidity sensors based on Heterostructure ZnO Nanorods/ WS_2 Nanosheets, Sensors Actuators B Chem. 310 (2020), https://doi.org/10.1016/j.snb.2020.127810, 127810.

[67] S. Kurosaki, The dielectric behavior of Sorbed water on silica gel, J. Phys. Chem. 58 (4) (1954) 320–324, https://doi.org/10.1021/j150514a009.

Index

Note: Page numbers followed by *f* indicate figures, *t* indicate tables, and *b* indicate boxes.

CPI Antony Rowe
Eastbourne, UK
December 01, 2022